STUDIES IN LOGIC

and

THE FOUNDATIONS OF MATHEMATICS

VOLUME 114

NORTH-HOLLAND
AMSTERDAM • NEW YORK • OXFORD • TOKYO

LOGIC, METHODOLOGY AND PHILOSOPHY OF SCIENCE VII

PROCEEDINGS OF THE SEVENTH INTERNATIONAL CONGRESS OF LOGIC, METHODOLOGY AND PHILOSOPHY OF SCIENCE, SALZBURG, 1983

Edited by

Ruth BARCAN MARCUS
Yale University, New Haven

Georg J.W. DORN
Institute for Philosophy, Salzburg

Paul WEINGARTNER
Institute for Philosophy, Salzburg

1986

NORTH-HOLLAND
AMSTERDAM • NEW YORK • OXFORD • TOKYO

ISBN: 0 444 87656 1

Published by:
ELSEVIER SCIENCE PUBLISHERS B.V.
P.O. Box 1991
1000 BZ Amsterdam
The Netherlands

Sole distributors for the U.S.A. and Canada:
ELSEVIER SCIENCE PUBLISHING COMPANY, INC.
52 Vanderbilt Avenue
New York, N.Y. 10017
U.S.A.

Library of Congress Cataloging in Publication Data

International Congress of Logic, Methodology, and
Philosophy of Science (7th : 1983 : Salzburg, Austria)
Logic, methodology, and philosophy of science, VII.

(Studies in logic and the foundations of mathematics;
v. 114)
Bibliography: p.
Includes index.
1. Science--Philosophy--Congresses. 2. Science--
Methodology--Congresses. 3. Logic, Symbolic and
mathematical--Congresses. I. Barcan Marcus, Ruth.
II. Dorn, Georg. III. Weingartner, Paul. IV. Title.
V. Series.
Q174.I58 1983 501 85-1640
ISBN 0-444-87656-1

PRINTED IN THE NETHERLANDS

PREFACE

This volume constitutes the Proceedings of the Seventh International Congress of Logic, Methodology and Philosophy of Science. The Congress was held at the University of Salzburg, Austria, from July 11 to July 16, 1983, under the auspices of the Division of Logic, Methodology and Philosophy of Science of the International Union of History and Philosophy of Science. The Congress took place under the patronage of Dr. Heinz Fischer, Minister for Science and Research in Austria, Dr. Wilfried Haslauer, Governor of the Province of Salzburg, and Professor Dr. Wolfgang Beilner, Rector of the University of Salzburg. The Congress was sponsored by the Austrian Ministry of Science and Research (the subvention was granted by the former minister Dr. Hertha Firnberg), by the Austrian National Bank, by the Österreichische Forschungsgemeinschaft, by the Province, the Town and the University of Salzburg. The Congress was organized by its Local Organizing Committee in close cooperation with its Programme Committee and the Executive Committee of the Division of Logic, Methodology and Philosophy of Science. The scientific programme of the Congress was drawn up by its Programme Committee together with 14 Advisory Committees, which corresponded to the 14 Sections of the Congress. (A list of the members of the various committees is appended to this preface.) The 14 Sections of the Congress were as follows:

1. Proof Theory and Foundations of Mathematics
2. Model Theory and its Applications
3. Recursion Theory and Theory of Computation
4. Axiomatic Set Theory
5. Philosophical Logic
6. General Methodology of Science
7. Foundations of Probability and Induction
8. Foundations and Philosophy of the Physical Sciences
9. Foundations and Philosophy of Biology
10. Foundations and Philosophy of Psychology
11. Foundations and Philosophy of the Social Sciences
12. Foundations and Philosophy of Linguistics

13. History of Logic, Methodology and Philosophy of Science
14. Fundamental Principles of the Ethics of Science

In each Section, three or four invited addresses were given. Every Section with the exception of Section 14 also contained contributed papers. Symposia were held in Section 5 ("A Linguistic Turn: New Directions in Logic" under the chairmanship of Charles D. Parsons, USA), in Section 6 ("The Structure of Theories" under the chairmanship of Wolfgang Stegmüller, West Germany), in Section 13 ("Life and Work of Kurt Gödel" under the chairmanship of John W. Dawson, Jr., USA), and in Section 14 ("Ethics of Medicine" under the chairmanship of Dag Prawitz, Norway, and "Scientific and Ethical Rationality" under the chairmanship of Evandro Agazzi, Switzerland). These Proceedings comprise the invited addresses only. A list of the contributed papers is given at the end of this volume. We should like to thank the authors and Elsevier Science Publishers B.V. for their support of our editorial work.

New Haven and Salzburg
March 1984

Ruth BARCAN MARCUS
Georg J.W. DORN
Paul WEINGARTNER

Appendix to the Preface

List of the members of the Executive Committee of the Division of Logic, Methodology and Philosophy of Science of the International Union of History and Philosophy of Science in 1983:

L. Jonathan Cohen	England	Secretary
Jens Erik Fenstad	Norway	Treasurer
Jerzy Łoś	Poland	President
Gert H. Müller	West Germany	2nd Vice-President
Wesley C. Salmon	USA	1st Vice-President

List of the members of the Programme Committee of the 7th International Congress of Logic, Methodology and Philosophy of Science:

Ruth Barcan Marcus	USA	Chairperson
Azriel Levy	Israel	
John J.C. Smart	Australia	
Vladimir A. Smirnov	USSR	
Paul Weingartner	Austria	
Natuhiko Yosida	Japan	

List of the members of the Sectional Programme Committees:

Section 1:

Solomon Feferman	USA	Chairperson
Per Martin-Löf	Sweden	
Dana Scott	USA	
A.S. Troelstra	The Netherlands	

Section 2:

Paul C. Eklof	USA	Chairperson
Wilfrid A. Hodges	England	
Alistair Lachlan	Canada	
Michael Morley	USA	

Section 3:

Jens Erik Fenstad	Norway	Chairperson
Peter G. Hinman	USA	
John C. Shepherdson	England	

Section 4:

Andras Hajnal	Hungary	
Menachim Magidor	Israel	
Donald Martin	USA	Chairperson

Section 5:

Hans Kamp	England	
Charles D. Parsons	USA	Chairperson
Bas van Fraassen	USA	
Ryszard Wojcicki	Poland	

Section 6:

Peter Gärdenfors	Sweden	
Carlos U. Moulines	Mexico	
Alan E. Musgrave	New Zealand	
Wolfgang Stegmüller	West Germany	Chairperson

Section 7:

Richard C. Jeffrey	USA	Chairperson
Henry Kyburg, Jr.	USA	
Hugh Mellor	England	
Ilkka Niiniluoto	Finland	

Section 8:

Alberto J. Coffa	USA	
Adolf Grünbaum	USA	Chairperson
Erhard Scheibe	West Germany	
Abner Shimony	USA	

Section 9:

Morton Beckner	USA	
I.T. Frolov	USSR	Chairperson
Regina S. Karpinsky	USSR	
Stephan A. Pastushny	USSR	

Section 10:

Gerhard H. Fischer	Austria	
Duncan Luce	USA	
Patrick Suppes	USA	Chairperson

Section 11:		
Bengt Hannson	Sweden	
J. Martin Hollis	England	Chairperson
Raimo Tuomela	Finland	
Section 12:		
Max J. Cresswell	New Zealand	Chairperson
Barbara Partee	USA	
Armin von Stechow	West Germany	
Section 13:		
Robert Butts	Canada	
Ernan McMullin	USA	Chairperson
Jürgen Mittelstrass	West Germany	
Section 14:		
Evandro Agazzi	Switzerland	Chairperson
Georg von Wright	Finland	

List of the members of the Local Organizing Committee:

Curt Christian	Vienna	
Johannes Czermak	Salzburg	
Georg J.W. Dorn	Salzburg	Secretary
Gerhard Frey	Innsbruck	
Rudolf Haller	Graz	
Edgar Morscher	Salzburg	
Christine Pühringer	Salzburg	
Wolfgang Stegmüller	Munich	
Paul Weingartner	Salzburg	Chairperson
Gerhard Zecha	Salzburg	

CONTENTS

5. PHILOSOPHICAL LOGIC

6. GENERAL METHODOLOGY OF SCIENCE

7. FOUNDATIONS OF PROBABILITY AND INDUCTION

8. FOUNDATIONS AND PHILOSOPHY OF THE PHYSICAL SCIENCES

9. FOUNDATIONS AND PHILOSOPHY OF BIOLOGY

10. FOUNDATIONS AND PHILOSOPHY OF PSYCHOLOGY

11. FOUNDATIONS AND PHILOSOPHY OF THE SOCIAL SCIENCES

12. FOUNDATIONS AND PHILOSOPHY OF LINGUISTICS

13. HISTORY OF LOGIC, METHODOLOGY AND PHILOSOPHY OF SCIENCE

14. FUNDAMENTAL PRINCIPLES OF THE ETHICS OF SCIENCE

0. Inaugural Address

Barcan Marcus et al., eds., Logic, Methodology and Philosophy of Science VII

ON SCIENTIFIC INFORMATION, EXPLANATION AND PROGRESS

STEPHAN KÖRNER

Yale University, New Haven, U.S.A.

The main purpose of this essay is to consider some aspects of the informative and explanatory function of science and their relevance to our understanding of scientific change and progress. The essay begins by considering the general notion of progress (Section 1). There follows a discussion of the informative function of scientific theories, especially their capacity to yield information not only to experts, but also to semiexperts and laymen (Section 2); and of the explanatory function of scientific theories, especially its dependence on certain supreme principles which govern a person's thinking (Section 3). In the light of this discussion the nature of informative and of explanatory progress and of possible conflicts between them are considered (Section 4). The concluding section briefly compares the views, outlined in this essay, with some influential, contemporary doctrines on the nature of scientific change (Section 5).

1. On the general notion of progress

The notion of any progressive process presupposes its division into phases and a respect in which of any two phases one surpasses the other, is surpassed by it or does not differ from it. A weak concept of progress, which on the whole will be sufficient for our purpose, can be defined as follows: (a) To every phase — except the last, if any — there exists at least one succeeding phase which surpasses all predecessors; (b) the last phase, if any, surpasses all its predecessors. This notion of progress may be strengthened by excluding the possibility of regressive phases, i.e. of phases which are surpassed by any of their predecessors. It may be further strengthened by excluding phases of stagnation, i.e. by requiring that every phase be surpassed by its immediate successor. To require the former strengthening, is to require that the phases of a progressive process be linearly ordered, to require in addition also the latter strengthening, is to require that the linear ordering be strict.

In order to believe that a process is progressive in any of the mentioned senses of the term, it is not necessary to have a clear knowledge of its division into phases, the respect in which the phases are compared with each other or the relation by means of which the comparison is made. Indeed some historically important conceptions of progress involve only a very fragmentary knowledge of these features. Thus a believer in divine providence may from the fact of social change, from his belief in an infinitely benevolent and powerful creator and from the inability of human beings to understand his ways, infer that the process which is human history is progressive, that, consequently, it is somehow divided into phases, that there exists a relation in terms of which the succession of the phases is progressive, but that this relation is either unknowable or as yet not known. Again, a biologist who accepts an early version of Darwin's theory of natural selection may similarly believe in the existence of progress, defined by an ordering relation which is far from clear. He may, for example, hold that "as natural selection works solely by and for the good of each being, all corporeal and mental endowment will tend to progress towards perfection"[1]. One of the difficulties of understanding "a progress towards perfection" — whether guaranteed by theology or biology — lies in the manner in which progressive or regressive changes in different dimensions are combined into one linear sequence of phases. Even the problem of so combining such changes in two dimensions, e.g. the dimensions of information and explanation, may admit only highly artificial solutions (and will not be discussed).

The process to be considered in this essay is scientific change, the phases of which are periods of time during which scientific experts in various fields of science accept certain theories. The separation of two neighbouring phases is marked by the empirical fact that at least one theory which in one phase is accepted by some experts, is in the other not accepted by any experts. There is no need to make the unrealistic assumption of an obvious and sharp demarcation line between scientific theories and other systems of belief, between experts and laymen or between the acceptance and abandonment of a theory by an expert. Such an assumption would be

[1] CHARLES DARWIN, *The Origin of Species* (1st edition, London 1859) quoted from 6th edition (London, 1894) Chapter XV, p. 428.

equally unrealistic in other fields of human endeavour, e.g. in the law where it is readily admitted that no sharp lines can be drawn, between legal rules and other social rules of conduct, between legal experts and laymen in any branch of the law or between a valid law and one that has become obsolete.

2. On the informative function of scientific theories

Even a highly technical and specialized theory may give new information not only to the experts who have grasped the theory's logico-mathematical structure and conceptual net, but also to others who have little or no understanding of these features of the theory. An expert nuclear physicist, his theoretically semiexpert assistant and a layman advocating or opposing unilateral, nuclear disarmament may share information which was unavailable before the advent of the theory. That this is so, follows from the relation which holds between on the one hand the expert's theoretical thinking, which makes use of the theory and the language in which it is expressed, and on the other hand commonsense thinking, expressed in the common language shared by experts and others. The nature of this relation constitutes an old and persistent problem. Among the proposed solutions are Plato's theory of the μέθεξις of the perceptual world, which "tumbles about between being and not being", in the world of Forms, which is reality; Descartes' opposition of "the fluctuating trust in the senses" to the allegedly achieved certainty in mathematics, which by the method of doubt can be extended to other regions of thinking; more recently, Frege's attempt at "purifying" ordinary thought and language and thereby making them scientific; and Wittgenstein's condemnation of his attempt as wholly misguided. For our purpose it is necessary to characterize, however briefly, the gap between commonsense and specialized theoretical thinking and the manner in which it is bridged when "grey theory" is applied to "life's green tree" and, in being applied, increases our knowledge about it[2].

Commonsense thinking admits of concepts which are inexact in the sense of having border-line cases and is governed by principles of deducibility and consistency which lack the precision of the corresponding principles of a formal logic. As against this, theories, especially those which are

[2] For a full account see *Experience and Theory* (London, 1966), for a brief synopsis see "Science and the Organization of Belief" in *Science, Belief and Behaviour*, Essays in honour of R.B. Braithwaite, edited by D.H. MELLOR (Cambridge, 1980).

mathematically formulated, are embedded in a logic which, as Frege insisted, excludes inexact concepts and comprises precise rules of consistency and deducibility or, at least, rules which are more precise than the corresponding rules of commonsense thinking. Theories are normally embedded in a version of classical logic. However, even in the comparatively rare cases in which they are embedded in a version of a constructivistic, quantum or some other logic, this logic requires exactness of concepts and conformity to principles of deducibility which diverge from the corresponding principles of commonsense.

If a theory is mathematically formulated, then the mathematical apparatus needed for this formulation imposes additional constraints, by which the theory is even further removed from commonsense thinking. It normally involves the introduction of numbers — in the case of physical theories usually a system of real numbers — and thereby a concept of infinity which is absent from commonsense thinking. Different theories, e.g. an economic and a physical theory, may be subject to the same logico-mathematical constraints.

The transition from commonsense to theoretical thinking — and the gap between them — is not only due to the imposition of logico-mathematical constraints. It also results from modifications of the commonsense conceptual net, which may be divided into 'deductive abstraction' and 'theoretical innovation'. Deductive abstraction is the elimination of concepts from the commonsense conceptual net because they are unnecessary for the purposes of the theory or because they (or concepts resembling them) are definable in terms of the remaining concepts — a procedure exemplified in classical particle dynamics and implied by Locke's distinction between primary and secondary qualities. Theoretical innovation is the introduction into a theory of concepts which are not contained in the commonsense conceptual net, e.g. of the concepts of momentary acceleration into classical particle dynamics or of entropy into thermodynamics.

To distinguish between commonsense and theoretical thinking is not to deny their interaction, in particular the influence of the latter on the former. Thus, there can be little doubt that post-Galilean commonsense has been influenced by Galilean physics, that post-Einsteinian commonsense has been influenced by Einsteinian physics or that quantum-mechanical concepts may not in some thinned out version seep into the commonsense thinking of future generations. Yet, however this may be, the concepts of a theory are not identical, but merely *for certain purposes and in certain contexts* identifiable with certain commonsense concepts. A simple example of such an as-if-identification is the replacement of the empirical concept of a triangle and of its instances by the Euclidean concept of a triangle and its instances.

Let us, for the sake of brevity, call '(theoretical) idealization' the replacement of commonsense concepts, particulars and propositions by the concepts, particulars and propositions of a theory with which they are — for the purpose and in the context of the theory's application — identifiable. And let us call the converse replacement 'de-idealization'. The procedure of increasing commonsense information, expressed in common language, by means of a theory, which may be highly specialized and technical, can then, somewhat schematically, be described as consisting in the following steps: (i) the description of a state of affairs by means of commonsense concepts and propositions; (ii) the theoretical idealization of this description; (iii) intratheoretical — deductive or probabilistic reasoning, leading from the theoretical, idealized state of affairs to another idealized state of affairs; (iv) de-idealization of this state of affairs into a state of affairs which is again described by means of commonsense concepts and propositions; (v) statement of the connection between the original state and the final state of affairs by means of commonsense concepts and propositions. The last mentioned statement is intelligible not only to the theoretical expert (e.g. a nuclear physicist), but also to somebody who has a limited knowledge of the theory (e.g. the physicist's technical assistant) or none at all (e.g. a proponent of nuclear disarmament who knows no theoretical physics). To put it more precisely, the connection between the original and the final commonsense proposition, though dependent on the context and purpose of the intervening steps, does not depend on their being taken and does not depend on understanding or accepting the theory.

Because of the gap between commonsense and theory — a gap bridged by idealization and de-idealization — one must distinguish between on the one hand the theory and its logical consequences and on the other hand the information which, though provided by the theory, is logically independent of it and may survive its rejection (e.g. because of the continued reliance on certain observations or experiments). It follows that whereas the set of the logical consequences of two inconsistent theories (together with suitable initial conditions) is necessarily inconsistent, the commonsense information, provided by them may be consistent. This is so, for example, if the inconsistency of the theories is due to the application of concepts which do not occur in the commonsense information, provided by the theories.

The trustworthiness of the information which a scientific theory provides to an expert who accepts it and a layman who trusts the expert, may be supported by a variety of reasons, in particular by appeals to arguments, put forward by philosophers of science holding an inductivist, deductivist or other view of scientific thinking. In accordance with these reasons the store of science-provided, commonsense information which at any time is being trusted by a person or group of persons, will vary in extent. In attempting to reduce it to a reasonable minimum, one might well recall the sceptical doctrine to the effect that even a sceptic must base his actions on some undogmatic belief. And one might be helped by the remark of Sextus Empiricus that a true sceptic "adhering to appearances" lives "in accordance with the normal rules of life, undogmatically seeing that we cannot remain wholly inactive"[3]; or by the remark of Hume that "none but a fool or madman will ever pretend to dispute the authority of experience, or to reject that great guide of human life ..."[4].

It does not matter for our present purpose which method is chosen for the demarcation of trustworthy and trusted commonsense information or, more specifically, of its science-provided component. Nor does it matter whether agreement about a chosen method would guarantee agreement about the result of following it. What is important and needs to be emphasized, is that whatever method is chosen and whatever result is achieved by applying it, science-provided commonsense information forms part of trustworthy and trusted commonsense information and that the development of science contributes to changes in it. Thus even a follower of Hume or of Sextus Empiricus is likely to act on the commonsense information, provided by nuclear physics, e.g. by avoiding a place on which an atomic bomb is about to be dropped.

3. On the explanatory function of scientific theories

Scientific thinking is subject not only to specific constraints, which distinguish it from commonsense thinking, but also to constraints, to which all thinking about the public world of intersubjective experience is subject. For the present purpose it will be sufficient to give a brief indication of these general constraints and of the possible variety of principles conforming to them[5]. Before doing so, it seems worthwhile to emphasize that while

---

[3] *Outlines of Pyrrhonism* I, 23.

[4] *An Enquiry Concerning Human Understanding*, Sect. IV, Part II, 32.

[5] *See Categorial Frameworks* (Oxford, 1970); for a brief synopsis see the paper referred to in Footnote 2.

a statement to the effect that a certain principle is accepted by some or all human beings, expresses an anthropological and, hence, an empirical fact, it does not follow that the accepted principle itself is empirical. That (almost) all human beings accept (a version of) the principle of non-contradiction, is no less an empirical statement than the statement that some human beings reject the principle of excluded middle. Yet neither principle is empirical.

All thinking about the world of intersubjective experience involves a distinction between particulars and attributes, a distinction between consistent and inconsistent propositions (and, hence, the acknowledgment of a relation of logical deducibility) and a distinction between the merely subjectively and the intersubjectively given. In order to be brief and to avoid repetition of what I have elsewhere argued at length, I shall set down my views on the nature of these distinctions by expressing *qualified* agreement with seminal thinkers who attached great importance to them: I accept on the whole Frege's distinction between concept and object[6], but disagree with his claim that it can be correctly drawn in one way only. I accept on the whole what Aristotle says about consistency and deducibility, but disagree with his claim and the claim of most of his successors that there is only one logic. (I do hold that there is a minimal logic containing the weak law of non-contradiction, according to which not every proposition is true.) I accept on the whole what Kant says about the conferment of intersubjectivity or objectivity on subjective experience by the application of Categories, but disagree with his claim that there is only one set of them.

A further distinction, though perhaps less universal than those mentioned so far, is the distinction between dependent particulars, the existence of which is dependent on the existence of other particulars and independent particulars, the existence of which is not so dependent. To define independent particulars in this fashion is to agree on the whole with Aristotle's definition of 'primary substance' (see *Categories* II). Yet this does not mean, and is here not intended to mean, that only Aristotle's way of distinguishing between dependent and independent particulars is legitimate. From the point of view of this essay one must, for example, acknowledge two contrasting, general conceptions of substance — one, according to which being a substance implies not being subject to change, the other, according to which transsubstantiation is possible.

In adopting a specific way of distinguishing between particulars and

[6] See "Begriff und Gegenstand", *Vierteljsch. f. Wissch. Philosophie* (vol. 16, 1892).

attributes and of making the other distinctions which have been mentioned as involved in thinking, a person *ipso facto* accepts a set of principles which together determine what, in some agreement with philosophical tradition, may be called his 'categorial framework'. It can be characterized as a differentiation of experience into particulars and attributes; a deductive logic containing principles determining consistency and deducibility; a set of intersubjectivity concepts (the application of which confers intersubjectivity on what is subjectively given); and a categorization of the intersubjective particulars into maximal kinds, with each of which are associated constitutive principles demarcating its membership and individuating principles for the identification of its distinct members. A categorial framework may in addition contain principles defining independent, as opposed to dependent, particulars, as well as principles admitting the use of various auxiliary concepts, which (like e.g. mathematical or theoretical idealizations) serve the application of intersubjectivity concepts.

The logical and non-logical principles which determine a person's categorial framework are for him supreme, in the sense that he requires all his beliefs to be consistent with them. A proposition which (by the principles of his logic) is incompatible with his supreme logical principles, is for him logically meaningless; a proposition which (by the principles of his logic) is incompatible with his supreme non-logical principles is for him 'metaphysically meaningless' or lacking in explanatory power. The criteria of metaphysical meaningfulness or explanatory power may differ from person to person and, hence, from one group of scientists to another. For it is an anthropological fact that a principle which is accepted as supreme by one person, and thus involved in the constitution or individuation of at least one maximal kind of his categorial framework (e.g. things, things and events, things and persons, atoms, monads etc.), need not be so accepted by everybody else. Thus for Kant, but not also for Einstein, a belief that spatial relations are non-Euclidean, would render a physical theory metaphysically meaningless and deprive it of any explanatory power. Again, Leibniz, Kant and Einstein, but not also Bohr and Born, would reject any physical theory which implied the physical possibility of discontinuous change as incapable of explaining natural phenomena — though they might admit that such a theory could yield important information about them. (While in this essay 'being metaphysically meaningful' is used as a necessary and sufficient condition of 'having explanatory power', it could, without affecting the main argument, be used as a merely necessary condition — in addition to other such conditions, e.g. 'being simple' or 'being aesthetically satisfying'.)

A person may be more or less aware of the supreme principles which determine his categorial framework and more or less able to formulate them. He may adhere to them with more or less conviction and even waver between the acceptance or rejection of a supreme principle. The situation is no different in the case of a person's grammar or his morality. Although the principles determining a person's categorial framework express some of his beliefs, they have, like other beliefs, also a practical or regulative function. A person who believes that something is the case will behave as if it were the case. Thus a scientist who believes that *natura non facit saltus* will try to construct theories which are consistent with this principle.

A person may, of course, also act and, indeed prudently act, *as if* he believed something which he does not believe. Thus it may in some circumstances be prudent for the driver of a motorcar to drive as if all other drivers were drunk, even if he does not believe this. And it may be prudent and praiseworthy for a scientist who believes that *natura non facit saltus* to make contributions to a theory which, though providing new information is based on the opposite assumption and to conduct empirical and thought-experiments as if he believed it. Although it is important to distinguish between the purely regulative function of a supreme principle, which is used but not believed, and the cognitively grounded regulative function of a supreme principle, which is used because it is believed, one may for some limited purpose be justified in examining the regulative function of supreme principles as such, i.e. without raising the question of their being cognitively grounded.[7]

4. On informative and explanatory progress

The preceding discussion of the general notion of progress, of the phases of scientific change, of the commonsense information provided by scientific theories and of their explanatory function, has some fairly obvious consequences for a discussion of informative and explanatory progress. A phase in the change of science-provided commonsense information has been characterized by the theories which the scientific experts accept in that phase, insofar as these theories provide commonsense information

[7] The distinction between cognitively grounded and purely regulative supreme principles is anticipated by Kant's doctrine of the (purely) regulative function of the Ideas. For a discussion of the regulative function of metaphysical principles in physics, see 'On Philosophical Arguments in Physics' in *Observation and Interpretation* (Bristol, 1957).

which is capable of being shared by experts and laymen. A historian of science or more specifically, a historian inquiring into the effects of science on society, might reasonably wish to work with a more specific concept of information which during a phase of scientific development has been considered trustworthy. Among the questions which he would have to face is the problem of dealing with mutually inconsistent informative propositions belonging to the same phase and the problem of dealing with mutually inconsistent informative propositions belonging to two different phases. As regards the former problem, it would seem reasonable to eliminate any two mutually inconsistent propositions belonging to the same phase. The decision is more difficult, if two mutually incompatible informative propositions belong to different phases. An overall, if somewhat crude, solution of such conflicts would be to eliminate the proposition belonging to the earlier phase, if the proposition belonging to the later phase is accepted by all the experts in that phase, and otherwise to eliminate both propositions.

Whatever procedure is adopted by a historian for whom scientific change coincides with change of theory-provided, commonsense information, it would be highly surprising if he did not come to the conclusion that scientific change has been progressive and that — provided that science will continue to be trusted — it is likely to make progress in the future. This judgment is supported by the point, made earlier, that even logically inconsistent theories may provide logically consistent commonsense information. In this connection it seems worthwhile to distinguish between three kinds of informative progress from one phase to another, namely theoretically unified progress, where in the later phase one of a set of information-providing theoreies logically implies the others, theoretically consistent progress, where the set of the theories is internally consistent, and theoretically conflicting progress where at least two of the theories are inconsistent with each other.

That the possibility of theoretically conflicting informative progress is by no means far fetched, is borne out by the contemporary coexistence of general relativity theory and of quantum mechanics in its dominant interpretation. In trying to exhibit this fact, one can hardly do better than to quote the concluding comment from a lecture given in 1982 on "Max Born and the Statistical Interpretation of Quantum Mechanics" by A. Pais[8]. The comment is itself a quotation of a remark by Norbert Wiener,

[8] *Science*, vol. 218, 17 Dec. 1982.

published in 1956: "It has been well said that the modern physicist is a quantum theorist on Monday, Wednesday and Friday" (when he assumes that *not* all laws of nature are expressible as differentiable functions) and "a student of gravitational relativity on Tuesday, Thursday and Saturday" (when he makes the opposite assumption) while "on Sunday he is neither, but is praying to his God that someone, preferably himself, will find the reconciliation between these two views".

"Other things being equal", theoretically unified informative change is preferred to theoretically consistent change, which in turn is preferred to a theoretically conflicting one. Yet there are differences of opinion as to the things whose equality matters. Thus, a radical positivist, who regards any distinction between the explanatory and informative function of scientific theories as spurious, will require that the preference for theoretically unified information does not involve any loss of theory-provided information. Again, a philosopher, who identifies theoretical unification with theoretical explanation, might in some cases be willing to aquiesce in such a loss, provided that he can regard it as provisional. Lastly, a scientist who distinguishes between information provided by a theory which has explanatory power, in the sense of conforming to his categorial framework, and a theory which lacks explanatory power, is likely to prefer a theoretically merely consistent — or even a theoretically conflicting — informative change which is brought about by theories possessing explanatory power, to a theoretically unifying change brought about by a theory lacking it.

Explanatory progress in science differs from theory-provided, informative progress, which does not depend on the acceptance or even the understanding of the theories by which it is provided, in that it not only involves an understanding of the theories but also of their conformity to a categorial framework. A person's information, provided by a theory, has explanatory power for him if, and only if, he understands the theory and correctly judges it to conform to the supreme cognitive principles determining his categorial framework. If Norbert Wiener's expert physicist, who uses both quantum theory and general relativity, is one of Einstein's followers, then he will regard only the latter theory as explanatory and will pray for a unifying theory conforming to the supreme principles to which that theory conforms. If he is one of Bohr's followers, then he will regard only the former theory as explanatory and will adjust his prayer accordingly.

Another way of characterizing this contrast is to say that each of these two physicists acts partly in accordance with his own supreme principles and, hence, the rules of conduct grounded in them; partly in accordance with the other's supreme principles, used in a purely regulative fashion. The possibility of a purely regulative use of principles was clearly recognized by Leibniz and exploited in his mathematical and physical thinking, in which his notion of well-founded fictions plays an important role. They enabled him to make contributions to the mechanistic physics of his day, while denying it any explanatory power[9]. Theory-provided, informative progress may involve explanatory stagnation and even regress. And one person's explanatory progress may be another's explanatory regress.

The conflict engendered by acquiring new information through a new theory which does not conform to one's categorial framework may be solved either by continuing to search for a theory which, while providing no less information than the new theory, conforms to the old categorial framework; or else by abandoning the old categorial framework in favour of a new one to which the new theory conforms. A scientist who adopts the former solution hopes for, and sometimes achieves, informative progress within his old categorial framework or, briefly, intracategorial progress. A scientist who adopts the latter solution hopes for and, sometimes achieves, informative progress within the new categorial framework or briefly transcategorial progress. Einstein believed until his death in intracategorial progress. He was convinced that "*one should start all over again*, as it were, and endeavour to obtain the quantum theory as a by-product of general relativistic theory or a generalization thereof"[10].

The contrast between intracategorial and transcategorial informative progress is clearly analogical to the contrast between orderly or non-revolutionary political progress within the constraints of a country's constitution and revolutionary political progress, involving the abandonment of a country's constitution. And just as a political change which is non-revolutionary in one country, may be revolutionary in another, so one scientist's intracategorial or orderly informative progress, i.e. progress within the constraints of his categorial framework, may be another scientist's transcategorial or revolutionary progress, i.e. progress involving the replacement of his old categorial framework by a new one. Thus Einstein "deprecated the idea that relativity is revolutionary and stressed that his theory was the natural completion of the work of Faraday, Maxwell and Lorentz" while "other physicists will quite reasonably object that the

---

[9] See e.g. Sections 80, 81 of the *Monadology* and the letter to Varignon of 20 June, 1702, *Mathem. Schriften* edited by C.I. GERHARDT, vol. 4, pp. 106f.

[10] See A. PAIS, *Subtle is the Lord* — The Science and the Life of Albert Einstein (Oxford Univ. Press, 1982) p. 461.

abandonment of absolute simultaneity and of absolute space are revolutionary steps"[11]. A Socratic dialogue, conducted with physicists belonging to these two groups would have shown that before those belonging to the first group became converted to the general theory of relativity, they accepted a loose version of the Newtonian (or Kantian) categorial framework; whereas before those belonging to the second group became converted to the theory, their categorial framework was more rigidly or closely Newtonian.

5. Some brief remarks comparing the position of this essay with other views

In order to understand the use of scientific theories in providing information, shared by experts, semi-experts and laymen, it is not sufficient to acknowledge that science involves idealization. What must in addition be shown, is how, in the case of a given theory, the idealization involves a transition from the logico-mathematical structure and conceptual net of commonsense or semi-expert thinking to the different logico-mathematical structure and conceptual net of the theory; and how the gap which is thereby created is bridged by as-if identifications within the limits of the context and purpose of the theory's employment. I used to object to the neglect by Karl Popper (and other deductivist philosophers of science) of the role played by idealization and de-idealization in the application of theories[12]. It therefore gave me some satisfaction to read in Popper's *Replies to My Critics*[13] his announcement "of an important principle", namely that "*all good science consists, and all good philosophy consists, of lucky oversimplification* or, if you prefer the term, *idealization*" (italics in the text). While one can no longer object to Popper's neglect of the important role, played by idealization in the construction and application of scientific theories, one must, I believe, still object to the vagueness of his account of it.

For Stegmüller and his 'structuralist' collaborators the gap between the logico-mathematical structure and conceptual net of a theory and the information which the theory provides to experts, semi-experts and non-experts alike, is not bridged by as-if identifications in the context and

[11] *Op. cit.* in A. Pais, see Footnote 10.

[12] See Section 2 and Chapter XII, *op. cit.* in Footnote 2.

[13] *The Philosophy of Karl Popper*, vol. II (La Salle, Illinois, 1974), p. 976.

for the purpose of the theory's application. It is bridged by supplementing a set-theoretical axiomatization of the theory in the manner of Suppes, by an informal semantics in the manner of Sneed. This semantics has the task of supplying the theory with "an open set I of *numerous, partly overlapping, intended applications*, all of which are 'anchored' in a paradigmatic subset I_0 of I"[14]. Stegmüller is cautious enough to expect "in the near future only a very fragmentary realization of his programme" (*op. cit.* p. 28), which he calls the "Suppes–Sneed programme" (*op. cit.* p. 84). At the moment I can see no reason for abandoning my account of the relation between a theory as formulated by an expert and the theory-provided information shared by experts, semi-experts and non-experts, in favour of an account in accordance with the Suppes–Sneed programme. Whereas the latter account assumes that only one language is involved, the former involves two languages, an exact theoretical and an inexact non-theoretical one, each of which has its own syntax and semantics. It may be worth noting that Suppes too is skeptical of anyone's providing an informal semantics as conceived by the structuralists, because he finds a very large gap "between the theoretical literature in any developed branch of science and the untraversable thicket of technical language used in the corresponding experimental work"[15].

Just as an adequate account of scientific information and informative progress must not neglect the role of the idealization of concepts and propositions, so an adequate account of scientific explanation and explanatory progress must not neglect the role of the supreme principles defining a person's categorial framework. Towards the end of his important *The Structure of Scientific Revolutions*, Thomas Kuhn remarks that in order to make the transition from Newton's to Einstein's universe "the whole conceptual web whose strands are space, time, matter, force, and so on, had to be shifted and laid down again on nature whole"[16]. This description — with its various elaborations — is incomplete in that it fails to account for the division of scientists and historians of science into two groups: one, to which Kuhn belongs, which regards the transition as revolutionary; the other, to which Einstein belongs, which regards it as normal progress. (See Section 4 and Footnote 10). The division can be explained by reference to the categorial frameworks, accepted by the scientists before and after their abandoning Newton's for Einstein's Universe. The transition, as has been

---

[14] Wolfgang STEGMÜLLER, *The Structuralist View of Theories* (Berlin, 1979), p. 27.

[15] *Profiles — Patrick Suppes*, edited by R.J. BOGDAN (Dordrecht, 1979) p. 208.

[16] *Op. cit.* (Chicago, 1963) p. 148.

argued earlier, was transcategorial for the first group and intracategorial for the second.

From what has been said about the informative and explanatory function of scientific theories, it follows that any attempt at defining scientific progress solely in terms of informative progress, is based on a misunderstanding of the relation between the two functions. An example of such a misunderstanding is Lakatos' conception of research programmes and of the manner in which they are, or should be, ranked. A research programme is essentially a set of — cognitively grounded or pure — regulative principles for the construction of theories. Of any two research programmes one is superior to the other if, and only if, the theories conforming to it have, compared with theories conforming to its competitor, "corroborated excess empirical content", i.e. provide more trustworthy information in a special sense of the terms which need not be elaborated here. It must, however, be emphasized that this special sense does not include any reference or appeal to a theory's explanatory power, as distinguished from the information provided by it[17]. Lakatos' ranking of scientific theories might be defensible if scientists were *only* aiming at information and if their regulative principles for the construction of theories were chosen *only* for their capacity to lead to the construction of theories providing ever increasing information. But, as e.g. Einstein's attitude to quantum mechanics shows, at least some scientists search for theories which, in addititon to being informative, have explanatory power, in the sense of conforming to the supreme principles constituting the scientists' categorial frameworks.

---

[17] See Imre LAKATOS, "Methodology of Scientific Research Programmes", in: *Criticism and the Growth of Science* edited by Lakatos and Musgrave (Cambridge Univ. Press, 1970), p. 116 and *passim*.

1. Proof Theory and Foundations of Mathematics

Barcan Marcus et al., eds., Logic, Methodology and Philosophy of Science VII

THE TYPE THEORETIC INTERPRETATION OF CONSTRUCTIVE SET THEORY: INDUCTIVE DEFINITIONS

PETER ACZEL

Dept. of Mathematics, Manchester Univ., England

Introduction

This is the third paper on the type theoretic interpretation of constructive set theory. The previous two are [1] and [2].

Constructive set theory originated with MYHILL'S paper [7]. It is a possible framework for the formalisation of constructive mathematics as practised by BISHOP and his co-workers (see [3], [4] and [8]). In [1] I formulated a system CZF + DC of constructive set theory which is essentially an extension of MYHILL'S system CST. I also gave an interpretation of CZF + DC in an extension of the framework of intuitionistic type theory as presented in [5]. In [2] the interpretation was reworked in the modified framework of type theory presented in [6]. In addition I showed that further axioms, the choice principles $\Pi\Sigma I$-AC and $\Pi\Sigma I$-PA are valid in the interpretation.

The main aim of the present paper is to formulate a new axiom for constructive set theory and show that it is valid in the interpretation. This new axiom I call the regular extension axiom, REA for short. With this axiom it is possible to show in constructive set theory that various inductively defined classes are actually sets. For example if A is a set and B_a is a set for each $a \in A$ then there is a smallest class $W = W_{a\in A} B_a$ such that if $a \in A$ and f: $B_a \rightarrow W$ then $\langle a, f\rangle \in W$. Assuming REA $W_{a\in A} B_a$ is a set. If this W operation is added to the Π and Σ operations used in formulating $\Pi\Sigma I$-AC and $\Pi\Sigma I$-PA then we obtain the axioms $\Pi\Sigma WI$-AC and $\Pi\Sigma WI$-PA which are also shown to be valid in the interpretation.

The proof of the validity of these new axioms depends essentially on the rules for the form of type $(Wx \in A)B(x)$ that were introduced in [6]. This form of type is simply a type theoretic version of the set theoretical W operation described above. $(Wx \in A)B(x)$ is a type W having the introduction rule

$$\frac{a \in A \qquad \begin{array}{c}(y \in B(a))\\ f(y) \in W\end{array}}{\sup(a, f) \in W}$$

The elimination rule for this type provides for definition by transfinite recursion on W and expresses that the elements of W are inductively generated using the introduction rule. Just as with the other basic type forming operations such as Π and Σ the type forming W operation should be reflected in the type of small types. So there is a rule

$$\frac{A \in U \qquad \begin{array}{c}(x \in A)\\ B(x) \in U\end{array}}{(Wx \in A)B(x) \in U}\ .$$

In Section 1 I review the formal system CZF + DC and its type theoretic interpretation $\underset{\sim}{V}$. I also take the opportunity to discuss informally the motivation for the interpretation. In fact it may be understood as a constructive version of the iterative notion of set used in explaining the meaning of classical set theory. The power set axiom and the full separation scheme are not theorems of CZF + DC and are not expected to be valid in $\underset{\sim}{V}$ unless strongly impredicative principles are added to type theory. I end the section with Theorem 1.2 which is proved in Appendix 1. This theorem provides rules that could be added to type theory so that the validity of the powerset axiom and the full separation scheme could then be obtained. There is no intended suggestion that these rules make any constructive sense. Nevertheless they do encapsulate in a convenient way the idea behind impredicativity. They seem to resemble Russell's axiom of reducibility. The type Ω should be compared with the subobject classifier of topos theory.

Section 2 consists of a review and some examples of the inductive definition of classes in CZF. The examples will play an essential role in the later sections.

The notion of base is needed to formulate the choice principles shown to be valid in $\underset{\sim}{V}$. In Section 3 this notion is discussed in some detail as it seems to be a fundamental notion for constructive set theory. The section ends with a result showing that the choice principles are equivalent to the slightly simpler axioms $\Pi\Sigma$-AC and $\Pi\Sigma$-PA.

The purpose of Section 4 is to show that in CZF + DC + $\Pi\Sigma$-AC an inner model of CZF + DC + $\Pi\Sigma$-PA may be defined. This result gives an approach to the interpretation of the latter system that avoids the somewhat inconvenient rule for type theory of definition by transfinite recursion on the type of small types.

The formulation of the axioms REA, $\Pi\Sigma WI$-AC and $\Pi\Sigma WI$-PA is given in Section 5. The proofs that these axioms are valid in $\underline{V}$ may be found in Appendix 2. Also in Section 5 may be found a definition of the notion of a bounded inductive definition and a proof assuming REA that such definitions always define sets. Several of the examples of inductive definitions that were considered in Section 2 turn out to be bounded.

It has been my intention to made the body of this paper readable by someone who is not too familiar with all the details of [2]. So work which relies on such details has been confined to the two appendices.

The notion of an inductive definition plays a prominent role in this paper. I end this introduction with an informal review of the notion.

An inductive definition usually involves the characterisation of a collection of objects as the smallest collection satisfying certain closure conditions. Such a characterisation can be made explicit in one of at least two ways. The first way is to define the collection as the intersection of all collections that satisfy the closure conditions. Such an explicit definition is thoroughly impredicative in that the collection is defined using quantification over all collections. The second way is to build up the collection from below as the union of a hierarchy of stages. These stages of the inductive definition are indexed using some suitable notion of 'ordinal'. In case the inductive definition is finitary these 'ordinals' can be the natural numbers. But in general transfinite 'ordinals' are needed. In order for this approach to work the 'ordinals' must satisfy suitable closure conditions and must themselves be inductively generated.

It would seem therefore that if one wants to make sense of infinitary inductive definitions it is necessary at some point to make use of some impredicative definitions. This is certainly the way that inductive definitions are legitimated in classical mathematics. But there is something unsatisfying with this conclusion. Many inductive definitions can be intuitively understood directly in their own terms and impredicative definitions are only required in order to represent them within a particular framework such as classical set theory.

The paradigm for a direct understanding of an inductive definition is that for the collection of natural numbers, which is characterised as the smallest collection containing zero and closed under the successor function. In constructive mathematics the natural numbers are viewed as objects *constructed* according to the following two rules:

(1) 0 is a natural number.

(2) If n is a natural number then so is $s(n)$.

What a natural number is is an object constructed according to these rules. As an infinitary example of an inductive definition we may take a constructive version of the second number class of ordinals. These ordinals are constructed according to the rules:

(1) $\bar{0}$ is an ordinal.

(2) If α is an ordinal then so is $\bar{s}(\alpha)$.

(3) If $\alpha_0, \alpha_1, \ldots$ is an infinite sequence of ordinals then $\sup(\alpha_n \mid n = 0, 1, \ldots)$ is an ordinal.

It is implicit here, as in the rules for the natural numbers, that each ordinal uniquely determines the combination of rules that is used in its construction. In classical set theory the above notion of ordinal could be represented as the set that is the intersection of all sets X such that $\bar{0} \in X$, $\bar{s}(\alpha) \in X$ if $\alpha \in X$ and $\sup(\alpha_n \mid n = 0, 1, \ldots) \in X$ if $\alpha_0, \alpha_1, \ldots \in X$, where we could define $\bar{0} = \emptyset$, $\bar{s}(\alpha) = \{\alpha\}$ and $\sup(\alpha_n \mid n = 0, 1, \ldots) = \{\langle n, \alpha_n \rangle \mid n = 0, 1, \ldots\}$. In constructive set theory this set can be shown to exist using the new axiom REA introduced in Section 5. This axiom avoids the impredicativity of the power set axiom and the full separation scheme but is strong enough to entail the existence of sets such as the above. It is in type theory where certain inductive definitions are treated directly and it is by using the inductively defined form of type $(Wx \in A)B(x)$ that the axiom REA is shown to be valid in the type theoretical interpretation of constructive set theory.

1. CZF + DC and its type theoretic interpretation

The axiom system CZF + DC

For our purposes here we shall use the standard first order language for set theory having "$\in$" as the only non logical symbol. The system is based on intuitionistic first order logic with equality. The non logical axioms of CZF are extensionality, pairing, union and infinity, all formulated as usual. The axiom schemes of CZF are set induction, restricted separation, strong collection and subset collection. I use DC for the strong form of dependent choices scheme that has sometimes been called relative dependent choices. In detail these schemes are as follows:

Set induction

$$\forall x(\forall y \in x\, \phi(y) \supset \phi(x)) \supset \forall x\, \phi(x)$$

for all formulae $\phi(x)$.

Restricted separation

$$\forall a\ \exists b\ \forall x(x \in b \equiv x \in a\ \&\ \phi(x))$$

for all restricted formulae $\phi(x)$, where a formula is restricted if all its quantifiers are restricted, i.e. have one of the forms $\forall x \in y$ or $\exists x \in y$ where these abbreviate $\forall x(x \in y \supset$ or $\exists x(x \in y\ \&$ respectively.

Strong collection

$$\forall a(\forall x \in a\ \exists y\ \phi(x, y) \supset \exists b\ \phi'(a, b))$$

for all formulae $\phi(x, y)$, where $\phi'(a, b)$ abbreviates

$$\forall x \in a\ \exists y \in b\ \phi(x, y)\ \&\ \forall y \in b\ \exists x \in a\ \phi(x, y).$$

Subset collection

$$\forall a\ \forall a'\ \exists c\ \forall u(\forall x \in a\ \exists y \in a'\ \phi(x, y) \supset \exists b \in c\ \phi'(a, b))$$

for all formulae $\phi(x, y)$, where it should be stressed that u may be free in $\phi(x, y)$.

Dependent choices (DC)

$$\forall x(\theta(x) \supset \exists y\ (\theta(y)\ \&\ \phi(x, y))) \ \supset\ \forall x(\theta(x) \supset \exists z\ \psi(x, z))$$

for all formulae $\theta(x)$ and $\phi(x, y)$, where $\psi(x, z)$ expresses that z is a function, whose domain is the set of natural numbers, such that $z(0) = x$ and for every natural number n $\theta(z(n))\ \&\ \phi(z(n), z(n+1))$ holds.

Below I list some basic facts concerning CZF which may help the reader. Some more details may be found in [1].

(1) CZF with classical logic has the same theorems as ZF.

(2) The strong collection scheme is needed to derive replacement. The ordinary collection scheme would not appear to suffice when only restricted separation is available.

(3) The power set axiom is not an axiom of CZF and in fact it is not proveable. In its place is the subset collection scheme. This scheme is in fact equivalent to the single instance where $\phi(x, y)$ expresses that $\langle x, y \rangle \in u$. Using subset collection Myhill's exponentiation axiom can be derived. This axiom expresses that for any two sets a and b the set b^a of functions from a to b exists. Assuming PA (see Section 3) subset collection is equivalent to the exponentiation axiom. In CZF the full power set axiom can be derived from the assumption that $\{\emptyset\}$ has a power set.

(4) Many aspects of the informal development of classical set theory still apply when working informally in CZF. For example natural numbers, ordered pairs, relations and functions can be defined just as in classical set theory. The notion of class and the notation associated with it is a convenient tool in informal classical set theory which also carries over without difficulty to the informal development of CZF.

The iterative notion of set: A constructive version

The type theoretic interpretation of constructive set theory may be found in detail in [2]. Without assuming a full familiarity with the type theoretic framework I wish here to discuss informally the idea behind the interpretation. The classical iterative notion of set has been used to seek to explain the meaning of classical set theory and so to give an interpretation to ZFC. The idea is to seek a constructive version of the iterative notion of set, i.e. that notion that arises by iterating the notion 'set of' to get sets, sets of sets, sets of sets of sets, etc. Assuming that we had a general notion of 'set of objects', applicable to an arbitrarily given domain of objects, then the universe of iterative sets might be viewed as that domain of objects that is inductively generated by the single rule:

if A is a *set of* iterative sets then A is an iterative set.

The logical approach to the notion 'set of objects' is to treat sets as classes, i.e. extensions $\{x \mid \phi(x)\}$ of predicates ϕ where x ranges over the domain of objects involved. Because of Russell's paradox we know that in case the domain of objects is to be the universe of iterative sets then x cannot be allowed to range over the whole domain. The modern view is to use a cumulative transfinite version of Russell's theory of types. This view requires the notion of cumulative level. Iterative sets are arranged in such levels and each set of a given level has its elements at lower levels. The levels are cumulative in the sense that a set is at a given level whenever it occurs at any lower level. The class $\{x \mid \phi(x)\}$ is now only accepted as a set at a given level if x is understood to range over the sets at lower levels.

This conception of a universe of sets arranged in cumulative levels has been used to give a somewhat plausible interpretation for ZFC. But the interpretation is on the face of it highly non-constructive. In forming sets $\{x \mid \phi(x)\}$ the predicate ϕ may be defined using quantification over the universe. Such impredicativity makes it impossible to view the rule given earlier for generating iterative sets as a rule for constructing them. Also the notion 'set of' being used in the rule involves the notion of cumulative level of the universe and it is not clear how to understand this from the constructive point of view.

There is an alternative approach to the above iterative notion of set. This approach takes the rule, given previously as a rule for generating iterative sets, as a rule for *constructing* iterative sets. What is needed is a suitable notion of 'set of objects' for an arbitrarily given domain of objects. The logical treatment of sets as classes $\{x \mid \phi(x)\}$ will not do. Notice that in order to grasp such a class it is necessary to survey in some sense each object x in the domain of objects and determine if it satisfies the predicate. But for example the set of natural numbers $\{0, 2, 5\}$ is naturally grasped to be the result of combining into a whole the selection of the natural numbers 0, 2 and 5. The natural numbers not in $\{0, 2, 5\}$ do not need to be surveyed in order to grasp the set. This suggests that in grasping a set only those objects selected to be in the set should need to be surveyed. In general let us take a set of objects from some domain to be the result of combining into a whole the selection of those objects from the domain that are to be the elements of the set. The set may be written $\{a_i\}_i$ where the a_i's are understood to be the selected elements of the set. Sets are to be treated extensionally. Two sets $\{a_i\}_i$ and $\{b_j\}_j$ are extensionally equal if every a_i is equal to some b_j and every b_j is equal to some a_i. Notice that we have used a variable i to index the selections of the elements a_i of the set $\{a_i\}_i$. What can be the range of i? This needs careful consideration if we are to avoid circularity. It is no good stipulating that i can range over any set. An independent notion is needed. Fortunately there is a suitable notion available. This is the notion of type from the intuitionistic theory of types. So our answer is that i can range over a type I and the set $\{a_i\}_i$ should be written more explicitly as $\{a_i\}_{i \in I}$.

The iterative sets are now inductively generated using the rule that for each type I

$$\text{if } a_i \text{ is an iterative set for } i \in I \text{ then } \{a_i\}_{i \in I} \text{ is an iterative set.}$$

This rule would seem to be acceptable as a rule of construction. But in order to use it in the type theoretic framework so as to give an interpretation of the set theoretical language it is necessary to have a type of iterative sets. If all types I are allowed in forming iterative sets then the iterative sets themselves cannot be expected to form a type. Instead if I is required to a *small* type in forming the iterative set then we obtain a relativised notion of iterative set over the type U of small types, and we can have the type V of iterative sets over U.

The type U of small types is obtained by reflection on the basic forms of type. As presented in [2] these are N_0, N, $(\Pi x \in A)F(x)$, $(\Sigma x \in A)F(x)$, $A + B$ and $I(A, b, c)$ where A and B are types, F is a family of types over A and $b, c \in A$. The rules for forming small types stipulate that the above types are small provided that A is small and $F(x)$ is small for $x \in A$.

The rule that inductively specifies the type V may be given by the following scheme

$$\frac{I \in U \qquad a_i \in V(i \in I)}{\{a_i\}_{i \in I} \in V}\,.$$

It is this type V with the above introduction rule and a corresponding elimination rule of transfinite recursion on V that has been used in [1] and [2] to give the interpretation of constructive set theory. (Note that in [2] the letters U and V are interchanged and that $\{a_i \mid i \in I\}$ is used instead of $\{a_i\}_{i \in I}$. In [1] I used $(\sup i \in I)a_i$ for $\{a_i\}_{i \in I}$.)

In order to use V to interpret the language of set theory it is necessary to have relations on V of extensional equality and membership $=_{\text{ext}}$ and $\in_{\text{ext}}$. These are defined so that if $\alpha = \{a_i\}_{i \in I}$ and $\beta = \{b_j\}_{j \in J}$ then

$$\left(\alpha \underset{\text{ext}}{=} \beta\right) = \left[\forall i \in I\, \exists j \in J\left(a_i \underset{\text{ext}}{=} b_j\right) \mathbin{\&} \forall j \in J\, \exists i \in I\left(a_i \underset{\text{ext}}{=} b_j\right)\right]$$

and

$$\left(\alpha \underset{\text{ext}}{\in} \beta\right) = \left[\exists j \in J\, \alpha \underset{\text{ext}}{=} b_j\right].$$

I will use $\underline{V}$ for the interpretation of the language of set theory where the variables are taken to range over the type V, $=_{\text{ext}}$ and $\in_{\text{ext}}$ are used to interprete '$=$' and '$\in$' and the 'propositions as types' interpretation is used for the logical operations (see [2]).

THEOREM 1.1. ([2]) *$\underline{V}$ models* CZF + DC.

The following result is proved in Appendix 1 and for those familiar with type theory it goes some way to explaining why $\underline{V}$ does not model the power set axiom and the full separation scheme.

By the *absolute* separation scheme I mean that scheme concerning the type V which expresses that the set $\{x \in \alpha \mid \phi(x)\}$ can be formed for every set α and every extensional predicate ϕ that can be defined in the type theoretic framework (not only those predicates definable in the first order language of set theory).

THEOREM 1.2. *Working in the framework of type theory the following are equivalent*

(i) $\underline{V}$ *models the power set axiom and the absolute separation scheme.*

(ii) *There is a small type* Ω *and a predicate* T *on* Ω *such that* $T(a)$ *is small for* $a \in \Omega$ *and for each type* A,

$$(\exists a \in \Omega)(T(a) \equiv A)$$

is true.

2. Inductive definitions of classes

As we shall see below it is often natural to introduce a class by an inductive definition. In classical set theory inductive definitions are usually dealt with using transfinite recursion on ordinals. But a direct treatment is possible and even convenient. Moreover in constructive set theory the ordinals are not at all as well behaved as they are classically. For example if we define an ordinal to be a transitive set of transitive sets (as seems necessary if every set is to have an ordinal rank) then every subset of $\{\emptyset\}$ is an ordinal and without the power set axiom they do not form a set of ordinals.

In this section we shall work informally in CZF. The following definition and result may be found in 4.2 of [2].

DEFINITION 2.1. For any class Φ the class X is *Φ-closed* if $A \subseteq X$ implies $a \in X$ for every ordered pair $\langle a, A\rangle \in \Phi$.

THEOREM 2.2. *For any class* Φ *there is a smallest* Φ*-closed class* $I(\Phi)$.

$I(\Phi)$ is the class *inductively defined* by Φ. Elements of $I(\Phi)$ are called *Φ-generated.* Usually the notion of Φ-closed class is defined directly in terms of a system of rules. It is then routine to extract from the system of rules the class Φ involved. For example the class of natural numbers may be characterised as the smallest class ω such that

(i) $\emptyset \in \omega$,

(ii) $a \cup \{a\} \in \omega$ if $a \in \omega$.

This can be rephrased as $\omega = I(\Phi)$ where

$$\Phi = \{\langle\emptyset, \emptyset\rangle\} \cup \{\langle a \cup \{a\}, \{a\}\rangle \mid a \in V\}.$$

Here V is the universal class of sets. Note that the axiom of infinity can be taken to assert that ω is a set.

Examples

EXAMPLE 2.3. Let A and R be classes such that $R \subseteq A \times A$ and $R_a = \{x \mid \langle x, a\rangle \in R\}$ is a set for each $a \in A$. Then let $Wf(A, R)$ be the smallest class X such that for $a \in A$

$$R_a \subseteq X \quad \text{implies } a \in X.$$

Then $Wf(A, R) = I(\Phi)$ where $\Phi = \{\langle a, R_a\rangle \mid a \in A\}$. $Wf(A, R)$ is the *well-founded part* of A with respect to R. Note that Φ and hence $Wf(A, R)$ could not be formed without the assumption that each R_a is a set.

EXAMPLE 2.4. Let A be a class. Let $H(A)$ be the smallest class X such that for $a \in A$

$$f \in X^a \quad \text{implies } \operatorname{ran} f \in X.$$

Here X^a is the class of functions from a to X and $\operatorname{ran} f$ is the range of the function f. $H(A)$ is the class of sets *hereditarily an image of a set in* A, where b is an *image* of a if there is a function from a onto b. For example $H(\omega)$ is the class of hereditarily finite sets and $H(\omega \cup \{\omega\})$ is the class of hereditarily countable sets where countable sets are taken to be those sets that are images of sets in $\omega \cup \{\omega\}$.

In general $H(A)$ can be characterised as the unique class H such that

$$H = \bigcup_{a \in A} \{\operatorname{ran} f \mid f \in H^a\}.$$

EXAMPLE 2.5. Let A be a class and let B_a be a set for each $a \in A$. Let $W_{a\in A} B_a$ be the smallest class X such that for $a \in A$

$$f \in X^{B_a} \quad \text{implies } \langle a, f\rangle \in X.$$

$W_{a\in A} B_a$ is a class of well-founded trees. If $\langle a, f\rangle \in W_{a\in A} B_a$ then $\langle a, f\rangle$ is a tree having $\langle a, f\rangle$ at the root and a node $f(x)$ immediately above it for each $x \in B_a$. Each $f(x)$ is itself a tree in $W_{a\in A} B_a$. For example if $A = \{0, 1, 2\}$, $B_0 = \emptyset$, $B_1 = \{0\}$ and $B_2 = \omega$ then $\mathcal{O} = W_{a\in A} B_a$ is a version of the constructive second number class. $\langle 0, \emptyset\rangle$ is the zero element of $\mathcal{O}$, $\langle 1, \{\langle 0, \alpha\rangle\}\rangle$ is the successor of $\alpha \in \mathcal{O}$ and if $f \in \mathcal{O}^\omega$ then $\langle 2, f\rangle$ is the supremum of f in $\mathcal{O}$.

In general $W_{a\in A} B_a$ can be characterised as the unique class W such that

$$W = \bigcup_{a \in A} \{a\} \times W^{B_a}.$$

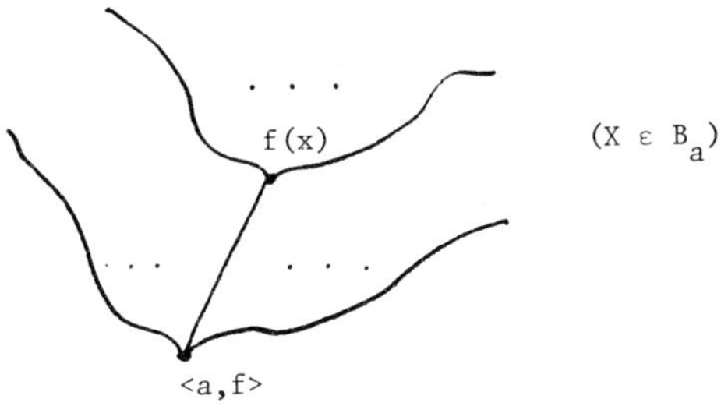

Fig. 1.

EXAMPLE 2.6. The next example is taken from 4.6 of [2]. Define a class X to be *ΠΣI-closed* if

(i) $\omega \in X$.

(ii) $\prod_{a\in A} B_a \in X$ and $\sum_{a\in A} B_a \in X$ whenever $A \in X$ and $B_a \in X$ for all $a \in A$.

(iii) $I(b, c) \in X$ for all b, $c \in A$ if $A \in X$.

In the above the cartesian product $\prod_{a\in A} B_a$ is the set of functions f with domain A such that $f(a) \in B_a$ for all $a \in A$ and the disjoint union $\sum_{a\in A} B_a$ is the set of pairs $\langle a, b\rangle$ such that $a \in A$ and $b \in B_a$. Also $I(b, c) = \{z \in \{\emptyset\} \mid b = c\}$. In [2] an appropriate class $\Pi\Sigma I$ of ordered pairs is defined. The class $I(\Pi\Sigma I)$ of $\Pi\Sigma I$-generated sets played a fundamental role in [2]. This will be examined in the next section where the class will be shown to be replaceable by the simpler class of $\Pi\Sigma$-generated sets. This class is defined by leaving out (iii) and adding to (i) $\emptyset \in X$.

3. The notion of a base

The notion of base was used in [1] and [2] to formulate the presentation axiom and other axioms shown to be modelled by $\underline{V}$ in [2]. Here I shall review the previous work and obtain some further results.

DEFINITION 3.1. The set A is a *base* if whenever for each $a \in A$ B_a is a set having an element then $\prod_{a\in A} B_a$ has an element.

Note that AC asserts that every set is a base. DC implies countable choice which asserts that ω is a base.

The following application of strong collection will sometimes be useful. As in most of this section we are working informally in CZF + DC.

THEOREM 3.2. *For any base A if $\forall x \in A\ \exists y\ \phi(x, y)$ then there is a function f with domain A such that $\forall x \in A\ \phi(x, f(x))$.*

PROOF. Under the assumption, $\forall x \in A\ \ \exists z\ \ \psi(x, z)$ where $\psi(x, z)$ is $\exists y(z = \langle x, y\rangle\ \&\ \phi(x, y))$. Hence by strong collection there is a set B such that

$$\forall x \in A\ \exists z \in B\ \psi(x, z)\ \&\ \forall z \in B\ \exists x \in A\ \psi(x, z).$$

Then for each $x \in A$ the set $B_x = \{y \mid \langle x, y\rangle \in B\}$ has an element so that, because A is a base, there is a function $f \in \prod_{a \in A} B_a$. Then $\forall x \in A$ $\phi(x, f(x))$ as desired.

Full AC is not constructively acceptable in constructive set theory as it implies unacceptable instances of excluded middle. But a general form of AC is available in type theory. For example see 1.15 of [2]. The intuition behind the results in [2] is that the notion of base is a set theoretical version of the notion of small type. By examining the notion of small type we are led to consider some new axioms concerning bases. Before reviewing these axioms let us list some closure properties of bases that can be easily derived in CZF + DC.

3.3. Base closure properties

(1) Each natural number $n = \{x \in \omega \mid x < n\}$ is a base and ω is a base.

(2) If A is a base and B_a is a base for each $a \in A$ then $\sum_{a \in A} B_a$ is a base. In particular if A and B are bases then so are $A \times B$ and $A + B$ $(= \{0\} \times A \cup \{1\} \times B)$.

(3) Any set in one–one correspondence with a base is a base.

(4) Any decidable subset of a base is a base.

By examining the rules for forming small types we are led to formulate the following axioms

BCA_{Π}: If A is a base and B_a is a base for each $a \in A$ then $\prod_{a \in A} B_a$ is a base.

BCA_I: If A is a base then $I(b, c) = \{z \in \{\emptyset\} \mid b = c\}$ is a base for all $b, c \in A$.

The following result gives alternative versions of these axioms.

THEOREM 3.4. (1) BCA_I *is equivalent to* $\mathrm{BCA}_{\mathrm{eq}}$: *If $f, g\colon A \to B$ where A and B are bases then the equaliser $\{x \in A \mid f(x) = g(x)\}$ is a base.*

(2) *Assuming* $\mathrm{BCA}_{\mathrm{eq}}$, BCA_{Π} *is equivalent to* $\mathrm{BCA}_{\mathrm{exp}}$: *If A and B are bases then so is B^A.*

PROOF. (1) First note that BCA_I is the special case of $\mathrm{BCA}_{\mathrm{eq}}$ when $A = \{\emptyset\}$. For the converse implication let $f, g: A \to B$ where A and B are bases. Their equaliser is clearly in one–one correspondence with $\Sigma_{a \in A} I(f(a), g(a))$. Hence by BCA_I, 3.3(2) and 3.3(3) the equaliser is a base.

(2) First note that $\mathrm{BCA}_{\mathrm{exp}}$ is a special case of BCA_{Π}. For the converse implication let A be a base and B_a be a base for each $a \in A$. Then $B = \Sigma_{a \in A} B_a$ is a base by 3.3(2). Note that $\Pi_{a \in A} B_a$ is in one–one correspondence with $C = \{f \in B^A \mid p(f(x)) = x \text{ for all } x \in A\}$, where $p: B \to A$ is given by

$$p(\langle a, b \rangle) = a \quad \text{for } \langle a, b \rangle \in B. \qquad \square$$

Define $F, G: B^A \to A^A$ by

$$F(f)(a) = p(f(a))$$
$$G(f)(a) = a$$

for $f \in B^A$, $a \in A$.

Then C is the equaliser of F and G. But by $\mathrm{BCA}_{\mathrm{exp}}$ the sets B^A and A^A are bases so that by $\mathrm{BCA}_{\mathrm{eq}}$ the set C is a base. Hence by 3.3(3) the set $\Pi_{a \in A} B_a$ is a base.

The base closure axioms BCA_{Π} and BCA_I were motivated by considering the rules for forming elements of the type U. The presentation axiom PA is motivated by considering the rule for forming elements of the type V. Each element of V has the form $\{a_i\}_{i \in I}$ where I is a small type. It is in a certain sense an image of the small type I. If small types are to be represented by bases in set theory then we are led to formulate the following axiom of set theory.

Presentation Axiom (PA). Every set is an image of a base.

In [2] I did not work directly with the axioms BCA_{Π}, BCA_I and PA, but rather with the following ones.

$\Pi\Sigma I$-axiom of choice ($\Pi\Sigma I$-AC). Every $\Pi\Sigma I$-generated set is a base.

$\Pi\Sigma I$-presentation axiom ($\Pi\Sigma I$-PA). Every $\Pi\Sigma I$-generated set is a base and every set is an image of a $\Pi\Sigma I$-generated set.

The following result is proved in [2].

THEOREM 3.5. $\underline{V}$ *models not only* CZF + DC *but also* $\Pi\Sigma I$-AC. *Moreover assuming definition by transfinite recursion on the type U of small types* $\underline{V}$ *also models* $\Pi\Sigma I$-PA.

Definition by transfinite recursion on U is the rule formulated in 1.10 of [2]. It expresses that the small types are inductively generated by the rules explicitly listed for forming the small types, i.e. one rule for each of the basic forms of type that U is reflecting. It is natural to keep the type U open to reflect any additional forms of type that can arise in the future. An example is the form of type $(Wx \in A)B(x)$ which was introduced in [6] and will be used in Section 5 and Appendix 2 of this paper. So the rule of definition by transfinite recursion on U is somewhat unnatural and needs to be modified each time a new form of type is to be reflected in U. In Section 4 a result is obtained which gives an alternative approach to modeling CZF + DC + $\Pi\Sigma I$-PA which avoids transfinite recursion on U.

Note that $\Pi\Sigma I$-PA is a strengthening of $\Pi\Sigma I$-AC which implies PA. It also implies BCA_Π and BCA_I. To see this observe that by 4.8 of [2] the class of bases is the class of those sets that are in one–one correspondence with a $\Pi\Sigma I$-generated set. It follows without undue difficulty that the class of bases is $\Pi\Sigma I$-closed and hence BCA_Π and BCA_I hold.

As a conclusion we get the result

THEOREM 3.6. $\underline{V}$ *models* CZF + DC + BCA_Π + BCA_I + PA, *assuming definition by transfinite recursion on U.*

Recall that the class of $\Pi\Sigma$-generated sets is defined like the class of $\Pi\Sigma I$-generated sets except that the rule involving I is left out and ϕ is explicitly put in. The axioms $\Pi\Sigma$-AC and $\Pi\Sigma$-PA are formulated in the obvious way. The remainder of this section is devoted to proving

THEOREM 3.7.

$$\Pi\Sigma\text{-AC} \equiv \Pi\Sigma I\text{-AC}, \qquad \Pi\Sigma\text{-PA} \equiv \Pi\Sigma I\text{-PA}.$$

This is easily seen to be an immediate consequence of the

LEMMA. *Assuming* $\Pi\Sigma$-AC, *every* $\Pi\Sigma I$*-generated set is in one–one correspondence with a* $\Pi\Sigma$*-generated set.*

PROOF. Let X be the class of those sets that are in one-one correspondence with a $\Pi\Sigma$-generated set. It suffices to show that X is $\Pi\Sigma I$-closed. Obviously $\omega \in X$. That X is closed under Π and Σ is not quite obvious. So let $A \in X$ and $B_a \in X$ for all $a \in A$. Then there is a $\Pi\Sigma$-generated set $\bar{A}$ and $f: A \approx \bar{A}$ (f is a one–one correspondence from A to $\bar{A}$). For each $x \in \bar{A}$, $B_{f^{-1}(x)} \in X$ so that there is a $\Pi\Sigma$-generated set $\bar{B}$ and $g: B_{f^{-1}(x)} \approx \bar{B}$. By $\Pi\Sigma$-AC the set $\bar{A}$ is a base. Hence by 3.2 there are functions assigning to each $x \in \bar{A}$ a $\Pi\Sigma$-generated set $\bar{B}_x$ and $g_x: B_{f^{-1}(x)} \approx \bar{B}_x$. Combining the one–one correspondences f and the g_x's in a straightforward way we see that $\prod_{a \in A} B_a \approx \prod_{x \in A} \bar{B}_x$ and $\sum_{a \in A} B_a \approx \sum_{x \in \bar{A}} \bar{B}_x$. As $\prod_{x \in \bar{A}} \bar{B}_x$ and $\sum_{x \in \bar{A}} \bar{B}_x$ are $\Pi\Sigma$-generated it follows that $\prod_{a \in A} B_a \in X$ and $\sum_{a \in A} B_a \in X$.

It remains to show that if $A \in X$ and $b, c \in A$ then $I(b, c) \in X$. This follows from the

Sublemma. For every $\Pi\Sigma$-generated set $\bar{A}$ if $x, y \in \bar{A}$ then $I(x, y) \in X$.

For if $A \in X$ then there is a $\Pi\Sigma$-generated set $\bar{A}$ and $f: A \approx \bar{A}$. If $b, c \in A$ then by the Sublemma $I(f(b), f(c)) \in X$. But as f is one–one $I(b, c) = I(f(b), f(c))$.

Proof of Sublemma. Let Y be the class of those $\Pi\Sigma$-generated sets $\bar{A}$ such that if $x, y \in \bar{A}$ then $I(x, y) \in X$. It suffices to show that Y is $\Pi\Sigma$-closed. Trivially $\emptyset \in Y$ and $\omega \in Y$ because if $n, m \in \omega$ then $I(n, m) = \{\emptyset\}$ if $n = m$ and $\emptyset$ if $n \neq m$. In either case $I(n, m) \in X$.

Now suppose that $A \in Y$ and $B_a \in Y$ for each $a \in A$. Hence $I(b, c) \in X$ for $b, c \in A$ and also for $b, c \in B_a$ where $a \in A$. We must show that $\prod_{a \in A} B_a \in Y$ and $\sum_{a \in A} B_a \in Y$. Now if $f, g \in \prod_{a \in A} B_a$ then $I(f, g)$ is easily seen to be in one–one correspondence with $\prod_{a \in A} I(f(a), g(a))$. As X is closed under Π, $A \in X$ and $I(f(a), g(a)) \in X$ for all $a \in A$ it follows that $\prod_{a \in A} I(f(a), g(a)) \in X$. Hence $I(f, g) \in X$. Thus $\prod_{a \in A} B_a \in Y$. If $\langle a, b \rangle$, $\langle a', b' \rangle \in \sum_{a \in A} B_a$ then $I(\langle a, b \rangle, \langle a', b' \rangle)$ is easily seen to be in one–one correspondence with $\sum_{z \in I(a,a')} I(b, b')$. As $a, a' \in A$, $I(a, a') \in X$. If $z \in I(a, a')$ then $a = a'$ and $b, b' \in B_a$ so that $I(b, b') \in X$. As X is closed under Σ it follows that $I(\langle a, b \rangle, \langle a', b' \rangle) \in X$. Hence $\sum_{a \in A} B_a \in Y$.

4. An inner model construction

In this section we work informally in CZF + DC.

DEFINITION 4.1. The class A is *regular* if it is transitive, i.e. every element of A is a subset of A, and for every $a \in A$ and set $R \subseteq a \times A$ if $\forall x \in a\ \exists y\ \langle x, y\rangle \in R$ then there is a set $b \in A$ such that

$$\forall x \in a\ \exists y \in b\ \langle x, y\rangle \in R\ \&\ \forall y \in b\ \exists x \in a\ \langle x, y\rangle \in R.$$

In particular if $R: a \to A$ then $\operatorname{ran} R \in A$.

Note that if A is regular then $H(A) = A$. One of the main aims of this section is to prove the following result.

THEOREM 4.2. *Assuming* $\Pi\Sigma$-AC, *there is a class M such that M is the smallest regular model of* CZF. *Moreover M is also a model of* DC + $\Pi\Sigma$-PA.

Note that a class M is a model of a set theoretical sentence ϕ if ϕ is true when all quantifiers are restricted to M. When this holds we write $M \models \phi$.

Using the above theorem and the results of Section 3 a model of $\mathrm{CZF} + \mathrm{DC} + \mathrm{BCA}_{\Pi} + \mathrm{BCA}_{I} + \mathrm{PA}$ can be obtained as follows. First use $\underline{V}$ to interpret CZF + DC + $\Pi\Sigma$-AC and then in this interpretation take the smallest regular model of CZF. This is an alternative to the exclusive use of $\underline{V}$ which requires the use of transfinite recursion on the type U.

The following lemma gives a method for constructing regular classes.

LEMMA 4.3. *If* Y *is a class of bases then* $H(Y)$ *is a regular class.*

PROOF. $H(Y)$ is transitive because every element of $H(Y)$ has the form $\operatorname{ran} f$ for some $f: y \to H(Y)$ for some $y \in Y$. But $\operatorname{ran} f \subseteq H(Y)$.

Now let $a \in H(Y)$ and let $R \subseteq a \times H(Y)$ be a set such that $\forall x \in a\ \exists z\ \langle x, z\rangle \in R$. Then for some $y \in Y$ and some $f: y \to H(Y)$ $a = \operatorname{ran} f$, so that

$$\forall x \in y\ \exists z \in H(Y)\ \langle f(x), z\rangle \in R.$$

As y is a base there is a function $g: y \to H(Y)$ such that

$$\forall x \in y\ \langle f(x), g(x)\rangle \in R.$$

If $b = \operatorname{ran} g$ then $b \in H(Y)$ and

$$\forall x \in a\ \exists z \in b\ \langle x, y\rangle \in R\ \&\ \forall z \in b\ \exists x \in a\ \langle x, y\rangle \in R. \qquad \square$$

The next lemma will be needed to verify the restricted separation scheme in our inner models.

LEMMA 4.4. *If* Y *is a* $\Pi\Sigma$*-closed class of bases then for each restricted* *sentence ϕ with parameters in $H(Y)$ there is a set $c \in Y$ such that*

$$\phi \equiv \exists x\,(x \in c).$$

PROOF. Let us use $!c$ to abbreviate $\exists x(x \in c)$. The following facts are easy to check if A, B are sets and B_a is a set for each $a \in A$.

(i) $$!A \;\&\; !B \equiv !(A \times B),$$

(ii) $$!A \vee !B \equiv !(A + B),$$

(iii) $$!A \supset !B \equiv !B^A, \text{ if } A \text{ is a base,}$$

(iv) $$(\exists x \in A)!B_x \equiv !\sum_{x \in A} B_x,$$

(v) $$(\forall x \in A)!B_x \equiv !\prod_{x \in A} B_x, \text{ if } A \text{ is a base.}$$

The lemma will be proved by induction on the way that the restricted sentence ϕ is built up. For atomic ϕ we need to prove the following claim.

Claim. For all $a, b \in H(Y)$ there is $c \in Y$ such that $(a = b) \equiv !c$.

This claim will be proved by a double set induction on $a, b \in H(Y)$. So as induction hypothesis, we may assume that

$$\forall x \in a\, \forall y \in b\, \exists z \in Y\, (x = y \equiv !z).$$

As $a, b \in H(Y)$ there are $a_0, b_0 \in Y$ and $f\colon a_0 \to a$, $g\colon b_0 \to b$ that are surjective. So we get

$$\forall x \in a_0\, \forall y \in b_0\, \exists z \in Y\, (f(x) = g(y) \equiv !z).$$

As a_0 and b_0 are bases in Y, so is $a_0 \times b_0$ so that by 3.2 there is a function $h\colon a_0 \times b_0 \to Y$ such that

$$\forall x \in a_0\, \forall y \in b_0\, (f(x) = g(y) \equiv !h(\langle x, y\rangle)).$$

Now if

$$c = \prod_{x \in a_0} \sum_{y \in b_0} h(\langle x, y\rangle) \times \prod_{y \in b_0} \sum_{x \in a_0} h(\langle x, y\rangle)$$

then $c \in Y$, as Y is $\Pi\Sigma$-closed, and by (i), (iv), (v) above

$$(a = b) \equiv !c.$$

This completes the proof of the claim and hence the lemma in the case where ϕ has the form $a = b$. $\square$

If ϕ has the form $a \in b$ where $a, b \in H(Y)$ then choose $b_0 \in Y$ and surjective $g: b_0 \to b$. By the claim

$$\forall y \in b_0 \, \exists z \in Y \, (a = g(y) \equiv !z).$$

As b_0 is a base, by 3.2 there is $h: b_0 \to Y$ such that

$$\forall y \in b_0 \, (a = g(y) \equiv !h(y)).$$

Hence

$$\begin{aligned} a \in b &\equiv \exists y \in b_0 \; a = g(y) \\ &\equiv \exists y \in b_0 \; !h(y) \\ &\equiv !c \end{aligned}$$

where

$$c = \sum_{y \in b_0} h(y) \in Y.$$

The final case of an atomic sentence is when ϕ is $\perp$. But if c is the empty set then $\phi \equiv !c$.

Now suppose that ϕ has one of the forms $\phi_1 \,\&\, \phi_2$, $\phi_1 \vee \phi_2$, $\phi_1 \supset \phi_2$, and that by the induction hypothesis there are $c_1, c_2 \in Y$ such that

$$\phi_1 \equiv !c_1 \quad \text{and} \quad \phi_2 \equiv !c_2.$$

Then by (i), (ii), (iii) above

$$\phi \equiv !c$$

where c has one of the forms $c_1 \times c_2$, $c_1 + c_2$, $c_2^{c_1}$. As Y is $\Pi\Sigma$-closed $c \in Y$.

Finally suppose that ϕ has one of the forms $(\forall x \in a)\phi_1(x)$ or $(\exists x \in a)\phi_1(x)$, where $a \in H(Y)$, and by induction hypothesis

$$\forall x \in a \; \exists z \in Y \, (\phi_1(x) \equiv !z).$$

As $a \in H(Y)$ choose $a_0 \in Y$ and surjective $f: a_0 \to a$. Then

$$\forall x \in a_0 \; \exists z \in Y \, (\phi_1(f(x)) \equiv !z).$$

As a_0 is a base there is $h: a_0 \to Y$ such that

$$\forall x \in a_0 \, (\phi_1(f(x)) \equiv !h(x)).$$

Hence by (iv), (v)

$$\phi \equiv !c$$

where c has one of the forms $\prod_{x\in a_0} h(x)$ or $\sum_{x\in a_0} h(x)$. In either case $c\in Y$ as Y is $\Pi\Sigma$-closed.

We can now prove the following result.

THEOREM 4.5. *If $M = H(Y)$ where Y is a $\Pi\Sigma$-closed class of bases then M is a regular model of* CZF + DC + $\Pi\Sigma$-AC.

PROOF. By 4.3 M is regular. We consider each axiom and scheme of CZF + DC + $\Pi\Sigma$-AC in turn. M models the *extensionality axiom* because it is transitive. To see that it models the *pairing axiom* let $a, b\in M$. Define $f\colon \omega\to M$ by putting $f(0) = a$ and $f(n+1) = b$ for $n\in\omega$. Then $\{a, b\} = \operatorname{ran} f$ so that, as $\omega\in Y$, $\{a, b\}\in M$. For the *union axiom* let $a\in M$. Choose $a_0\in Y$ and surjective $f\colon a_0\to a$. As $a\subseteq M$, if $x\in a_0$ then $f(x)\in M$ so that there is $y\in Y$ and surjective $g\colon y\to f(x)$. As a_0 is a base there is a function $b\colon a_0\to Y$ and a function g with domain a_0 such that for all $x\in a_0$ $g(x)\colon b(x)\to f(x)$ is surjective. As Y is $\Pi\Sigma$-closed $c = \sum_{x\in a_0} b(x)\in Y$. Now we can define $h\colon c\to M$ by

$$h(\langle x, y\rangle) = g(x)(y)$$

for $x\in a_0$, $y\in b(x)$. So

$$\begin{aligned} z\in \bigcup a &\equiv \exists x\in a\ (z\in x)\\ &\equiv \exists x\in a_0\ (z\in f(x))\\ &\equiv \exists x\in a_0\ \exists y\in b(x)\ (z = g(x)(y))\\ &\equiv z\in \operatorname{ran} h. \end{aligned}$$

Hence $h\colon c\to \bigcup a$ is surjective so that $\bigcup a\in M$. For the *infinity axiom* we first show that each natural number is in M. As $\emptyset\in Y$ and $\emptyset : \emptyset\to\emptyset$ is surjective it follows that $\emptyset\in M$. As M models the pairing and union axioms if $a\in M$ then $a\cup\{a\}\in M$. Hence by mathematical induction $\omega\subseteq M$. As $\omega\in Y$ it follows that $\omega\in M$.

The *set induction scheme* is easily seen to be modeled by any class. For the *restricted separation scheme* let $a\in M$ and let $\phi(x)$ be a restricted formula with parameters in M and having x as the only free variable. We need to prove that $\{x\in a \mid \phi(x)\}\in M$. By Lemma 4.4

$$\forall x\in a\ \exists z\in Y\ (\phi(x)\equiv\, !z).$$

As $a\in M$ we may choose $a_0\in Y$ and surjective $f\colon a_0\to a$. As a_0 is a base there is a function $h\colon a_0\to Y$ such that

$$\forall x\in a_0 (\phi(f(x))\equiv\, !h(x)).$$

Let $b_0 = \Sigma_{x \in a_0} h(x) \in Y$, and define $g\colon b_0 \to M$ so that $g(\langle x, y\rangle) = f(x)$ for all $x \in a_0$, $y \in h(x)$. Then $\operatorname{ran} g \in M$ and

$$\begin{aligned} z \in \operatorname{ran} g &\equiv \exists x \in a_0\ \exists y \in h(x)\ (z = g(\langle x, y\rangle)) \\ &\equiv \forall x \in a_0\ \exists y \in h(x)\ (z = f(x)) \\ &\equiv \exists x \in a_0\ (!h(x)\ \&\ z = f(x)) \\ &\equiv \exists x \in a_0\ (\phi(f(x))\ \&\ z = f(x) \\ &\equiv z \in a\ \&\ \phi(z) \\ &\equiv z \in \{x \in a \mid \phi(x)\}. \end{aligned}$$

Hence $\{x \in a \mid \phi(x)\} = \operatorname{ran} g \in M$.

For the *strong collection scheme* let $a \in M$ such that $M \models \forall x \in a\ \exists y \phi(x, y)$ where $\phi(x, y)$ is a formula having parameters in M and at most the variables x, y occuring free. We must find $b \in M$ such that $M \models \phi'(a, b)$. By strong collection there is a set $R \subseteq a \times M$ such that $\forall x \in a\ \exists y\ \langle x, y\rangle \in R$ and $M \models \phi(x, y)$ whenever $\langle x, y\rangle \in R$. As M is regular there is $b \in M$ such that

$$\forall x \in a\ \exists y \in b\ \langle x, y\rangle \in R\ \&\ \forall y \in b\ \exists x \in a\ \langle x, y\rangle \in R.$$

It follows that $M \models \phi'(a, b)$.

For the *subset collection scheme* let $a, a' \in M$. Choose $a_0, a'_0 \in Y$ and surjective $f\colon a_0 \to a$, $f'\colon a'_0 \to a'$. As Y is $\Pi\Sigma$-closed $a_0'^{a_0} \in Y$ and if $h \in a_0'^{a_0}$ then $F(h) \in M$ where $F(h) = \{f'(h(x)) \mid x \in a\}$. Hence if $c = \operatorname{ran} F$ then $c \in M$. Now suppose that $M \models \forall x \in a\ \exists y \in a' \phi(x, y)$ where $\phi(x, y)$ is a formula having parameters in M and at most the variables x, y free. (Note that c was defined independently of $\phi(x, y)$.) Then

$$\forall x \in a_0\ \exists y \in a'_0\ M \models \phi(f(x), f'(y)).$$

As a_0 is a base there is $h\colon a_0 \to a'_0$ such that

$$\forall x \in a_0\ M \models \phi(f(x), f'(h(x))).$$

If $b = F(h)$ then $b \in c$ and $M \models \phi'(a, b)$. Hence $M \models \exists b \in c\ \phi'(a, b)$.

We next consider the *dependent choices scheme.* So assume that

$$M \models \forall x(\theta(x) \supset \exists y(\theta(y)\ \&\ \phi(x, y))),$$

and let $a \in M$ such that $M \models \theta(a)$. By DC there is $f\colon \omega \to M$ such that $f(0) = a$ and for all $n \in \omega$

$$M \models \theta(f(n))\ \&\ \phi(f(n), f(n+1)).$$

But $f = \{\langle n, f(n)\rangle \mid n \in \omega\} \subseteq M$ and $\omega \in Y$ so that $f \in M$.

Finally we show that M models $\Pi\Sigma$-AC. We must show that if $a \in M$ such that $M \models$ "a is $\Pi\Sigma$-generated" then $M \models$ "a is a base". As we have already seen that M is a regular model of CZF, if $a \in M$ such that $M \models$ "a is $\Pi\Sigma$-generated" then by Lemma 4.6 below a really is $\Pi\Sigma$-generated. Hence by $\Pi\Sigma$-AC it follows that a is a base and hence it is easily seen that $M \models$ "a is a base".

Lemma 4.6. *If M is a regular model of* CZF *then M is $\Pi\Sigma$-closed and for $a \in M$*

$$(a \text{ is } \Sigma\Pi\text{-generated}) \equiv M \models \text{“}a \text{ is } \Pi\Sigma\text{-generated”}.$$

The conclusion of this result may be formulated as stating that the class of $\Pi\Sigma$-generated sets is absolute over M. In order to prove this it is necessary to review a definition of the class of $\Pi\Sigma$-generated sets and check that each part of the definition is absolute for M. The class is given by an inductive definition which can be replaced by an explicit definition as in the proof of Theorem 2.2 that is to be found in 4.2 of [2]. The details of the absoluteness proof are straightforward but somewhat long, so the proof of the lemma will not be presented here.

Proof of Theorem 4.2. Assume $\Pi\Sigma$-AC and let $M = H(Y)$ where Y is the class of $\Pi\Sigma$-generated sets. As Y is a $\Pi\Sigma$-closed class of bases it follows from Theorem 4.5 that M is a regular model of CZF + DC + $\Pi\Sigma$-AC.

We also need to show that M is a model of $\Pi\Sigma$-PA. So let $a \in M$. Choose $a_0 \in Y$ and surjective $f\colon a_0 \to a$. As M is $\Pi\Sigma$-closed $Y \subseteq M$ so that $a_0 \in M$. By Lemma 4.6 $M \models$ "a_0 is $\Pi\Sigma$-generated". Note also that $f \in M$. Hence $M \models$ "a is an image of a $\Pi\Sigma$-generated set".

Finally suppose that M' is a regular model of CZF. We must show that $M \subseteq M'$. By Lemma 4.6 M' is $\Pi\Sigma$-closed. It follows that $Y \subseteq M'$ and hence $M = H(Y) \subseteq H(M')$. Hence $M \subseteq M'$ as M' is regular. □

5. The regular extension axiom

The form of type $(Wx \in A)B(x)$

Up till now the interpretation $\underline{V}$ of constructive set theory has been based on type theory with rules for the following forms of type: N_0, N,

$(\Pi x \in A)B(x)$, $(\Pi x \in A)B(x)$, $A + B$, $I(A, b, c)$, U and of course the type V of iterative sets over U. In this section we consider the effect on $\underline{V}$ of adding the form of type $(Wx \in A)B(x)$ and rules for it. This new form of type was first introduced by MARTIN-LÖF in [6]. If B is a family of types over the type A then $(Wx \in A)B(x)$ is a type W having the introduction rule

$$\frac{a \in A \qquad \begin{array}{c}(y \in B(a))\\ f(y) \in W\end{array}}{\sup(a, f) \in W}\,.$$

There are also rules for definition by transfinite recursion on W which express that the elements of W are inductively generated using the above introduction rule. Notice that the type V and its rules correspond exactly to the type $(Wx \in U)x$ and its rules. Also notice that the inductively defined class $W_{x \in A} B_x$ of 2.5 is a set theoretical version of the type $(Wx \in A)B(x)$.

In addition to the already mentioned rules for the new form of type there is a rule for reflectings $(Wx \in A)B(x)$ in the type U:

$$\frac{A \in U \qquad \begin{array}{c}(x \in A)\\ B(x) \in U\end{array}}{(Wx \in A)B(x) \in U}\,.$$

When this rule is added to the other rules for forming small types then the rule for definition by transfinite recursion (if it is to be used at all) has to be modified to allow for the new form of small type.

Inductive definitions of sets

In Section 2 we considered inductive definitions Φ of class $I(\Phi)$ in constructive set theory. Under what conditions on Φ will the class $I(\Phi)$ be a set? One might expect from classical set theory that $I(\Phi)$ should certainly be a set when Φ itself is a set. In fact by considering classical examples such as the set of hereditarily countable sets one might expect $I(\Phi)$ to be a set for certain classes Φ which are not sets. This is the case but even when Φ is a set a new axiom of constructive set theory seems to be needed. The notion of a regular class was defined in 4.1.

The Regular Extension Axiom (*REA*). Every set is a subset of a regular set.

We shall see that $\underline{V}$ models this axiom in the context of type theory with the W-form of type.

DEFINITION 5.1. An inductive definition Φ is *bounded* if

(i) for each set A the class Φ_A is a set where

$$\Phi_A = \{a \mid \langle a, A\rangle \in \Phi\},$$

(ii) there is a set B such that if $\langle a, A\rangle \in \Phi$ then A is an image of a set in B. The set B is called a *bound* for Φ.

First notice that if Φ is a set then Φ is bounded with bound the set $\{A \mid \exists a\ \langle a, A\rangle \in \Phi\}$. In particular if A and R are sets with $R \subseteq A \times A$ then the inductive definition in 2.3 of $Wf(A, R)$ is bounded.

A simple example of a bounded inductive definition that is not a set is the inductive definition

$$\Phi = \{\langle \emptyset, \emptyset\rangle\} \cup \{\langle a \cup \{a\}, \{a\}\rangle \mid a \in V\}$$

of the class of natural numbers. It has bound $\{\emptyset, \{\emptyset\}\}$.

If A is a set then the class $H(A)$ defined in 2.4 is inductively defined by

$$\Phi = \{\langle \operatorname{ran} f, \operatorname{ran} f\rangle \mid f \in V^a \text{ for some } a \in A\}.$$

This is bounded with bound A.

As a final example if A is a set and B_a is a set for each $a \in A$ then the class $W_{a\in A} B_a$, defined in 2.5, has inductive definition

$$\Phi = \{\langle\langle a, f\rangle, \operatorname{ran} f\rangle \mid f \in V^{B_a} \text{ and } a \in A\}.$$

This is bounded with bound $\{B_a \mid a \in A\}$.

THEOREM 5.2 (CZF + REA). *Every bounded inductive definition inductively defines a set.*

COROLLARY 5.3 (CZF + REA). (i) *If A is a set and $R \subseteq A \times A$ is a set then $Wf(A, R)$ is a set.*

(ii) *If A is a set then $H(A)$ is a set.*

(iii) *If A is a set and B_a is a set for $a \in A$ then $W_{a\in A} B_a$ is a set.*

The classical theory of inductive definitions is usually presented in terms of transfinite iterations of a monotone operator. If Φ is an inductive definition for each set x let

$$\Gamma(x) = \{a \mid \langle a, A\rangle \in \Phi \text{ for some } A \subseteq x\}.$$

In general $\Gamma(x)$ is a class. Note that $x \subseteq y$ implies $\Gamma(x) \subseteq \Gamma(y)$ and for any class X

$$[X \text{ is } \Phi\text{-closed}] \equiv [\Gamma(x) \subseteq X \text{ for all } x \subseteq X].$$

LEMMA 5.4 (CZF). *If Φ is a bounded inductive definition then*

(i) *$\Gamma(x)$ is a set for each set x.*

(ii) *There is an assignment of a set Γ^a to each set a such that*

$$\Gamma^a = \Gamma(\bigcup\{\Gamma^y \mid y \in a\}).$$

(iii) $I(\Phi) = \bigcup\{\Gamma^a \mid a \in V\}$.

PROOF. (i) Let B be a bound for Φ. Then $\langle a, A\rangle \in \Phi$ implies that there is $b \in B$ and surjective $f\colon b \to A$. Hence for each set x

$$\Gamma(x) = \bigcup\{\Phi_{\operatorname{ran} f} \mid f \in C\}$$

where $C = \bigcup\{x^b \mid b \in B\}$. By the exponentiation, replacement and union axioms C is a set. As Φ is bounded $\Phi_{\operatorname{ran} f}$ is a set for all $f \in C$. Hence by the replacement and union axioms $\Gamma(x)$ is a set.

(ii) Let X be the smallest class such that if

$$\forall y \in a\ \exists z \in b\ \langle y, z\rangle \in X\ \&\ \forall z \in b\ \exists y \in a\ \langle y, z\rangle \in X$$

then

$$\langle a, \Gamma(\bigcup b)\rangle \in X.$$

This definition can be put in the form of an inductive definition coming under the scope of Theorem 2.2. By set induction one can easily prove that for each set a there is a unique set x such that $\langle a, x\rangle \in X$, and if this unique x is written Γ^a then

$$\Gamma^a = \Gamma(\bigcup\{\Gamma^y \mid y \in a\}).$$

(iii) First note that if $\Gamma^y \subseteq I(\Phi)$ for all $y \in a$ then $\bigcup\{\Gamma^y \mid y \in a\} \subseteq I(\Phi)$ and hence $\Gamma^a \subseteq I(\Phi)$ as $I(\Phi)$ is Φ-closed. Hence by set induction $\Gamma^a \subseteq I(\Phi)$ for all sets a so that $\bigcup\{\Gamma^a \mid a \in V\} \subseteq I(\Phi)$. For the converse inclusion it suffices to show that $\bigcup\{\Gamma^a \mid a \in V\}$ is Φ-closed. So let x be a set such that $x \subseteq \bigcup\{\Gamma^a \mid a \in V\}$. Then

$$\forall y \in x\ \exists a\ y \in \Gamma^a.$$

By collection there is a set b such that

$$\forall y \in x\ \exists a \in b\ y \in \Gamma^a.$$

It follows that $x \subseteq \bigcup\{\Gamma^a \mid a \in b\}$ and hence $\Gamma(x) \subseteq \Gamma^b \subseteq \bigcup\{\Gamma^a \mid a \in V\}$.

PROOF OF THEOREM 5.2. Let Φ be a bounded inductive definition with bound the set A. By REA we may assume without loss that A is a regular set. Let

$$I = \bigcup\{\Gamma^a \mid a \in A\}.$$

By the replacement and union axioms I is a set. By (iii) of the lemma $I \subseteq I(\Phi)$. Hence it suffices to show that I is Φ-closed, because then $I(\Phi) \subseteq I$ so that $I(\Phi) = I$ is a set.

So let $\langle y, Y\rangle \in \Phi$ with $Y \subseteq I$. We must show that $y \in I$. As Φ has bound A there is $a \in A$ and surjective $f\colon a \to Y$. Hence

$$\forall x \in a\ f(x) \in I,$$

so that

$$\forall x \in a\ \exists z \in A\ f(x) \in \Gamma^z.$$

As A is regular there is $b \in A$ such that

$$\forall x \in a\ \exists z \in b\ f(x) \in \Gamma^z.$$

It follows that $Y \subseteq \bigcup\{\Gamma^z \mid z \in b\}$ so that $y \in \Gamma^b \subseteq I$.

Assuming the presentation axiom REA has several equivalents.

THEOREM 5.5 (CZF + PA). *The following are equivalent.*

(i) REA.
(ii) $I(\Phi)$ *is a set for every bounded* Φ.
(iii) $H(A)$ *is a set for every set* A.
(iv) $H(A)$ *is a set for every set* A *of bases.*
(v) $W_{a\in A} B_a$ *is a set for every set* A *and sets* B_a *for* $a \in A$.

PROOF. The implications

$$\text{(i)} \to \text{(ii)} \begin{array}{l} \nearrow \text{(iii)} \to \text{(iv)} \\ \searrow \text{(v)} \end{array}$$

are clear from the work above. We shall complete the circle by showing that (v) → (iv) → (i).

(v) → (iv). Let A be a set of bases. Let $W = W_{a\in A}\, a$. Assuming (v) W is a set. It is the smallest class such that if $a \in A$ and $f\colon a \to W$ then

$\langle a, f\rangle \in W$. There is a function $F\colon W \to V$ such that

$$F(\langle a, f\rangle) = \{F(f(x)) \mid x \in a\}$$

for $a \in A$ and $f\colon a \to W$. In fact F may be defined inductively as the smallest class such that if $a \in A$, $f\colon a \to W$ and $g\colon a \to V$ such that $\forall x \in a$ $\langle f(x), g(x)\rangle \in F$ then $\langle\langle a, f\rangle, \operatorname{ran} g\rangle \in F$. Let $H = \operatorname{ran} F$. As H is a set it suffices to show that $H(A) = H$.

To show that $H(A) \subseteq H$ it suffices to show that if $b \subseteq H$ is an image of a set in A then $b \in H$. So let $b = \operatorname{ran} f$ where $f\colon a \to H$ with $a \in A$. Then

$$\forall x \in a\ \exists y \in W\ (f(x) = F(y)).$$

As a is a base there is $g\colon a \to W$ such that

$$\forall x \in a\ f(x) = F(g(x)).$$

Hence $b = \operatorname{ran} f = F(\langle a, g\rangle) \in H$. For $H \subseteq H(A)$ we prove that $b \in H$ implies $b \in H(A)$ by set induction. So assume that for all $y \in b$ if $y \in H$ then $y \in H(A)$ and let $b \in H$. Then $b = F(\langle a, g\rangle)$ for some $a \in A$ and $g\colon a \to W$. So $b = \operatorname{ran} f$ where $f(x) = F(g(x)) \in H(A)$ for $x \in A$. Hence $b \in H(A)$.

(iv)$\to$(i). If A is a set we must find a regular set $H \supseteq A$. Without loss we may assume that A is transitive. By PA there is a surjective function $\pi\colon B \to A$ where B is a base. By PA again, for each $b \in B$ there is a surjective function $\pi'\colon B' \to \pi(b)$ where B' is a base. As B is a base there is a function that assigns to each $b \in B$ a surjective function $\pi\colon B_b \to \pi(b)$ where B_b is a base. Now let $H = H(\{B_b \mid b \in B\})$. By Lemma 4 H is a regular class. By (iv) H is a set. It only remains to show that $A \subseteq H$. We show that $a \in A$ implies $a \in H$ by set induction. So assume that for all $x \in a$ if $x \in A$ then $x \in H$ and let $a \in A$. As A is transitive $a \subseteq H$. Choose $b \in B$ such that $\pi(b) = a$. Then as $\pi_b\colon B_b \to a$ is surjective it follows that $a \in H$ by the definition of H. □

The interpretation of CZF + DC + REA + BCA_{Π} + BCA_W + BCA_I + PA

Recall that the base closure axioms BCA_{Π} and BCA_I were formulated by considering the rules of type theory for forming small types. With the new form of small type $(Wx \in A)B(x)$ we are led to the base closure axiom for W.

BCA_W: If A is a base and B_a is a base for each $a \in A$ then $W_{a\in A} B_a$ is a base.

The discussion and proofs in Section 3 and Section 4 can be reworked so as to include the axioms REA and BCA_W. The notion of $\Pi\Sigma I$-generated set needs to be modified to the notion of $\Pi\Sigma WI$-generated set by incorporating the W-operation in the obvious way. The axioms $\Pi\Sigma WI$-AC and $\Pi\Sigma WI$-PA are formulated in the obvious way. The following modification of Theorem 3.5 will be proved in Appendix 2 using the W-rules of type theory.

THEOREM 5.6. *$\underline{V}$ models* CZF + DC + REA + $\Pi\Sigma WI$-AC. *Moreover assuming definition by transfinite recursion on the type U of small types $\underline{V}$ also models* $\Pi\Sigma WI$-PA.

The following result is proved exactly as in the proof of Theorem 4.8 of [2].

THEOREM 5.7 (CZF + REA + $\Pi\Sigma WI$-PA). *A set is a base if and only if it is in one–one correspondence with a $\Pi\Sigma WI$-generated set.*

Hence, as in the discussion leading up to 3.6 we get

THEOREM 5.8. *$\underline{V}$ models* CZF + DC + REA + BCA_Π + BCA_W + BCA_I + PA, *assuming definition by transfinite recursion on U.*

If the notion of $\Pi\Sigma W$-generated set and the axioms $\Pi\Sigma W$-AC and $\Pi\Sigma W$-PA are defined in the obvious way then the following result can be proved along the same lines as Theorem 3.7.

THEOREM 5.9 (CZF + REA).

$$\Pi\Sigma W\text{-AC} \equiv \Pi\Sigma\text{-AC}, \qquad \Pi\Sigma W\text{-PA} \equiv \Pi\Sigma\text{-PA}.$$

Finally by reworking Section 4 we can get the following result.

THEOREM 5.10 (CZF + DC + REA + $\Pi\Sigma W$-AC). *There is a class M such that M is a minimal regular model of* CZF + REA. *Moreover M is also a model of* DC + $\Pi\Sigma W$-PA.

In fact $M = H(Y)$ where Y is the class of $\Pi\Sigma W$-generated sets. As in Section 4 this result gives an alternative approach to the interpretation of CZF + DC + REA + BCA_Π + BCA_W + BCA_I + PA to that in Theorem 5.8.

Appendices[1]

A1. Proof of Theorem 1.2

We shall work informally in type theory as in [2]. Let us first assume I to prove II. As $\underline{V}$ models the power set axiom there is an $\alpha \in V$ such that for $\beta \in V$

$$\beta \in \alpha \equiv \beta \subseteq \{\emptyset\},$$

i.e. α is the power set of $\{\emptyset\}$. Let $\Omega = \bar{\alpha}$ and let $T(x) = \overline{\tilde{\alpha}(x)}$ for $x \in A$. Then $\Omega \in U$ and $T(x) \in U$ for $x \in \Omega$. For any type A let θ be the extensional species on V where $\theta(x) = A$ for $x \in V$. By the absolute separation scheme there is a set $\beta \in V$ such that for $\gamma \in V$

$$\gamma \in \beta \equiv \gamma \in \{\emptyset\} \;\&\; \theta(\gamma) \equiv \gamma \doteq \emptyset \;\&\; A.$$

As $\beta \subseteq \{\emptyset\}$ $\beta \in \alpha$ so that for some $x \in \Omega$ $\beta \doteq \tilde{\alpha}(a)$. Hence

$$A \equiv \emptyset \doteq \emptyset \;\&\; A = \emptyset \in \beta \equiv \emptyset \in \tilde{\alpha}(x) \equiv \exists y \in T(x)(\emptyset \doteq \tilde{\alpha}(x)(y)).$$

But

$$\forall y \in T(x)(\emptyset \doteq \overline{\tilde{\alpha}(x)}(y)).$$

Hence

$$A \equiv \exists y \in T(x)(\emptyset \doteq \emptyset) \equiv T(x).$$

Thus $(\exists a \in \Omega)(T(a) \equiv A)$ and we have proved II.

Now let us assume II to prove I. Let $\alpha = (\sup x \in \Omega) f(x)$ where $f(x) = (\sup y \in T(x))\emptyset$ for $x \in \Omega$. Then $\alpha \in V$. We will show that α is the power set of $\{\emptyset\}$. First note that $f(x) \subseteq \{\emptyset\}$ for all $x \in \Omega$ so that α is a set of subsets of $\{\emptyset\}$. Now let $\beta \subseteq \{\emptyset\}$. Choose $x \in \Omega$ such that $T(x) \equiv (\emptyset \in \beta)$. This is possible by II. Then for $\gamma \in V$

$$\gamma \in \beta \equiv \gamma \doteq \emptyset \;\&\; \emptyset \in \beta \equiv \gamma \doteq \emptyset \;\&\; T(x) \equiv \exists y \in T(x)(\gamma \doteq \emptyset) \equiv \gamma \in f(x).$$

Hence $\beta \doteq f(x)$ so that $\beta \in \alpha$. Thus in $\underline{V}$ $\{\emptyset\}$ has a powerset. But, as shown in 2.3 of [1], this implies that $\underline{V}$ models the powerset axiom.

---

[1] In the appendices the same symbol '$\in$' has been used for both the fundamental type theoretic relation between an object and its type and the defined extensional membership relation on V, expressed by '$\in$'$_{\text{ext}}$ in Section 1. The context should always make clear which is intended.

Finally let $\delta \in V$ and let ϕ be an extensional species on V. By II

$$\forall y \in \bar{\delta}\ \exists x \in \Omega(T(x) \equiv \phi(\tilde{\delta}(y))).$$

Hence by AC for type theory (see 1.15 of [2]) there is $f \in \bar{\delta} \to \Omega$ such that

$$\forall y \in \bar{\delta}\ T(f(y)) \equiv \phi(\tilde{\delta}(y)).$$

Let $A = (\Sigma y \in \bar{\delta})T(f(y))$. Then $A \in U$ and if $\beta = (\sup z \in A)\tilde{\delta}(p(z))$ then $\beta \in V$. For $\gamma \in V$

$$\begin{aligned}\gamma \in \beta &\equiv \exists z \in A\ \gamma \doteq \tilde{\delta}(p(z))\\ &\equiv \exists y \in \bar{\delta}\ \exists u \in T(f(y))\ (\gamma \doteq \tilde{\delta}(y))\\ &\equiv \exists y \in \bar{\delta}\ (T(f(y))\ \&\ \gamma \doteq \tilde{\delta}(y))\\ &\equiv \exists y \in \bar{\delta}\ (\phi(\tilde{\delta}(y))\ \&\ \gamma \doteq \tilde{\delta}(y))\\ &\equiv \exists y \in \bar{\delta}\ (\phi(y)\ \&\ \gamma \doteq y)\\ &\equiv \gamma \in \bar{\delta}\ \&\ \phi(\gamma).\end{aligned}$$

Hence $\underline{V}$ models the absolute separation scheme and I is proved.

REMARKS. (1) $\underline{V}$ models the powerset axiom if and only if II′, where II′ is the weakening of II which only requires $(\exists a \in \Omega)(T(a) \equiv A)$ for small types A.

(2) $\underline{V}$ models the absolute separation scheme if and only if for each type A $(\exists a \in U)(a \equiv A)$ is true.

A2. Proof of Theorem 1.2

As in Section A1 we shall work informally in type theory. In addition to the rules of type theory used in [2] and Section A1 we shall use the rules for the form of type $(Wx \in A)B(x)$ as discussed in Section 5 and formulated in detail in [6]. In view of the previous work we need here to prove the following results.

A2.1 $\underline{V}$ models REA.

A2.2 $\underline{V}$ models $\Pi\Sigma WI$-AC.

A2.3 $\underline{V}$ models $\Pi\Sigma WI$-PA, if definition by transfinite recursion on the type U of small types is assumed.

Proof of A2.1

In CZF every set is a subset of a transitive set. Hence it suffices to show that if $\alpha_0 \in V$ is transitive then there is a regular set $\alpha \in V$ such that

$\alpha_0 \subseteq \alpha$. For the regularity of α it suffices to show that if $\beta \in V$ such that $\beta \in \alpha$ then $\beta \subseteq \alpha$ and for each species F on $V \times V$

$$\forall x \in \beta\ \exists y \in \alpha\ F(x, y) \supset \exists \beta' \in \alpha\ F'(\beta, \beta') \tag{*}$$

where $F'(\beta, \beta')$ is

$$\forall x \in \beta\ \exists y \in \beta'\ F(x, y)\ \&\ \forall y \in \beta'\ \exists x \in \beta\ F(x, y).$$

So let $\alpha_0 \in V$ be transitive. Let $A_0 = \bar{\alpha}_0$ and let $B_0 = (x)\tilde{\alpha}_0(x)$. Then $A_0 \in U$ and $B_0 \in A_0 \rightarrow U$ so that if $A = (Wx \in A_0)B_0(x)$ then $A \in U$. Now define $h \in A \rightarrow V$ by transfinite recursion on A so that

$$h(\sup(a, f)) = (\sup u \in B_0(a))h(f(u))$$

for $a \in A_0$, $f \in B_0(a) \rightarrow A$. Then $\alpha \in V$ where $\alpha = \sup(A, h)$.

LEMMA. *Let $\beta \in V$. If $\beta \doteq \gamma$ for some $\gamma \in V$ such that $\bar{\gamma} = B_0(a)$ for some $a \in A_0$ then (*) holds for each species F on $V \times V$.*

PROOF. Assume that $\forall x \in \beta\ \exists y \in \alpha\ F(x, y)$. Then by the assumptions on β

$$\forall x \in B_0(a)\ \exists y \in A\ F(\tilde{\gamma}(x), h(y)).$$

Hence by AC (1.15 of [2]) there is $f \in B_0(a) \rightarrow A$ such that

$$\forall x \in B_0(a)\ F(\tilde{\gamma}(x), h(f(x))).$$

So $\sup(a, f) \in A$ and if

$$\beta' = h(\sup(a, f)) = (\sup u \in B_0(a))h(f(u))$$

then $\beta' \in V$ such that $\beta' \in \alpha$ and

$$\forall x \in B_0(a)\ F(\tilde{\gamma}(x), \tilde{\beta}'(x))$$

so that by Theorem 2.9(i) of [2] $F'(\gamma, \beta')$. As $\beta \doteq \gamma$ we get $F'(\beta, \beta')$ as desired.

To show that α is regular let $\beta \in V$ such that $\beta \in \alpha$. Then $\beta \doteq h(c)$ for some $c \in A$. But $c = \sup(a, f)$ for some $a \in A_0$ and $f \in B_0(a) \rightarrow A$. Hence $\beta \doteq (\sup u \in B_0(a))h(f(u))$. As $h(f(u)) \in \alpha$ for $u \in B_0(a)$ it follows that $\beta \subseteq \alpha$. Finally let F be a species on $V \times V$. The assumptions of the lemma hold with $\gamma = h(c)$. Hence (*) holds as desired.

It remains to show that $\alpha_0 \subseteq \alpha$. We show that $\beta \in \alpha_0 \supset \beta \in \alpha$ by set induction on $\beta \in V$. So as induction hypothesis we assume that

$$\forall y \in \beta\ (y \in \alpha_0 \supset y \in \alpha).$$

Now if $\beta \in \alpha_0$ then $\beta \doteq \tilde{\alpha}_0(a)$ for some $a \in A_0$. So the assumptions of the lemma hold with $\gamma = \tilde{\alpha}_0(a)$. Hence (*) holds, where $F(x, y)$ is $x \doteq y$. As α_0 is transitive, by the induction hypothesis $\beta \subseteq \alpha$ so that

$$\forall x \in \beta\ \exists y \in \alpha (x \doteq y).$$

Hence by (*) there is $\beta' \in \alpha$ such that

$$\forall x \in \beta\ \exists y \in \beta'(x \doteq y)\ \&\ \forall y \in \beta'\ \exists x \in \beta(x \doteq y).$$

Hence $\beta \doteq \beta'$ so that $\beta \in \alpha$ as required.

Proof of A2.2

The proof of the validity of $\Pi\Sigma I$-AC was carried out in Section 6 of [2]. Here we shall only discuss the additional constructions that are needed to transform Section 6 into a proof of the new result. In 6.2 and 6.3 of [2] $\Pi(\alpha, \beta) \in V$ and $\Sigma(\alpha, \beta) \in V$ are defined for $\alpha, \beta \in V$ such that $\bar{\alpha} = \bar{\beta}$, and in 6.4 these are related to the set theoretical disjoint union and cartesian product operations when α is injectively presented and $\tilde{\beta}(x)$ is injectively presented for each $x \in \bar{\alpha}$. Moreover it is also shown that $\Pi(\alpha, \beta)$ and $\Sigma(\alpha, \beta)$ are injectively presented under these conditions. We need to carry out a similar construction for the set theoretical W-operation of 2.5. Once this has been done then theorem 6.7 of [2] can be strengthened to obtain the result that the class of strong bases is $\Pi\Sigma WI$-closed and hence the validity of $\Pi\Sigma WI$-AC.

So let $\alpha, \beta \in V$ with $\bar{\alpha} = \bar{\beta}$. We wish to define $W(\alpha, \beta) \in V$. First note that $E \in U$ where $E = (Wx \in \bar{\alpha})\overline{\tilde{\beta}(x)}$. Define $h \in E \to V$ by transfinite recursion on E so that for $x \in \bar{\alpha}$ and $f \in \overline{\tilde{\beta}(x)} \to E$

$$h(\sup(x, f)) = \langle \tilde{\alpha}(x), S(\tilde{\beta}(x), (\sup u \in \overline{\tilde{\beta}(x)})h(f(u)))\rangle.$$

Finally let $W(\alpha, \beta) = \sup(E, h)$.

LEMMA. *Let $\alpha \in V$ be injectively presented and let $\beta \in V$ such that $\bar{\beta} = \bar{\alpha}$ and $\tilde{\beta}(x)$ is injectively presented for all $x \in \bar{\alpha}$. Then*

(1) *If $\eta = W(\alpha, \beta)$ then*

If $\gamma \in \alpha$ and δ is a function from γ' to η, where $\langle \gamma, \gamma' \rangle \in S(\alpha, \beta)$, then $\langle \gamma, \delta \rangle \in \eta$. (**)

(2) *If $\eta \in V$ such that (*) then $W(\alpha, \beta) \subseteq \eta$.*

(3) *$W(\alpha, \beta)$ is injectively presented.*

PROOF. (1) Let $\eta = W(\alpha, \beta)$, $\gamma \in \alpha$ and let δ be a function from γ' to η where $\langle \gamma, \gamma' \rangle \in S(\alpha, \beta)$. To show that $\langle \gamma, \delta \rangle \in \eta$.

As $\gamma \in \alpha$, $\gamma \doteq \tilde{\alpha}(x)$ for some $x \in \bar{\alpha}$. Hence $\langle \tilde{\alpha}(x), \gamma' \rangle \in S(\alpha, \beta)$ so that by 5.3(i) of [2] $\gamma' \doteq \tilde{\beta}(x)$. It follows that δ is a function from $\tilde{\beta}(x)$ to η so that, by 5.3(ii), $\delta \doteq S(\tilde{\beta}(x), \delta')$ for some $\delta' \in V$ such that $\bar{\delta}' = \overline{\tilde{\beta}(x)}$. As δ is a function with values in η we may use 5.3(i) again to get $(\forall u \in \bar{\delta}')\tilde{\delta}'(u) \in \eta$. Hence $(\forall u \in \bar{\delta}')(\exists z \in E)(\tilde{\delta}'(u) \doteq h(z))$ and we may use the type theoretical AC to get an $f \in \bar{\delta}' \to E$ such that $(\forall u \in \bar{\delta}')(\tilde{\delta}'(u) \doteq h(f(u)))$. So $\delta' \doteq (\sup u \in \overline{\tilde{\beta}(x)})h(f(u))$ and by 5.3(iii)

$$\delta \doteq S(\tilde{\beta}(x), \delta') \doteq S(\tilde{\beta}(x), (\sup u \in \tilde{\beta}(x))h(f(u))).$$

Hence finally $\langle \gamma, \delta \rangle \doteq h(\sup(x, f)) \in \eta$.

(2) Assume that $\eta \in V$ such that (**). To show that $W(\alpha, \beta) \subseteq \eta$; i.e. $h(e) \in \eta$ for $e \in E$.

We will do this by transfinite induction on E. So if $e = \sup(x, f)$ where $x \in \bar{\alpha}$ and $f \in \overline{\tilde{\beta}(x)} \to E$ then we wish to prove that $h(e) \in \eta$ under the induction hypothesis that

$$(\forall u \in \overline{\tilde{\beta}(x)})(h(f(u)) \in \eta).$$

If $\delta' = (\sup u \in \overline{\tilde{\beta}(x)})h(f(u))$ then $\delta' \in V$ and $\bar{\delta}' = \overline{\tilde{\beta}(x)}$ so that if $\delta \doteq S(\tilde{\beta}(x), \delta')$ then by 5.3 of [2] and the induction hypothesis $\delta \in V$ is a function from $\tilde{\beta}(x)$ to η. Also $\langle \tilde{\alpha}(x), \tilde{\beta}(x) \rangle \in S(\alpha, \beta)$ so that $\langle \tilde{\alpha}(x), \delta \rangle \in \eta$ by (**). Hence $h(e) = h(\sup(x, f)) = \langle \tilde{\alpha}(x), \delta \rangle \in \eta$.

(3) We shall use transfinite induction on E. We shall use a binary version which can easily be derived from the usual formulation. So if $z_i = \sup(x_i, f_i)$ for $x_i \in \bar{\alpha}$, $f_i \in \overline{\tilde{\beta}(x_i)} \to E$ for $i = 1, 2$ we shall prove (3) under the induction hypothesis that for $u_1 \in \overline{\tilde{\beta}(x_1)}$ and $u_2 \in \overline{\tilde{\beta}(x_2)}$

$$h(f(u_1)) \doteq h(f_2(u_2)) \supset f_1(u_1) =_E f_2(u_2).$$

So assume that $h(z_1) \doteq h(z_2)$. Then by the definition of h this implies that

(i) $\tilde{\alpha}(x_1) \doteq \tilde{\alpha}(x_2)$, and

(ii) $S(\tilde{\beta}(x_1), (\sup u \in \overline{\tilde{\beta}(x_1)}h(f_1(u))) \doteq S(\tilde{\beta}(x_2), (\sup u \in \overline{\tilde{\beta}(x_2)}h(f_2(u)))$.

As α is injectively presented $x_1 = x_2$.

Hence by 5.3(iii) of [2]

$$h(f_1(u)) = h(f_2(u)) \quad \text{for } u \in \tilde{\beta}(x_1)$$

so that by the induction hypothesis

$$f_1(u) = f_2(u) \quad \text{for } u \in \tilde{\beta}(x_1)$$

so that $f_1 = f_2$ and hence $z_1 =_E z_2$.

Proof of A2.3

This follows the lines of Section 7 of [2] where the validity of $\Pi\Sigma I$-PA is shown. In the statement of Theorem 7.1 the extra equality

$$\tau(W(A,F)) = W(\tau(A),(\sup x \in A)\tau(F(x)))$$

should be added. In the proof of 7.1 the following extra equation in the definition of σ is needed:

$$\sigma(W(A,F)) = h,$$

where h is defined by transfinite recursion on $W(A,F)$ so that

$$h(\sup(x,f)) = \langle \sigma(A)(x), S(\sup(F(x),\sigma(F(x)))), (\sup u \in F(x))h(f(u))\rangle.$$

The remainder of Section 7 goes through with only trivial changes provided that $\Pi\Sigma I$ is replaced by $\Pi\Sigma WI$ everywhere, and in the modified 7.5 we get the validity of $\Pi\Sigma WI$-PA.

References

[1] ACZEL, P., 1978, *The type theoretic interpretation of constructive set theory*, in: Logic Colloquium '77, A. Macintyre, L. Pacholski and J. Paris, eds. (North-Holland, Amsterdam).

[2] ACZEL, P., 1982, *The type theoretic interpretation of constructive set theory*: Choice Principles, in [8].

[3] BISHOP, E., 1967, *Foundations of constructive analysis* (McGraw-Hill, New York).

[4] BRIDGES, D. S., 1979, *Constructive functional analysis*, Research Notes in Mathematics, Vol. 28 (Pitman, London).

[5] MARTIN-LÖF, P., 1975, *An intuitionistic theory of types: Predicative part*, in: Logic Colloquium '73, H.E. Rose and J.C. Shepherdson, eds. (North-Holland, Amsterdam).

[6] MARTIN-LÖF, P., 1979, *Constructive mathematics and computer programming*, in: Proceedings of the 6th International Congress for Logic, Methodology and Philosophy of Science (North-Holland, Amsterdam).

[7] MYHILL, J., 1975, *Constructive set theory*, Journal of Symbolic Logic, Vol. 40, pp. 347–382.

[8] TROELSTRA, A.S. and VAN DALEN, D., eds., 1982, *The L.E.J. Brouwer centenary symposium* (North-Holland, Amsterdam).

Barcan Marcus et al., eds., Logic, Methodology and Philosophy of Science VII

PROVING PROGRAMS AND PROGRAMMING PROOFS

M.J. BEESON

Dept. of Mathematics and Computer Science, San Jose State Univ., San Jose, CA 95192, U.S.A.

"Proving programs" is computer science; "programming proofs" is logic. The meeting ground of the two is that both depend on formal systems; those of logic are designed for proving, those of computer science are designed for programming. Future progress, in particular progress in applications to large-scale computer sytems, depends on the design of new formal systems. What should these be like and how might they be used? This paper raises some issues (in its first half) and makes (in its second half) a technical contribution by considering some theories of Feferman from the viewpoint of computer science, and comparing them with the theories of Martin-Löf. Perhaps the best introduction to the paper is a listing of the section headings:

1. Proving programs: command language and assertion language?

2. Programming proofs: mathematics as a high-level programming language.

3. It matters how you do it, not just what you do!

4. Mathematics is dynamic as well as static.

5. The role of constructive mathematics.

6. Programming proofs: a logician's view of automatic program generation.

7. Programming proofs: Automatic deduction in artificial intelligence.

8. Formal systems, new and used: will the next generation be cheaper and better?

9. Types and data structures: Martin-Löf's theories.

10. Logic of partial terms.

11. Flexible typing: Feferman's theories.

12. Proving programs and programming proofs, revisited.

1. Proving programs: command language and assertion language?

By "proving programs" is meant more explicitly: proving properties of programs. Or better still: proving properties of the execution of programs. People speak of the "correctness" of a program with respect to its "specifications"; by this they mean that if the program gets an input of the kind it is designed for, it will produce an output having certain specified relations with the input. This notion is divided into the two notions of "total correctness" (an output is always produced and it is correct) and "partial correctness" (if an output is produced then it is correct). Theoretically-minded people invented these notions after considering the practical problems of "software reliability": one wishes to have certainty that the programs used in air-traffic control, in the systems that monitor nuclear reactor safety, in telephone exchanges, banking, air defense, etc., do not contain hidden bugs that will show up tomorrow under unusual conditions and cause a disaster. Since these programs are very large and usually written by teams, it is hard to check their correctness. Theoretically, it should be possible to provide "comments" in a suitable formal language, which would describe what the programmer thought should be true as that part of the code is executed. Then the program could be run through a "verifier", which would find an inductive proof of the partial correctness of the program. Note that in practical situations partial correctness is more vital than total correctness: what we are worried about is wrong answers that we don't know are wrong.

Theoretical studies of the problem typically proceed by setting up a formal language in which one can express "conditions", and writing $\{\phi\}P\{\psi\}$ to express the partial correctness of program P with respect to "input conditions" ϕ and "output conditions" ψ; that is, if the input satisfies ϕ, then the output of P, if any, satisfies ψ. Manna uses the words *command language* and *assertion language:* ϕ and ψ are written in assertion language, and P is written in command language. This reflects a distinction between *dynamics* and *statics:* P is supposed to do something, and ϕ and ψ are supposed to express facts about the static situations before and after P does something. This distinction between dynamic and static is a recurrent theme in our work; it recurs in several different situations, and the question is always: what is the best way to handle the static and dynamic aspects of a situation and their relationships? Separating the two into two separate languages necessitates a third formalism (e.g. Hoare's logic) to discuss their relationships.[1]

---

[1] Another phrase in the literature is "specification language", which means the same as

The distinction between dynamics and statics summarizes the difference between the programming languages of computer science and the formal languages of logic: the former are for doing things, the latter for stating and proving things. Separating the two into two separate languages for program verification is better than leaving one of them out entirely; but it is still artificial.

The command language and the assertion language should be the same

Assertions should be allowed to mention programs as well as input and output. Programs (algorithms) in the "command language" should be allowed to work on assertions. The distinction between data and program has long been recognized as artificial; the distinction between propositions (assertions) and programs is equally artificial. This point may require a little elaboration. First of all, it is well-known that every program may be regarded as data. It is less well-known that data may often by regarded as programs. For example, even numbers written in various customary notations are best thought of as programs. For example, 10^{10} represents a program for computing a certain natural number. It is in fact distinguished from most of its neighbors in the natural-number sequence by having this very short program in the simple "command language" of expressions built up from symbols for addition, multiplication, and exponentiation. This way of looking at data was explained to me by Per Martin-Löf: each data type, when specified, will include a *canonical representation* for each of its members (e.g. natural numbers are canonically represented by tally marks); then an object of this type is in general a program which "evaluates" to canonical form. We shall have more to say about this point below, when Martin-Löf's ideas are discussed more fully.

2. Programming proofs: Mathematics as a high-level programming language

The formal systems of logic were created in order to be studied, not in order to be used. It is an interesting exercise to try to formalize (for example) Hardy and Wright's number theory book in Peano arithmetic (**PA**). Any logician will see that it can theoretically be done, but to do it in

"assertion language". For the state of the art in specification languages, see e.g. the chapter of BURSTALL and GOGUEN in BOYER–MOORE [1981]. For the basic theory of program correctness see DE BAKKER [1980].

practice is far too cumbersome, for some reasons which are touched upon in Section 3. This has not bothered logicians, who (at least since *Principia Mathematica*) have not been interested in actually formalizing anything, but only in the possibilities of so doing. Similarly, they are satisfied with Turing machines or combinatory logic as a theory of computability. Anybody who tries to program a Turing machine to do anything complicated will realize why Pascal and LISP are needed: to makes programs which are machine-readable, and also comprehensible (that is: writeable and readable) by humans. Similarly, formal systems are needed in which one can write machine-readable proofs that are still comprehensible by humans. There are at least six projects in progress (of which the author is aware) in which elaborate computer systems have been constructed with this (or a similar) aim. These projects and their theoretical backgrounds are surveyed in BEESON [1983].[2] One lesson the creators of all these systems have had to learn is that what goes for programs goes for proofs: to be readable, they must be *well-structured*. To state the point clearly:

> *The systems we want must be as great an improvement over traditional logical systems like* **ZF** *and* **PA** *set theory, as modern computer languages like Pascal and* LISP *are over Turing machine language.*

To put is as graphically as possible:

$$\frac{?}{\mathbf{PA}} = \frac{\text{LISP}}{\text{TM}}$$

3. It matters how you do it, not just what you do!

In order to bring out more clearly what we consider the defects of **PA** as a high-level programming language, we shall consider an example in some detail: Euclid's algorithm for finding the greatest common divisor (gcd) of two numbers. We shall consider the example in LISP and then in **PA**, in order to bring out the advantages and disadvantages of each language. The algorithm can be expressed in a few lines of LISP:

---

[2] Study of these systems was an essential phase of the development of the ideas in this paper, but space limitations preclude a discussion of them here. The projects are: AUTOMATH, under the direction of de Bruijn at Eindhoven; PRL and related projects under the direction of Constable at Cornell; FOL under the direction of Weyhrauch at Stanford; LCF at Edinburgh; Algos under the direction of Graves; and the language PROLOG which is widely used in artificial intelligence research. The list of references contains a trail that can be followed by the interested reader.

```
(DEFUN EUCLID (N M)
  (COND ((EQUAL N 0) M)
    ((EQUAL M 0) N)
    ((LT N M)(EUCLID M (REM M N)))
    (T (EUCLID N (REM N M)))))
```

Translated into English: define a function EUCLID of two arguments N and M, as follows: if $N = 0$ the answer is M; if $M = 0$ the answer is N; if $N < M$ the answer is EUCLID(M REM(M, N)); where REM(M, N) is the remainder of M after division by N; otherwise the answer is EUCLID(N, REM(N, M)). It is an accolade to LISP that the translation is harder to read than the algorithm.

Now consider Euclid's algorithm in Peano arithmetic **PA**. The most obvious difficulty is that Euclid's algorithm is defined by recursion, and **PA** has no direct facility for definitions by recursion. Nevertheless, since the 1930's we have known how to make recursive definitions in **PA**; Gödel showed us how to use the Chinese remainder theorem to construct a formula $R(u, m, i, x)$ which can be thought of as "u codes a sequence of length at least m, of which x is the i-th member". This formula R can then be used to replace the recursive definition of EUCLID by an explicit definition of the relation $E(n, m, y)$ which holds if $y = \text{EUCLID}(n, m)$: $E(n, m, y)$ holds if there is a double sequence u_{ij} coding up the values of $E(i, j)$ for all $i < n$ and $j < m$; that is, if we think of u_{ij} as the value of $E(i, j)$ then the recursion equations for E are satisfied, and $u_{nm} = y$.

There are two points to be made about the treatment of Euclid's algorithm in **PA**. First, were we to be presented with the formula defining $E(n, m, y)$ explicitly, without explanation, we would require a long time to understand that it had anything to do with greatest common divisors. This contrasts with the extreme readability of the LISP algorithm above.

Second, the formula mentions no algorithm. It is a mere statement of some relationships between numbers. To the extent that one may say there is an algorithm *implicit* in the formula, *it is the wrong algorithm!* The recursively-defined algorithm EUCLID has been replaced by an iterative algorithm, requiring us to compute $E(i, j)$ for all i and j less than the given arguments. The result, of course, is the same as the result of Euclid's algorithm, but the method is different. The distinction between "iterative" and "recursive" algorithms is made in first-year computer science courses, but entirely ignored in traditional logic. The fact that primitive recursion and searching can be used to define every general recursive function (the so-called Kleene normal form theorem) shows, to the satisfaction of the logician, that iteration and recursion are the same thing. What has been overlooked is that an algorithm cannot be identified with the function that it computes. Philosophers use the words "extensional equality" and in this connection: two algorithms are extensionally equal if they produce outputs for exactly the same inputs, and always produce the same output at a given input, no matter what the internal workings of the algorithm. Intensional equality is a less well-defined concept; it refers to two algorithms differing only in inessential respects. We may then summarize the defect of **PA** to which we have called attention as follows:

PA *does not allow the intensionally-correct representation of all number-theoretic algorithms in a natural way.*

A logician may object that the theory of Turing machines can be formalized in PA, and the proof of the recursion theorem for Turing machines can be formalized, and so one can find a number which is the index of a Turing machine which works recursively in a manner similar to the algorithm EUCLID, provably in **PA**. In fact, one may by suitable Gödel numberings even formalize the theory of LISP, so that there is a code number of a coded LISP interpreter and a code number of algorithm EUCLID. But all this is an artifice; the objection is that **PA** does not allow the intensionally correct representations of all number-theoretic algorithms in a *natural* way.

Lest it seem that we should just forget about **PA** and work in LISP, let us now consider the one point in **PA**'s favor: After having defined E, we can give a formal proof in **PA** of the facts that for each n and m, there is a unique y such that $E(n, m, y)$ and this y is the greatest common divisor of n and m. We may not have the algorithm, but we have the correctness proof; and there is no way to prove anything in LISP, whose only statements are commands. What we need is a language in which we can do both.

4. Mathematics is dynamic as well as static

All that the usual formal results on representability of recursive functions in PA show is that *every number-theoretic fact can be stated in* **PA**. Traditionally, this has been felt to be satisfactory. But,

It only formalizes the static aspect of number theory, ignoring the dynamic aspect.

LISP and Pascal, on the other hand, formalize only the dynamic aspect, neglecting the static aspect. Informal mathematics typically includes both aspects. People say, "Now take x to be any number larger than y". That phrase has connotations of action, but in formalization it gets translated to the static hypothesis $x > y$, which is tacked on to all subsequent formulae in the argument. Informal mathematics is made up of statements like, "if you perform the following constructions, the result will be such-and-such". After translation into traditional formal systems, the dynamic feature is erased, replaced by a function symbol or symbols combined into a term of the formal language. The *evaluation* of the terms is regarded as a part of metamathematics, not built into the formal system.

We have already seen one example of this point in the preceding section. Another interesting example is furnished by interactive symbol-manipulation systems such as MACSYMA, or its cousin **vaxima** with which the author has had some interesting experiences. The dynamic and static aspects of mathematics receive some explicit attention in **vaxima**: every function name has a noun form and a verb form. Using the verb form causes the function to be applied and the result evaluated; using the noun form causes it to be left in symbolic form, e.g. sin(0) instead of 0. The same distinction applies as well to operations of what the logician calls "higher type", such as the operation DIFF of taking the derivative.

Like LISP, **vaxima** is an interactive language; the user communicates with the **vaxima** interpreter. At any time, this communication takes place in an *environment*, in which certain variables have been assigned values (the values can be numbers or defined functions). The phrase "now take x to be 2" which you might find in a mathematics book (or more likely in a conversation) becomes the **vaxima** command $x:2$. Then x has the value 2 until you change it. There is nothing corresponding to this in traditional logical languages like **PA**. One can, of course, *substitute* 2 for x as a step in a formal derivation. But this is a process which has to be done *outside* the system itself. Indeed, one of the principal technical lemmas in the elementary metamathematics of **PA** is that formal derivations are capable of "reflecting" all Turing-machine computations. I refer to the fact that all recursive functions are representable in **PA**; it says that any computation can be replaced by another computation with consists in searching for a formal proof of a certain formula (if all we care about is the result of the computation). *One pushes the dynamics out of the formal system into the metamathematics. The entire nature of the interactive relationship between the user and such a program as* **vaxima** *is alien to the view of mathematics fostered by the study of static systems such as* **PA**.

This discussion makes it clear why traditional formal systems can't be used for the study of program verification: the very problem of program verification involves a dynamic aspect. Of course, we can state in **PA** the theorem $\forall m\, \forall n\, \exists y\, A(n, m, y)$ which says that the primitive recursive function corresponding to Euclid's algorithm always produces an output. But since every primitive recursive function always produces an output, this is a triviality. The content of the termination of Euclid's algorithm has vanished in the reduction of EUCLID to some primitive recursive function.

There is another reason why traditional formal systems are inadequate for program verification: they don't provide anything corresponding to "the environment". Suppose we have a program P which is supposed to transform input conditions ϕ to output conditions ψ. Suppose further that the above difficulties do not arise, and program P can be adequately and naturally described in formal system T, by a term t of T. Then we may express the partial correctness of P by

$$\forall x (\phi(x) \rightarrow \psi(t(x))).$$

But this doesn't allow for the changes that the execution of P might make in the environment, i.e. for the "side effects" of the execution. While this formalism might work for a "one-run" algorithm like Euclid's algorithm, it is ill-adapted to programs where the "side effects" are as important or even more important than the input and output. Indeed, many programs are designed to "run forever", e.g. operating systems, so that the *only* interesting aspects are the side effects.

The inability of traditional formal systems to represent the environment comes up again when one considers the problem of natural formalization of mathematics. Suppose one tries to formalize, for example, Hardy and Wright's well-known number theory text. The first page goes rather well in **PA**. On the second page, one encounters the convention that the letter "p" will always stand for a prime number. The traditional logician will not worry: we just remember to preface every theorem mentioning the letter p by the formula defining "p is prime":

$$\forall x\, \forall y\, (x \cdot y = p \rightarrow x = p \vee x = \mathrm{s}(0))$$

(where s is the successor function). However, this will get hopelessly awkward as convention after convention has to be unwound in this fashion.[3]

---

[3] This example was brought to my attention by Richard Wehyrauch, who pointed out to me that his system FOL doesn't suffer from this defect.

Moral of the above discussion: The defects of traditional formal systems are the same, whether one is interested in program verification, or in interactive computerized mathematics, or in formal languages for mathematics which are readable both by machines and by humans.

There is another lesson to be learned from experience with **vaxima**: the user of **vaxima** soon learns that system never tells you how it gets the answer; and since the program has a few bugs (like any piece of software developed and modified by teams) there is room for doubt. The apparatus for *justifying* an answer, which is central in systems like **PA**, is entirely absent in **vaxima**.

The systems of the future should be able to answer the question, "How do you know that?"

This applies not only to mathematical systems, but to systems in artificial intelligence and in data base management. "Data base" is static, "management" is dynamic. As management systems get more sophisticated, the problem of justifying the answers they give us becomes more crucial. This is related to, but more complicated than, "program verification".

The problem of treating both the dynamic and static aspects of information is fundamental and arises in all branches of information science.

The solution will necessitate the construction of languages which can treat *both* statics and dynamics.

5. The role of constructive mathematics

The phrase "mathematics as a high-level programming language" is due to Bishop, whose book *Foundations of Constructive Analysis* kindled a new interest in constructive mathematics. By *constructive mathematics* I mean mathematics in which "there exists" means "we can find explicitly". Bishop's view is that if mathematics is properly written, one should be able to extract what he called "numerical information" from the proof. Every mathematical proof boils down, according to this view, to the assertion that if such-and-such computations are performed on the positive integers, they will have such-and-such results. The parallel with the formulation $\{\phi\}P\{\psi\}$ is striking: Bishop says every mathematical theorem should have this form.

Hence the phrase "programming proofs": if we start with a constructive proof, we should be able to extract a program from it, which contains the computational information implicit in the proof. Thus "programming proofs" has the sense: *extracting programs from proofs.*

Space for this paper is very limited; otherwise I would devote several pages to examples of extracting algorithms from proofs. If the subject is new to you, begin by looking up the usual proof of the existence of a greatest common divisor of two numbers, and observing that no algorithm can be extracted from it. Now find a proof from which Euclid's algorithm can be extracted.

Next, consider the different proofs of the standard existence theorem for the differential equation $dy/dx = f(x, y)$. The proof by successive approximations furnishes an algorithm; the proof by Arzela–Ascoli does not, since it depends on finding a convergent sequence in a compact set, which we have no algorithm to do.

The fundamental theorem of algebra is fertile ground for experimentation: there are many different proofs of it, and many different algorithms for finding roots of polynomials. Try to extract an algorithm from the least constructive proof, the one by Liouville's theorem! You will first have to constructivize the proof; or more accurately, find a constructive proof based on the non-constructive one. The idea is to compute the winding number around some nested squares; but to compute the integral you need squares on whose boundaries the function is bounded away from zero. Thus to find zeroes you need non-zeroes. The details may be found in WEYL [1924]. The example illustrates Bishop's point: many a proof that seems non-constructive actually does have a numerical content if one looks for it.

6. Programming proofs: a logician's view of automatic program generation

The general problem of automatic program generation is to produce automatically a program meeting certain specifications when presented with the specifications. Clearly this is asking too much: the program generator has to be told *how* to generate the program. As always when you want to know how to compute something, the right question to ask is

What additional data do I need?

That is, "what data in addition to the specifications that the program is supposed to need will enable me to find such a program?" The least we could ask for is a proof that the thing the program is supposed to compute actually exist! So an automatic program generator can be viewed as a device for extracting programs from proofs.

For example, in practice one has programs which generate parsers automatically, when one is given a suitable grammar for the language to be parsed. (Such a program was used to generate the mathematical typesetting preprocessor that the author used to prepare the manuscript of this paper. It translates, for example, "*x* sub *i* sup 2" into instructions which cause the typesetter to print "x_i^2".) A grammar (of the right kind) is in fact a kind of existence proof for a parser; the automatic parser generator passes from such an existence proof to a parser as output. In this case, there is no formal language in which the existence proof has to be expressed, because the domain of applicability of such a program generator is extremely narrow, although quite useful.

There is reason to believe that the extraction of programs from proofs may eventually permit the construction of much more useful and general automatic program generators.

Logicians have spent considerable effort in studying how algorithms can be extracted from proofs. Their conclusions may be summarized as follows:

(i) One has to use *constructive* proofs if one hopes to extract algorithms from them. There is an elegant logic which corresponds to constructive proofs, and it has been thoroughly studied.

(ii) One can extract algorithms from constructive proofs in at least two ways: by realizability (and its variants), and by cut-elemination (or normalization).

The rather large body of formal results which are summarized in these two short statements is hardly known to most computer scientists, and computer science is hardly known to most of the logicians who have developed these results. Hence the potential power of these methods is as yet untapped.[4]

The challenge is to *implement* the logical theory. By using the word "implement" I do not mean to imply that only a programming task remains. On the contrary, the difficult part of the task, it seems to me, lies in the construction of suitable languages, whose structure (syntax) mirrors the structures we want to talk about in a natural way.

7. Programming proofs: automatic deduction in artificial intelligence

John McCarthy has said, "A reasoning program should express its knowledge in logical terminology, and then deduce or infer a suitable

---

[4] Although Goad has begun to do so, see e.g. GOAD [1980].

action, and then carry out this action". Of course, this presumes that the program has some goals in the light of which it will decide what action is suitable. We draw an analogy between the world of mathematics and the world in which the reasoning program is supposed to operate (think of a factory environment or the well-known "blocks world"). From a formal proof that the goal of the program is a possible state of the world, we should be able to extract an algorithm (suitable sequence of actions) for achieving the goal, just as we extract number-theoretic algorithms from constructive proofs in number theory.

8. Formal systems, new and used: will the next generation be cheaper and better?

The common thread of the above examples is this:

The choice of formal system is crucial!

One decision that has to be made before constructing new formal systems is whether one wants a *typed* system, in which every object is known to belong to a certain data structure (in computer science language), or type (in logician's language); or whether one does not want to have this restriction built in to the language. So far, typed systems have been more fully developed for computer implementation than untyped systems (e.g. AUTOMATH). The author thinks that untyped (or better: flexibly typed) systems should be considered as well. To open the discussion, let us consider only one of the reasons: It is natural to use terms even when one doesn't have any idea if they actually denote anything, let alone what type it might be. The linguists are fond of "The present king of France". An example closer to computer science is, "the output of this program I just wrote", when you haven't debugged it yet. One will have a difficult time formalizing mathematics naturally without the use of such terms as $\sum_{n=1}^{\infty} a_n$, when one hasn't yet proved the convergence of the series. Another example, which may show that the problem is not irrelevant even for the most applications-oriented computer scientist: "The first available flight from San Francisco to Madrid on next June 17". If there is no available flight then this expression does not denote. How should a computerized travel agent deal with it?

We have just discussed whether every term must have a type. There is a related question: suppose a term does have a type, must that type be unique? In other words, must every name make clear what kind of an object it is supposed to name? A system in which this is true is said to have *strict typing.* It means, for example, that 7 must have different names when considered as an object of the type of integers representable in eight bits than when considered as an object of the type "bignum". This is not traditional in mathematics, where people think that "Seven is seven is seven"; but evidently it has its uses in computer science. Strict typing prevents the formation of "subtypes" in the natural way; a positive real number is not a real number as in traditional mathematics, but a real number together with a lower bound or "witness" to its positivity.

In the next decades, very large and complex computer systems will be designed to deal with the ever-increasing need for and flow of information. These systems will involve major software engineering projects, and in some cases (e.g. the Japanese Fifth-Generation Computer Systems project, described in FUCHI [1983]) hardware development as well. These systems will have powerful methods of defining data types; their designers must resolve questions such as how to deal with untyped terms, and whether the system should be strictly typed or not. The design of these systems needs a theoretical basis. In BEESON [1983], the author has reviewed several precursors of such systems and their theoretical bases. In this paper, the emphasis is on the theoretical side; but the issues raised here arise when practical applications are considered. We think that systems developed by proof-theorists for other reasons may turn out to be useful; here we make studies which are still very theoretical, but moving in the direction of eventual applications. The proof-theorists alluded to are Martin-Löf and Feferman, who have each presented formal systems in a series of papers. The main purpose of the rest of the paper is to present a version of Feferman's systems chosen with an eye to applications in computer science, and compare it with Martin-Löf's systems. In the process we will return to the themes mentioned in the title of the paper.

9. Types and data structures: Martin-Löf's theories

These are often called "type theories", because the idea idea behind them is the principle of "strict typing" discussed above. They grew out of Martin-Löf's proof-theoretical studies, and were at first mainly studied by proof-theorists, though the relevance to computer systems soon became apparent. The fundamental statements of the theory (called "judgments") have four possible forms:

$$t\colon A \text{ (read, } t \text{ is of type } A\text{)},$$

$$A\ \textit{type} \text{ (read, } A \text{ is a type)}.$$

The third possible form of judgment is

$$s = t\colon A \text{ (read, } s \text{ and } t \text{ are equal as objects of type } A\text{).}$$

On Martin-Löf's conception of "type", each type comes equipped with its own natural notion of equality. It is this notion that is meant in this form of judgment, not some underlying notion of absolute identity.

The fourth form of judgment expresses the equality of two data types,

$$A = B.$$

Intuitively, this is *extensional* equality, i.e. $A = B$ if and only if the same objects have type A as have type B.

The system contains primitive type constructors which enable one to construct product and sum types, starting from the basic types $\mathbf{N}$ of the natural numbers and $\mathbf{N}_k$ of the natural numbers less than a fixed number k. We shall briefly describe these constructors. If A is a type, and for each $x\colon A$, $B(x)$ is a type, and if $x = y\colon A$ implies $B(x) = B(y)$, then B is a called a *family of types* over A. In that case the product type $(\Pi x\colon A)B(x)$ consists of those operations f such that $x\colon A$ implies $f(x)\colon B(x)$ and $x = y\colon A$ implies $f(x) = f(y)\colon B(x)$. Note that if the base type A is $\mathbf{N}_k$, then the product type is what in computer science would be called an "array of length k", with i-th entry from $B(i)$.

Another important special case of the product type is when $B(x)$ is independent of x, say is a constant type C. Then $(\Pi x\colon A)B(x)$ is written C^A; it is the type of all functions from A to C.

With A and B as above, the sum type $(\Sigma x\colon A)B(x)$ consists of those pairs (x, y) with $x\colon A$ and $y\colon B(x)$. Note that if $B(x)$ does not actually depend on x, say $B(x) = C$, then the sum type in question is just the Cartesian product $A \times C$. (This is responsible for a notational confusion: sometimes the term "product type" refers to a type built using Π, sometimes to a cartesian product, which is formally a sum.)

Martin-Löf's systems are what a proof-theorist would call "logic-free". That is, they do not provide for the building up of complex expressions by the usual logical operations, "and", "or", and so on. Instead, logic is indirectly embedded, or defined, using the propositions-as-types idea. According to this scheme, every proposition is associated with a certain type: intuitively, the type of all (constructive) proofs of the proposition. Thus, for example, the proposition $A \mathbin{\&} B$ is associated with the Cartesian product of the types associated to A and B: in order to prove $A \mathbin{\&} B$, we have to give a pair (x, y) where x is a proof of A and y is a proof of B. This idea is a fundamental one, which has its historical roots in KOLMOGOROV [1929] and was developed by Howard and Tait in the proof theory of the fifties and sixties. For a more leisurely introduction, see [BEESON, 1985] (Chapter XI) or [MARTIN-LÖF, 1982].

Martin-Löf's philosophy calls for a *strictly typed* system, i.e. one in which every object has a unique type. Thus e.g. 7 as an object of type $\mathbf{N}_{13}$ is different from 7 as an object of type $\mathbf{N}_{11}$. His [MARTIN-LÖF, 1975] system had the corresponding formal property that if t: A and t: B are both provable, so is $A = B$. His [1982] system would have it too, but for a minor technicality.[5]

Martin-Löf has extended the possibilities for constructing types in two directions. First, some of his theories contain symbols for "universes". In the simplest such theory, there is just one universe, represented by a constant symbol U. Intuitively, this is the type of all "small types". This might be just the types mentioned above, or it might include others; the exact meaning of U is to be left open. Hence no axioms for proof by induction on the construction of elements of U are included. (The computer scientist who will not be satisfied with incompletely specified data types, may complete the specification of U as desired.) The main axioms that are included about U are that it is closed under the formation of product and sum types, and contains $\mathbf{N}$ and each $\mathbf{N}_k$.

The second direction in which the basic theory has been extended is to include some axioms for inductively-generated types. These rules are rather complicated. They do address an important issue, however, and some sound theoretical basis for inductive definitions will have to be provided before these theories can be effectively applied to the design of useful computer systems.[6]

10. Logic of partial terms

The purpose of this section is to describe one convenient logic for dealing with "partial terms", i.e. terms that may not denote anything. We

---

[5] The technicality in question is that the same constant $\mathbf{r}$ is allowed to be of any type $I(A, a, a)$. To recover the strict typing property we have to write $\mathbf{r}(A)$ instead of $\mathbf{r}$.

[6] With this in mind the author has worked out how one uses Martin-Löf's rules to introduce the data type List which is fundamental to the programming language LISP. This is an interesting and instructive exercise, but it is omitted here for lack of space. The principal difficulty to be resolved is that the definition of "list" flagrantly violates the principle of strict typing (as already discussed), while that principle is basic to Martin-Löf's systems. In other words, LISP is a type-free system; how can it be imitated in Martin-Löf's strictly-typed system? The answer is that one has to "cheat" by changing the definition of the type List to conform to strict typing.

shall give such a logic, compare it briefly with other such systems, give a semantics for it, and state some theorems about it which generalize well-known theorems about the predicate calculus to this situation.

LPT (logic of partial terms) is a logic in the same sense as the predicate calculus. If we are given any collection of predicate symbols, function symbols, and constants as in the usual predicate calculus, there will be a language in LPT based on these symbols. The rules for forming terms are the same as in ordinary predicate calculus. Every atomic formula in the usual sense is still an atomic formula; but there is one more kind of atomic formula, namely: if t is a term then $t\downarrow$ is an atomic formula. This may be read "t is defined". It should be emphasized, however, that the intended meaning is that the term "t" denotes something. That is, one says of an object that it exists, of a term that it denotes or is defined. All objects exist, of course, so that to say something does not exist is a figure of speech; what is meant is that the term we have mentioned does not denote.[7]

In case equality is part of the language, we use $t \cong s$ to abbreviate $(t\downarrow \rightarrow t=s)$ & $(s\downarrow \rightarrow t=s)$. In words: if either t or s denotes anything, then they both denote the same thing. Note, however, that $\cong$ is not an official part of the language.

We shall use the notation $A[t/x]$ to mean the result of substituting t for the free occurrences of x in A. The customary inference from $\forall xA$ to $A[t/x]$ is not valid if t is a non-denoting term: "if everything exists then the king of France exists" is an invalid inference, since the antecedent is true but the consequent is false. We are now ready to set out a list of rules and axioms for making correct inferences in LPT. In this list, t and s are terms, while x and y are variables.

Axioms and rules of LPT

$$\frac{B \rightarrow A}{B \rightarrow \forall xA} \quad \text{if } x \text{ is not free in } B \tag{Q1}$$

$$\frac{A \rightarrow B}{\exists xA \rightarrow B} \quad \text{if } x \text{ is not free in } B \tag{Q2}$$

[7] It may seem that the above is too obvious to state, but there is an entire book devoted to the subject of "Non-existent objects". It is dedicated to "my parents, without whom I might have been one". What the author means is that without his parents, his name would have been a non-denoting term. Compare the famous Zen koan which asks for your original face, the one you had before your father and mother were born. Another non-denoting term; but part of the point of Zen is to break the confusion between words and reality; a word is only an approximate description of the reality it denotes.

$$\forall xA \;\&\; t\downarrow \rightarrow A[t/s] \tag{Q3}$$

$$A[t/x] \;\&\; t\downarrow \rightarrow \exists xA \tag{Q4}$$

$$x = x \;\&\; (x = y \rightarrow y = x) \tag{E1}$$

$$t \cong s \;\&\; \phi(t) \rightarrow \phi(s) \tag{E2}$$

$$t = s \rightarrow t\downarrow \;\&\; s\downarrow \tag{E3}$$

$$R(t_1, \ldots, t_n) \rightarrow t_1\downarrow \;\&\; \cdots \;\&\; t_n\downarrow \tag{S1}$$

$$c\downarrow \text{ for constants } c \tag{S2}$$

$$x\downarrow \text{ for variables } x \tag{S3}$$

Note that E3 is a special case of S1. Another special case of S1 worthy of special mention[8] is:

$$f(t_1, \ldots, t_n)\downarrow \rightarrow t_1\downarrow \;\&\; \cdots \;\&\; t_n\downarrow$$

Semantics of LPT

LPT has a natural semantics, both classically and intuitionistically. For simplicity we consider the classical semantics first. A *partial structure* is like a structure (that is, it consists of a set and some relations and functions to match the symbols of the language), except that the function symbols can be interpreted by *partial* functions, i.e. functions not necessarily everywhere defined. Note that the relations are treated as usual; there is no such thing as a "partial relation". Let M be a structure; we then wish to define Val(t) the value of t in M for each term t. This is done by induction: if $\hat{f}$ is the partial function which interprets the function symbol f in M, we set Val($f(t)$) to be $\hat{f}$(Val(t)), and similarly if f takes several arguments. This rule will assign values to certain terms t and leave Val(t) undefined for some terms t; to be precise, we are taking the least fixed point of this inductive definition. We then say that the formula $t\downarrow$ holds in M if and only if Val(t) is defined. The rest of the definition of satisfaction is the same as for the ordinary predicate calculus.

[8] This axiom may well turn out to be too strict for some future applications. It prevents, for instance, the possibility that "the throne of the king of France" might denote something even though there is no present king of France; or more practically, that "Seat 13B on the first available flight to Madrid on June 17" might denote something even if there is no such available flight. We take the view that this is correct: the phrase may well have a *meaning*, but that is more subtle and depends on the context of the phrase. It does not have a *denotation*. Its meaning, if any, is the *reference* in the sense of Frege.

Now we consider the intuitionistic semantics corresponding to Kripke models. A partial Kripke structure is like a Kripke model, except that the function symbols are to be interpreted by partial functions, subject to the restriction that if $f(x)$ is defined at one node of the model, it must also be defined at any higher node (and of course take the same value). Then for each node α there is a function Val_α such that $\mathrm{Val}_\alpha(t)$ is the thing denoted by t at node α, if any; and as above we say that $f(t)\downarrow$ holds at node α iff $\mathrm{Val}_\alpha(t)$ is defined.

Comparison to Scott's logic

Scott [1979] has given a logic similar to LPT, but with slightly more general aims and a different motivation. The result has at least one defect, in the author's opinion, in that $\forall xA$ is not equivalent to $A(x)$. We are so accustomed to being able to omit universal quantifiers when stating axioms or results that it is quite awkward to work in a logic where this is illegal. The root of the difficulty is the different conception underlying Scott's logic: he is thinking of models in which some objects "exist" and some do not. Thus what we write as $t\downarrow$, Scott would write as $E(t)$, which is to be read "t exists". Variables are to range over all objects, existing or not, and bound variables are to range only over existing objects. In other words, Scott treats existence like an ordinary predicate, a property of objects and not of terms. In certain contexts, this is not entirely unnatural: for example, in studying models of the λ-calculus, one may wish to make a model whose elements are all terms, and where only the normal terms "exist". It was such situations that led Scott to create his logic.

Scott's logic is more general than LPT in that is also deals with partial predicates and with descriptors. A descriptor is a term of the form "the x such that $\phi(x)$"; or even "some x such that $\phi(x)$". These will in general be partial terms, since there may not be any suitable object x. A systematic treatment of descriptors should be possible on the basis of LPT, but we have purposely not undertaken it here in order not to obscure the basic issues. Renardel [1982] gives an excellent survey of the literature on descriptors.

Translation of LPT *to ordinary predicate calculus*

It is possible to reduce LPT to ordinary predicate calculus in a straightforward way. Namely, to every function symbol f we associate a predicate symbol R_f to stand for the graph of f. We then assign to each term t a formula $A_t(x)$ with the intuitive idea that $A_t(x)$ should be true when x is the value of t. The definition of A_t is by induction on the complexity of the term t; there is one inductive clause corresponding to each function symbol f. If f is unary, that clause is

$$A_{f(t)}(x) \text{ is } \exists y(A_t(y) \mathbin{\&} R_f(y, x)).$$

The clause corresponding to a function symbol with more than one argument is similar. The base clause of the inductive definition is the case when t is a variable or constant. In that case we take $A_t(x)$ to be $x = t$; so $A_{f(x)}(x)$ comes out equivalent to $R_f(z, x)$, as it intuitively should.

Next we translate every formula of LPT into a corresponding formula of ordinary predicate calculus. Each atomic formula of the form $t\downarrow$ is translated to $\exists x A_t(x)$. An atomic formula of the form $R(t)$ is translated to $\exists x(A_t(x) \mathbin{\&} R(x))$. The translation commutes with the logical connectives and quantifiers. It is a sound translation in the following precise sense:

PROPOSITION. *The translation of every theorem of* LPT *can be derived in ordinary predicate calculus, supplemented by the axioms asserting that each* R_f *is the graph of a partial function. This is true for both the intuitionistic and classical versions of* LPT.

PROOF. The axioms mentioned in the theorem are $R_f(x, y) \mathbin{\&} R_f(x, z) \rightarrow y = z$. These permit one to prove $A_t(x) \mathbin{\&} A_t(y) \rightarrow x = y$ for each term t. Let B^0 denote the translation of B. One then proves by induction on the complexity of the formula B that

$$A_t(y) \mathbin{\&} B[t/x]^0 \rightarrow B^0[t/y].$$

This makes it easy to verify the translations of Q3 and Q4. The rest of the axioms are easily checked. One then proceeds by induction on the length of the proof of a theorem of LPT. □

REMARK. This translation is similar to the device used by FEFERMAN [1975, 1979] to avoid a logic of partial terms in his theories. However, it is not exactly the same. That is, Feferman uses $\phi(t)$ as an abbreviation expressing that t denotes and $\phi(y)$ is true, where y is what t denotes. With this convention, if we take $\phi(x)$ to be $\neg x\downarrow$, then $\phi[t/x]$ is false (no matter what term t is). In LPT, however, $\phi[t/x]$ might be true, if t is a term that does not denote.

The converse of the proposition is also true; that is, the translation of LPT into predicate calculus is *faithful* in the sense that if the translation of a formula A is provable in predicate calculus plus the axioms for the R_f, then A is provable in LPT. To see this, we just replace every atomic formula $R_f(t, s)$ by $f(t) = s$ and observe that theorems of the predicate calculus go over to theorems of LPT, as do the axioms $R_f(x, y)$ & $R_f(x, t) \rightarrow y = z$. This simple observation has an interesting corollary:

THEOREM. (*Completeness of* LPT): *If* LPT *does not prove* ϕ *then there is a model in which* ϕ *does not hold.*

REMARKS. If we take LPT with only intuitionistic logic, then "model" means "Kripke model". The completeness theorem itself is not constructive

PROOF. Suppose ϕ is unprovable in LPT. Then by the faithfulness of the translation into predicate calculus, its translation ϕ^* is unprovable in predicate calculus. By the completeness theorem for predicate calculus, there is a counter-model to ϕ^* in which each relation symbol R_f is interpreted in this model by the graph of a partial function $\hat{f}$. Using $\hat{f}$ to interpret the function symbol f, we get a model of LPT in which ϕ fails. □

11. Flexible typing: Feferman's theories

FEFERMAN [1975, 1979] introduced theories of "operations and classes" with the purpose of formalizing Bishop's constructive mathematics. We shall here formulate a minor variant of these theories with the needs of computer science in mind. Feferman's "classes" can be thought of as "data structures"; instead of reading $x \in A$ as "x is a member of the class A", we can read it as "x is an object of type A". In order to emphasize this reading, we shall write $x : A$ instead of $x \in A$, as we did for Martin-Löf's theories above. We shall call our version of Feferman's theories FT, which stands for "flexible typing". This name reflects our view about what is important and likely to be useful about these theories: they permit the formation of types, without requiring that every object has to have a type or that the type has to be unique. Theories with this property have sometimes been called "untyped", but that is a misnomer, since objects may very well have types. If a programming language were to make use of constructions like those in FT, we could certainly declare the types of variables if we chose to; so all the facilities available in a typed situation would be available here also — but not compulsory.

FT is a two-sorted theory; we use small letters for individuals, and capital letters for data types. The underlying logic is the logic of partial terms. This may be taken either classical or intuitionistic; for definiteness, and because we think it is the appropriate logic for theories of computation, we take the intuitionistic version. The underlying idea in representing data structures in Feferman's theories is that a data structure has two aspects: it is a *classifier*, as when we say that an object has type X. On the other hand, it is in turn a piece of data itself, as when we want to manipulate it, e.g. in using it to form a new data type. These two aspects should not be confused: for instance, if data types are treated as classifiers, equality should be extensional (two types X and Y should be equal if everything of type X is also of type Y and vice versa). On the other hand, if they are treated as pieces of data, there is every reason to want to distinguish between two differently-constructed types which happen to classify the same objects. In Feferman's theories, this distinction is easily made: a data type as a classifier is represented by a capital letter, and the same type as an object is represented by a small letter. The theory FT includes a function symbol E such that $E(x)$ is the extension of X, that is, the data structure *qua* classifier whose name is the object x. Since FT is based on the logic of partial terms, it will not matter if $E(x)$ is sometimes undefined, which it will be if x is not the name of a data structure.

One of the kinds of atomic formulae in FT is $x: A$, where x is an individual variable and A is a type variable. There are also atomic formulae $x = y$ and $X = Y$; that is, there are two kinds of equality. It is not allowed, however, as in some of Feferman's theories, to write $x = Y$. This is replaced by $E(x) = Y$ in FT, which we think represents a more carefully considered view of the relationship between individuals and data types.

In FT, it is not allowed to quantify over type variables. In this respect FT is not a theory in the ordinary logic of partial terms; with respect to its second sort of variables it is quantifier-free. To be precise, the definition of "formula" is given by the usual clauses, except that the clauses permitting quantification over capital-letter variables are omitted; and in stating the rules of inference, the quantifier rules and axioms involving type-variable quantifiers are omitted. This feature of FT is not absolutely essential, but it simplifies several definitions and the metatheory in general, and seems to correspond to a certain intuition that data structures as classifiers form too vague a universe to quantify over; individuals, on the other hand, correspond to things that can be stored in the computer, and are quite concrete. Note that we do not assume that every data structure has a name; nor would we be able to state that assumption in FT.

Feferman's theories are constructed so as to permit the definition of operations by recursion, i.e. to make the recursion theorem provable. In order to arrange this, one has to settle upon a basic theory of operations. Feferman chose to base his theories on combinatory logic. Recently HAYASHI [1983] has given a variant of Feferman's theories based on LISP. We think this is a step in the right direction (towards applicability), but there is no doubt that it complicates the metatheory. In order not to distract attention from the main issue here, which is the proper treatment of data types, we base FT on combinatory logic like Feferman's theories. In eventual applications, of course, this will have to be changed.[9]

We want to build in certain operations for the construction of data types. It is natural to take operations corresponding to the product and join constructions in Martin-Löf's theories. We arrange this by including constants Π and Σ. The idea is that if a is a name of A, and if $f(x)$ is of type $B(x)$ whenever x is of type A, and if $b(x)$ is a name of $B(x)$ whenever x is of type A, then $\Sigma(a, f)$ and $\Pi(a, f)$ are names of what Martin-Löf would call $(\Sigma x: A)B(x)$ and $(\Pi x: A)B(x)$, respectively. Note that Σ and Π are operations on data structures as objects, not on data structures as classifiers; for short we can say they operate on names of data structures. The meaning of Σ and Π would be determined by some compiler which would produce computer representations for new data types when given representations for the component types.

To start the process of type construction, we need some basic types. In practice one would want at least the types of lists, character strings, fixed-point numbers, etc. For the present theoretical purposes, it suffices to take only one basic type, that of the natural numbers. FT includes constants $\mathbf{N}$ for the natural numbers. Note that $\mathbf{N}$ is an individual constant, not a second-order constant, in spite of the fact that a capital letter is used for it. However, as a practical matter, it is convenient to write $x: \mathbf{N}$ instead of $x: E(\mathbf{N})$. No ambiguity is possible; since $\mathbf{N}$ is an individual constant, when it occurs on the right of the colon it must be abbreviating $E(\mathbf{N})$. We shall quite generally omit to write E in places where the restoration of E is obvious.

In FT, typing is not strict; that is, the same object may have several different types. The number 8, for example, is of the type of integers, the

---

[9] Exercise: work out the term in combinatory logic (even allowing the extra combination of FT) which denotes the algorithm EUCLID. This term is constructed by the fixed-point theorem as $\lambda x.t(xx)(\lambda x.t(xx))$ for a suitable term t expressing the recursion equations EUCLID is supposed to satisfy. These equations involve some arithmetic operations and a definition by cases. The term requires the better part of a handwritten page to write out in FT.

type of even integers, and the type of integers smaller than 256. This interpretation of typing permits the construction of subtypes in a straightforward way. Namely, if ϕ is a formula, and A is a type, named by a, then we can construct the subtype $c_\phi(a)$. The objects of this type are exactly those x of type A for which $\phi(x)$ is true. This will not be permitted for every formula ϕ of the system, but only for so-called *elementary* formulae.

DEFINITION. The formula ϕ is called *elementary* if only free variables occur on the right of the colon in ϕ; that is, if in every subformula $x: A$ of ϕ, A is a free variable.

Thus what is not allowed is terms containing E. For example, the formula $\neg x: E(x)$ is not elementary. The idea behind this restriction is that any parameters in ϕ should stand for already-constructed data types.

The exact "universe" of data types is not specified, even in the "intended model". It is purposely left open; perhaps it is just those data types which can be built up by terms of FT. Perhaps, on the other hand, it is much larger, consisting of all data types anyone might ever define. It turns out that the theory is consistent with the existence of the data type V of all (names of) data types. (The Russell paradox is blocked by the restriction to elementary formulae in the subtype construction.)

Language and axioms of FT

Constants. 0, $\mathbf{p}_{\mathbf{N}}$, $\mathbf{s}_{\mathbf{N}}$, $\mathbf{d}$, $\mathbf{N}$, Π, Σ, $\mathbf{k}$, $\mathbf{s}$, $\mathbf{p}_0$, $\mathbf{p}_1$, Dom, Ran. There is also a constant c_n for each integer n. If n is the Gödel number of an elementary formula $\phi(y, X)$ we write c_ϕ for c_n. (We are uninterested in the c_n for other n.)

Function symbol. There are two function symbols in FT: Ap and E. Ap takes individual arguments and individual values; E takes individual arguments and type values. We always abbreviate $Ap(t, s)$ by $t(s)$ or just ts. In longer terms, association is presumed to be to the left, e.g. xyz abbreviates $(xy)z$. Operations with several arguments are treated as usual in combinatory logic, e.g. $f(x, y)$ means $f\,xy$.

Type Construction Axioms.

$$\forall x: A(E(fx)\downarrow) \leftrightarrow E(\Sigma(A, f))\downarrow \qquad \textit{Join}$$

$$E(\Sigma(A, f))\downarrow \rightarrow (x: \Sigma(A, f) \leftrightarrow \mathbf{p}_0 x: A \;\&\; \mathbf{p}_1 x: E(fx)) \qquad \textit{Join}$$

$$\forall x: A(E(fx))\downarrow \leftrightarrow E(\Pi(A, f))\downarrow \qquad \textit{Product}$$

$$E(\Pi(A,f))\downarrow \to (x\colon \Pi(A,f) \leftrightarrow \forall z\colon A(xz\colon E(fz))) \qquad \textit{Product}$$

$$E(z)\downarrow \to E(\mathrm{Dom}(z))\downarrow \ \&\ (x\colon E(\mathrm{Dom}(z)) \leftrightarrow \exists y(\mathbf{p}xy\colon E(z))) \quad \textit{Domain}$$

$$E(z)\downarrow \to E(\mathrm{Ran}(z))\downarrow \ \&\ (y\colon E(\mathrm{Ran}(z)) \leftrightarrow \exists x(\mathbf{p}xy\colon E(z))) \quad \textit{Range}$$

$$E(a)\downarrow \ \&\ \ E(x)\downarrow \to (z\colon c_\phi(a,y,x) \leftrightarrow z\colon E(a)\ \ \&\ \ \phi(z,y,E(x)))$$

for ϕ elementary *Separation*

Here all the free variables of ϕ are shown; y and X may be lists of variables, and $E(X)$ abbreviates the conjunction of the $E(X_i)$.

Axioms for programs

$$\mathbf{k}xy = x$$

$$\mathbf{s}xy\downarrow \ \&\ \mathbf{s}xyz \cong xz(yz)$$

$$\mathbf{p}_0(\mathbf{p}xy) = x$$

$$\mathbf{p}_1(\mathbf{p}xy) = y$$

$$a\colon \mathbf{N} \to \mathbf{d}aaxy = x$$

$$a\colon \mathbf{N}\ \&\ b\colon \mathbf{N}\ \&\ \neg\, a = b \to \mathbf{d}abxy = y$$

Axioms for the natural numbers

$$0\colon \mathbf{N} \qquad \textit{zero}$$

$$x\colon \mathbf{N} \to \mathbf{s}_{\mathbf{N}}x\colon \mathbf{N} \qquad \textit{successor}$$

$$x\colon \mathbf{N} \to \mathbf{p}_{\mathbf{N}}(\mathbf{s}_{\mathbf{N}}x) = x \qquad \textit{predecessor}$$

$$x\colon \mathbf{N}\ \&\ \neg\, x = 0 \to \mathbf{s}_{\mathbf{N}}(\mathbf{p}_{\mathbf{N}}x) = x \qquad \textit{successor-onto}$$

$$0\colon A\ \&\ \forall x(x\colon \mathbf{N}\ \&\ x\colon A \to \mathbf{s}_{\mathbf{N}}x\colon A) \to \forall z(z\colon N \to z\colon A) \qquad \textit{induction}$$

Note that the induction axiom corresponds to what proof-theorists call "restricted induction": the only thing that can be proved by induction in FT is that something is of a given type. An arbitrary formula can be proved by induction in FT only if it can be shown that the formula defines a type. This is not particularly important from the point of view of applications, but to the proof-theorist it is important since it will determine the "proof-theoretic strength" of the theory. If we put in only restricted induction, the resulting theory FT has the strength of arithmetic; if we put in full induction, i.e. induction for arbitrary formulae, then the theory has the strength of Σ^1_1-AC. (See BEESON [1985], Chapter XII, for a survey of similar results on proof-theoretic strength.)

Relation of FT *to Feferman's formulations of his theories*

FT can be translated into Feferman's (two-sorted) versions of his theories by erasing "E" and replacing "x: A" by "$x \in A$". The reverse translation is not possible since FT does not permit quantifiers over class variables. We shall show that nevertheless, theories with class quantification are conservative over the corresponding theories without.

PROPOSITION. *Let T be* FT *or an extension of* FT. *Let S be the corresponding theory in Feferman's formulation, i.e. with class quantifiers allowed and without E. Then T proves the same elementary theorems as S.*

PROOF. Let ϕ be elementary and suppose S proves ϕ. Then for some finite list Γ of axioms of S, there is a proof p in predicate calculus (formulated Gentzen-style) of the sequent $\Gamma \vdash \phi$. By the cut-elimination theorem, we may suppose p is a cut-free proof. Since the language of T does not permit class quantification, no class quantifiers appear in Γ, and hence, by the subformula property of cut-free proofs, not in the proof p either. The proof p is not yet a proof of ϕ in FT, since FT is formulated in LPT and not in ordinary predicate calculus. But by adding E on the right of $\in$ (and replacing "$\in$" by ":"), we convert p to a proof of the translation of ϕ in predicate calculus. (Note that Feferman's predicate App(x, y, z) is just the R_f mentioned in the general translation of LPT into predicate calculus, when f is the application operator Ap of FT.) Since we have already seen that the translation is faithful, FT proves ϕ. □

Relation of FT *to Martin-Löf's theories*

We shall interpret Martin-Löf's simplest system $\mathbf{ML}_0$ in FT. For a description of $\mathbf{ML}_0$ see BEESON [1982] or BEESON [1985]. It is essentially the system of MARTIN-LÖF [1982] with no "universes". This system can be interpreted in intuitionistic arithmetic, as shown in BEESON [1982] and BEESON [1985]. So of course it can also be interpreted in the stronger theory FT. However, the interpretation used for this proof-theoretical result consists in formalizing the "recursive model" of $\mathbf{ML}_0$. Our interest in this paper is different; we are not just interested in the traditional concern of proof theory, the proof-theoretic strength. We are interested in whether there is a *meaning-preserving* translation between Feferman's and Martin-Löf's theories. Are they talking about the same kinds of data types? We think that in one direction at least, the answer is yes: the data types of Martin-Löf can be discussed in Feferman's theories.

We shall give what we claim is a meaning-preserving translation from $\mathbf{ML}_0$ to FT. The idea of the interpretation is to translate Martin-Löf's types as pairs (x, y) where $E(x)$ and $E(y)$ are both defined and $E(y)$ is an equivalence relation on $E(x)$. Since FT contains the product and join constructions basic to $\mathbf{ML}_0$, the only problem is to define suitable equivalence relations on product and join types. But this too is straightforward: first consider a join type $\Sigma(A, f)$. Two members (x, y) and (a, b) of $\Sigma(A, f)$ should be equal if and only if x and y are equal as objects of type A, and y and b are equal as objects of type $E(fx)$. Note that $E(fx)$ and $E(fa)$ have to be equal types in order that Martin-Löf will count f as a "family of types over A"; and only under that condition will he allow the formation of $\Sigma(A, f)$; so there will be no conflict: if y and b are equal as objects of type $E(fx)$, they will be equal as objects of type $f(a)$ too.

Now consider a product type $\Pi(A, f)$. We shall set u and v to be equal as members of $\Pi(A, f)$ if whenever x is of type A, ux and vx are equal as objects of type $E(fx)$.

This informal translation can be turned into a formal interpretation, under which each provable judgment A of $\mathbf{ML}_0$ is translated into a theorem of FT. The detailed definition and verification have been omitted for lack of space.

Now consider the converse. How could we interpret FT meaningfully in $\mathbf{ML}_0$? The difficulty is that FT is not strictly typed, while $\mathbf{ML}_0$ is. Note that when we interpret $\mathbf{ML}_0$ into FT, we need only a few types defined by elementary separation[10] domain, and range. It is these constructors, however, that permit us to construct subtypes. *We think that this difference between* FL *and* $\mathbf{ML}_0$ *is interesting and important, bearing as it does on the design of next-generation computer systems.*

One aspect of the difference between FT and $\mathbf{ML}_0$ is that FT permits the use of existential quantification in defining subtypes. For example, we can define the type of even integers as the type of integers n for which there exists an m with $2m = n$. In $\mathbf{ML}_0$, we can only define the type of pairs (m, n) with $2m = n$. Here m is an example of what is called a *witness*, in this case a witness to the even-ness of n. The type constructors of $\mathbf{ML}_0$ permit only the construction of "fully-presented" types, with all witnesses explicity present. This is good: all the information is carried along, and so is readily accessible if needed. It is also bad: all the information is carried along; it takes up space, distracts attention, and has to be manipulated.

---

[10] For technical purposes in interpreting e.g. the I-rules of Martin-Löf's theories.

The metamathematical technique known as *realizability* can be used to recover the "missing witnesses" when subtypes are defined using separation in FT. There is no space in this paper to explain realizability; we refer the reader to BEESON [1985] for a discussion of witnesses and realizability. The point is that to each type A is associated the type of pairs (x, y) for which x realizes $y: A$. This latter type is defined by a negative formula of FT, i.e. one without any existential quantifiers or disjunctions. This construction is similar to the method usually used to interpret (for example) arithmetic in Martin-Löf's theories. (See e.g. BEESON [1982].)

We shall not present purely formal interpretations of FT into $\mathbf{ML}_0$, since the point of interest is to compare the meanings of the theories; and the domain of discourse of FT is somewhat broader than that of $\mathbf{ML}_0$. There is, however, a natural translation of FT into Martin-Löf's theory with one universe, $\mathbf{ML}_1$, which interprets the type variables as ranging over the universe U, and the combinators as suitable terms built up by the operation of "abstraction" permitted in Martin-Löf's theories. The range of the individual variables may also be taken as U. One generalizes the interpretation of arithmetic in a straightforward way. Even full induction is soundly interpreted; which is pleasant, since FT plus full induction and $\mathbf{ML}_1$ have the same proof-theoretic strength.

Extensionality

There is another difference between FT and $\mathbf{ML}_0$: extensionality. MARTIN-LÖF [1982] has extensionality built in, in the sense that types come equipped with equality relations, and families of types are supposed to respect these equality relations, i.e. if x and y are equal objects of type A and f is a family of types over A, then fx and fy have to be equal types. This is not required in FT. It is also not required in the [1975] version of MARTIN-LÖF'S theories, and hence is not essential to that style of theory, although he now feels that it is necessary to a coherent philosophical explanation of his notions. There are some interesting formal results in connection with extensionality. For example, in BEESON [1982] it is shown that Church's thesis is refutable in Martin-Löf's theories; extensionality plays a key role in making Church's thesis mean something other than it would in the absence of extensionality. (Hence this result cannot be interpreted as an argument against the constructivity of Martin-Löf's theories.)

Another interesting formal result about extensionality is the theorem of Gordeev (which can be found in BEESON [1985] (Chapter X, section 11) or

[1982b]) that extensionality is inconsistent with Feferman's theories. In the context of Feferman's theories, by extensionality we mean

$$\forall x(x: A \leftrightarrow x: B) \rightarrow A = B) \qquad (Ext)$$

Gordeev's proof works for Feferman's versions of his theories, but it does not work for FT. If one tries to translate the proof into FT, one finds that one would need the principle

$$E(x) = E(y) \rightarrow x = y$$

to make the proof work. Otherwise put: Gordeev's proof shows that FT is inconsistent with (*Ext*) plus the principle just mentioned. In Feferman's formulation of his theories, there is no distinction between types as classifiers and types as names, so (*Ext*) carries the meaning that two types classifying the same objects have equal names. In FT, (*Ext*) only says that two types which classify the same objects are equal as classifiers. This seems to be much weaker and might even be taken as defining the intended meaning of equality of types. These considerations raise the formal question whether (*Ext*) is consistent with FT. We answer this question in the affirmative.

THEOREM. FT *is consistent with* (*Ext*).

REMARK. The proof applies as well if full induction is added to FT.

PROOF. Let M be the model of FT constructed by FEFERMAN'S [1975] method. That is, the universe of M (the range of the individual variables) is the natural numbers, and the operations Π and Σ, as well as the pairing functions, are interpreted by indices of some trivial functions, so that e.g. $\Pi(A, b)$ is interpreted as $\langle 1, A, b\rangle$. Similarly, we interpret $c_\phi(x, Y)$ as $\langle m, x, Y\rangle$, where m is the Gödel number of ϕ. We interpret the function E as the identity function on a certain inductively defined set CL which will serve to interpret the type variables. CL is defined simultaneously with the relation $M \models x: A$ as follows: the number chosen to interpret N is in CL, and if A has already been put in CL, and $M \models x: A$ implies $\{e\}(x)$ is in CL, then the interpretations of $\Pi(A, e)$ and $\Sigma(A, e)$ are in CL. If A and Y have already been put in CL, then the interpretation of $c_\phi(A, x, Y)$ is in CL. These clauses enable us to advance one stage in the inductive definition of CL if we know the relation $M \models x: A$ for A already in CL. To continue, we must be able to determine this relation for the A just added; but that can be done, since it only requires to determine whether M

satisfies some elementary formulae with type parameters already known to be in CL. To determine this, we need only know the "type of" relation restricted to those parameters; but that is already known, by induction hypothesis.

The reader who finds this proof too informal may find details in BEESON [1985]. So far, nothing new has been added to Feferman's original model construction. But now we come to the point: in FT, we have a separate equality relation for the type variables. We are free to define the conditions under which M will satisfy $A = B$. We define this to hold just in case M satisfies $x: A$ if and only if it satisfies $x: B$. Since equality between type variables cannot occur in an elementary formula, how we define this relation does not affect the fact that the model satisfies the non-logical axioms of FT. We only have to check that the equality axioms are satisfied. One proves by induction on the complexity of ϕ that if M satisfies $A = B$ and $\phi(A)$ then it satisfies $\phi(B)$. (Here A and B are parameters from M.) One basis case is when ϕ is $x: A$; in that case the conclusion holds by construction. The other basis case is when ϕ is $A = C$; we have to check that this equality relation is transitive and symmetric, which it is. □

The point is that the axioms of FT permit the equality relation on type variables to be any equivalence relation which refines extensional equality. This is pleasant: we have decoupled the role of types as classifiers from their role as names of classifiers.

12. Proving programs and programming proofs revisited

It may seem that our extended discussion of data types has brought us rather far afield from the initial topics, those mentioned in the title. This section is intended to remove this impression and show the usefulness of having a system that treats data types adequately. Consider how to formulate program correctness in FT: a program will be represented by a term t, the pre-conditions by a formula ϕ, and the post-conditions by a formula ψ. We then have

Partial correctness:

$$\phi(x)\ \&\ tx\downarrow\ \rightarrow \psi(x, tx).$$

Total correctness:

$$\phi(x) \rightarrow tx\downarrow\ \&\ \psi(x, tx).$$

The aim of having a single language combining "assertion language" and "command language" has been achieved. Note that the input x can be mentioned in the post-conditions without the artificial device of carrying it along as another output, as is necessary in the Hoare logic formalism.

Now consider "programming proofs"; given a formal proof p of $\phi(x) \rightarrow \exists y \psi(x, y)$ in FT, when can we extract an algorithm t from p that gets y from x? Not always, since ϕ may be defined using some existential quantifiers; for example, if $\phi(x)$ says that $x > 0$ and $\psi(x, y)$ says that y is a positive rational smaller than the real number x, then there is no hope of extracting y from x alone. However, if ϕ is "almost-negative", i.e. contains no $\exists$ or $\vee$, then various proof-theoretic tools, for example realizability, may be used to extract a term t from p that gets y from x. We shall make use of the standard theory of realizability (more precisely **q**-realizability; see e.g. BEESON [1985]) to prove:

THEOREM (Correctness of extracted algorithms). *Let ϕ be almost-negative. Suppose* FT *proves $\phi(x) \rightarrow \exists y \psi(x, y)$. Then a term t can be found such that* FT *proves $\phi(x) \rightarrow tx \downarrow \ \& \ \psi(x, tx)$.*

PROOF. We assume the reader is familiar with formalized realizability. In order to avoid having to make any hypothesis on the formula ψ, we use a variant of realizability known as **q**-realizability; thus in $e \mathbf{r} A$, the **r** means **q**-realizability. Since ϕ is almost-negative, we can find a term j such that

$$\mathrm{FT} \vdash \phi(x) \rightarrow jx \, \mathbf{r} \, \phi(x). \tag{1}$$

By hypothesis, we have

$$\mathrm{FT} \vdash \phi(x) \rightarrow \exists y \psi(x, y). \tag{2}$$

By the soundness of realizability, we have some term q such that

$$\mathrm{FT} \vdash q \, \mathbf{r} \, [\phi(x) \rightarrow \exists y \psi(x, y)]. \tag{3}$$

By (1) and (3), we have

$$\mathrm{FT} \vdash \phi(x) \rightarrow q(jx) \downarrow \ \& \ q(jx) \mathbf{r} \exists y \psi(x, y). \tag{4}$$

Take t to be $\lambda x . \mathbf{p}_0(q(jx))$. Then

$$\mathrm{FT} \vdash \phi(x) \rightarrow tx \downarrow \ \& \ \psi(x, tx) \ \& \ \mathbf{p}_1(q(jx)) \, \mathbf{r} \, \psi(x, tx). \tag{5}$$

Dropping the last conjunct, we have

$$\mathrm{FT} \vdash \phi(x) \rightarrow tx \downarrow \ \& \ \psi(x, tx) \tag{6}$$

as claimed. □

The proof is very simple; the point is not that a complicated or deep result has been proved, but that a result of interest in computer science has been proved by a simple application of standard methods of proof theory. We have given a single theory in which programs can be written and their correctness stated; we have shown how to extract programs from proofs and how to prove the programs so extracted.

References

DE BAKKER, J., 1980, *Mathematical theory of program correctness* (Prentice-Hall, Englewood Cliffs, NJ).

BATES, J. and CONSTABLE, R., 1982, *Programs as proofs*. Technical Report TR 82-532, November, Department of Computer Science, Cornell University, Ithaca, New York.

BEESON, M., 1982, *Recursive models for constructive set theories*, Annals of Math. Logic 23, pp. 127–178.

BEESON, M., 1982, *Problematic principles in constructive mathematics*, in: VAN DALEN, D., LASCOM, D. and SMILEY, T.J. (eds.), *Logic Colloquium '80*, (North-Holland, Amsterdam) pp. 11–56.

BEESON, M., 1983, *Designing intelligent information systems: some issues and approaches*, Language of Data Project, Los Altos, CA.

BEESON, M., 1985, *Foundations of constructive mathematics: Metamathematical studies* (Springer, Berlin).

BISHOP, E., 1967, *Foundations of constructive analysis* (McGraw-Hill, New York).

BOYER, R.S. and MOORE, J.S., 1981, *The correctness problem in computer science* (Academic Press, London).

DE BRUIJN, N.G., 1980, *A survey of the project AUTOMATH*, in: SELDIN, J.P. and HINDLEY, J.R. (eds.), *To H.B. Curry: Essays on Combinatory Logic, Lambda Calculus and Formalism* (Academic Press, New York).

CLARK, K.L. and TÄRNLUND, S.-A., 1982, *Logic programming* (Academic Press, London).

CLOCKSIN and MELLISH, 1981, *Programming in PROLOG* (Springer, Berlin).

CONSTABLE, R., 1971, *Constructive mathematics and automatic program writers*, Proc. of IFIP Congress, Ljubjana 1971, pp. 229–233.

CONSTABLE, R., 1982, *Programs as proofs*, Technical Report 82–532, November, Department of Computer Science, Cornell University, Ithaca, New York.

CONSTABLE, R. and O'DONNEL, M., 1978, *A programming logic* (Winthrop, Cambridge).

DOLBY, J., 1982, *The language of data*, Language of Data Project, Los Altos, CA.

FEFERMAN, S., 1975, *A language and axioms for explicit mathematics*, in: *Algebra and Logic*, Lecture Notes in Mathematics 450, pp. 87–139 (Springer, Berlin).

FEFERMAN, S., 1979, *Constructive theories of functions and classes*, in: BOFFA, M., VAN DALEN, D. and MCALOON, K. (eds.), *Logic Colloquium '78: Proceedings of the Logic Colloquium at Mons, 1978*, pp. 159–224 (North-Holland, Amsterdam).

FEFERMAN, S., 1982, *Inductively presented systems and the formalization of meta-mathematics*, in: VAN DALEN, D., LASCOM, D. and SMILEY, T.J. (eds.), *Logic Colloquium '80*, pp. 95–128 (North-Holland, Amsterdam).

FEFERMAN, S., *Towards useful type-free theories, I* (to appear).

FUCHI, K., 1983, *The direction the Fifth Generation Computer System project will take*, New Generation Computing 1, pp. 3–9.

GOAD, C., 1980, *Proofs as descriptions of computation*, in: BIBEL, W. and KOWALSKI, R. (eds.), *5th Conference on Automated Deduction, Les Arcs, France, 1980*, pp. 39–52 (Springer, Berlin).

GORDON, M., MILNER, R. and WADSWORTH, C., 1979, *Edinburgh LCF: A Mechanized Logic of Computation*, Lecture Notes in Computer Science 78 (Springer, Berlin).

GRAVES, H., 1983, *The Algos system*, Language of Data Project report, Los Altos, CA.

HAYASHI, S., 1983, *Extracting Lisp programs from constructive proofs: a formal theory of constructive mathematics based on Lisp*, Publications of the Research Institute for Mathematical Sciences, Kyoto University 19, pp. 161–191.

VAN HEIJENOORT, J. (ed.), 1967, *From Frege to Gödel: A Source Book in Mathematical Logic, 1879–1931* (Cambridge University Press, Cambridge, MA).

KOLMOGOROV, A.N., 1925, *O principe tertium non datur* (On the principle of tertium non datur), Math. Sb. 32, pp. 646–667 (Russian); English translation in: [VAN HEIJENOORT, 1967] pp. 414–437.

MARTIN-LÖF, P., 1975, *An intuitionistic theory of types: predicative part*, in: ROSE, H.E. and SHEPHERDSON, J.C., *Logic Colloquium '73*, pp. 73–118 (North-Holland, Amsterdam).

MARTIN-LÖF, P., 1982, *Constructive mathematics and computer programming*, in: COHEN, L.J., LÖS, J., PFEIFFER, H. and PODEWSKI, K.P., *Logic, Methodology, and Philosophy of Science VI*, pp. 153–179 (North-Holland, Amsterdam).

SCOTT, D.S, 1975, *Identity and existence in intuitionistic logic*, in: FOURMAN, M.P., MULVEY, C.J. and SCOTT, D.S. (eds.), *Applications of Sheaves*, Lecture Notes in Mathematics 753, pp. 660–696 (Springer, Berlin).

WEYHRAUCH, R., 1980, *Prologomena to a theory of mechanized formal reasoning*, Artificial Intelligence 13, pp. 133–170.

WEYL, H., (1924), *Randbemerkungen zu Hauptproblemen der Mathematik*, Math. Zeitschrift 20, pp. 131–150.

Barcan Marcus et al., eds., Logic, Methodology and Philosophy of Science VII

THE USE OF ORDINALS IN THE CONSTRUCTIVE FOUNDATIONS OF MATHEMATICS

WILLIAM HOWARD

Dept. of Mathematics, Univ. of Illinois, Chicago, U.S.A.

Introduction

In [4] and [5] GENTZEN introduced a method of analyzing formal theories by the constructive use of ordinals, and his approach has been developed extensively since that time. Our purpose will be to consider certain aspects of this line of development from the viewpoint of the constructive foundations of mathematics. Also, in Section 4, we will describe a method of measuring functionals of finite type by ordinals in such a way that operations such as composition and primitive recursion are reflected by corresponding functions on the measures. This allows one to see in a direct way what ordinals can be expected to be associated with a given family of functionals. If the functionals arise from Gödel's functional interpretation of a theory C, this indicates what ordinals will be associated with C.

1. Constructive foundations and reductive proof theory

By a foundations of mathematics is meant a framework of ideas, principles, concepts, definitions, and axioms within which some part, or the whole, of mathematics can be developed. An example is Dedekind's set-theoretic foundations of the real number system which provides a basis for the differential and integral calculus.

Some of the ideas of constructive foundations are as follows. The notion of an effective process is taken to be fundamental. This includes functions on natural numbers. Depending on one's philosophy, one may regard an effective process as occurring in the physical world (for example, as a computation) or as a constructional activity of the mind (Brouwer). The classical notion of truth is replaced by a notion of proof. When a natural number with a given property is proved to exist, an effective means must be given for producing (or 'constructing') it. A species of mathematical objects is not regarded as existing as a completed totality; rather, the objects in the species are to be constructed.

Since set theory is widely accepted as a foundations of mathematics at the present time, it may be asked, "What is the interest in constructive foundations?" A reply to this is as follows. A goal of foundational research is to obtain or develop answers to the question, "What is the relation between mathematics and: knowledge and experience as a whole?" In thinking about this question it is obviously worthwhile to consider constructive ideas, set-theoretic ideas, or any other ideas which appear to be fundamental to mathematical reasoning. This can be illustrated by the following example. What is an irrational number? The standard answer is that it is a certain kind of set. This leads to problems concerning the nature of sets. It may be argued that mathematical intuition shows us that sets exist and that the axioms of set theory are true [6, p. 271]. In reply, one might claim that it is properties, rather than sets, which are fundamental. The language of set theory might be regarded as merely providing a geometric imagery which helps us talk about properties. Thus the set-theoretic approach leads to various questions. On the other hand, in considering how irrational numbers such as $\sqrt{2}$ and $\bar{\pi}$ are used by machinists and astronomers, one might decide that an irrational number should be regarded as an approximation process; in other words, a rule of calculation. This leads to the theory of computable real numbers, which has its own problems. Another method of attempting to base the theory of real numbers on constructive ideas is provided by *reductive proof theory*, which we shall now consider.

In the program of reductive proof theory, first various areas of ordinary mathematics are formalized; then the resulting formal theories are analyzed by constructive methods. This approach is, of course, an outgrowth of Hilbert's Program. Hilbert's idea was to prove the consistency of various formal theories by means of a particularly elementary form of constructive reasoning which he called *finitistic*. As Gödel's Second Incompleteness Theorem shows, the methods which Hilbert and his followers were using are not strong enough for the required consistency proofs. As Gentzen showed, some progress in reductive proof theory can be made by appealing to a constructive principle of transfinite induction.

2. Extensions of Skolem arithmetic by "transfinite induction"

In [4] and [5] GENTZEN proved the consistency of Peano arithmetic by means of an appeal to a constructive principle of transfinite induction over the ordinals less than Cantor's first epsilon number ε_0. Specifically, in the 1938 paper he attached ordinal notations $\mathrm{ord}(d)$ to derivations d and gave a reduction procedure f such that if d is a derivation of an inconsistency, then $f(d)$ is also a derivation of an inconsistency and, moreover, $\mathrm{ord}(f(d)) < \mathrm{ord}(d)$. He then concluded "by transfinite induction" that d cannot be the derivation of an inconsistency. He emphasized that, except for this appeal to transfinite induction, the reasoning used in the consistency of proof is of an elementary constructive nature. The purpose of the present section is to consider formulations of 'transfinite induction' which are appropriate for this situation. We will emply Skolem, free variable, primitive recursive arithmetic PRA, which provides a useful formulation of an elementary part of constructive reasoning.

2.1. Rule of transfinite induction

If the ordinal notations less than ε_0 are numbered in the usual way, then the order relations of the ordinals is reflected by a relation $x < y$ on numbers, and Gentzen's consistency proof can be carried out in PRA extended by the following free variable rule of transfinite induction which was formulated by KREISEL [11, p. 47; 12, p. 322]: from $\neg\, h(x) < x \rightarrow B(x)$ and $B(h(x)) \rightarrow B(x)$, infer $B(t)$ for an arbitrary term t. In general, supposing that a set of ordinal notations for at least the ordinal notations less than ξ has been numbered, let PRA(ξ) denote PRA extended by the free variable rule to transfinite induction restricted to the ordinals less than ξ.

2.2. Ordinal recursion

A second method of extending PRA consists of using the following scheme for introducing function φ on the basis of given functions f, g, and h by ordinal recursion [1; 12, p. 322]:

$$h(x) < x \rightarrow \varphi(x) = g(x, \varphi(h(x)))$$

$$\neg\, h(x) < x \rightarrow \varphi(x) = f(x).$$

Let REC(ξ) be a PRA extended in this way for the restriction of the ordering $x < y$ to numbers corresponding to the ordinals less than ξ. As KREISEL has shown [12, p. 323], the free variable rule of transfinite induction for ordinals less than ξ is a derived rule of REC(ξ). Thus REC(ξ) is at least as strong as PRA(ξ). On the other hand, it is not much stronger (relative to the formulas of PRA) since the consistency of REC(ξ) can be proved in PRA($(\xi + 1)^{\omega}\omega$), as was shown by TAIT [20].

2.3. *Descending chain principle*

Returning to the account of Gentzen's consistency proof given at the beginning of the present section, we note that the role of ordinals consists in their use in showing that a certain process terminates. From this point of view it is the descending chain principle which is fundamental. A corresponding extension of PRA can be obtained by introducing function variables and a λ-operator for numerical variables (alternately, the appropriate combinators of level 2) together with a sign E for a functional of type 2 and the axiom scheme

$$f(0) < \xi \rightarrow \neg f(E(f)+1) < f(E(f))$$

for terms f of type 1. Let CHN(ξ) denote the resulting extension of PRA. By means of the methods of [9] or [19] it is not hard to show that, after replacing the function variables by closed terms, CHN(ξ) can be interpreted in REC(ξ) and vice versa.

2.4. *Use of the three extensions of* PRA

The extensions 2.1–2.3 provide an implementation of Gentzen's idea of taking the metamathematical reasoning to be elementary except for an appeal to a constructive form of 'transfinite induction'. Moreover, one way of measuring the strength of a formal theory is to take ξ to be the strength of the theory T if the Π^0_1 theorems of T are the same as those of $\bigcup\{\text{PRA}(\alpha)\colon \alpha < \xi\}$; cf. POHLERS [16, p. 124].

The theories REC(ξ) and CHN(ξ) serve as rather minimal extensions of PRA which allow the use of constructive versions of Π^0_2 and Π^1_1 statements, respectively. Thus, for example, REC(ξ) is appropriate for proving that a computation process terminates, and CHN(ξ) is suitable for proving the well-foundedness of trees. To put the matter another way: the ordinal recursive functions (respectively, the descending chain functional) provide a means for handling the notion of the *termination of a process* (respectively, the termination of a non-deterministic process), which, of course, is a basic constructive idea.

3. Infinite derivations

The constructive ω-rule and the corresponding use of infinite derivations were emphasized by Schütte and have subsequently played an important role in proof theory. The proof-theoretic ordinals arise as the lengths of the trees. From the viewpoint of constructive foundations a question arises as to the nature of these trees. One answer is that they are inductively generated objects, but another answer has been emerging which will be described below.

In SCHÜTTE'S work [17] a derivation tree can be understood as being provided by a constructive function f such that, for a finite sequence $\mathcal{N}$ of natural numbers, the value of $f(\mathcal{N})$ indicates whether $\mathcal{N}$ is a node of the tree, whether $\mathcal{N}$ is a terminal node, and what formula is attached to $\mathcal{N}$. Thus Schütte's metamathematics must be able to talk about functions as objects.

It has become customary to encode infinite derivation trees by natural numbers in a manner similar to the encoding of the Church–Kleene constructive ordinals. One thing this achieves is that it makes the metamathematics more elementary because now the domain of individuals in the metamathematics consists of natural numbers. The main emphasis, however, has been in the use of such an encoding as a technical tool [18].

The use of encodings as just described has led to an awareness of the continuity of the syntactical transformations employed in the process of cut-elimination. Such continuity can be inferred in a general manner as follows. Since the mappings of the codes are defined by Kleene's recursion theorem, they are defined on all trees rather than merely on well-founded trees. Moreover these transformations can be extended to all recursive functions by employing an extensional effective operation which maps an arbitrary recursive function into a function that defines a tree. Hence by [13] these transformations are continuous.

When infinite derivations are viewed as inductively generated objects, it is natural to define the syntactical transformations by transfinite recursion, hence successively from the terminal nodes back to the principal node. On the other hand, since the transformations are continuous, they can be defined by starting from the principal node and proceeding out to the terminal nodes. Thus the transformations are also defined on non-well-founded trees [14].

Another interesting aspect of these developments is the use of primitive recursive codes [18], [15]. The syntactical transformations are then represented by primitive recursive functions on the codes. As is well known, the possibility of characterizing the notion of computable function depends on the fact that if a computable function f is defined by means of a computation process whose steps are given by another computable function, then f can also be defined by means of a computation process whose steps are elementary (for example, primitive recursive). Perhaps the phenomena concerning continuity and primitive recursive codes, above, have an analogous significance.

These considerations lead to a metamathematical treatment which is very much in harmony with the idea, expressed in Section 2, that all steps in the metamathematics can be taken to be elementary except for the use of the descending chain principle to show that a process terminates.

4. Functionals

An approach to reductive proof theory which has interesting points of contact with the Gentzen line of development is provided by GÖDEL'S functional interpretation [7]. First a given classical theory C is mapped into an intuitionistic theory I by means of the 'negative translation', then Gödel's functional interpretation is applied to the theory I. In this way, one obtains a collection of terms for functionals of finite type over the natural numbers. A constructive analysis of C will be obtained if it can be shown, by constructive means, that the terms of type 0 are computable. Also, an ordinal analysis of the functionals yields an ordinal analysis of C. One approach to the ordinal analysis of functionals of finite type is provided by the use of infinite terms in analogy with SCHÜTTE'S use of infinite derivations [21]. In the present section we will describe another approach, which uses ordinals to measure the lengths of computation trees.

4.1. Types

Natural numbers have type 0. If σ and τ are types, then so is $\sigma \to \tau$, where the level of $\sigma \to \tau$ is the maximum of $1 + \text{level}(\sigma)$ and $\text{level}(\tau)$. The level of the type 0 is zero. We denote $\sigma_1 \to (\sigma_2 \to \cdots (\sigma_p \to \tau) \cdots)$ by $(\sigma_1, \ldots, \sigma_p) \to \tau$ to indicate that a functional of this type can be regarded as a function with argument places of types $\sigma_1, \ldots, \sigma_p$ and values of type σ. If $\emptyset$ denotes the empty set, a functional of type σ can be represented as a functional of type $\emptyset \to \sigma$; in other words, p is 0 in the above. Hence by putting ρ_r equal to $\emptyset$ in Section 4.3, below, application is expressed as a special case of composition.

4.2. Evaluation trees

The notion of an evaluation tree for a functional of level not exceeding 2 can be explained by considering the case in which F has type $(\sigma_1, \sigma_2, 0) \to 0$, where σ_1 is $(0,0) \to 0$ and σ_2 is $0 \to 0$. The evaluation of $F(\alpha, \beta, x)$ proceeds by pursuit of a path through the tree. When a node is reached, numbers n and m are given, and one of the following three questions is asked. "What is the value of $\alpha(n, m)$?" "What is the value of $\beta(n)$?" "What is the value of x?" The answer determines which branch is to be taken. When a terminal node is reached, the value of $F(\alpha, \beta, x)$ is given.

When the determination and value just mentioned are given by a computable function, then the evaluation tree is called a computation tree. To say that, for every α, β, and x, the value of $F(\alpha, \beta, x)$ *exists* is just to say that the evaluation tree is well founded. It is easy to see that a functional has an evaluation tree if and only if it has a Kleene-associate.

If ordinals are assigned to the nodes of a tree in such a way that the ordinal assigned to a node $\mathcal{N}$ is greater than the ordinals assigned to the immediate successor nodes of $\mathcal{N}$, and if this assigns the ordinal b to the principal node, then we say that the tree has length b. If a functional F of level not greater than 2 has an evaluation tree with length b, say that F has measure b and write $\text{meas}(F) \leq b$.

4.3. *Composition*

Let F be a functional of type $(\sigma_1, \ldots, \sigma_j) \to 0$. If a list $\mathcal{L}$ of indices $r \leq j$ is selected, and if G_r is a functional of type $\rho_r \to \sigma_r$ for every r in $\mathcal{L}$, and if $\boldsymbol{G}$ is the corresponding list of functionals G_r, then $F \circ \boldsymbol{G}$ denotes the result of composing F with the functionals G_r at the corresponding argument places in F. Thus $F \circ \boldsymbol{G}$ is a functional of type $(\xi_1, \ldots, \xi_j) \to 0$, where ξ_r is ρ_r or σ_r depending on whether r is in $\mathcal{L}$. We say that the composition is *uniform* if the types σ_r with r in $\mathcal{L}$ all have the same level. We say that *the composition is of the first kind* if it is uniform and if the level of $F \circ \boldsymbol{G}$ is less than the level of F. Otherwise the composition is said to be *of the second kind*. Note that a composition of the first kind must involve all the argument places of maximum level in F.

THEOREM 4.1. *Suppose F and $G_1, \ldots, G_j$ are functionals with level not exceeding 2 and with measures c and $b_1, \ldots, b_j$, respectively. Let $\boldsymbol{G}$ denote the list $G_1, \ldots, G_j$. Then $F \circ \boldsymbol{G}$ has the following measures:*

(i) $(1 + \max\{b_r\})(c + 1)$ *in general, and*

(ii) $c + \sum_r b_r$ *if $G_1, \ldots, G_j$ have values of type* 0.

PROOF. By hypothesis F and $G_1, \ldots, G_j$ have evaluation trees σ and $\tau_1, \ldots, \tau_j$ with lengths c and $b_1, \ldots, b_j$, respectively. Thus there are assignments $\text{ord}(\mathcal{M})$ and $\text{ord}(\mathcal{N})$ of ordinals to the nodes of σ and $\tau_1, \ldots, \tau_j$ respectively. The evaluation of $F \circ \boldsymbol{G}$ proceeds by going from state to state, where a state is labelled by a pair $(\mathcal{M}, \mathcal{N})$. For fixed $\mathcal{M}$, the value of some

$G_r(n_1,\ldots,n_k)$ is being sought, and this involves the pursuit of a sequence of nodes $\mathcal{N}, \mathcal{N}', \mathcal{N}'', \ldots$ in the tree τ_r. When this value is found, then we go to a successor of $\mathcal{M}$. Thus the nodes of the evaluation tree of $F \circ \mathbf{G}$ can be taken to be finite sequences of pairs of nodes $(\mathcal{M}, \mathcal{N})$ which satisfy the predecessor relation obtained by ordering these pairs lexicographically (but pairs $(\mathcal{M}, \mathcal{N})$ with $\mathcal{M}$ terminal are not included). Let b denote the maximum of $b_1, \ldots, b_j$, and take $\text{ord}(\mathcal{M}, \mathcal{N})$ to be $(1+b)\text{ord}(\mathcal{M}) + 1 + \text{ord}(\mathcal{N})$.

In case (ii), first evaluate $G_j, G_{j-1}, \ldots, G_1$, getting numbers n_j, n_{j-1}, n_1, then evaluate $F(n_1,\ldots,n_j)$. In the state $(r, \mathcal{N})$ the value of G_r is being sought. Take $\text{ord}(r, \mathcal{N})$ to be $c + b_1 + \cdots + b_{r-1} + \text{ord}(\mathcal{N})$.

In Theorem 3.3, p. 95 of [9], replace $(b+1)f$ by $(1+b)(f+1)$. □

The preceding ideas suffice for the ordinal analysis of primitive recursive functionals of level not exceeding 2. These functionals are generated by starting with zero, successor, and projection functionals, and applying the definition schemes for composition and primitive recursion; namely,

$$H(Z) = F(G_1(Z), \ldots, G_j(Z))$$

and

$$H(Z, 0) = G(Z),$$

$$H(Z, n+1) = F(Z, n, H(Z, n)) \quad \text{for } n = 0, 1, 2, \ldots .$$

The *level* of the primitive recursion is 1 plus the level of $H(Z, n)$. Let $H_n(Z)$ denote $H(n, Z)$. Supose F and G have measures c and b, respectively. We wish to find a measure h for H. Suppose H_n has measure h_n.

Primitive recursion on level 1

By case (ii) of Theorem 4.1 we can take h_{n+1} to be $c + h_n$. Hence $h_n = cn + b$ by induction. Hence $h \leqslant (\max\{b, c\})\omega$. Thus primitive recursion on level 1 is reflected by multiplication of the measures by ω. By case (ii) of Theorem 4.1, composition is reflected by, essentially, multiplication of the measures. The starting functionals have measure 2. Thus the functionals generated by composition and primitive recursion on level 1 have measures less than ω^{ω}

Primitive recursion on level 2

In this case, use (i) of Theorem 4.1 to infer that h_n can be taken to be $(1+b)(1+c)^n$. Thus $h \leqslant (1+b)(1+c)^{\omega}$. Alternatively, $h \leqslant (\max\{b, c\})^{\omega}$ so long as b and c are greater than 1. Thus primitive recursion on level 2 has the effect of raising the measures to the power ω. Hence, in the light of Theorem

4.1, what is required for the ordinal analysis of the primitive recursive functionals of level not exceeding 2 is a (non-trivial) set of ordinals closed under addition, multiplication, and raising to the power ω. Thus the ordinals less than ω^d, where $d = \omega^\omega$, will do.

4.4. Extension of measure to higher types

It is natural to consider the following notion of measure function. A functional F of level 3 is said to have a measure function f if, for all compositions of the first kind (Section 4.3) with lists $\boldsymbol{D}$ of functionals of level 2: if meas$(\boldsymbol{D}) \leq d$, then meas$(F \circ \boldsymbol{D}) \leqq f(d)$.

In composition involving functionals of level 3, the type levels can have various combinations. For example, composition of a functional F of level 3 with functionals $G_1, \ldots, G_j$ of levels 2 or 3 at argument places of level 2 yields a functional $F \circ \boldsymbol{G}$ whose level may be either 3 or less than 3. To handle all these combinations at once it is convenient to think of a functional G of level less than 3 as represented by a functional G^* of level 3, where $G = G^*(B)$ for some trivial functional with measure 2. Hence we say that G has measure function g *for height 3* if G has measure $g(2)$. Also, if a functional F of level 3 has a measure function f, then we say F has measure function f for height 3.

With this understanding, and assuming all measure functions are strictly monotone increasing, it is easy to prove, for (uniform) composition at some of the argument places of level 2 in F: if F and $\boldsymbol{G}$ have measure functions f and g, respectively, for height 3, then $F \circ \boldsymbol{G}$ has measure function $f \circ g$ for height 3.

For the functionals being analyzed, it may be that a suitable supply of measure functions for the functionals of level 3 is provided by some family h_c parametrized by ordinals c. For example, to analyze the primitive recursive functionals of finite type we use $h_c(x) = x^c$. To analyze the functionals of finite type generated by bar recursion of type 0, we use the Bachmann functions $\bar{\varphi}_c$, $c < \varepsilon_{\Omega+1}$, [8].

For the analysis of functionals of level 3, if we have such a parametric family h_c, then: if F has a measure function h_c for height 3, we say F has measure c for height 3. These ideas can be extended to functionals of level greater than 3 in a manner which we will illustrate for the case of primitive recursive functionals.

4.5. Primitive recursive functionals of finite type

By induction on n we define: $e(0, b) = b$ and $e(n+1, b) = 2^{e(n,b)}$. If a functional H of level i has measure $e(n, c)$, then H is said to have measure c for height $j+n$, where $j = \max\{i, 2\}$.

Suppose a functional F of level $s+1>2$ has type $(\sigma_1, \ldots, \sigma_j) \to 0$ and consider compositions $F \circ \boldsymbol{D}$ of the first kind. We say that F has measure c if, for all such compositions, $F \circ \boldsymbol{D}$ has measure d^c for height s whenever $\mathrm{meas}(\boldsymbol{D}) \leq d$.

THEOREM 4.2. *Suppose F and all functionals in the list $\boldsymbol{G}$ have level not exceeding $s+1$, where $s>1$. If F and $\boldsymbol{G}$ have measure c and b for height $s+1$, respectively, then: if F is composed with $\boldsymbol{G}$ at some of the argument places of level k in F, the resulting functional $F \circ \boldsymbol{G}$ has the following measure for height $s+1$:*

(i) *bc if $k = s$,*

(ii) *$b + c$ if $2 \leq k < s$,*

(iii) *$b + c + i - 1$ if $k < 2$, where i is the number of functionals in the list $\boldsymbol{G}$.*

This theorem can be proved by induction on s by the method of proof of Lemma 2.1 of [8, p. 110].

Using this theorem and proceeding essentially as in Section 4.5, it is easy to show that every primitive recursive functional of finite type has a measure less than ε_0.

4.6. *The metamathematics*

The discussion in Sections 4.4 and 4.5 suppose some notion of functional of finite type. In order to reduce the discussion to the metamathematics considered in Section 2, one would first take a model which can be discussed in arithmetic; for example, a term model in a λ-calculus. Then one might modify the discussion so that it could be carried out directly in one of the theories in Section 2. Alternately, a reduction to one of the theories in Section 2 might be obtained by use of an intermediate metamathematical theory; for example, Peano arithmetic extended by transfinite induction.

5. Accomplishments, difficulties

Gentzen hoped eventually to obtain a consistency proof for classical analysis; that is, second order arithmetic with a comprehension axiom with respect to arbitrary formulas. This is the theory which has customarily been regarded as appropriate for formalizing elementary calculus. Although the goal of obtaining a consistency proof for classical analysis has not been attained, the program has been carried out for certain subtheories: Π_1^1 comprehension being a landmark [22], and the Σ_2^1 axiom of choice plus bar induction being the strongest theory handled so far. Summaries of this work are given in [3] and [16]. The following two difficulties have been encountered.

(1) There is the purely mathematical difficulty of discovering suitable systems of ordinal notations. The problem is to name the ordinals belonging to sufficiently large segments of the second number class. The best solution so far is based on ideas of Veblen and Bachmann. At the present time it appears that the Veblen–Bachmann approach has reached a fairly natural stopping point [16, p. 134]. To get significantly larger notational systems, some new ideas will be needed.

(2) Supposing a formal theory to have been analyzed by use of a constructive form of transfinite induction with respect to some system of ordinal notations, the question arises, "Are we to take transfinite induction with respect to these notations as a fundamental constructive principle?" If not, then, in pursuing the program of reductive proof theory (Section 1), it becomes necessary to give a constructive proof of the principle of transfinite induction for the system of notations used. GENTZEN himself felt this had to be done for the notations less than ε_0. His proof is based on the concept of accessibility: see 15.4 and 16.11 of [4]. Is the concept of accessibility to be taken as a basic constructive idea? BROUWER [2] gives a proof of transfinite induction up to ε_0 based on ideas about inductive generation which presumably are to be regarded as more fundamental than the idea of accessibility. FEFERMAN [3, p. 81] has formulated a constructive theory T_0, based on an axiom of accessibility, which is strong enough to prove transfinite induction for every proper lower segment of the large system of ordinal notations mentioned in (1). Should Feferman's theory T_0 be taken as fundamental or should it be analyzed on the basis of more fundamental ideas?

The ideas about constructive reasoning which we have at present are fragmentary. It is an open problem to find some basic constructive principles upon which a coherent system of constructive reasoning may be built. The most extensive system of constructive ideas we have at present consists of those developed by Brouwer; but in Brouwer's system a basic role is played by an abstract notion of proof (or, more generally, construction) which needs to be clarified.

References

[1] Ackermann, W., 1940, *Zur Widerspruchsfreiheit der Zahlentheorie*, Mathematische Annalen 117, pp. 162–194.

[2] Brouwer, L.E.J., 1926, *Zur Begrundung der intuitionistische Mathematik III*, Mathematische Annalen 96, pp. 451–488.

[3] Buchholz, W., S. Feferman, W. Pohlers and W. Sieg, 1981, *Iterated inductive definitions and subsystems of analysis: recent proof-theoretic studies*, Lecture Notes in Mathematics 897 (Springer-Verlag).

[4] Gentzen, G., 1936, *Die Widerspruchsfreiheit der reinen Zahlentheorie*, Mathematische Annalen 112, pp. 493–565.

[5] Gentzen, G., 1938, *Neue Fassung des Widerspruchsfreitheitsbeweises für die reine Zahlentheorie*. Forschungen zur Logik und zur Grundlegung der exakten Wissenschaften, New Series, No. 4, pp. 19–44 (Hirzel).

[6] Gödel, K., 1964, *What is Cantor's continuum problem*?, revised version in: P. Benacerraf and H. Putnam (eds). Philosophy of Mathematics: Selected Readings, pp. 258–273 (Prentice-Hall).

[7] Gödel, K., 1958, *Über eine bisher noch nicht benützte Erweiterung des finiten Standpunktes*, Dialectica 12, pp. 280–287.

[8] Howard, W., 1981, *Ordinal analysis of bar recursion of type zero*, Compositio Mathematica 42, 105–119.

[9] Howard, W., 1981, *Computability of ordinal recursion of type level two*, in: F. Richman (ed.), Constructive mathematics, Lecture Notes in Mathematics 893, pp. 87–104 (Srpinger).

[10] Jäger, G. and Pohlers, W., 1982, *Eine beweistheoretische Untersuchung von* $(\Delta^1_2\text{-CA})+$ (BI) *und verwandter Systeme*, Sitzungsberichten der Bayerische Akademie der Wissenschaften, pp. 1–28.

[11] Kreisel, G., 1952, *On the interpretation of non-finitist proofs II*, Journal of Symbolic Logic 17, pp. 43–58.

[12] Kreisel, G., 1959, *Proof by transfinite induction and definition by transfinite induction in quantifier-free systems*, Journal of Symbolic Logic 24, pp. 322–323.

[13] Kreisel, G., Lacombe, D. and Schoenfield, J., 1959, *Partial recursive functionals and effective operations*, in: A. Heyting (ed.), *Constructivity in Mathematics*, pp. 290–297 (North-Holland).

[14] Kreisel, G., Mints, G. and Simpson, S., 1975, *The use of abstract language in elementary metamathematics: some pedagogic examples*, in: R. Parikh (ed.), Logic Colloquium, Lecture Notes in Mathematics 453, pp. 38–131 (Springer).

[15] López-Escobar, E.G.K., 1976, *On an extremely restricted* ω*-rule*, Fundamenta Mathematicae 90, pp. 159–172.

[16] Pohlers, W., 1982, *Admissibility in proof theory; a survey*, in: Studies in Logic and the Foundations of Mathematics 104, pp. 123–139 (North-Holland).

[17] Schütte, K., 1960, *Beweistheorie* (Springer).

[18] Schwichtenberg, H., 1977, *Proof theory: some applications of cut-elimination*, in: Handbook of Mathematical Logic, pp. 867–895 (North-Holland).

[19] Tait, W., 1961, *Nested recursion*, Mathematische Annalen 143, pp. 236–250.

[20] Tait, W., 1965, *Functionals defined by transfinite recursion*, Journal of Symbolic Logic 30, pp. 155–174.

[21] Tait, W., 1965, *Infinitely long terms of transfinite type*, in: J. Crossley and M. Dummet (eds.), Formal Systems and Recursive Functions, pp. 176–185 (North-Holland).

[22] Takeuti, G., 1967, *Consistency proofs of subsystems of analysis*, Annals of Mathematics 86, pp. 299–348.

Barcan Marcus et al., eds., Logic, Methodology and Philosophy of Science VII
© Elsevier Science Publishers B.V. (1986) 95–98

APPLICATIONS OF PROOF-THEORETIC TRANSFORMATION (ABSTRACT)

G.E. MINC

Leningrad, U.S.S.R.

We present here three applications in mathematical logic, one in algebra and two in computer science.

1. Conservativity of $(\mathrm{AC}+\mathrm{RDC})^{\omega}$ over Heyting arithmetic (HA)

Finite types are constructed from 0 by $\rightarrow$. Terms of finite types are defined from constants (including 0, +, S, $\cdot$) by application. Atomic formulas are equations of terms of the same type, and formulas are built up by &, $\supset$ and quantifiers for all finite types. The formulation of HA^{ω} in terms of sequents $A_1, \ldots, A_n \rightarrow B$ has as postulates intuitionistic natural deduction rules (like $A, X \rightarrow B / X \rightarrow (A \supset B)$) modified to take account of many-sorted language and usual arithmetic axioms. AC, RDC denote axiom scheme of choice and relativized dependent choice respectively. The conservativity of these schemata for all finite types (cf. [14]) over HA is proved in three steps.

(a) Infinitary natural deduction system H_{∞} using formulas-as-types notation and ordinals $< \varepsilon_0$ is built up and $\mathrm{HA}^{\omega}+\mathrm{AC}^{\omega}+\mathrm{RDC}^{\omega}$ is embedded into H_{∞}.

(b) Normalization theorem for H_{∞} is established (in PRA).

(c) Normal derivation of any HA-formula in H_{∞} is transformed in one containing no types except 0 and then into HA-derivation (using reflection).

2. Proof of Novikov's hypothesis

Familiar Gödel–Tarski translation of the intuitionistic logic into nodal one is simply the result of prefixing the necessity sign $\Box$ to any subformula occurrence. It evidently preserves derivability. The proof of McKinsey and

Tarski of its faithfullness for the propositional calculus was extended to the predicate case by Rasiowa and Sikorski. Novikov [2] conjectured extension of this result to his formulation of HA with partial recursive functions. This turned out to be false in general but we were able to prove faithfulness in [3] for all formulas containing no symbols for partial functions. The scheme of the proof is the same as in Section 1.

3. Normalization theorem for predicate logic implying one for arithmetic

To simplify notation consider Gentzen-type L-formulation (with the rules for introduction in antecedent and succedent) of the classical predicate calculus in the language $\forall$, $\neg$, &. In fact everything extends to the language with other connectives as well as to the intuitionistic case.

An $(\forall\rightarrow)$-inference $A[t], \forall xA, \Gamma \rightarrow \Sigma / \forall xA, \Gamma \rightarrow \Sigma$ is *reducible* if $A[t]$ is derivable. Corresponding reduction consists in replacing this inference by a cut on $A[t]$. A derivation is irreducible if it is cut-free, contains no reducible inferences and its free individual variables are exactly free variables of the endsequent and eigenvariables of $(\rightarrow\forall)$-inferences.

Theorem. *Any derivation is reducible to an irreducible derivation of the same endsequent.*

Scheme of the proof. Given derivation is transformed in the infinitary one like in [4] and normalized. The normal derivation is pruned and induction up to ε_0 shows the resulting figure to be required (finite) irreducible derivation.

It would be interesting to obtain simple model-theoretic proof of the corresponding normal form theorem.

Irreducible derivation of a sequent of the form $I \rightarrow A$ where I is a conjunction of arithmetic axioms is easily (primitive recursively) transformed into normal arithmetic derivation of A.

4. Pruning

In this section we mean by pruning the deletion of obviously superfluous parts of a derivation like replacing

$$\frac{d_0\colon X \rightarrow A \vee B \qquad d_1\colon A, X \rightarrow C \qquad d_2\colon B, X \rightarrow C}{X \rightarrow C}$$

by $d_2^-: X \to C$ if the rightmost derivation d_2 does not use the assumption B.

This transformation was discovered by KLEENE [5] and independently by SHANIN [6] and applied by GOAD [7] to computer program optimization in the framework of mixed computations [8].

5. Coherence theorems

These are theorems of category theory of the form: all diagrams commute (under suitable conditions). The applications of proof theory outlined by LAMBEK [9] and developed in detail by MACLANE [10] are based on the correspondence between canonical morphisms in categories with additional structure S and (equivalence classes of) derivations in suitable non-classical calculus C_S. The most popular example is S = (cartesian closed) and C_S = (intuitionistic proportional calculus). In this case the (reformulation of) coherence theorem takes the form: if any variable occurs no more than twice in $A \to B$, then any two derivations of $A \to B$ are equivalent (modulo standard normalization steps for natural deductions). This was proved by SOLOVJOV and BABAEV [11] by rather long arguments and is proved now by much shorter argument using pruning.

6. Program synthesis

Standard approach to program synthesis from proofs (see [12] for example) suggests to apply some realizability interpretation to a given deduction d of a sentence $\forall x \exists y A(x, y)$ to obtain (a term describing) program π_d such that $\forall x A(x, \pi_d(x))$ holds. This requires either construction of d by a man (as was done in Goad's experiment) or some proof-search program. The latter is impractical for most decidable theories, so the problem of finding suitable efficiently decidable subclasses arises. Very close relation between intuitionistic propositional logic and working program-synthesis system PRIZ [13] was discovered during investigation of planner (program synthesis) module of this system. The planning of PRIZ turned out to be sound and complete P-SPACE proof search algorithm for the implication-conjunction intuitionistic propositional calculus. Other standard intuitionistic propositional connectives can be eliminated preserving deductive equality. First falsity $\perp$ is replaced by conjunction of all proposition variables present plus a new one. After this depth-reducing transformations (due to Wajsberg and Jaskowski) are applied leaving occurrence of $\vee$ only in the form $(A \mathbin{\&} (x \rightarrow y \vee z) \mathbin{\&} B) \rightarrow u$ and then occurrences of $x \rightarrow y \vee z$ are replaced by

$$\& \; (((y \rightarrow v) \mathbin{\&} (z \rightarrow v)) \rightarrow (x \rightarrow v))$$

with conjunction taken over all variables v. Efficiently decidable (and most useful in practice) class is one with implication nesting $\leqslant 2$.

References

[1] MINC, G.E., 1978, J. Soviet Math. 10, pp. 548–596.

[2] NOVIKOV, P.S., 1977, *Constructive mathematics from the viewpoint of the classical one* (Russian) (Moscow, Nauka).

[3] MINC, G.E., 1978, *On Novikov's Hypothesis.* Modal and Intensional logics, 102–106 (Russian) (Moscow).

[4] MINC, G.E., 1975, *Proc. 6th Internat. Congress Logic, Methodology and Philosophy of Science* (North-Holland, Amsterdam).

[5] KLEENE, S.C., 1952, Mem. Amer. Math. Soc. 10, pp. 1–26.

[6] SHANIN, M.A., et al. 1965, *An algorithm for computer search of a natural logical deduction in the propositional calculus* (Leningrad).

[7] GOAD, C., 1980, Lecture Notes Comput. Sci. 87, pp. 39–52.

[8] ERSHOV, A.P., 1977, Inform. Process. Lett. 6, pp. 38–41.

[9] LAMBEK, J., 1958, Math. Syst. Theory 2, pp. 287–318.

[10] KELLY, G. and MACLANE, S., 1971, J. Pure Appl. Alg. 1, pp. 97–140.

[11] BABAEV, A. and SOLOVJOV, S., 1979, Zap. Naučn. Sem. Leningrad Otdel Math. Inst. Steklov 88, pp. 3–29. (In Russian).

[12] KREISEL, G., 1977, Colloq. Intern. Log. (Paris) pp. 123–134.

[13] KAHRO, M., KALJA, A. and TOUGU, E. 1981, Instrumental programming system PRIZ, (Russian) (Moscow).

[14] GOODMAN, N., 1984, J. Symbolic Logic 49, pp. 192–203.

2. Model Theory and its Applications

Barcan Marcus et al., eds., Logic, Methodology and Philosophy of Science VII

ASPECTS OF $\aleph_0$-CATEGORICITY

GREGORY CHERLIN

Dept. of Mathematics, Rutgers Univ., New Brunswick, NJ 08903, U.S.A.

Prologue

I have always found the classification of structures blessed with an unusual degree of symmetry, such as the finite simple groups, an extremely attractive subject, even when there are no immediate prospects of success (finite projective planes). As far as the study of $\aleph_0$-categorical structures is concerned, we cannot realistically expect any very explicit classification, as simple examples show. The situation is analogous in algebra, where nilpotent groups or rings are intrinsically unclassifiable. The most one might aim at in the context of general $\aleph_0$-categorical theories is a good notion of "nilpotent radical". I do not as yet have a sufficiently well-developed case of megalomania to aim at this, and I will confine myself to a discussion of some specific problems whose solution might be expected to advance the field.

Before we begin I should comment on my title. I avoid the word "survey" for various reasons, one of which is worth mentioning explictly: my neglect of work on ordered structures, for which one may consult papers of SCHMERL [1]. A survey would necessarily have to deal systematically with this work as well.

The convention will be in force throughout the discussion that all structures are *countable*, although not invariably infinite. This allows us in particular to give the following succinct formulations of $\aleph_0$-categoricity. A structure M is *$\aleph_0$-categorical* if any structure M' elementarily equivalent to M is in fact isomorphic with it. Equivalently [3]:

$$M^n/\mathrm{Aut}\, M \text{ is finite for each } n. \tag{1}$$

Here M^n is the space of n-tuples from M, on which the automorphism group Aut M acts naturally, and $M^n/\mathrm{Aut}\, M$ is the corresponding space of orbits (the model theorist's n-types). Someone has coined the term "almost n-transitive" for the property exhibited in (1).

This elegant characterization makes it possible to discuss the subject intelligently with algebraists, and in any case entirely supplants the original definition in practice. As a first consequence, immediate but fundamental, any $\aleph_0$-categorical structure M is *uniformly locally finite*, that is we have an *a priori* estimate on the size of the substructure of M generated by an arbitrary finite subset A, which is of the form:

$$\operatorname{card}\langle A\rangle \leqslant f(\operatorname{card} A)$$

where indeed we may take $f(n) = |M^{n+1}/\operatorname{Aut} M|$.

I shall divide my subject into two parts. We will consider classical algebraic theories in the first part, and then pass to topics of purely model-theoretic interest.

I. ALGEBRAIC THEORIES

We will consider $\aleph_0$-categorical theories of modules, rings, or groups.

A. Modules

An explicit algebraic characterization of $\aleph_0$-categorical modules as direct sums of *finitely* many modules of the form

$$E^{(\alpha)} \quad \left(\text{that is } \bigoplus_{i<\alpha} E_i \text{ with } E_i \simeq E\right)$$

with E finite and indecomposable, $\alpha \leqslant \infty$, was given by BAUR [4]. A curious feature of his argument is that he finds it necessary to make explicit use of the *stability* of theories of modules, more specifically of indiscernible sets, within a purely algebraic analysis. As a purely model-theoretic consequence of his analysis, $\aleph_0$-categorical modules are in fact $\aleph_0$-stable of finite Morley rank.

As it stands, Baur's result fits naturally into the general model theoretic structure theory of modules, which is reorganized and developed in a forthcoming article by ZIEGLER [5]. But for our purposes it would be desirable to have a very different description of the $\aleph_0$-categorical modules, as follows.

PROBLEM 1. Describe $\aleph_0$-categorical modules in terms of their transitive constituents.

By this I mean the following. If M is an $\aleph_0$-categorical structure, a *transitive constituent* of M is an orbit in M under Aut M, equipped with the structure inherited from M. Thus if P is a transitive constituent of M, Aut P is the restriction to P of Aut M. The problem is

(a) Describe the transitive constituents of $\aleph_0$-categorical modules.

(b) Describe the linkages among the transitive constituents that allow us to reconstruct the module from them.

As described in Part II, theory predicts the form of the solution to (a), but it is not clear to me how explicit such an analysis can be. So far part (b) has been ignored completely by the general theory, so we don't even know precisely what we are looking for.

There are any number of problems in a related vein, such as:

(c) Describe the abstract group Aut M explictly.

(d) If Aut $M \simeq$ Aut X where M is an $\aleph_0$-categorical module and X is an $\aleph_0$-categorical structure, what can one say about X? This last question is admittedly bizarre, and certainly lies off the main line, but it seems intriguing.

B. Rings

Here the situation is more complex. If R is an $\aleph_0$-categorical ring then both the Jacobson radical J and the semisimple quotient $\bar{R} = R/J$ are $\aleph_0$-categorical, as J is 0-definable in R. Since we have our hands full with the study of the possible radicals J and the quotients $\bar{R}$, we are not going to confront the extension problem (reconstruction of R from J, $\bar{R}$). What is known about J and $\bar{R}$ runs as follows.

(1) If $\bar{R}$ is an $\aleph_0$-categorical *biregular* ring (meaning that ideals are generated by central idempotents), in particular if $\bar{R}$ is $\aleph_0$-categorical, semisimple, and *commutative*, then it has a very explicit representation as a so-called *filtered Boolean power*, in symbols:

$$\bar{R} \simeq C(\mathfrak{X}, \mathfrak{A}).$$

Here $C(\mathfrak{X}, \mathfrak{A})$ denotes the ring of all *locally constant* functions $f: \mathfrak{X} \to \mathfrak{A}$ where:

$$\mathfrak{X} = (X; X_1, \ldots, X_k)$$

is an augmented Boolean space (X is Boolean and the $X_i \subseteq X$ are closed subsets), and

$$\mathfrak{A} = (A; A_1, \ldots, A_k)$$

is a finite augmented ring (the $A_i \leq A$ are subrings), where in fact A is a matrix ring over a finite field. Here $f: \mathfrak{X} \to \mathfrak{A}$ signifies that $f[X_i] \subseteq A_i$.

This representation is given by MIRAVAGLIA [6].

(2) J is nilpotent. This was proved in [8].

(3) SARACINO and WOOD [9] have given $2^{\aleph_0}$ examples of radicals J which are not only $\aleph_0$-categorical, but are commutative, nilpotent of exponent 3 ($xyz = 0$), of any odd prime characteristic, and more. This seems to be the end of the line for classifiers, but a vestige of legitimate doubt remains (see the end of Part II).

Scholia

I intend to comment on these three items at length, under two headings.

1. Filtered Boolean powers

In the course of the last decade filtered Boolean powers were extensively studied by universal algebraists, with incursions by model theorists (compare the bibliography in Johnstone's *Stone Spaces*, 1983). To my way of thinking the subject remains elusive, mainly because the representation of a given structure as a filtered Boolean power is not generally canonical. In the case at hand, this makes it difficult to determine whether a given ring is $\aleph_0$-categorical.

PROBLEM 2. When is a filtered Boolean power $\aleph_0$-categorical?

We know that the following are equivalent:

(i) $\mathfrak{X}$ is $\aleph_0$-categorical (by Stone duality $\mathfrak{X}$ corresponds to a Boolean algebra with distinguished ideals; whence a notion of $\aleph_0$-categoricity for $\mathfrak{X}$);

(ii) The (dual) Heyting algebra H of closed sets generated by $X_1, \ldots, X_k$ in X is of "finite type", that is H is finite, and every element of H has finitely many isolated points;

(iii) The Boolean closure algebra generated by $X_1, \ldots, X_k$ in X (taking Boolean operations plus the closure operation) is of finite type in the sense above.

Here I combine MACINTYRE–ROSENSTEIN [7] with recent comments of

APPS [17]. The point of all this is that these conditions imply that $C(\mathfrak{X}, \mathfrak{A})$ is $\aleph_0$-categorical if $\mathfrak{A}$ is finite. What is missing is some sort of converse.

There is of course a more obvious open question.

PROBLEM 3. Analyze the general semisimple $\aleph_0$-categorical ring $\bar{R}$.

We know so little about this problem in general that we cannot even tell whether it should be difficult. For all I know these rings may be always biregular! Conceivably we can steal some of the ideas that have been used successfully to analyze $\aleph_0$-categorical groups (see below).

2. *Calculemus*

There are connections between $\aleph_0$-categoricity on the one hand, and the notions of model-completeness and QE (quantifier elimination) on the other, which in favorable circumstances can be exploited to reduce the question of the existence of (some, many) $\aleph_0$-categorical structures of a given type to a fairly elementary, if tedious, calculation.

Recall that a structure M is model complete if to each formula ϕ in the language of M we can associate an *existential* formula:

$$\phi^* = \exists \bar{x} \phi_0^* \quad (\phi_0^* \text{ quantifier-free})$$

with the same free variables, so that ϕ and ϕ^* define the same relation on M. If ϕ^* can always be taken to be quantifier-free, we say M is QE (queuey). We have gradually learned to appreciate the following connections.

Le côté Fraissé

> Uniformly locally finite QE structures can be mass-produced economically. (1)

Since all such structures are automatically $\aleph_0$-categorical, this has become the primary source of examples. The method for producing uniformly locally finite structures was described explicitly on a theoretical level by FRAISSÉ [10]; one takes a suitable supply of finite structures and amalgamates them all together into a huge ratatouille. Two decades later ASH, EHRENFEUCHT, GLASSMIRE, and HENSON [11] suddenly gave recipes for $2^{\aleph_0}$ distinct ratatouilles.

You get $2^{\aleph_0}$ $\aleph_0$-categorical digraphs this way, which is frankly embarassing. We have been converted from cooks to sorcerer's apprentices. This situation will be reexamined at the end of Part II; what matters here is that this idea eventually led SARACINO and WOOD [9] to the construction of $2^{\aleph_0}$ uniformly locally finite QE commutative rings (and an analogous class of groups).

QE structures can be analyzed *a priori*. (2)

I don't know whether anyone tried to manufacture QE groups and rings *ad nauseam* in the early seventies. In the event, Saracino and Wood's choice of ingredients is quite subtle. They work with certain very special finite commutative rings of prime characteristic satisfying:

$$xyz = 0 \quad \text{all } x, y, z, \tag{QE1}$$

$$x^2 = 0 \Rightarrow xy = 0 \quad \text{all } x, y. \tag{QE2}$$

Work of BOFFA–POINT–MACINTYRE [12] on the one hand and BERLINE and CHERLIN [13] on the other showed that one *can't* build very many QE rings of prime characteristic out of anything else! The combination of (1) and (2) is irresistible, though practitioners of the art have been known to grumble at the amount of calculation involved, both in the preliminary analysis and in the actual constructions.

Before abandoning this topic, I offer a political slogan.

$$\aleph_0\text{-categoricity} = \lim_{L\to\infty} \mathrm{QE}(L; \mathrm{ULF}). \tag{3}$$

Here L is a variable finite language, and ULF means "uniformly locally finite". The stock of examples provided by (1, 2) applied to various languages is already quite rich.

Robinson's way

Let T be an inductive first order theory, that is we assume that the class of models of T is closed under increasing unions. The example to bear in mind here is the theory of nil rings of exponent n which are not nilpotent.

SARACINO observed [14]:

If T has an $\aleph_0$-categorical model, then it has a model-complete $\aleph_0$-categorical model. (4)

I want to argue that this is a significant observation, by showing that it leads to a proof that $\aleph_0$-categorical nil rings are nilpotent, at least in the commutative case.

Reineke suggested looking for a commutative counterexample J of characteristic 2 and exponent 2. Bearing in mind (4), take it to be existentially complete (that is make as many existential sentences

$\exists \bar{x} \phi(\bar{a}, \bar{x})$ true in J as possible, when $\bar{a} \in J$). This guarantees that J is not nilpotent, and is the simplest way to aim at model-completeness. J is our prime candidate to refute the theorem. (To make the last two points really convincing requires more background than I want to go into here.)

Now we know exactly what to do: compute the existential n-types for each n, and see if there are finitely many. The point is that as we deal only with existential formulas, this really is *just* a computation (for fixed n). This situation being rather murky, we try $n = 2$. The computation is short, and the list of 2-types is finite.

As the situation is still murky, we try $n = 3$. In the fullness of time it becomes evident that this involves an infinite computation. A typical existential formula F_k in three variables a, b, c is found in Exhibit A.

$$\begin{gathered}(\exists x_1 \cdots x_k\ ax_1 = bx_2 \ \&\ ax_2 = bx_3 \ \&\ \cdots \ \&\ ax_{k-1} = bx_k \ \&\ acx_1 \neq 0] \\ \& \\ [\exists y_1 \cdots y_k\ ay_1 = bc \ \&\ ay_2 = by_1 \ \&\ \cdots \ \&\ ay_k = by_{k-1} \ \&\ by_k = 0].\end{gathered} \tag{A}$$

It is extremely easy to see that as k varies the formulas $F_k(a, b, c)$ are pairwise contradictory, and that they are all satisfied in J. So $J^{(3)}/\mathrm{Aut}\, J$ is infinite.

Reverse engines. For any $\aleph_0$-categorical commutative ring J and large enough k, J *omits* F_k, that is: no triple in J satisfies F_k. The next point is a bit subtle, and involves some more computations: if the commutative ring J omits F_k, then it satisfies:

$$\forall z_1 \cdots z_{2k+1} [z_1^2 = \cdots = z_{2k+1}^2 = 0 \ \Rightarrow\ z_1 \cdots z_{2k+1} = 0].$$

From here it is downhill all the way.

I think it is pretty clear how this sort of thing flows out of (4). One specializes the problem to make the computations practical and afterwards jettisons any accidental features of the result. Can one do this in other contexts?

I published this proof together with some heuristic remarks in the foregoing vein in a (very) short article [8]. I wish I could give a similar treatment of the noncommutative case, but it is completely *ad hoc*. One exploits identities coming from $x^n \equiv 0$ to arrive ultimately at

$$x^{n-1} y^{n-1} = (-yx)^{n-1},$$

a relation enough like commutativity to permit the previous argument to be used. This leads to $x^{n-1} \equiv 0$ and so on. Observe that for $n = 2$ one has the desired identity as an immediate consequence of $x^2 \equiv 0$, and our law is just as manageable as the commutative law, while for $n > 2$ we have to reduce first to the characteristically simple case to get our identity, and tinker with the subsequent argument. I insist on these minor points because we will want something similar in the next section, in a more subtle context — something which we perhaps cannot get.

C. Groups

Notice at the outset that there are many $\aleph_0$-categorical nilpotent groups of class 2. QE technology fails us slightly, since most QE nilpotent groups are of exponent 4, which is not very satisfactory. I think one can repair this by taking QE groups in a language with a predicate for the center, and this may even be implicit in other work of Saracino and Wood, but I have not checked. In any case there is a "Mal'cev correspondence" between rings and the corresponding upper triangular unipotent (1's on the diagonal) $n \times n$ matrices (e.g. $n = 3$) which preserves $\aleph_0$-categoricity (but not QE of course).

But I want to comment on the work of WILSON [16] in the direction of a classification. Wilson has surveyed his work and the recent work of Apps in the proceedings of the St. Andrews group theory conference, and APPS' papers are in course of publication in the usual British journals [17].

Let G be an $\aleph_0$-categorical group. It has of course a finite characteristic (i.e. 0-definable) series:

$$1 = G_0 < G_1 < \cdots < G_n = G.$$

The quotients G_{i+1}/G_i are $\aleph_0$-categorical, and essentially independent of the series chosen. Apps looks a bit at the extension problem, but I will not, so let us take G itself to be characteristically simple. There are three possibilities:

(1) G is elementary abelian.

(2) G is a Boolean power (unfiltered) of a finite simple group.

(3) G is a perfect ($[G, G] = G$) p-group for some prime p.

This uses the classification of the finite simple groups and is due to Wilson, using results of Kargapolov and Higman. As far as I know the details can only be found in a paper by Apps (1983?). Is there a similar result available for rings by similar methods?

PROBLEM 4. Prove that $\aleph_0$-categorical p-groups are nilpotent.

As Apps notes, we can try locally nilpotent lie rings first, if we aim to generalize the treatment of nil rings described previously.

Wilson has proved that $\aleph_0$-categorical *solvable* locally nilpotent groups are nilpotent, and observes that this is essentially the analog of my result on $\aleph_0$-categorical nil rings in the commutative case. One reduces first to the case of metabelian groups, that is solvable groups of class 2. This means essentially that an abelian group B acts on an abelian group A, so that we have:

$$B \to \mathrm{End}(A), \qquad b \to \gamma_b$$

where $\gamma_b(a) = [b, a] = (a^{-1})^b a$. If we let J be the subring of End(A) generated by (γ_b: $b \in B$) then our objective is (precisely) to show that J is nilpotent. Now J is easily seen to be commutative, and (unexpectedly, using commutativity) also $\aleph_0$-categorical, so that my result applies. Details are in Wilson's survey article [16].

Summing up we have the following apparent analogies.

Ring	*Group*
Nil	Locally nilpotent (formally: right Engel)
Nilpotent	Nilpotent
Commutative	Solvable class 2
$x^2 = 0$	$[y, x, \ldots, x] = 1$
$x^2 = 0 \Rightarrow xy = -yx$	$[y, x, x] \equiv 1 \Rightarrow$ nilpotent class 3
$x^{n-1}y^{n-1} = (-yx)^{n-1}$	? ?

In the penultimate line, we have some relations that don't depend on $\aleph_0$-categoricity (the one on the right is nontrivial). The one on the left has a useful generalization in the characteristically simple case. Actually it is unfortunate that $[y, x, x] \equiv 1$ implies nilpotence outright, since it makes it impossible to translate the ring-theoretic analysis sensibly into the group-theoretic setting. I think it would be useful to have a purely group theoretic proof of Wilson's theorem, based on the type structure in a suitable sort of "existentially complete" group.

II. GENERAL MODEL THEORY

I will discuss stable $\aleph_0$-categorical structures, and QE structures for small (microscopic) languages.

A. Stability

My own favorite question in the area of $\aleph_0$-categoricity is the following.

PROBLEM 5. Show that any stable $\aleph_0$-categorical structure is $\aleph_0$-stable.

We know a great deal about $\aleph_0$-categorical $\aleph_0$-stable structures, and this leads to various equivalent formulations of Problem 5.

I will describe five properties of $\aleph_0$-categorical $\aleph_0$-stable structures, combining work of ZIL'BER [18, 19] with work of HARRINGTON, LACHLAN, and CHERLIN [20]. The first fact seems technically central. Throughout M denotes an $\aleph_0$-categorical $\aleph_0$-stable structure.

1. Coordinatization

I need to speak of geometries, coordinate systems, and grassmannians. A geometry is either an affine or projective geometry of infinite dimension over a finite field, or (degenerate case) an infinite set with no additional structure. A coordinate system is, roughly speaking, a disjoint sum of finitely many isomorphic geometries. If $\mathcal{H}$ is a coordinate system and p is an orbit in $\mathcal{H}$ of some finite algebraically closed set under Aut $\mathcal{H}$ then the "grassmannian" $\mathrm{Gr}(p, \mathcal{H})$ is the structure whose underlying set is p and whose automorphism group is Aut $\mathcal{H}$ with the natural action. (This way of putting things may seem abstract, but I don't know a better one.)

A *coordinatization* of our structure M is an isomorphism $M/E \to \mathrm{Gr}(p, \mathcal{H})$ between the quotient M/E of M by a 0-definable equivalence relation, and a grassmannian structure.

The first significant property of our structure M is that it admits a coordinatization.

2. $\mathrm{rank}(M)$ is finite (in the sense of Morley rank).

3. M has the finite submodel property: any sentence ϕ true of M is true of a finite submodel of M.

4. Definable sets in M are Boolean combinations of sets definable from single parameters.

5. The definable sets in M are "flat" in the following sense. If $\mathcal{F}$ is a definable family of sets of constant rank r and $\mathcal{F}$ covers M then

$$\mathrm{rank}\, \mathcal{F} + r = \mathrm{rank}\, M,$$

assuming that the sets in $\mathcal{F}$ are normalized in the following sense:

$$A, B \in \mathcal{F} \text{ distinct } \Rightarrow \mathrm{rank}(A \cap B) < r.$$

I think the fourth property is suggestive in connection with Problem 5, if one knows Shelah's local ranks (Δ-rk). There is another line of attack suggested ten years ago by LACHLAN [21]. He calls a combinatorial geometry $(P, L; I)$ (that is: points, lines; incidence) a *pseudoplane* if *each* point or line is incident with infinitely many lines or points respectively, while no two share infinitely many partners. Lachlan suggested

PROBLEM 5′. Prove that there is no $\aleph_0$-categorical stable pseudoplane.

He showed that this contains Problem 5; any $\aleph_0$-categorical stable but not $\aleph_0$-stable structure involves a pseudoplane. Now we know that the two problems are equivalent, because there is no $\aleph_0$-categorical $\aleph_0$-stable pseudoplane. (If $(P, L; I)$ is such, we can easily assume the rank r of the lines as subsets of P is constant, and apply property 5 above with $\mathscr{F} = L$ to conclude:

$$r + \operatorname{rank} L = \operatorname{rank} P.$$

Then compute rank(I) two ways for a contradiction.)

Is Problem 5′ a good reformulation of Problem 5? The results on $\aleph_0$-categorical $\aleph_0$-stable structures are all essentially equivalent with the nonexistence of $\aleph_0$-categorical $\aleph_0$-stable pseudoplanes, and can be obtained by two separate methods: an application of the classification of the finite simple groups to determine the structure of strongly minimal sets, or a direct attack (ZIL'BER [19]) on the pseudoplane problem, making heavy use of the fact that only pseudoplanes of rank 2 need be considered. Neither approach is very plausible as a way of analyzing $\aleph_0$-categorical stable structures, but of the two Zil'ber's is slightly more promising.

However Lachlan actually suggested we might prove:

PROBLEM 6. Show that there is no $\aleph_0$-categorical pseudoplane.

My impression is that he had no very pronounced opinion as to the veracity of this assertion, but it is known as Lachlan's Pseudoplane Conjecture. I have come to believe that it is true, and in fact I believe something stronger:

CONJECTURE. There is no uniformly locally finite pseudoplane.

I must of course give you the correct definition of "subplane generated by A", so that the notion of uniform local finiteness will be sensibly defined. We take:

$$\langle A \rangle = \operatorname{acl}^{\mathrm{ex}}(A),$$

the algebraic closure of A relative to existential formulas. (Cf. Saracino's principle, part IB.) I would very much like to know whether for each n, e.g. $n=5$, there is a uniformly n-finite pseudoplane. Conceivably there is an easy construction for each n separately, but if there seems to be no uniformly 5-finite pseudoplane, then that would bring the whole issue comparatively down to earth.

B. QE structures

If L is a finite relational language I denote by QE(L) and QE(L; st) the classes of QE or QE and stable structures, respectively. Since our language contains no function symbols, all structures are uniformly locally finite, and all sorts of phenomena are going to be simplified. Lachlan and co. have been looking at these two classes for some time, with interesting results. Bear in mind that QE(L) contains finite structures.

QE(L, st)

EXAMPLE. The stable QE graphs are of three types:

(A) $m \cdot K_n$ or its complement ($m, n \leqslant \infty$),

(B) C_5,

(C) K_3^2, with an edge between (i, j) and (k, l) if $\{i, j\} \cap (k, l) \neq \emptyset$.

K_n is the complete graph on n points, C_5 is the 5-cycle. This example generalizes as follows.

THEOREM. *With L fixed,* QE(L, st) *decomposes into finitely many families $\mathcal{F}_1, \ldots, \mathcal{F}_k$ such that within a given family $\mathcal{F}_i$, each structure M is determined up to isomorphism by its* dimensions $d_{ij}(M)$, *which are defined as the dimensions of the geometries involved in coordinatizations of M and suitable substructures by grassmannians.*

This is imprecise in a number of ways — one must allow finite geometries, and only the degenerate ones are relevant — but it does capture the way this theorem fits into the line of Section A. This formulation of the theorem incorporates a technical result proved recently by Lachlan and myself. Lachlan uses a rank function designed to make sense on finite structures and satisfying:

$$\mathrm{QE}(L, st) = \bigcup_{n<\infty} \mathrm{QE}(L, n)$$

for $QE(L, n) = \{M \in QE(L): \text{rank } M \leq n\}$. For technical reasons he proves his theorem for each $QE(L, n)$ separately, but as he suspected:

THEOREM. *For each L there is an n with*

$$QE(L, st) = QE(L, n).$$

What is really at issue here is the following:

COORDINATIZATION LEMMA. *For L fixed there is an m so that for any $M \in QE(L, st)$ and any maximal 0-definable equivalent relation E on M with M/E finite of cardinality at least m:*

$$M/E \simeq \textit{some grassmannian.}$$

This is a theorem about finite permutation groups, and the recent literature abounds in relevant information.

I find it quite interesting that the theory of $\aleph_0$-categorical $\aleph_0$-stable structures closely resembles that of $QE(L; st)$, after one allows nondegenerate geometries.

QE(L)

It is very hard to get a *complete* classification of the QE structures for even the simplest languages, but Lachlan — in part with Woodrow — has had considerable success. I don't feel I can go into the methods here, but let me indicate the current situation.

QE *graphs:* (2 symmetric 2-types)
Stable: listed above.
Unstable: $\neg K_{n+1}$-generic (containing every graph not embedding K_{n+1}), or the complementary graph,
or generic (universal).

QE *tournaments:* (2 asymmetric 2-types)
1 (one point),
$\vec{C}_3$ (oriented triangle),
$\mathbb{Q}$ (rational order),
$\mathbb{Q}^*$ (Skolem's circular order-points at rational angles on the unit circle; arrows in the positive direction up to half-way around).
Generic.

If we take a stock of three 2-types (with one or all symmetric) there are $2^{\aleph_0}$ examples by the ratatouille method. This leads to a very interesting problem, which should be attributed to Lachlan, unless he disowns it.

PROBLEM 7. Find all QE simple digraphs.

A solution to this problem would provide the first explicit *classification* of a *natural* uncountable family of $\aleph_0$-categorical structures. For the record, here are the ones I know of.

Some QE simple digraphs

(1) $\vec{C}_4$ (oriented square).

(2) SL(2, 3): the points are the eight nonzero vectors in the plane over F_3. The automorphism group is SL(2, 3); this determines the digraph.

(3) $m \cdot X$ or $X[m]$, X one of the five QE tournaments, in the empty graph on m points, $m \leqslant \infty$.

(4) The generic partial ordering.

(5) The generic digraph for which "not joined by an edge" is an equivalence relation, with m classes ($m \leqslant \infty$).

(6) $\hat{\mathbb{Q}}$, $\hat{T}^{\infty}$ where T^{∞} is the generic tournament and the operation "^" applied to the tournament T does the following:

(a) adds a point o to T to form T_o, with $o \to T$;

(b) creates a second copy T'_o of T_o;

(c) joins a in T_o to b' in T'_o by an arrow the "wrong" way ($a \neq b$);

(d) leaves a, a' unlinked.

($\hat{1} = \vec{C}_4$, $\hat{\vec{C}}_3 = \mathrm{SL}(2,3)$).

(7) $\mathbb{Q}_3$: this is the digraph defined on the points $r = 1$, $\theta \in \mathbb{Q}$ (in polar coordinates) where $a \to b$ means b lies less than one third of the way around the circle in the positive direction.

(8) Generic omitting m (m is the digraph on m points with no arrows).

(9) Generic omitting a fixed set X of tournaments (where X is closed upward, and otherwise arbitrary).

Perhaps this list is complete as it stands.

Notes added in proof

(1) For the classification of characteristically simple $\aleph_0$-categorical groups (part C), compare also R. GILMAN, J. Symbolic Logic 49 (1984), pp. 900–907.

(2) I have classified all imprimitive homogeneous digraphs (cf. Problem 7). It turns out that there is one additional variant of number (5) for $m = \infty$. I have not learned of any other primitive examples.

(3) Peter Neumann showed that if $\operatorname{Aut} X \simeq \operatorname{Aut} \mathbb{Q}$ with X countable, then X is interpretable in $\mathbb{Q}$. Compare Problem 1(d).

References

[1] SCHMERL, J., 1980, *Decidability and $\aleph_0$-categoricity of theories of partially ordered sets*, J. Symbolic Logic 45, pp. 585–611.

[2] ROSENSTEIN, J., 1969, *$\aleph_0$-categoricity of linear orderings*, Fund. Math. 44, pp. 1–5.

[3] ENGELER, E., 1959, *A characterization of ...*, Notices Amer. Math. Soc. 6, p. 161; RYLL-NARDZEWSKI, C., 1959, ... *categoricity in power* $\leqslant \aleph_0$ ..., Bull. Acad. Pol. Sci. 7, pp. 545–548. SVENONIUS, L., 1959, ... *in first-order predicate calculus*, Theoria 25, pp. 82–94.

[4] BAUR, W., 1975, *$\aleph_0$-categorical modules*, J. Symbolic Logic 40, pp. 213–226.

[5] ZIEGLER, M. 1984, *Model theory of modules*, Ann. Pure Appl. Logic 26, pp. 149–213.

[6] MIRAVAGLIA, F., 1977, *On $\aleph_0$-categorical biregular rings*, Thesis, Yale.

[7] MACINTYRE, A. and J. ROSENSTEIN, 1976, *$\aleph_0$-categoricity for rings without nilpotent elements and for Boolean structures*, J. Algebra 43, pp. 129–154.

[8] CHERLIN, G., 1980, *On $\aleph_0$-categorical nil rings* I, II. Algebra Universalis 10, pp. 27–30 and J. Symbolic Logic 45, pp. 291-301.

[9] SARACINO, D. and C. WOOD, 1984, *QE commutative nilrings*, J. Symbolic Logic 49, pp. 644–651.

[10] FRAISSÉ, R., *Sur certaines ratatouilles qui généralisent l'ordre des nombres rationnels*, C.R. Acad. Sci. 237, pp. 540–542.

[11] GLASSMIRE, W., 1971, *There are* $2^{\aleph_0}$ ..., Bull. Acad. Pol. Sci. 19, pp. 185–190. ASH, C., 1971, ... *undecidable* ..., Amer. Math. Soc. Notices 18, p. 423. HENSON, C.W., 1972, ... *countable homogeneous relational structures and* ... J. Symbolic Logic 37, pp. 494–500. EHRENFEUCHT, C.W., 1972, *$\aleph_0$-categorical theories*, Bull. Acad. Pol. Sci. 20, pp. 425–427.

[12] BOFFA, M., A. MACINTYRE and F. POINT, 1980, *The quantifier elimination problem for rings without nilpotent elements and for semisimple rings*, in: PACHOLSKI et al., eds., Set theory and hierarchy theory, Lecture Notes in Math. 834 (Springer, New York) pp. 20–30.

[13] BERLINE, C. and G. CHERLIN, 1981, *QE nilrings of prime characteristic*, Bull. Soc. Math. Belg. Sér. B 33, pp. 3–17.

[14] SARACINO, D., 1973, *Model companions for $\aleph_0$-categorical theories*, Proc. Amer. Math. Soc. 39, pp. 591–598.

[15] SARACINO, D. and C. WOOD, 1983, *QE nil-2 groups of exponent 4*, J. Algebra 76, pp. 337–352.

[16] WILSON, J., 1952, *The algebraic structure of $\aleph_0$-categorical groups*, CAMPBELL and ROBERTSON, eds., Groups — St. Andrews 1981, London Math. Soc. Lecture Notes Ser. 71 (*Cambridge*).

[17] APPS, A., 1983, *On the structure of $\aleph_0$-categorical groups*, J. Algebra 81, pp. 320–339 (related work in Math. Proc. Camb. Phil. Soc. 91 (1982) and PLMS 47 (1983)).

[18] ZIL'BER, B.I., 1980, *Totally categorical theories: structural properties and the non-finite axiomatizability*, in: Model theory of algebra and arithmetic, Karpacz 1979, Lecture Notes in Math. 834 (Springer, Berlin) pp. 381–410.

[19] ZIL'BER, B.I., 1980, 1984, *Strongly minimal $\aleph_0$-categorical structures* I–III (Russian), Sibirsk Mat. Z.: Part I: 21, pp. 98–112; Part II, III: to appear.
[20] CHERLIN, G., L. HARRINGTON and A. LACHLAN, 1985, *$\aleph_0$-categorical $\aleph_0$-stable structures*, Ann. Pure Appl. Logic 28, pp. 103–135.
[21] LACHLAN, A., 1974, *Two conjectures regarding the stability of $\aleph_0$-categorical theories*, Fund. Math. 81, pp. 133–145.
[22] LACHLAN, A., 1984, *On countable stable structures homogeneous for a finite relational language*, IJM 49, pp. 69–153.
[23] CHERLIN, G. and A. LACHLAN, *Stable finitely homogeneous structures*, submitted.
[24] CAMERON, P., 1981, *Finite permutation groups and finite simple groups*, Bull. London Math. Soc. 13, pp. 1–22.
[25] LACHLAN, A. and R. WOODROW, 1980, *Countable ultrahomogeneous graphs*, Trans. Amer. Math. Soc. 262, pp. 51–94.
[26] LACHLAN, A., 1984, *Countable homogeneous tournaments*, Trans. Amer. Math. Soc. 284, pp. 431–461.

Barcan Marcus et al., eds., Logic, Methodology and Philosophy of Science VII
© Elsevier Science Publishers B.V. (1986) 115–128

STRUCTURAL PROPERTIES OF MODELS OF $\aleph_1$-CATEGORICAL THEORIES

B.I. ZIL'BER

Kemerovo University, Kemerovo 43, 650043 USSR

1.

The structural theory of categoricity (in uncountable powers) began with the works of BALDWIN [1972] and BALDWIN & LACHLAN [1971], in which the notions of a strongly minimal set and algebraic closure were introduced and it was shown that the structure of a strongly minimal set with respect to algebraic closure (acl) affects essentially the structure of the model itself.

The structure of a strongly minimal set S with respect to the closure operator acl can be essentially characterized by the *geometry* associated with S. The geometry associated with S over a subset A is given by its points, which are the sets of the form $\mathrm{acl}(a, A)$ for $a \in S\text{-}\mathrm{acl}(A)$, and its n-dimensional subspaces, which are $\mathrm{acl}(a_0, \dots, a_n, A)$, where $a_0, \dots, a_n$ are algebraically independent over A. We omit "over A", if $A = \emptyset$.

If the geometry associated with S over any non-algebraic element is isomorphic to a geometry of a projective space over a division ring then the geometry associated with S is called *locally projective.*

If the division ring in the definition is finite, then the main result of DOYEN & HUBAUT [1971] describes the locally projective geometry as an affine or projective geometry over the division ring.

Call a strongly minimal structure S disintegrated if $\mathrm{acl}(X \cup Y) = \mathrm{acl}(X) \cup \mathrm{acl}(Y)$ for every $X, Y \subseteq S$. This is equivalent to the degeneracy of the geometry associated with S (i.e. all subsets of the geometry are subspaces).

Natural examples of strongly minimal structures with projective geometries are strongly minimal abelian groups and, more generally, modules. Affine spaces over division rings have locally projective geometries which are not projective. The natural numbers with the successor operation is a typical example of a strongly minimal disintegrated structure.

On the other hand such strongly minimal structures as algebraically closed fields can hardly be characterized in terms of their geometries. More adequate in this situation seems the following notion introduced by LACHLAN [1973/74].

A pseudoplane is a triple $\langle P, L, I\rangle$, where P is a set of "points", L is a set of "lines" and $I \subseteq P \times L$ is an incidence relation satisfying the following:

(1) every line is incident to an infinite set of points;

(2) every point is incident to an infinite set of lines;

(3) any two distinct points are incident in common to at most finite number of lines;

(4) any two distinct lines are incident in common to at most finite number of points.

CONJECTURE. For any uncountably categorical pseudoplane there is an algebraically closed field such that the field is definable in the pseudoplane and the pseudoplane is definable in the field.

In the paper the following theorem will be proved:

TRICHOTOMY THEOREM. *For an uncountably categorical structure M one and only one of the following holds:*

(1) *An uncountably categorical pseudoplane is definable in M.*

(2) *For every strongly minimal structure S definable in M the geometry associated with S is locally projective.*

(3) *Every strongly minimal structure definable in M is disintegrated.*

In the connection with the Trichotomy Theorem the following theorem is of special interest.

THEOREM 2. *There is no totally categorical pseudoplane* (*i.e. one the complete theory of which is categorical in all infinite powers*).

Theorem 2 was proved independently by CHERLIN et al. [1981] and the author [1977] (the complete proof is to appear in Sibirsk. M.Ž.). The proofs are quite different, that of CERLIN et al. [1981] relies on the classification of all finite simple groups. The proof of the author is rather long but does not use any deep results outside model theory.

As was shown in ZIL'BER [1980a] the global properties of an uncountably categorical structure M depend essentially on the structure of groups definable in M. Therefore the following theorems, which will be proved in the paper, are of much importance for the structural theory.

THEOREM 3. *Let M be an uncountably categorical structure satisfying* (2) *of the Trichotomy Theorem and G a group definable in M. Then*

(i) *G is abelian-by-finite.*

(ii) *If G is infinite and has no proper infinite definable subgroup, then G is strongly minimal.*

THEOREM 4. *If M is an uncountably categorical structure satisfying* (3) *of the Trichotomy Theorem, then no infinite group is definable in M. It follows from this that M is almost strongly minimal.*

Note that Theorem 3(i) contains the known theorem of BAUR, CHERLIN and MACINTYRE [1979] which states that totally categorical groups are abelian-by-finite.

2. Proofs

An *incidence sytem* is a triple $\langle P, L, T\rangle$, where P is a set of "points", L is a set of "lines" and $I \subseteq P \times L$ is an arbitrary relation called an incidence relation.

For a binary relation R and an element x we denote

$$xR = \{y: xRy\}, \qquad Rx = \{y: yRx\}.$$

Thus, for $p_0 \in P$, $l_0 \in L$

$$p_0I = \{l \in L: p_0Il\}, \qquad Il_0 = \{p \in P: pIl_0\}.$$

Let $A \subseteq M^n$ be an X-definable subset of a structure M and E be an X-definable in M equivalence relation on A. Sets of the form A/E are called *X-definable sets in M.* Definable means X-definable for some $X \subseteq M$.

An *X-definable structure in M* is an X-definable set with X-definable relations.

A natural construction considered in SHELAH [1978, III, §6], ZIL'BER [1980a], CHERLIN et al. [1981] allows us to treat definable sets in M as definable subsets of some larger structure M^* which contains M and preserves categoricity, ranks and definability.

Now we begin with the proof of the Trichotomy Theorem. From now on M is an uncountable categorical structure.

LEMMA 1. *Let $\langle P, L, I\rangle$ be an incidence system 0-definable in M,*

$$\langle p_0, l_0\rangle \in I,$$

$$\operatorname{rank}(l_0, \emptyset) = \operatorname{rank}(L), \qquad \operatorname{rank}(p_0, \emptyset) = \operatorname{rank}(P),$$

$$\operatorname{rank}(p_0, \{l_0\}) = \operatorname{rank}(Il_0), \qquad \operatorname{rank}(l_0, \{p_0\}) = \operatorname{rank}(p_0 I).$$

Then there exist an 0*-definable incidence system* $\langle P', L', I'\rangle$ *in* M, $l_0' \in L'$ *and a mapping* $m: L' \to L$ *such that*

$$P' = P, \qquad \operatorname{rank}(L') = \operatorname{rank}(L),$$

$$m^{-1}(l) \textit{ is finite} \quad \textit{for all } l \in L, \qquad m(l_0') = l_0,$$

$$\operatorname{rank}(I'l_0') = \operatorname{rank}(Il_0), \qquad \deg(I'l_0') = 1,$$

$$\textit{for all } p \in P \quad m(pI') \subseteq pI, \quad \operatorname{rank}(pI') = \operatorname{rank}(pI).$$

PROOF. By the Finite Equivalence Relation Theorem in SHELAH [1978, III, T2.28] there is a two-variable formula E_{l_0} with constant l_0, which defines an equivalence relation on Il_0 with finite number of classes, each of the classes having degree 1 or rank less than $r_0 = \operatorname{rank}(Il_0)$. Let the number of classes be k_0. Put

$$L_1 = \{l \in L : E_l \text{ is an equivalence relation on } Il \text{ with } k_0 \text{ classes}\}.$$

Evidently, L_1 is 0-definable, $l_0 \in L_1$, therefore $\operatorname{rank}(L_1) = \operatorname{rank}(L)$.

Define an equivalence relation E on $I \cap (P \times L_1)$:

$$\langle p, l\rangle E \langle p', l'\rangle \quad \text{iff} \quad l = l' \;\&\; p E_l p',$$

and put

$$L' = I \cap (P \times L_1)/E.$$

It is easy to see that for every $l \in L_1$ there are precisely k_0 elements $l' \in L'$ of the form $l' = \langle p, l\rangle E$ for some $p \in P$. Define $m(l') = l$ in this case. Evidently, $l' \in \operatorname{acl}(l)$, therefore, in particular, $\operatorname{rank}(L') = \operatorname{rank}(L)$. Put

$$pI'l' \quad \text{iff} \quad l' = \langle p, l\rangle E \;\&\; \operatorname{rank}(\langle p, l\rangle E) = r_0.$$

Note that the last condition is definable in M since M is uncountably categorical. Put

$$l_0' = \langle p_0, l_0\rangle E.$$

It is clear that $I'l_0' \subseteq Il_0$, $I'l_0'$, is an E_{l_0}-equivalence class and $p_0 \in I'l_0'$, therefore $\operatorname{rank}(I'l_0') = r_0$, $\deg(I'l_0') = 1$.

LEMMA 2. *Let* M *be a strongly minimal structure. If there are elements* a_1, a_2,

b_1, b_2, c in M, every four of which are algebraically independent, $c \in \operatorname{acl}(a_1, a_2, b_1, b_2)$ and $\operatorname{acl}(a_1, a_2, c) \cap \operatorname{acl}(b_1, b_2, c) = \operatorname{acl}(c)$, then there is an incidence system $\langle P, L, I \rangle$ which is 0-definable in M and:

$$\operatorname{rank}(P) = 2, \qquad \operatorname{rank}(L) \geqslant 2, \qquad \deg(P) = 1;$$

$$\operatorname{rank}(Il) = 1 \quad \textit{for every } l \in L;$$

$$\textit{if } l_1, l_2 \in L,\ l_1 \neq l_2, \quad \textit{then } \operatorname{rank}(Il_1 \cap Il_2) = 0.$$

PROOF. Let $P_0 = M \times M$, $L_0 = M \times M \times M$ and $I_0 \subseteq P_0 \times L_0$ be an arbitrary 0-definable relation such that

$$\langle b_1, b_2 \rangle\, I_0 \langle a_1, a_2, c \rangle$$

and

$$\langle x_1, x_2 \rangle\, I_0 \langle y_1, y_2, z \rangle \rightarrow z \in \operatorname{acl}(x_1, x_2, y_1, y_2).$$

It is easy to check that putting $p_0 = \langle b_1, b_2 \rangle$, $l_0 = \langle a_1, a_2, c \rangle$ we have all the assumptions of Lemma 1 satisfied. Hence for some L_0', $l_0' \in L_0$, I_0' we have

$$\operatorname{rank}(L_0') = \operatorname{rank}(L_0) = 3, \qquad l_0' \in \operatorname{acl}(l_0), \quad l_0 \in \operatorname{acl}(l_0'),$$

$$\operatorname{rank}(I_0'l_0') = \operatorname{rank}(I_0 l_0) = 1, \qquad \deg(I_0'l_0') = 1.$$

Put

$$L_1 = \{l_1 \in L_0' \colon \operatorname{rank}(I_0'l_1) = 1\ \&\ (\forall l_2 \in L_0')$$
$$(\operatorname{rank}(I_0'l_1 \cap I_0'l_2) > 0 \rightarrow \operatorname{rank}(I_0'l_1 - I_0'l_2) = 0).$$

Since $I_0'l_0'$ is strongly minimal, hence $l_0' \in L_1$, therefore

$$\operatorname{rank}(L_1) = \operatorname{rank}(l_0', \emptyset) = \operatorname{rank}(L_0').$$

Define an equivalence relation E on L_1:

$$l_1\, E\, l_2 \quad \text{iff} \quad \operatorname{rank}(I_0'l_1 \dot{-} I_0'l_2) = 0.$$

Now put $P = P_0$, $L = L_1/E$ and for $p \in P$, $l \in L_1$

$$pI(lE) \quad \text{iff} \quad \operatorname{rank}(pI_0' - lE) < \operatorname{rank}(lE).$$

It follows from Proposition 1.5 of ZIL'BER [1980a] that for every l of L_1 there is p of $I_0'l$ such that $pI(lE)$ (consider $\gamma = I_1 l$, $\varphi = lE$, $\psi = I$). Moreover it follows from the same proposition that $pI(lE)$ holds for almost all p of $I_0'l$, i.e.

$$\operatorname{rank}(I(lE) \dot{-} I_0'l) = 0, \qquad \operatorname{rank}(I(lE)) = 1.$$

In particular for our p_0 and l_0, if we put $\bar{l}_0 = l_0'E$ we get the strong minimality of $I\bar{l}_0$ and $p_0I\bar{l}_0$.

If $l_1, l_2 \in L_1$, $\bar{l}_1 = l_1E$, $\bar{l}_2 = l_2E$, $\operatorname{rank}(I\bar{l}_1 \cap I\bar{l}_2) > 0$, then

$$\operatorname{rank}(I_0'l_1 \cap I_0'l_2) > 0 \quad \text{and} \quad \operatorname{rank}(I_0'l_1 \dot{-} I_0'l_2) = 0,$$

which follows from the definition of L_1, hence $\bar{l}_1 = \bar{l}_2$.

We show now that $\operatorname{rank}(\bar{l}_0, \emptyset) \geq 2$ and therefore $\operatorname{rank}(L) \geq 2$.

Suppose $\operatorname{rank}(\bar{l}_0, \emptyset) \leq 1$. Then, since $\operatorname{rank}(p_0, \{\bar{l}_0\}) = 1 < \operatorname{rank}(p_0, \emptyset)$, $\operatorname{rank}(\bar{l}_0, \{p_0\}) < \operatorname{rank}(\bar{l}_0, \emptyset) \leq 1$. Thus $\bar{l}_0 \in \operatorname{acl}(p_0) = \operatorname{acl}(b_1, b_2)$. Evidently $c \notin \operatorname{acl}(\bar{l}_0)$, therefore there is c' of M such that

$$t(\langle c, c'\rangle, \{\bar{l}_0\}) = t(\langle b_1, b_2\rangle, \{\bar{l}_0\}).$$

Since $\operatorname{rank}(\langle c, c'\rangle, \{\bar{l}_0\}) = 1$,

$$c' \in \operatorname{acl}(\bar{l}_0, c) \subseteq \operatorname{acl}(b_1, b_2, c) \cap \operatorname{acl}(a_1, a_2, c).$$

By the definition $c' \notin \operatorname{acl}(c)$. This contradicts the assumptions of the lemma. Hence, $\operatorname{rank}(\bar{l}_0, \emptyset) \geq 2$.

LEMMA 3. *Let for an incidence system* $\langle P, L, I\rangle$: $p_0 \in P$, $\operatorname{rank}(p_0, \emptyset) \geq \operatorname{rank}(L)$, $\operatorname{rank}(p_0I) > 0$ *and if* $p_1, p_2 \in P$, $p_1 \neq p_2$, *then* $\operatorname{rank}(p_1I \cap p_2I) = 0$.

Then there is $l_0 \in L$ *such that*

$$\langle p_0, l_0\rangle \in I, \qquad \operatorname{rank}(p_0, \{l_0\}) > 0,$$

$$\operatorname{rank}(l_0, \emptyset) > \operatorname{rank}(p_0I), \qquad \operatorname{rank}(Il_0) > 0.$$

PROOF. It follows from the assumptions of the lemma that there is no 0-definable subset L' of L such that $L' \supseteq p_0I$, $\operatorname{rank}(L') = \operatorname{rank}(p_0I)$. Thus, by the Compactness Theorem there is

$$l_0 \in p_0I - \operatorname{acl}(p_0), \qquad \operatorname{rank}(l_0, \emptyset) > \operatorname{rank}(p_0I).$$

Now counting

$$\begin{aligned} \operatorname{rank}(\langle p_0, l_0\rangle, \emptyset) &= \operatorname{rank}(p_0, \{l_0\}) + \operatorname{rank}(l_0, \emptyset) \\ &= \operatorname{rank}(l_0, \{p_0\}) + \operatorname{rank}(p_0, \emptyset), \end{aligned}$$

we have

$$\begin{aligned} \operatorname{rank}(p_0, \{l_0\}) &= \operatorname{rank}(l_0, \{p_0\}) + \operatorname{rank}(p_0, \emptyset) \\ &\quad - \operatorname{rank}(l_0, \emptyset) \geq \operatorname{rank}(l_0, \{p_0\}) > 0. \end{aligned}$$

Hence, in particular, $\operatorname{rank}(Il_0) > 0$.

LEMMA 4. *If all the assumptions of Lemma 2 hold then an uncountably categorical pseudoplane $\langle P, L, I\rangle$ is definable in M with* $\text{rank}(P) = \text{rank}(L) = 2$, $\deg(P) = \deg(L) = 1$.

PROOF. Using Lemma 2 and the symmetry of the definition we get an incidence system $\langle P_0, L_0, I_0\rangle$ definable in M such that the following hold:

(i) $\text{rank}(P_0) \geqslant 2, \quad \text{rank}(L_0) = 2, \quad \deg(L_0) = 1;$

(ii) $\text{rank}(pI_0) = 1$ for all p of P_0;

(iii) if $p_1, p_2 \in P_0$, $p_1 \neq p_2$, then $\text{rank}(p_1I_0 \cap p_2I_0) = 0$.

Considering a definable subset of rank 2 degree 1 of P_0 instead of P_0 and taking an inessential expansion of M we preserve (i), (ii), (iii) having $\text{rank}(P_0) = 2$ and the incidence structure 0-definable in M.

Apply now Lemma 3 to find $\langle p_0, l_0\rangle \in I_0$ such that

$$\text{rank}(p_0, \emptyset) = \text{rank}(l_0, \emptyset) = 2,$$

$$\text{rank}(p_0I) = 1 = \text{rank}(l_0, \{p_0\}),$$

$$\text{rank}(Il_0) = 1 = \text{rank}(p_0, \{l_0\}).$$

Now by Lemma 1 we get an incidence system $\langle P_0, L_0', I_0'\rangle$ and $l_0' \in L_0'$. $I_0'l_0'$ is strongly minimal in the system. In addition for different p_1, p_2, of P_0 the set $p_1I_0' \cap p_2I_0'$ is finite, since it lies in $m^{-1}(p_1I_0 \cap p_2I_0)$.

Put as in the proof of Lemma 2

$$L_1 = \{l_1 \in L_0' \colon \text{rank}(I_0'l_1) = 1 \;\&\; (\forall l_2 \in L_0')$$
$$(\text{rank}(I_0'l_1 \cap I_0'l_2) > 0 \rightarrow \text{rank}(I_0'l_1 - I_0'l_2) = 0)\};$$

$$l_1 \mathrel{E} l_2 \quad \text{iff} \quad \text{rank}(I_0'l_1 \mathbin{\dot{-}} I_0'l_2) = 0;$$

$$P = P_0, \qquad L = L_1/E,$$

$$pI(lE) \quad \text{iff} \quad \text{rank}(pI_0' - lE) < \text{rank}(lE).$$

Observe that every class lE is finite, since if $l \mathrel{E} l_1$, then there are $p_1, p_2 \in I_0'l \cap I_0'l_1$,

$$\text{rank}(p_1, \{l\}) = \text{rank}(p_1, \{l, l_1\}) = 1,$$

$$\text{rank}(p_2, \{p_1, l\}) = \text{rank}(p_2, \{p_1, l, l_1\}) = 1,$$

and by the reciprocity principle

$$\text{rank}(l_1, \{l\}) = \text{rank}(l_1, \{l, p_1\}),$$

$$\operatorname{rank}(l_1, \{l, p_1\}) = \operatorname{rank}(l_1, \{l, p_1, p_2\}),$$

thus

$$\operatorname{rank}(l_1, \{l\}) \leq \operatorname{rank}(l_1, \{p_1, p_2\}) = 0$$

(since $l_1 \in p_1 I'_0 \cap p_2 I'_0$), i.e.

$$l_1 \in \operatorname{acl}(l).$$

Granting the finiteness of lE,

$$pI(lE) \quad \text{iff} \quad lE \subseteq pI'_0.$$

Hence

$$pI \subseteq pI'_0/E,$$

$$\operatorname{rank}(p_1 I \cap p_2 I) = 0 \quad \text{for distinct } p_1, p_2 \in P.$$

As is shown in the proof of Lemma 2 for distinct l_1, l_2 of L

$$\operatorname{rank}(Il_1 \cap Il_2) = 0, \qquad \operatorname{rank}(Il_1) = 1.$$

To get $\operatorname{rank} pI = 1$ for all $p \in P$ remove all $p \in P$ with $\operatorname{rank} pI = 0$ from P. Since $\operatorname{rank}(p_0, \emptyset) = 2$, $\operatorname{rank}(p_0 I) = 1$ and $\operatorname{rank}(P) = 2$, $\deg(P) = 1$, the set of the points removed has rank not greater than 1, therefore removing these points we diminsh the rank of only a finite number of Il, $l \in L$. Remove these lines too, denote by the same letters P, L the new sets, and the construction of the pseudoplane $\langle P, L, I\rangle$ is finished. In Zil'ber [1980b, Proposition 11] it is proved that $\langle P, L, I\rangle$ is an uncountably categorical pseudoplane.

Proposition 5. *If no uncountably catagorical pseudoplane is definable in M then for every strongly minimal structure S definable in M the geometry associated with S is locally projective or degenerate.*

The proof of the proposition follows from Lemma 4 as is shown in Zil'ber [1980b, Section 2].

Proposition 6. *If a pseudoplane is definable in M, $A \subseteq M$, then for every A-definable in M strongly minimal structure S the geometry associated with S over a is neither projective nor degenerate.*

Proof. Let $\langle P, L, I\rangle$ be a C-definable in M incidence system satisfying the following:

(i) $$\operatorname{rank}(Il_1 \cap Il_2) = 0, \qquad \operatorname{rank}(p_1 I \cap p_2 I) = 0$$

for $l_1, l_2 \in L$, $p_1, p_2 \in P$, $l_1 \neq l_2$, $p_1 \neq p_2$.

(ii) For some pair $\langle p_0, l_0 \rangle \in I$

$$\operatorname{rank}(l_0, C) = \operatorname{rank}(L), \qquad \operatorname{rank}(p_0, C) = \operatorname{rank}(P),$$

$$\operatorname{rank}(l_0, \{p_0\} \cup C) > 0, \qquad \operatorname{rank}(Il_0) > 0, \qquad \operatorname{rank}(p_0 I) > 0.$$

(iii) The four-tuple $\langle \operatorname{rank}(P), \deg(P), \operatorname{rank}(L), \deg(L) \rangle$ is lexicographically minimal among such four-tuples for every $C \subseteq M$ and systems $\langle P, L, I \rangle$ satisfying (i) and (ii).

Observe that it follows from Lemma 3 and the minimality condition that

$$\operatorname{rank}(P) = \operatorname{rank}(L), \qquad \deg(P) = 1, \qquad \deg(L) = 1.$$

Condition (ii) implies the existence of such C-definable P', L', I', $p_0 \in P' \subseteq P$, $l_0 \in L' \subseteq L$, $\langle p_0, l_0 \rangle \in I' \subseteq I$ that for every $p \in P'$, $l \in L'$

(iv) $\operatorname{rank}(pI') > 0, \qquad \operatorname{rank}(I'l) > 0.$

We assume that $P' = P$, $L' = L$, $I' = I$. Put $r = \operatorname{rank}(P) = \operatorname{rank}(L)$.

Observe also that any extension of C preserves conditions (i)–(iv), thus we may assume C contains all the parameters required in what follows.

Let S be a C-definable in M set and $\psi(x, y)$ a formula with parameters in C which is a stratification of L over S of rank less than r, i.e.:

(v) For every $l \in L$ there is an $s \in S$ such that $\psi(s, l)$.

(vi) For every $s \in S$, $\operatorname{rank}(\psi(s, M)) < r$.

The stratification ψ exists if C is sufficiently large, ZIL'BER [1974] (see also another version of the statement in SHELAH [1978, V. 6.1]).

Let us prove

(vii) For every $s \in S$, $\operatorname{rank}(p_0 I \cap \psi(s, M)) = 0$. Indeed, otherwise, putting

$$L_0 = \psi(s, M) \cap L, \qquad I_0 = I \cap (P \times L_0)$$

we get

$$\operatorname{rank}(p_0, C \cup \{s\}) \geq r - 1 \geq \operatorname{rank}(L_0), \qquad \operatorname{rank}(p_0 I_0) > 0.$$

Evidently $\langle P, L_0, I_0 \rangle$ over $C \cup \{s\}$ satisfies (i), and (ii) follows from Lemma 3. This contradicts the minimality of $\langle P, L, I \rangle$.

Observe again that if we take

$$P' = \{p \in P \colon \forall s \in S, \operatorname{rank}(pI \cap \psi(s, M)) = 0\},$$
$$I = I \cap (P' \times L),$$

we get system $\langle P', L, I' \rangle$ satisfying (i)–(iii), therefore it can be assumed $P' = P$.

Let Q be a definable subset of $P \cup L$ of the rank maximal among all such Q that

$$Q \subseteq \operatorname{acl}(S \cup C'), \quad C' \supseteq C, \quad C' \text{ is finite.}$$

Since C was assumed to be sufficiently large, hence $C' = C$. Also $Q \neq \emptyset$, since $Q \supseteq (P \cup L) \cap C$. Let

$$\operatorname{rank}(Q \cap L) \geqslant \operatorname{rank}(Q \cap P),$$

$$P^* = P \cap Q, \qquad L^* = \bigcup\{pI \colon p \in Q \cap P\}.$$

For every l from L^* there are $p \in P^*$ and $s \in S$ such that $l \in pI \cap \psi(s, M)$. Since the last set is of rank 0,

$$l \in \operatorname{acl}(p, s, C); \qquad L^* \subseteq \operatorname{acl}(S \cup C),$$

i.e. $\operatorname{rank}(L^*) \leqslant \operatorname{rank}(P^*)$. Choose $p_0^* \in P^*$ so that $\operatorname{rank}(p_0^*, C) \geqslant \operatorname{rank}(L^*)$ and by Lemma 3 we get (ii) for $\langle P^*, L^*, I \cap (P^* \times L^*)\rangle$. Since (i) for this system follows from that of $\langle P, L, I\rangle$,

$$\operatorname{rank}(P^*) = \operatorname{rank}(L^*) = r.$$

We will assume $P^* = P$, $L^* = L$, i.e. $P \cup L \subseteq \operatorname{acl}(S \cup C)$.

It can be easily proved by induction on k, that for every C-definable set Q of rank k, if $Q \subseteq \operatorname{acl}(S \cup C)$, then C can be extended so that for every $q \in Q$ there are $d_1, \ldots, d_k \in S$

$$\operatorname{acl}(q, C) = \operatorname{acl}(d_1, \ldots, d_k, C).$$

Let now Q be $P \cup L$. Assume for simplicity $C = \emptyset$. Then

$$\operatorname{acl}(p_0) = \operatorname{acl}(s_1, \ldots, s_r), \qquad \operatorname{acl}(l_0) = \operatorname{acl}(t_1, \ldots, t_r),$$

for $s_1, \ldots, s_r, t_1, \ldots, t_r \in S$. Since $\operatorname{rank}(p_0, \emptyset) = r$, $\operatorname{rank}(l_0, \emptyset) = r$,

$$\dim(s_1, \ldots, s_r) = r = \dim(t_1, \ldots, t_r).$$

It follows from (i) that

$$\operatorname{rank}(p_0, \{l_0\}) < r,$$

therefore

$$\dim(s_1, \ldots, s_r, t_1, \ldots, t_r) < 2r.$$

If the geometry associated with S over $\emptyset$ is degenerate or projective, then the last condition implies the existence of $u_0 \in S - \operatorname{acl}(\emptyset)$

$$u_0 \in \operatorname{acl}(s_1, \ldots, s_r) \cap \operatorname{acl}(t_1, \ldots, t_r),$$

i.e. $u_0 \in \operatorname{acl}(p_0) \cap \operatorname{acl}(l_0)$.

It is easy to get, using the last fact, a formula $\psi(x, y)$ without parameters such that

$$\psi(u_0, l_0) \quad \text{and} \quad \text{rank}(\psi(u_0, M)) < r.$$

Such a formula can be easily touched up so that (v) and (vi) be satisfied. Therefore $p_0 I \cap \psi(u_0, M)$ is finite. This set contains l_0, hence

$$l_0 \in \text{acl}(p_0, u_0) = \text{acl}(p_0).$$

This contradicts condition (ii). Thus the geometry associated with S over C is neither projective nor degenerate. Since $A \subseteq C$, the proposition is proved.

Proof of the Trichotomy Theorem. By Propositions 5 and 6 the non-definability of pseudoplanes in M is equivalent to the fact that all strongly minimal structures in M have locally projective or degenerate geometries. Adding a new constant for any locally projective strongly minimal structure we can assume that the locally projective structure is projective. Since every two strongly minimal sets in an uncountably categorical structure are nonorthogonal, their geometries are isomorphic provided they are projective or degenerate, as is proved in CHERLIN et al. [1981, 2.8].

Now we begin with the proof of Theorem 3. M is the structure which does not satisfy (i) of the Trichotomy Theorem, G is the group definable in M.

Proof of Theorem 3(ii). We will prove that if G has no proper infinite definable subgroups and Q is its strongly minimal subset, then $G - Q$ is finite. Let

$$H = \{h \in G\colon Qh \dot{-} Q \text{ is finite}\},$$

$$H' = \{h \in G\colon hQ \dot{-} Q \text{ is finite}\}.$$

It is known from ZIL'BER [1977, Lemma 10] that H and H' are definable subgroups of G and for some $g, g' \in Q$, $\text{rank}(gH - Q) < \text{rank}(H)$, $\text{rank}(H'g' - Q) < \text{rank}(H)$. By our assumptions H and H' are finite or equal to G. If $H = G$ or $H' = G$, then $G - Q$ is finite. Thus we may assume that H and H' are finite.

Now by the definition $Q \dot{-} H'QH$ is finite, we assume $Q = H'QH$. Put

$$P = \{gH'\colon g \in G\}, \qquad L = \{Hg\colon g \in G\},$$

$$I = \{\langle g'H', Hg\rangle\colon g' \in Qg\}.$$

If H and H' are finite, then all the axioms of a pseudoplane are satisfied by $\langle P, L, I\rangle$, which contradicts the assumptions of the theorem. Theorem 3(ii) is proved.

LEMMA 7. *Let $J \subseteq U \times G$ be a binary definable relation such that for every $u \in U$ the set uJ is a strongly minimal subgroup of G and for any distinct $u_1, u_2 \in U$, $u_1 J \neq u_2 J$. Then U is finite.*

PROOF. Suppose not. Then we may assume U is strongly minimal. Put

$$H = \{h \in G\colon Jh \text{ is infinite}\},$$

$$P = \{Hg\colon g \in G\}, \qquad L = \{g \cdot uJ\colon g \in G, u \in U\},$$

$$I = \{\langle gH, g \cdot uJ\rangle\colon g \in G, u \in U\}.$$

H is finite for otherwise, since for any $h_1, \ldots, h_k \in H$ $Jh_1 \cap \cdots \cap Jh_k$ is infinite we can find distinct $u_1, u_2 \in U$ such that $u_1 J \cap u_2 J$ contains at least k elements $h_1, \ldots, h_k$, therefore distinct u_1, u_2 can be found with $u_1 J \cap u_2 J$ infinite, which contradicts assumptions of the lemma.

H is a subgroup of G, since for $h_1, h_2 \in H$, $Jh_1 \cdot h_2^{-1} \supseteq Jh_1 \cap Jh_2$ is infinite. Now it can be directly verified that $\langle P, L, I\rangle$ is a pseudoplane, which is a contradiction.

LEMMA 8. *G possesses a definable normal nilpotent subgroup of finite index.*

PROOF. We may assume that G is connected (i.e. has no proper definable subgroup of finite index, see CHERLIN [1979] or ZIL'BER [1977]). Then $G \times G$ is also connected.

Let H be a strongly minimal subgroup of G, which exist by Theorem 3(ii), if $G \neq 1$. Let $gI\langle h, h'\rangle$ mean $h \in H$ & $h' = g^{-1}hg$.

Clearly for every $g \in G$ the set gI is a subgroup of $G \times G$ isomorphic to H, i.e. gI is strongly minimal.

Let $g_1 E g_2$ denote $g_1 I = g_2 I$. Then, by Lemma 7, G/E is finite. This means that the centralizer $C(H)$ of H in G has a finite index. Since G is connected, $C(H) = G$ and thus H lies in the center of G. It follows by induction on rank(G) that G is nilpotent.

LEMMA 9. *G possesses a definable normal abelian subgroup of finite index.*

PROOF. Now we may assume G is connected and nilpotent of class 2 (i.e. $G/C(G)$ is abelian). It is sufficient to prove that $G = C(G)$.

Denote $\bar{G}=G/C(G)$ and supposing $\bar{G}\neq 1$ we get by the connectedness of G that $\bar{G}$ is infinite and by Theorem 3(ii) $\bar{G}$ has a strongly minimal subgroup H. Put

$$gI\langle h,h'\rangle \quad \text{iff} \quad h,h'\in H \ \&\ g\in G \ \&\ h'=hgh^{-1}g^{-1},$$

$$g_1 E g_2 \quad \text{iff} \quad g_1 I = g_2 I.$$

It is evident that gI is a strongly minimal subgroup of $\bar{G}\times C(G)$. By Lemma 7, G/E is finite, in other words the subgroup

$$\{g\in G: \forall h\in H,\ hgh^{-1}=g\}$$

has a finite index in G and thus coincides with G. This means $H=1$ in $\bar{G}$, contradiction.

This proves the lemma and concludes the proof of Theorem 3.

Proof of Theorem 4. First we suppose $G\subseteq \operatorname{acl}(S)$ for some strongly minimal set S. We shall prove that G is finite.

Let $\operatorname{rank}(G)=k$. It is easy to prove for any set $G\subseteq \operatorname{acl}(S)$ by induction on k that there is a finite $A\subseteq M$ such that for every $g\in G$ there are $s_1,\dots,s_k\in S$ with

$$\operatorname{acl}(g,A)=\operatorname{acl}(s_1,\dots,s_k,A).$$

Assume for simplicity $A=\emptyset$ and choose $g,h\in G$ independent over $\emptyset$ with $\operatorname{rank}(g,\emptyset)=\operatorname{rank}(h,\emptyset)=k$. We have

$$\operatorname{acl}(g)=\operatorname{acl}(s_1,\dots,s_k), \qquad \operatorname{acl}(h)=\operatorname{acl}(t_1,\dots,t_k)$$

for some $s_1,\dots,s_k,t_1,\dots,t_k\in S$. It follows from the independence of g and h that

$$\operatorname{acl}(s_1,\dots,s_k)\cap \operatorname{acl}(t_1,\dots,t_k)=\operatorname{acl}(\emptyset).$$

Let $g\cdot h=f$, $\operatorname{acl}(f)=\operatorname{acl}(u_1,\dots,u_k)$, $u_1,\dots,u_k\in S$. Since $f\in \operatorname{acl}(g,h)$, $h\in\operatorname{acl}(g,f)$, $g\in \operatorname{acl}(h,f)$ we have, granting S is disintegrated,

$$\operatorname{acl}(s_1,\dots,s_k)\cup \operatorname{acl}(t_1,\dots,t_k)\supseteq \operatorname{acl}(u_1,\dots,u_k),$$

$$\operatorname{acl}(s_1,\dots,s_k)\cup \operatorname{acl}(u_1,\dots,u_k)\supseteq \operatorname{acl}(t_1,\dots,t_k),$$

$$\operatorname{acl}(t_1,\dots,t_k)\cup \operatorname{acl}(u_1,\dots,u_k)\supseteq \operatorname{acl}(s_1,\dots,s_k).$$

This is possible only if all the sets lie in $\operatorname{acl}(\emptyset)$. Thus $k=0$ and G is finite.

Now if $M\not\subseteq \operatorname{acl}(S)$, then an infinite group G is definable in M with $G\subseteq \operatorname{acl}(S)$, as is shown in ZIL'BER [1980a, Proposition 4.3]. This is impossible as was shown above, and it follows that $G\subseteq \operatorname{acl}(S)$ for every group G definable in M and G is finite.

References

BALDWIN, J.T., 1972, *Almost strongly minimal theories*, J. Symbolic Logic 37, pp. 481–493.

BALDWIN, J.T. and LACHLAN, A.H., 1971, *On strongly minimal sets*, J. Symbolic Logic 36, pp. 79–96.

BAUR, W., CHERLIN, G. and MACINTYRE, A., 1979, *Totally categorical groups and rings*, J. Algebra 57, pp. 407–440.

CHERLIN, G., 1979, *Groups of small Morley rank*, Ann. Math. Logic 17, pp. 1–28.

CHERLIN, G., HARRINGTON, L. and LACHLAN, A.H., 1981, *$\aleph_0$-categorical $\aleph_0$-stable structures*, Preprint, to appear in Ann. Pure Appl. Logic.

DOYEN, J. and HUBAUT, X., 1971, *Finite regular locally projective spaces*, Math. Z. 119, pp. 83–88.

LACHLAN, A.H., 1973/74, *Two conjectures regarding the stability of ω-categorical theories*, Fund. Math. 81, pp. 133–145.

SHELAH, S., 1978, *Classification Theory and the Number of Nonisomorphic Models* (North-Holland, Amsterdam).

ZIL'BER, B.I., 1980, *Totally categorical theories: structural properties and non-finite axiomatizability*, in: Model Theory of Algebra and Arithmetic, PACHOLSKI et al. eds., Lecture Notes in Math. 834 (Springer, Berlin), pp. 381–410.

ZIL'BER, B.I., 1977, *Gruppy i kol'ca, teorii kotorych kategoričny*, Fund. Math. 95, pp. 173–188.

ZIL'BER, B.I., 1980a, *Sil'no minimal'nye sčetno kategoričnye teorii*, Sib. Mat. Žurn. 21, pp. 98–112.

ZIL'BER, B.I., 1980b, *O range transcendentnosti formul $\aleph_1$-kategoričnych toriĭ*, Mat. Zam. 15, pp. 321–329.

ZIL'BER, B.I., 1981, *Total'no kategoričnye struktury i kombinatornye geometrii*, DAN SSSR 259, pp. 1039–1041.

3. Recursion Theory and Theory of Computation

Barcan Marcus et al., eds., Logic, Methodology and Philosophy of Science VII

AN INTRODUCTION TO THE ADMISSIBILITY SPECTRUM

SY D. FRIEDMAN*

M.I.T., Cambridge, MA 02139, U.S.A.

The admissibility spectrum provides a useful invariant for studying definability properties of reals. An ordinal α is *R-admissible* if $L_\alpha(R)$ obeys Σ_1 replacement. If R is a subset of ω, let $\Lambda(R)$ denote the class of all R-admissible ordinals greater than ω. Then $\Lambda(R)$ is a proper class containing all $L(R)$-cardinals. The least element of $\Lambda(R)$ is precisely ω_1^R, the least non-R-recursive ordinal.

The ordinal ω_1^R has received a great deal of attention in the literature. It can be characterized in many equivalent ways: the least R-admissible greater than ω, the least non-R-recursive ordinal, the closure ordinal for R-arithmetical positive inductive definitions, the least α such that the logic $\mathscr{L}_A$, $A = L_\alpha(R)$, is Σ_1 compact. A beautiful relationship between ω_1^R and the hyperdegree of R was discovered by Spector.

SPECTOR CRITERION. $\omega_1^R > \omega_1^{ck}$ iff $\mathcal{O} \leq_h R$ (where $\mathcal{O}$ is Kleene's complete Π_1^1 set of integers and $\leq_h$ is hyperarithmetic reducibility).

It is reasonable to expect that other elements of the admissibility spectrum $\Lambda(R)$ would provide further information concerning definability properties of R. This is illustrated below; in particular there is a natural generalization of Spector's Criterion which relates $\Lambda(R)$ to the L-degree of R.

1. Early results

Work of SACKS [1976] and JENSEN [1972] characterizes the countable sets which can occur as an initial segment of $\Lambda(R)$ for some real R. We present

* This research was supported by NSF Contract #MCS 7906084.

proofs of these results in this section which are somewhat simpler than the original ones. (SACKS [1976] actually proves a result stronger than what we consider here. See the discussion at the end of this section.)

THEOREM 1 (Sacks). *If $\alpha > \omega$ is admissible and countable then there is a real R such that $\omega_1^R = \alpha$.*

PROOF (Almost Disjoint Forcing). We can assume that α is a limit of admissibles as otherwise if $\beta = \sup(\alpha \cap \mathrm{Adm})$, we can force over L_α with finite conditions from ω into β; this produces a generic real R so that α is the least R-admissible greater than ω. (Admissibility is always preserved when forcing with a set of conditions which is an element of the ground model.)

Now the desired real R is obtained in two steps.

Step 1. Find $A \subseteq \alpha$ so that $\beta \in \mathrm{Adm} \cap \alpha \to L_\beta[A]$ is inadmissible.

Step 2. "Code" A by a real R so that $\beta \in \mathrm{Adm} \cap \alpha \to A \cap \beta$ is $\Delta_1(L_\beta(R))$.

In both steps we of course want to preserve the admissibility of α.

To accomplish Step 1 first force $A_0 \subseteq \alpha$ so that $L_\alpha[A_0]$ is *locally countable*; i.e., $L_\alpha[A_0] \models$ "Every set is countable". This can be done by forcing with finite conditions p from $\alpha \times \omega$ into α with the property that $p(\beta, n) < \beta$. Note that if $\mathcal{P}_0$ denotes this forcing and $\beta \in \mathrm{Adm} \cap \alpha$ then any maximal antichain M for $\mathcal{P}_0^\beta = \mathcal{P}_0 \cap L_\beta$ is also a maximal antichain for $\mathcal{P}_0$. It follows that the $\mathcal{P}_0$-forcing relation is Σ_1 when restricted to ranked sentences and that given p such that $p \Vdash \exists\beta\phi$, $\phi\Delta_0$, one can effectively produce a maximal antichain M below p so that $M \in L_\alpha$ and $q \in M \to q \Vdash \phi(\beta_q)$ for some β_q. These facts imply that if A_0 is $\mathcal{P}_0$-generic over L_α then $L_\alpha[A_0]$ is admissible.

Second, we add $A_1 \subseteq \alpha$ so that $L_\alpha[A_0, A_1]$ is admissible but $\beta \in \mathrm{Adm} \cap \alpha \to L_\beta[A_0, A_1]$ is inadmissible. This is done by forcing with $\mathcal{P}_1$ consisting of all conditions $p: \beta_p \to 2$ in $L_\alpha[A_0]$ so that $\beta \in \mathrm{Adm} \cap (\beta_p + 1) \to L_\beta[A_0, p]$ is inadmissible. Using the fact that $L_\alpha[A_0]$ is locally countable it is easy to see that $p \in \mathcal{P}_1$, $\beta < \alpha \to \exists q \leq p$, $\beta_q \geq \beta$. It is easy to see that the forcing relation is Σ_1 when restricted to pairs (p, ϕ), ϕ a ranked sentence of rank $< \beta_p$, as in this case $p \Vdash \phi$ iff $L_{\beta_p}[p] \models \phi$. Lastly if $\langle D_i \mid i < \omega\rangle$ is a uniformly $\Sigma_1(L_\alpha[A_0])$ sequence of dense open sets, $p \in \mathcal{P}_1$ then we can effectively define $p = p_0 \geq p_1 \geq \cdots$ so that $p_{i+1} \in D_i$ and $\langle p_i \mid i < \omega\rangle$ is $\Sigma_1(L_\beta[A_0])$, $\beta = \bigcup\{\beta_{p_i} \mid i < \omega\}$. Thus $L_\beta[A_0]$ is inadmissible and $p = \bigcup\{p_i \mid i < \omega\}$ is a condition. This form of distributivity suffices to show that if A_1 is $\mathcal{P}_1$-generic over $L_\alpha[A_0]$ then $L_\alpha[A_0, A_1]$ is admissible. To complete Step 1 define $A = A_0 \vee A_1$.

Step 2 is accomplished using almost disjoint forcing. We assign a real R_β to each $\beta < \alpha$ so that R_β is definable over $L_\beta[A]$ uniformly in β. Note that for any $\beta < \alpha$ there must be an $L_\beta[A]$-definable bijection of ω and $L_\beta[A]$ as the least counterexample $L_\beta[A]$ to this assertion would have to be admissible, contrary to hypothesis. Thus we can in fact choose R_β to be Cohen generic over $L_\beta[A]$ as well, say for all $\Sigma_2(L_\beta[A])$ dense sets.

A condition in the forcing $\mathscr{P}$ for coding A is a pair $(r, \bar{r})$ where r is a finite subset of ω and $\bar{r}$ is a finite subset of $\{R^*_\beta \mid \beta \in A\} \cup \{r^* \mid r$ a finite subset of $\omega\}$. Here we make use of the canonical operation $R \mapsto R^* = \{\text{Code}(R \restriction n) \mid n < \omega\} \subseteq \omega$ for converting distinct subsets of ω into almost disjoint ones. Write $(r', \bar{r}') \leq (r, \bar{r})$ if $r \subseteq r'$, $\bar{r} \subseteq \bar{r}'$ and $b \in \bar{r} \to b \cap r' \subseteq b \cap r$. Thus generically we produce a real R so that $\beta \in A$ iff R, R^*_β are almost disjoint. Also note that as each R_β is uniformly definable over $L_\beta[A]$ we obtain that $A \cap \beta$ is uniformly $\Delta_1(L_\beta(R))$, by induction on β. (To define $A \cap (\beta + 1)$ we need to know $A \cap \beta$ and R_β; but the latter is definable over $L_\beta[A \cap \beta] = L_\beta[A]$.)

We need only show that $\mathscr{P}$ preserves the admissibility of $L_\alpha[A]$. As in the first part of Step 1 it suffices to argue that if $M \subseteq \mathscr{P}^\beta = \mathscr{P} \cap L_\beta[A]$ is a maximal $\mathscr{P}^\beta$-antichain and Σ_1-definable over $L_\beta[A]$ then M is a maximal antichain in $\mathscr{P}$. It is for the proof of this assertion that we chose R_β to be Cohen generic over $L_\beta[A]$. Indeed suppose $(r, \bar{r}_0 \cup \bar{r}_1)$ were incompatible with each element of M, where $\bar{r}_0 \subseteq L_\beta[A]$, $\bar{r}_1 \cap L_\beta[A] = \emptyset$. Note that the reals $\bar{r}_1 \subseteq \{R_{\beta'} \mid R^*_{\beta'} \in \bar{r}_1\}$ are *mutually* Cohen generic over $L_\beta[A]$ as if $\beta_1 < \beta_2 < \cdots < \beta_k$ then R_{β_1} is Cohen generic over $L_{\beta_1}[A]$, R_{β_2} is Cohen generic over $L_{\beta_2}[A] \supseteq L_{\beta_1}[A][R_{\beta_1}], \ldots$ and we use the product lemma. So in fact the preceding assertion about $(r, \bar{r}_0 \cup \bar{r}_1)$ is forced by a Cohen condition c on $\bar{r}_1$. But then $(r, \bar{r}_0 \cup \{s^* \mid s \in c\}) \in L_\beta[A]$ would be incompatible with each element of M, contradicting the maximality of M. This completes the proof of Theorem 1. $\square$

To be sure, there are many published proofs of the preceding result. We have included the above proof here, however, to serve as a model for the following proof of Jensen's result, as yet unpublished. To save notation we introduce:

CONVENTION. When writing $L_\alpha[X_1, \ldots, X_n]$ we refer to the structure $\langle L_\alpha[X_1, \ldots, X_n], X_1, \ldots, X_n \rangle$.

THEOREM 2 (Jensen). *Suppose X is a countable set of countable admissibles greater than ω and $\alpha \in X \to L_\alpha[X]$ is admissible. Then for some real R, X is an initial segment of $\Lambda(R)$.*

PROOF. We can assume that X has a greatest element α. As in the proof of Theorem 1 we proceed in two steps.

Step 1. Find $A \subseteq \alpha$ so that $\beta < \alpha \to L_\beta[A]$ is admissible iff $\beta \in X$, $L_\beta[A]$ is not recursively Mahlo.

Step 2. Code A by a real R so that $\beta < \alpha \to A \cap \beta$ is $\Delta_1(L_\beta(R))$. In both steps we want to preserve the admissibility of the elements of X.

To accomplish Step 1 first add $A_0 \subseteq \alpha$ so that $L_\alpha[A_0]$ is locally countable, as in the proof of Theorem 1 except over the ground model $L_\alpha[X]$. Then $L_\beta[X]$ admissible $\to L_\beta[X, A_0]$ admissible for all $\beta \leqslant \alpha$. (To see this note that if A_0 is $\mathcal{P}_0$-generic then $A_0 \cap \beta$ is $\mathcal{P}_0^\beta$-generic.) Also if $\hat{\beta}$ = least p.r. closed ordinal greater than β then $\beta < \alpha \to \beta$ is countable in $L_{\hat{\beta}}[X, A_0]$. Second, add $A_1 \subseteq \alpha$ so that $L_\beta[X, A_0, A_1]$ is not recursively Mahlo for all $\beta \leqslant \alpha$. The collection of conditions Q_0 for doing this consists of all $p: \beta_p \to 2$ so that

(i) $\beta \leqslant \beta_p \to p \restriction \beta \in L_{\hat{\beta}}[X, A_0]$,

(ii) $\beta \leqslant \beta_p$, $L_\beta[X, A_0]$ admissible $\to L_\beta[X, A_0, p]$ admissible,

(iii) $\beta \leqslant \beta_p \to L_\beta[X, A_0]$ is not recursively Mahlo.

We must show that $p \in Q_0$, $\alpha > \beta > \beta_p \to$ there is a $q \leqslant p$, $\beta_q \geqslant \beta$. Then the argument of the second part to Step 1 in the proof of Theorem 1 shows that Q_0 is sufficiently distributive so as to preserve the admissibility of $L_\alpha[X, A_0]$.

The extendibility assertion is proved by induction on β. If β is a successor ordinal then the result is clear. If β is a limit ordinal but $L_\beta[X, A_0]$ is inadmissible then the construction of q is easy by induction, using the fact that β is countable in $L_{\hat{\beta}}[X, A_0]$. If $L_\beta[X, A_0]$ is admissible then first we force with $Q_0^\beta = Q_0 \cap L_\beta[X, A_0]$ to obtain q': $\beta \to 2$ so that $q' \in L_{\hat{\beta}}[X, A_0]$ and $q' \supseteq p$. (Note that $p \in Q_0^\beta$.) Then $L_\beta[X, A_0, q']$ is admissible as Q_0^β preserves admissibility just as does Q_0. We must arrange that $L_\beta[X, A_0, q]$ is not recursively Mahlo. This requires one further forcing. Let Q_1^β consist of all closed $p \subseteq \beta$, $|p| = \max(p) \in p$ so that $p \in L_\beta[X, A_0, q']$ and

(i) $\beta' \leqslant |p| \to p \cap \beta' \in L_{\hat{\beta}'}[X, A_0, q']$,

(ii) $\beta' \leqslant |p|$, $L_{\beta'}[X, A_0, q']$ admissible $\to L_{\beta'}[X, A_0, p, q']$ admissible,

(iii) $\beta' \in p \to L_{\beta'}[X, A_0, q']$ inadmissible.

(Note that (ii) is actually redundant due to (i), (iii) and the fact that p is closed.) Now force q'' to be Q_1^β-generic, $q'' \in L_{\hat{\beta}}[X, A_0]$. Then Q_1^β can be shown to preserve admissibility much as could Q_0^β. Clearly $L_\beta[X, A_0, q', q'']$ is not recursively Mahlo as q'' provides a closed unbounded set of $\beta' < \beta$ such that $L_{\beta'}[X, A_0, q']$ is inadmissible. Finally we define $q \leqslant p$ so as to code q', q''. Then $q \in Q_0$, $\beta_q = \beta$.

We now have that $L_\beta[X]$ admissible $\to L_\beta[X, A_0, A_1]$ admissible, $\beta < \alpha \to L_\beta[X, A_0, A_1]$ is not recursively Mahlo. In particular $\beta < \alpha \to$ there is an $L_\beta[X, A_0, A_1]$-definable bijection of ω and $L_\beta[X, A_0, A_1]$. At last we now complete Step 1. We add $A_2 \subseteq \alpha$ so that $\beta < \alpha \to L_\beta[X, A_0, A_1, A_2]$ is admissible iff $\beta \in X$. The collection of conditions $\mathcal{P}_1$ for doing this consists of all $p: \beta_p \to 2$ in $L_\beta[X, A_0, A_1]$ so that

(i) $\beta \leq \beta_p \to L_\beta[X, A_0, A_1, p]$ is admissible iff $\beta \in X$,

(ii) $\beta \leq \beta_p \to p \restriction \beta \in L_{\hat\beta}[X, A_0, A_1]$.

We must show that for all $p \in \mathcal{P}_1$, $\beta < \alpha$ there exists $q \leq p$, $\beta_q \geq \beta$. Once this is accomplished we have completed Step 1 as the argument that $\mathcal{P}_1$ preserves admissibility is much like that for Q_0.

The extendibility assertion is proved by induction on β. As before the nontrivial case is where $\beta \in X$. Then the desired $q \leq p$ is obtained by forcing with $\mathcal{P}_1^\beta = \mathcal{P}_1 \cap L_\beta[X, A_0, A_1]$. Such a q can be found in $L_{\hat\beta}[X, A_0, A_1]$. And, $\mathcal{P}_1^\beta$ preserves admissibility just as did Q_0^β. This completes Step 1: let $A = X \vee A_0 \vee A_1 \vee A_2$ where A_2 is $\mathcal{P}_1$-generic.

Step 2 is precisely as in the proof of Theorem 1. Note that we can choose R_β to be definable over $L_\beta[A]$, as in that proof, since $\beta < \alpha \to$ there is an $L_\beta[A]$-definable bijection of ω and $L_\beta[A]$. Lastly note that $\beta \in X$, R $\mathcal{P}$-generic over $L_\alpha[A] \to R$ $\mathcal{P}^\beta$-generic over $L_\beta[A]$ (for Σ_2 definable dense sets) so it follows that $L_\beta[R]$ is admissible. □

As we mentioned earlier, SACKS [1976] establishes a result somewhat stronger than Theorem 1: If $\alpha > \omega$ is a countable admissible ordinal then $\alpha = \omega_1^R$ for some real R such that $S <_h R \to \omega_1^S < \omega_1^R$, where $\leq_h$ refers to hyperarithmetic reducibility. Sacks uses pointed perfected forcing and in addition, when L_α is not locally countable, perfect trees of Lévy collapsing maps.

Recently, R. LUBARSKY [1984] has established a version of the preceding result in the context of Jensen's theorem. He shows that, assuming X as in Jensen's theorem has a greatest element α and in addition that $X \cap \beta$ is uniformly definable over L_β for $\beta \in X$, that there is a real R so that X is an initial segment of $\Lambda(R)$ and in addition, $S \in L_\alpha(R) \to R \in L_\alpha(S)$ or X is not an initial segment of $\Lambda(S)$. Lubarsky's proof is a significant extension of Sacks'; the key difference is that $\alpha = \omega_1^S \to L_\alpha(S)$ is locally countable, however $\beta \in \Lambda(S) \not\to L_\beta(S)$ is locally countable. Thus when establishing minimality for R, Lubarsky must consider that for $S \in L_\alpha(R)$ one need not have the local countability of $L_\beta(S)$ for $\beta \in X$ (though $L_\beta(R)$ is locally countable for $\beta \in X$). A new argument is required to rule out the possibility that such an S may obey "X is an initial segment of $\Lambda(S)$".

2. The full spectrum-limitations

The theme of this section is that $\Lambda(R)$ is a useful invariant for detecting the set-theoretic complexity of R. Let Λ denote $\Lambda(0)=$ all admissibles greater than ω.

THEOREM 3. *Suppose $R\in L$. Then $\Lambda(R)$ contains $\Lambda-\beta$ for some $\beta<\aleph_1^L$.*

PROOF. Choose β so that $R\in L_\beta$, $\beta<\aleph_1^L$. □

Thus it follows from Theorem 2 that if $V=L$ then the possible admissibility spectra $\Lambda(R)$ can be completely characterized: they are of the form $X\cup(\Lambda-\alpha)$ where X is as in Jensen's theorem, $X\in L_\alpha$.

Note that if R is a Sacks real (R is generic for perfect set forcing over L) then a density argument shows that the conclusion of Theorem 3 fails. However we have the following.

THEOREM 4. *Suppose R is set-generic over L (R belongs to $L(G)$ where G is $\mathcal{P}$-generic over L, $\mathcal{P}\in L$). Then:*

(a) *$\Lambda(R)\supseteq\Lambda-\beta$ for some β.*

(b) *For any $\alpha<\aleph_1$ there exist $\beta,\gamma<\aleph_1$ such that $\Lambda\cap(\beta,\gamma)$ has ordertype $\geqslant\alpha$ and is contained in $\Lambda(R)$.*

PROOF. (a) Choose β so that $\mathcal{P}\in L_\beta$ where $R\in L_\beta(G)$, G is $\mathcal{P}$-generic over L. If $\alpha>\beta$ is admissible then $L_\alpha(G)$ is admissible as forcing with a set of conditions preserves admissibility. Thus $L_\alpha(R)$ is admissible since $R\in L_\beta(G)\subseteq L_\alpha(G)$.

(b) By the result of (a) we know that there exist $\beta,\gamma\in\mathrm{ORD}$ such that $\Lambda\cap(\beta,\gamma)$ has ordertype $\geqslant\alpha$ and is contained in $\Lambda(R)$. But HC $=$ (the hereditarily countable sets) is a Σ_1 elementary substructure of V. So there must exist such β, γ which are countable. □

The preceding result imposes severe restrictions on which admissibility spectra can be obtained via set-forcing over L. It implies that even when restricting to *countable* admissible ordinals, simple spectra such as $\{\alpha_{2i}\mid i\in\mathrm{ORD}\}=$ (even admissibles) cannot be realized by $\Lambda(R)$ for set-generic R (where $\alpha_0<\alpha_1<\cdots$ is the increasing enumeration of Λ).

The next result implies that certain spectra cannot be realized without the use of large cardinals.

THEOREM 5 (Silver). *Suppose $\Lambda(R)-\beta$ is contained in the class of all L-cardinals for some β. Then $0^{\#} \leq_L R$.*

PROOF. Let κ be a singular cardinal greater than β. Then $(\kappa^+)^{L(R)} > (\kappa^+)^L$ since there are R-admissible ordinals between κ and $(\kappa^+)^{L(R)}$. By Jensen's Covering Theorem (see DEVLIN–JENSEN [1974]), $0^{\#} \in L(R)$. □

This result can in fact be strengthened to provide a natural generalization of Spector's Criterion, in the context of L-degrees.

DEFINTION. $X \subseteq \mathrm{ORD}$ is *Σ_1-complete* if whenever $Y \subseteq \mathrm{ORD}$ is $\Sigma_1(L)$, Y is $\Delta_1\langle L[X], X\rangle$.

THEOREM 6. *$\Lambda(R)$ is Σ_1-complete iff $0^{\#} \leq_L R$.*

PROOF. X is Σ_1-complete whenever X is unbounded and $X \subseteq L\text{-Card} = \{\alpha \mid \alpha \text{ is an } L\text{-cardinal}\}$, as if Y is $\Sigma_1(L)$ with defining formula $\phi(y)$ then $y \notin Y$ iff $\exists \alpha \in X\ (L_\alpha \models \sim \phi(y)$ and $y, p \in L_\alpha)$ where p is the parameter in ϕ. (We are using the fact that α an L-cardinal $\rightarrow L_\alpha <_{\Sigma_1} L$; i.e., α is stable.) Thus $\Lambda(R)$ is Σ_1-complete whenever $0^{\#} \leq_L R$ as $\Lambda(0^{\#}) \subseteq L$-Card. Conversely if $\Lambda(R)$ is Σ_1-complete then L-Card is $\Sigma_1(L(R))$ and as in the proof of Theorem 5, $(\kappa^+)^{L(R)} > (\kappa^+)^L$ for sufficiently large singular κ. (We are using the R-stability of $(\kappa^+)^{L(R)}$.) By the Covering Theorem, $0^{\#} \leq_L R$. □

Theorem 6 has the consequence that certain spectra X are ruled out entirely, even though the Jensen criterion ($\alpha \in X \rightarrow \langle L_\alpha[X], X\rangle$ is admissible) is satisfied.

COROLLARY. *There is no real R obeying any of the following:*

(a) *$\Lambda(R) = \Sigma_2$-admissible L-cardinals,*

(b) *$\Lambda(R) = \Sigma_2$-admissible stables,*

(c) *R is generic over L via an amenable class forcing, $\Lambda(R) \subseteq$ stables.*

PROOF. (a), (b) are clear, using Theorem 6. (c) follows from the fact that the condition on R contradicts $0^{\#} \leq_L R$ (see Beller–Jensen–Welch [1982], p. 157). □

We have left open the possibility of solutions to spectrum equations $\Lambda(R) = X$, where X is not Σ_1-complete. We discuss this in the next section.

3. The full spectrum-positive results

The results of Section 2 imply that a real R satisfying $\Lambda(R)=$(even admissibles) cannot be set-generic and cannot construct $0^{\#}$ (i.e., $0^{\#} \leq_L R$). Thus such reals are entirely ruled out by the following conjecture of Solovay.

SOLOVAY'S CONJECTURE. $0^{\#} \leq_L R \rightarrow R$ is set-generic (over L).

Fortunately for our purposes, Solovay's conjecture is false. This was shown by Jensen (see BELLER–JENSEN–WELCH [1982]).

THEOREM 7 (Jensen). *If $A \subseteq$ ORD then there is an $\langle L[A], A\rangle$-definable forcing for extending $L[A]$ to $L(R)$, $R \subseteq \omega$ so that $L(R) \models$ ZFC and A is definable over $L(R)$.*

COROLLARY. *The negation of Solovay's conjecture is consistent.*

PROOF. Choose $A \subseteq$ ORD to be amenable but not L-definable. By Jensen's theorem we can get $R \subseteq \omega$ so that A is definable over $L(R)$. Then R cannot be set-generic over L as otherwise there is a condition $p \in \mathcal{P}$ and a formula ϕ (where $R \in L(G)$, G $\mathcal{P}$-generic over L) such that for unboundedly many $\alpha \in$ ORD, $p \Vdash \overline{A \cap \alpha}$ is an initial segment of $\{\beta \mid \phi(\beta)\}$. Thus $\beta \in A$ iff $\exists x \in L$ $(p \Vdash \beta \in x$, x an initial segment of $\{\beta \mid \phi(\beta)\})$. $\square$

As it turns out the technique used to prove Theorem 7, Jensen's coding method, suffices to get the first example of a nontrivial spectrum.

THEOREM 8 (David, Friedman). *There is an L-definable forcing for producing a real R so that $L(R) \models$ ZFC and $\Lambda(R) \subseteq$ (even admissibles).*

IDEA OF PROOF. The desired forcing is made up of certain "building blocks", which are not difficult to describe. Jensen coding is used to put these building blocks together.

We wish to arrange that α R-admissible $\rightarrow \alpha$ is an even admissible. Suppose that we have $D \subseteq \aleph_1$ so that: $L_\alpha[D]$ admissible $\rightarrow \alpha$ is even. Then we could hope to choose R so as to code D and satisfy the desired property.

The problem is that if we code D by R in the usual way (with almost disjoint forcing) we only obtain the following: For all α, $D \cap (\aleph_1)^{L_\alpha}$ is $\Delta_1(L_\alpha(R))$. The reason is that to decode D from R we need to know the almost disjoint coding reals R_β and it is only for $\beta < (\aleph_1)^{L_\alpha}$ that we have $R_\beta \in L_\alpha$. Thus the recovery of D from R is not "fast enough". On the other hand we are in great shape if D has the following stronger properties:

$$L_\alpha(D \cap \xi) \text{ admissible}, L_\alpha(D \cap \xi) \models \xi = \aleph_1 \rightarrow \alpha \text{ is even.} \tag{*}$$

$$L_\alpha[D] \text{ admissible and locally countable} \rightarrow \alpha \text{ is even.} \tag{**}$$

For then we need only recover $D \cap (\aleph_1)^{L_\alpha}$ inside $L_\alpha(R)$ to guarantee that α is even (or inadmissible relative to R), a recovery that can be made.

The question is how to obtain $D \subseteq \aleph_1$ obeying (*), (**). The natural thing to do is force with conditions d which are initial segments of $\aleph_1$ obeying (*), (**) for $\xi \leqslant \sup(d)$. We now come to the heart of the argument, which is contained in the following two observations:

(1) Extendibility for this forcing is trivial because given d and $\xi > \sup(d)$ we are free to extend d to length ξ by *killing all admissibles* between $\sup(d)$ and ξ. It is crucial for this argument that we are only concerned with killing admissibility, not in preserving it.

(2) Distributivity for this forcing is easily established assuming the following (!): There exists $D' \subseteq \aleph_2$ such that:

$$L_\alpha(D' \cap \xi) \text{ admissible}, L_\alpha(D' \cap \xi) \models \xi = \aleph_2 \rightarrow \alpha \text{ is even} \tag{*'}$$

$$L_\alpha[D] \text{ admissible}, L_\alpha[D] = \forall x\, (\text{card}(x) \leqslant \aleph_1) \rightarrow \alpha \text{ is even.} \tag{**'}$$

Thus we are faced with the original problem, but one cardinal higher!

Proof by induction does not look promising. However note that we need not already "have" all of D' before we can "start building" D; thus the idea of the proof (as in all Jensen coding constructions) is to build $R, D, D', D'', \ldots$ simultaneously and check distributivity for any final segment of the forcing. □

A proof of the preceding result will appear in DAVID [1984]. In that paper the above ideas are combined with some ideas from "strong coding" (mentioned below) to improve the conclusion of Theorem 8 to: $\Lambda(R) \subseteq \{\alpha \mid L \models \phi(\alpha)\}$, where ϕ is Σ_1 and $L \models \phi(\kappa)$ for all cardinals κ.

The next step in the study of admissibility spectra is to introduce the requirement of admissibility preservation into the above. Thus for example we wish to obtain solutions to the equation $\Lambda(R) = $ (even admissibles). This requires the method of strong coding.

THEOREM 9. *There is a $\Delta_1(L)$-definable forcing $\mathscr{P}$ for producing a real R so that $L(R) \models$ ZFC and $\Lambda(R) =$ (even admissibles).*

IDEA OF PROOF. We approach the problem as in Theorem 8. Of course now the extendibility property is much more difficult (distributivity is the same). Indeed the desired extension of d to d' of length $\geqslant \xi$ must be made generically, so as to preserve even admissibles. Thus we see that our conditions must be constructed out of generic sets for "local" versions of the very same forcing. Thus in fact we construct a strong coding $\mathscr{P}^\beta \subseteq L_\beta$ at each admissible β and then inductively build $\mathscr{P}^\beta$ out of generic sets for various $\mathscr{P}^{\beta'}$, $\beta' < \beta$.

The main difficulty is in showing that the desired generic sets actually exist; note that we want a $\mathscr{P}^\beta$-generic over L_β where β may indeed be uncountable. The proof of generic existence is by a simultaneous induction with the proofs of extendibility, distributivity and requires use of the critical projecta of FRIEDMAN [1982]. (These projecta are closely related to Jensen's notion of dependency in the theory of higher-gap morasses.)

The other difficulty in the extendibility argument is the conflict between the genericity requirement and the need to "avoid" the almost disjoint codes R^*_β: Recall that in almost disjoint forcing, $(r', \bar{r}') \leqslant (r, \bar{r})$ iff $r' \supseteq r$, $\bar{r}' \supseteq \bar{r}$ and $b \in \bar{r} \rightarrow b \cap r' \subseteq r$. This last requirement causes difficulty with the need for making r' generic. Solving this requires the construction of special "supergeneric" codes R_β. These codes will not be Cohen generic but instead generic for a suitable forcing, defined inductively. □

4. Recent work

A complete characterization of those $A \subseteq$ ORD which can be realized as admissibility spectra $\Lambda(R)$ is not known. However some hints as to the nature of such a characterization are hinted at by the following examples.

(a) Suppose $A = L$-Card, the class of L-cardinals. Then A cannot be of the form $\Lambda(R)$ as A fails to satisfy: $\alpha \in A \rightarrow L_\alpha[A]$ is admissible.

(b) Suppose $A =$ (all α such that $L_\alpha \models$ Power set). Then A cannot be of the form $\Lambda(R)$ as then the $L(R)$-cardinal successor to $\aleph_\omega$ would be greater than the L-cardinal successor to $\aleph_\omega$, hence $0^\# \in L(R)$; but then $\Lambda(R) - \beta \subseteq L$-Card for some β.

(c) Suppose $A = \{\alpha \mid \alpha$ a successor admissible, L-Card(α) a successor L-Cardinal$\} \cup \{\alpha \mid \alpha$ recursively inaccessible, L-Card(α) a limit L-cardinal$\}$. Then A cannot be of the form $\Lambda(R)$ else L-Card is $\Delta_1(L(R))$ and thus $0^\# \in L(R)$; this is a contradiction as in (b).

(d) Suppose A = nonprojectibles = $\{\alpha \mid \Sigma_1 \text{ projectum}(\alpha) = \alpha\}$. Then A cannot be of the form $\Lambda(R)$ for then the least R-admissible α greater than $\aleph_1$ would have cofinality ω, but this is false since $\text{cof}(\alpha) = \text{cof}(\Sigma_1 \text{ projectum } \alpha \text{ relative to } R) = \aleph_1$.

Also note the following: If $A = \Lambda(R)$ then A is $\Delta_1(L(R))$ and hence A "collapses to itself" when transitively collapsing Σ_1^R Skolem hulls. More precisely, for any $x \in \mathcal{P}_{\omega_1}(\text{ORD}) = \{x \subseteq \text{ORD} \mid x \text{ is countable}\}$ let π_x be the unique order-preserving function from x onto ordertype(x). Then in $L(R)$, $A^* = \{x \in \mathcal{P}_{\omega_1}(\text{ORD}) \mid \pi_x[A] \text{ is an initial segment of } A\}$ contains a closed unbounded class (namely $\{x \in \mathcal{P}_{\omega_1}(\text{ORD}) \mid x <_{\Sigma_1} L(R)\}$). Thus $\langle L[A], A\rangle \models A^*$ is stationary in $\mathcal{P}_{\omega_1}(\text{ORD})$, assuming $(\aleph_1)^{L[A]} = (\aleph_1)^{L(R)}$.

The above considerations lead us to conjecture what the situation is in a very special case of the general problem. Namely suppose α = (least α such that $L_\alpha \models \text{KP}$ and $\aleph_2$ exists). We conjecture the following.

(*) Suppose $A \subseteq \alpha$ is amenable and $\langle L_\alpha, A\rangle$ is admissible. Then there is a real R such that $A = \Lambda(R) \cap \alpha$, $L_\alpha(R) \models \text{KP} + \aleph_2$ exists iff:

(i) $\aleph_1^{L_\alpha}, \aleph_2^{L_\alpha} \in A$,

(ii) $\beta \in A \rightarrow \langle L_\beta[A], A \cap \beta\rangle$ is admissible,

(iii) $\aleph_1^{L_\alpha} < \beta$, β a successor element of $A \rightarrow L_\alpha \models \text{cof}(\beta) = \aleph_1$,

(iv) $\langle L_\alpha, A\rangle \models A^*$ is stationary on $\mathcal{P}_{\omega_1}(\text{ORD})$.

The key step in establishing this conjecture should be to obtain an $(\omega_1, 1)$-morass of A-preserving maps, using property (iv) to show that the natural forcing for doing this is ω-distributive.

References

BELLER–JENSEN–WELCH, 1982, *Coding the Universe*, London Math. Soc. Lecture Notes (Cambridge Univ. Press, Cambridge).

DAVID, 1984, *A functorial Π_2^1-singleton*, Advances in Math., to appear.

DEVLIN–JENSEN, 1974, *Marginalia to a theorem of Silver*, Kiel Logic Conference, Lecture Notes in Math. 499 (Springer, Berlin).

FRIEDMAN, 1982, *Uncountable admissibles I: Forcing*, Trans. AMS.

JENSEN, 1972, *Forcing over admissible sets*, Handwritten Notes.

LUBARSKY, 1984, Thesis, MIT.

SACKS, 1976, *Countable admissible ordinals and hyperdegrees*, Advances in Math.

Barcan Marcus et al., eds., Logic, Methodology and Philosophy of Science VII

ARE RECURSION THEORETIC ARGUMENTS USEFUL IN COMPLEXITY THEORY?

WOLFGANG MAASS*

Dept. of Mathematics and Computer Science Division, Univ. of California, Berkeley, CA 94720, U.S.A.

1. Introduction

Recursion theory is that area of mathematical logic where one studies the *qualitative* aspects of computability. Here one is only interested in the question whether a computation converges at all, i.e. yields a result after finitely many computation steps. In complexity theory, which is part of computer science, one studies in addition *quantitative* aspects of computations. For example one studies for computations on a mathematical computer model the computation time as a function of the size of the input.

Over the last few decades a number of quite powerful techniques have been developed in recursion theory — most of them so-called priority arguments — that finally allowed to solve a number of difficult open recursion theoretic problems (see SOARE [25]). In complexity theory, on the other hand, a variety of concepts and methods have been introduced but many basic and important problems remain open. We analyze and survey in this paper some of our recent research in the light of the question whether arguments from recursion theory are useful in complexity theory. We arrive at the conclusion that recursion theoretic techniques are in fact useful in complexity theory, although in general only in combination with arguments about algorithms for concrete problems or with arguments about concrete computer models.

Many problems in complexity theory deal with the question whether certain mathematical problems can be solved by computations whose computation time is polynomially related to the size of the input. It is

* During the preparation of this paper the author has been supported by the Heisenberg Programm of the Deutsche Forschungsgemeinschaft, Bonn.

Permanent address (after Fall 84): Dept. of Mathematics, Statistics and Computer Science, University of Illinois at Chicago.

tempting to view such quantitative questions as qualitative questions in a new generalized recursion theory where one interprets the basic concept of "finite" as "of polynomial size in the considered parameters" and "recursive function" as "in polynomial time computable function." It is well known that many arguments from recursion theory can be transferred to generalized recursion theories, where the basic notions of "finite" and "recursive function" are substituted by other notions (see e.g. FENSTAD [5]). We look in Section 2 of this paper at a number of open problems about the structure of NP where one can *prove* that even under the assumption $P \neq NP$ recursion theoretic arguments will not suffice. Ironically our proof uses a recursion theoretic argument.

In Sections 3 and 4, on the other hand, we exhibit examples from complexity theory where a strategy that is very reminiscent of a well-known strategy from priority arguments in recursion theory is used in combination with concrete arguments about algorithms (Section 3) resp. computer models (Section 4). In Section 3 we construct polynomial time approximation schemes for some strongly NP-complete problems that arise e.g. in robotics. In Section 4 we survey a proof of optimal lower bounds for two tapes versus one on deterministic and nondeterministic Turing machines. We further get results that show a substantial superiority of nondeterminism over determinism resp. co-nondeterminism over nondeterminism for one-tape Turing machines (which have an additional one-way input tape). We show that both in Section 3 and in Section 4 one can view the proof of the desired result as the construction of a winning strategy for a two-person game. Further the winning strategy that we give employs a tactic that is familiar from modern priority arguments. Our winning strategy consists of a system of different strategies which have the property that the failure of one strategy (which after all tells us a little bit about the opponent) increases the chances of the other strategies to beat the opponent. Such tactic is actually used quite often in complexity theory, although it usually remains hidden in the combinatorics. We believe that it is worthwhile to make this feature more explicit because its full power has not yet been exploited. It is quite plausible that the proofs of many theorems in complexity theory have not yet been found for the same reasons that delayed the solution of several problems in recursion theory. One tends to insist on winning strategies that try to reach their goal too uniformly, i.e. besides the *outcome* of the game they also want to prescribe *how* the game is won (which is unnecessary and often impossible). The previously sketched tactic leaves it open *which* strategy in our system will overcome the opponent. Thus it offers a way to exploit the power of inconstructive mathematics. It further appears that similarly as in recursion theory the description of lower bound proofs as games makes it possible to keep track of increasingly complex situations (with nested strategies, etc.).

There are many interesting interactions between recursion theory and complexity theory that we do not even touch in this paper. We refer to SOARE [24] for a recent survey concerning the qualitative theory of complexity measures (it turns out that in this area one also finds applications of concepts from complexity theory to recursion theory, see also MAASS [17]). Additional results and references can be found in HARTMANIS and HOPCROFT [9] and JOSEPH [16].

The previously indicated possibility to view polynomial time computable functions as the "recursive" functions of a generalized recursion theory is made explicit in forthcoming work by Moschovakis.

We do not assume in this paper any knowledge from complexity theory. In particular we try to define and illustrate all concepts from complexity theory that we use.

2. On the limits of recursion theoretic arguments in complexity theory

We assume that the reader is familiar with the standard definition of a Turing machine (abbreviated: TM). A set of binary strings is in the class P if its characteristic function can be computed by a deterministic TM in time $p(n)$ for some polynomial p (n is the length of the input for the considered computation). The only new feature of a nondeterministic TM N is that its transition function is multiple-valued. Thus for every input w one has instead of one computation a tree of many different computations of TM N on this input. One says that N accepts input w if one of the branches in the tree ends with an accepting final state (assume that all final states of N have been partitioned into accepting and nonaccepting states). N accepts w in time t if there is at least one such branch of length $\leq t$ (or one can demand that every accepting branch has length $\leq t$ — it does not make a difference in the following). Finally one says that a set of binary strings is in the class NP if there is a nondeterministic TM that accepts exactly the strings in this set and further accepts each string of length n in time $p(n)$ for some polynomial p. Notice that for sets in NP there is an asymmetry between being *in* the set and being *out* of the set, similarly as for recursively enumerable sets.

It is tempting to view the classes P and NP as downward projections of the classes of recursive and recursively enumerable sets. Note that one may view the elements of a recursively enumerable set $f[\mathbb{N}]$, where f is some total recursive function, as those elements w that are accepted by a nondeterministic TM that tries in each computation branch a different argument x and halts at the end of the branch in an accepting state iff $f(x) = w$.

Unfortunately so far one cannot answer even the most basic questions about this downward projected recursion theory (e.g. P = NP?). BAKER et al. [2] have shown that the situation is even worse. They consider relativizations P^O and NP^O of P and NP where the attached "oracle" O is some set of binary strings. One can use e.g. oracle-TM's like in recursion theory to define such relativized complexity classes. An oracle-TM may ask its attached oracle O at any time and as often as it likes during the computation whether the string u that it has currently written on its special oracle-tape is in the set O or not. The oracle O gives in one step the correct answer. General experience says that every recursion theoretic argument "relativizes", i.e. remains valid if one attaches the same oracle O everywhere in the argument (for an arbitrarily chosen set O). This relativized argument proves then an accordingly relativized theorem. BAKER et al. [2] show that it is impossible to prove P = NP or P ≠ NP by an argument that relativizes. They do this by constructing via simple diagonalization sets A and B s.t. $\mathrm{P}^A = \mathrm{NP}^A$ and $\mathrm{P}^B \neq \mathrm{NP}^B$.

This result leaves the possibility open that one can get under the assumption P ≠ NP via recursion theoretic arguments a clear picture of the structure of the classes P and NP (following the standard tradition in logic to take as an axiom what one cannot prove). The following result shows that there are also limitations to this program.

THEOREM 2.1 (HOMER and MAASS [14]). *The following statements S are "independent" from the assumption* P ≠ NP *in the sense that there are recursive sets A and B s.t.* $\mathrm{P}^A \neq \mathrm{NP}^A$ *and* S^A *but* $\mathrm{P}^B \neq \mathrm{NP}^B$ *and* $\neg S^B$:

(1) *every infinite set in* NP *has an infinite subset that is in* P,

(2) *there are simple elements in the lattice of* NP-*sets* (*with set theoretic inclusion*),

(3) *there is a set U in* NP *that is universal for* P, *i.e.* $\mathrm{P} = \{\{v \mid \langle v, w\rangle \in U\} \mid w \text{ a binary string}\}$ *for some standard pairing operation* $\langle\cdot,\cdot\rangle$.

To prove Theorem 2.1 one splits for each statement S the desired properties of A resp. B into infinitely many requirements. One constructs A and B in stages s.t. gradually all requirements become satisfied. This construction is somewhat delicate because there arise conflicts between requirements of different types. One possible way to solve such conflicts is to use a finite injury priority construction. Alternatively — since one has in these constructions a recursive a priori bound on the stages where an earlier attempt might be injured — one can eliminate with some additional work all injuries. On the other hand, one encounters usually still delays of the activities for a given requirement and in order to show that each requirement is only finitely long delayed one has to argue like in a finite injury priority argument. In general it may be appropriate to view a delay of a requirement in the restricted world of constructions of recursive sets (instead of recursively enumerable sets) as a form of injury.

Following Theorem 2.1 a large number of similar "independence" results has been found (see references in JOSEPH [16]).

What methods remain that might possibly answer the mentioned questions from complexity theory if recursion theoretic arguments (actually more generally: arguments from mathematical logic) do not suffice? We would like to mention two possible escapes. If one proves (by any argument) that a *concrete* NP-complete problem (say HAMILTONICITY) is not in P then this proof of $P \neq NP$ does not relativize. There is not even a natural definition of HAMILTONICITY^O for an oracle O. Second one might analyze more closely the *concrete* structure of computations on a specific computation model. In general such arguments do not remain valid if one adds an oracle tape to the computation model. Thus in any case it appears to be unavoidable that the recursion theorist gets "his fingers dirty".

3. Approximation algorithms

In this section we apply a strategy that is familiar from recursion theory in order to design approximation algorithms.

The following computational problem arises in the context of motion planning and positioning of robots:

Given: n points in Euclidean space (e.g. spots that have to be welded by a robot) and some type of industrial robot.

Wanted: a minimal number k of positions for the base of the robot s.t. each of the n points can be reached by the arm of the robot from one of these k positions.

We look first at the 1- and 2-dimensional versions of this problem. Assume that all given points lie in a fixed horizontal plane. Assume that from any fixed base position the arm of the robot can reach any point that has a distance between r and $r+w$ from the (vertical axis through the) base of the robot, where r and w depend on the flexibility of the arm of the considered type of robot. Thus we arrive at the mathematical problem of covering n given points in the Euclidean plane by a minimal number of rings with inner radius r and outer radius $r+w$. Unfortunately the following result suggests that no computer is able to solve this problem (for nontrivial sizes of n).

THEOREM 3.1 (FOWLER, PATERSON and TAMIMOTO [6]). *The problem whether n given points in the Euclidean plane can be covered by k rings of inner radius r and outer radius $r+w$ is strongly* NP-*complete* (*even if we fix $r=0$, i.e. consider only discs*).

We would like to explain briefly to those readers that are not familiar with nondetermininistic computations what this means. It is easy to see that the considered problem (which we identify with the *set* $\{\langle\langle x_1, y_1\rangle, \ldots, \langle x_n, y_n\rangle, r, w, k\rangle \mid$ the n points with coordinates $\langle x_i, y_i\rangle$ can be covered by k rings of inner radius r and outer radius $r+w$; all numbers are rational$\}$) lies in the class NP. A nondeterministic Turing machine (see definition in Section 2) just guesses the positions of up to k rings and checks whether all points are covered by these rings. If a tuple $\langle\langle x_1, y_1\rangle, \ldots, \langle x_n, y_n\rangle, r, w, k\rangle$ is in the considered set then along some branch of the computation tree of the nondeterministic computation the TM guesses k ring positions that cover all points and therefore it halts at the end of this branch in an accepting state. Since it takes only polynomially many steps (in the length of the considered tuple) to write down k guessed ring positions and to verify that all n points are covered (assume that one can compute in one step the distance between two points), this accepting branch is of polynomial length in the size of the input. Thus the problem is in NP. To say that the problem is NP-complete means that it is in NP and that every other problem in NP can be reduced to it by a deterministic polynomial time computable function (like in many-one reducibility). This implies that the problem is not in P unless P = NP. *Strong* NP-completeness means that this holds even if we write down the data of the problem in unary code (which makes the size of the input much longer).

Notice that if we cannot compute in polynomial time the minimal number k of rings that are needed for a covering, we certainly cannot compute an optimal covering in polynomial time.

We refer to GAREY and JOHNSON [7] for further information about NP-completeness.

Usually one can escape NP-completeness in geometric location problems by looking only at special cases that are essentially 1-dimensional. In our case one might want to consider given points on a straight line (or on a fixed number of parallel lines). Notice that the intersection of a ring with a straight line is a pair of closed intervals. Unfortunately our problem is quite obnoxious.

THEOREM 3.2 (MAASS [18]). *The problem whether n given points on the line can be covered by k pairs of closed intervals of length w and distance 2r is strongly* NP-*complete.*

Does NP-completeness imply that it is hopeless to attack these problems on a computer? No, because even NP-complete problems may have good approximation algorithms (another possibility would be to look at randomized algorithms, a third possibility would be to show that P = NP). Instead of a minimal number of robot positions an approximation algorithm for the considered problem computes a nearly minimal number of robot positions from which all points can be reached. If for an instance I of our problem OPT(I) is an optimal solution and A(I) is an approximate solution that is produced by approximation algorithm A one calls

$$\frac{\big|\,|\mathrm{A}(I)| - |\mathrm{OPT}(I)|\,\big|}{|\mathrm{OPT}(I)|}$$

the error of A on instance I ($|\mathrm{OPT}(I)|$, $|\mathrm{A}(I)|$ are the numbers of robot positions that are used in the respective solutions). One calls A a *polynomial time approximation scheme* for some combinatorial optimization problem Π if the scheme A provides for every given $\varepsilon > 0$ a polynomial time approximation algorithm A_ε that has error $\leqslant \varepsilon$ for all instances I of Π.

Not all NP-complete problems have good approximation algorithms. In particular polynomial time approximation schemes for strongly NP-complete problems are very rare (see [7]).

We sketch in the rest of this section the construction of polynomial time approximation schemes for the considered strongly NP-complete problems. We will also point out how one can view these constructions as the construction of winning strategies in certain 2-person games. Our winning strategy employs a system of complementary strategies with the properties that we described in Section 1.

THEOREM 3.3 (HOCHBAUM and MAASS [13]). *For every finite dimension d the problem of computing for n given points in d-dimensional Euclidean space positions for a minimal number of d-dimensional balls with radius w that cover all n points has a polynomial time approximation scheme* (*this problem is strongly* NP-*complete for* $d \geq 2$).

PROOF. It is sufficient to illustrate the idea for $d = 2$. For a given $\varepsilon > 0$ we describe a polynomial time approximation algorithm A_ε. Fix a natural number l s.t. $(1 + 1/l)^2 \leq 1 + \varepsilon$. Cut the 2-dimensional Euclidean plane into vertical strips of width $l \cdot 2w$. We use the divide-and-conquer method and solve the covering problem separately in each strip. We then take the union of all discs that we use for the coverings in the strips and get a covering of all n given points. The problem with this approach is that it may cause an error $\gg \varepsilon$. This occurs in particular if most of the given points happen to lie close to a cut line.

To view algorithm design as a 2-person game one imagines that player I ("we") wants to produce an algorithm with the desired properties and player II ("opponent") wants to construct an instance of the considered problem where player I's algorithm fails. In the preceding situation player II can win by placing most of the n given points in suitable positions close (i.e. in distance $\leq 2w$) to a cut line of player I's algorithm. Player I is now more clever and uses instead of one strategy S for cutting the plane into strips of width $l \cdot 2w$, l different strategies $S_1, \ldots, S_l$ where $S_1 \equiv S$ and S_{i+1} results from S_i by shifting all cut lines of strategy S_i over a distance $2w$ to the right. The rationale behind this approach is that if player II decides to place e.g. most of the n points close to the cut lines of strategy S_1, he must place accordingly fewer points close to the cut lines of the other strategies S_i. This implies that one of the strategies S_i causes a relatively small error. More precisely each disc of a fixed optimal global solution is cut by a cut line of at most one strategy S_i. Thus for some S_i the cut lines of S_i cut at most $1/l$ of these discs. Further the number of additional discs (compared with the fixed optimal solution) that the divide-and-conquer strategy S_i uses can be bounded by the number of discs in the fixed optimal solution that are cut by a cut line of this strategy S_i. Therefore some strategy S_i causes an error $\leq 1/l$.

So far we have assumed that each strategy S_i computes an optimal covering in each of its strips of width $l \cdot 2w$. Since we do not know how to do this in polynomial time, we use again for each strip an approximation algorithm. We cut now the considered strip by horizontal lines in distance $l \cdot 2w$ and apply again the divide-and-conquer method. In each resulting

$l \cdot 2w \times l \cdot 2w$ square we can afford to compute an optimal covering by exhaustive search because this takes only polynomially in m many steps (where m is the number of points in that square). We use here the fact that at most $2l^2$ discs are needed for an optimal covering of such square. Further every disc in an optimal covering that covers more than one of the given points has w.l.o.g. at least two of the given points on its periphery and these two points determine its position up to two possibilities.

Of course we may again produce an error $\geqslant \varepsilon$ by this divide-and-conquer method for the considered strip. Therefore we try l different substrategies $T_1, \ldots, T_l$ for cutting this strip into $l \cdot 2w \times l \cdot 2w$ squares, where T_{i+1} results from T_i by moving all cut lines of T_i upward over distance $2w$. Some T_i is guaranteed to cause in this strip an error $\leqslant 1/l$ (same argument as before).

Altogether approximation algorithm A_ε proceeds as follows. It tries successively l strategies $S_1, \ldots, S_l$ for cutting the plane into vertical strips of width $l \cdot 2w$. Separately for each strip that arises in some S_i it tries successively l substrategies $T_1, \ldots, T_l$ for cutting it into $l \cdot 2w \times l \cdot 2w$ squares. For each resulting square it computes an optimal covering by exhaustive search. It returns with the resulting covering from that substrategy T_j which uses the fewest discs. Finally A_ε outputs the covering of the n given points which arises from that strategy S_i which uses the fewest discs.

By the preceding A_ε uses at most $(1 + 1/l)^2 \cdot |\mathrm{OPT}(I)|$ discs. It is easy to verify that the running time of A_ε is polynomial in n and w.

Compared with a "supermind" which *knows* immediately the best cutting strategy the previous algorithm A_ε has to try various guesses at the opponent's strategy. A_ε has to pay for this lack of knowledge with a time penalty: a factor of l^2 in the time bound for A_ε. These delays correspond to the injuries in a finite injury priority construction.

Concerning the problem from Theorem 3.1 one gets in the same way a polynomial time approximation scheme for each fixed bound on the "nonconvexity" measure r/w of the covering rings. For the 1-dimensional problem there is a more subtle approach that allows to eliminate the term r/w from the exponent of the time bounds. This yields the following result.

THEOREM 3.4 (HOCHBAUM and MAASS [12]). *There is a polynomial time approximation scheme for the strongly* NP-*complete problem of Theorem* 3.2.

One can improve the time bounds of the previous approximation algorithms considerably by using insight into the combinatorial resp. geometrical structure of an optimal local covering (see Sections 5 and 6 in [12]).

Of course one gets in the same way approximation schemes for covering with objects of various other shapes. A nice application is the problem of covering given points with a minimal number of squares, which comes up in image processing [26]. Also the same methods happen to provide polynomial time approximation schemes for NP-complete packing problems where one wants to pack without overlap a maximal number of objects of a given size and shape into a given area (JOHNSON [15] describes how such problems arise in the context of VLSI-design).

4. Lower bounds for Turing machines

The generic question of machine-based complexity theory is the following. Given are two classes T_1 and T_2 of mathematical models for computers, where models of type T_2 appear to be more powerful than those of type T_1. Find the slowest growing function S_{T_1,T_2} s.t. any model of type T_2 whose time bound is $t(n)$ (for some function $t(n)$) can be simulated by a model of type T_1 with time bound $O(S_{T_1,T_2}(t(n)))$ (simulation just means that the same output is produced on the same input). Nonlinear lower bounds for S_{T_1,T_2} tell us that models of type T_2 are in fact more powerful and the precise growth rate of S_{T_1,T_2} provides a quantitative measure for the superiority of models of type T_2 over models of type T_1.

Questions of this form arise quite frequently in computer science, e.g. if one wants to make an intelligent choice between several competing designs for hardware or software. Such questions also arise in more theoretical considerations where one wants to classify the inherent computational difficulty of mathematical problems (which often can be determined only for a special type of computer model, e.g. only for nondeterministic machines).

Unfortunately questions of the considered type have only been solved for very few classes T_1 and T_2. The most prominent open problem is the instance where T_1 is the class of deterministic Turing machines and T_2 is the class of nondeterministic Turing machines (P $\stackrel{?}{=}$ NP problem, S_{T_1,T_2} is nonlinear by [22]). Many other open problems of the considered type are not related to nondeterminism. This suggests that there is not just a single "trick" missing (the one that shows P $\neq$ NP). Rather a new mathematical area has to be developed that provides techniques for sharp lower bound results.

We want to report in this section about some new results in this area, that rely on the method of playing simultaneously several strategies against the opponent in a 2-person game. We will describe primarily those aspects that are relevant to this aspect and refer to MAASS [19] for all missing details.

The first problem that we consider is the instance where T_1 is the class of 1-tape deterministic Turing machines and T_2 is the class of 2-tape deterministic Turing machines. We assume that every Turing machine (TM) possesses besides its work tapes (whose number we indicate) an additional one-way input tape (one-way means that the associated head can move only in one direction). Further one head is associated which each tape. All heads may move simultaneously.

The problem of comparing these two classes T_1 and T_2 is actually quite old. Traditionally only 1-tape TM's have been considered. 2-Tape TM's emerged right at the beginning of machine-based complexity theory because one can write for these machines programs that run substantially faster than all known programs for 1-tape TM's. Unfortunately although similarly fast programs have not been found, one was neither able to prove that they do not exist. The obvious disadvantage of a 1-tape TM is the fact that it needs $\Omega(l \cdot d)$ steps to move on its work tape a string of l symbols over a distance of d cells, while a 2-tape TM can do this in time $O(l + d)$. This observation allows to prove easily quadratic lower bounds for a weak form of 1-tape TM's that do not have an extra input tape (they receive the input on the work tape), see HENNIE [11]. E.g. such machine cannot compute for any $\delta > 0$ in $O(n^{2-\delta})$ steps whether a string $x_1 \cdots x_n y_1 \cdots y_n$ is a "palindrome", i.e. for all $i : y_i = x_{n+1-i}$. The 1-tape TM with an extra one-way input tape — this is the model that is usually studied in the current lower bound literature — is quite a bit more powerful and can e.g. recognize palindromes in linear time. In addition for more difficult problems such 1-tape TM has the option to choose a clever "datastructure" for the representation of the input on its work tape which makes it unnecessary to perform during the computation a lot of time-consuming copying operations. In particular, the machine can use several "tracks" on its single work tape and it may also write immediately each input symbol that it reads from the input tape at a number of different locations on the work tape. In order to get strong lower bound results for the function S_{T_1,T_2} in question one has to show that all these tricks cannot help. On the other hand, there are related situations where the use of clever datastructures helps very well. For example one can simulate a k-tape TM with time bound $t(n)$ by a 2-tape TM without a severe time loss in time $O(t(n) \cdot \log t(n))$, for any $k > 2$ (HENNIE and STEARNS [10]).

The best known upper bound for the function S_{T_1,T_2} in question is $S_{T_1,T_2}(m) = O(m^2)$ (HARTMANIS and STEARNS [8]). The best known lower bound result shows that not $S_{T_1,T_2}(m) = O(m \cdot \log\log m)$ (DURIS et al. [4]; one should also mention related earlier work by RABIN [23], AANDERAA [1] and PAUL [21]).

THEOREM 4.1. *For no* $\delta > 0$, $S_{T_1,T_2}(m) = O(m^{2-\delta})$.

Another open problem of the considered type (see DURIS et al. [4] for a recent list of open problems, we solve here 1. and 7.) deals with the classes T_1^N of nondeterministic 1-tape TM's and T_2^N of nondeterministic 2-tape TM's. The HARTMANIS and STEARNS simulation [8] provides again the best upper bound $S_{T_1^N,T_2^N}(m) = O(m^2)$ and the best lower bound result shows that not $S_{T_1^N,T_2^N}(m) = O(m \cdot \log\log m)$ (DURIS et al. [4]).

Strong lower bounds for *nondeterministic* 1-tape TM's are a bit more difficult. Notice that these machines accept e.g. some NP-complete problems like 3-COLORABILITY in linear time. Further in terms of the previously discussed possibilities a nondeterministic 1-tape TM has an important additional tool. In order to simulate a 2-tape TM without significant time loss it can choose for each input an "individualized" data-structure on its work tape, which facilitates the particular computation that is performed on this particular input. In addition BOOK et al. [3] have proved that for any $k > 2$ one can simulate a nondeterministic k-tape TM by a nondeterministic 2-tape TM without any increase in computation time. Furthermore for alternating TM's (which iterate nondeterminism) PAUL et al. [20] have shown that for any $k > 1$ one can simulate a k-tape alternating TM by a 1-tape alternating TM without any increase in computation time.

THEOREM 4.2. *For no* $\delta > 0$, $S_{T_1^N,T_2^N}(m) = O(m^{2-\delta})$.

So far we have compared classes that have the same control structure but different storage facilities. We consider now pairs of classes which have the same storage facilities (one work tape besides the one-way input tape) but different control structures. We write $\text{DTIME}_1(t(n))$ and $\text{NTIME}_1(t(n))$ for the classes of sets that are accepted by deterministic resp. nondeterministic 1-tape TM's (always with an additional one-way input tape). We write CO-NTIME$_1(t(n))$ for the class of sets whose complement is in NTIME$_1(t(n))$.

THEOREM 4.3. $\mathrm{NTIME}_1(n) \not\subseteq \bigcup_{\delta>0} \mathrm{DTIME}_1(n^{2-\delta})$.

THEOREM 4.4. $\text{CO-NTIME}_1(n) \not\subseteq \bigcup_{\delta>0} \mathrm{NTIME}_1(n^{2-\delta})$.

Notice that Theorem 4.4 implies Theorem 4.3. In a somewhat related result PAUL et al. [22] have shown that

$$\mathrm{NTIME}_2(n) \not\subseteq \bigcup_{k\geq 1} \mathrm{DTIME}_k\,(n \cdot (\log^* n)^{1/4}).$$

Concerning stronger separation results the authors of [22] point out that their method might yield at best an $n \cdot \log n$ lower bound. We use here a different type of argument (analysis of the structure of computations for concrete languages) which seems to have no a priori limitations. We construct a language L_1 that satisfies the following lemmata (which obviously imply Theorems 4.1–4.4).

LEMMA 4.5. *L_1 is accepted by a deterministic* 2-*tape* TM *in linear* (*even real*) *time.*

LEMMA 4.6. *The complement of L_1 is accepted by a nondeterministic* 1-*tape* TM *in linear* (*even real*) *time.*

LEMMA 4.7 (Main Lemma). *There is no $\delta > 0$ s.t. L_1 is accepted by a non-deterministic* 1-*tape* TM *in time* $O(n^{2-\delta})$.

The language L_1 consists of finite sequences of symbols 0, 1, 2, 3, 4. We interpret these symbols as commands that tell a deterministic 2-tape TM M' to perform certain operations and tests. We assume that initially M' is always in "writing mode". In this mode M' copies the initial segment of its input Y from left to right on both work tapes until it encounters in the input a first symbol $z \notin \{0, 1\}$. M' rejects the input unless $z = 4$. M' changes now into the "testing mode" (it never changes back to the writing mode). M' always interprets the symbol 4 as the command to change the direction of movement for both of its work heads. M' interprets 2(3) as the command to move work head 1(2) one cell in the currently required direction. M' in testing mode interprets a symbol $y \in \{0, 1\}$ as the command to test whether the work head that moved last reads currently the symbol y. We put a string Y in L_1 iff all these test that M' performs for input Y have a positive outcome. With this definition of L_1 we have proved simultaneously Lemma 4.5. The proof of Lemma 4.6 is also quite obvious.

As an example for words in L_1 we note that a binary string $x_1 \cdots x_n y_1 \cdots y_n$ is a palindrome iff the string $x_1 \cdots x_n 42 y_1 2 y_2 \cdots 2 y_n$ is in L_1. For the lower bound argument we will consider words in L_1 of the following structure. Let $X = x_1 \cdots x_n$ be a binary string and let $L = l_1, l_2, \ldots$ and $R = r_1, r_2, \ldots$ be two sets of subsequences of consecutive bits ("blocks") from X. We assume that the blocks in L and R are listed in the order of their occurrence from right to left in X. Let $l_{i,1} \cdots l_{i,p}$ and $r_{i,1} \cdots r_{i,p}$ be the symbols of block l_i resp. r_i in the order of their occurrence in X from right to left. Let $d_l(i)$ ($d_r(i)$) be the number of bits between blocks l_i and l_{i+1} (r_i and r_{i+1}) in X. Further let $d_l(0)$ ($d_r(0)$) be the number of bits in X to the right of block l_1 (r_1). Then the following string is in L_1:

$$x_1 \cdots x_n 4 \underbrace{2 \cdots 2}_{d_l(0) \text{ times}} 2 l_{1,1} 2 l_{1,2} \cdots 2 l_{1,p} \underbrace{3 \cdots 3}_{d_r(0) \text{ times}} 3 r_{1,1} 3 r_{1,2} \cdots 3 r_{1,p} \underbrace{2 \cdots 2}_{d_l(1) \text{ times}} 2 l_{2,1} 2 l_{2,2} \cdots 2 l_{2,p}$$

$$\underbrace{3 \cdots 3}_{d_r(1) \text{ times}} 3 r_{2,1} 3 r_{2,2} \cdots 3 r_{2,p} \cdots \text{ (etc., alternating through all blocks of } L \text{ and } R).$$

We view the proof of Lemma 4.7 as a 2-person game where player I ("we") wants to prove the claimed lower bound and player II ("opponent") claims to have a counterexample. The opponent starts the game by choosing a nondeterministic 1-tape TM M and constants $\delta, K > 0$. He claims that M accepts L_1 in time $K \cdot n^{2-\delta}$. Player I continues the game by choosing an input $X^\frown Z$ in L_1 on which he tests M. $X^\frown Z$ is chosen as follows.

We assume that some canonical way of coding TM's $\tilde{M}$ by binary strings has been fixed. We write $|\tilde{M}|$ for the length of the binary string that codes $\tilde{M}$. The first part $X = x_1 \cdots x_n$ of the input is a binary string s.t. $K(X) \geq n \gg |M|$. Here the Kolmogorov complexity $K(X)$ is defined as

$$K(X) := \min\{|\tilde{M}| \mid \tilde{M} \text{ is a TM which produces (for the empty input) output } X\}.$$

The notion of Kolmogorov complexity has been introduced into complexity theory by PAUL (see [2]). Its advantage is that if $K(X) \geq |X| \gg |M|$ we can be sure that TM M has nearly no special knowledge about X (X looks like a random string to M).

We define for the rest of this section $\tilde{n} := n^{1-\delta/3}$. Note that (for large n) $\tilde{n}^2$ is bigger than the time bound for M.

To motivate the choice of the second part Z of the input we first give a result that holds for any Z.

LEMMA 4.8 ("Desert Lemma"). *Assume that C is an accepting computation of* TM M *on input* $X^\frown Z$ *with no more than* $K(10n\cdot \log n)^{2-\delta}$ *steps* (n *is the length of string* X). *Then for large enough* n *there is an interval* D ("*desert*") *of* $\tilde{n}$ *cells on the work tape of* M *and there are two sets* $\tilde{L}$ *and* $\tilde{R}$ *s.t. both* $\tilde{L}$ *and* $\tilde{R}$ *contain exactly* $\tilde{n}/2-2n^{1-\delta/2}$ *blocks* B *from* X *with* $|B| = n^{\delta/3}$ *for each* B *and s.t. in computation* C *the work head of* M *is always left* (*right*) *of* D *while its input head reads from a block* B *in* X *that belongs to* $\tilde{L}$ ($\tilde{R}$).

The *proof* of Lemma 4.8 requires a lengthy combinatorial argument which we cannot give here. One uses in particular that among any $\tilde{n}$ cells on the work tape of M there is one which is visited during at most $\tilde{n}$ steps. This may be viewed as playing $\tilde{n}$ substrategies against the opponent — one of which is guaranteed to win.

If we put ourselves for a moment in the easier situation of the proof of Theorem 4.1 where the opponent's 1-tape TM M is a deterministic machine, we are after Lemma 4.8 already quite close to the completion of the proof. In this case the first part of the computation C of M on input $X^\frown Z$ until the step t_0 where M's input head moves onto the first symbol of Z does not depend on Z. Therefore we need not specify Z before step t_0. Lemma 4.8 deals only with the part of C before step t_0. Thus we can use the sets $\tilde{L}$ and $\tilde{R}$ that are provided by Lemma 4.8 for the definition of Z. From $\tilde{L}$ and $\tilde{R}$ we define Z as in the example right after the definition of L_1, with $p := n^{\delta/3}$, $L := \tilde{L}$, $R := \tilde{R}$. Then we can complete the proof by using Lemma 4.10 below (call every subsequence of Z an $\tilde{L}-\tilde{R}$ pair that consists of the commands to check a block from $\tilde{L}$ and to check in immediate succession a block from $\tilde{R}$).

When we return now to the proof of Lemma 4.7 (the nondeterministic case) we see that our strategic situation is much weaker. In this case the first part of computation C until step t_0 depends already on the second part Z of the input (e.g. M may choose a representation of X on its work tape that facilitates the particular test sequence Z; technically M can guess Z while reading X and verify its guesses later while reading Z). But if we define already Z before the beginning of the computation, with some arbitrarily chosen sets L, R in the way of our previous example, we can hardly expect that the opponent is so kind to arrange C s.t. the sets $\tilde{L}$, $\tilde{R}$ that come out of Lemma 4.8 are the same — or even similar — to the sets L, R we started with. Therefore we use a system of several different strategies against the opponent. We use in our first strategy a guess L_1, R_1 at the future $\tilde{L}$, $\tilde{R}$ that may be totally wrong. But if this is the case we learn at least something about the opponent and the second guess L_2, R_2 that we use in our second strategy is designed to approximate any $\tilde{L}$, $\tilde{R}$ that are totally different from L_1, R_1. Analogously L_3, R_3 is designed to approximate any $\tilde{L}$, $\tilde{R}$ that are totally different from L_1, R_1 and L_2, R_2. Altogether we design a system of $\log \tilde{n}$ "guesses" L_i, R_i and we use L_i, R_i to define the ith section Z_i of Z. Z_1 is defined from L_1, R_1 exactly as the string in our previous example had been defined from sets L, R. Z_2 is a similar command sequence that tells the 2-tape TM M' to check in alternation the blocks in L_2 and R_2, the first ones with head 1, the second ones with head 2. This is done on M' during one sweep from left to right of both heads. Z_3 uses like Z_1 a sweep from right to left to check in alternation the blocks in L_3, R_3.

We partition X into $\tilde{n}$ blocks of length $n^{\delta/3}$. We number these blocks in X from left to right by binary sequences of length $\log \tilde{n}$ (assume w.l.o.g. that $\log \tilde{n}$ is a natural number). We say that two blocks are i-connected if their associated binary sequences differ exactly at the i-last bit. If two blocks are i-connected we put the left one into L_i and the right one into R_i.

Finally we define $Z := Z_1{}^\frown \cdots {}^\frown Z_{\log \tilde{n}}$. Notice that any two blocks from X that are i-connected for some i are tested in immediate succession somewhere in command sequence Z. It is obvious that $X^\frown Z \in L_1$.

We have now specified the complete input $X^\frown Z$ and Lemma 4.8 provides for this input a "desert" D and two sets $\tilde{L}$, $\tilde{R}$ of $\tilde{n}/2 - 2n^{1-\delta/2}$ blocks each. We call a subsequence of Z an $\tilde{L}$–$\tilde{R}$ pair if it consists of the commands to check in immediate succession two blocks b_1, b_2 from X s.t. one belongs to $\tilde{L}$ and the other to $\tilde{R}$.

LEMMA 4.9. *Assume that the $\tilde{n}$ blocks of X have been partitioned into any three sets $\tilde{L}$, $\tilde{R}$, G (G consists of those blocks that are neither in $\tilde{L}$ nor in $\tilde{R}$). Then there are at least* $\min\{|\tilde{L}|, |\tilde{R}|\} - |G| \log \tilde{n}$ *$\tilde{L}$–$\tilde{R}$ pairs in the previously defined sequence Z.*

PROOF OF LEMMA 4.9. We verify now that our previously described tactic — where we play a system of $\log \tilde{n}$ strategies against the opponent — is successful. Assume for simplicity that $G = \emptyset$ and $|\tilde{L}| = |\tilde{R}| = \tilde{n}/2$. We view the partition into $\tilde{L}$, $\tilde{R}$ as a coloring of the blocks in X. Consider the case where our first strategy fails completely and Z_1 contains no $\tilde{L}$–$\tilde{R}$ pair. This implies (by the definition of L_1, R_1 respectively the definition of "1-connected") that the first and second block in X have received the same color, the third and fourth block in X have received the same color, etc. Assume in addition that the second strategy fails completely and the second section Z_2 of Z contains also no $\tilde{L}-\tilde{R}$ pair. Together with the previous information this implies that the first through fourth block in X have the same color, the fifth through eighth block in X have the same color, etc. Apparently, this cannot go on for all sections $Z_1, \ldots, Z_{\log \tilde{n}}$ of Z because otherwise all blocks in X would have received the same color, a contradiction to $|\tilde{L}| = |\tilde{R}| = \tilde{n}/2$.

It is not difficult to fill in the precise proof of Lemma 4.9, which proceeds by induction on $\log \tilde{n}$.

Lemma 4.9 implies that for the two sets $\tilde{L}$, $\tilde{R}$ that have been provided by Lemma 4.8 there are at least $\tilde{n}/2 - 6n^{1-\delta/2} \log \tilde{n}$ $\tilde{L}-\tilde{R}$ pairs in Z, which is more than $\tilde{n}/4$ for large n. The final knockout is delivered by the following lemma.

LEMMA 4.10. *For at least* 1/3 *of the* $\tilde{L}-\tilde{R}$ *pairs in* Z *the work head of* M *crosses the* $\tilde{n}/3$ *cells in the middle of desert* D *during those steps where its input head reads from that* $\tilde{L}-\tilde{R}$ *pair in* Z.

The *proof* of Lemma 4.10 requires a lengthy combinatorial argument. The intuition is that M cannot too often check blocks from X (as demanded by Z) without moving its work head close to the area where it had written notes about this block while reading the corresponding part of X. Of course one has to be aware that M may have written down each block at several locations and it may also have spread information about each block to other areas during its later head movements.

Lemma 4.10 implies that the work head of M crosses (for large n) at least $1/3 \cdot \tilde{n}/4$ often the $\tilde{n}/3$ cells in the middle of desert D. This takes at least $\tilde{n}^2/36$ steps, which exceeds for large n the time bound of $K(10n \cdot \log n)^{2-\delta}$ steps for machine M on input $X^\frown Z$. This finishes the proof of Lemma 4.7.

References

[1] AANDERAA, S.O., 1974, *On k-tape versus* $(k-1)$*-tape real time computations*, in: Complexity of Computation, R.M. Karp, ed., SIAM-AMS Proceedings, Vol. 7 (AMS, Providence), pp. 75–96.

[2] BAKER, T., GILL, J. and SOLOVAY, R., 1975, *Relativizations of the* P $\stackrel{?}{=}$ NP *question*, SIAM J. Comput 4 (4), pp. 431–442.

[3] Book, R.V., Greibach, S.A. and Wegbreit, B., 1970, *Time and tape bounded Turing acceptors and AFL's*, J. Comput. Syst. Sci. 4, pp. 606–621.

[4] Duris, P., Galil, Z., Paul, W. and Reischuk, R., 1983, *Two nonlinear lower bounds*, Proceedings of the STOC Conference of the ACM, pp. 127–132.

[5] Fenstad, J.E., 1980, *General Recursion Theory: An Axiomatic Approach* (Springer, Berlin).

[6] Fowler, R.J., Paterson, M.S. and Tamimoto, S.L., 1981, *Optimal packing and covering in the plane are NP-complete*, Inform. Process. Lett. 12, pp. 133–137.

[7] Garey, M.R. and Johnson, D.S., 1979, *Computers and Intractability* (Freeman, San Francisco).

[8] Hartmanis, J. and Stearns, R.E., 1965, *On the computational complexity of algorithms*, Trans. AMS 117, pp. 285–306.

[9] Hartmanis, J. and Hopcroft, J., 1971, *An overview of the theory of computational complexity*, J. ACM 18, pp. 444–475.

[10] Hennie, F.C. and Stearns, R.E., 1966, *Two-tape simulation of multitape Turing machines*, J. ACM 13, pp. 533–546.

[11] Hennie, F.C., 1965, *One-tape, off-line Turing machine computations*, Information and Control 8, pp. 553–578.

[12] Hochbaum, D.S. and Maass, W., *Fast approximation algorithms for a nonconvex covering problem*, to appear.

[13] Hochbaum, D.S. and Maass, W., 1985, *Approximation algorithms for covering and packing problems in image processing and VLSI*, J. ACM 32, pp. 130–136.

[14] Homer, S. and Maass, W., 1983, *Oracle dependent properties of the lattice of NP-sets*, Theoret. Comput. Sci. 24, pp. 279–289.

[15] Johnson, D.S., 1982, *The NP-completeness column: an ongoing guide*, J. Algorithms 3, pp. 182–195.

[16] Joseph, D., 1983, *Three proof techniques in complexity theory*, to appear in Proceedings of a Conference on Computational Complexity Theory in Santa Barbara (March 1983).

[17] Maass, W., 1983, *Characterization of recursively enumerable sets with supersets effectively isomorphic to all recursively enumerable sets*, Trans. AMS 279, pp. 311–336.

[18] Maass, W., *On the complexity of nonconvex covering*, SIAM J. Comput., to appear.

[19] Maass, W., 1984, *Quadratic lower bounds for deterministic and nondeterministic one-tape Turing machines*, Proc. STOC Conf. ACM, pp. 401–408.

[20] Paul, W.J., Prauss, E.J. and Reischuk, R., 1980, *On alternation*, Acta Informatica 14, pp. 243–255.

[21] Paul, W.J., 1982, *On-line simulation of $k+1$ tapes by k tapes requires nonlinear time*, Proceedings of the 23rd IEEE FOCS Conference, pp. 53–56.

[22] Paul, W.J., Pippenger, N., Szemeredi, E. and Trotter W.T., *On determinism versus nondeterminism and related problems*, Proceedings of the 24th IEEE FOCS Conference.

[23] Rabin, M.O., 1963, *Real time computation*, Israel J. Math. 1, pp. 203–211.

[24] Soare, R.I., 1981, *Computational complexity and recursively enumerable sets*, to appear in Proceedings of the Workshop on Recursion Theoretic Approaches to Computer Science (Purdue, May).

[25] Soare, R.I., 1984, *Recursively Enumerable Sets and Degrees: the Study of Computable Functions and Computably Generated Sets* (Springer, Berlin).

Added in proof. Some improvements and detailed proofs of the results in Section 4 can be found in: Maass, W., Combinatorial lower bound arguments for deterministic and nondeterministic Turing machines, Trans. AMS, to appear.

4. Axiomatic Set Theory

Barcan Marcus et al., eds., Logic, Methodology and Philosophy of Science VII
© Elsevier Science Publishers B.V. (1986) 159–169

REALS AND POSITIVE PARTITION RELATIONS

STEVO TODORČEVIĆ

Dept. of Mathematics, Univ. of California, Berkeley, CA 94720, U.S.A.

The purpose of this note is to give several remarks and informations about the partition relation

$$2^{\aleph_0} \to (2^{\aleph_0}, \alpha)^2 \quad \text{for all } \alpha < \omega_1$$

as the strongest positive ordinary partition relation not refuted by the well-known Sierpiński partition $2^{\aleph_0} \not\to (\aleph_1, \aleph_1)^2$. One of the first informations about this relation was given by K. KUNEN [6] who showed that

If κ is real-valued measurable, then
$\kappa \to (\kappa, \alpha)^2$ for all $\alpha < \omega_1$.

Let us note that R. SOLOVAY [10] had previously shown that if κ is a measurable cardinal and if $\mathscr{P}$ adds a number of random reals, then $\mathscr{P}$ forces κ is real-valued measurable. Thus $\Vdash_{\mathscr{P}} \kappa \to (\kappa, \alpha)^2$ for all $\alpha < \omega_1$.

Later R. LAVER [8] defined a new saturation property of κ-ideals, i.e., a (λ, μ, ν)-saturation property of κ-ideals, and showed for example that

If κ is $(\kappa, \kappa, \aleph_0)$-saturated, then
$\kappa \to (\kappa, \alpha)^2$ for all $\alpha < \omega_1$.

In this paper we give some further information about this partition by proving the following preservation theorem for uncountable cardinals κ.

THEOREM 1. *If $\kappa \to (\kappa, \kappa)^2$ and if $\mathscr{P}$ is any of the standard posets for adding a number of independent reals, then $\Vdash_{\mathscr{P}} \kappa \to (\kappa, \alpha)^2$ for all $\alpha < \omega_1$.*

By "standard poset for adding a number of independent reals" we mean any of the standard posets for adding side-by-side a number of, say, Cohen, random, Sacks, Silver, $\cdots$ reals. However, an examination of the proof of Theorem 1 will show that $\mathscr{P}$ can be any product of small posets with small supports which preserves ω_1. We do not know which of the weaker positive partition relations on κ are also preserved in this sense under some reasonable forcing-real extensions. However, we know that the relation $\kappa \to (\kappa, \alpha)^2$ where $\alpha < \omega_1$, in general, is not preserved under such extensions. For example, $\omega_1 \to (\omega_1, \omega : 2)^2$ fails if one Cohen or one random real is added. On the other hand,

$$\omega_1 \to (\omega_1, \alpha)^2 \quad \text{for all } \alpha < \omega_1$$

is consistent relative only to the consistency of ZF ([11]).

The proof of Theorem 1 readily generalizes to higher levels of cardinal exponentiation. So, for example, if $\mathscr{P}$ adds κ many Cohen subsets of ω_1, then $\mathscr{P}$ forces

$$2^{\aleph_1} \to (2^{\aleph_1}, \alpha)^2 \quad \text{for all } \alpha < \omega_2.$$

Our second result gives a counterexample to the partition relation $2^{\aleph_0} \to (2^{\aleph_0}, \omega + 2)^2$ which can be considered as the weakest positive partition relation not provable in ZFC. The first result in this direction is due to A. HAJNAL [4] who showed that CH implies $2^{\aleph_0} \not\to (2^{\aleph_0}, (\omega : 2))^2$. Later R. LAVER [7] showed that $\mathrm{MA} + 2^{\aleph_0} = \aleph_2$ implies $2^{\aleph_0} \not\to (2^{\aleph_0}, (\omega : 2))^2$. Thus

$$\mathrm{MA} + 2^{\aleph_0} \leq \aleph_2 \quad \text{implies} \quad 2^{\aleph_0} \not\to (2^{\aleph_0}, (\omega : 2))^2.$$

It is interesting to note that this result cannot be generalized further, i.e., that $\mathrm{MA} + 2^{\aleph_0} = \aleph_3$ does not imply $2^{\aleph_0} \not\to (2^{\aleph_0}, (\omega : 2))^2$ (compare this with [3; p. 271]). This can be proved easily using the methods of Section 2. In [1], J. BAUMGARTNER proved the consistency of $2^{\aleph_0} \not\to (2^{\aleph_0}, (\omega : 2))^2$ where $2^{\aleph_0}$ is the successor of an arbitrary regular cardinal. He also proved the consistency of $2^{\aleph_0} \to (2^{\aleph_0}, (\omega : 2^{\aleph_0}))^2$ without using large cardinals. Subsequently R. LAVER [7] showed the consistency of $2^{\aleph_0} \not\to (2^{\aleph_0}, (\omega : 2))^2$ where $2^{\aleph_0}$ is a weakly Mahlo cardinal. In this note we shall prove the following.

THEOREM 2. *If* $\operatorname{cf} \kappa > \omega$ *then there is a ccc poset* $\mathscr{P}$ *of size* κ *which adds at least* κ *new reals such that* $\Vdash_{\mathscr{P}} \kappa \not\to (\kappa, \omega + 2)^2$.

Thus if $\kappa^{\aleph_0} = \kappa$ then $\Vdash_{\mathscr{P}} 2^{\aleph_0} \not\to (2^{\aleph_0}, \omega + 2)^2$. If, moreover, κ is a measurable cardinal, then since $\mathscr{P}$ is a ccc poset, we have that $\mathscr{P}$ forces

$$\text{there is a } \sigma\text{-saturated } \kappa\text{-ideal ([9]) but } \kappa \not\to (\kappa, \omega + 2)^2.$$

This shows that the fine saturation properties used by Kunen and Laver for getting $\kappa \to (\kappa, \alpha)^2$ are in a sense necessary. This also shows that in Theorem 1 $\mathscr{P}$ cannot be an arbitrary ω_1 preserving (or even ccc) poset.

Note also that this shows that $2^{\aleph_0} \to (2^{\aleph_0}, (\omega : 2))^2$ is strictly weaker than $2^{\aleph_0} \to (2^{\aleph_0}, \omega + 2)^2$ since the existence of a σ-saturated κ-ideal easily implies $\kappa \to (\kappa, (\omega : 2))^2$.

The proof of Theorem 2 is given in Section 1. In Section 2 we show that for certain kinds of reals (e.g., Sacks reals) Theorem 1 holds in a much stronger form. In Section 3 we finish the proof of Theorem 1.

1. $2^{\aleph_0} \not\to (2^{\aleph_0}, \omega + 2)^2$

Let $p \in \mathcal{P}$ iff $p = \langle D_p, f_p \rangle$ where D_p is a finite subset of κ and $f_p \colon [D_p]^2 \to 2$. Before defining the order on $\mathcal{P}$ we need some definitions. For $\alpha < \beta$ in D_p let

$$\mathcal{A}_p(\alpha, \beta) = \{A \subseteq D_p \colon A < \alpha \text{ and } [A]^2 \cup (A \otimes \{\alpha, \beta\}) \subseteq f_p^{-1}(1)\}.$$

We order $\mathcal{A}_p(\alpha, \beta)$ by: $A \lessdot B$ iff A is an initial part of B. Let $\mathrm{rk}_p(\alpha, \beta) \colon \mathcal{A}_p(\alpha, \beta) \to \omega$ be the (unique) rank function on $\mathcal{A}_p(\alpha, \beta)$, i.e.,

$$\mathrm{rk}_p(\alpha, \beta)(A) = \max\{\mathrm{rk}_p(\alpha, \beta)(B) + 1 \colon A \lessdot B \in \mathcal{A}_p(\alpha, \beta)\}.$$

Finally, the ordering on $\mathcal{P}$ is defined by: $p \leqslant q$ iff

$$D_p \supseteq D_q,\ f_p \supseteq f_q \quad \text{and} \quad \mathrm{rk}_p(\alpha, \beta) \supseteq \mathrm{rk}_q(\alpha, \beta) \quad \text{for all } \alpha < \beta \text{ in } D_q.$$

CLAIM 1. *$\mathcal{P}$ is a ccc poset.*

PROOF. Suppose that p and q are isomorphic conditions and that the isomorphism is an identity on $\Delta = D_p \cap D_q$. It suffices to show that p and q are compatible. Let $D_r = D_p \cup D_q$, and let f_r be the extension of f_p and f_q such that $f_r(\{\alpha, \beta\}) = 0$ for $\{\alpha, \beta\} \notin [D_p]^2 \cup [D_q]^2$. The following facts show that $r \leqslant p, q$:

(i) If $\{\alpha, \beta\} \in [D_p]^2 \setminus [D_q]^2$, then $\mathcal{A}_r(\alpha, \beta) = \mathcal{A}_p(\alpha, \beta)$, and so $\mathrm{rk}_r(\alpha, \beta) = \mathrm{rk}_p(\alpha, \beta)$.

(ii) If $\{\alpha, \beta\} \in [D_q]^2 \setminus [D_p]^2$, then $\mathcal{A}_r(\alpha, \beta) = \mathcal{A}_q(\alpha, \beta)$, and so $\mathrm{rk}_r(\alpha, \beta) = \mathrm{rk}_q(\alpha, \beta)$.

(iii) If $\{\alpha, \beta\} \in [\Delta]^2$, then $\mathcal{A}_r(\alpha, \beta) = \mathcal{A}_p(\alpha, \beta) \cup \mathcal{A}_q(\alpha, \beta)$ and $\mathrm{rk}_r(\alpha, \beta) = \mathrm{rk}_p(\alpha, \beta) \cup \mathrm{rk}_q(\alpha, \beta)$.

The facts (i) and (ii) follow immediately from the way f_r is defined. The same is true for the first conclusion of (iii). The second conclusion of (iii) follows easily from the first using the fact that $\mathrm{rk}_p(\alpha, \beta)(A) = \mathrm{rk}_q(\alpha, \beta)(A)$ for $A \in \mathcal{A}_p(\alpha, \beta) \cap \mathcal{A}_q(\alpha, \beta)$ which follows from the isomorphism condition of p and q.

CLAIM 2. $\Vdash_{\mathscr{P}} \kappa \not\rightarrow (\kappa, \omega+2)^2$.

PROOF. The fact that the generic partition has no 1-homogeneous $\omega+2$ follows directly from the existence of the generic ranking function on $\mathscr{A}_G(\alpha, \beta)$. To show that the generic partition has no cofinal in κ 0-homogeneous set it suffices to show the following: Suppose p and q are isomorphic conditions with the isomorphism identity on $\Delta = D_p \cap D_q$. Let $\alpha_0 \in D_p$ be such that $\Delta < \alpha_0$, let $\beta_0 \in D_q$ correspond to α_0 in the isomorphism, and let $\alpha_0 < \beta_0$. Let $D_r = D_p \cup D_q$ and let f_r be the extension of f_p and f_q such that $f_r(\{\alpha_0, \beta_0\}) = 1$ and $f_r(\{\alpha, \beta\}) = 0$ for all $\{\alpha, \beta\} \in [D_r]^2 \setminus [D_q]^2 \cup [D_q]^2$ with $\{\alpha, \beta\} \neq \{\alpha_0, \beta_0\}$. Then $r \leq p, q$.

Again it suffices to prove the facts (i)–(iii) from the proof of Claim 2. The fact (iii) follows as in the previous case since $f_r(\{\alpha_0, \beta_0\}) = 1$ has no effect on $\mathscr{A}_r(\alpha, \beta)$ for $\{\alpha, \beta\} \in [\Delta]^2$. Since (i) and (ii) are similar we shall prove only (ii). Assume (ii) is false, and pick $\{\alpha, \beta\} \in [D_q]^2 \setminus [D_p]^2$ and $A \in \mathscr{A}_r(\alpha, \beta)$ such that $A \notin \mathscr{A}_q(\alpha, \beta)$. Pick $\xi \in A \setminus D_q$.

Case I: $\xi \neq \alpha_0$. Let $\eta \in \{\alpha, \beta\}$ be such that $\eta \notin D_p$. Then by the definition of f_r we have $f_r(\{\xi, \eta\}) = 0$, a contradiction.

Case II: $\xi = \alpha_0$. Hence $\Delta < \alpha_0 < \{\alpha, \beta\}$, and so $\{\alpha, \beta\} \cap D_p = \emptyset$. Pick $\eta \in \{\alpha, \beta\}$ such that $\eta \neq \beta_0$. Then by the definition of f_r we have $f_r(\{\xi, \eta\}) = 0$, a contradiction.

This finishes the proof of Claim 2 and also the proof of Theorem 2.

2. Sacks reals and $2^{\aleph_0} \rightarrow (2^{\aleph_0}, \alpha)^2$

In this section a stronger form of Theorem 1 will be proved for a certain class of reals, typical members of which are Sacks and Silver reals. For notational convenience we shall work only with Sacks reals, but it should not be any problem in defining a class of reals for which the same argument works.

Let κ be a fixed uncountable cardinal such that $\kappa \rightarrow (\kappa, \kappa)^2$. For $A, B \subseteq \kappa$ and $K \subseteq [\kappa]^2$ we define

$$A \otimes B = \{\{\alpha, \beta\}: \alpha \in A,\ \beta \in B \text{ and } \alpha \neq \beta\},$$

$$A / B = \{\{\alpha, \beta\}: \alpha \in A,\ \beta \in B \text{ and } \alpha < \beta\},$$

$$K(A) = \{\beta < \kappa : A \otimes \{\beta\} \subseteq K\}.$$

Let $\mathscr{S}_\theta$ denote the countable-support product of θ copies of the perfect-set poset, i.e., the standard poset for adding side-by-side θ Sacks reals. Then

$\mathcal{S}_\theta$ is a $(2^{\aleph_0})^+$-cc poset with the property that any countable set of ordinals from the extension is contained in a countable set of ordinals from the ground model. This fact will be used later in this section.

Let $\lambda < \kappa$ and let $[\kappa]^2 = \bigcup_{i<\lambda} \dot{K}_i$ be a given partition in $V^{\mathcal{S}_\theta}$ with no 0-homogeneous set of size κ. Let $\dot{W}$ be an $\mathcal{S}_\theta$-name for a member of $[\kappa]^\kappa$. Pick $A \in [\kappa]^\kappa$ and for each $\alpha \in A$, a $q_\alpha \in \mathcal{S}_\theta$ such that $q_\alpha \Vdash \alpha \in \dot{W}$ and such that the q_α's form a Δ-system. Define H: $[A]^2 \to \lambda$ by

$$H(\{\alpha,\beta\}) = \begin{cases} i & \text{if } i \text{ is the minimal } j \text{ such } p \Vdash \{\alpha,\beta\} \in \dot{K}_j \\ & \text{for some } p \leq q_\alpha, q_\beta, \text{ if such a } j \text{ exists,} \\ 0 & \text{otherwise.} \end{cases}$$

By our assumption on the partition and by $\kappa \to (\kappa)^2_2$, we may assume that for some $1 \leq i < \lambda$, $H''[A]^2 = \{i\}$. Let $\langle p_{\alpha\beta} : \{\alpha,\beta\} \in [A]^2 \rangle$ be a fixed sequence of conditions such that $p_{\alpha\beta} \leq q_\alpha, q_\beta$ and $p_{\alpha\beta} \Vdash \{\alpha,\beta\} \in \dot{K}_i$.

Now for $\alpha < \beta < \gamma$ in A we define $H_0(\{\alpha,\beta,\gamma\})$ to be a pair (c,d) where c codes $p_{\alpha\beta}$ and $p_{\alpha\gamma}$ as structures as well as all relations between the ordinals of $\operatorname{dom} p_{\alpha\beta}$ and $\operatorname{dom} p_{\alpha\gamma}$, and where d does the same thing for $p_{\alpha\gamma}$ and $p_{\beta\gamma}$. Since there are only $2^{\aleph_0}$ such pairs and since $\kappa \to (\kappa)^3_{2^{\aleph_0}}$ holds there exist $B \in [A]^\kappa$ and (c,d) such that $H_0''[B]^3 = \{(c,d)\}$. It is now easily seen that for each $\alpha \in B$, $\langle p_{\alpha\beta} : \beta \in B \setminus (\alpha+1) \rangle$ forms a Δ-system with root p^0_α $(\leq q_\alpha)$ and that for each $\gamma \in B$, $\langle p_{\beta\gamma} : \beta \in B \cap \gamma \rangle$ forms a Δ-system with root p^1_γ $(\leq q_\gamma)$. Moreover, p^0_α's and p^1_γ's form Δ-systems with roots p^0 and p^1, respectively. We shall call $\langle p_{\alpha\beta} : \{\alpha,\beta\} \in B/B \rangle$ a *double Δ-system* with root $p^0 \cup p^1$.

A more economical way of getting such B would be by using the canonical partition relations of ERDÖS–RADO [2]. Namely, one first defines a natural set of $\aleph_1$ functions from $[A]^2$ into V (e.g., $f(\{\alpha,\beta\} = \operatorname{tp} \operatorname{dom} p_{\alpha\beta}$, $g_\xi(\{\alpha,\beta\}) =$ the ξth member of $\operatorname{dom} p_{\alpha\beta}$, $h_\xi(\{\alpha,\beta\}) = p_{\alpha\beta}(g_\xi(\{\alpha,\beta\})))$, and then refine a set which is canonical with respect to each of those functions to obtain a B with the above properties.

Let $\dot{X}$ be the set of all $\alpha \in B$ such that $p^0_\alpha \in G_{\mathcal{S}_\theta}$ and let $\dot{Y}$ be the set of all $\gamma \in B$ such that $p^1_\gamma \in G_{\mathcal{S}_\theta}$. Then clearly

$$p^0 \cup p^1 \Vdash \dot{X},\ \dot{Y} \in [\dot{W}]^\kappa,$$

and we shall show that $p^0 \cup p^1$ forces the following fact about $\dot{X}$ and $\dot{Y}$.

(1) For all $C \in [\dot{X}]^\kappa$ and $D \in [\dot{Y}]^\kappa$ there exists $\delta < \kappa$ such that $|K_i(E) \cap D| = \kappa$ for all $E \in [C \setminus \delta]^{\aleph_0}$.

Assume that $p^0 \cup p^1$ does not force (1). Then we can find $\dot{D} \in [\dot{Y}]^\kappa$ and for each $\gamma \in \dot{D}$ an $\dot{E}_\gamma \in [\dot{X} \setminus \gamma]^{\aleph_0}$ such that

$$p^0 \cup p^1 \Vdash \dot{E}_\gamma < \delta \text{ and } \delta \notin \dot{K}_i(\dot{E}_\gamma) \text{ for all } \gamma < \delta \text{ in } \dot{D}.$$

Working in V, we can pick $B_0 \in [B]^\kappa$ such that for each $\gamma \in B_0$ we can find $r_\gamma \leq p^1_\gamma$ and $F_\gamma \in [B \setminus \gamma]^{\aleph_0}$ such that $r_\gamma \Vdash \gamma \in \dot{D}$ and $\dot{E}_\gamma \subseteq F_\gamma$. We may assume that r_γ's form a Δ-system with root $\leq p^0 \cup p^1$, that $r_\gamma \leq p^0_\beta$ for all $\beta \in F_\gamma$, and that $F_\gamma < \delta$ for all $\gamma < \delta$ in B_0. Since $\langle p_{\alpha\beta} : \beta \in B \setminus (\alpha+1)\rangle$ forms a Δ-system, we may also assume that

$$\operatorname{dom} r_\gamma \cap \operatorname{dom}(p_{\beta\delta} \setminus p^0_\beta) = \emptyset \quad \text{for all } \gamma < \delta \text{ in } B_0 \text{ and } \beta \in F_\gamma.$$

Pick $\delta \in B_0$ such that $B_0 \cap \delta$ is uncountable. Since $\langle p_{\beta\delta} : \beta \in B \cap \delta\rangle$ forms a Δ-system with root p^1_δ and since $\operatorname{dom} r_\delta$ is countable, only for countably many $\gamma \in B_0 \cap \delta$ there exists $\beta \in F_\gamma$ such that $\operatorname{dom}(p_{\beta\delta} \setminus p^1_\delta) \cap \operatorname{dom} r_\delta \neq \emptyset$. So pick a $\gamma \in B_0 \cap \delta$ such that

$$\operatorname{dom}(p_{\beta\delta} \setminus p^1_\delta) \cap \operatorname{dom} r_\delta = \emptyset \quad \text{for all } \beta \in F_\gamma.$$

Define $r \in \mathcal{S}_\theta$ as follows:

$$\operatorname{dom} r = \operatorname{dom} r_\gamma \cup \operatorname{dom} r_\delta \cup \bigcup_{\beta \in F_\gamma} \operatorname{dom}(p_{\beta\delta} \setminus p^1_\delta),$$

$$r \restriction \operatorname{dom}(r_\gamma \cup r_\delta) = r_\gamma \cup r_\delta,$$

and

$$r(\xi) = p_{\beta\delta}(\xi) \quad \text{for } \xi \in \operatorname{dom} p_{\beta\delta} \setminus \operatorname{dom}(r_\gamma \cup r_\delta).$$

Then r is a well-defined condition with the property that $r \leq r_\gamma$, r_δ and $r_{\beta\delta}$ for all $\beta \in F_\gamma$. So r forces that $\gamma < \delta$ are members of $\dot{D}$ and that $\delta \in \dot{K}_i(\dot{E}_\gamma)$ which is a contradiction.

A completely analogous proof shows that $p^0 \cup p^1$ forces the following property of $\dot{X}$ and $\dot{Y}$.

(2) For all $C \in [\dot{X}]^\kappa$ and $D \in [\dot{Y}]^\kappa$ there exists $\delta < \kappa$ such that $\dot{K}_i(E) \cap C \cap \delta \neq \emptyset$ for all $E \in [D \setminus \delta]^{\aleph_0}$.

Thus we have proved that the following combinatorial property of κ is forced by $\mathcal{S}_\theta$.

(3) If $\lambda < \kappa$ and if $[\kappa]^2 = \bigcup_{i<\lambda} K_i$ is a given partition, then either $[A]^2 \subseteq K_0$ for some $A \in [\kappa]^\kappa$, or else for every $B \in [\kappa]^\kappa$ there exist $i \geq 1$ and $C, D \in [B]^\kappa$ such that for all $C_0 \in [C]^\kappa$ and $D_0 \in [D]^\kappa$ there exists $\delta < \kappa$ such that $|K_i(E) \cap D_0| = \kappa$ and $K_i(F) \cap C_0 \cap \delta \neq \emptyset$ for all $E \in [C_0 \setminus \delta]^{\aleph_0}$ and $F \in [D_0 \setminus \delta]^{\aleph_0}$.

CLAIM 3. (3) *implies* $\kappa \to (\kappa, \alpha)^2$ *for all* $\alpha < \omega_1$.

PROOF. Assume $[\kappa]^2 = K_0 \cup K_1$ is a given partition with no 0-homogeneous κ. By induction on $\alpha < \omega_1$ we shall show that every $B \in [\kappa]^\kappa$ contains a 1-homogeneous set of order type α. So let $\alpha < \omega_1$ have the property that for all $B \in [\kappa]^\kappa$ and $\beta < \alpha$ there exists $C \in [B]^\beta$ such that $[C]^2 \subseteq K_1$. Note that this together with (3) implies that

(4) For all $B \in [\kappa]^\kappa$ and $\beta < \alpha$ there exists $C \in [B]^\beta$ such that $[C]^2 \subseteq K_1$ and $|K_1(C) \cap B| = \kappa$.

So we assume that α is a limit ordinal. Let $\langle \alpha_n : n < \omega \rangle$ be a sequence of smaller ordinals such that $\alpha = \sum_{n<\omega} \alpha_n$, and let $B \in [\kappa]^\kappa$ be fixed. An easy induction on $n < \omega$ using (4) shows that we can construct sequences $\langle C_n : n < \omega \rangle$ and $\langle B_n : n < \omega \rangle$ such that $B_0 = B$ and

(a) $$C_n \subseteq B_n, \quad B_{n+1} \in [B_n]^\kappa \quad \text{and} \quad C_n < B_{n+1}.$$

(b) $$\operatorname{tp} C_n = \alpha_n \quad \text{and} \quad [C_n]^2 \cup (C_n \otimes B_n) \subseteq K_1.$$

Let $C = \bigcup_{n<\omega} C_n$. Then $C \subseteq B$, $\operatorname{tp} C = \alpha$ and $[C]^2 \subseteq K_1$. This finishes the proof.

In [1], BAUMGARTNER shows how to generalize notions of Sacks and Silver reals to uncountable regular cardinals σ (see also [5]). Using the above arguments one can prove an analogue of (3) for such generalizations. In particular, if one adds κ Sacks subsets of ω_1 then in the extension $2^{\aleph_1} = \kappa$ and $(3)_{\aleph_1}$ holds. Moreover, the extension satisfies $\lambda^{\aleph_0} < \kappa$ for all $\lambda < \kappa$. Using the methods of the proof of Claim 3 one easily proves

CLAIM 4. *Assume* $2^{\aleph_1} = \kappa$ *and* $\lambda^{\aleph_0} < \kappa$ *for all* $\lambda < \kappa$. *Then* $(3)_{\aleph_1}$ *implies* $2^{\aleph_1} \to (2^{\aleph_1}, \alpha)^2$ *for all* $\alpha < \omega_2$.

Let us conclude this section with the remark that some weak versions of (3) hold in the extension by $\mathcal{S}_\theta$ assuming, for example, only that $\kappa = (2^{2^{\aleph_0}})^+$.

3. Cohen reals and $2^{\aleph_0} \to (2^{\aleph_0}, \alpha)^2$

In this section we shall finish the proof of Theorem 1. Since the statement (3) from Section 2 is not true in all forcing-real extensions, we shall need to carry here a more delicate argument in constructing long 1-homogeneous sets. But the following weak form of (3) is still true in any forcing-real extension and will be useful in our proof of $\kappa \to (\kappa, \alpha)^2$.

(5) If $[\kappa]^2 = K_0 \cup K_1$ is given, then either $[A]^2 \subseteq K_0$ for some $A \in [\kappa]^\kappa$, or else for every $B \in [\kappa]^\kappa$ there exist $C, D \in [B]^\kappa$ such that:

(i) For all $C_0 \in [C]^\kappa$ and $D_0 \in [D]^\kappa$ there exists $\delta < \kappa$ such that $|K_1(E) \cap D_0| = \kappa$ for all finite $E \subseteq C_0 \setminus \delta$.

(ii) For all $C_0 \in [C]^\kappa$ and $D_0 \in [D]^\kappa$ there exists $\delta < \kappa$ such that $K_1(E) \cap C_0 \cap \delta \neq \emptyset$ for all finite $E \subseteq D_0 \setminus \delta$.

Actually, we shall have to improve (5) slightly, and in order to do this let us, for notational convenience, assume that we are working with the standard poset $\mathscr{C}_\theta$ for adding θ Cohen reals. We shall prove that $\mathscr{C}_\theta$ forces the following.

(6) If $[\kappa]^2 = K_0 \cup K_1$ is given, then either $[A]^2 \subseteq K_0$ for some $A \in [\kappa]^\kappa$, or else for each $\alpha < \omega_1$ there is a sequence $\langle C_\xi : \xi < \alpha \rangle$ of subsets of κ of size κ such that if $\xi < \alpha$ and if $D_\xi = \bigcup_{\xi < \eta < \alpha} C_\eta$, then the pair $\langle C_\xi, D_\xi \rangle$ satisfies (5)(i) and (ii).

CLAIM 4. (6) *implies* $\kappa \to (\kappa, \alpha)^2$ *for all* $\alpha < \omega_1$.

PROOF. Let $[\kappa]^2 = K_0 \cup K_1$ be given and let us assume that there are no 0-homogeneous sets of size κ. Let $\omega \leq \alpha < \omega_1$ and let $\langle C_\xi : \xi < \alpha \rangle$ satisfy (6). Let $\{N_\xi : \xi \leq \alpha \rangle$ be a strictly increasing continuous sequence of elementary submodels of H_{κ^+} of size $< \kappa$ such that K_0, K_1, $\langle C_\xi : \xi < \alpha \rangle \in N_0$ and $\kappa(\xi) = N_\xi \cap \kappa \in \kappa$ for all $\xi \leq \alpha$. Let $\langle \xi_n : n < \omega \rangle$ be a 1–1 enumeration of α. By induction on $n < \omega$ we shall construct a sequence $\langle \alpha_n : n < \omega \rangle$ of ordinals from κ such that:

(a) $\alpha_n \in C_{\xi_n} \cap (\kappa(\xi_n + 1) \setminus \kappa(\xi_n))$,

(b) $[\{\alpha_i : i < n\}]^2 \subseteq K_1$,

(c) If $I \subseteq n$ and if $\xi < \alpha$ is $> \xi_i$ for all $i \in I$, then

$$|K_1(\{\alpha_i : i \in I\}) \cap C_\xi| = \kappa.$$

Clearly, this will finish the proof of Claim 4 since $\{\alpha_n : n < \omega\}$ is a 1-homogeneous set of type α. Let α_0 be any member of $C_{\xi_0} \cap (\kappa(\xi_0 + 1) \setminus \kappa(\xi_0))$ with the property (c). Such an α_0 exists since $\langle C_{\xi_0}, D_{\xi_0} \rangle$ satisfies (5)(i) and since N_{ξ_0+1} is an elementary submodel of H_{κ^+}. Assume $\alpha_0, \ldots, \alpha_{n-1}$ have been defined so that (a)–(c) hold. Let $I = \{i < n : \xi_i < \xi_n\}$ and let $J = n \setminus I$. Let

$$C = C_{\xi_n} \cap K_1(\{\alpha_i : i \in I\}).$$

Then by the inductive hypothesis, $C \in [\kappa]^\kappa \cap N_{\xi_n}$. Let

$$C^0 = \{\gamma \in C : \text{for all } \xi > \xi_n, |K_1(\{\alpha_i : i \in I\} \cup \{\gamma\}) \cap C_\xi| = \kappa\}.$$

Then by the property (5)(i) of (C_{ξ_n}, D_{ξ_n}), $|C \setminus C^0| < \kappa$. Since clearly $C^0 \in N_{\xi_n+1}$, and since (C_{ξ_n}, D_{ξ_n}) satisfies (5)(ii), there exists $\alpha_n \in C^0 \cap (\kappa(\xi_n+1) \setminus \kappa(\xi_n))$ such that $\alpha_n \in K_1(\{\alpha_j : j \in J\})$. It remains to show that $\alpha_0, \ldots, \alpha_n$ still satisfies (a)–(c). Only (c) is nontrivial and follows easily from the fact that $\langle C_\xi : \xi < \alpha \rangle$ satisfies (6) and by the fact that α_i's are separated by the elementary submodels N_ξ's. This finishes the proof of Claim 4.

So we are left with the proof of $\Vdash_{\mathscr{C}_\theta}$ (6). First we need some definitions. Let $[\kappa]^2 = \dot{K}_0 \cup \dot{K}_1$ be a fixed partition in $V^{\mathscr{C}_\theta}$ with no 0-homogeneous κ. Recall that a sequence $\boldsymbol{p} = \langle p_{\alpha\beta} : \{\alpha, \beta\} \in A/B \rangle$, where $A, B \in [\kappa]^\kappa$ is a *double Δ-system* with root $p^0 \cup p^1$ iff

(d) For each $\alpha \in A$, $\langle p_{\alpha\beta} : \beta \in B \setminus (\alpha+1) \rangle$ is a Δ-system with root p^0_α.

(e) For each $\beta \in B$, $\langle p_{\alpha\beta} : \alpha \in A \cap \beta \rangle$ is a Δ-system with root p^1_β.

(f) $\langle p^0_\alpha : \alpha \in A \rangle$ and $\langle p^1_\beta : \beta \in B \rangle$ are Δ-systems with roots p^0 and p^1, respectively.

(g) For each $\{\alpha, \beta\} \in A/B$, $p_{\alpha\beta} \Vdash \{\alpha, \beta\} \in \dot{K}_1$.

Let

$$X_{\boldsymbol{p}} = \{\alpha \in A : p^0_\alpha \in G_{\mathscr{C}_\theta}\} \quad \text{and} \quad Y_{\boldsymbol{p}} = \{\beta \in B : p^1_\beta \in G_{\mathscr{C}_\theta}\}.$$

Two double Δ-systems

$$\boldsymbol{p} = \langle p_{\alpha\beta} : \{\alpha, \beta\} \in A/B \rangle \quad \text{and} \quad \boldsymbol{q} = \langle q_{\alpha\beta} : \{\alpha, \beta\} \in C/D \rangle$$

are *consistent* iff $p^0 \cup p^1$ and $q^0 \cup q^1$ are compatible and for $\{\alpha, \beta\} \in A/B$ and $\{\gamma, \delta\} \in C/D$ with $\{\alpha, \gamma\} < \{\beta, \delta\}$ and $\beta \neq \delta$, we have that

$$\operatorname{dom}(p_{\alpha\beta} \setminus p^0_\alpha) \cap \operatorname{dom}(q_{\gamma\delta} \setminus q^0_\gamma) = \emptyset.$$

Clearly, the methods of Section 2 show that if $\boldsymbol{p}$ and $\boldsymbol{q}$ are consistent then $(p^0 \cup p^1) \wedge (q^0 \cup q^1)$ forces that $\langle \dot{X}_{\boldsymbol{p}} \cap \dot{X}_{\boldsymbol{q}}, \dot{Y}_{\boldsymbol{p}} \cup \dot{Y}_{\boldsymbol{q}} \rangle$ satisfies (5)(i) and (ii). So in order to prove that $\mathscr{C}_\theta$ forces (6) it suffices to show, in $V^{\mathscr{C}_\theta}$, that for every $\bar{\alpha} < \omega_1$ there exists a sequence $\langle C_\xi : \xi < \bar{\alpha} \rangle$ of elements of $[\kappa]^\kappa$ such that:

(h) For each $\xi < \eta < \bar{\alpha}$ there is (in V) a double Δ-system $\boldsymbol{p}(\xi, \eta)$ such that $C_\xi \subseteq X_{\boldsymbol{p}(\xi,\eta)}$ and $C_\eta \subseteq Y_{\boldsymbol{p}(\xi,\eta)}$.

(i) For each $\xi < \eta, \eta'$, $\boldsymbol{p}(\xi, \eta)$ and $\boldsymbol{p}(\xi, \eta')$ are consistent.

We say that B is a club subset of $A \in [\kappa]^\kappa$ iff $B = i''_A C$ for some closed and unbounded $C \subseteq \kappa$, where $i_A : \kappa \to A$ is the increasing enumeration of A. Let $\bar{\alpha} < \omega_1$ and suppose that for each $\bar{\beta} < \bar{\alpha}$ and for each finite sequence

$$\boldsymbol{q}(0) = \langle q_{\alpha\beta}(0) : \{\alpha, \beta\} \in A_0/B_0 \rangle, \ldots, \boldsymbol{q}(k) = \langle q_{\alpha\beta}(k) : \{\alpha, \beta\} \in A_k/B_k \rangle$$

of double Δ-systems with roots in $G_{\mathscr{C}_\theta}$ and for every

$$X \subseteq \bigcap_{l \le k} X_{q(l)}$$

of cardinality κ there is a sequence $\langle C_\xi : \xi < \bar{\beta} \rangle$ which satisfies (h) and (i) and club sets $\bar{B}_l \subseteq B_l$ for $l \le k$ such that:

(j) For all $\xi < \bar{\beta}$, $C_\xi \subseteq X$.

(k) For all $\xi < \eta < \bar{\beta}$ and $l \le k$, $\boldsymbol{p}(\xi, \eta)$ and $\langle q_{\alpha\beta}(l) : \{\alpha, \beta\} \in A_l / \bar{B}_l \rangle$ are consistent.

Assume first that $\bar{\alpha} = \bar{\beta} + 1$. Using the methods of Section 2 we can easily select a double Δ-system $\boldsymbol{r} = \langle r_{\alpha\beta} : \{\alpha, \beta\} \in E/F \rangle$ such that $X_r, Y_r \in [X]^\kappa$. Let C be the set of all limits $\sigma < \kappa$ with the following property: For all $l \le k$, all $\alpha \in E$ and $\gamma \in A_l$ such that $i_E(\alpha)$, $i_{A_l}(\gamma) < \sigma$ and for all $\beta \in F$ and $\delta \in B_l$ with the property $\mathrm{dom}(r_{\alpha\beta} \setminus r^0_\alpha) \cap \mathrm{dom}(q_{\gamma\delta}(l) \setminus q^0_\gamma(l)) \neq \emptyset$, we have that $i_F(\beta) < \sigma$ iff $i_{B_l}(\delta) < \sigma$. Clearly C is a club in κ. Let C' be the set of all limit points of C and let

$$F' = i''_F C \setminus C' \quad \text{and} \quad B'_l = i''_{B_l} C' \quad \text{for } l \le k.$$

By the induction hypothesis there exists a sequence $\langle C_\xi : \xi < \bar{\beta} \rangle$ of elements of $[X_r]^\kappa$ and club sets $\bar{F} \subseteq F'$ and $\bar{B}_l \subseteq B'_l$ such that for all $\xi < \eta < \bar{\beta}$, $\boldsymbol{p}(\xi, \eta)$ is consistent with each of the systems $\boldsymbol{s} = \langle r_{\alpha\beta} : \{\alpha, \beta\} \in E/\bar{F} \rangle$ and $\langle q_{\alpha\beta}(l) : \{\alpha, \beta\} \in A_l / \bar{B}_l \rangle$ for $l \le k$. Then $\langle C_\xi : \xi < \bar{\beta} \rangle^\frown \langle Y_s \rangle$ satisfies (j) and (k).

Assume now that $\bar{\alpha}$ is limit. Pick an increasing sequence $\langle \bar{\alpha}_n : n < \omega \rangle$ which converge to $\bar{\alpha}$.

Pick a double Δ-system $\boldsymbol{r}(0) = \langle r_{\alpha\beta}(0) : \{\alpha, \beta\} \in E_0/F_0 \rangle$ such that for some club sets $B'_l \subseteq B_l$, $l \le k$, $\boldsymbol{r}(0)$ is consistent with each of $\langle q_{\alpha\beta}(l) : \{\alpha, \beta\} \in A_l / B'_l \rangle$ and such that $X_{r(0)}, Y_{r(0)} \in [X]^\kappa$. By the induction hypothesis there exist $\langle C_\xi : \xi < \bar{\alpha}_0 \rangle$ in $X_{r(0)}$ and club sets $\bar{F}_0 \subseteq F_0$ and $\bar{B}^0_l \subseteq B'_l$, $l \le k$ such that for each $\xi < \eta < \bar{\alpha}_0$, $\boldsymbol{p}(\xi, \eta)$ is consistent with each of $\boldsymbol{s}(0) = \langle r_{\alpha\beta}(0) : \{\alpha, \beta\} \in E_0/\bar{F}_0 \rangle$ and $\langle q_{\alpha\beta}(l) : \{\alpha, \beta\} \in A_l / \bar{B}^0_l \rangle$, $l \le k$. Now pick a double Δ-system $\boldsymbol{r}(1)$ and club sets $B''_l \subseteq \bar{B}^0_l$, $l \le k$ such that $\boldsymbol{r}(1)$ is consistent with each of $\langle q_{\alpha\beta}(l) : \{\alpha, \beta\} \in A_l / B''_l \rangle$, $l \le k$ and such that $X_{r(1)}$, $Y_{r(1)} \in [Y_{s(0)}]^\kappa$. Using the induction hypothesis pick $\langle C_\xi : \bar{\alpha}_0 \le \xi < \bar{\alpha}_1 \rangle$ in $X_{r(1)}$ and clubs $\bar{B}^1_l \subseteq B''_l$, $l \le k$ and $\bar{F}_1 \subseteq F_1$ such that for each $\bar{\alpha}_0 \le \xi$, $\eta < \bar{\alpha}_1$, $\boldsymbol{p}(\xi, \eta)$ is consistent with each of $\boldsymbol{s}(1) = \langle r_{\alpha\beta}(1) : \{\alpha, \beta\} \in E_1/\bar{F}_1 \rangle$ and $\langle q_{\alpha\beta}(l) : \{\alpha, \beta\} \in A_l / \bar{B}^1_l \rangle$, $l \le k$, and so on. Proceeding in this way we construct for each $n < \omega$, $\langle C_\xi : \bar{\alpha}_{n-1} \le \xi < \bar{\alpha}_n \rangle$ and club sets $\bar{B}^n_l \subseteq B_l$ for $l \le k$ $(\bar{\alpha}_{-1} = 0)$ so that if we let $\bar{B}_l = \bigcap_{n<\omega} \bar{B}^n_l$ for $l \le k$ then $\langle C_\xi : \xi < \bar{\alpha} \rangle$ and $\bar{B}_l$, $l \le k$ satisfy (j) and (k). This finishes the induction step $\lim(\bar{\alpha})$, and so the proof of $\Vdash_{\mathscr{C}_\theta}$ (6) is completed.

The above proof can be generalized to higher levels of cardinal exponentiation without much difficulty. So, for example, if $\mathcal{P}$ is the standard poset for adding κ Cohen subsets of ω_1, then

$$\Vdash_{\mathcal{P}} 2^{\aleph_1} \rightarrow (2^{\aleph_1}, \alpha)^2 \quad \text{for all } \alpha < \omega_2.$$

The only nontrivial point in such generalization is to show that $\mathcal{P}$ forces the $\aleph_1$-analogue of the statement (6). So, in the construction of long sequences $\langle C_\xi : \xi < \bar{\alpha} \rangle$ with properties (h) and (i), we have to be able to deal with induction stages $\lim(\bar{\alpha})$ where $\bar{\alpha}$ is not necessarily of cofinality ω. The crucial point in this is to provide that the sequence $\boldsymbol{r}(0), \boldsymbol{r}(1), \ldots$ from the above proof does not vanish after some countable number of steps. This can be done by using a κ-complete ultrafilter over $\mathcal{P}(\kappa) \cap M$ where M is a substructure of some large enough H_ρ which has size κ and which is closed under $< \kappa$ sequences.

References

[1] BAUMGARTNER, J.E., 1976, *Almost disjoint sets, the dense set problem, and the partition calculus*, Ann. Math. Logic 10, pp. 401–439.

[2] ERDÖS, P. and RADO, R., 1950, *A combinatorial theorem*, J. London Math. Soc. 25, pp. 249–255.

[3] ERDÖS, P. and HAJNAL, A., 1974, *Unsolved and solved problems in set theory*, Proc. Symp. in Pure Math. 25 (Amer. Math. Soc., Providence, RI), pp. 269–278.

[4] HAJNAL, A., 1960, *Some results and problems in set theory*, Acta Math. Acad. Sci. Hungar. 11, pp. 277–298.

[5] KANAMORI, A., 1980, *Perfect-set forcing for uncountable cardinals*, Ann. Math. Logic 19, pp. 97–114.

[6] KUNEN, K., 1971, *A partition theorem*, Notices Amer. Math. Soc. 18, pp. 425.

[7] LAVER, R., 1975, *Partition relations for uncountable cardinals* $\leq 2^{\aleph_0}$, Coll. Math. Soc. János Bolyai 10, Infinite and Finite Sets, Keszthely, Hungary (1973), (North-Holland, Amsterdam), pp. 1029–1042.

[8] LAVER, R., 1978, *A saturation property on ideals*, Comp. Math. 36, pp. 233–242.

[9] PRIKRY, K., 1970, *Changing measurable into accessible cardinals*, Dissertationes Math. 68.

[10] SOLOVAY, R., 1971, *Real valued measurable cardinals*, Proc. Symp. Pure Math. 13 (Amer. Math. Soc., Providence, RI), pp. 397–418.

[11] TODORČEVIĆ, S., 1983, *Forcing positive partition relations*, Trans. Amer. Math. Soc. 280, pp. 703–720.

Barcan Marcus et al., eds., Logic, Methodology and Philosophy of Science VII
© Elsevier Science Publishers B.V. (1986) 171–181

ASPECTS OF DETERMINACY

HUGH WOODIN

Dept. of Mathematics, Caltech, Pasadena, CA 91125, U.S.A.

Introduction

The purpose of this paper is essentially of an expository nature though some new results will be presented. The goal is to review some recent developments that indicate an unexpected relationship between descriptive set theory within the context of determinacy and conventional combinatorial set theory.

Before reviewing the basic concepts we are compelled to point out that one of the fundamental and more subtle relationships will be ignored, specifically that of determinacy and large cardinals.

At a simple level the difference between set theory and descriptive set theory is that between the class of (arbitrary) sets and the class of definable sets. For our purposes we shall be concerned only with sets of real numbers, thus this distinction can be made a little more precise. The set theorist is concerned with arbitrary sets of reals, here, for example, the continuum hypothesis is of natural interest. The descriptive set theorist is intrigued by definable sets of reals, i.e. borel sets, projective sets and sets of reals in natural hierarchies beyond the projective hierarchy. The continuum problem has several manifestations to the descriptive set theorist, whether uncountable (definable) sets of reals contain perfect subsets and computing the lengths of (definable) prewellorderings of the reals.

For the study of descriptive set theory it is convenient to view real numbers as infinite sequences of natural numbers so that the space of real numbers is regarded as $\mathbb{N}^{\mathbb{N}}$ which from a logical point of view is denoted by ω^ω. The precise relationship between ω^ω and the Euclidean space, $\mathbb{R}$, is clarified from a topological perspective, i.e. considering ω as a discrete space endows ω^ω with the natural product topology. With this topology ω^ω is homeomorphic to the space of irrationals, $\mathbb{R}\setminus\mathbb{Q}$.

The space ω^ω accommodates more naturally the notion of a game. Suppose $A \subseteq \omega^\omega$ is a set of reals. Associated to the set A is a game G_A involving two players:

$$\begin{array}{cc} \text{I} & \text{II} \\ n_0 & m_0 \\ n_1 & m_1 \\ \vdots & \vdots \\ \hline x & y \end{array} \qquad \begin{array}{l} \text{Player I wins iff } x * y \in A \\ x * y = \langle n_0, m_0, \ldots \rangle \end{array}$$

The players alternate selecting natural numbers with player I selecting first. After infinitely many moves they have collaborated in constructing an element of ω^ω,

$$x * y = \langle n_0, m_0, n_1, m_1, \ldots \rangle.$$

Player I wins provided $x * y$ belongs to A.

The game G_A is determined (in this case it is customary to say that A is determined) if there is a winning strategy for one of the players. More precisely, a strategy (for player I) is simply an algorithm by which to play, formally it is a function that assigns to finite sequences of natural numbers (of even length), natural numbers. The function is interpreted as providing a natural number to play given the sequence of plays so far. A strategy is a winning strategy for player I if in following the strategy player I wins regardless of the play by player II.

The axiom of determinacy (abbreviated AD) asserts that for every set of reals, A, the corresponding game, G_A, is determined. An unfortunate consequence of the axiom of choice is that there is a set of reals which is not determined. However the axiom of choice seems essential in constructing a nondetermined set, there is no known proof of the existence of such a set assuming only ZF, further such sets cannot be borel as Martin has proved that all borel sets are determined.

Hence to assume AD requires banishment to a world without choice. There are several metamathematical solutions to this. The prevalent one presently is to assume that AD holds in some fragment of the universe, typically $L(\mathbb{R})$ the smallest inner model of V satisfying ZF containing all the reals and all the ordinals. A gratifying consequence of this point of view is the development of a complete structure theory for $L(\mathbb{R})$. Of course the assumption of $V = L$ also provides a comprehensive theory for $L(\mathbb{R})$. The point is that AD provides a rich structure theory while allowing for desirable regularity results such as every set of reals (in $L(\mathbb{R})$) is Lebesque measurable.

The concept of determinacy illuminates a key difference between the descriptive set theorist and the set theorist, in that the descriptive set theorist is free to assume (additional) axioms inconsistent for the set theorist. Specifically to aid in the study of the definable sets of reals the descriptive set theorist will often want to assume the axiom of (definable) determinacy, i.e. that all definable sets of reals are determined. It is our convention that a definable set of reals is a set of reals that is first-order definable in V, the universe of sets, allowing ordinal and real parameters. Hence there is an inner model of V, $\text{HOD}_{\mathbb{R}}$, satisfying ZF + DC whose sets of reals are precisely the definable sets of reals. Thus, working in ZFC under the assumption of definable determinacy leads naturally to working in ZF + DC + AD. This position is strengthened (or weakened, depending on one's bias) by the equiconsistency of "ZFC + definable determinacy" and "ZF + DC + AD".

The concept of a tree and the related notion of a Souslin set of reals has been the focal point for developing the structure theory of $L(\mathbb{R})$ within the context of AD. These concepts will be the focus of this paper.

1. Trees

Suppose D is a set. A D-tree is simply a set of finite sequences with members from D, that is closed under initial segments. Let $D^{<\omega}$ denote the set of all finite sequences from D, hence $D^{<\omega}$ is a D-tree and any D-tree is a subtree of $D^{<\omega}$.

Suppose T is a D-tree. Then let $[T]$ denote the set of (infinite) "branches" through T, i.e.

$$[T] = \{f \in D^{\omega} \mid f \upharpoonright n \in T \text{ for all } n \in \omega\}.$$

The tree, T, is well founded if and only if $[T] = \emptyset$. Typically D will be of the form $\omega \times \kappa$ for some cardinal κ. In this situation it is convenient to view elements of a D-tree as pairs of finite sequences of the same length. Thus, if T is an $\omega \times \kappa$-tree then,

$$T \subseteq \{(s, t) \mid (s, t) \in \omega^{<\omega} \times \kappa^{<\omega} \text{ and } l(s) = l(t)\}.$$

We adopt a similar convention for the set of branches through T,

$$[T] = \{(f, g) \mid f \in \omega^{\omega}, g \in \omega^{\omega} \text{ and } (f \upharpoonright n, g \upharpoonright n) \in T \text{ for all } n \in \omega\}.$$

Suppose T is an $\omega \times \kappa$-tree. For $s \in \omega^{<\omega}$ let

$$T(s) = \{t \in \kappa^{<\omega} \mid (s, t) \in T\}.$$

Similarly for $x \in \omega^{\omega}$ let $T(x) = \{t \in \kappa^{<\omega} \mid (x \restriction n, t) \in T$ for some n, i.e. $n = l(t)\}$. Hence for $s \in \omega^{<\omega}$, $T(s) \subseteq \kappa^{l(s)}$ and for $x \in \omega^{\omega}$, $T(x)$ is a κ-tree.

Finally, let

$$p[T] = \{x \in \omega^{\omega} \mid (x, g) \in [T] \text{ for some } g \in \kappa^{\omega}\}.$$

Note

$$p[T] = \{x \in \omega^{\omega} \mid [T(x)] \neq \emptyset \text{ i.e. } T(x) \text{ is not well founded}\}.$$

A set of reals $B \subseteq \omega^{\omega}$ is Souslin if for some cardinal, κ, there is an $\omega \times \kappa$-tree, T, with $B = p[T]$. In this case we say that B is κ-Souslin.

We recall the definition of projective sets of reals. The projective sets are generated by closing the borel sets under continuous images and complements. Thus the projective sets form naturally a hierarchy, $\boldsymbol{\Sigma}^1_1$ denotes those that are continuous images of borel sets (i.e. analytic sets), a set is $\boldsymbol{\Pi}^1_1$ (i.e. coanalytic) if it is the complement of a $\boldsymbol{\Sigma}^1_1$ set, a set is $\boldsymbol{\Sigma}^1_2$ if it is the continuous image of a $\boldsymbol{\Pi}^1_1$ set, etc.

Of course the projective sets are precisely those sets of reals that are first-order definable with parameters over the structure of the reals viewed as a model of second-order number theory. This view has the distinct advantage in that it allows for a "lightface" (i.e. parameter-free) version of the projective hierarchy.

Assuming the axiom of choice it is easily seen that every set of reals is Souslin. The problem facing the descriptive set theorist is that of which definable sets are definably Souslin (i.e. if $B \subseteq \omega^{\omega}$ is definable, is there a definable tree, T, with $B = p[T]$?). This is analogous to the problem of finding Souslin representations of sets of reals without appealing to the axiom of choice. Without the axiom of choice the problem can become nontrivial. For instance, the property of being Souslin can be related to other natural properties of the set. The following theorem is implicit in SOLOVAY [12].

THEOREM 1 (Solovay), (ZF). *Assume there is no uncountable sequence of distinct reals. Then every Souslin set is Lebesque measurable and has the property of Baire.*

Here begins a pattern we shall follow. The version of Theorem 1 of interest to a descriptive set theorist is

THEOREM 2 (ZFC). *Assume there is no definable uncountable sequence of distinct reals. Then every set of reals that is definably Souslin is Lebesque measurable and has the property of Baire.*

For the sake of exposition we are translating considerations of interest to the descriptive set theorist to a context of ZF without (assuming) the axiom of choice. At a naive level this is simply a change in perspective. One can work in V (i.e. ZFC) and study the definable sets or equivalently one can work in the inner model, $\mathrm{HOD}_{\mathbb{R}}$, and thus study arbitrary sets, though in the restrictive context of ZF + DC.

Though there is a parallel between the efforts of a descriptive set theorist and those of a set theorist attempting to work in ZF without AC, the deeper aspects of descriptive set theory are lost in the translation, for example that ZF proves $\boldsymbol{\Sigma}^1_2$ sets are Souslin ignores the more subtle aspects, that Σ^1_2 sets admit Σ^1_2 scales (for the relevant definitions see MOSCHOVAKIS [9]).

A great deal of research in descriptive set theory has been devoted to the question of which sets of reals are definably Souslin. Classical results offer hope of nontrivial answers, consider the following theorem of SHOENFIELD [10].

THEOREM (Shoenfield) (ZF). *$\boldsymbol{\Sigma}^1_2$ sets are Souslin.*

2. Homogeneous and weakly homogeneous trees

Our intent is to identify a property for sets of reals which within ZFC will play a role analogous to the property of being Souslin in ZF, in particular we seek a version of Theorem 1 within the context of the axiom of choice.

Suppose D is a set. A measure on D is a countably complete ultrafilter on (the boolean algebra) $P(D)$. If μ is a measure and B is a set we write $\mu[B]=1$ to indicate that the set B belongs to the ultrafiter μ, i.e. that B has μ-measure 1.

Suppose D is a set and that $\langle \mu_i : 0<i<\omega\rangle$ is a sequence of measures on $D^{<\omega}$ with the ith measure concentrating on sequences of length i, i.e. for each i, $\mu_i[D^i]=1$. The sequence of measures $\langle \mu_i : 0<i<\omega\rangle$ defines a tower, more precisely the measures cohere, if for $i<j$ and $B \subseteq D^i$ with $\mu_i[B]=1$, $\mu_j[B^*]=1$ for $B^*=\{s\in D^j \mid s\restriction i \in B\}$. In this case (assuming coherence) $\mu_i[B]=1$ if and only if $\mu_j[B^*]=1$ so that in fact μ_i is the (appropriate) projection of μ_j.

The tower of measures $\langle \mu_i : 0 < i < \omega \rangle$ is countably complete if for any sequence $\langle B_i : 0 < i < \omega \rangle$ of subsets of $D^{<\omega}$ with the property that $\mu_i[B_i] = 1$ for all i, there is an infinite sequence $g \in D^\omega$, such that for every index i, $g \upharpoonright i \in B_i$ (which is to say the suitable tree determined by $\bigcup B_i$ has a branch). The main point here is that the tower of measures $\langle \mu_i : 0 < i < \omega \rangle$ naturally defines through an inverse limit a filter of subsets of D^ω. The tower is countably complete just in case this filter is countably complete, in the sense that any countable intersection of sets in the filter is nonempty.

We now define the notion of a homogeneous tree, the original idea is implicit in work of KUNEN [4] and MARTIN [6], see KECHRIS [2]. We adopt for the duration of this paper the convention of considering only those $\omega \times \kappa$-trees, T, for which $T(s) \neq \emptyset$ for all $s \in \omega^{<\omega}$.

DEFINITION 3. Suppose T is an $\omega \times \kappa$-tree. The tree, T, is homogeneous if there is a function, ν, defined on $\omega^{<\omega}$ such that:

(1) For $s \in \omega^{<\omega}$, with $s \neq \emptyset$, $\nu(s)$ is a measure on $\kappa^{<\omega}$ with $\nu(s)[T(s)] = 1$.

(2) For $x \in \omega^\omega$, the sequence of measures $\langle \nu(x \upharpoonright i) : 0 < i < \omega \rangle$ defines a tower which is countably complete if $x \in p[T]$.

It actually follows from condition (1) that if $x \in \omega^\omega$ and if the tower of measures $\langle \nu(x \upharpoonright i) : 0 < i < \omega \rangle$ is countably complete then $x \in p[T]$, simply consider the sequence of measure 1 sets $\langle T(x \upharpoonright i) : 0 < i < \omega \rangle$. Thus $x \in p[T]$ if and only if the tower of measures $\langle \nu(x \upharpoonright i) : 0 < i < \omega \rangle$ is countably complete.

The homogeneity of an $\omega \times \kappa$-tree, T, has rather pleasant consequences for its projection, $p[T]$. In fact the concept of an homogeneous tree originated in part with the following theorem in mind.

THEOREM (Kunen, Martin), (ZF + DC). *Suppose T is an $\omega \times \kappa$-tree that is homogeneous. Then the set of reals, $A = p[T]$, is determined.*

Current proofs of determinacy from large cardinal hypotheses tend to work through homogeneous trees. For example Martin's proof of the determinacy of $\boldsymbol{\Pi}^1_1$ sets from the existence of a measurable cardinal is easily reformulated as

THEOREM (Martin), (ZFC). *Assume there is a measurable cardinal. Suppose $A \subseteq \omega^\omega$ is $\boldsymbol{\Pi}^1_1$. Then there is an homogeneous tree T with $A = p[T]$.*

Similarly Martin's proof of the determinacy of $\boldsymbol{\Pi}^1_2$ sets from the existence of the appropriately large cardinal (now referred to as a Martin cardinal) succeeds via the construction of an homogeneous tree with projection the desired $\boldsymbol{\Pi}^1_2$ set. For further details see MARTIN [7].

The concept of homogeneity is rather a strong condition on a tree as it is evident that many trees are not homogeneous. We now recall the less restrictive condition of weak homogeneity. Again KECHRIS [2] is a good reference. In the following definition we continue to regard $(\omega \times \omega)^{<\omega}$ as the set consisting of pairs of finite sequences of natural numbers, of the same length.

DEFINITION 4. Suppose T is an $\omega \times \kappa$-tree. The tree, T, is weakly homogeneous if there is a function, ν, defined on $(\omega \times \omega)^{<\omega}$ such that:

(1) For $(s,t) \in (\omega \times \omega)^{<\omega}$, with $s \neq \emptyset$, $\nu(s,t)$ is a measure on $\kappa^{<\omega}$ with $\nu(s,t)[T(s)] = 1$.

(2) For $x \in \omega^\omega$ and $y \in \omega^\omega$ the sequence of measures $\langle \nu(x \restriction i, y \restriction i) \colon 0 < i < \omega \rangle$ defines a tower and if $x \in p[T]$ then for some $y \in \omega^\omega$ the tower, $\langle \nu[x \restriction i, y \restriction i) \colon 0 < i < \omega \rangle$, is countably complete.

As in the case of homogeneity it follows from condition (1) that the latter part of condition (2) is an equivalence more precisely, $x \in p[T]$ if and only if for some $y \in \omega^\omega$ the tower of measures given by $\langle \nu(x \mid i, y \mid i) \colon 0 < i < \omega \rangle$ is countably complete.

The generalizations of homogeneity and weak homogeneity to trees on $\omega^k \times \kappa$ are immediate given the desire (for homogeneity ν has domain $(\omega^k)^{<\omega}$ and for weak homogeneity ν has domain $(\omega^k \times \omega)^{<\omega}$).

Suppose B is a subset of the plane, i.e. $B \subseteq \omega^\omega \times \omega^\omega$. We denote by $\exists^{\mathbb{R}} B$ the set of reals defined by the projection of B in the first coordinate, $\exists^{\mathbb{R}} B = \{x \in \omega^\omega \mid (x,y) \in B \text{ for some } y \in \omega^\omega\}$.

It is an immediate consequence of the definition that if T is an $\omega \times \kappa$-tree and T is weakly homogeneous then there is an $\omega^2 \times \kappa$-tree, T^*, such that T^* is homogeneous and $p[T] = \exists^{\mathbb{R}} p[T^*]$. The converse is also easily verified.

An easy classical result regarding $\boldsymbol{\Sigma}^1_2$ sets of reals is that they are projections of $\boldsymbol{\Pi}^1_1$ subsets of the plane. Hence:

THEOREM (ZFC). *Assume there is a measurable cardinal. Suppose $A \subseteq \omega^\omega$ is $\boldsymbol{\Sigma}^1_2$ then there is a weakly homogeneous tree, T, with $A = p[T]$.*

The following theorem is in some sense a generalization of the principal result in SOLOVAY [12] that if there is a measurable cardinal then $\boldsymbol{\Sigma}^1_2$ sets of reals are Lebesque measurable and have the property of Baire.

THEOREM 5 (ZFC). *Assume B is a set of reals such that $B = p[T]$ for some weakly homogeneous tree, T. Then B is Lebesque measurable and has the property of Baire.*

This theorem is similar to Theorem 1. We contend that weakly homogeneous trees provide the structural representations for sets of reals, in the context of ZFC, that approximate Souslin representations in ZF (e.g. the remarks at the beginning of this section).

Of course one could observe that Theorem 5 is also true for homogeneous trees. However weak homogeneity is a far less restrictive condition on trees in fact, in some sense, almost any tree is weakly homogeneous as the following theorems indicate. We betray a bias by scarcely noting that for nontrivial weakly homogeneous trees to exist at all, there must be a measurable cardinal.

Suppose λ is a cardinal. Let $\mathbb{P}_\lambda$ denote the Lévy conditions for collapsing λ to ω, i.e. $\mathbb{P}_\lambda \sim \lambda^{<\omega}$, the order given by extension. Let $\mathbb{Q}_\lambda$ denote the Lévy conditions for collapsing all smaller cardinals to ω, i.e. $\mathbb{Q}_\lambda \sim \prod_{\delta<\lambda} \mathbb{P}_\delta$, the product computed with finite support.

Suppose κ is strongly inaccessible and that $G \subseteq \mathbb{Q}_\kappa$ is generic. In the generic extension, $V[G]$, one can define the Solovay model, more precisely the inner model of $V[G]$ consisting of those sets hereditarily definable in parameters from $V \cup \mathbb{R}$. This of course is simply the constructible extension of V determined by the set of new reals, hence we denote the Solovay model by $V(\mathbb{R})$. For more information on Solovay's construction see SOLOVAY [11].

We take for granted the definition of a supercompact cardinal (see KANAMORI and MAGIDOR [1]).

THEOREM 6 (ZFC). *Suppose κ is supercompact and that T is an $\omega \times \lambda$-tree for some λ. Then there is a cardinal $\delta < \kappa$ such that if $G \subseteq \mathbb{P}_\delta$ is generic, $V[G] \models T$ is weakly homogeneous.*

Thus in the presence of a supercompact cardinal any tree, modulo some small forcing, is weakly homogeneous. This is the best that one can hope for, assuming the axiom of choice (e.g. Theorem 7).

The proof of Theorem 6 actually reveals a slightly stronger result in that the measures witnessing weak homogeneity can be chosen to be κ-complete. Hence as a corollary to Theorem 6 one can show the following.

THEOREM 7 (ZFC). *Suppose κ is supercompact and that $G \subseteq \mathbb{Q}_\kappa$ is generic. Let $V(\mathbb{R}) \subseteq V[G]$ be the Solovay model obtained, $V(\mathbb{R}) \models$ Every $\omega \times \lambda$-tree is weakly homogeneous, for any λ.*

In particular Theorem 7 shows that granting the consistency of the existence of a supercompact cardinal, ZF + DC is consistent with the proposition that every $\omega \times \lambda$-tree is weakly homogeneous. Theorem 6 and its corollary, Theorem 7, were inspired by the result of MARTIN [8] that assuming $AD_{\mathbb{R}}$ (the axiom of determinacy for real games) every tree is weakly homogeneous.

Working only in ZF + DC it follows from the results of KUNEN [4] and MARTIN [6] that if a set of reals is the projection of a weakly homogeneous tree then its complement is Souslin. Hence given that every $\omega \times \lambda$-tree is weakly homogeneous it follows that Souslin sets are closed under complements. This coupled with the obvious fact that the Souslin sets are closed under $\exists^{\mathbb{R}}$ yields that assuming every tree is weakly homogeneous, every projective set is the projection of a weakly homogeneous tree.

THEOREM 8 (ZFC). *Assume κ is supercompact. There is a cardinal $\delta < \kappa$ such that if $G \subseteq \mathbb{P}_\delta$ is generic then $V[G] \models$ Every projective set of reals is the projection of a weakly homogeneous tree.*

3. Some applications

Let HC denote the collection of sets with countable transitive closure, i.e. HC is the set of hereditarily countable sets. We need to fix a coding of elements of HC by reals that is reasonably effective. Suppose $b \in \mathrm{HC}$ and let TC(b) denote the transitive closure of B. The set b is easily coded by coding the countable structure $\langle \mathrm{TC}(b), b, \in \rangle$. Thus natural codes for the set b are structures $\langle \omega, B, E \rangle$ (where $B \subseteq \omega$, $E \subseteq \omega \times \omega$) that are isomorphic to $\langle \mathrm{TC}(b), b, \in \rangle$. These codes of b are in essence elements of $P(\omega) \times P(\omega \times \omega)$ which via a recursive map can be viewed as reals.

This defines a partial map, $\pi: \omega^\omega \to \mathrm{HC}$, that is Δ_1 definable over the structure $\langle \mathrm{HC}, \in \rangle$. The domain of π is easily computed to be a Π^1_1 set of reals. These features, abstractly, are what we require, fix such a coding map, π.

Suppose $S \subseteq \omega_1$. Let S^* denote the set of reals coding initial segments of S, $S^* = \{x \in \omega^\omega \mid \pi(x) = S \cap \alpha \text{ for some } \alpha < \omega_1\}$.

The following theorem of KECHRIS [3] demonstrates that if one strengthens the failure of the axiom of choice in the hypothesis of Theorem 1 to the assumption that ω_1 is measurable then Souslin sets of reals have even nicer properties.

A set $S \subseteq \omega_1$ is constructible from a real if for some $x \in \omega^\omega$, $S \in L[x]$. It is a well-known theorem of Solovay that assuming AD, every subset of ω_1 is constructible from a real.

THEOREM 9 (Kechris), (ZF + DC). *Assume ω_1 is measurable. Then $S \subseteq \omega_1$ is constructible from a real if and only if S^* is a Souslin set of reals.*

Observe that if $S \subseteq \omega_1$ and $S \in L[x]$ for some real, x, for which $x^\#$ exists then S^* is $\boldsymbol{\Pi}^1_1$ and hence Souslin.

There is of course a ZFC version of Theorem 9 and it is also due to Kechris. The ideas involved in proving these theorems are very similar.

THEOREM 10 (Kechris) (ZFC). *Assume there is a measurable cardinal. Then $S \subseteq \omega_1$ is constructible from a real if and only if S^* is the projection of a weakly homogeneous tree.*

We now turn toward an application of these ideas. The problem we shall consider is that of the possible saturation of the nonstationary ideal on ω_1. For a discussion of this problem and the related problems of saturated ideals in general see [1].

Let NS denote the ideal of nonstationary subsets of ω_1. The problem is simply, within ZFC can $P(\omega_1)/\mathrm{NS}$ satisfy the ω_2 chain condition (contain no antichains of size ω_2)?

The first related consistency result was the following theorem of STEEL and VAN WESEP [13]. $\mathbb{R}$-AC indicates the axiom of choice for families indexed by the reals.

THEOREM (Steel, Van Wesep). *Assume "ZF + AD + $\mathbb{R}$-AC" is consistent. Then so is "ZFC + the nonstationary ideal on ω_1 is ω_2 saturated".*

In WOODIN [14] the hypothesis is refined to the consistency of ZF + AD. However using the ideas implicit in the proofs of Theorem 9 and Theorem 10, and the techniques of WOODIN [14] one can prove the following theorems.

THEOREM 11. *Assume "ZF + DC + ω_1 is measurable + every set of reals in $L(\mathbb{R})$ is Souslin" is consistent. Then so is "ZFC + the nonstationary ideal on ω_1 is ω_2 saturated".*

THEOREM 12 (ZFC). *Assume every set of reals in $L(\mathbb{R})$ is the projection of a weakly homogeneous tree. Then "ZFC + the nonstationary ideal on ω_1 is ω_2 saturated" is consistent.*

In both theorems the relevant set of reals is just the complete $\Sigma_1^2(L(\mathbb{R}))$ set. The reason for consistency in Theorem 12 as opposed to relative consistency in Theorem 11 is simply that assuming the axiom of choice if there is a measurable cardinal then $L(R)^{\#}$ exists.

As in Woodin [14] one actually obtains the consistency of a combinatorial condition apparently stronger than that of the saturation of the nonstationary ideal, the principle $*$ (see WOODIN [14] for details).

For more details of the proofs of the theorems stated in this paper see the forthcoming WOODIN [15].

References

[1] KANAMORI, A. and MAGIDOR, M., Survey paper.

[2] KECHRIS, A.S., 1981, *Homogeneous trees and projective scales*, in: Cabal Seminar 77–79, Lecture Notes in Mathematics 839 (Springer, Berlin), pp. 33–34.

[3] KECHRIS, A.S., private communication.

[4] KUNEN, K., 1971, *On $\boldsymbol{\delta}_3^1$*, circulated note.

[5] MARTIN, D.A., 1975, *Borel determinacy*, Annals of Math. 102, pp. 363–371.

[6] MARTIN, D.A., January, 1977, *On subsets of $\boldsymbol{\delta}_5^1$*, circulated note.

[7] MARTIN, D.A., *Borel and projective games*, monograph in preparation.

[8] MARTIN, D.A., private communication.

[9] MOSCHOVAKIS, Y.N., 1980, *Descriptive Set Theory* (North-Holland, Amsterdam).

[10] SHOENFIELD, J.R., 1961, *The problem of predicativity*, Essays on the Foundations of Mathematics (Magnes Press, Hebrew Univ. Jerusalem), pp. 132–139.

[11] SOLOVAY, R., 1970, *A model of set theory in which every set is Lebesque measurable*, Annals of Math. 92, pp. 1–56.

[12] SOLOVAY, R., 1969, *On the cardinality of Σ_2^1 sets of reals*, in: Foundation of Mathematics, Bullof et al., eds. (Springer, Berlin), pp. 38–73.

[13] STEEL, J.R. and VAN WESEP, *Two consequences of determinacy consistent with choice*, Trans. Amer. Math. Soc.

[14] WOODIN, H., 1983, *Some consistency results in ZFC using AD*, in: Cabal Seminar 79–81, Lecture Notes in Mathematics 1019 (Springer, Berlin), pp. 172–198.

[15] WOODIN, H., in preparation.

5. Philosophical Logic

Barcan Marcus et al., eds., Logic, Methodology and Philosophy of Science VII

THE SITUATION IN LOGIC — I

JON BARWISE

Dept. of Philosophy, Stanford Univ., Stanford, CA 94305, U.S.A.

This paper argues for a broader conception of what logic is all about than prevails among logicians. In particular, it claims that ordinary usage of the words *logic*, *inference*, *information*, and *meaning* defines a natural subject matter that is broader than logic as presently studied. More specifically, I argue that logic should seek to understand meaning and inference within a general theory of information, one that takes us outside the realm of sentences and relations between sentences of any language, natural or formal. I also want to suggest that the theory of situations and situation types developed with John Perry provides a tool with which one can begin to study some of the neglected aspects of logic.

1. The commonsense view of logic

1.1. Logic versus first-order logic

Logic, we tell our students, has to do with arguments, good and bad, sound and unsound, valid and invalid, verbal and written, but with arguments in some language. This motivates us to consider the relation of one sentence as a "logical consequence" of other sentences. This relation is given a semantic explication in terms of truth in models, a syntactic explication in terms of proof, or both kinds of analyses connected by a completeness theorem. Somewhere along the way, in an attempt to account for the logical structure of certain kinds of arguments, certain linguistic items are singled out as "logical constants," with the rest relegated to the nonlogical. In its most extreme form, this leads to what I have elsewhere called the *first-order thesis*, that is, the claim that logic is the study of the properties of *and*, *or*, *not*, *implies*, *every*, *some*, and *identity*[1] and that anything that cannot be defined in terms of these is outside the domain of logic.

The reasons for the widespread, often uncritical, acceptance of the first-order thesis are numerous. Partly, it grew out of interest in and hopes for Hilbert's Program in the foundations of mathematics. Partly, it was spawned by the great success in the formalization of parts of mathematics in first-order theories like Zermelo-Fraenkel set theory. And, partly, it grew out of a pervasive philosophical nominalism in the mid-20th century, led by Quine, among others. As late as 1953, QUINE wrote in his book *From a Logical Point of View:*

> The bulk of logical reasoning takes place on a level which does not presuppose abstract entities. Such reasoning proceeds mostly by quantification theory, the laws of which can be represented through schemata involving no quantification over class variables. Much of what is commonly formulated in terms of classes, relations, and even number, can easily be reformulated schematically within quantification theory plus perhaps identity theory. (p. 116)

By now we know that this is false, that there is a wealth of notions used in mathematics, science, and everyday life that take us outside the realm of first-order logic. As logicians, we do our subject a disservice by convincing others that logic is first-order logic and then convincing them that almost none of the concepts of modern mathematics or everyday life can be captured in first-order logic.

For the person in the street, the term *logic* connotes much more than can be captured in first-order logic or, indeed, in any other theory that restricts itself to the study of relations between sentences of some language. Logic has to do with valid forms of reasoning, from the most mundane uses in our day-to-day lives to the most sophisticated uses in science and mathematics. If you and I are discussing some topic, like fixing the roof, a law of genetics, or the solution to some partial differential equation, and I say, "The logic of that escapes me," what I mean is that I do not see how the conclusion you have come to follows from our shared assumptions and concepts, including the conception of the task at hand. How does it follow from the properties of this roof we are on, or the laws of genetics that we both accept, or the concepts involved in differential equations, and the problem before us? Similarly, when we say that a certain process or activity like perception, conversation, golf, or winning a marathon has its own logic, we

[1] Among past and present adherents to this view there is a difference of opinion as to whether identity counts as a logical constant.

use the word *logic* in a perfectly ordinary way. However, this use of the word is not one that can be reduced to the notion of logic that we, we logicians, have explicated.

In the commonsense view of logic, all the concepts we use to cope with and organize our world have their own logic. As logicians, we are perfectly entitled to delve into their logic. Within metamathematics, this view has led to the study of so-called *strong logics*, that is, formal languages that attempt to capture the logic of various mathematical notions that lie outside the domain of first-order logic. Within philosophy, it has led to the study of formal languages that attempt to capture the logic of knowledge, belief, time, and various modal notions. And within computer science, it has led to the study of formal languages about the logic of programs.

1.2. Logic and language

This is all well and good, but what I want to suggest is that an understanding of logic and the related notions of meaning and inference requires us to look beyond language. Inference, for example, is an activity that attempts to use facts about the world to extract additional information, information implicit in the facts. A sound inference, I will argue, is one that has the logical structure necessary to serve as a link in an informational chain but that need not use language at all. For example, I may correctly infer roughly how cold it is going to feel when I go outside by looking at the people walking past my office, and this inference supports the logic of my leaving my hat and gloves behind. Even when language is used, the connection between sound argument and sound inference is less than straightforward. A sound argument leads to a sound inference on someone's part only if the argument is also successful. On the other hand, unsound arguments often lead to sound inferences. For example, even if the logic of my paper is faulty and its argument is unsound, you will successfully and soundly infer many facts about me and my views from it. To explicate the commonsense view of logic, we must develop accounts of information and inference that do not presuppose language and so limit us to inferences based on language.

1.3. Overview of this paper

The rest of this paper is divided into three parts. In the next and longest section, I take up a question of strategy, the strategy that has rested behind traditional approaches in logic to the study of semantics, or meaning. I want to question the "divide and conquer" approach to meaning that prevails today and that has virtually defined logic's subject matter over the past 50 years. I will argue that even if what you want is a mathematical theory of linguistic meaning, this theory must be set within a more general theory of meaning, inference, and information if it is to succeed. We need a theory that can do all the usual things we expect of model theory,[2] but one that can also underwrite a theory of information flow of the kind discussed in DRETSKE's book *Knowledge and the Flow of Information* (1981).

In the third section, I turn from strategy to tactics. Having argued for the need for a more general theory, I will outline one attempt to provide this kind of theory, the approach that John Perry and I call *situation semantics*. In attempting to work out situation semantics as a theory of linguistic meaning, we found that we were forced to develop a more general theory. It now seems that this was inevitable but that this fact can be separated from the details of our own theory. Not only can they be separated, but I think they must be if comparison of competing theories is not going to get hopelessly confused. We need to distinguish between what I would call "in house" criticisms, that is, those that accept the basic strategy but find fault with our tactics, from criticisms of the basic strategy. In the final subsection, I return to examine briefly what the study of logic and inference might look like within situation semantics.[3]

2. Meaning and logic

In this section, I take one of the words that fall under the purview of logic, the word *meaning*, and argue that it must be understood within a general theory of information. I could equally well have taken *logic* or *inference* to make the point. First, however, comes a little reassurance.

---

[2] A minor terminological point. I will use "model theory" for mathematical approaches to linguistic meaning of the traditonal sort, especially as they have been developed for natural languages, that is, theories that view the subject as a question of the relation of language to the world. By contrast, I will refer to the more general theories of the kind I claim we need as "theories of meaning and information." Briefly put, the claim of Section 2 is that a successful model theory has to be set within a more general theory of meaning and information.

[3] Much of this paper was taken from a larger paper I wrote, called "Loss of Meaning," to be read at the University of Wisconsin and at Princeton University, and the discussions after these presentations were quite helpful. This is the first in a projected series of papers on logic and situation semantics. A draft of this paper was prepared for a symposium held in Salzburg. When circumstances prevented my attending the symposium, Hans Kamp graciously agreed to read the paper for me. He also provided me with very helpful comments on that draft.

2.1. The inventor's paradox

There is a phenomenon in mathematics that often strikes the uninitiated as paradoxical. It sometimes happens that you want to prove some theorem T but are unable to do so. However, it turns out that it is possible (even easy) to prove some *stronger* theorem T'. For example, on noticing that $1+3+5=9$ and $1+3+5+7+9=25$, you might conjecture that the sum of consecutive odd numbers $1+3+\cdots+(2k+1)$ is always a perfect square. If you attempt to prove this, the proof breaks down because the induction hypothesis does not contain enough information to get the required conclusion. However, if you try to prove something stronger, that the sum of the first k consecutive odd numbers is exactly k^2 (and hence a perfect square), the proof is easy. You prove it by induction, and you have just the right information needed at the inductive step to keep the proof going. This phenomenon is pedagogically so important that Polya has called it the *inventor's paradox*. To prove what you want, you may have to prove more than you want, simply to get the flow of information to work out properly.

What I want to suggest, in this first part, is that we logicians have suffered from the inventor's paradox. That is, in investigating the semantics of ordinary language, we have been trying to do too little and so have not been able to do even that. We have been concerned solely with the truth conditions of sentences, the conditions under which a sentence can be truly asserted. We have not been concerned with the more general problem of accounting for how sentences can be used to convey information and, as a result, have not been able to get even the truth conditions right.

There have been more than enough critics of model-theoretic semantics to point out our failings. Frequently, however, they give the impression that model theory is trying to do the impossible. What I want to suggest is that, just as in the case of the sum of odd numbers, we have been trying to do too little. It is not attention to truth conditions that I want to call into question but the attempt to develop a theory of truth conditions or some other model-theoretic analysis of logic, inference, and linguistic meaning isolated from the flow of information.

What is it that has prompted us to partition meaning the way we do into various kinds of meaning? I suppose it is the feeling, which I had for years, that the word *meaning* conflates many different things. For example, the meaning of life is often taken to be its purpose.[4] The meaning of a name is

---

[4] The meaning of life is often taken to be its purpose, but this is a confusion. Many things have meaning that do not have purpose. Smoke means fire but that is not its purpose. In fact,

often taken to be the individual it names. The meaning of a declarative sentence is often taken to be its sense or its truth conditions. The meanings of mental states and events are often taken to be the role they play in psychology. Even physical events and states of affairs are said to have meaning, as when we say that the smoke in the distance means fire or that the position of the hands on my watch means that it is 4 p.m.

At first glance, these do seem like vastly different sorts of things that are deemed meaningful: lives, names, sentences, mental states and events, and physical states and events. So, too, the meanings that they are full of appear to be quite different sorts of things: purposes, people, senses or truth conditions, psychological roles, and other external events. This apparent hodgepodge of meaningful items and meanings attributed to them gives the impression that the word *meaning* conflates things that must be kept distinct, and some distinctions as to different kinds of meanings have emerged.

2.2. Situation-type meaning and situation meaning

You can always sell a philosopher a distinction, so let me try to sell you one, a distinction between two ways that the word *means* is used in ordinary English. If I say, "That hair in the butter means that the cat has been on the table," I am talking about the meaning of a particular situation. However, if I say, "Cat hair's being in the butter always means that a cat has been on the table," I am not talking about the meaning of a particular situation but about the meaning of a certain type of situation. Similarly, if I say that John's saying "I'll meet your at the airport" meant that he would meet me at the airport, I am referring to the meaning of a particular utterance. By contrast, the meaning of the disembodied English sentence

I'll meet you at the airport. (1)

whatever your theory says it is — be it sense, truth conditions, or whatever — is not that John would meet me at the airport.

This same dichotomy appears across the board. If I point to my watch and say, "That means it's 4 p.m." I am using *meaning* in the first sense. If I

---

as far as one can tell, smoke has no purpose. It is the wrong kind of thing to have a purpose. And even if something has both meaning and purpose, they need not be the same. My saying that it is 4 p.m. may mean that it is 4 p.m.; however, the purpose of my saying that is not that it is 4 p.m. but, rather, to tell you the time. While it may be that life has both meaning and purpose, they are probably not the same.

say, "The big hand pointing at 12 and the little hand pointing at 4 always means that it is 4 o'clock," I am using *means* in the second sense. This distinction is one that must be borne in mind. Perry and I use the phrase *situation meaning* in the first case, as contrasted to *situation-type meaning* in the second.[5] I will rely heavily on this distinction in what follows. When it is particularly crucial, I use the subscripts "s" for situation meaning and "t" for situation-type meaning.

I am not claiming that meaning_s and meaning_t are unrelated, any more than I would say that a token of some type is unrelated to the type. It will be one task of a theory of meaning to spell out the relationship between the two. For now, I simply want to point out that there are these two ways we use the word *means*, so that there is a possibility for confusion when we talk about what something means.

2.3. Natural and conventional meaning

A more familiar distinction among different kinds of meaning is Grice's distinction between natural meaning and nonnatural, or conventional, meaning. This parallels Peirce's distinction between sign and symbol. Smoke is a sign of fire but the sentence *My house is on fire* is a symbol. The former has natural meaning; the latter, conventional meaning. It means what it does because of the conventions that make up the English language.

I think this is a sound distinction, one that should find its place in any theory of meaning. The fact that conventions can be violated gives conventional carriers of meaning an importantly different property from natural carriers of meaning. Some utterances of the sentence

My house is on fire. (2)

do not really mean that the speaker's house is on fire, not with the same reliability that smoke pouring from his windows does. The speaker can be wrong in a way that smoke cannot. Any theory of meaning has to have room for this dichotomy.

---

[5] This name, of course, slightly prejudices the case, since it points to our analysis of the distinction, but I will stick with it here. Actually, it is what we called "event meaning" and "event-type meaning" in our book *Situations and Attitudes* (BARWISE and PERRY, 1983). The phrase *situation type* picked up a certain technical meaning in the early days of situation semantics. It was the name we chose for certain abstract objects that we use to represent uniformities across states of affairs, that is, static situations. Later, we realized we needed a more powerful way to represent uniformities. In our book, we called these "event types," since "situation types" was used up. Here, I am using *situation type* for the more powerful notion, since it accords more closely with the informal notion of type of situation.

This distinction is one of species within a single genus, however, and there are infinitely subtle variations from one to the other. Consider the most straightforward of exchanges. You ask for the time. I glance at my watch and say, "It's 4 p.m." Your infer that you are late for a seminar and so dash off. What is it that makes my utterance of "It's 4 p.m." mean that it is 4 p.m.? This, I take it, is a question we must answer if our theory of linguistic meaning is to account for the significance of language.

I am asking here about the meaning$_s$ of my utterance, that is, what about that particular utterance made it mean what it did and what let you draw the inferences you did from it. Part of what is involved is the meaning$_t$ of the English sentence used — but only part. I want to argue that in order to account for what that utterance meant we must look well beyond conventional meaning and, hence, well beyond the linguistic meaning of

It's 4 p.m. (3)

Here is a schematic diagram of the flow of information illustrated in this simple example:

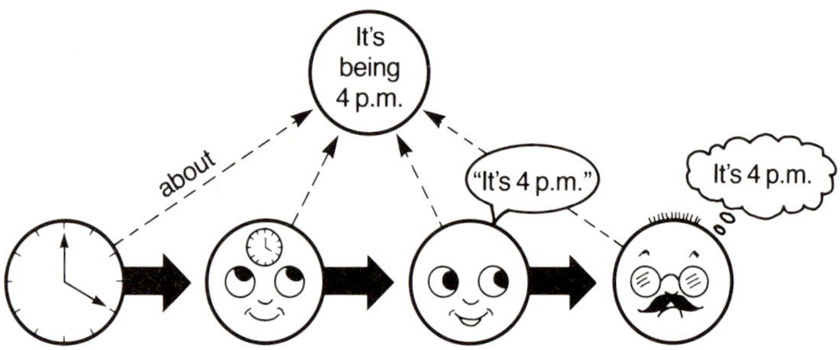

Fig. 1

When we ask about the meaning$_s$ of my utterance, that is, what made it mean that it was 4 p.m., we are focusing on the third situation in the informational chain above.

Recalling that the issue for the time being is the viability of using the distinction between natural and conventional to define a subject matter, let us imagine for a while that we are obsessed not with the logic of utterances but with the logic of telling time. Thus, instead of focusing on linguistic meaning, let us focus on chronometric meaning, the meaning of clocks and watches and what they say to us when we look at them. That is, let us think about the first situation in the informational chain.

Is chronometric meaning natural meaning or conventional meaning? Are watches part of the natural order or the conventional order? It seems clear that they are not part of either one alone but of both. For a theory of chronometric meaning, the state of the watch does not fall cleanly into either category of sign or symbol. Natural laws about quartz, for example, are exploited in the design of the watch and are responsible for its pointing to 4 p.m. at 4 p.m. and not at noon. On the other hand, the system of time measurement that the watch is measuring is conventional. Every time we go onto or off of daylight saving time or fly from one time zone to another we are changing the conventions and so have to change the clocks to keep them from telling the wrong time. Timepieces, like people, can tell the wrong time, if they do not fit the conventions within which they are placed.

The theory of chronometric meaning would also have to have more room in it for error than just change of convention. Normally, if my clock points to 4 p.m., that means it is 4 p.m. If we are near the end of April or October, or have just taken a trip, there is some room for doubt because of shifting conventions, but otherwise we do not have to worry about that. Still, things *can* go wrong. The battery in my watch may run down, or I might attempt to use it under conditions for which it was not designed, say, under water. It is only under the proper, normal conditions that the watch means what it says. In other circumstances, it loses its meaning. My grandparents' old, broken clock always points to 4 o'clock, but that fact never means that it is 4 o'clock. It no longer means that it is 4 p.m. even when it is 4 p.m.

Besides the fact that our theory of chronometric meaning will not lie squarely within a theory of natural meaning or of conventional meaning, there are a couple of secondary points here. First, the "truth conditions" on watches are that they point to *t* only at time *t*. However, a watch may point to 4 o'clock without that meaning that it really is 4 o'clock, for a variety of reasons, even if it is, in fact, 4 p.m. Second, and a consequence of the first point, we see that the meaning_s of the watch "utterance" does not depend solely on the internal state of the watch but also on how and where the watch sits in the wider world. Finally, the state of my watch on my arm does not have a single meaning but many. For example, besides meaning that it is 4 p.m., it means that someone made it and that I remembered to put it on this morning.

The points made here have ramifications for the theory of linguistic meaning in general and for what it is that makes my assertion "It's 4 p.m." mean that it *is* 4 p.m. First, just as the theory of chronometric meaning does not lie neatly and solely in either the natural or the conventional order, neither does linguistic meaning. What made my telling you what time it was mean that it was 4 p.m. was not *just* the conventions of English, even augmented with the facts that we both speak English and are in the same time zone. Also involved are things like natural laws about quartz, my knowing how to tell time, and the fact that the speed of sound is great enough that we can ignore the time it takes for you to hear my utterance. If we are to account for the meaning$_s$ of particular utterances, that is, for the information they carry, we have to set them within a general theory of meaning that accounts for the flow of information.

The secondary points have their analogues, too. First, a point about truth conditions. The truth conditions on sentence (3) are that a statement made with it is true if and only if it is 4 p.m. at the time and place of utterance. However, my statement of "It is 4 p.m." does not necessarily mean that it *is* 4 p.m., *even if it is true.* I might go around saying that all the time, just like my grandparents' old clock. Second, my mental state alone does not determine the meaning$_s$ of my utterance, any more than the watch's did. If exactly the same state of mind had been produced by a hallucination at midnight, my utterance would not have meant that it was 4 p.m. Or, for another example, no matter how firmly I believe I am Bobby Fischer, nothing can make my utterance of "I'm Bobby Fischer and a great chess player" mean that I am Bobby Fischer or even a moderately good chess player. And, finally, an utterance of "It's 4 p.m." does not mean$_s$ just one thing, that it is 4 p.m. It may well also mean that I know how to tell time, that I speak English, and so forth. You can soundly infer any of these things from my utterance, in the appropriate circumstances.

Let me repeat that I do not doubt that the natural versus the conventional meaning is a genuine distinction. I do doubt that the meaning of very much can be determined by looking at just natural meaning or just conventional meaning. Rather, it will be a complex combination of various kinds of meaning that makes almost *anything* have the situation meaning (or meanings) it does.

2.4. External and internal meaning

Another distinction is often made between external and internal (or psychological) meaning or significance: meaning "in the world" versus meaning "in the head." The distinction is not universally accepted, since there are people who doubt the very existence of one or the other form of meaning. Let us be generous for a moment, though, and see if we can find external and internal meaning in the episode above of time telling.

Presumably, if the state of my watch meant_s that it was 4 p.m. when I looked at it, then one minute earlier, when it said it was 3 : 59, that must have meant_s that is was 3 : 59, even if no one was looking at it. Wherever that meaning is, it must be external. It would seem, then, that the analogous element of external meaning must have been present at 4 p.m. It would certainly be odd if my looking at my watch could make it lose its external significance.

On the other hand, the meaning_s of my internal representation of the time, illustrated by the second situation in the informational chain above, is presumably psychological, if anything is. It follows that if exactly the same mental state had involved something else, say, a faulty watch, it would have had the same psychological significance. For example, it would have caused me to say the same thing.

Notice that even if my state were totally spurious, caused, say, by a hallucination, understanding the psychological significance of my state would require an understanding of its setting in the normal course of events, where that state is caused by my looking at my good, old, reliable watch. That is, understanding the meaning_s of a particular mental state requires an understanding of the meaning_t of that *type* of state, as it normally functions in the external life of the agent. For that matter, the only way we have of describing my mental state is with reference to the way that type of state functions in the normal course of events. Saying something like "an image of a watch pointing at 4" uses talk of watches to describe types of mental states, the type of state normally produced by a watch pointing at 4.

Just as the natural/conventional distinction will have its place in any overarching theory of meaning, so, too, this distinction between external and internal will have to find its place in such a theory. More than that, no theory of either external or internal significance can succeed in isolation. On the one hand, as we have just seen, describing and understanding the psychological significance of any particular mental state requires an understanding of the meaning of that type of state under normal circumstances. On the other hand, part of wat makes the external significance of my telling you the time what it is, is the psychological significance of my mental state. For my state is an informational intermediary between my watch and my utterance. It has to carry part of the burden of the information in an information chain that goes from my watch to you.

This point is made better with a different example. I was born in Missouri. That is a fact. Now consider the meaning_s of that little utterance of mine, my saying "I was born in Missouri." At least part of its situation meaning is that I was born in Missouri, a piece of information about an external event, my birth. However, my psychological state, the fact that I know where I was born, is crucial to my utterance's having the meaning$_s$ it did. No account of the information contained in my utterance can escape the fact that the information was stored in me and my psychological state.

2.5. Meaning and information

Part of what makes a particular state of affairs or event mean$_s$ what it does, that is, have the situation meaning it does, is how it fits into the general flow of information. This includes signs like fire and symbols like the sentence *My house is on fire.* Conversely, what information a signal or situation contains depends in part on what it means. Even if we are interested only in chronometric meaning, linguistic meaning, or psychological meaning, we are going to have to see it within the general flow of information to which DRETSKE (1981) has drawn our attention.

The strategy I propose, then, is that in pursuing the study of logic in general, and of developing model theory in particular, we pursue it as part of a general theory of information flow.

Having come to this conclusion, it is unsettling to find that Dretske, in his book, goes to some pains to disassociate information from meaning.

> Although information, as ordinarily understood, may be a semantic concept, this does not mean that we must assimilate it to the concept of *meaning*. For, on the face of it, there is no reason to think that every meaningful sign must carry information or, if it does, that the information it carries must be identical to its meaning. (p. 42)

Dretske goes on to give several examples of the difference between meaning and information and concludes:

> The information embodied in a signal (linguistic or otherwise) is only incidentally related to the meaning (if any) of that signal. (p. 44)

If Dretske is right, we are ill advised to look toward information in developing a mathematical theory of meaning. However, I think that on this point Dretske is mistaken. In terms of the overall theme of his book, it amounts to a quibble, but in terms of the theme of this paper, namely, the strategy for developing a mathematical theory of meaning, it becomes a major point. Dretske admits, in a footnote, that he uses "meaning" to denote conventional meaning, which, of course, we have already rejected. However, even within the realm of conventional meaning, he rejects the identification of meaning and information.

Dretske's rejection of a direct correlation between meaning and information stems from a genuine problem. As Dretske correctly (I believe) insists, a situation or signal cannot contain the information that it is 4 p.m. unless it is indeed 4 p.m.; by contrast, a speaker may mean what he says but be wrong. However, I think that the correct response to this problem can be found by carefully distinguishing between meaning_s and meaning_t.

Since this point is of some importance to the whole endeavor, let us look at two of Dretske's examples that involve conventional meaning. Dretske's first example contrasts the meaning of a conventional bid at bridge and the information it conveyed. His partner bid 5 clubs using the Blackwood convention, which meant, in that context, that she had no aces.[6] Dretske argues that he is required by the rules of duplicate bridge to tell his opponents the meaning of the bid but not to tell them what information is conveyed to him. Another way of putting this is that he is required to tell them the meaning_t of the bid, that is, what it means in general, but not the meaning_s of the bid, that is, what the bid meant_s to him. He is required to explain the general convention, but he does not have to explain that, in these particular circumstances, the bid meant that his partner had no aces. What the example shows is that the information conveyed is not the meaning_t of the convention, but, it seems to me, the information conveyed *is* the situation meaning_s of the bid.

In another example, Dretske argues that there is no tight connection between meaning and information in language.

> If I do *not* have a toothache, my saying that I have a toothache fails to convey the information that I have a toothache. The words I utter, "I have a toothache," are meaningful. They mean that I have a toothache. Yet ... this is not the information they convey. (pp. 43–44)

Dretske slips too easily from the meaning_t of the sentence to the meaning_s of his utterance. Under the circumstances he describes, we normally would not say that his utterance meant that he had a toothache any more than we would that the state of my grandparents' clock at midnight means that it is 4 p.m.

Dretske's sentence is misleading, since in discussion it tends to raise a spurious issue, namely, whether there is a matter of fact about aches and pains above and beyond the convictions of the individual. A sentence like

[6] The convention works as follows. After a suit has been determined, one partner bids "4 No-trump." The other partner bids 5 clubs if he has 0 or 4 aces, 5 diamonds if he has one ace, 5 hearts if he has 2 aces, and 5 spades if he has 3 aces. Since Dretske had some aces in his hand, he could tell that the bid of 5 clubs meant that his partner had no aces.

It is 4 p.m. avoids this issue. No amount of sincere conviction alone can conceivably make it 4 p.m. at 4:30, much as we would all like it to be otherwise when we are late.

Let us suppose that, as a result of looking at his faulty watch, Ed becomes convinced that it is 4 p.m. and so says, "It is 4 p.m." at 4:30. While we can truly report that

Ed means what he says.

we can also truly report that

Ed's statement does not mean that it is 4 p.m.

The reason for this nearly, but not quite, paradoxical state of affairs, it seems to me, is that different uses are being made of the verb *means*. In the first case, the use is clearly related to means_t: Ed's mental state is a *type* of state that usually has the external significance of its being 4 p.m. — normally, that is, but not in this instance. That is why Ed used a sentence that means_t what it does. On the other hand, the second use of *means* is clearly $\textit{means}_s$. Statements can mean_s this or that, but they cannot mean_t. Sentences do that.

What all of Dretske's examples show is that the information in a particular situation cannot be identified with the meaning_t of some meaningful type of signal or situation. The examples do not show that meaning_t is not intimately connected with the flow of information. So they do not preclude the existence of a theory of meaning that, among other things, underwrites the flow of information. It is just that such a theory will have more to do with the ordinary use of the word *meaning* than with the use that identifies meaning with linguistic or conventional meaning.

It might seem from all this that I want to identify situation meaning with information. It might also seem that we have lost sight of logic and inference. To explain why these impressions are not right, I turn to situation semantics.

3. Situation semantics

So much for general strategy. I have discussed elsewhere the assumptions built into standard approaches to model theory that make it inappropriate for the kind of theory we need.[7] I will restrict myself here to a discussion of the situation semantics approach.

[7] See my commentary on Dretske's book in BARWISE (1983, p. 65).

Above and beyond the general strategy, the driving force behind situation semantics is a commitment to a form of realism, to be specific, to the claim that meaning does not reside in the head or in some mysterious realm but in the interaction of real, living things and their actual environment. This is not an original idea. What is new is the attempt to use it as a basis for a formal theory of meaning and information.

3.1. Logic and constraints on types of situations

As Perry and I develop this idea, it becomes the claim that meaning, be it natural or conventional, linguistic or cultural, resides in systematic relations of a special sort between different types of situations — relations to which agents are attuned. Systematic constraints between types of situations are what allows one situation to contain information about another situation. An agent's attunement to such constraints is what allows the agent to infer soundly from the one's being the case to the other's being the case. Said in different words, meaning_t is what allows an event of a particular type to have meaning_s. Attunement to meaning_t is what allows an agent with information about the first to soundly infer what it means_s.

Let me give a few examples.

EXAMPLE 1. There is a systematic relation between types of situations where a watch points at 4 and those states of affairs where it is 4 p.m. It is not a simple relation, but it is systematic, and it is what allows my watch's pointing at 4 to mean that it is 4 p.m. Anyone who is attuned to the relation can use the watch's state to infer the time. Under the appropriate circumstances, this inference will be sound.

EXAMPLE 2. That I have an image of a watch pointing at 4 is a certain state of affairs, or situation. The image is a complex situation, probably physiological. There is a systematic relation between types of images, those that are of watches pointing at 4, and other situations, actual watches pointing at 4. This relation is a consequence of the laws embodied in the logic of perception and accounts for why a particular image can mean that my watch is pointing at 4.

EXAMPLE 3. There is also a systematic relation between utterances of a certain type, those in which someone uses the sentence

It's 4 p.m. (3)

and the state of affairs of its being 4 p.m. That is why you can infer that it is 4 p.m. from my telling you. This relation is just what we try to capture in discussing the truth conditions of sentence (3).

There are many other constraints on the use of the sentence (3). For example, there is a conventional constraint that one should believe what one says. There is a certain type S of belief that is associated with sincere utterances of (3), and having a belief of type S is, in turn, systematically (but not unconditionally) related to its being 4 p.m. It is not part of the truth conditions of the sentence, but it is one of the constraints English puts on the use of the sentence, that one have a belief of type S when one asserts (3). So, from a psychological point of view, this relation between utterances of the sentence and certain mental states is also part of the meaning of the sentence.

In all these cases, we see that the reason a particular state of affairs s has a particular meaning_s is that it is of some type S, and that type of situation involves there being a situation s' of a second type S'. There are many ways one type S can involve a second type S'. One can cause the other or be characteristically caused by the other, for example. On the other hand, it can be a conventional relation, as when s is an utterance describing s'. What is common, though, is that actual situations of type S involve there being actual situations of type S'. We describe this state of affairs, that is, this relation between types of situations, by writing

$$S \Rightarrow S'.$$

So, what we are claiming is that a situation s has the situation meaning that there is a situation of type S' if there is an actual constraint $S \Rightarrow S'$ such that s is of type S.

When we search for the "logic" of some activity, what we are after is the collection of constraints $S \Rightarrow S'$ that govern this activity. For example, the logic of perception consists of the set of constraints that govern perception.

Among the various states of affairs and events in the world are mental (or brain) states of affairs and events, and these types of situations are also systematically linked to other situations, external and internal. The type of image of a clock pointing at 4 normally involves there being a clock pointing at 4. Let us use $\#S$, $\#S'$, etc., to vary over situation types of mental situations. Fix a given agent a. In order that a be able to discriminate situations of type S, there must be a type $\#S$ of mental state that, under normal conditions, means that there is a situation of type S. That is, if a is in a mental state of type $\#S$, then, under normal circumstances, there is a situation of type S. In symbols, what is required is that $\#S \Rightarrow S$. Under other conditions, usually more stringent, one also has the converse, $S \Rightarrow \#S$.

We can now say roughly what it is for an agent a to be attuned to a constraint C, say, $S \Rightarrow S'$. What we need is for the agent to be able to discriminate these types of situations and for the types of mental situations involved in the discrimination to be suitably linked:

$$\begin{array}{ccc} S & \Rightarrow & S' \\ \Downarrow & & \Downarrow \\ \#S & \Rightarrow & \#S' \end{array} \tag{4}$$

For example, a's being in a visual state of type $\#S$ involves there being an actual scene s of type S, with such a scene involving there being a situation s' of type S'. Going around the other way, a's being in a visual state of type $\#S$ causes a to go into a belief state of type $\#S'$. And that is systematically linked to states of type S'. That the agent a is attuned to the constraint C amounts to the constraint $\#C$: $\#S \Rightarrow \#S'$ being actual. The second constraint is what accounts for the agent's being able to make the inferences that keep his mental state "in synch" with reality.

We can now see, in a very general way, what the study of commonsense logic should give us. External states, processes, and events have their logic; mental states, processes, and events have their own logic. The latter processes are, in general, "inferences" about the former. Sound inferences are those that meet the conditions necessary to ensure that the resulting mental states contain information concerning the external situations they are about. The problem for our subject, then, is to get clear about these conditions necessary to preserve information. It is not an easy matter.

3.2. The multiplicity of things a situation can mean

If situation meaning is a consequence of situation-type meaning, why (some have asked) make all the fuss about the distinction between the situation meaning and situation-type meaning in the previous section? First, of course, the distinction was needed simply to sort out how we use the word *means*, to motivate the kind of theory I am proposing. But, more important, it is crucial if we are to understand how it is that one state of affairs or event can mean so many things. A given situation, be it ever so simple, is going to be of more than one type. And each of these types can play a role in one or more constraints. Suppose, for example, that s is of both types $S1$ and $S2$. And let us suppose that there are constraints $C1$ and

C1′ involving situations of type *S1* and *C2* and *C2′* involving situations of type *S2*. A given agent could extract information from *s* if it were attuned to any nonempty subset of these four constraints, so we have a total of 15 possible sets of constraints. Any such set will generate a situation meaning from *s*. When we speak of *the* meaning$_s$ of any given situation, we are speaking relative to some set of constraints.

Nowhere is this more important to remember than in the study of linguistic meaning. A given utterance of "It's 4 p.m." can mean all kinds of things, besides just that it is 4 p., since the utterance is of many different types. In our book, Perry and I concentrate on what the utterance means about the described situation, because we are primarily interested in the straightforward use of language to convey information. Thus, we identify the linguistic meaning of declarative sentences with the specific constraints that holds between utterances of the sentence and situations described by such utterances. By picking out one constraint, the one that holds between the types of utterances of sentence (3) and the types of situations it describes, and calling that the linguistic meaning of (3), we are betraying the fact that our primary concern is for the subject matter of the sentence, for finding out what it is talking about and what it is saying about that subject matter. This does not preclude our accounting for the fact that such utterances also mean other things, that the sentence has psychological meaning or even meaning as to the state of our vocal chords, in virtue of utterances being of many different types at once, types that can play a role in other constraints to which people can be attuned. My utterance of (3) meant that it was 4 p.m. relative to certain constraints. But to an onlooker with different concerns and abilities, someone who is an expert at placing people by their accents, it might mean that I am from the American Midwest. So, too, in the case of the bid of 5 clubs in bridge. To Dretske, holding some aces, it meant that his partner had no aces. To his opponents, it may have meant that his partner had 0 or 4 aces.

This only scratches the surface of the complexity involved. Consider an utterance of

My watch says that it's 4 p.m. but it is slow. (5)

Such an utterance would not, under normal conditions, mean that it is 4 p.m. If anything, it would mean that it is after 4 p.m. Yet it contains a subutterance of sentence (3).

The conventional constraints of English relate the meaning$_t$ of (3) with that of (5). More generally, the rules of English relate the meaning$_t$ of complex expressions to those of its constituent expressions. That is what allows us to generate expressions that have never before been uttered and use them to convey information. It also allows us, however, to generate utterances that do not mean what they say. Consider:

$$\textit{The earth is flat.} \tag{6}$$

The sentence has a meaning$_t$ as projected by the rules of English, but no utterance of it can ever mean$_s$ that the world is flat. It can, of course, relative to certain other constraints, mean$_s$ that the speaker believes that the world is flat.

So, to repeat, in our theory, meaning is a product of constraints that hold between types of situations, constraints to which an agent is attuned. To work out a theory based on this picture requires us to develop (a) a theory of situations, (b) a theory of situation types, and (c) a theory of constraints that hold between types of situations. The logic of any particular natural activity will be the set of constraints that govern that activity. To understand inference, which has its own logic, we need to recognize that there are, in general, two parallel sets of constraints, one on some activity A and the other on cognitive activity *about* A, and we need to understand the relations that enable cognitive activity to adequately "track" the activity it is about.

Perry and I have tried to initiate this development in our book, but we have said next to nothing about inference or logic. We start with real situations. For life to be possible, there must be similarities or uniformities across various distinct situations. Among these are individuals, properties, relations, and spatiotemporal locations. We then represent situations in terms of individuals having properties and standing in relations at various spatiotemporal locations. We next develop a theory of situation types (actually, we called them event types in the book), so that we can say what it means for a situation to be of a given type. We use situation types and schemata (sets of situation types) to spell out a theory of constraints as relations between types of situations (and schemata). The relation is that of one type S of situation involving another type (or schema) S' of situation. Within this theory of constraints we can make the distinction between various kinds of constraints, including that between natural and conventional constraints, as well as account for both external and psychological significance.

We can now spell out the difference between situation meaning and information. Both have the same general form: A situation s contains the information that it is 4 p.m., say, if there is an actual constraint $S \Rightarrow S'$ so that s is of type S and every situation of type S' is one in which it is 4 p.m.

What else is required for s to mean that it is 4 p.m.? It seems that we do not talk about meaning unless there are agents that are attuned to the constraint $S \Rightarrow S$. The situations we perceive presumably contain much more information than they do meaning, because no living creatures are attuned to the appropriate constraints. In this sense, science is a search for meaning.

3.3. *Inference*

Central to the commonsense notion of logic are the interlocking concepts of meaning, inference, and information. Resting behind these is the idea of a law, pattern, or constraint, as we have come to call them. Inference is an activity that attempts to use facts about the world to extract additional information, information implicit in the facts. As we saw above, a sound inference is one that has the logical structure necessary to serve as a link in an informational chain.

This definition of sound inference, placing it within the context of the flow of information, does not exclude the traditional one, in terms of truth-preserving rules in a language; rather, it expands on it. One sentence's being a logical consequence of others is but one way that we extract information out of our environment by making sound inferences.

Looked at within the theory of information flow, however, an important shortcoming of the tradition within the semantics of natural language emerges; it has no notion of subject matter, of what in particular an assertion or argument is *about*. The emphasis on situations in situation semantics can be seen as an attempt to work out a notion of the particular subject matter of a statement. Statements are about situations; informative statements contain information about the situations they are about.

Statements, like arguments and inferences, are particular kinds of situations themselves, events in which someone says something. The meaning of a sentence is a conventional constraint between those types of situations where the sentence could be informatively asserted and those it would thereby describe. Traditional approaches to inference through the notion of logical consequence examine one very narrow sort of conventional constraint, those involving the meanings of the so-called logical constants. A genuine explication of the commonsense notion of logic must examine these constraints in the context of all the other constraints that enable cognitive agents to pick up information from their environment and to serve as links in informational chains, transmitting information from one to another. Turning this view into a genuine mathematical theory is an exciting but enormous challenge. In a sequel to this paper, I hope to use the technical machinery sketched in Perry's and my book to develop the view of logic set out here in more formal detail.

References

BARWISE, J., 1983, *Information and semantics*, Behavioral and Brain Sciences 6, p. 65.

BARWISE, J., and PERRY, J., 1983, *Situations and Attitudes* (Bradford Books, MIT Press, Boston).

DRETSKE, F., 1981, *Knowledge and the Flow of Information* (Bradford Books, MIT Press, Boston).

QUINE, W.V.O., 1953, *From a Logic Point of View* (Harvard Press, Cambridge, MA).

Barcan Marcus et al., eds., Logic, Methodology and Philosophy of Science VII

A LINGUISTIC TURN: NEW DIRECTIONS IN LOGIC

JOHAN VAN BENTHEM

Centrale Interfaculteit, Rijksuniversiteit Groningen, The Netherlands

1. Formal semantics

Modern logic has derived its main inspiration from the science of mathematics, with important and well-known results. The time has come now to reconsider a more traditional source of inspiration, emanating from natural language. Some new directions in logical research on the common border with modern linguistics will be presented here.

Over the past decade, there has been a vigorous development of logical semantics of natural language — in the spirit, if not always according to the letter of Richard Montague's pioneering approach (cf. MONTAGUE 1974). Thus, theories arise aiming at a combination of a linguistically faithful grammar with a logically precise model-theoretic interpretation. As in all healthy disciplines, two main types of question are generated in this way. One is more empirical, concerning the 'how' of particular linguistic phenomena, another is more theoretical, explaining the 'why' of the matter. Especially in the latter direction, there lies a lot of logic.

General developments in the research program of Montague semantics show various phases. Originally, there was the concern with a general and systematic 'fit' between grammatical constructions of natural language and their denotational counterparts in semantic models. That this was at all possible is a major insight from this early phase. Nevertheless, this very global concern as such has not inspired much original logical research. Two exceptions are JANSSEN 1983, containing a universal algebraic investigation of the Montague framework, and VAN BENTHEM 1984c, studying a 'semantic hierarchy' arising by imposing various constraints on the components of the Montague format. Thus, one might arrive at a discipline of mathematical semantics, comparable in aims and generality to the existing field of mathematical linguistics.

In a second phase, however, questions of linguistic 'fine-structure' have come to the fore. Many of Montague's original proposals have been modified, and extended. Thus, interesting formal analyses have appeared describing the logic of various natural language constructions beyond the usual province of logic. Montague's own analysis of tense and modality is a case in point, though still derived from traditional philosophical logic. Less traditional examples of clear logical interest may be found in the logic of tense and aspect (cf. KAMP 1979), the study of conditionals (cf. KRATZER 1981), or the analysis of comparatives provided in KLEIN 1982, VAN BENTHEM 1982.

But also, in between global questions of categorial fit and detailed description of single lexical items, there lies an intermediate level of logical interest. Given a certain category of expression in natural language and its corresponding semantic type, there is a multitude of possible denotations, only a fragment of which is realized by actual expressions. Thus, there arises a systematic study of plausible constraints, bearing some resemblance to Montague's 'meaning postulates', in order to focus upon the essential denotations. For the case of noun phrases and determiner expressions, this type of study was initiated in BARWISE and COOPER 1981, KEENAN and STAVI 1982, which have sparked off many subsequent publications. It is this theme which will be elaborated in this paper. We shall give a survey of the theory at its present stage, while adding several new results to the existing literature.

But already a fourth phase of development may be discerned. The Montagovian mode of semantic description is concerned with already established interpretative ties between linguistic items and their denotations, rather than with the dynamics of an utterance in context acquiring its meaning. One richer perspective is found in the discourse representation theory of KAMP 1981, which adds a level of mental constructs 'mediating' between language and model. Evidently, there are fundamental logical questions at issue here, since truth now becomes a more complex relation between sentences, models and discourse representations. Moreover, a study of inference would now seem to presuppose processing of such representations.

Another notable development belonging to this fourth, more dynamical phase is the situation semantics of BARWISE and PERRY 1983. Within the present scheme, its contribution may be described as a systematic charting of all logical aspects implicit in actual situations of utterance. Again, a significant widening of logical research is being proposed here, not only as regards varieties of inference, but even as to the very mechanism of concept formation. At the moment, the logical research lines inspired by this fourth phase of semantic theory still represent promises rather than achievements. For this reason, the present exposition will be confined to the third phase, where at least a body of results can be shown to justify further expectations. Eventually, however, the same logical themes can be pursued in a fourth phase setting too.

Finally, despite the above linguistic motivation, this is a paper on logic, not primarily on linguistics. We shall be concerned with general patterns, rather than with all concrete empirical details of actual usage. Beyond a certain borderline, the latter will hamper, rather than stimulate logical research. But our linguistic turn does show that this border lies much closer to the realities of natural language than modern logical orthodoxy has suggested for too long a time.

2. Generalized quantifiers

The basic pattern of natural language sentences is encapsuled in the following two grammatical rules:

$$\text{sentence} \Rightarrow \text{noun phrase} + \text{verb phrase},$$

$$\text{noun phrase} \Rightarrow \text{determiner} + \text{noun}.$$

It is the determiner expressions which provide the logical 'glue', relating the subject term with the predicate term; as in the basic pattern of traditional logic:

$$(D\,X)\;Y.$$

Examples are the usual quantifiers *all*, *some*, *three*, but also, e.g., *most*, *many*, *enough*, *all but two*, and even expressions such as *too few*, *some but not all*, *Mary's*. Here is where much of the logical action occurs, and accordingly, we shall study the linguistic category of determiners.

One very convenient approach to these matters employs the notion of a 'generalized quantifier', implicit in Montague's work, whose linguistic significance was brought out explicitly in BARWISE and COOPER 1981. (In mathematical logic, the idea dates back to MOSTOWSKI 1957.) Disregarding intensional phenomena, any noun phrase may be regarded as denoting a set of sets of individuals, viz. the denotations of those verb phrases to which it applies. Thus, e.g., for any given universe E,

$$[\![\mathit{all}\ X]\!] = \{B \subseteq E \mid [\![X]\!] \subseteq B\},$$

$$[\![\mathit{some}\ X]\!] = \{B \subseteq E \mid [\![X]\!] \cap B \neq \emptyset\},$$

$$[\![\mathit{most}\ X]\!] = \{B \subseteq E \mid |[\![X]\!] \cap B| > |[\![X]\!] - B|\}.$$

These denotations exhibit various mathematical structures which can be used in classifying linguistic items and formulating hypotheses about their behaviour.

For our purposes, it is profitable to 'flatten' the above analysis, viewing determiners as denoting relations between predicate extensions:

$$[\![all]\!]\, AB \quad \text{iff} \quad A \subseteq B,$$

$$[\![some]\!]\, AB \quad \text{iff} \quad A \cap B \neq \emptyset,$$

$$[\![most]\!]\, AB \quad \text{iff} \quad |A \cap B| > |A - B|.$$

For a convenient visualization of this view, Venn diagrams may be used, as in Fig. 1.

In certain cases, the determiner denotation may depend on the universe E; as with (*relatively*) *many*, meaning that the proportion of B in A exceeds that of B in the universe E. Hence, in general, a *generalized quantifier* Q now becomes a functor assigning, to each universe E, some binary relation Q_E between subsets of E.

Evidently, there are many more possible generalized quantifiers than are realized in actual determiner expressions of natural language. Thus, one major query becomes the search for plausible constraints. Some interesting results in this direction have been obtained in KEENAN and STAVI 1982, WESTERSTÅHL 1982, 1984 and VAN BENTHEM 1983a, 1984b. We shall mention a few of them in due course, while adding various new notions and theorems.

One conspicuous constraint must be mentioned at the outset, as it makes the whole topic rather different from earlier logical investigations of generalized quantifiers, as found in abstract model theory. Attention will be restricted to *finite universes E*. These are the proper sphere for a linguistically oriented semantics, where our intuitions feel at home — with the extrapolation to the infinite realm occurring only afterwards, if at all.

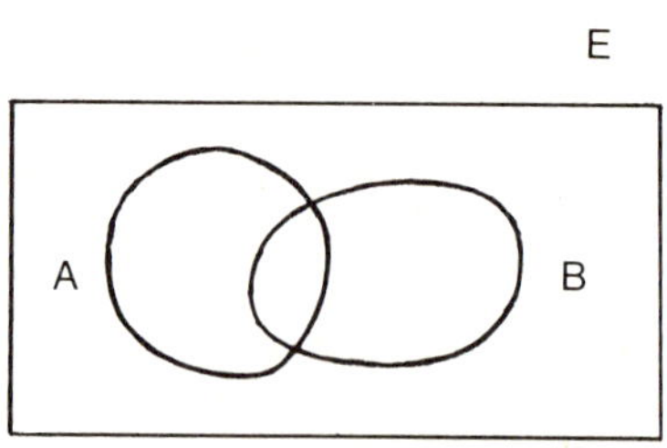

Fig. 1.

The only thing lost in this way seem to be some esoteric set-theoretic questions, while the gains are considerable.

After these preliminaries, we may now turn to our first concrete theme of logical investigation.

3. Logical quantifiers

Usually, the existence of a limited set of logical constants is taken as an ultimate fact. But, why are there just these? Such metaphysical questions seem to be beyond the realm of logic itself. As it turns out, however, exact answers may be obtained here by formulating plausible intuitive constraints on logicality of generalized quantifiers.

General constraints

A first general condition ensures the priority of the first argument in a generalized quantifier relation, setting the scene as it were:

CONS For all E, $A, B \subseteq E$,
$Q_E AB$ iff $Q_E A(B \cap A)$.

This requirement is called *Conservativity*; and it has been defended for determiners in general in KEENAN and STAVI 1982. Further motivation for this and the following conditions may be found also in VAN BENTHEM 1984b.

More specific for logical quantifiers is their 'topic-neutrality': no individual plays a distinguished role. Equivalently, it is only the numbers of objects in the four slots of the earlier Venn diagram which determine the truth of the statement that $Q_E AB$:

QUANT For all E, $A, B \subseteq E$, and all bijections π defined on E,
$Q_E AB$ iff $Q_{\pi[E]} \pi[A] \pi[B]$.

In particular, this condition of *Quantity* requires that Q be invariant under permutations of the universe E. As soon as distinguished individuals do play a role, this invariance may disappear, as in the case of the determiner expression *Mary's*. (Cf. VAN BENTHEM 1983a.) And indeed, eventually, our study can be generalized so as to do without Quantity (cf. Section 6).

Moreover, logical quantifiers are 'context-neutral' (unlike the earlier mentioned sense of *many*):

EXT For all E, $A, B \subseteq E \subseteq E'$,
$Q_E AB$ iff $Q_{E'} AB$.

This postulate of *Extension* will enable us to drop the subscript 'E' where convenient.

With these three conditions, logical quantifiers may be represented as subsets of the *tree of numbers* (a, b) $(a = |A - B|, b = |A \cap B|)$, depicted in Fig. 2. The quantifier Q is given as the set of couples at which it holds. Further conditions on quantifiers will now translate into geometrical constraints on such sets. (Notice that the adequacy of this tree representation, conversely, implies the earlier three postulates.)

Finally, if the relation Q_E is really to depend on its second argument; alternatively, if the relevant logical constant is to do some work for us, it ought to exhibit some *Variety* of behaviour:

VAR For all $\emptyset \neq A \subseteq E$, there exist $B, B' \subseteq E$ such that Q_EAB, not Q_EAB'.

This requirement, though not essential, will facilitate the exposition considerably. Its geometrical effect in the tree is this: every horizontal row must contain both + and − cases of the quantifier. (Notice that this rules out quantifiers such as *at least two*, *at most three*. But then, Variety may be dropped later on — be it at a combinatorial price.)

Against the background of these general constraints, various classes of actual logical quantifiers may now be analyzed in terms of special properties.

Special conditions: monotonicity

One of the most common notions in this area is stability of the relation under increases or decreases of its arguments. This phenomenon of *Monotonicity* comes in various kinds, indicated by the following notation:

$$Q \text{ is MON}\uparrow \quad \text{if} \quad Q_EAB,\ B \subseteq B' \quad \text{implies} \quad Q_EAB',$$

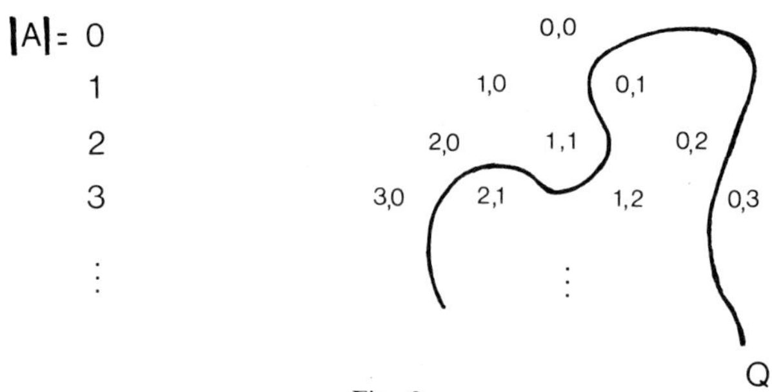

Fig. 2.

Q is MON$\downarrow$ if $Q_E AB$, $B' \subseteq B$ implies $Q_E AB'$,

Q is $\uparrow$MON if $Q_E AB$, $A \subseteq A'$ implies $Q_E A'B$,

Q is $\downarrow$MON if $Q_E AB$, $A' \subseteq A$ implies $Q_E A'B$.

For instance, *all* is $\downarrow$MON$\uparrow$, *some* is $\uparrow$MON$\uparrow$, while *most* is merely MON$\uparrow$. Various uses of these notions are found in BARWISE and COOPER 1981, who call the right-hand form 'monotonicity' and the left-hand form 'persistence'.

The latter property turns out to be the defining characteristic of the four quantifiers in the classical Square of Opposition: *all, some, no, not all.*

3.1. THEOREM. *The Square of Opposition consists of precisely the persistent logical quantifiers.*

PROOF. That all these quantifiers are persistent follows by inspection of their monotonicity types. (In addition to the above observations, *no* has $\downarrow$MON$\downarrow$ and *not all* $\uparrow$MON$\downarrow$.)

Conversely, consider any persistent logical quantifier in the tree. There are only four possible top triangles in the tree (by VAR, the second row must already have $+-$ or $-+$). Each of these determines one quantifier in the tree, through the following observation on the geometric effect of persistence:

— $\uparrow$MON means that the whole downward subtree generated by a point in the quantifier belongs to that quantifier,

— $\downarrow$MON expresses the upward mirror image property of $\uparrow$MON. Now, e.g., a top triangle $\overset{+}{+\,-}$ already violates $\uparrow$MON; whence the corresponding quantifier must be $\downarrow$MON. This again implies that $+$ can only occur on the left edge of the tree (otherwise, the indicated $-$ would have to be $+$); where indeed it must occur by VAR. Thus, this case determines the quantifier *no.* The remaining three cases are similar. □

Even without Quantity, this characterization obtains, but now involving both sides of monotonicity.

3.2. THEOREM. *Without Quantity, the Square of Opposition consists of precisely the doubly monotone quantifiers.*

PROOF. Here is a sample argument. Suppose that Q is doubly monotone, of type $\downarrow$MON$\downarrow$. It will now be shown that Q must be the quantifier *no.*

(1) Suppose that $A \cap B = \emptyset$. Choose $\emptyset \neq A' \supseteq A$. For some X, $QA'X$ (VAR), and hence $QA\emptyset$ ($\downarrow$MON$\downarrow$). It follows that QAB (CONS).

(2) Suppose that QAB. Then $Q(A \cap B)B$ ($\downarrow$MON), and so $Q(A \cap B)(A \cap B)$ (CONS). By MON$\downarrow$, $Q(A \cap B)X$ for all X, and hence $A \cap B = \emptyset$ (VAR). □

Without right-monotonicity, however, there exist non-quantitative persistent quantifiers outside of the Square of Opposition.

EXAMPLE. Fix any object a. Set QAB if

$$\begin{cases} A \cap B = \{a\}, & a \in A, \\ A \cap B = \emptyset, & a \notin A. \end{cases}$$

This quantifier satisfies all general postulates, except Quantity. Moreover, it is downward persistent (though not monotone either way).

Special conditions: continuity

Monotonicity rules out such ordinary quantifiers as *one* or *all but one*, whose tree patterns fail to obey the relevant geometric constraints. But, these are still definable as Boolean *combinations* of monotone ones. For instance, *one* is equivalent to *some and at most one* (where the latter quantifier has monotonicity type $\downarrow$MON$\downarrow$).

Another, less restrictive condition with a similar flavour is *Continuity*. Logical constants should be 'continuous' in the sense that, given any situation Q_EAB, increasing or decreasing the relevant sets can change the truth value only once. In more formal terms,

CONT (1) If $B_1 \subseteq B \subseteq B_2$ and Q_EAB_1, Q_EAB_2, then Q_EAB; and likewise for $\neg Q_E$,
(2, 3) Similar principles for varying the sets $A - B$ (with fixed $A \cap B$) and $A \cap B$ (with fixed $A - B$).

In terms of the tree of numbers, looking in the three main directions $\leftrightarrow$, $\swarrow\!\!\nearrow$, $\nwarrow\!\!\searrow$ reveals at most one change of truth value. For instance, in any horizontal row, this amounts to a choice between MON$\uparrow$ or MON$\downarrow$.

Again, there is a simple geometric meaning to this condition. At every row, a continuous quantifier consists of a + segment and a − segment. Across the rows, corresponding segments lie on the same side. Moreover, the 'transition point' in the + segment can only shift one position towards the left or right at a time, when moving down to the next row. Thus, the continuous quantifiers are given by the patterns of Fig. 3. (The exact combinatorial argument is omitted here.) It follows that there are already $2^{\aleph_0}$ of them, in contrast to Theorem 3.1.

There exists a natural connection between this notion and a speculation in BARWISE and COOPER 1981 concerning the complexity of establishing the quantifier relation.

EXAMPLE. On a universe with n elements, it takes at least one inspection to falsify a universally quantified statement, and at least n to verify it. But, e.g., *exactly one* is of a higher count complexity, requiring at least two inspections for falsification and n for verification ($n \geqslant 2$).

The idea would be that natural language has a preference for basic lexical determiners of 'minimal complexity'. To make this precise, VAN BENTHEM 1984c defines the logical quantifiers of *minimal count complexity* to be those Q for which, on each universe with n elements, there exists a minimal refutation pair (i, j) $(i + j \leqslant n)$ and a confirmation pair (k, l) $(k + l \leqslant n)$ such that every pair (r, s) with $r + s = n$ is determined by them:

$$(r, s) \in Q \quad \text{if } k \leqslant r \text{ and } l \leqslant s,$$

$$(r, s) \notin Q \quad \text{if } i \leqslant r \text{ and } j \leqslant s.$$

It is easily seen that, in this case, $i + j + k + l = n + 1$. I.e., *all* (and indeed all quantifiers in the Square of Opposition) are of minimal count complexity; but so are *most*, *least* and many others.

In the tree of numbers, minimal count complexity means that each row consists of two adjacent + and − parts forming the base of complete + and − triangles. By a geometrical argument, we now have a surprising connection.

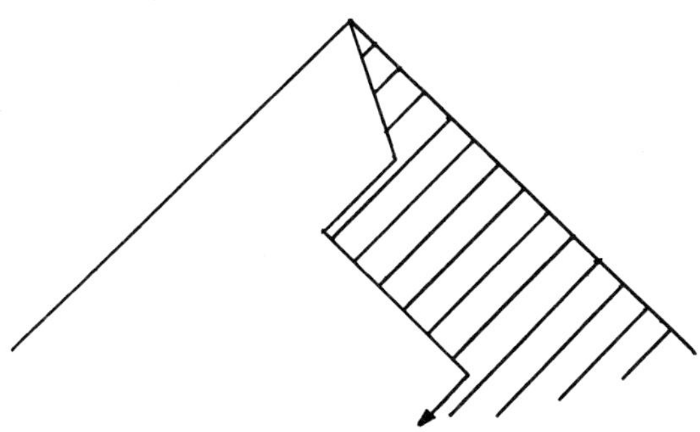

Fig. 3.

3.3. THEOREM. *The continuous quantifiers are precisely those of minimal count complexity.*

PROOF. It suffices to note that the two mentioned geometrical descriptions are equivalent. □

Special conditions: uniformity

The general framework of our postulates is also an excellent laboratory for testing out more volatile intuitions. For instance, there is a strong intuitive conviction that logical constants should behave 'uniformly' across various situations. In its strongest form, this becomes

UNIF Equal truth values in the tree generate equal (downward) tree patterns.

The effect of this *Uniformity* condition is recorded in the next result.

3.4. THEOREM. *The uniform logical quantifiers are precisely the following: no, an even number of, all, all but an even number of, some, an odd number of, not all, all but an odd number of.*

PROOF. By Variety, there are only four possible top triangles. Each of these splits up into two possibilities for the third row, after which the patterns have become fixed. □

An obvious weakening of this condition would allow a *finite variety* of tree patterns for each truth value: UNIF*. This lets in quite a few additional quantifiers, many of them of a well-known kind.

3.5. THEOREM. *Assuming Continuity, the* UNIF* *logical quantifiers are precisely those which are first-order definable in a monadic first-order language (with vocabulary* X, Y, $=$*).*

PROOF. By Fraïssé's characterization of first-order definability in terms of insensitivity for back-and-forth connections of suitably high complexity, the hallmark of first-order definability for Q is this:

There exists a number n such that, whenever $A \cap B =_n A' \cap B'$, $A - B =_n A' - B'$, then QAB iff $QA'B'$;
where $U =_n V$ if $|U| = |V| < n$ or $|U|, |V| \geq n$.

In particular, for such quantifiers, there exists a row $a + b = 2n$ in the tree whose truth values 'propagate', as in Fig. 4. (Notice the 'characteristic triangle' below (n, n).)

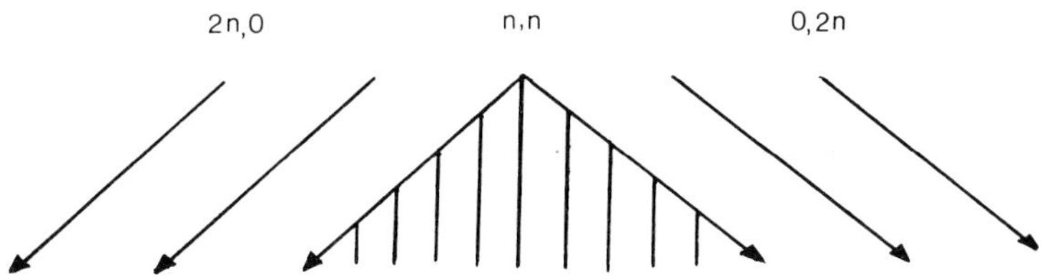

Fig. 4.

Evidently, quantifiers with such a pattern satisfy UNIF*: at some finite stage, they calm down. But also conversely, continuous quantifiers satisfying UNIF* must reach such a stage. For instance, suppose the tree pattern is as in Fig. 3, with + values on the right-hand side. If no point on the right edge of the tree generates a downward triangle which is completely +, then, by UNIF*, no arbitrarily large south-east + sequences can emanate from this edge, and hence there is some upper limit to these sequences. Thus, the tree either contains some downward triangle which is completely +, or one which is completely −. Additional arguments about the diagonals adjoining such trees in uppermost position will then establish the existence of a 'stable row' as described above, and hence of first-order definability. □

The preceding results illustrate two important types of question:

— *definability:* how do quantifiers in the above classes relate to those definable in the standard logical languages?

— *hierarchy:* how can these conditions be used to classify quantifiers in a natural way?

Perhaps the best example of a hierarchy arises with a slightly different perspective upon uniformity. Continuous quantifiers had the property that their true/false boundary can only shift one step at a time in going down the tree. This leads to a natural classification by degrees of uniformity, counting from the top position:

one-step shift pattern repeats itself: *all, no, some, not all,*

two-step shift pattern repeats itself: add *most, least, not most, not least,*

etcetera.

Thus, one sifts out the better-behaved higher-order quantifiers stage by stage.

To conclude, the enterprise of this section has been to explore the nature of logicality, searching for 'natural kinds' of quantifier. Nevertheless, our results do not indicate *why* natural language, or the human mind, should care to entertain these: deeper explanations are always possible. We have made it intelligible, however, why, on reasonable assumptions, a small number of patterns must recur — rather in the spirit of the catalogue of forms occurring in René Thom's account of morphogenesis. Even a logical treatment of this kind of question already constitutes a significant widening of the traditional subject matter of logic.

Appendix: some further topics of current research

As these sections are intended to give an outline of main ideas, rather than an exhaustive survey of ongoing research, a few further special themes will be appended briefly.

(1) *Closure conditions.* One may also study classes of quantifiers from an inductive point of view, admitting basic cases, and then closing under certain constructions — notably Boolean operations. For the interplay of this point of view with the present one, approaching from the outside by accumulating restrictive conditions, cf. KEENAN and STAVI 1982, VAN BENTHEM 1983a. (Keenan and Stavi raise several interesting questions about definability of determiner denotations, in an algebraic setting related to the present one. Notably, the important class of conservative determiners has an inductive definition 'locally', though not 'uniformly'.)

(2) *Other special postulates.* Various other versions of continuity and uniformity, as well as other conditions may be found in VAN BENTHEM 1984b, 1983a.

(3) *Definability results.* Deeper results than Theorem 3.5 exist. Two examples are the following:

— even without VAR, left-monotonicity implies first-order definability (WESTERSTÅHL 1984),

— even without VAR and CONS, double monotonicity implies first-order definability (this follows from the proof of Theorem 4.2.1 in VAN BENTHEM 1984b).

(4) *Counting problems.* On a universe with n elements, there are

— 2^{4^n} general quantifiers,

— 2^{3^n} CONS general quantifiers (KEENAN and STAVI 1982),

— $2^{\binom{n+3}{2}}$ QUANT general quantifiers (HIGGINBOTHAM and MAY 1981),

— $2^{(n+1)(n+2)/2}$ CONS, QUANT general quantifiers (VAN BENTHEM 1984b).

Many more results on the numerical effects of proposed conditions are in THIJSSE 1983 — where familiar mathematical notions turn out to be relevant in this area. For instance, the Fibonacci series is needed in counting left monotone continuous quantifiers. Perhaps the most important open counting problem, with a mathematical history, is the following:

What is the number of CONS, MON$\uparrow$ general quantifiers?

(5) *Computational complexity of denotations.* The above quantifiers can also be viewed as procedures for computing truth values. For instance, the UNIF* ones are those representable by *finite state automata.* This viewpoint is developed in VAN BENTHEM 1985.

4. Patterns of inference

When viewed as binary relations, quantifiers exhibit many familiar properties: reflexivity, transitivity, symmetry, linearity, etcetera.

EXAMPLE 1. *All* is reflexive and transitive,
not all is irreflexive and linear,
some is symmetric and quasi-reflexive ($\forall xy(Rxy \to Rxx)$),
no is symmetric and quasi-universal ($\forall xy(Rxx \to Rxy)$).

These properties were used in formulating semantic universals in ZWARTS 1983, 1985. For instance, studying actual lists of natural language expressions, he noted systematic gaps such as the following: "no human language has asymmetric determiner expressions".

One motive in starting the earlier general study of quantifiers has been the wish to evaluate such semantic universals. For instance, in this particular case, one can prove that the statement is true for all logical quantifiers in the earlier sense (cf. VAN BENTHEM 1984b). A short repetition of the relevant argument will show how one can argue about such matters.

EXAMPLE 2. There are no non-trivial asymmetric logical quantifiers.

PROOF. Let Q be any non-trivial quantifier; i.e., QAB for some A, B. Consider $A, A \cap B$ only — and add a new individuals, to form A^* as in Fig. 5. Then we have: QAB, $QA(B \cap A)$ (CONS), $QA(A^* \cap A)$, QAA^* (CONS) and hence QA^*A (QUANT). Thus, Q cannot have been asymmetric. □

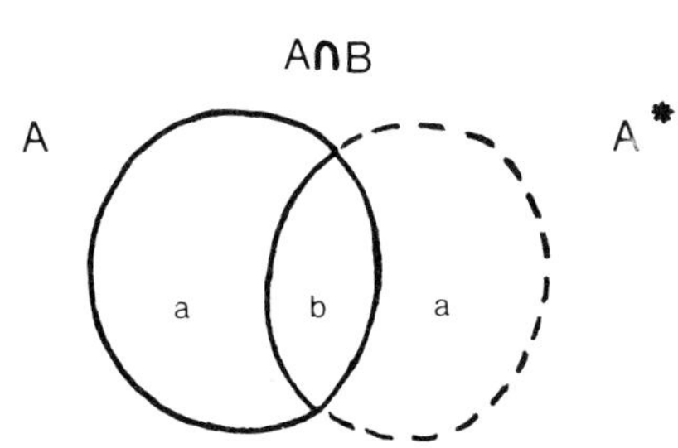

Fig. 5.

A further discussion of the status of this and other semantic universals, and the role of logical and mathematical considerations in evaluating them, is found in VAN BENTHEM 1984c.

(Evidently, not all proposed semantic universals will be a priori true, as in this particular case.)

Sometimes also, these relational conditions characterize a particular quantifier uniquely.

EXAMPLE 3. Modulo Variety, *all* is the only reflexive, transitive quantifier.

PROOF. For the non-trivial direction, let Q be reflexive and transitive. (1) If $A \subseteq B$, then $A \cap B = A$, and we have QAA (reflexivity), QAB (CONS). (2) If QAB, then also $Q(A - B)A$ (as in (1)), and hence $Q(A - B)B$ (transitivity). Then also $Q(A - B)\emptyset$ (CONS), and so $A - B = \emptyset$ (VAR); i.e., $A \subseteq B$. □

From a more general logical point of view, these relational conditions are nothing but basic *patterns of inference*. In that light, the present query amounts to reversing the usual logical investigation. Instead of giving the logical constants, and determining their valid inferences, one is giving the inference patterns and asking for the range of constants (if any) validating them. This new type of '*inverse logic*' may by explored quite systematically, for all possible inferential theories.

For instance, for syllogisms mostly negative results are found, as in the above: no plausible inference patterns have remained unused in the tradition. A prominent example of a 'failed syllogism' is in VAN BENTHEM 1984b:

4.1. THEOREM. *There are no non-trivial circular quantifiers, satisfying the inference pattern QXY, QYZ/QZX.*

But there are also pleasant characterizations, such as the following result from the same paper:

4.2. THEOREM. *The quantifiers in the Square of Opposition are each completely determined by the conditions mentioned in the first example.*

It may be instructive to compare this result with existing attempts at proof-theoretic characterization of the logical constants (cf. ZUCKER 1978).

Without the assumption of Variety, classification results become more complex combinatorially. Nevertheless, henceforth, Variety will be dropped — and the only assumption remaining is that quantifiers are to be non-trivial: i.e., they are neither empty nor universal. For, after all, there is a counteracting consideration which may compensate for this.

In traditional syllogistic, several quantifiers may interact in the same inference, unlike in the earlier pure cases. Thus, the above questions of existence and characterization also arise with respect to several quantifiers at once. For instance, in addition to the above properties of *all, some*, there is also their combined inference

$$all\ XY,\ some\ XZ / some\ YZ.$$

Even so, there are usually whole ranges of solutions. One telling result is the following; allowing both statements QXY and their negations in inference patterns.

4.3. THEOREM. *The complete syllogistic theory of 'some' and 'all' is satisfied by precisely all couples* $\langle$*at least n X are Y*; *there are at most* $n-1$ *X or all X are Y*$\rangle$, *with* $n = 1, 2, \ldots$

Thus, in a sense, classical logic fails to enforce its intended interpretation.

PROOF. The argument proceeds in two stages. First, the class of possible solutions is narrowed down to the above list. Then, it is shown that all of these validate the same syllogistic inference patterns.

First, the earlier relational conditions on $Q_1 = some$, $Q_2 = all$ already restrict the possibilities to the following (cf. WESTERSTÅHL 1984):

— Q_1 must be *at least m X are Y*, for some $m \geqslant 0$.

By way of illustration, notice that, wherever Q_1 has a + position, its entire north-east/south-west diagonal will be contained in Q_1: symmetric quantifiers only depend on their b-values. Moreover, once such a diagonal occurs, quasi-reflexivity will bring in the whole further right edge of the tree — and again all its parallel diagonals.

— Q_2 must be *there are at most* $n-1$ *X, or all X are Y*, for some $n \geqslant 1$.

The added effect of the above combined inference is this. If, at any row, $Q_2X\emptyset$, and also Q_1XY, then $Q_1\emptyset Y$- and hence $Q_1\emptyset\emptyset$. By the above observations then, Q_1 would comprise the whole tree. It follows that the picture must be as in Fig. 6.

Finally, the combined inference $\neg\, some\ XX / all\ XY$ implies that $m = n$.

Next, to prove that all remaining possibilities are on a par, an auxiliary result is needed.

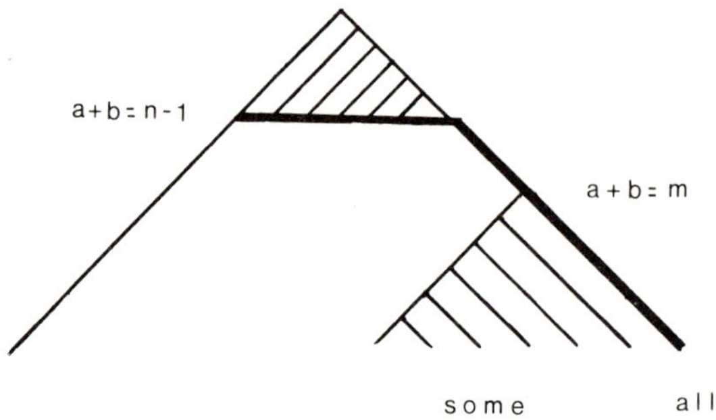

Fig. 6.

LEMMA. *Any set of (negated) statements* $(\neg)Q_{(1,2)}X_1Y_1, \ldots, (\neg)Q_{(1,2)}X_kY_k$ *(with* $Q_{(1,2)} = Q_1$ *or* Q_2*) is consistent in one of the above readings if and only if it is consistent for* $Q_1 = $ *some,* $Q_2 = $ *all.*

PROOF. Let $A_1, B_1, \ldots, A_k, B_k$ be any choice of sets verifying the above statements for $Q_1 =$ *some*, $Q_2 =$ *all* (i.e., the n-reading with $n = 1$). This choice can be turned into a verification for any n-reading ($n \geqslant 2$) by adding new individuals as follows. Consider any non-empty intersection $A \cap B$, and add n new objects $d_1^{A,B}, \ldots, d_n^{A,B}$ to both A, B and all sets containing them both. It may be checked that, for this new choice of sets $A_1^+, B_1^+, \ldots, A_k^+, A_k^+$ (where C^+ may equal C, if undisturbed in the above process), all statements in the above list retain their original truth value:

all UV: $U^+ \subseteq V^+$ (by construction),

not *all UV*: $U^+ \not\subseteq V^+$ (by construction),

some UV: $U^+ \cap V^+ \neq \emptyset$ (by construction),

not *some UV*: Suppose that $U^+ \cap V^+$ has become non-empty. Then, by the construction, $U \supseteq A, B$, $V \supseteq A, B$, and so: $U \cap V \neq \emptyset$.

A similar process works for cases $A \not\subseteq B$. Repeating this procedure until all intersections and non-inclusions have been treated, one obtains a verification for the original list of statements in which every intersection has at least n members, while cases of non-inclusion only occur for sets of size at least n.

REMARK. This part of the proof may also be established much more simply, by adding n new objects to each non-empty 'state description' $(-)A_1 \cap (-)B_1 \cap \cdots \cap (-)A_k \cap (-)B_k$. The resulting structure is elementarily equivalent to the original one in monadic predicate logic without identity — and hence it still verifies the original list in its first reading. But, unlike the earlier one, this type of argument does not work in the next part of the proof.

Conversely, suppose that the above list is verified by some n-reading. This time, an ordinary reading is required. Instead of removing objects from intersections, one first adds *one new* element in '*some*' and 'not *all*' situations, like above. (Again, it has to be checked that no truth values are disturbed in this way.) Then, *all old* individuals are omitted — and it may be seen that the list is still verified: under the standard reading ($n = 1$). □

From the Lemma, the assertion about coincidence of valid syllogistic inference for the above family of readings follows at once — and the theorem has been proven. □

Imposing stronger inferential conditions than those expressed by the present syllogistic forms might still lead to a unique solution. The latter would be a completeness theorem characterizing the usual quantifiers as the only ones with their particular inferential behaviour. (Some partial advances along this line are reported in WESTERSTÅHL 1986.)

Another important type of query is the study of the effects of imposing the above inferential conditions upon a prescribed range of 'truth definitions' for general quantifiers, say those expressible in first-order logic. In that case, e.g., a search arises for 'preservation results' for predicate logical sentences satisfying the above properties.

This section has introduced a second new direction in logic, viz. the systematic exploration of 'realizable' inference patterns. Of course, examples of 'inverse logic' have been at stake whenever some particular syntactic 'logic' is being modelled. But this time, we are after the general phenomenon.

5. Other linguistic categories

The investigation up till now has been restricted to the linguistic category of determiners. But, the previous techniques can be brought to bear upon arbitrary grammatical types: adjectives, connectives, adverbs, etcetera. To be definite, a categorial grammar will be considered with basic types e ('entity') and t ('truth value'), allowing formation of functional types (a, b) ('from a-denotations to b-denotations').

One prime example of a condition which generalizes to all such types at once is *Quantity:*

Let π be any permutation of the universe $E = D_e$. Setting π equal to the identity on D_t, this function may be lifted inductively to all universes D_a, by stipulating for $f \in D_{(a,b)}$,

$$\pi(f) = \{\langle \pi(x), \pi(y)\rangle \mid \langle x, y\rangle \in f\}.$$

Now, an object x is *quantitative* if $\pi(x) = x$ for all permutations π. For determiners (type $((e, t), ((e, t), t))$), this amounts to the earlier formulation.

To see the power of this constraint, notice that, for individual binary relations (type $(e, (t, t))$), Quantity leaves only *identity* and its Boolean compounds as logical constants — as ought to be the case.

As a further illustration of this type of enquiry, here are some more extensive reflections on logical constants.

Connectives

In natural language, the usual Boolean connectives occur as operations on predicates in the first place; with the sentential connections as a less frequent case. (Eventually, one comes to see them as operations on almost arbitrary types; cf. KEENAN and STAVI 1982.) For instance, the former view was already prior in the formulation of Conservativity:

$$QXY \quad \text{iff} \quad QX(Y \textit{ and } X).$$

Also, Monotonicity has obvious Boolean formulations of this kind:

$$QXY/Q(X \textit{ and } Z)Y \ (\downarrow \text{MON}); \qquad QX(Y \textit{ and } Z)/QXY \ (\text{MON} \uparrow).$$

The effects of earlier general conditions on denotations (Section 3) may be gauged for such set operations. For instance, consider an arbitrary k-ary set connective in the light of the above basic requirement (VAN BENTHEM 1983a).

5.1. THEOREM. *For any choice of k arguments, Quantity leaves exactly those values which can be described by some Boolean polynomial.*

Other conditions become less plausible in this case. E.g., the complement operation is universe-dependent; whence it fails to satisfy *Extension.* But then, new conditions may be forthcoming for special categories. For instance, for many operations C, a principle of *Restriction* is plausible:

REST If $E' \subseteq E$, then, for $A_1, \ldots, A_k \subseteq E$,
$C_{E'}(A_1 \cap E', \ldots, A_k \cap E') = C_E(A_1, \ldots, A_k) \cap E'$.

This condition holds for k-ary Boolean set connectives, and this is no coincidence (VAN BENTHEM 1983a):

5.2. THEOREM. QUANT, REST *characterize the Boolean operations uniformly.*

Not only the topics of Section 3 return in this wider setting. The inferential concerns of Section 4 still apply as well.

QUERY. Which triples of QUANT operations on sets satisfy the complete set of Boolean identities in $\neg$, $\wedge$, $\vee$?

Actually, the earlier phenomenon of 'multiple solutions' has been known for a long time in this particular area, as 'duality' of $\wedge$, $\vee$. By way of illustration, here is a special case.

5.3. THEOREM. *In propositional truth value semantics, the complete set of Boolean identities has exactly two solutions:*

$$\neg, \wedge, \vee \quad \textit{and} \quad \neg, \vee, \wedge .$$

Notice that *identities* are necessary here because of the operational perspective upon connectives. (A sentential inference like 'X *and* Y/X' does not make sense right now.) Therefore, the earlier mentioned connection with related proof-theoretic concerns is certainly not straightforward.

PROOF. As with Theorem 4.3, the argument proceeds in two stages.

There are at most these solutions. Various valid identities will narrow down the range of possible solutions to just these two:

$\neg\neg X = X$: $\neg$ can only be identity or value reversal.

$X \wedge X = X$: $\wedge(0,0) = 0$, $\wedge(1,1) = 1$.

$X \wedge Y = Y \wedge X$: $\wedge(0,1) = \wedge(1,0)$.

$(X \wedge \neg X) \wedge Y = X \wedge \neg X$: If $\neg$ were the identity, then $X \wedge Y = X$ would be valid — and a contradiction arises:

— $X = 0$, $Y = 1$: $\wedge(0,1) = 0$

— $X = 1$, $Y = 0$: $\wedge(1,0) = 1$.

So, $\neg$ denotes value reversal.

For disjunction, similar observations imply $\vee(0,0) = 0$, $\vee(1,1) = 1$, $\vee(0,1) = \vee(1,0)$. Moreover, because of the mixed principle $X \wedge (Y \vee X) = X$: $\vee(0,1) \neq \wedge(0,1)$. (Otherwise,

— if $\vee(0,1) = \wedge(0,1) = 0$, set $X = 1$, $Y = 0$: $\wedge(1, \vee(0,1)) = \wedge(1,0) = 0 \neq 1$ (the X-value),

— if $\vee(0,1) = \wedge(0,1) = 1$, set $X = 0$, $Y = 1$: likewise.

That there are at least these two solutions follows by any one of the familiar duality arguments. □

Connectives and quantifiers

The two main kinds of logical constant may also be combined. For instance, consider the logic of *all* and *and*. This time, Conservativity may be formulated as a combined inference pattern, rather than a background postulate. Moreover, of course, there were the relational properties of reflexivity and transitivity for *all* alone, while *and* satisfies the obvious Boolean identities. These principles already imply several other basic facts:

MON↑:

$$\dfrac{\textit{all } X(Y \textit{ and } Z) \qquad \dfrac{\dfrac{\textit{all } (Y \textit{ and } Z)(Y \textit{ and } Z)}{\textit{all } (Y \textit{ and } Z)(Y \textit{ and } (Y \textit{ and } Z))}}{\textit{all } (Y \textit{ and } Z)Y}\,\text{CONS}}{\textit{all } XY}\,\text{trans}$$

↓MON:

$$\dfrac{\textit{all } (X \textit{ and } Z)X \text{ (as above)} \qquad \textit{all } XY}{\textit{all } (X \textit{ and } Z)Y}$$

CONJ:

$$\dfrac{\dfrac{\textit{all } XY}{\textit{all } X(X \textit{ and } Y)}\,\text{CONS} \qquad \dfrac{\dfrac{\dfrac{\textit{all } XZ}{\textit{all } (X \textit{ and } Y)Z}\,\text{MON}}{\textit{all } (X \textit{ and } Y)(X \textit{ and } Y \textit{ and } Z))}\,\text{CONS}}{\textit{all } (X \textit{ and } Y)(Y \textit{ and } Z)}\,\text{MON}}{\textit{all } X(Y \textit{ and } Z)}\,\text{trans}$$

Are there again multiple solutions (within the constraint of Quantity) to the above mixed inferential theory? One would expect that addition of connectives will restrict the range of possibilities for quantifiers left open by Theorem 4.3.

QUESTION. Modulo Quantity, does the logic of *all*, *some*, *and*, *or* and *not* fix the interpretation of these logical constants uniquely?

Quantifiers revisited

Within a general categorial framework, quantifiers may also be regarded differently, as *reducers of argument places.* Thus, e.g., the quantifier phrase 'someone' turns the binary relation 'loves' into the unary property 'love someone'. This point of view is the basis of predicate logic as developed in QUINE 1966. Its major operations on predicates are 'permutation', 'identification' and 'projection'; of which the latter two are relevant here:

$$\mathrm{id}(R) =_{\mathrm{def}} \{x \mid \langle x, x\rangle \in R\},$$

$$\mathrm{proj}(R) =_{\mathrm{def}} \{x \mid \exists y\, \langle x, y\rangle \in R\}.$$

This scarcity of argument drop mechanisms (also known from linguistics) may be understood in our framework.

Let us restrict attention to relation-reducing functions sending binary relations to subsets of their domains. The above examples satisfy Quantity as a general postulate. Moreover, their special distinguishing feature turns out to be a well-known mathematical continuity condition:

UNION For all families $\{R_i \mid i \in I\}$,

$$f\Big(\bigcup_i R_i\Big) = \bigcup_i f(R_i).$$

5.4. THEOREM. *The Quine operations* id, proj *are essentially the only relation-reducing functions satisfying* QUANT *and UNION.*

PROOF. Let E be any universe, with a binary relation R on E. Any function f satisfying QUANT, UNION will map R to $\bigcup\{f(\{\langle d, e\rangle\}) \mid \langle d, e\rangle \in R\}$. So, it is to be determined what can be assigned in these singleton cases.

(1) $\langle d, d\rangle \in R$: f assigns either $\emptyset$ (1.1) or $\{d\}$ (1.2).

By QUANT, if case (1.2) occurs for any object d, it will occur for all $d' \in R$. (Consider a permutation interchanging d, d', while leaving all other objects fixed.)

(2) $\langle d, e\rangle \in R$, $d \neq e$: f assigns either $\emptyset$ (2.1) or $\{d\}$ (2.2).

Again by QUANT, case (2.2) either occurs for all doublets, or for none.

Summing up, there are four possible cases:

(1.1, 2.1) $f(R) = \emptyset$,

(1.1, 2.2) $f(R) = \text{proj}(R) - \text{id}(R)$,
(1.2, 2.1) $f(R) = \text{id}(R)$,
(1.2, 2.2) $f(R) = \text{proj}(R)$. □

This second view of quantifiers will also arise in the earlier perspective of Section 2, once determiners are considered, not in subject, but in direct *object* position ('Mary loved every lamb'). In this position, quantifiers no longer stand for relations between sets, but for functions on sets and binary relations, yielding sets of individuals. E.g.,

$$[\![\text{every}]\!]([\![\text{love}]\!], [\![\text{lamb}]\!]) = \{e \in E \mid [\![\text{lamb}]\!] \subseteq \{d \mid \langle e, d\rangle \in [\![\text{love}]\!]\}\}.$$

In a relational perspective, there is a ternary relation here between an individual, a binary relation and a property:

$$Q_E(e, R, A).$$

What we want, then, is a reduction to the earlier treatment of quantification in Sections 2, 3. One strategy is to use a principle of *Locality:*

$$Q_E(e, R, A) \quad \text{iff} \quad Q_E(e', R', A), \quad \text{whenever } R_e = R'_{e'};$$

with $R_e =_{\text{def}} \{d \mid \langle e, d\rangle \in R\}$.

Then, given suitable versions of QUANT, CONS, etcetera, for the above ternary relations, the binary relation

$$\{(A, R_e) \mid Q_E(e, R, A)\}$$

can be treated exactly as before.

These few examples will have illustrated how the study of logical constants can be extended, from the original heartland, to cover all types of categorial grammar.

There is a more general significance to this move, in the light of our main theme. After an initial predominance of transformationally enriched phrase structure grammars, flexible categorial grammars are becoming an attractive alternative nowadays as a model for linguistic description. But also from a logical point of view, these grammars are of interest, being fragments of a certain theory of types. To illustrate this development, and its attendant logical questions, a brief introduction to a flexible version of the above categorial grammar concludes this section (cf. LAMBEK 1958, ZWARTS 1985, VAN BENTHEM 1983d).

Traditional categorial grammar, with fixed types assigned to linguistic items, has usually been regarded as an attractive, but inadequate form of grammatical analysis. (Notably, 'action at a distance' between various constituents of an expression has turned out difficult to describe.) But then, there is another, more flexible way of using this grammar, which has a much greater descriptive potential.

Many expressions of natural language do not stay in one category, but have a certain freedom of choice, as the linguistic context requires. Several examples of this phenomenon have occurred already in the preceding sections. *Not* can be both sentence negation (type (t, t)) and predicate negation (type $((e, t), (e, t))$); *every lamb* can have the NP type $((e, t), t)$, but also the direct object type $((e, (e, t)), (e, t))$, as we have seen. The general rule here would seem to be this:

(1) *occurrences in type* (a, b) *can also be in type* $((c, a), (c, b))$.

Another case is the Montagovian raising of proper names (type e) to the type $((e, t), t)$, in order for them to fit the NP VP scheme of Section 2:

(2) *occurrences in type* a *can also be in type* $((a, b), b)$.

Thus, a family arises of flexible categorial grammars with systematic rules of 'type raising', which can handle many more phenomena than their rigid ancestor. Ironically, an elegant grammar of this kind had already been proposed in LAMBEK 1958, a paper which was ignored in the development of transformational generative grammar. The connection with logic is particularly striking in Lambek's grammar, of which a *rough* version runs as follows.

Expressions are evaluated by operating on their associated string of basic types, allowing all type changes of the form

$$\mathbf{a} \Rightarrow b \qquad (\mathbf{a} = a_1; \ldots ; a_n)$$

generated by the following '*calculus of sequents*':

$$\textit{function-elimination:} \quad \begin{array}{l} (a, b);\ a \Rightarrow b \\ a;\ (a, b) \Rightarrow b \end{array}$$

$$\textit{function-introduction:} \quad \frac{\mathbf{a};\ a \Rightarrow b}{\mathbf{a} \Rightarrow (a, b)}$$

EXAMPLE 1.

$$\frac{\dfrac{(a, b);\ (c, a);\ c \Rightarrow (a, b);\ a \Rightarrow b}{(a,b);\ (c, a) \Rightarrow (c, b)}}{(a, b) \Rightarrow ((c, a), (c, b))}$$

EXAMPLE 2.

$$\frac{a;\ (a, b) \Rightarrow b}{a \Rightarrow ((a, b), b)}$$

These grammars are finding ever more linguistic applications these days, even to languages lacking the basic SVO syntax of Section 2.

Another, slightly different motivation for investigating these flexible categorial grammars is found in VAN BENTHEM 1984c. Here the view is proposed that our notions of syntactic *grammaticality* and semantic *interpretability* are largely independent from one another; and categorial grammar is used in providing a syntax-free account of the latter notion. Afterwards, some examples are given of 'completeness results' relating the two notions, on the scheme

INTERPRETABLE = TRANSFORMATIONS (GRAMMATICAL).

This is but one instance of a logical question generated by this newer kind of categorial grammar. Lambek himself considered the more traditional issue whether his calculus of type change is *decidable*. (The answer is positive; by the standard proof-theoretic method of Cut Elimination.) Another obvious question, this time one of theoretical linguistics, concerns generative power: in particular, are the languages recognized by the above flexible grammar still *context-free*?

But, these grammars also introduce new types of query. For instance, in VAN BENTHEM 1984c, the issue is studied which *variety* of raised types exists for any single given type. There is no total freedom here: certain 'invariants' turn out to be preserved by the above rules; but much is possible. In particular, the obvious conjecture that an expression evaluates to a family of types 'generated' from its basic type by the above rules (1), (2) turns out to be false.

Perhaps the most important new question of all is the following. The above rules of type raising are more or less syntactic in nature. Can a sensible independent *semantics* be given for 'admissible' type transitions? If so, the linguist would feel safer, having some semantic constraints on the exuberance of the above scheme — while the logician might wish to prove a completeness theorem matching the two concepts of type change. Such results are indeed found in VAN BENTHEM 1983d, in terms of 'lambda-recipes'.

Even from this brief sketch, it will have become clear that the 'logic of categorial grammar' contains various interesting questions, and may well become a chapter of its own in our linguistic turn.

Appendix: further topics

(1) *Determining logical constants in arbitrary categories.* Cf. WESTERSTÅHL 1982.

(2) *Formulating conditions applicable to all categories.* Cf. WESTERSTÅHL 1982, VAN BENTHEM 1983a.

(3) *Case studies of particular categories of logical or linguistic interest.* For instance, determiners with more than two arguments (*more X than Y are Z*) have been studied in KEENAN and MOSS 1985.

(4) *From ordinary to flexible categorial grammar.* There remains the matter of the connection between the final part of this section and the preceding investigation. What is the mechanism for connecting our theory of Sections 3, 4 and the beginning of 5, as it applies to the simplest categorization of a linguistic item, with that for its higher occurrences?

(5) *The logic of categorial grammar.* Various mathematical results on Lambek grammar may be found in BUSZKOWSKI 1982, especially concerning generative power (an area where many open questions remain). Buszkowki also presents an 'algebraic' semantics, for which a set-theoretic possible worlds variant is found in VAN BENTHEM 1986, chapter 7. Actually, there is a whole spectrum of Lambek grammars, which can be studied via transcription of their derivations into type-theoretic lambda-terms (VAN BENTHEM 1983d contains several applications of this idea).

(6) *Operations across all categories.* Finally, to extend the present analysis to full type theory, an account would be needed of 'transcategorial' operations, such as *lambda abstraction.* Such a global categorical perspective is in line with recent semantic attempts at capturing broad semantic mechanisms operative across natural language. A tentative survey of this area is found in VAN BENTHEM 1986, chapter 3. (See also KEENAN and FALTZ 1985.)

6. Variations

Returning to the starting point of this investigation, even the basic constraints on admissible determiner denotations can be varied as the need arises. For instance, the realm of *infinite* universes may be studied after all — nothing in the preceding actually prevents this.

The two most important variations from a linguistic point of view concern the postulates of *Extension* and *Quantity.* Extension was our reason in the preceding to disregard phenomena of context dependence. But eventually, if the linguistic dynamics of context change are to be studied, an account will have to be provided of relations between various universes, and possible accompanying changes in denotations. (Indeed, a study of *growth* in the finite realm would also be our preference over speculation about the infinite.) Most urgent, however, is a liberalization of Quantity; as this principle breaks down in several cases of immediate logical interest.

EXAMPLE 1 (genitives). The statement 'Lucia's curls are genuine' need not be preserved under arbitrary permutations of the universe: it is sensitive to the underlying 'possession structure' among individuals. (Cf. VAN BENTHEM 1983b.)

EXAMPLE 2 (conditionals). Statements of the form '*if X, then Y*' need not be preserved under arbitrary permutations of the universe of possible worlds in intensional semantics: the 'similarity pattern' matters. (Cf. VAN BENTHEM 1984a.)

What is needed here is a transition from Quantity to *Quality:* invariance only occurs with respect to some group of E-automorphisms respecting some relevant structure among individuals in E. Actually, for any quantifier Q, its associated permutation group

$$G_Q(E) = \{\pi \mid \pi \text{ is a permutation of } E \text{ and, for all } A, B \subseteq E, \\ Q_E AB \text{ iff } Q_E \pi[A]\pi[B]\}$$

may be represented (locally in E) as an automorphism group. It suffices to select a suitable one of its invariant relations among individuals. But of course, the point is to find interesting invariants across all universes.

A paradigmatic example of qualitative behaviour in logic is displayed by *conditionals*; treated extensively in VAN BENTHEM 1984a. A conditional statement of the form

$$if\ XY$$

involves a generalized quantifier relation between the set $[\![X]\!]$ of antecedent occasions and the set $[\![Y]\!]$ of consequent occasions. And this relation is sensitive, in general, to intensional 'hidden structure' among such occasions: some are more 'relevant' than others.

Two concrete examples of conditionals will illustrate these points about qualitative analysis, while also providing a connection with the earlier theory.

EXAMPLE 3 (classical entailment). The logic of this conditional contains reflexivity and transitivity, as well as Conservativity. By earlier observations (Section 3), its monotonicity type is $\downarrow$ MON $\uparrow$. (Notice that QUANT was not used in the relevant argument.) Moreover, assuming Variety, this conditional must be inclusion (*all*; cf. Theorem 3.2). Thus, classical entailment must be quantitative after all.

EXAMPLE 4 (counterfactual implication). The basic (counterfactual) conditional logic of LEWIS 1973 has a much weaker inferential theory. Semantically, its corresponding intuition is that some antecedent occasions are more important than others — as stated above. A natural 'hierarchical' relation between individual occasions to consider, then, is the following:

$$x \leqslant y \quad \text{iff} \quad if\ \{x, y\}\{y\}.$$

It can be shown that this preference pattern indeed determines the behaviour of the conditional.

6.1. THEOREM. *A conditional generalized quantifier relation 'if' satisfies the counterfactual base logic if and only if*

(1) *its induced relation* $\leqslant$ *is a partial order,*

(2) *if* XY *is defined by* $\forall x \in X\ \exists y \in X \cap Y\ x \leqslant y$.

PROOF. 'Only if': all transitions in the following argument are valid in Lewis' semantics.

(1) Reflexivity: *if* $\{x\}\{x\}$.

Transitivity:

$$\dfrac{\dfrac{\dfrac{if\ \{x, y\}\{y\} \qquad if\ \{y, z\}\{z\}}{if\ \{x, y, z\}\{y, z\} \qquad if\ \{x, y, z\}\{x, z\}}}{if\ \{x, y, z\}\{z\}}}{if\ \{x, z\}\{z\}.}$$

Antisymmetry:

$$\dfrac{if\ \{x, y\}\{x\} \qquad if\ \{x, y\}\{y\}}{if\ \{x, y\}\{x\} \cap \{y\}}$$

Now, if $\{x\} \cap \{y\} = \emptyset$, then *if* $\{x, y\}\emptyset$. In Lewis' semantics, this only holds when $\{x, y\}$ is empty. It follows that, on the contrary, $\{x\} \cap \{y\} \neq \emptyset$: i.e., $x = y$.

(2) First, suppose that *if* XY. Let $x \in X$. Either $x \in X \cap Y$ (and $x \leqslant x$: we are done), or $x \in X - Y$. In the latter case, *if* $X(Y \cap X)$, and hence *if* $(\{x\} \cup (Y \cap X))(Y \cap X)$. Now, an auxiliary observation is needed, which is valid in Lewis' semantics:

Claim. If $Y_1 \cap Y_2 = \emptyset$, $x \notin Y_1 \cup Y_2$, then *if* $(\{x\} \cup Y_1 \cup Y_2)(Y_1 \cup Y_2)$ implies *if* $(\{x\} \cup Y_1)Y_1$ or *if* $(\{x\} \cup Y_2)Y_2$.

By repeated applications of this assertion, it follows that *if* $(\{x\} \cup \{y\})\{y\}$ (i.e., $x \leq y$) for at least one $y \in X \cap Y$.

Conversely, suppose that, for all $x \in X$, $x \leq y_x$ for some $y_x \in X \cap Y$. Then, *if* $\{x, y_x\}\{y_x\}$ (all $x \in X$), and therefore *if* $\{x, y_x\}Y$ (all $x \in X$; by MON↑), and *if* XY follows by Disjunction of antecedents.

'If': This time, it is to be checked that all principles of the basic counterfactual logic follow from the above explanation. It suffices to inspect the principles of Lewis' complete axiomatization. By way of illustration, here is the case of Conjunction.

Suppose that *if* XY, *if* XZ. Let $x \in X$. Choose some $\leq$-maximal successor x^+ of x in X. (Here the fact is used that $\leq$ is a partial order on a *finite* universe.) For some $y \in X \cap Y$, $x^+ \leq y$: it follows that $x^+ = y$, and $x^+ \in Y$. Again, for some $z \in X \cap Z$: $x^+ \leq z$, and hence $z = x^+$, and $x^+ \in Z$. □

The above qualitative analysis of the counterfactual conditional explains the genesis of 'hidden patterns' (usually merely postulated) among possible worlds in intensional semantics. Many further results in this vein are found in VELTMAN 1985.

In this connection, other themes from the preceding sections are relevant too. For instance, the earlier questions of definability now amount to asking for a range of admissible *truth definitions* for a given logic. Especially, in the realm of first-order truth definitions, the earlier conditions on denotations translate into familiar model-theoretic preservation properties. For instance, a standard argument establishes the following characteristic result concerning 'transmitting' conditions (type ↓MON↑).

6.2. THEOREM. *A first-order sentence in* $R^{(2)}$, $=$, $X^{(1)}$, $Y^{(1)}$ *satisfies* CONS, EXT, ↓MON↑ *if and only if it is equivalent to a conjunction of the forms*

$$\forall^A x_1 \cdots \forall^A x_n \left(\rho(x_1, \ldots, x_n) \rightarrow \sum_{i \in I} Bx_i\right);$$

where ρ *is a quantifier-free condition on* R, $=$, *and* $I \subseteq \{1, \ldots, n\}$.

Special cases are

— $\forall^A x\, \forall^A y(Rxy \rightarrow By)$ (as in modal logic),

— $\forall^A x\, \forall^A y(x \neq y \rightarrow Bx \vee By)$ (as in the classification of QUANT, ↓MON↑ quantifiers given in VAN BENTHEM 1984b).

Thus, the present enquiry also leads to a systematic background study of semantic modellings in current possible worlds semantics and related intensional theories.

Appendix: further topics

(1) *Infinite models.* The prospects of the present theory in infinite models are explored in THIJSSE 1983, VAN DEEMTER 1985.

(2) *Genitives and other cases.* The treatment of genitives in the present perspective has been started in VAN BENTHEM 1983b, THIJSSE 1983. (Cf. also PARTEE 1983.)

(3) *Permutation groups G_O and their invariants.* A general theory of the emergence of relations among individuals, in order to account for selective invariance behaviour of generalized quantifiers, remains to be developed.

Several people have suggested that the above kind of analysis would be particularly useful in the study of *temporal* connectives, and their influence upon the underlying conception of the structure of Time. Indeed, the analysis of Theorem 6.1 can be transferred quite easily to the 'temporal conjunction' connective *whenever XY*, representing it as $\forall t\,(Xt \rightarrow \exists t' \geqslant t\; Yt')$, for some reflexive and transitive temporal order $\geqslant$. (Interestingly, *whenever* does not satisfy VAR, or even CONS.)

(4) *Preservation theorems in qualitative settings.* Several results of this kind are proven in VAN BENTHEM 1983c, 1984a. One typical outcome: for normal modal logics, the truth definition scheme $\Box p \mapsto \forall y\,(\rho(x, y) \rightarrow Py)$ is the only possible one.

7. Natural logic

This section is devoted to an old direction in logic. The traditional subject matter of logic has been the systematic description of valid inference. Now, the present more linguistic perspective throws some new light upon this old task as well. The ideal division of labour would be one in which the logician could borrow the linguist's grammatical analysis in his account of inference, instead of having to set up his own shop for producing 'logical forms'. Some strands in the present enquiry might be useful in creating such a 'natural logic', witness the following tentative discussion.

The relational perspective of Section 2 is quite congenial to the 'two-term theory' of classical pre-Fregean logic, which stayed close to the surface forms of natural language. And there are more specific analogies. For instance, downward monotonicity of arguments in a quantified statement QXY (cf. Section 3) turns out to be equivalent to what was called 'distributed' occurrence in traditional syllogistic. Likewise, upward monotonicity reflects a classical type of argument called the 'Dictum de Omni':

"what is true of every X is true of what is X".

In our terminology, if every X is Y, and 'X' occurs in upward monotone position in some statement $\ldots X \ldots$, then that same statement holds for Y:$\ldots Y \ldots$. The Dictum de Omni has been regarded as the principle par excellence governing syllogistic inference. Besides, it even extends beyond the latter into the logic of relations. (For instance, De Morgan's famous non-syllogistic relational example 'All horses are animals. Therefore, all horse tails are animal tails' can be subsumed under it, by considering the true statement 'All horse tails are *horse* tails'.) Thus, monotonicity, and perhaps other conditions in the above would seem to be promising notions for a natural logic reviving classical ideals.

Now, one classical ideal was that inference should be studied in harmony with grammatical form. Another important idea, implicit in the formulation of the Dictum de Omni, is that inference rules can be *global*, operating at statement level, without presupposing any proof-theoretic fine-structure analysis. At least in an intuitive psychological sense, this idea seems realistic. Whether these ideals can be realized in a workable theory of inference matching the power of Fregean predicate logic remains to be seen — but we shall outline a program.

A possible set-up for a natural logic comes in several phases.

(1) *Grammatical rules.* For purposes of illustration, one may think here of a simple phrase structure grammar, with rules such as the following: S ⇒ NP VP (*S*entence, *N*oun *P*hrase, *V*erb *P*hrase), NP ⇒ Det N (*Determiner* *N*oun), NP ⇒ NP *and* NP, VP ⇒ V (NP) (*V*erb), V ⇒ V *and* V, Det ⇒ *all, some, no.*

(2) *Inferentially sensitive positions.* The Dictum de Omni needs the following concepts for its precise formulation:

— an occurrence of an expression X in an expression $\alpha(X) = \cdots X \cdots$ is *positive* if $[\![X]\!] \leqslant [\![X']\!]$ implies $[\![\cdots X \cdots]\!] \leqslant [\![\cdots X' \cdots]\!]$ (*notation:* $\cdots \overset{+}{X} \cdots$),

— an occurrence is *negative* when the inverse correspondence holds (*notation:* $\cdots X \cdots$).

Comments. (a) This formulation assumes the existence of a suitable inclusion order $\leqslant$ on all relevant denotations. E.g., for unary predicates, $\leqslant$ is just inclusion — for truth value expressions, it is the usual order on $\{0, 1\}$. Etcetera.

(b) When read in dynamic terms, positive occurrence of X in α means both:

— increasing the denotation of X will at most increase that of α,

— decreasing the denotation of X will at most decrease that of α.

So, the main point of the distinction $+/-$ is that of direct versus inverse correlation.

Now, in order to exploit positive and negative occurrences in inference, they must be syntactically recognizable. Thus, this information must have been built in during the very process of sentence construction.

(3) *Inference marking.* Marking inferentially sensitive positions requires several phases. First, the 'logic' of certain lexical items consists (partly) in their $+/-$ effects on their linguistic environment. Then, the $+/-$ effects of various phrase structure rules have to be taken into account. And finally, the $+/-$ effects of combined rules are to be established by some rule of calculation. In this way, sentences are constructed (or understood) with marked inferentially sensitive positions of key items.

This process may be described as an algorithm on any phrase structure tree. But here, we shall only display a few examples.

EXAMPLE 1 ($+/-$ effects of phrase structure rules). The rule $S \Rightarrow NP\,VP$ gets the marking $\overset{+}{NP}\,VP$. The reason lies in the nature of NP-denotations, as given in Section 2: enlarging these will only make the sentence 'more true'. There is no VP-marking, however: both increasing [[VP]] and decreasing it may change a truth value from 1 to 0, depending on the NP.

The general principle is this: in 'functional' phrase structure rules, the functor gets a + value. (Perhaps a tricky case: the correct rule for $VP \Rightarrow V\,NP$ turns out to be $V\,\overset{+}{NP}$.)

EXAMPLE 2 ($+/-$ effects of specific lexical items). The determiners have the double monotonicity effects described in Section 3. E.g., for (*all* N) VP, the marking is as in Fig. 7a; while 7b gives the pattern for V (*all* N). Connectives also have their expected behaviour. E.g., *and* behaves as in Fig. 7c.

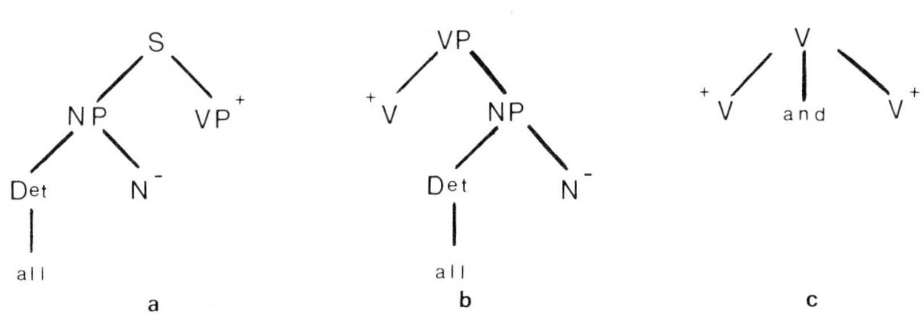

Fig. 7.

Example 3 (+/− effects of nestings). The rule of overall calculation is simply algebraic multiplication:

$$+ + = +, \qquad + - = -, \qquad - + = -, \qquad - - = +.$$

Using these tools, concrete sentences can be inferentially marked.

Example 4. "No girl loves a cat" has a structure as in Fig. 8, resulting in the marking

$$\text{No } \overset{-}{\text{girl}} \ \overset{-}{\text{loves}} \text{ a } \overset{-}{\text{cat}}.$$

By considering some examples, these predictions are borne out. The sentence indeed implies: no *pretty* girl loves a cat, no girl *loves and kisses* a cat, no girl loves a *curly* cat.

That there are still some surprises in store is shown by

Example 5. "No girl loves a cat and no dog". In this case, the two final determiners disagree on the marking of the main verb, and hence the overall pattern becomes just

$$\text{No } \overset{-}{\text{girl}} \text{ loves a } \overset{-}{\text{cat}} \text{ and no } \overset{+}{\text{dog}}.$$

Even in this sketchy form, the mechanism of a modest natural logic will have become clear. Now, as it stands, the above logic only accounts for monotonicity inferences — an important class, but by no means the only one, even in direct practice. (For instance, monotonicity alone cannot

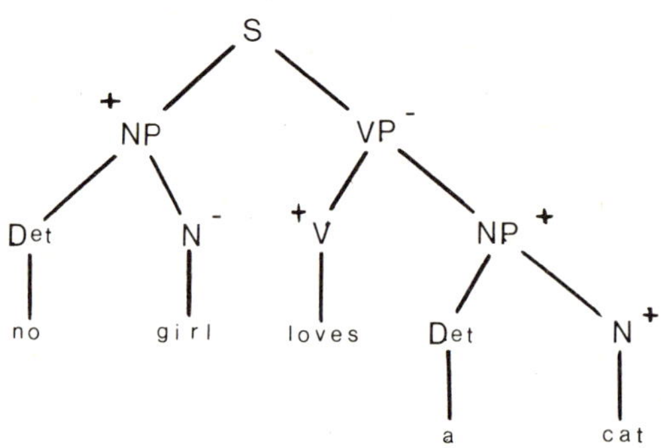

Fig. 8.

account for the characteristic inference of Conjunction. Cf. Section 5.) Thus, additions will be necessary. Staying with the earlier general conditions, a good candidate for the natural logic of a fragment like the above would be *Monotonicity*, as indicated, together with *Conservativity*, as well as several algebraic 'working rules' for the connectives.

EXAMPLE 6. Transitivity follows by Monotonicity:

$$\frac{all\ X\overset{+}{Y} \qquad all\ YZ}{all\ XZ}$$

(Notice that marking the second premise will not do. Thus, there is some basis for the traditional distinction between a 'major' and a 'minor' premise in syllogistic inference.)

EXAMPLE 7. Conjunction follows by Monotonicity and Conservativity:

$$\frac{\dfrac{all\ XY}{all\ X(Y\ and\ \overset{+}{X})} \qquad all\ XZ}{all\ X(Y\ and\ Z)}$$

What the above approach provides us with, for fragments of natural language, is a theory of logical inference directly based upon grammatical form. On the other hand, of course, such fragments also have a 'Fregean logic', through the usual transcriptions into predicate-logical form. Thus, the vexed controversy between 'classical' and 'modern' logic can be given an exact content: how do the two logics compare for various fragments of natural language? In particular, is the natural logic of the above fragment *complete* with respect to its Fregean rival? In this way, a philosophical quarrel becomes a matter of research.

Appendix: further topics

(1) *Extensions of the above inferentially marked grammar.* At least one reasonably realistic fragment ought to be treated in the above manner. A category which should be covered at least is that of relative clauses — as these are the natural means for obtaining higher nestings of quantifiers in language.

(2) *Comparisons with predicate logic for specific phenomena.* Although the above theory can handle both syllogistic and relational kinds of inference, it seems especially weak in capturing involved anaphoric relations. Therefore, it is at this point that a more detailed comparison with Fregean logic will be useful.

(3) *Connections with attempts at reviving classical logic.* There exist various attempts at rehabilitating pre-Fregean views, often condemned by modern orthodoxy. The present analysis might change this verdict in some cases. Notably, the very suggestive work of SOMMERS 1982 ought to be reviewed in this light. For other types of connection with classical logic, see VAN BENTHEM 1986, chapter 6, on the Square of Opposition.

(4) *Further theoretical issues.* In addition to the above-mentioned completeness questions, there are also more intrinsic questions for a natural logic. For instance, what is the precise relation between the semantic and syntactic accounts of positive and negative occurrence, as presented above?

(5) *Choice of an optimal grammatical paradigm.* Although a phrase structure grammar has some didactic virtues, a categorial grammar often becomes preferable when generality and elegance are to be achieved. (Cf. the general considerations at the end of Section 5.) ZWARTS 1985 contains a theory of this kind, using the system of LAMBEK 1958. One important topic then becomes the systematic *interplay* between the calculi of type change and of inference.

8. Further logical prospects

Several new directions of logical research have been introduced in this paper. Many others had to be omitted. For instance, even just quantification in natural language has many other logical aspects not considered here. One notable example of philosophical interest is the pervasive duality between its *discrete* uses (as in this paper) and the *continuous* ones (*no* milk, *some* tea, *much* wine). One further challenge is to develop a unified quantifier theory covering both uses. (Cf. VAN BENTHEM 1983a for a first attempt.) This extension, as well as other ones, is considered in the monograph VAN BENTHEM 1986, which gives the full story behind the present paper.

There are also aspects of our subject matter that will probably remain beyond the resources of this framework. A case in point are the subtle differences in universal quantification between *all*, *every*, *each*, *any* — with the attendant differences in plural and singular nouns. More sensitive approaches from the 'fourth phase' (cf. Section 1) may be needed here.

Nevertheless, the present 'third phase' developments do provide a kind of laboratory for themes to be developed in the wider semantics. For instance, a systematic development of *constraints* on semantic structures will also be useful both in discourse representation theory and situation semantics. The enterprise of a *natural logic* has already become a concern for both these theories as well. And finally, our more sensitive denotational analyses may conceivably lead to a study of semantic complexity and *learnability*.

But the new directions of this paper can also stand on their own. We have found a whole realm of logical questions about the basics of our subject, where many used to see mere metaphysics (if anything at all). Moreover, interesting new insights were forthcoming by elementary means: a rather pleasant form of economy. It is a comforting thought that, even at an elementary level, our discipline remains full of surprises.

References

BARWISE, J. and COOPER, R., 1981, *Generalized quantifiers and natural language*, Linguistics and Philosophy 4, pp. 159–219.

BARWISE, J. and PERRY, J., 1983, *Situations and Attitudes* (Bradford Books, Montgomery).

VAN BENTHEM, J., 1982, *The Logic of Time*, Synthese Library vol. 156 (Reidel, Dordrecht).

VAN BENTHEM, J., 1983a, *Determiners and Logic*, Linguistics and Philosophy 6, pp. 447–478.

VAN BENTHEM, J., 1983b, *Five easy pieces*, in: A. ter Meulen, ed., Studies in Modeltheoretic Semantics, GRASS series vol. 1 (Foris, Dordrecht), pp. 1–17.

VAN BENTHEM, J., 1983c, *Possible Worlds Semantics: a research program that cannot fail?*, Report 83–29, Dept. of Math., Simon Fraser Univ., Burnaby. (Also in Studia Logica 43, pp. 379–393.)

VAN BENTHEM, J., 1983d, *The Semantics of Variety in Categorial Grammar*, Report 83–26, Dept. of Math., Simon Fraser Univ., Burnaby. (To appear in: W. BUSZKOWSKI et al., eds., Categorial Grammar (Benjamin, Amsterdam).)

VAN BENTHEM, J., 1984a, *Foundations of conditional logic*, J. Philosophical Logic 13, pp. 303–349.

VAN BENTHEM, J., 1984b, *Questions about quantifiers*, J. Symbolic Logic 49, pp. 443–466.

VAN BENTHEM, J., 1984c, *The logic of semantics:* in F. Landman and F. Veltman, eds., Varieties of Formal Semantics, GRASS series vol. 3 (Foris, Dordrecht), pp. 55–80.

VAN BENTHEM, J., 1985, *Semantic Automata*, Report 85-27, Center for the Study of Language and Information, Stanford. (To appear in: D. de Jongh et al., eds., Information, Interpretation and Inference, GRASS series vol. 5 (Foris, Dordrecht).)

VAN BENTHEM, J., 1986, *Essays in Logical Semantics*, Synthese Language Library (Reidel, Dordrecht).

BUSZKOWSKI, W., 1982, *Lambek's Categorial Grammar*, Instytut Matematyki, Uniwersytet Im. Adama Mickiewicza, Poznań.

VAN DEEMTER, K., 1985, *Generalized quantifiers — finite versus infinite*, in: J. van Benthem and A. ter Meulen, eds., Generalized Quantifiers in Natural Language, GRASS series vol. 4 (Foris, Dordrecht), pp. 145–159.

HIGGINBOTHAM, J. and MAY, R., 1981, *Questions, quantifiers and crossing*, The Linguistic Review 1, pp. 4–80.

JANSSEN, T., 1983, *Foundations and Applications of Montague Grammar* (Mathematical Centre, Amsterdam).

KAMP, H., 1979, *Instants, events and temporal discourse*, in: R. Bäuerle et al., eds., Semantics from Different Points of View (Springer, Berlin), pp. 376–417.

KAMP, H., 1981, *A theory of truth and semantic representation*, in: J. Groenendijk et al., eds., Formal Methods in the Study of Language (Mathematical Centre, Amsterdam). (Reprinted in J. Groenendijk et al. eds., Truth, Interpretation and Information, GRASS series vol. 2 (Foris, Dordrecht), pp. 1–41.)

KEENAN, E. and FALTZ, L., 1985, *Boolean Semantics for Natural Language*, Synthese Language Library 23 (Reidel, Dordrecht).

KEENAN, E. and MOSS, L., 1985, *Generalized quantifiers and the expressive power of natural language*, in: J. van Benthem and A. ter Meulen, eds., Generalized Quantifiers in Natural Language, GRASS series vol. 4 (Foris, Dordrecht), pp. 73–124.

KEENAN, E. and STAVI, Y., 1982, *A semantic characterization of natural language determiners*, Linguistics and Philosophy, to appear.

KLEIN, E., 1982, *The interpretation of adjectival comparatives*, J. Linguistics 18, pp. 113–136.

KRATZER, A., 1981, *The notional category of modality*, in: H.-J. Eikmever and H. Rieser, eds., Words, Worlds and Contexts (W. de Gruyter, Berlin).

LAMBEK, J., 1958, *The mathematics of sentence structure*, Amer. Math. Monthly 65, pp. 154–170.

LEWIS, D., 1973, *Counterfactuals* (Blackwell, Oxford).

MONTAGUE, R., 1974, *Formal Philosophy* (Yale Univ. Press, New Haven, CT).

MOSTOWSKI, A., 1957, *On a generalization of quantifiers*, Fund. Math. 44, pp. 12–36.

PARTEE, B., 1984, *Compositionality*, in: F. Landman and F. Veltmans, eds., Varieties of Formal Semantics, GRASS series vol. 3 (Foris, Dordrecht), pp. 281–311.

QUINE, W., 1966, *Variables explained away*, in: Selected Logic Papers (Random House, New York).

SOMMERS, F., 1982, *The Logic of Natural Language* (Clarendon, Oxford).

THIJSSE, E., 1983, *Laws of Language*, Master's Thesis, Department of Philosophy, Rijksuniversiteit, Groningen.

VELTMAN, F., 1985, *The Logic of Conditionals*, Dissertation, Filosofisch Instituut, Universiteit van Amsterdam.

WESTERSTÅHL, D., 1982, *Logical constants in quantifier languages*, Linguistics and Philosophy, to appear (1985).

WESTERSTÅHL, D., 1984, *Some results on quantifiers*, Notre Dame J. Formal Logic 25, pp. 152–170.

WESTERSTÅHL, D., 1986, *Quantifiers*, in: D. Gabbay and F. Guenthner, eds., Handbook of Philosophical Logic, vol. 4 (Reidel, Dordrecht).

ZUCKER, J., 1978, *The adequacy problem for classical logic*, J. Philosophical Logic 7, pp. 517–535.

ZWARTS, F., 1983, *Determiners: a relational perspective*, in: A. ter Meulen, ed., Studies in Modeltheoretic Semantics, GRASS series vol. 1 (Foris, Dordrecht), pp. 37–62.

ZWARTS, F., 1985, *Modeltheoretic Semantics and Natural Language: a case study in modern Dutch*, Dissertation, Nederlands Instituut, Rijksuniversiteit, Groningen.

Barcan Marcus et al., eds., Logic, Methodology and Philosophy of Science VII

THE RELEVANCE OF QUANTUM LOGIC IN THE DOMAIN OF NON-CLASSICAL LOGICS

MARIA LUISA DALLA CHIARA

Istituto di Filosofia dell' Università di Firenze, Italy

The research for general classification criteria in the universe of logics is a well known crucial question of the logical investigations of recent years. This research is mainly responsible for a shift, in the contemporary studies about the foundations of logic, from a *metalogical* to a *metametalogical* approach, in the same way as the logical work of the Twenties and of the Thirties was characterized by a shift from a *logical* to a *metalogical* attitude.

I will refer to a very general notion of *logic*. Roughly, any logic L may be intended as a theory for a *consequence-relation* $\vDash_L$ (or more generally for a system of consequence-relations) which may hold between sets of formulas and formulas (or sets of formulas and sets of formulas) of a given language, where a consequence-relation is intended to be characterized either in a *proof-theoretical* or in a *model-theoretical* way, and is supposed to satisfy some obvious minimal conditions (like for instance closure under substitution).

For historical reasons, adopting a kind of "ptolemaic description" of the logics which have been studied so far, one may recognize within this universe a singular point represented by classical logic (CL): the sublogics of CL are usually termed *weak*, whereas the superlogics of CL are called *strong*. Strangely enough, the weak logics are often classified in the literature as *philosophical logics*, whereas many strong logics (for instance all the strong logics which are studied in *abstract model theory*) are usually recognized as *mathematical logics*.

Quantum logic (QL) is a weak logic. Since most of the significant peculiarities of this logic appear already at the sentential level, for the sake of simplicity, in this paper, I will mainly refer to sentential QL, which I will suppose formalized in a standard language with sentential letters

$(p_1, p_2, \ldots)$ and the usual connectives $(\neg, \wedge, \vee, \rightarrow)$. The conditional connective will not appear in all cases that I will consider.

As is well known, two kinds of semantical approaches have turned out to be particularly successful in the attempt to characterize the elements of the subuniverse of weak logics: the *algebraic* and the *Kripke-semantics.* From an intuitive point of view, the algebraic semantics is founded on the following basic idea: to interpret a language consists essentially in associating a *truth-value* (or a *truth-degree*) to any sentence of the language. On the contrary, in a Kripkian semantical approach, one assumes that to interpret a language consists in determining the *situations* (or the *worlds*) where any sentence of the language holds. Whereas the algebraic approach seems to be founded on a somewhat *universal method,* the range of applicability and the limits of the Kripkian semantics are still an object of discussion.

In this paper, I will refer to a Kripkian characterization of QL [6], [9], [2]. Generally, the notion of *Kripke-realization* for a sentential language can be described as a system

$$\mathcal{K} = \langle I, R_1, \ldots, R_n, \Pi, \rho \rangle$$

consisting of a set of *worlds* I, a sequence of *world-relations* $R_1, \ldots, R_n$, a set of *propositions* Π (which contains a privileged proposition — possibly empty — called the *absurd proposition*), and an interpretation-function ρ, which associates to any sentence of the language a proposition in Π (representing, intuitively, the meaning of the sentence). In the different logics ρ will satisfy different sets of conditions C.

The general semantical definitions I will refer to are the usual ones:

DEFINITION 1. A world i is said to *verify* a sentence α $(i \models \alpha)$ iff $i \in \rho(\alpha)$.

α is called *true* in a realization $\mathcal{K}$ $(\models_{\mathcal{K}} \alpha)$ iff for any world i of $\mathcal{K}$: $i \models \alpha$.

α is called a *consequence in* $\mathcal{K}$ of a set of sentences T $(T \models_{\mathcal{K}} \alpha)$ iff for any world i of $\mathcal{K}$: if for any $\beta \in T$, $i \models \beta$ then $i \models \alpha$.

The subsystem of a realization $\mathcal{K}$ consisting of the set of worlds, of the world-relations and of the set of the propositions is called the *frame* of $\mathcal{K}$; the part of the frame containing only the set of worlds and the world-relations will be called the *semiframe* of $\mathcal{K}$. If $\mathcal{F}$ is a class of similar frames and C a set of conditions, we will indicate by $\mathrm{Real}^C(\mathcal{F})$ the class of all possible realizations whose frame is in $\mathcal{F}$ and that satisfy C. Let L be a logic with a corresponding consequence-relation $\models_L$; we will say that a class $\mathrm{Real}^C(\mathcal{F})$ *characterizes* (*strongly*) the logic L iff the following condition holds: for any set of sentences T and any sentence α,

$$T \models_L \alpha \quad \text{iff} \quad \text{for any } \mathcal{K} \in \mathrm{Real}^C(\mathcal{F}): \quad T \models_{\mathcal{K}} \alpha.$$

An interesting case is represented by the class of Kripke-frames with a single world-relation R (which is usually called *accessibility-relation*). I will speak in such cases of *simple* frames. It seems very natural to distinguish a class of simple frames, where the accessibility-relation is at least reflexive and transitive (i.e. it is a *pre-order*) and a class of frames where the accessibility-relation is at least reflexive and symmetrical (i.e. it is a *similarity-relation*). I will use the symbol $\leqslant$ for pre-orders, and the symbol $\not\perp$ for similarity-relations. The first class gives rise, in a natural way, to a class of logics, which for obvious reaons may be called *epistemic* (intuitionistic logic and the so called *intermediate logics* belong to this class); the second class gives rise to logics, which may be called *similarity-logics*. Accordingly, I will speak in the two different cases respectively of *epistemic* and *similarity*-frames (and semiframes).

From an intuitive point of view, one can understand pretty well the reasons why similarity-logics may find interesting applications in the logical analysis of physical theories. Indeed, in the case of physical theories, what is generally interesting to describe is not the "possible evolutions of states of knowledge with respect to a constant world"; rather "sets of physical situations which may be similar with respect to which states of knowledge must single out some invariants".

When considering a simple semiframe, one can define in terms of the accessibility-relation R the notion of *possible proposition* of the semiframe:

DEFINITION 2. Let $\langle I, R\rangle$ be a semiframe and $X \subseteq I$. X is called a *possible proposition* ($X \in \Pi^{\mathrm{P}}$) of $\langle I, R\rangle$ iff

$$\text{om}\, i\, \{i \in X \text{ iff om}\, j\, [Rij \Rightarrow \text{ex}\, k\, (k \in X \text{ and } Rkj)]\}^1.$$

In other words, a possible proposition is a kind of *maximal* set of worlds, which contains all and only the worlds whose accessible worlds are not unaccessible to the whole set.

For the epistemic semiframes (but not for the similarity-frames) one can prove:

[1] om i (ex i) is a metalinguistic abbreviation for "for any i" ("for at least one i").

THEOREM 1. *A set of worlds X is a possible proposition of the semiframe $\langle I, R\rangle$ iff it is R-closed (i.e. it satisfies the following condition:* om i, j $(i \in X$ *and* $Rij \Rightarrow j \in X))$.

As a consequence, one may conclude that the notion of possible proposition represents a good generalization for the intuitive concept of *possible meaning* of a sentence, which seems to adequately behave both in the case of epistemic and similarity-frames.

Let $\langle I, R\rangle$ be either an epistemic or a similarity-semiframe; we may define, on the power-set of the set of worlds I, a 1-ary operation ⦹ which we will call *propositional complement:*

DEFINITION 3. For any $X \subseteq I$, $X^{⦹} = [i \mid \text{om}\, j\, (Rij \text{ and } j \in X \Rightarrow j \in J)]$ where J is the absurd proposition of $\langle I, R\rangle$.

In other words, a world i belongs to the propositional complement of X iff any world j accessible to i, which belongs to X, belongs to the absurd proposition.

THEOREM 2. *In the epistemic and similarity-semiframes, there holds*:

(1) $X \subseteq I \Rightarrow X^{⦹} \in \Pi^{\mathrm{P}}$.

(2) $X, Y \in \Pi^{\mathrm{P}} \Rightarrow X \cap X \in \Pi^{\mathrm{P}}$.

In the epistemic (but generally not in the similarity-semiframes) there holds:

(3) $X, Y \in \Pi^{\mathrm{P}} \Rightarrow X \cup Y \in \Pi^{\mathrm{P}}$.

In the following, for the sake of simplicity, I will usually consider only *scotian* similarity-semiframes, where the absurd proposition J is empty.

On this basis, we can now define the two notions of *epistemic* and *similarity-(Kripkian) realization* for a sentential language.

DEFINITION 4. An *epistemic realization* is a system $\mathscr{K} = \langle I, R, \Pi, \rho\rangle$ where

(1) $\langle I, R, \Pi\rangle$ is an *epistemic frame*, that means: $\langle I, R\rangle$ is an epistemic semiframe and Π is a set of possible propositions closed at least under ⦹, $\cap$, $\cup$ and containing I.

(2) $\rho(p) \in \Pi$,
$\rho(\neg\beta) = \rho(\beta)^{⦹}$,
$\rho(\beta \wedge \gamma) = \rho(\beta) \cap \rho(\gamma)$,
$\rho(\beta \vee \gamma) = \rho(\beta) \cup \rho(\gamma)$.

DEFINITION 5. *A similarity-realization* is a system $\mathscr{K} = \langle I, R, \Pi, \rho\rangle$ where

(1) $\langle I, R, \Pi\rangle$ is a *similarity-frame*, that means: $\langle I, R\rangle$ is a similarity-

semiframe and Π is a set of possible propositions closed at least under $^{\oplus}$, $\cap$ and containing I.

(2) $\rho(p) \in \Pi$,
$\rho(\neg\beta) = \rho(\beta)^{\oplus}$;
$\rho(\beta \wedge \gamma) = \rho(\beta) \cap \rho(\gamma)$,
$\rho(\beta \vee \gamma) = (\rho(\beta)^{\oplus} \cap \rho(\gamma)^{\oplus})^{\oplus}$.

DEFINITION 6. A realization $\mathcal{K} = \langle I, R, \Pi, \rho \rangle$ is called *standard* iff the set of propositions Π coincides with the set of all possible propositions Π^{P}.

The class of all similarity-realizations characterizes a weak form of QL, which has been called *minimal quantum logic* (MQL) (or also *orthologic*). This logic satisfies, among others, the following principles[2]:

$$\models_{\mathrm{MQL}} \alpha \vee \neg\alpha \qquad (\textit{tertium non datur}),$$

$$\neg\neg\alpha \models_{\mathrm{MQL}} \alpha \qquad (\text{strong double negation}),$$

$$(\alpha \wedge \beta) \vee (\alpha \wedge \gamma) \models_{\mathrm{MQL}} \alpha \wedge (\beta \vee \gamma) \qquad (\text{weak distributivity}).$$

But the strong distributivity breaks down:

$$\alpha \wedge (\beta \vee \gamma) \not\models_{\mathrm{MQL}} (\alpha \wedge \beta) \vee (\alpha \wedge \gamma).$$

A stronger form of QL may be obtained by requiring that the set of propositions, in any realization $\mathcal{K}$, satisfies the *orthomodular property*, which can be defined equivalently by each of the following conditions:

(O1) For any propositions X, Y of $\mathcal{K}$:

$$X \subseteq Y \quad \text{iff} \quad X \cap (X \cap Y)^{\oplus} = \emptyset.$$

(O2) For any propositions X, Y of $\mathcal{K}$:

$$X \cap [(X \cap (X \cap Y)^{\oplus})]^{\oplus} \subseteq Y.$$

The logic characterized by the class of all orthomodular similarity-realizations has been called *orthomodular quantum logic* (OQL); and just this logic has found successful applications to the logical analysis of quantum mechanics. An interesting feature of OQL is represented by the fact that in this logic one can define a well-behaved conditional connective in terms of negation and conjunction:

$$\alpha \to \beta \overset{\text{def}}{=} \neg(\alpha \wedge \neg(\alpha \wedge \beta)).$$

[2] Given a logic L, the notation $\models_L \alpha$ ($\alpha \models_L \beta$) is used as an abbreviation for $\emptyset \models_L \alpha$ ($\{\alpha\} \models_L \beta$).

This connective, first proposed by Mittelstaedt and Finch, and which is often referred to as *Sasaki-hook*, turns out to be a well-behaved conditional, in the sense that it satisfies the following conditions, which have been often proposed in the literature as minimal conditions to be required for a binary connective in order to be accepted as a "good" conditional:

(I) $\models \alpha \to \alpha$ (identity-principle),

(II) $\alpha \wedge (\alpha \to \beta) \models \beta$ (Modus Ponens principle).

Our second formulation of the orthomodular property (O2) shows immediately that the Sasaki-hook will not generally satisfy the Modus Ponens principle in a non orthomodular realization; hence it cannot represent a "good" conditional for MQL.

An important theorem, which has been proved by Goldblatt, states that: orthomodularity is not generally describable as an elementary property of the accessibility-relation. On this basis Goldblatt concludes that: "this is further evidence of the intractability of OQL. Since it is perhaps the first example of a natural and significant logic that leaves the usual methods defeated" [10].

Both MQL and OQL admit of a number of different axiomatizations and soundness and completeness proofs with respect to the Kripke-semantics. Significantly enough, the canonical model which is constructed in the completeness proof for OQL turns out to be a nonstandard realization (in the sense of Definition 6) [4].

Quantum logics give rise to some characteristic *metalogical anomalies*, which lead to conjecture that the distinction between epistemic and similarity-logics represents a highly significant dividing line within the universe of weak logics. Let L be any logic characterized by a Kripkian semantics; for the sake of simplicity, I will refer here to scotian logics, but the restriction is not essential. One may define the following pairs of semantical concepts (where α, β represent any sentences):

DEFINITION 7. (1) α is called *realizable* (Real α) iff α has a *quasi-model* (i.e. a realization, where at least one world verifies α).

(2) α is called *verifiable* (Ver α) iff α has a *model* (i.e. a realization, where any world verifies α).

DEFINITION 8. (1) β is a *weak consequence* of α ($\alpha \mid\equiv \beta$) iff any model of α is a model of β.

(2) β is a *consequence* of α ($\alpha \models \beta$) iff for any realization $\mathcal{K}$ ($\alpha \models_{\mathcal{K}} \beta$).

These definitions may be naturally extended also to sets of sentences.

Now, in the case of most familiar epistemic logics, one can prove that the two members of both pairs collapse into the same concept. Namely:

$$\text{Ver}\,\alpha \quad \text{iff} \quad \text{Real}\,\alpha,$$

$$\alpha \models \beta \quad \text{iff} \quad \alpha \mid\!\equiv \beta.$$

On the contrary, in the case of similarity-logics, generally there holds only:

$$\text{Ver}\,\alpha \quad \Rightarrow \quad \text{Real}\,\alpha,$$

$$\alpha \models \beta \quad \Rightarrow \quad \alpha \mid\!\equiv \beta.$$

All this recalls an analogous situation which arises for open formulas in classical first-order logic, where — as is well known — one may properly distinguish between verifiability and realizability, and between strong and weak logical consequence. On this ground, one may conclude that, strangely enough, sentences in similarity-logics turn out to share some characteristic properties that in classical logic (and also in most familiar epistemic logics) hold only for open formulas.

A significant example of a sentence, which is realizable but not verifiable in QL [3] is the following α (which asserts a particular negation of the distributive law in terms of negation, conjunction and disjunction):

$$\alpha = \neg\{[\beta \wedge (\gamma \vee \delta)] \wedge [(\beta \wedge \gamma) \vee (\beta \wedge \delta)]\} \wedge$$
$$\neg\{\neg[\beta \wedge (\gamma \vee \delta)] \wedge \neg[(\beta \wedge \gamma) \vee (\beta \wedge \delta)]\}.$$

Since, using the soundness and the completeness theorem of QL (with respect to a given axiomatization) one can prove that:

$$\text{Real}\,\alpha \quad \text{iff} \quad \alpha \nvdash_{\text{QL}} \beta \wedge \neg\beta \quad \text{for any } \beta;$$

one may conclude that in QL there are non contradictory sentences that do not admit any model.

As an interesting consequence, one obtains a violation of the *Lindenbaum-property* and of a strong variant of the *Robinson-property*. The Lindenbaum-property is violated, because there are non-contradictory sentences (for instance the just constructed α) which cannot be extended to any non contradictory and *complete* set of sentences T (such that for any sentence β, either $\beta \in T$ or $\neg\beta \in T$). Indeed, one can easily show that any non contradictory and complete set of sentences has trivially a model; and we already know that our sentence α cannot have any model. Further, a strong variant of the Robinson-property is violated, for the following equivalence breaks down:

$$\text{Ctr } T_1 \cup T_2 \quad \text{iff} \quad \text{for at least one sentence } \alpha:$$

$$T_1 \vdash \alpha \quad \text{and} \quad T_2 \vdash \neg\alpha.$$

Indeed, one can easily show that in QL the Robinson-property (in this formulation) implies the Lindenbaum-property, and — as we already know — the Lindenbaum-property breaks down.

An interesting question is the following: can we characterize by means of a Kripke-semantics any significant "very weak" logic, which shares at the same time some characteristic aspects of both epistemic and similarity-logics? From an intuitive point of view, a somewhat natural approach to this question might be the following: it is well known that MQL admits of a modal interpretation in the (classical) modal system **B** [9], [2]. Let us now take, instead of classical **B**, an intuitionistic version (**IB**) of **B** and let us consider the same linguistic translation τ which works in the case of MQL and classical **B** (the language of **IB** is supposed to contain the intuitionistic connectives $\sim$, $\cdot$, $\curlyvee$ and the modal operators L, M):

$$\tau(p) = LMp \quad \text{for any atomic sentence } p,$$

$$\tau(\neg\beta) = L \sim \tau(\beta),$$

$$\tau(\beta \wedge \gamma) = \tau(\beta) \cdot \tau(\gamma),$$

$$\tau(\beta \vee \gamma) = LM(\tau(\beta) \curlyvee \tau(\gamma)).$$

Now, if we next interpret intuitionistic logic in the classical modal system S_4 (according to the well known Gödel's translation) we will obtain an overall-tranlation τ^* of the language of QL into a *bimodal* classical logic (whose language contains the classical connectives $-$, &, $\dot{\vee}$, the **B**-operators L, M and the S_4-operators $\Box$, $\Diamond$):

$$\tau^*(p) = LM\Box p \quad \text{for any atomic sentence } p,$$

$$\tau^*(\neg\beta) = L\Box - \tau^*(\beta),$$

$$\tau^*(\beta \wedge \gamma) = \tau^*(\beta) \,\&\, \tau^*(\gamma),$$

$$\tau^*(\beta \vee \gamma) = LM(\tau^*(\beta) \dot{\vee} \tau^*(\gamma)).$$

I will use this bimodal translation of the language of QL only as a heuristic guide, which suggests a Kripkian characterization of a weak logic, which I will call *epistemic quantum logic* (EQL)[3].

[3] EQL is deeply close to a weak logic that in an other context Mittelstaedt and Stachow have called *effective quantum logic*.

EQL is characterized by the the class of all *epistemic-similarity realizations*, where the notion of epistemic-similarity realization is defined as follows:

DEFINITION 9. An *epistemic-similarity realization* is a system $\mathcal{K} = \langle I, \not\perp, \leqslant, \Pi, \rho\rangle$ where

(1) $\langle I, \not\perp, \leqslant\rangle$ is an *epistemic-similarity semiframe*, consisting of a set of worlds I, and two world-relations: a similarity-relation $\not\perp$ and a pre-order relation $\leqslant$, correlated by the following conditions:

(C1) $i \leqslant j$ and $j \not\perp k \Rightarrow i \not\perp k$;

(C2) $i \not\perp j$ and $j \leqslant k \Rightarrow$ ex h ($h \not\perp k$ and $h \leqslant i$).

(2) Π contains all (and only) the possible propositions of $\mathcal{K}$, that means all the subsets of I which are at the same time possible propositions with respect to $\not\perp$ and $\leqslant$ (in the sense of Definition 2).[4]

(3) $\rho(p) \in \Pi$,

$$\rho(\neg\beta) = [i \mid \text{om}\, j \not\perp i \ \text{om}\, k \geqslant j\ (k \notin \rho(\beta))],$$
$$\rho(\beta \wedge \gamma) = \rho(\beta) \cap \rho(\gamma),$$
$$\rho(\beta \vee \gamma) = [i \mid \text{om}\, j \not\perp i\ \text{ex}\, k \not\perp j\ (k \in \rho(\beta) \text{ or } k \in \rho(\gamma))].$$

The definition is correct, since one can prove that for any α, $\rho(\alpha) \in \Pi$.

EQL turns out to satisfy, among others, the following principles:

$\alpha \models_{\text{EQL}} \neg\neg\alpha$ (weak double negation),

$\neg\neg\neg\alpha \models_{\text{EQL}} \neg\alpha$ (Brouwer),

$\alpha \wedge \neg\alpha \models_{\text{EQL}} \beta$ (Duns Scoto),

$(\alpha \wedge \beta) \vee (\alpha \wedge \gamma) \models_{\text{EQL}} \alpha \wedge (\beta \vee \gamma)$ (weak distributivity).

But the *tertium non datur*, the strong double negation and the strong distributivity break down:

$$\not\models_{\text{EQL}} \alpha \vee \neg\alpha,$$

$$\neg\neg\alpha \not\models_{\text{EQL}} \alpha,$$

$$\alpha \wedge (\beta \vee \gamma) \not\models_{\text{EQL}} (\alpha \wedge \beta) \vee (\alpha \wedge \gamma).$$

EQL admits of a class of interesting physical models. In order to understand them, it will be expedient first to recall the form of certain

[4] For the sake of simplicity, we have required here that $\mathcal{K}$ be standard; however, this restriction may be omitted, by supposing convenient closure-conditions on Π.

privileged physical models of OQL. In such models, the set of worlds consists of all *pure states* of a quantum physical system (where any pure state is identified with a particular vector in a Hilbert space); hence the somewhat metaphysical notion of *possible world* acquires here a very concrete physical meaning. The accessibility-relation is the *non-orthogonality* relation between vectors, the set of propositions is identified with the set of all *closed subspaces* of the Hilbert space. Using a fundamental axiom of quantum mechanics (von Neumann's *projection postulate*) one can prove that: two pure states i and j are accessible iff i can be transformed into j after the performance of a measurement (concerning a physical quantity) in the physical system represented by i. It turns out also that in this case the accessibility-relation represents a kind of *weak physical indiscernibility* relation, in the sense that: i and j are accessible iff there is no proposition X such that i satisfies X and j satisfies the propositional complement $X^{\perp}$. As a consequence one obtains, for instance, that the quantum-logical negation acquires the following physical meaning: a physical state i verifies the negation of β iff i cannot be transformed (after the performance of a measurement) into a state which verifies β [5].

Let us now try and characterize similarly a class of natural physical models of EQL. For such models, we will take as a set of worlds, instead of the set of all pure states, the set of all *statistical operators* associated to a quantum physical system. As is well known, according to the orthodox interpretation of quantum mechanics, whereas pure states represent *intrinsic uncertainties* of physical systems, statistical operators, on the contrary, represent also our *ignorance* about the systems. Any statistical operator i is naturally associated to a subspace X^i: intuitively, X^i represents the smallest subspace containing all the pure states in which the physical system represented by the statistical operator i might be with probability different from 0. As to the world-relations, the similarity-relation $\not\perp$ is identified (analogously to the previous case) with the non-orthogonality relation between the corresponding subspaces ($i \not\perp j$ iff X^i is not orthogonal to X^j); the (epistemic) pre-order $\leqslant$ is identified with the inverse of the inclusion-relation between the corresponding subspaces ($i \leqslant j$ iff $X^i \supseteq X^j$). From an intuitive point of view, our epistemic pre-order corresponds to an *increasing of information.* Indeed, a statistical operator j gives more information than a statistical operator i iff the subspace associated to j is a subset of the subspace associated to i. Pure states (associated to unidimensional subspaces) represent a *maximum of information*; hence, in case of pure states, we will have that the epistemic pre-order relation collapses into identity. Finally, the set of propositions Π

is identified with the set of all *principal quasi-ideals* in the algebra of the subspaces (where a set of subspaces S is called a *principal quasi-ideal* iff there exists a subspace Z, such that S contains all and only the non-null subspaces of Z).

In these particular physical models, the connective *not* will acquire an interesting epistemic meaning. It turns out that a physical state verifies $\neg\beta$ iff any other physical state which is physically indiscernible with respect to our original state cannot be extended to a more informational state which verifies β.

As a conclusion, let me now try and sum up the most significant aspects (in my opinion) of the relevance of QL within the universe of weak logics. First of all, as we have seen, quantum logics transfer to sentences characteristic properties that in other logics hold only for open formulas. However, whereas from an intuitive point of view, the origin of the semantical ambiguity of open formulas is clearly intelligible, what might be the cause of a similar semantical ambiguity of sentences in quantum logics?

At the same time, quantum logics transfer to very deep semantical levels characteristic properties which in other logics hold only at syntactical levels. The failure of the Lindenbaum-property represents a significant event which may lead to the following general observation: one can distinguish at least three different degrees of validity of the *tertium-non datur property* in the different logics. At the first degree, the *tertium non datur* holds at the semantical level (for the concept of *truth*) but generally *not* at the syntactical level (for the concept of *theorem*). To this degree belong obviously classical logic and most strong logics. As is well known, by restricting through convenient effectivity-conditions the concept of theorem, one obtains that some "very important" scientific theories are intrinsically incompatible with a syntactical *tertium-non datur* property. At the second degree, the *tertium non datur* generally breaks down simultaneously at the syntactical and at the semantical level. This happens, for instance, to most epistemic logics. However, the validity of the Lindenbaum-property still warrants that any non contradictory theory has at least one model with respect to which every problem expressed in the language of the theory is semantically decided. Metaphorically, we may say: every problem can be semantically decided, at least *in mente Dei.* Finally, at the third level, where even the Lindenbaum-property breaks down, we get a situation of very strong semantical undecidability: there are theories, which are intrinsically incompatible with the semantical *tertium-non datur.* Going on with our metaphor, we might say: God cannot avoid playing dice (theoretically)!

A similar situation arises in connection with the general concept of *logical individual.* As is well known, in any logic theories are not generally required to be *syntactically rich* (in the sense that an existential sentence $\exists x \alpha$ is a theorem iff there is an individual (closed) term t such that $\alpha(t)$ is a theorem). In spite of this, at the semantical level, the notion of truth is usually constructed in such a way that a *semantical richness* property holds: $\exists x \alpha$ is true iff there exists at least one individual of the domain which satisfies α. However, this property breaks down in first-order QL, where it may even happen that: $\exists ! x \alpha$ (there exists exactly one x which is α) is true, while there is no precise individual in the domain that satisfies the property expressed by α! This situation renders, of course, very problematic the possibility of a reasonable description-theory in QL. But, in spite of its apparent logical disagreeability, such a behaviour of the concept of individual seems to fit very well with characteristic features of particular objects in microphyscis (for instance, the particles that are governed by the Bose–Einstein statistics).

On this basis, we may conclude that similarity-logics seem to bring about deep changes concerning some very basic and ancient questions of the foundations of logic, like for instance: what is a logical object? What is a name? What is a property? What is the meaning of identity and of truth?

References

[1] BUGAJSKI, S., 1983, *Languages of similarity*, J. Philosophical Logic 12, pp. 1–8.

[2] DALLA CHIARA, M.L., 1977, *Quantum logic and physical modalities*, J. Philosophical Logic 6, pp. 391–404.

[3] DALLA CHIARA, M.L., 1981, *Some metalogical pathologies of quantum logic*, in: E. BELTRAMETTI and B. VAN FRAASSEN, eds., Current Issues in Quantum Logic (Plenum Press, New York), pp. 147–159.

[4] DALLA CHIARA, M.L., *Quantum logic*, to be published in: Handbook of Philosophical Logic (Reidel, Dordrecht).

[5] DALLA CHIARA, M.L., 1983, *Physical implications in a Kripkian semantical approach to physical theories*, in: Logic in the 20th Century, Scientia, pp. 37–52.

[6] DISHKANT, H., 1972, *Semantics for the minimal logic of quantum mechanics*, Studia Logica 30, pp. 23–36.

[7] FINCH, P.D., 1972, *Quantum logic as an implication algebra*, Bull. Austral. Math. Soc. 2, pp. 101–106.

[8] FISCHER SERVI, G., 1981, *Semantics for a class of intuitionistic modal calcuuli*, in: M.L. DALLA CHIARA, ed., Italian Studies in the Philosophy of Science (Reidel, Dorrecht), pp. 59–72.

[9] GOLDBLATT, R.H., 1974, *Semantic analysis of orthologic*, J. Philosophical Logic 3, pp. 19–35.

[10] GOLDBLATT, R.H., 1984, *Orthomodularity is not elementary*, J. Symbolic Logic 49, pp. 401–404.

[11] HARDEGREE, G.M., 1976, *The conditional in quantum logic*, in: P.A. SUPPES, ed., Logic and Probability in Quantum Mechanics (Reidel, Dordrecht), pp. 55–72.
[12] MITTELSTAEDT, P., 1972, *On the interpretation of the lattice of subspaces of Hilbert space as a propositional calculus*, Z. für Naturforschung 27a, pp. 1358–62.
[13] MITTELSTAEDT, P., 1978, *Quantum Logic* (Reidel, Dordrecht).
[14] STACHOW, E.W., 1975, *Dissertation* (Köln).

6. General Methodology of Science

Barcan Marcus et al., eds., Logic, Methodology and Philosophy of Science VII
 255–289

THEORIES, APPROXIMATIONS, AND IDEALIZATIONS

ILKKA NIINILUOTO

Dept. of Philosophy, Univ. of Helsinki, Finland

1. Introduction

Approximation and idealization are two important aspects of scientific knowledge formation. Many scientific laws, and the empirical evidence which supports them, are stated in an approximate form. Most of the philosophically interesting relations into which scientific theories enter — the relations that theories have to the world, to empirical data, to experimental laws, and to other theories (see Fig. 1) — have more or less approximate character. The method of idealization helps to create exact laws and theories, but — instead of describing the actual world directly — they tell how physical and social systems would behave under idealized counterfactual conditions. Therefore, the applications of such laws to concrete actual situations have to be based upon approximations.

A logical reconstruction of the structure of scientific theories should help us to understand the role of approximation and idealization within scientific theorizing. The so-called Received View of theories (cf. SUPPE, 1974) did not pay very much attention to these problems, however. The notions of truth, confirmation, explanation, and reduction, as developed by Tarski, Carnap, Hempel, and Nagel from the 1930's to the 50's, are primarily intended for non-approximate cases involving non-idealized theories.[1] Still, some of the limitations of this approach were at least recognized. For example, SCRIVEN (1961) argued that the "key property" of physical laws is their "inaccuracy". Duhem had already in 1906 pointed out that Kepler's laws and Newton's theory in fact strictly speaking contradict

[1] For brief discussions about idealization, see HEMPEL (1965), pp. 160–171, RUDNER (1966), BARR (1974), and SCHWARTZ (1978). See also SUPPES (1962), BUNGE (1970), and SUPPE (1974), pp. 42–45.

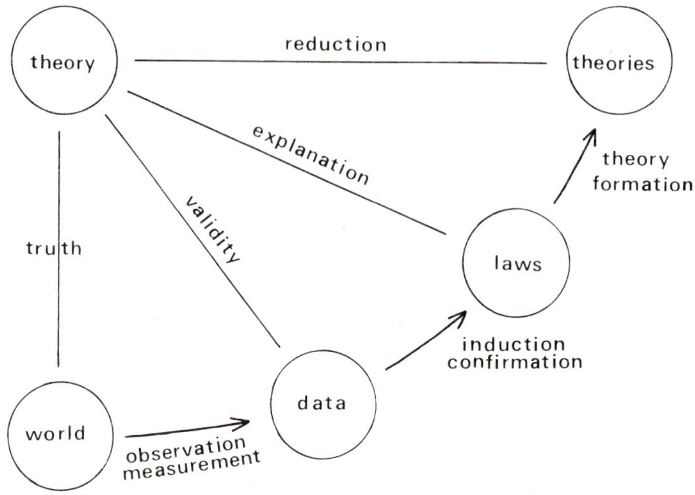

Fig. 1.

each other, so that the standard types of inductive and deductive relations cannot hold between them (DUHEM, 1954, pp. 193–195). This observation was repeated by FEYERABEND (1962), HEMPEL (1965), p. 344, and POPPER (1972), pp. 198–202, and it served as a stimulus for new attempts to outline philosophical accounts of *approximate truth*[2], *approximate explanation*[3], and *approximate reduction*[4]. Other relevant developments include the theory of measurement[5] in the 60's and the semantics of vague, inexact, or fuzzy concepts[6] in the 70's.

Within the latest ten years, approximation and idealization have been widely recognized as belonging to the most central problems of the contemporary philosophy of science. The nature of idealization, and its role within the natural and social sciences, has been the focus of the Poznan school in Poland[7]. The concept of truthlikeness, and the possibility of its

---

[2] Popper proposed his theory of verisimilitude (truthlikeness) in 1960 (cf. POPPER, 1972; NIINILUOTO, 1978). See also SCRIVEN (1961), WOJCICKI (1976), and KRAJEWSKI (1977).

[3] See HEMPEL (1965), p. 344, SCHEIBE (1973), BARR (1974), SCHWARTZ (1978), and TUOMELA (1979).

[4] See SCHAFFNER (1967), POST (1971), NICKLES (1973), STEGMÜLLER (1979), MOULINES (1980). For the related views of Sellars, see PITT (1981).

[5] See SUPPES and ZINNES (1963). Cf. also KUHN (1961) and SNEED (1979).

[6] See the special issues of *Synthese*, Vol. 30, Nos. 3/4 (1975), and Vol. 33, Nos. 2/3/4 (1976).

[7] See NOWAK (1972, 1980) and works by Nowak, Nowakova, Patryas, and others, published in 1975–79 in the *Poznan Studies in the Philosophy of the Sciences and the Humanities*. While the earlier work of the Poznan school was associated with radical conventionalism

definition, has become a hot issue among logicians.[8] Most recently, the collaborators of the Ludwig-approach and the structuralist Suppes–Sneed–Stegmüller-approach to scientific theories have analyzed approximations within science by means of the Bourbaki-notion of uniformity,[9] while Rantala and Pearce have used the tools of non-standard analysis for the same purpose.[10]

In spite of the impressive work done recently in this field, there is at this moment no unified treatment of approximations and idealizations in science. For example, while the treatment of the Poznan school is primarily syntactical, the structuralist school defines approximate reduction essentially as a relation between classes of structures. The diversity of these approaches — which have many obvious connections as well — is at least partly due to the existence of different accounts of the structure of scientific theories. The traditional logistic approach, which views theories as deductively closed sets of sentences in some language, has by now several alternatives: Beth's and VAN FRAASSEN's (1970, 1980) state space conception, LUDWIG's (1978) and SCHEIBE's (1979) Bourbaki-style species-of-structures approach, SNEED's (1979) and STEGMÜLLER's (1979) 'structuralist' or 'non-statement' view, and PEARCE's and RANTALA's (1983) sophisticated treatment in terms of abstract logic.[11]

In this paper, I try to show how one can apply to idealized scientific laws and theories such notions as approximate truth, approximate validity, approximate counterpart, approximate deduction, approximate application, approximate explanation, and approximate prediction. I am primarily interested in quantitative theories which can be formulated by equations between quantities. The relevant notion of a scientific theory is thus closely associated with the state space conception which, I argue, turns out to be a quantitative variant of an emended Carnapian statement view for qualitative theories. My notion of approximation is the standard metric concept

---

(Ajdukiewicz) and the history of instrumentalism (Giedymin), L. Nowak has combined the method of idealization with scientific realism. Nowak has primarily discussed this method of idealization in connection with Marx's economical and social theories. KRAJEWSKI (1977) develops the theme of idealization in the natural sciences.

[8] See NIINILUOTO (1978, 1982a, 1982b, 1982c, 1983a), ODDIE (1981), KUIPERS (1982). See also PRZELECKI (1976), ADAMS (1982) and LAYMON (1980, 1982).

[9] See LUDWIG (1978, 1981), MOULINES (1976, 1980), MAYR (1981a, b), MAJER (1981).

[10] See RANTALA (1979), PEARCE and RANTALA (1983b, c).

[11] For arguments against the tenability or desirability of a sharp distinction between a 'statement view' and a 'non-statement view' of scientific theories, see NIINILUOTO (1981), PEARCE (1982), PEARCE and RANTALA (1983a).

from the mathematical theory of approximation (cf. Rice, 1964, 1969). But before going to technical details (Sections 4–6), I have to outline briefly some the philosophical issues which are related to the problems of approximation and idealization (Sections 2 and 3).

2. Theories and truth

Contemporary philosophers of science can be divided in several groups by reference to issues regarding the notion of truth and its role within science. First, the *semantic realists* support some version of the correspondence theory of truth, while the *semantic anti-realists* replace the realist notion of truth by some epistemic surrogate (e.g., warranted assertability, limit of inquiry). Secondly, the semantic realists can be divided in *scientific realists* who think that all scientific statements (including laws and theories) have a truth value and *scientific instrumentalists* who assign a truth value at most to empirical scientific statements but not to theoretical ones. Scientific realists in turn include *methodological realists* who take truth (usually together with information or systematic power) to be an important aim of scientific inquiry and *methodological non-realists* who replace truth as an aim of science by some methodological surrogate (e.g., successful prediction, empirical adequacy, problem-solving ability). Finally, methodological realism may be *naive* (truths are easily obtained and accumulated) or *critical* (science makes gradual progress towards the truth by finding new theories which have a better correspondence with larger fragments of reality than the old theories).

In terms of these distinctions (see Fig. 2), Dummett and Putnam (1981) are typical semantic anti-realists, Duhem a scientific instrumentalist, Kuhn and Laudan methodological non-realists, while Peirce, Engels, Popper, and Sellars are critical realists. Sneed's and Stegmüller's position has a strongly instrumentalist flavour: theories do not have truth values, but they can be used to make 'empirical claims'. However, it has been suggested that the structuralist view — and Laudan's (1977) related treatment of the problem-solving ability of theories — can be reconstrued in the fashion of critical realism.[12] van Fraassen's 'constructive empiricism' seems to be a variant of methodological non-realism: a theory is capable of having a truth value, but this is irrelevant, since its acceptance involves only the belief that

[12] See Niiniluoto (1981). For Sneed's own recent evaluation of his position, see Sneed (1983).

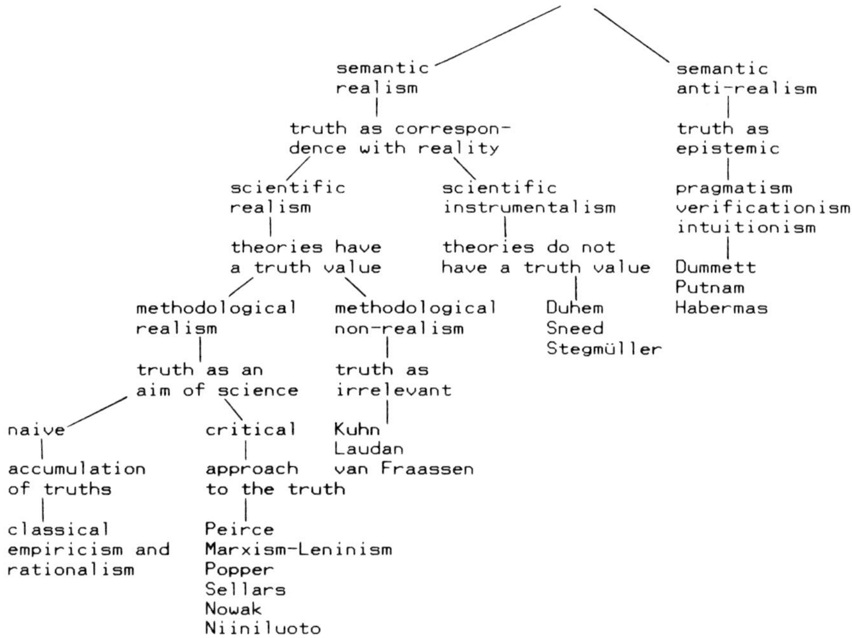

Fig. 2.

it is 'empirically adequate', i.e., what it says about the observable phenomena is true.[13]

A common assumption shared by the methodological realists and most semantic anti-realists is that truth (as *they* define it) is the central aim of science; this distinguishes them from the scientific instrumentalists and the methodological non-realists. On the other hand, there is a common motivation for the redefinition of truth by the semantic anti-realists and for the reformulation of the aim of science by the methodological non-realists, viz. the fear that truth in the realist sense would be a utopian goal for science and the conviction that the search for unreachable goals is irrational (cf. PUTNAM, 1981; LAUDAN, 1981). Already St. Augustine argued *contra academicos* that a man cannot be happy if he only seeks for the truth without ever reaching it.

VAN FRAASSEN (1980), p. 8, defines scientific realism as the doctrine that science aims at "a literally true story of what the world is like" and that the acceptance of a scientific theory "involves the belief that it is true". I think

[13] See VAN FRAASSEN (1980, 1981), and comments by BOYD (1983).

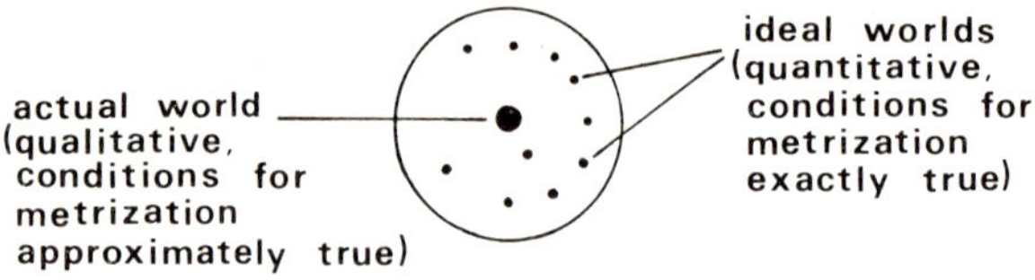

Fig. 3.

qualitative relations can be metricized, i.e., represented (uniquely or up to some class of transformations) by real numbers. For example, conditions guaranteeing the existence of subjective probability distributions and utility functions have been given in various representation theorems. Such conditions are idealizations — they are exactly satisfied only by ideally rational agents. Nevertheless, such conditions may be *approximately true* of agents in real life.

In such situations, quantitative descriptions of reality are 'fuzzy' in the sense that the actual world corresponds to a *class* of structures for the quantitative language, not to a single structure (cf. PRZELECKI, 1976, 1978). However, at the same time the concept of approximate truth should allow us to say the following: even if quantitative subjective probabilities do not 'really' exist in the actual world, there are *ideal worlds arbitrarily close to the actual world* where such quantities do exist. A similar situation may hold for physical quantities. In such cases, the use of quantities in our theories may be justified by the cognitive aim of science: not in instrumentalist terms concerning empirical adequacy and predictive power, but realistically in terms of information about the world and explanatory power. In other words, in the case of Fig. 3, the most efficient way of making progress towards the ultimate best theory about the qualitative actual world may proceed through ideal worlds which contain quantities.

4. Quantitative laws and theories

Carnap's later work in inductive logic[16] was based upon *monadic conceptual systems*, i.e., monadic first-order languages where the non-logical one-place predicates are collected into families

$$\mathcal{F}_1 = \{M_{11}, \ldots, M_{1k_1}\},$$
$$\vdots$$
$$\mathcal{F}_n = \{M_{n1}, \ldots, M_{nk_n}\}.$$

[16] See CARNAP (1971, 1980) and NIINILUOTO (1978).

The predicates within each family are assumed to be mutually exclusive and jointly exhaustive:

$$\vdash \bigvee_{j=1}^{k_i} M_{ij}(x) \qquad (i = 1, \ldots, n)$$

$$\vdash \sim (M_{ij}(x) \;\&\; M_{im}(x)) \qquad (m \neq j)\,(i = 1, \ldots, n).$$

The n families $\mathscr{F}_i$, $i = 1, \ldots, n$, generate an n-dimensional conceptual space $\mathbf{Q}$, where the basic elements are the *Q-predicates* defined by conjunctions

$$M_{1j_1}(x) \;\&\; M_{2j_2}(x) \;\&\; \cdots \;\&\; M_{nj_n}(x), \tag{1}$$

where $1 \leqslant j_1 \leqslant k_1, \ldots, 1 \leqslant j_n \leqslant k_n$.

The Q-predicates constitute a classification system with $q = k_1 \cdot k_2 \cdot \cdots \cdot k_n$ cells: each individual belongs to one and only one cell. Further, Carnap assumed that the distance between the predicates within each family $\mathscr{F}_i$ is defined by a function d_i, and that the distance between two Q-predicates can be defined by the Euclidean measure

$$d(M_{1j_1}(x) \;\&\; \cdots \;\&\; M_{nj_n}(x),\, M_{1m_1}(x) \;\&\; \cdots \;\&\; M_{nm_n}(x))$$

$$= \sqrt{\sum_{i=1}^{n} d_i\,(M_{ij_i}, M_{im_i})^2}. \tag{2}$$

If follows that d is a metric on the space $\mathbf{Q}$ of Q-predicates, i.e., the pair $(\mathbf{Q}, d)$ is a metric space.

Carnap allowed for the possibility that a single family of predicates may be obtained from a countable partition of the value space of a quantity. For example, if we are interested in classifying objects with respect to their length, we might use the family $\mathscr{F}_i$ defined by

$$M_{ij}(x) = \text{the length of } x \text{ is between } j-1 \text{ and } j \text{ cm.}$$

Carnap wished to avoid uncountable partitions, because that would generate an uncountable number of Q-predicates as well. However, if we replace each $\mathscr{F}_i$ simply by the real axis, so that the elements of a family correspond to real numbers, then the Q-predicates correspond to n-tuples of real numbers and the pair $(\mathbf{Q}, d)$ is the n-dimensional Euclidean space R^n. In this case, $\mathbf{Q}$ is called the *state space*. For some applications, $\mathbf{Q}$ may also be a finite subspace of R^n, an infinite-dimensional Hilbert space, or some other normed linear space.

Let L be a first-order language with predicates Q_j $(j = 1, \ldots, q)$ designating the cells in $\mathbf{Q}$. Then sentences of the form

$$\bigwedge_{j=1}^{q} (\pm)(\exists x)Q_j(x), \tag{3}$$

where $(\pm)$ is replaced by the negation sign $\sim$ or by nothing, are called *constituents*. Another way of writing the constituent (3) is to denote by $\mathbf{CT} \subseteq \mathbf{Q}$ those Q-predicates that it claims to be instantiated:

$$\bigwedge_{Q_j \in \mathbf{CT}} (\exists x)Q_j(x) \;\&\; (x)\Big[\bigvee_{Q_j \in \mathbf{CT}} Q_j(x)\Big]. \tag{4}$$

Constituents are the strongest generalizations in L: each quantificational sentence in L can be expressed as a disjunction of some constituents.

While a constituent tells which kinds of individuals there exists in the world, a *nomic constituent*[17] tells what kinds of individuals possibly exist in the world. If $\Diamond$ is the operator for physical (nomic) possibility and $\Box$ for physical necessity, then a nomic constituent has the form

$$\bigwedge_{Q_j \in \mathbf{CT}} \Diamond(\exists x)Q_j(x) \;\&\; \Box(x)\Big[\bigvee_{Q_j \in \mathbf{CT}} Q_j(x)\Big]. \tag{5}$$

Thus, remembering the definition (1) of Q-predicates, a nomic constituent of the form (5) tells *which combinations of properties are physically possible* and which are not. They are the strongest *laws of coexistence* expressible in language $L(\Box)$ (i.e., L enriched with the operators $\Diamond$ and $\Box$). *Probabilistic* laws of coexistence are obtained by replacing the operator $\Diamond$ with a probability measure P which expresses degrees of physical possibility.

To express *laws of succession* relative to the conceptual space $\mathbf{Q}$, we have to relativize statements of the form $Q_j(x)$ to time t, i.e., we write $Q_j^t(x)$, and then replace the Q-predicates $Q_j(x)$ in constituent (4) by conjunctions of the form $Q_j^t(x) \;\&\; Q_m^{t+1}(x)$ (cf. UCHII, 1977). If for $Q_j^t(x)$ only $Q_m^{t+1}(x)$ is physically possible as the next state, then we have deterministic transitions of the form

$$\Box(Q_m^{t+1}(x) \text{ given } Q_j^t(x)). \tag{6}$$

Probabilistic laws of succession are obtained from (6) by replacing the operator $\Box$ with a probability measure P which defines *transition probabilities* of the form $P(Q_m^{t+1}/Q_j^t) = p_m$ (where $p_1 + \cdots + p_q = 1$).

If the families $\mathscr{F}_i$ are defined by real-valued quantities h_i $(i = 1, \ldots, n)$, so that $\mathbf{Q}$ is the n-dimensional Euclidean state space, nomic constituents (5) correspond to quantitative statements which specify the regions of

[17] Cf. UCHII's (1977) notion of 'a basic causal law'. See also NIINILUOTO (1983a).

physically possible states. Typical examples are those laws of coexistence which can be expressed by equations in the space $\mathbf{Q}$:

$$f(h_1(x), \ldots, h_n(x)) = 0. \tag{7}$$

Here f is a real-valued function $f\colon R^n \to R$ of n arguments which expresses the connection between the physically possible values of the quantities $h_1, \ldots, h_n$. For example, Boyle's law has the form

$$\frac{p(x)V(x)}{R} - T(x) = 0, \tag{8}$$

where $p(x)$ is the pressure of x, $V(x)$ the volume of x, $T(x)$ the absolute temperature of x, and R is a constant. If the equation (7) can be solved with respect to $h_1(x)$, i.e., there is a function g which gives the value of $h_1(x)$ in terms of the values $h_2(x), \ldots, h_n(x)$, then the equation

$$h_1(x) = g(h_2(x), \ldots, h_n(x))$$

entails the following: it is physically necessary that $h_1(a) = g(r_2, \ldots, r_n)$ given $h_2(a) = r_2, \ldots, h_n(a) = r_n$.

If discrete time is replaced by the continuous time variable t in transformations (6), deterministic laws of succession (6) will correspond to dynamical equations which express the state of the system x at time t as a function of time t and some initial state at time t_0:

$$(h_1(x, t), \ldots, h_n(x, t)) = F(t, h_1(x, t_0), \ldots, h_n(x, t_0)). \tag{9}$$

Equations of the form (9) are usually obtained as solutions of systems of equations which involve the state variables h_i and their derivatives $\mathrm{d}h_i/\mathrm{d}t$ with respect to time t. Function F in (9) is a transformation function $F\colon R \times \mathbf{Q} \to \mathbf{Q}$ which describes all *physically possible trajectories* relative to space $\mathbf{Q}$.

For example, if a ball with mass m is thrown at time 0 and at point $(0, 0)$ with the velocity v_0 to the direction which forms the angle α with the x-axis, if the gravitation of the earth gives to the ball a constant acceleration g downward the y-axis, and if no other forces are operative, then its position $(s_x(t), s_y(t))$ at time t is given by the ballistic equations

$$s_x(t) = tv_0 \cos \alpha \tag{10}$$

$$s_y(t) = tv_0 \sin \alpha - gt^2/2.$$

By choosing $\alpha = -\pi/2$ and $v_0 = 0$, equation (10) gives as a special case Galileo's law of free fall

$$s_y(t) = -gt^2/2. \tag{11}$$

Probabilistic laws of coexistence and laws of succession relative to the state space **Q** can be defined by specifying the physical probabilities of states or the probabilities of trajectories.

What we have achieved so far is a straightforward derivation of van Fraassen's state space conception of physical theories[18] as a generalization of nomic constituents (5) and transformations (6) relative to discrete conceptual spaces. To outline a semantics for this notion of a theory, let us first write the law (7) in a more complete form

$$(x)(Cx \Rightarrow f(h_1(x), \ldots, h_n(x)) = 0), \tag{12}$$

where Cx says that x is a (physical or social) object or system of a certain kind, and $\Rightarrow$ is an intensional if-then-connective. (Sentence '$p \Rightarrow q$' is read: if it were the case that p, then it would be the case that q.) Then (12) may express a true counterfactual about worlds which do not contain any individuals satisfying the condition C. Thus, (12) has *models* without individuals of type C. On the other hand, law (12) directly asserts that its consequent holds in those possible worlds which satisfy its antecedent. Hence, the *intended models* of (12) are structures of the form

$$\mathcal{U} = \langle D, (h_1(x))_{x \in D}, \ldots, (h_n(x))_{x \in D} \rangle, \tag{13}$$

where D is a non-empty domain of actual or possible objects, Cx is true for all x in D and, for each $x \in D$, the values $h_1(x), \ldots, h_n(x)$ satisfy the equation (7). In particular, a model of (12) may have only one individual a in its domain:

$$\mathcal{U}_a = \langle \{a\}, h_1(a), \ldots, h_n(a) \rangle.^{19} \tag{14}$$

To obtain the structuralist conception of a theory, let M_p be the class of all structures of the type (13), let $M \subseteq M_p$ be the class of models for the equation (7), and let $J \subseteq M_p$ be the class of those structures to which the scientific community intends to apply the equation (7). Then $K = \langle M_p, M \rangle$ is the *core* of a Sneedian theory-element $\langle K, J \rangle$, and J is its set of *intended*

[18] See VAN FRAASSEN (1970, 1972, 1980), SUPPE (1974, 1976), and DALLA CHIARA (1983). This conception of a theory has been applied to many physical theories, such as classical mechanics, quantum mechanics, and thermodynamics. It is also applicable to many theories from the social sciences. For a formulation of quantum statistical mechanics in a 'fuzzy phase space', see PRUGOVEČKI (1979).

[19] Note that we have restricted the discussion here to monadic quantities which attribute quantitative properties to one individual. If the state space **Q** involves dyadic quantities (such as the distance between x and y or the force by which x attracts y), then the intended models of a theory relative to **Q** typically should contain at least two individuals.

applications.[20] The *empirical claim* of $\langle K, J\rangle$ is the statement that all intended applications are models:

$$J \subseteq M. \tag{15}$$

It is further assumed that J is defined through paradigmatic exemplars: J includes all elements of a set J_0 and other 'sufficiently similar' structures. If J can be defined by the condition

$$(x)((\mathcal{U}_x \in J) \equiv Cx), \tag{16}$$

then the claim (15) is equivalent to the statement (12).

We have not yet in this section paid any attention to the idealizational nature of quantitative laws and theories. As the state space formulation of theories turns out to be very convenient for the study of approximations, we shall take up this topic first and postpone the treatment of idealization to Section 6.

5. Approximation

Let **Q** be a state space generated by the real-valued quantities $h_1, \ldots, h_n$, and let d be a metric on **Q**.[21] Let $Q(a)$ (or $Q(a, t)$) be a statement to the effect that the properties of individual a (at time t) satisfy the point Q in **Q**. Then the *distance* between the statements $Q_1(a)$ and $Q_2(a)$ is simply $d(Q_1, Q_2)$. Further, $Q_1(a)$ is *closer to* $Q_2(a)$ than to $Q_3(a)$ if and only if $d(Q_1, Q_2) < d(Q_1, Q_3)$. (See Fig. 4.) Statements $Q_1(a)$ and $Q_2(a)$ are

---

[20] This sketch of the structuralist conception is simplified, since it does not take into account the distinction between theoretical and non-theoretical functions (cf. SNEED, 1979; BALZER and MOULINES, 1980) or Sneed's notion of constraint. While Sneed and Stegmüller take the intended applications of a theory to be non-theoretical structures, I argued in NIINILUOTO (1981) that the intended application in J could be taken to be structures with theoretical functions. BLAZER (1982, 1983) has recently suggested that J could include any structures $\mathcal{U}$ which are substructures of a theoretical structure $\mathcal{U}'$ in the sense that $\mathrm{dom}(\mathcal{U}) \subseteq \mathrm{dom}(\mathcal{U}')$ and $\mathcal{U}$ contains some of the functions of $\mathcal{U}'$. As $\mathcal{U}'$ itself is a substructure of $\mathcal{U}'$ in this sense, Balzer's proposal includes my formulation as a special case: J may consist of theoretical structures. Balzer also mentions another, extremely empiricist requirement: the intended applications in J consist only of individuals and functions which have been "actually observed, identified and measured". In this reconstruction of the structuralist programme, theories would not be applied to reality but to finite bodies of empirical data. (For such a tendency in Ludwig's conception of a physical theory, see KAMLAH, 1981.)

[21] For a theory of approximation in a qualitative conceptual system **Q**, see NIINILUOTO (1978, 1983a). The latter paper contains my answer to what KUIPERS (1982) calls the problem of 'theoretical verisimilitude'.

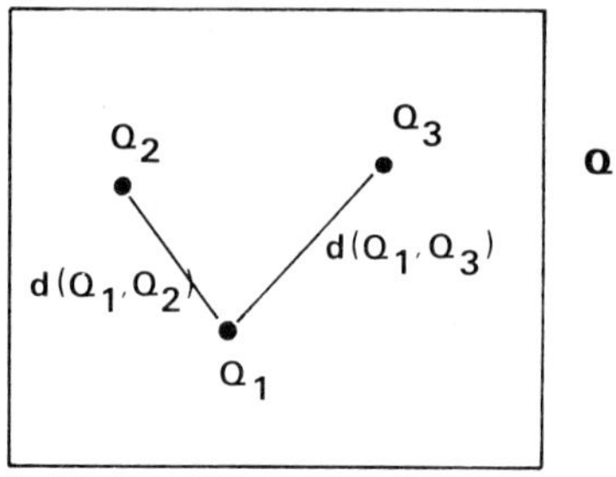

Fig. 4.

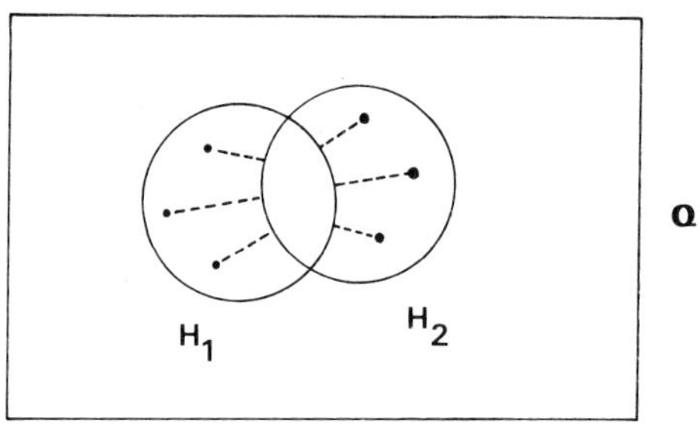

Fig. 5.

δ-*close* if and only if $d(Q_1, Q_2) < \delta$ (where $\delta > 0$ is a constant). Statements $Q_1(a)$ and $Q_2(a)$ are *approximate counterparts* to each other if and only if they are δ-close for sufficiently small δ.

These definitions can be immediately generalized to arbitrary singular sentences about one individual a. Such sentences have the form $H(a)$ (or $H(a, t)$), where H is a region in space $\mathbf{Q}$. The distance between two singular sentences $H_1(a)$ and $H_2(a)$ can be defined by extending the metric d to the subsets of $\mathbf{Q}$. One way of doing this is to use the weighted symmetric difference between H_1 and H_2:

$$d(H_1, H_2) = \frac{1}{2} \sum_{Q \in H_1} \min_{Q' \in H_2} d(Q, Q') + \frac{1}{2} \sum_{Q \in H_2} \min_{Q' \in H_1} d(Q, Q'). \qquad (17)$$

(See Fig. 5.)[22] Hence, in particular,

$$d(H_1, \{Q_2\}) = \frac{1}{2} \sum_{Q \in H_1} d(Q, Q_2) + \frac{1}{2} \min_{Q \in H_1} d(Q, Q_2), \qquad (18)$$

[22] Definition (17) is based on the proposal for the distance between theories in NIINILUOTO (1978). It is not the only possibility for defining $d(H_1, H_2)$. For the special case where H_1 is an interval of real numbers and H_2 is a real number, see NIINILUOTO (1982b).

$$d(\{Q_1\},\{Q_2\}) = d(Q_1, Q_2). \tag{19}$$

These notions allow us to define a number of methodologically interesting concepts. For example, sentence $H(a)$ is *approximately deducible* from premises Σ if and only if an approximate counterpart of $H(a)$ is deducible from Σ. A theory Σ *approximately explains* $H(a)$ if and only if Σ explains an approximate counterpart of $H(a)$. A sequence of sentences $H_1(a), H_2(a), \ldots$ is *convergent* if and only if for all $\varepsilon > 0$ there is a $n_0 > 0$ such that $d(H_m(a), H_n(a)) < \varepsilon$ for all $m \geqslant n_0$, $n \geqslant n_0$. Sequence $H_1(a), H_2(a), \ldots$ *converges to* $H(a)$ if and only if $d(H_n(a), H(a)) \rightarrow 0$, when $n \rightarrow \infty$.

Assume that the truth about the location of individual a in space $\mathbf{Q}$ is expressed by sentence $Q_*(a)$. Then the *degree of truthlikeness* of sentence $H(a)$ is defined by $1/(1 + d(H, \{Q_*\}))$ (cf. formula (18)). Sentence $H(a)$ is *approximately true* if and only if $d(H, \{Q_*\})$ is sufficiently small.[23] Sentence $H_1(a)$ is *closer to the truth* than sentence $H_2(a)$ if and only if $d(H_1, \{Q_*\}) < d(H_2, \{Q_*\})$. A sequence of sentences $H_1(a), H_2(a), \ldots$ *converges to the truth* if it converges to $Q_*(a)$.

If some of the quantities $h_1, \ldots, h_n$ defining $\mathbf{Q}$ are semantically indeterminate, so that the truth about individual a has to be represented by the sentence $H_*(a)$, where H_* is a region in $\mathbf{Q}$ (cf. Section 3), then the degree of truthlikeness of $H(a)$ is defined by $1/(1 + d(H, H_*))$ (cf. formula (17)).

Two quantitative laws are approximate counterparts to each other if their distance is sufficiently small. Thus, all the concepts defined above can be generalized to laws as soon as we have introduced a way of measuring the distance between laws.

Let us consider first laws of coexistence of the form (7), where the equation $f(h_1(x), \ldots, h_n(x)) = 0$ can be solved with respect to the first argument:

$$h_1(x) = g(h_2(x), \ldots, h_n(x)). \tag{20}$$

Equation (20) defines a surface in space $\mathbf{Q}$ if the function $g\colon R^{n-1} \rightarrow R$ is continuous. (If $n = 2$, this surface reduces to a curve in R^2.) Let g_1 and g_2 be two such continuous functions which define surfaces in Q. Then the distance between the corresponding laws can be measured by the L_p-

[23] This notion — which in the fashion of Popper's verisimilitude combines the ideas of truth and information — should be distinguished from the notion of being 'false but almost true'. Sentence $H(a)$ may be said to be *almost true* if $\min_{Q \in H} d(Q, \{Q_*\})$ is sufficiently small (but greater than 0). For example, if it is true that $\theta = 2.5$, then the claim $\theta \geqslant 2.6$ is almost true, but its degree of truthlikeness is not very high.

metrics for function spaces:

$$L_p(g_1, g_2) = \left(\int_{R^{n-1}} |g_1(z) - g_2(z)|^p \, dz \right)^{1/p}. \tag{21}$$

As special cases of these Minkowski metrics, we obtain the *city-block metric* ($p = 1$), the *Euclidean metric* ($p = 2$), and the *Tchebycheff metric* ($p = \infty$):

$$L_1(g_1, g_2) = \int_{R^{n-1}} |g_1(z) - g_2(z)| \, dz \tag{22}$$

$$L_2(g_1, g_2) = \sqrt{\int_{R^{n-1}} (g_1(z) - g_2(z))^2 \, dz}$$

$$L_\infty(g_1, g_2) = \sup_{z \in R^{n-1}} |g_1(z) - g_2(z)|.$$

According to L_1, two laws are close to each other if the *volume* between the corresponding surfaces is small. On the other hand, L_∞ requires that the *maximum* distance between these surfaces must be small.[24] (See Figs. 6 and 7.)

Deterministic laws of succession correspond to functions of the form $F: R \times \mathbf{Q} \rightarrow \mathbf{Q}$ (cf. (9)). The distance between two such laws can be defined by applying the L_p-metrics and the metric d on $\mathbf{Q}$:

$$L_p(F_1, F_2) = \left(\int_R \int_{\mathbf{Q}} d(F_1(t, Q), F_2(t, Q))^p \, dt \, dQ \right)^{1/p}. \tag{23}$$

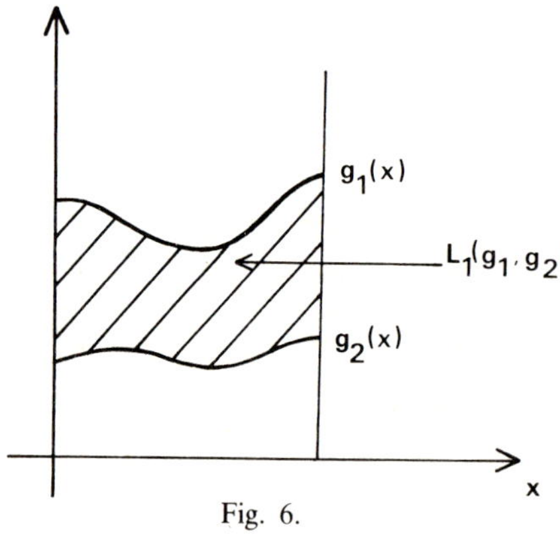

Fig. 6.

[24] Note that L_1 and L_2 give finite values only if the functions g_1 and g_2 are restricted to a finite subspace of R^{n-1} — or some suitable normalization is used. L_3 has the advantage that such qualifications are unnecessary.

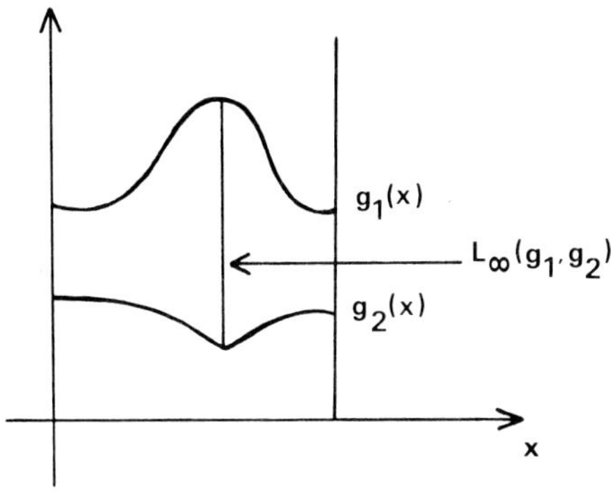

Fig. 7.

For probabilistic laws, we have to define the distance between two probability measures. This can be done, e.g., by using Jeffreys' notion of divergence.[25]

We can now apply these ideas in the following definitions. A theory T *approximately explains* a law if and only if T explains an approximate counterpart of the law. A law of coexistence of the form (20) is *approximately true* if the L_p-distance of g from the true g_* is sufficiently small.[26] A law of succession of the form (9) is approximately true if the L_p-distance of F from the true F_* is sufficiently small. Similarly, the notions *closer to the truth* and *convergence to the truth* can be defined for sequences of quantitative laws.

To see how the approximate truth of a law is reflected on the level of its intended models, assume that $\mathcal{U} = \langle D, (h_1(x))_{x \in D}, \ldots, (h_n(x))_{x \in D} \rangle$ is a model of the true law $h_1(x) = g_*(h_2(x), \ldots, h_n(x))$. If the law $h_1(x) = g(h_2(x), \ldots, h_n(x))$ is approximately true, then it has a model $\mathcal{U}' = \langle D, (g(h_2(x), \ldots, h_n(x))_{x \in D}, (h_2(x))_{x \in D}, \ldots, (h_n(x))_{x \in D} \rangle$ which is 'close' to $\mathcal{U}$. For example, if $\langle \{a\}, 1/5, 5 \rangle$ is a model of

$$h_1(x) = 1/h_2(x),$$

then $\langle \{a\}, 10/51, 5 \rangle$ is a model of

$$h_1(x) = 1/(h_2(x) + 1/10).$$

Thus, to any model $\mathcal{U}$ of the true law there is a model $\mathcal{U}'$ of the approximately true law such that $\mathcal{U}$ and $\mathcal{U}'$ are close to each other. The

[25] Cf. ROSENKRANTZ (1980) and NIINILUOTO (1982b).

[26] For the case where such connection does not exist in **Q**, see Section 6.

distance between two structures of the form

$$\mathcal{U} = \langle D, (u_1(x))_{x\in D}, \ldots, (u_n(x))_{x\in D}\rangle,$$
$$\mathcal{U}' = \langle D, (v_1(x))_{x\in D}, \ldots, (v_n(x))_{x\in D}\rangle$$

can be measured here by

$$d(\mathcal{U}, \mathcal{U}') = \sum_{i=1}^{n} L_p(u_i, v_i), \tag{24}$$

where

$$L_p(u_i, v_i) = \Big(\sum_{x\in D} |u_i(x) - v_i(x)|^p\Big)^{1/p}.$$

If $\mathcal{U}$ and $\mathcal{U}'$ have different but intersecting domains D and D', respectively, then we may take the sum in the definition of $L_p(u_i, v_i)$ over $x \in D \cap D'$.

If J and J' are two classes of structures, let us write $J \sim J'$ if for each $\mathcal{U} \in J$ there is $\mathcal{U}' \in J'$ such that $d(U, U')$ is small and for each $\mathcal{U}' \in J'$ there is $\mathcal{U} \in J$ such that $d(U, U')$ is small. (Cf. MOULINES, 1976.) Then we have the result:

> If Mod(T) and Mod(T') are the classes of models of two laws T and T', respectively, then Mod(T) ~ Mod(T') if and only if T and T' are approximate counterparts. (25)

A law T *applies approximately* to a structure $\mathcal{U}'$ if and only if T is true in a structure $\mathcal{U}$ which is close to $\mathcal{U}'$. By (25), this is equivalent to saying that there is an approximate counterpart T' of T such that T' is true in $\mathcal{U}'$ (see Fig. 8).

These results can be compared to the alternative ways of 'blurring' the empirical claims of Sneedian theory-elements in MOULINES (1976). Three ways of replacing the claim $J \subseteq M$ (cf. (15)) by an 'approximate' claim are the following:

$$\exists J'(J \sim J' \ \&\ J' \subseteq M), \tag{26}$$

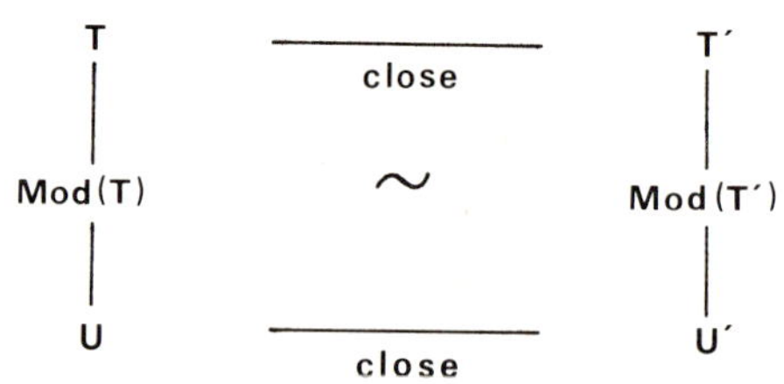

Fig. 8.

$$\exists M'(M \sim M' \mathbin{\&} J \subseteq M'), \tag{27}$$

$$\exists J' \exists M'(J \sim J' \mathbin{\&} M \sim M' \mathbin{\&} J' \subseteq M'). \tag{28}$$

For example, if law (20) is expressed as the empirical claim of a Sneedian theory-element $\langle K, J\rangle$, where the set of the intended applications J has the form

$$J = \{\langle\{a\}, h_1(a), \ldots, h_n(a)\rangle \mid a \in D\},$$

then (26)–(28) correspond to the statements:

$$\forall x \in D\, \exists r_1 \cdots \exists r_n\, (r_1 \approx h_1(x) \mathbin{\&} \cdots \mathbin{\&} r_n \approx h_n(x) \mathbin{\&} r_1 = g(r_2, \ldots, r_n)) \tag{29}$$

$$\forall x \in D\, (h_1(x) \approx g(h_2(x), \ldots, h_n(x)) \tag{30}$$

$$\forall x \in D\, \exists r_1 \cdots \exists r_n\, (r_1 \approx h_1(x) \mathbin{\&} \cdots \mathbin{\&} r_n \approx h_n(x) \mathbin{\&} r_1 \approx g(r_2, \ldots, r_n)). \tag{31}$$

Here (29) says that law (20) applies approximately to the cases J, (30) says some approximate counterpart of (20) applies exactly to the cases J, and (31) says that some approximate counterpart of (20) applies approximately to the cases D.

Thus, (26) and (27) correspond to the Fig. 8 with $\mathcal{U}$ replaced by the set J and $\mathcal{U}'$ replaced by the set J'. Hence, (26) and (27) are equivalent, and both entail (28).[27]

As further methodological notions for quantitative laws, we define approximate validity and approximate prediction. For simplicity, only laws of coexistence will be discussed, but everything we say can be immediately generalized to laws of succession.

Let $H_k = \{\langle h_1(a_i), \ldots, h_n(a_i)\rangle \mid i = 1, \ldots, k\}$ be a finite set of points in space $\mathbf{Q}$. We may think that the elements of H_k represent the results of k ideal observations or measurements of individuals $a_1, \ldots, a_k$ (or averages within classes of such measurements). Then the distance of the law (20) (with function g) from set H_k is defined as a specialization of (21):

$$L_p(g, H_k) = \left(\sum_{i=1}^{k} |g(h_2(a_i), \ldots, h_n(a_i)) - h_1(a_i)|^p \right)^{1/p} \tag{32}$$

The choice $p = 2$ in (32) gives the traditional formula for the *least square difference.* The case $p = \infty$ has been discussed by PATRYAS (1977). (Cf. also KRAJEWSKI, 1977.) Then law (20) is said to be *approximately valid* relative to data H_k if and only if $L_p(g, H_k)$ is sufficiently small.

Let G be a class of functions $g: R^{n-1} \to R$ which has been chosen on the

[27] This conclusion differs from Moulines's claim about the logical relations of (26)–(28). Cf. also BALZER (1981).

basis of *simplicity* considerations. For example, G may consist of linear functions or quadratic functions. Then we may formulate a *rule of acceptance:*

Accept law $g \in G$ on the basis of H_k if and only if g minimizes the distance $L_p(g, H_k)$ among the elements of G. (33)

Rule (33) says that, among the 'simple' hypotheses in G, the one with the highest degree of aproximate validity should be accepted.[28] Instead of aiming at the "simplest useful approximation" (SCRIVEN, 1961, p. 100), it recommends us to find the closest approximation within the class of simple laws.

From the exact law (20) one can make *exact predictions* only if the precise values $h_2(a) = r_2, \ldots, h_n(a) = r_n$ are known for some individual a. From these initial conditions and law (20), we can strictly deduce the prediction $h_1(a) = g(r_2, \ldots, r_n)$ and approximately deduce statements $h_1(a) = r$ where $r \approx g(r_2, \ldots, r_n)$ (i.e., the difference $|r - g(r_2, \ldots, r_n)|$ is small). Moreover, we have the following general result (cf. NIINILUOTO, 1982b):

Assume that the initial conditions $h_i(a) = r_i$ are true $(i = 1, \ldots, n)$. If the law $h_1(x) = g(h_2(x), \ldots, h_n(x))$ is approximately true in the sense of the L_∞-norm, then the prediction $h_n(a) = g(r_2, \ldots, r_n)$ is close to the truth. (34)

In contrast, predictions from true initial conditions based on an approximately true law in the sense of L_1- and L_2-norms are close to the truth in the average. On the other hand, if the initial conditions are not known precisely, the predictions deducible from (20) are likewise imprecise. Thus, the *approximate predictions* of law (20) can be based on the following consequence of (20):

If $h_2(x) \in [r_{20}, r_{21}], \ldots, h_n(x) \in [r_{n0}, r_{n1}]$, then $h_1(x) \in [r_{10}, r_{11}]$, where

$$r_{10} = \min_{\substack{h_2(x) \in [r_{20}, r_{21}] \\ \vdots \\ h_n(x) \in [r_{n0}, r_{n1}]}} g(h_2(x), \ldots, h_n(x)),$$

$$r_{11} = \max_{\substack{h_2(x) \in [r_{20}, r_{21}] \\ \vdots \\ h_n(x) \in [r_{n0}, r_{n1}]}} g(h_2(x), \ldots, h_n(x)). \tag{35}$$

[28] The mathematical theory of approximation gives solutions precisely to such minimatization problems for L_p-metrics (see RICE, 1964, 1969).

However, small changes in the values of $h_2(x), \ldots, h_n(x)$ may induce large errors in the value of $h_1(x)$.[29] Therefore, the following principle is not generally a valid consequence of law (20):

$$\text{If } h_2(x) \approx r_2, \ldots, h_n(x) \approx r_n, \text{ then } h_1(x) \approx g(r_1, \ldots, r_n). \qquad (36)$$

In contrast with the result (34), the predictions based on an exactly true law and on approximately true initial conditions need not be approximately true.[30]

6. Idealization and concretization

In Sections 4 and 5, we have ignored the fact that most laws in science are idealized. For example, Boyle's law (8) holds exactly only for 'ideal gases', and Galileo's law (11) describes exactly only the behaviour of a perfectly spherical body which falls with constant acceleration without disturbing forces in a vacuum near the surface of the perfectly spherical earth. In other words, scientific laws tell what the physically possible behaviour of objects and systems would be under certain counterfactual assumptions.

The problem of idealization is intimately connected with the notion of lawlikeness: laws are often distinguished from merely accidental generalizations by their ability to support counterfactuals. Following HEMPEL (1965), one might then suggest that idealized laws are 'theoretical idealizations' in the sense that they can be derived as special cases of highly lawlike scientific theories. However, this cannot be the whole story, since — as Nowak and Krajewski point out — the general theories (such as Newton's mechanics) may themselves be idealizations. It nevertheless has an important consequence: scientific theories typically have both factual and counterfactual intended applications.

A law of coexistence of the form

$$(x)(Cx \Rightarrow h_1(x) = g(h_2(x), \ldots, h_n(x))) \qquad (37)$$

---

[29] For example, if $h_1(x) = 1000000 h_2(x)$, then an error of the size 1 in the value of $h_2(x)$ generates an error of the size 1000000 in the value of $h_1(x)$.

[30] These results answer the question raised by LAUDAN (1981) about the preservation of approximate truth in deduction. NEWTON-SMITH's (1981) attempt to explicate verisimilitude is based upon a mistaken adequacy condition: "If a theory T_2 has greater verisimilitude than a theory T_1, T_2 is likely to have greater observation success than T_1" (p. 198). This is wrong, since a theory may have a high degree of truthlikeness even if it does not have any observational consequences (cf. NIINILUOTO, 1983b).

may fail to give a realistic description of the actual world, since there is no function g which would express the connection between the quantities $h_1, \ldots, h_n$ in all factual intended applications. In such a case, the state space $\mathbf{Q}$ based upon quantities $h_1, \ldots, h_n$ is insufficient, and $\mathbf{Q}$ has to be extended to a new state space $\mathbf{Q}^k$ by taking into account new factors $w_1, \ldots, w_k$. Then we may hope to find the desired factually true law in $\mathbf{Q}^k$:

$$(x)(Cx \Rightarrow h_1(x) = g_k(h_2(x), \ldots, h_n(x), w_1(x), \ldots, w_k(x))). \tag{38}$$

Law (38) is *factually true* if and only if it includes all the factors which could influence the values of the function $h_1(x)$ in the actual world and states correctly the dependence of $h_1(x)$ of these factors. As (38) is a law, there are then no physically possible worlds where $h_1(x)$ depends on still further factors. But, as we have noted above, law (38) has consequences about counterfactual cases where some of the factors $w_1(x), \ldots, w_k(x)$ do not have any influence on $h_1(x)$.

In the Hegelian terminology, laws involving more factors are more 'concrete' and less 'abstract' than laws involving less factors. The process of arising from laws relative to $\mathbf{Q}$ towards laws relative to $\mathbf{Q}^k$ is called *concretization* by NOWAK (1980) and *factualization* by KRAJEWSKI (1977). Nowak argues that this method of idealization and concretization is a common element of the natural and the social sciences — and that it gives an explication to the 'ascent from the abstract to the concrete' in Marx's *Capital*. He describes this method through the following steps:

$$(x)(Cx \supset h_1(x) = g(h_2(x), \ldots, h_n(x))) \tag{39}$$

$$(x)(Cx \,\&\, w_1(x) = 0 \,\&\, \cdots \,\&\, w_k(x) = 0 \supset h_1(x) = g(h_2(x), \ldots, h_n(x))) \tag{40_0}$$

$$(x)(Cx \,\&\, w_1(x) \neq 0 \,\&\, w_2(x) = 0 \,\&\, \cdots \,\&\, w_k(x) = 0 \supset h_1(x) = g_1(h_2(x), \ldots, h_n(x), w_1(x))) \tag{40_1}$$

$$\vdots$$

$$(x)(Cx \,\&\, w_1(x) \neq 0 \,\&\, \cdots \,\&\, w_k(x) \neq 0 \supset h_1(x) = g_k(h_2(x), \ldots, h_n(x), w_1(x), \ldots, w_k(x))). \tag{40_k}$$

Here $(40)_0$ modifies the original law (39) by explicity introducing the idealized assumptions $w_1(x) = 0, \ldots, w_k(x) = 0$ which claim that the quantities $w_1, \ldots, w_k$ have no influence upon the values of h_1. In the step from $(40)_0$ to $(40)_1$, the first of these assumptions is removed by replacing it with $w_1(x) \neq 0$ and by introducing $w_1(x)$ as a new factor in the equation. The same process is repeated until we finally have removed all the idealizing assumptions and obtained $(40)_k$.[31] Nowak assumes further that each $g_j(h_2(x), \ldots, h_n(x), w_1(x), \ldots, w_j(x))$ can be expressed as a function of $g_{j-1}(h_2(x), \ldots, h_n(x), w_1(x), \ldots, w_{j-1}(x))$ and some function of $w_j(x)$, but this requirement seems to be unnecessarily restrictive. KRAJEWSKI (1977) requires that the process of factualization satisfies the *Principle of Correspondence:*

$$g_1(h_1(x), \ldots, h_n(x), z) \to g(h_2(x), \ldots, h_n(x)), \quad \text{when } z \to 0.$$

$$\ldots$$

$$g_k(h_2(x), \ldots, h_n(x), w_1(x), \ldots, w_{k-1}(x), z)$$
$$\to g_{k-1}(h_2(x), \ldots, h_n(x), w_1(x), \ldots, w_{k-1}(x)) \quad \text{when } z \to 0.$$

Nowak assumes further that the factors $w_1, \ldots, w_k$ are introduced in the process of concretization in the order of their *significance.*[32] If $h_2, \ldots, h_n$ are the most important factors in the determination of h_1 (*principal factors*), then $(40)_0$ expresses the *essence* of the facts about h_1. Law $(40)_0$ describes the essential or "internal structure" of a fact, and the sequence $(40)_0, \ldots, (40)_k$ brings us closer to its "manifest structure". From this perspective, idealizational laws are not insignificant side steps in the progress of science, only practically necessary because we lack knowledge of factual laws, but rather it is an important aim of science to discover the internal structure of facts in their 'pure' form — without 'disturbing' factors.

If the assumptions $w_i(x) = 0$ $(i = 1, \ldots, k)$ are counterfactual, then the use of the material implication $\supset$ in the laws (40) would give the undesirable result that all idealizational laws — independently of their consequents — are true. Nowak concludes that this is a compelling reason for revising the "classical definition of truth" (NOWAK, 1980, pp. 134–135), but a more natural move is to interpret the if-then-connective as the conditional $\Rightarrow$ (cf. Section 4). Moreover, the laws (39) and (40) are written so that they are logically independent of each other — except that (39) entails $(40)_0$ when material implications are used. For these reasons, the process of concretization should be rewritten in the following way:

$$(x)(Cx \Rightarrow E_0(x)), \qquad (T)$$

---

[31] PATRYAS (1975) suggests that laws (39) and (40) should contain a *ceteris paribus* condition as well, but I shall not deal with this issue here. Cf. CARTWRIGHT (1980).

[32] This concept might be defined as follows: w_1 is more significant than w_1' for h_1 (relative to $h_2, \ldots, h_n$) if the concretized function $g_1(h_2(x), \ldots, h_n(x), w_1(x))$ differs more from the original function $g(h_2(x), \ldots, h_n(x))$ than the concretized function $g_1'(h_2(x), \ldots, h_n(x), w_1'(x))$.

$$(x)(Cx \,\&\, w_1(x)=0 \,\&\, \cdots \,\&\, w_k(x)=0 \Rightarrow E_0(x)), \tag{T_0}$$

$$(x)(Cx \,\&\, w_2(x)=0 \,\&\, \cdots \,\&\, w_k(x)=0 \Rightarrow E_1(x)), \tag{T_1}$$

$$\ldots$$

$$(x)(Cx \Rightarrow E_k(x)) \tag{T_k}$$

where

$$h_1(x)=g_0(h_2(x),\ldots,h_n(x)), \tag{$E_0(x)$}$$

$$h_1(x)=g_1(h_2(x),\ldots,h_n(x),w_1(x)), \tag{$E_1(x)$}$$

$$\ldots$$

$$h_1(x)=g_k(h_2(x),\ldots,h_n(x),w_1(x),\ldots,w_k(x)). \tag{$E_k(x)$}$$

Moreover, we assume that the functions $w_1,\ldots,w_k$ have non-zero values in the actual world:

$$(x)(w_1(x)\neq 0 \,\&\, \cdots \,\&\, w_k(x)\neq 0).$$

Then $T_0,\ldots,T_{k-1}$, which explicitly mention counterfactual assumptions, are *idealizational laws*; T and T_k are *factual laws*. A factual law may nevertheless be *idealized* in the sense that it fails to mention some of the actually relevant factors. T is an idealized factual law in this sense, and T_k may be one as well.

If the Principle of Correspondence holds, i.e.,

$$\text{For } j=1,\ldots,k,$$

$$\lim_{z\to 0} g_j(h_2(x),\ldots,h_n(x),w_1(x),\ldots,w_{j-1}(x),z)$$
$$= g_{j-1}(h_2(x),\ldots,h_n(x),w_1(x),\ldots,w_{j-1}(x)), \tag{41}$$

then each law in the list $T_0,\ldots,T_k$ entails the preceding ones. Thus, T_j entails $T_0,\ldots,T_{j-1}$ for all $j=1,\ldots,k$.[33] Moreover, if the function g_1 depends on the argument $w_1(x)\neq 0$, then all the laws $T_1,\ldots,T_k$ are inconsistent with the original law T.[34]

Let us give an example of concretization. One of the idealizations involved in the ballistic equations (10) is the assumption that the resistance

[33] This is similar to KRAJEWSKI's (1977) "renewed implicative version" of correspondence, but instead of saying that (a) T_k entails T_0, (b) $E_k(x)$ entails $w_1(x)=0 \,\&\, \cdots \,\&\, w_k(x)=0 \Rightarrow E_0(x)$, (c) $E_k(x) \,\&\, w_1(x)=0 \,\&\, \cdots \,\&\, w_k(x)=0$ entails $E_0(x)$, he claims that (d) $E_k(x) \,\&\, w_1(x)=0 \,\&\, \cdots \,\&\, w_k(x)=0$ entails $w_1(x)=0 \,\&\, \cdots \,\&\, w_k(x)=0 \supset E_0(x)$.

[34] We exclude here the possibility that, e.g., $g_2(h_2(x), w_1(x), w_2(x)) = g(h_2(w)) + w_1(x) + w_2(x)$ and $w_1(x) = -w_2(x)$ for all x.

of air is zero. If we add to the derivation of (10) the condition that the force due to the resistance of air is proportional to the velocity v of the projectile, i.e., $-\beta v$ where $\beta > 0$ is a constant, then we obtain

$$s_x(t) = \frac{m v_0 \cos \alpha}{\beta}\left(1 - e^{-\beta t/m}\right)$$

$$s_y(t) = -\frac{mg}{\beta}t + \left(\frac{m^2 g}{\beta^2} + \frac{m v_0 \sin \alpha}{\beta}\right)\left(1 - e^{-\beta t/m}\right). \tag{42}$$

If the constant β approaches the limit 0, then the equations (42) approximate more and more closely the equations (10).[35]

In this example, concretization is achieved by means of a general theory: the subsequent concretizations of, e.g., Galileo's laws are derived from Newton's mechanics. If the relation of concretization is denoted by $\dashv$ (cf. NOWAK, 1980), then the sequence

$$T \rightsquigarrow T_0 \dashv T_1 \dashv \cdots \dashv T_k \tag{43}$$

does not adequately describe here the order of discovery, but should be replaced by

(44)

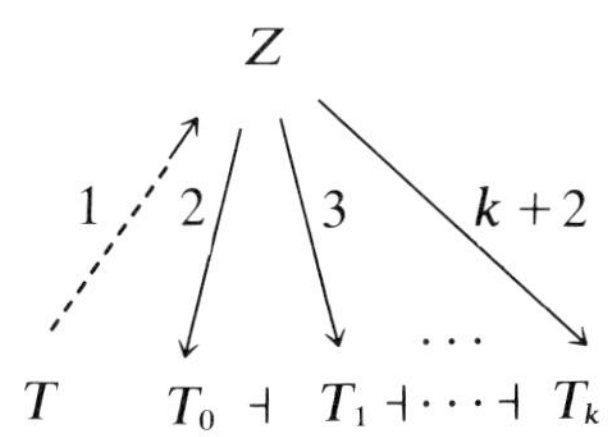

where $\rightarrow$ denotes entailment and numbers indicate the order of the inferential steps. If the theory Z in schema (44) is itself idealized, then the last concretization T_k is likewise idealized in the same respects. However, examples of sequences of type (43) can also be found in many fields of science.

Let us now consider the process of concretization in more detail — in particular, how it is reflected on the level of the structures satisfying the laws. The first thing to notice is that all the laws $T_0, \ldots, T_k$ are expressed in the same state space $\mathbf{Q}^k$. But while T_k describes a surface in $\mathbf{Q}^k$, the

---

[35] In this example, we may think that the concretization is achieved either by introducing the constant function $w_1(x) = \beta > 0$ or the function $w_1(x) =$ the resistance of air on x. In the latter case, we put up the separate assumption $w_1(x) = -\beta v(x)$, but this function does not occur explicitly in the equations (42) any more.

idealizational laws $T_{k-1}, \ldots, T_0$ describe surfaces in more and more restricted subspaces of $\mathbf{Q}^k$. But if the relation of correspondence holds between T_{j-1} and T_j ($j = 1, \ldots, k$), then all these small surfaces are simply parts of the whole surface defined in $\mathbf{Q}^k$ by T_k. On the other hand, the original law T describes a surface in $\mathbf{Q}$, so that the step from T to $T_0, \ldots, T_k$ involves conceptual enrichment. Within the enlarged space $\mathbf{Q}^k$, law T defines a surface which coincides with the surface defined by T_k in the subspace of $\mathbf{Q}^k$ with $w_0 = 0, \ldots, w_k = 0$, but deviates from this surface outside this subspace. As all the laws $T, T_0, \ldots, T_k$ correspond to regions in $\mathbf{Q}^k$, the ideas and methods of Section 5 can be used to measure their distance. For example, the distance of T from a surface in $\mathbf{Q}^k$ can be defined either by the maximum or by the average distance from T to the surface. These suggestions lead to the following result for $j = 1, \ldots, k$:

$$T_j \text{ is closer to } T_k \text{ than } T_{j-1} \text{ is.} \tag{45}$$

But there does not seem to be any general result concerning the relative distances of T and T_j to T_k: as T is more informative than $T_0, \ldots, T_{k-1}$ in making a claim for all values $w_1(x), \ldots, w_k(x)$, it follows that T may in some cases be closer to T_k than $T_0, \ldots, T_j$ for some j. This is still true if T_j is replaced by its *approximate version* AT_j (cf. NOWAK, 1980) which is less restricted but also less precise than T_j:

$$\begin{aligned}(x)(Cx \,\&\, w_{j+1}(x) \leq \alpha_{j+1} \,\&\, \cdots \,\&\, w_k(x) \leq \alpha_k \\ \Rightarrow h_1(x) \approx g_j(h_2(x), \ldots, h_n(x), w_1(x), \ldots, w_j(x))).\end{aligned} \tag{46}$$

Secondly, as T_j entails T_{j-1}, all the models of T_j are also models of T_{j-1}:

$$\mathrm{Mod}(T_k) \subseteq \mathrm{Mod}(T_{k-1}) \subseteq \cdots \subseteq \mathrm{Mod}(T_0).$$

This means that if T_k is a factually true statement, then its consequences $T_{k-1}, \ldots, T_0$ are true counterfactuals,[36] and T is a false factual statement. If T_k is approximately true, then the claims of $T_1, \ldots, T_{k-1}$ for the counterfactual cases are close to the truth in the sense of (34).

To compare the laws $T, T_0, \ldots, T_k$, it is most instructive to consider their intended models (cf. Section 4): for $0 \leq j < k$, T_j is then interpreted as a

[36] There is some unclarity about this point in Nowak. In arguing that idealizational laws T_i cannot be interpretative systems in Hempel's sense, he argues that their idealizing assumptions should be fulfilled in all the models of T_i. (NOWAK, 1980, p. 62.) However, counterfactuals can be true in the actual world. Krajewski's formulation is equally misleading: "The ideal law is fulfilled only in the ideal models". (KRAJEWSKI, 1977, p. 23.)

statement to the effect that the equation $E_j(x)$ holds for those structures which satisfy the idealizing condition $w_{j+1}(x) = \cdots = w_k(x) = 0$. Let I_j be the class of such structures for $\mathbf{Q}^k$, i.e.,

$$I_j = \mathrm{Mod}((x)(w_{j+1}(x) = 0 \;\&\; \cdots \;\&\; w_k(x) = 0)). \tag{47}$$

(See Fig. 9.) Then typical structures $\mathcal{U}_j \in I_j - I_{j-1}$ which satisfy the equation E_j look as follows:

$$\mathcal{U}_0 = \langle \{a\}, g_0(h_2(a), \ldots, h_n(a)), h_2(a), \ldots, h_n(a), 0, 0, \ldots, 0 \rangle,$$

$$\mathcal{U}_1 = \langle \{a\}, g_1(h_2(a), \ldots, h_n(a), w_1(a)), h_2(a), \ldots, h_n(a), w_1(a), 0, \ldots, 0 \rangle,$$

$$\ldots$$

$$\mathcal{U}_k = \langle \{a\}, g_k(h_2(a), \ldots, h_n(a), w_1(a), \ldots, w_k(a)),$$
$$h_2(a), \ldots, h_n(a), w_1(a), \ldots, w_k(a) \rangle. \tag{48}$$

Then all the *factual structures* belong to the class $I_k - I_{k-1}$, and the *idealized structures* to the classes $I_0, \ldots, I_{k-1}$. It is important to notice, however, that the factual law T_k has as its intended models structures from all the classes $I_0, \ldots, I_k$. Similarly, law T_j has intended models in all the classes $I_0, \ldots, I_j$, but not in $I_{j+1}, \ldots, I_k$. The situation is asymmetric in the following sense: idealizational laws do not have factual structures as intended models, but factual laws have idealized (and factual, of course) structures as intended models.

A typical structure $\mathcal{U}$ for $\mathbf{Q}$ satisfying T looks as follows:

$$\mathcal{U} = \langle \{a\}, g_0(h_2(a), \ldots, h_n(a)), h_2(a), \ldots, h_n(a) \rangle. \tag{49}$$

Structure (49) corresponds to an infinite class of structures for $\mathbf{Q}^k$ which are

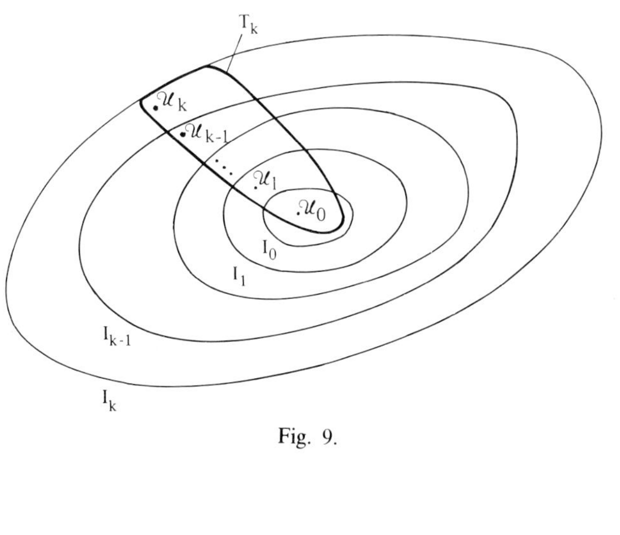

Fig. 9.

also models of T:

$$\{\langle\{a\}, g_0(h_2(a),\ldots,h_n(a)), h_2(a),\ldots, h_n(a), w_1(a),\ldots,w_k(a)\rangle \mid w_1(a)\in R,\ldots,w_k(a)\in R\}. \quad (50)$$

This class (50) contains structures which have arbitrarily great distances from the model $\mathfrak{A}_0$ of T_0. The minimum distance of the elements of class (50) from structure $\mathfrak{A}_k$ depends on the absolute difference between the values $g_0(h_2(a),\ldots,h_n(a))$ and $g_k(h_2(a),\ldots,h_n(a), w_1(a),\ldots,w_k(a))$.

Class I_0 contains the most idealized structures which are, in an obvious sense, most 'distant' from the factual structures in $I_k - I_{k-1}$. On the other hand, it need not always be the case that the distances between structures $\mathfrak{A}_0,\ldots,\mathfrak{A}_k$, as defined by (24), perfectly match with these degrees of idealization: whether the condition

$$d(\mathfrak{A}_0, \mathfrak{A}_k) > d(\mathfrak{A}_1, \mathfrak{A}_k) > \cdots > d(\mathfrak{A}_{k-1}, \mathfrak{A}_k)$$

holds depends essentially on the size and direction of corrections to the value of $h_1(a)$ that the new factors $w_1(a),\ldots,w_k(a)$ make. Nevertheless, the principle of correspondence (41) guarantees the following result:

> For each model $\mathfrak{A}_{j-1}$ of T_{j-1} in $I_{j-1} - I_{j-2}$ there is a sequence of models $\mathfrak{A}_j^m$ of T_j in $I_j - I_{j-1}$, $m = 1, 2, \ldots$, such that $d(\mathfrak{A}_{j-1}, \mathfrak{A}_j^m) \to 0$, when $m \to \infty$. (51)

For example, for a projectile satisfying the parabolic ballistic equations (10) there is a sequence of cases of projectiles with smaller and smaller resistance of air (cf. (42)) which indefinitely approximates the given parabolic case. In this sense, it may be said that theory T_{j-1} is *approximately reducible* to theory T_j if the relation of correspondence holds between T_{j-1} and T_j.[37] This relation of approximate reduction need not hold between the factual theories T and T_k.

Is it the case, as Nowak and Krajeski suggest, that the process of concretization defines sequences of theories which converge towards the truth? The answer to this question of course depends essentially on the theories $T_0,\ldots,T_k$, since concretization can be made in very mistaken ways. If T_k is completely mistaken and therefore has a low degree of truthlikeness, it need not be the case that sequence $T_0,\ldots,T_k$ brings us

---

[37] Cf. the results of RANTALA (1979), MOULINES (1980, 1981), and MAYR (1981a, b) concerning the approximate reduction of the classical mechanics to the relativistic mechanics or Kepler's laws to Newton's theory.

closer to the truth at all. However, if T_k is factually true, then $T_0, \ldots, T_k$ is a sequence of more and more informative true theories — and any reasonable theory of truthlikeness (cf. NIINILUOTO, 1978, 1982c; *pace* ODDIE, 1981) gives the result that the degrees of truthlikeness of $T_0, \ldots, T_k$ increase (cf. also result (45)). However, it depends on the case which of the statements T, T_j, or AT_j has the highest degree of truthlikeness.

Let us conclude this section with some remarks about the testability of idealizational laws and their role in scientific explanation and prediction. If we are dealing with a sequence of type (44), then we may test an idealizational law T_j by testing the theory Z from which it is deducible — and the best way of doing this is to test factual laws (such as T_k) derivable from Z. The use of law T_j for the purposes of explanation and prediction can be replaced by the use of theory Z: to explain the behaviour of a cannon ball we may use Newton's mechanics rather than the idealized ballistic equations. RUDNER (1966) in fact argues that this is the only role that idealizational laws may play in explanation. If we explain a regularity by deriving its concretization from theory Z (e.g., the explanation of Kepler's laws by Newton's theory), then this explanation is not only approximate but also *corrective*, since it shows the original formulation of the regularity to be imprecise.

If we are dealing with a sequence of type (43), then the move to a higher theory Z is not available. If an idealization law T_j is used as a premise of a deductive argument (cf. BARR, 1974, CARTWRIGHT, 1980), its conclusion H_j may approximate the given factual explanadum H, but there is no reason to regard this approximate explanation as corrective (see Fig. 10). Therefore, it is more appropriate to replace T_j by its factual concretization T_k — or, if T_k is not available, by an approximation AT_m of its least idealized known concretization T_m $(j < m < k)$ which is sufficiently broad to include the appropriate initial conditions (see Fig. 11). (Cf. NOWAK, 1980.) These remarks apply to predictions as well.

To test an idealized law T_j in this case, we may either try to calculate what our data about the given case a would have been if the idealizing assumptions $w_{j+1}(a) = 0, \ldots, w_k(a) = 0$ had been valid (SUPPES, 1962) or to

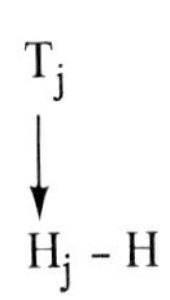

Fig. 10.

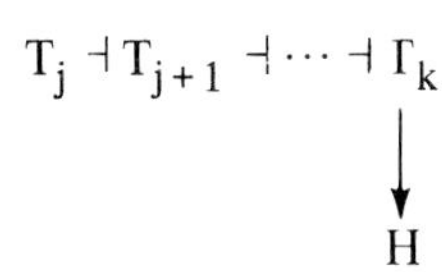

Fig. 11.

imitate experimentally the idealizing conditions of T_j as closely as possible (cf. SUCH, 1978). In the latter case, we are in effect testing the approximate version AT_j of T_j (cf. NOWAK, 1980).[38]

7. Concluding remarks

The treatment of approximation and idealization that has been developed in this paper gives rise to a number of general conclusions.

(1) The notion of metric is less general than Bourbaki's concept of uniformity. However, uniformities can easily be defined by using metric concepts — and, indeed, this seems typically to have been the case also in the methodological applications of this notion. Therefore, it is not evident that there is any real loss in generality if the nature of approximation is analysed by metric distance functions rather than by entourages of uniform spaces. But there is a clear gain in having explicitly defined distances which are based upon a rich and well-developed mathematical theory.

(2) In principle, it seems possible to apply metrics in the study of approximation in connection with any conception of scientific theories. For example, we can use them in Ludwig's approach and in the Sneed–Stegmüller approach by introducing, through a 'micrological' treatment, a metric in the class of structures. On the other hand, the state space conception of theories is especially suitable for the study of approximation: by treating statements as corresponding to regions in the state space, the metric structure of that space can be used in the definition of approximations between statements — and these in turn are reflected in approximations between the models of statements. A sophisticated statement view of theories is able to acount for approximation and also the role of idealizations in science.

(3) A successful analysis of approximations between statements (or between structures) gives us a systematic basis for explicating many important methodological concepts — such as approximate explanation, approximate prediction, approximate reduction, approximate validity, and approximate application. If the notion of truth is well-defined, we can include in this list also the concept of approximate truth. Thereby we obtain an indispensable tool for developing the philosophical basis of

[38] LAYMON (1982) suggests that a scientific theory is confirmed if it can be shown that using more realistic initial conditions will lead to correspondingly more accurate predictions. In our terms, this means the following: an idealized theory T_j is confirmed if we find a concretization T_{j+1} of T_j which has greater approximate validity than T_j.

critical scientific realism. In this sort of realism, the idea of increasing truthlikeness is not a "teleological myth", as Stegmüller has suspected, but rather an exactly definable property that successful sequences of scientific theories possess.

(4) Idealizational laws in science are counterfactuals which are deducible from more 'concrete' factual laws. This analysis implies that as a rule the intended models of a scientific theory include ideal structures, i.e., domains of objects with counterfactual properties. This is especially clear if the class of intended models is assumed to contain a subset of 'exemplars' or 'paradigmatic' applications in Kuhn's sense, as the structuralist reconstruction of 'Kuhn-theories' requires. This is the case, because such exemplars typically consist of the early successful applications of the theory which are repeated as examples and exercises in textbooks — and surely the exemplars for Newton's mechanics, for example, include such strange entities as perfectly spherical projectiles moving in a vacuum or balls rolling with a constant acceleration on a frictionless plane. Still, this dimension of ideal models is lost, if it is thought that theories are always applied to "chunks of empirical reality" (Sneed) or to actually observed and measured "data". Moreover, to accept a theory is definitely more than just to believe that "what a theory says about what is observable is true" (van Fraassen) — it is also to believe that what the theory says about the unobservable, even nonexistent ideal cases is at least approximately true.

(5) The ability of scientific theories to give approximately true descriptions of ideal cases gives us an argument for critical scientific realism. Methodological non-realists have sometimes suggested that there is no need to explain the success of current scientific theories: this is no miracle, since "only the successful theories survive" in the "fierce competition" (VAN FRAASSEN, 1980, p. 40). But this is not sufficient: if theories are chosen on the basis of their observable success, why do they continue to be successful in similar situations? How do they successfully make predictions about unobserved and even counterfactual ideal situations? These are genuine puzzles which need to be answered — and for a realist the only plausible answer is to suppose that the best current theories are in fact close to the truth in the relevant respects.

Bibliography

ADAMS, E., 1982, *Approximate generalizations and their idealization*, in: ASQUITH and NICKLES, 1982, pp. 199–207.

ASQUITH, P.D. and NICKLES, T., eds., 1982, *PSA 1982*, vol. 1 (Philosophy of Science Association, East Lansing).

BALZER, W., 1981, *Sneed's theory concept and vagueness*, in: HARTKÄMPER and SCHMIDT, 1981, pp. 147–163.

BALZER, W., 1982, *Empirical claims in microeconomics*, in: W. Balzer, W. Spohn and W. Stegmüller, eds., Philosophy of Economics (Springer, Berlin), pp. 16–40.

BALZER, W., 1983, *Theory and measurement*, Erkenntnis 19, pp. 3–25.

BALZER, W. and MOULINES, C.U., 1980, *On theoreticity*, Synthese 44, pp. 467–494.

BARR, W.F., 1974, *A pragmatic analysis of idealizations in physics*, Philosophy of Science 41, pp. 48–64.

BOYD, R., 1983, *On the current status of the issue of scientific realism*, Erkenntnis 19, pp. 45–90.

BUNGE, M., 1970, *Theory meets experience*, in: H.E. Kiefer and M.K. Munitz, eds., Mind, Science, and History (State Univ. of New York Press, Albany), pp. 138–165.

CARNAP, R., 1971, *A basic system of inductive logic, Part I*, in: R. Carnap and R. Jeffrey, eds., Studies in Inductive Logic and Probability, vol. I (Univ. of California Press, Berkeley), pp. 33–165.

CARNAP, R., 1980, *A basic system of inductive logic, Part II*, in: R. Jeffrey, ed., Studies in Inductive Logic and Probability, vol. II (Univ. of California Press, Berkeley), pp. 7–155.

CARTWRIGHT, N., 1980, *The truth doesn't explain much*, Amer. Philosophical Quart. 17, pp. 159–163.

DALLA CHIARA, M.L., 1983, *Physical implications in a Kripkian semantical approach to physical theories*, Scientia: Logic in the 20th Century (Milano), pp. 37–52.

DUHEM, P., 1954, *The Aim and Structure of Physical Theory* (Princeton Univ. Press, Princeton).

FEYERABEND, P., 1962, *Explanation, reduction, and empiricism*, in: H. Feigl and G. Maxwell, eds., Minnesota Studies in the Philosophy of Science, vol. III (Univ. of Minnesota Press, Minneapolis), pp. 28–97.

VAN FRAASSEN, B., 1970, *On the extension of Beth's semantics of physical theories*, Philosophy of Science 37, pp. 325–339.

VAN FRAASSEN, B., 1972, *A formal approach to the philosophy of science*, in: R. Colodny, ed., Paradigms and Paradoxes: The Philosophical Challenge of the Quantum Domain (Univ. of Pittsburgh Press, Pittsburgh), pp. 303–366.

VAN FRAASSEN, B., 1980, *The Scientific Image* (Clarendon Press, Oxford).

VAN FRAASSEN, B., 1981, *Theory construction and experiment: an empiricist view*, in: P.D. Asquith and R.N. Giere, eds., *PSA 1980*, vol. 2 (Philosophy of Science Association, East Lansing), pp. 663–677.

GURWITSCH, A., 1967, *Galilean physics in the light of Husserl's phenomenology*, in: E. McMullin, ed.; Galileo, Man of Science (Basic Books, New York), pp. 388–401.

HARTKÄMPER, A. and SCHMIDT, H.-J., eds., 1981, *Structure and Approximation in Physical Theories* (Plenum Press, New York).

HEMPEL, C.G., 1965, *Aspects of Scientific Explanation* (The Free Press, New York).

KAMLAH, A., 1981, *G. Ludwig's positivistic reconstruction of the physical world and his rejection of theoretical concepts*, in: HARTKÄMPER and SCHMIDT, 1981, pp. 71–90.

KRAJEWSKI, W., 1977, *Correspondence Principle and the Growth of Knowledge* (Reidel, Dordrecht).

KUHN, T.S., 1961, *The functions of measurement in modern physical science*, Isis 52, pp. 161–193.

KUIPERS, T., 1982, *Approaching descriptive and theoretical truth*, Erkenntnis 18, pp. 343–378.

LAUDAN, L., 1977, *Progress and Its Problems* (Routledge and Kegan Paul, London).

LAUDAN, L., 1981, *A confutation of convergent realism*, Philosophy of Science 48, pp. 19–49.

LAYMON, R., 1980, *Idealization, explanation, and confirmation*, in: P.D. Asquith and R.N.

Giere, eds., *PSA 1980*, vol. 1 (Philosophy of Science Association, East Lansing), pp. 336–350.

LAYMON, R., 1982, *Scientific realism and the hierarchical counterfactual path from data to theory*, in: ASQUITH and NICKLES, 1982, pp. 107–121.

LUDWIG, D., 1978, *Die Grundstrukturen einer physikalischen Theorie* (Springer, Berlin).

LUDWIG, G., 1981, *Imprecision in physics*, in: HARTKÄMPER and SCHMIDT, 1981, pp. 7–19.

MAJER, U., 1981, *Abstraction, idealization and approximation*, in: HARTKÄMPER and SCHMIDT, 1981, pp. 91–111.

MAYR, D., 1981a, *Investigations of the concept of reduction II*, Erkenntnis 16, pp. 109–129.

MAYR, D., 1982b, *Approximative reduction by completion of empirical uniformities*, in: HARTKÄMPER and SCHMIDT, 1981, pp. 55–70.

MOULINES, C.U., 1976, *Approximative application of empirical theories: a general explication*, Erkenntnis 10, pp. 201–227.

MOULINES, C.U., 1980, *Intertheoretic approximation: the Kepler–Newton case*, Synthese 45, pp. 387–412.

MOULINES, C.U., 1981, *A general scheme for intertheoretic approximation*, in: HARTKÄMPER and SCHMIDT, 1981, pp. 123–146.

NEWTON-SMITH, W.H., 1981, *The Rationality of Science* (Routlege and Kegan Paul, Boston).

NICKLES, T., 1973, *Two concepts of intertheoretic reduction*, J. Philosophy 70, pp. 181–201.

NIINILUOTO, I., 1978, *Truthlikeness: comments on recent discussion*, Synthese 38, pp. 281–329.

NIINILUOTO, I., 1980, *Scientific progress*, Synthese 45, pp. 427–462.

NIINILUOTO, I., 1981, *The growth of theories: comments on the structuralist approach*, in: J. Hintikka et al., eds., Proc. 1978 Pisa Conference on the History and Philosophy of Science, vol. 1 (Reidel, Dordrecht), pp. 3–47.

NIINILUOTO, I., 1982a, *What shall we do with verisimilitude?*, Philosophy of Science 49, pp. 181–197.

NIINILUOTO, I., 1982b, *Truthlikeness for quantitative statements*, in: ASQUITH and NICKLES, 1982, pp. 208–216.

NIINILUOTO, I., 1982c, *On explicating verisimilitude: a reply to Oddie*, British J. Philosophy of Science 33, pp. 290–296.

NIINILUOTO, I., 1983a, *Verisimilitude and legisimilitude*, Studia Logica 42, pp. 315–329.

NIINILUOTO, I., 1983b, *Truthlikeness, realism, and progressive theory-change*, in: J. Pitt, ed., Proc. Fourth International Conference on History and Philosophy of Science, Blacksburg, 1982 (Reidel, Dordrecht).

NIINILUOTO, I., 1984, *Is Science Progressive?* (Reidel, Dordrecht).

NIINILUOTO, I. and TUOMELA, R., eds., 1979, *The Logic and Epistemology of Scientific Change*, Acta Philosophica Fennica 30 (North-Holland, Amsterdam).

NOWAK, L., 1972, *Laws of science, theory, measurement*, Philosophy of Science 39.

NOWAK, L., 1980, *The Structure of Idealization: Towards a Systematic Interpretation of the Marxian Idea of Science* (Reidel, Dordrecht).

NOWAKOVA, I., 1975, *Idealization and the problem of correspondence*, Poznan Studies in the Philosophy of the Sciences and the Humanities 1, pp. 65–70.

ODDIE, G., 1981, *Verisimilitude reviewed*, British J. Philosophy of Science 32, pp. 237–265.

PATRYAS, W., 1975, *An Analysis of the "Caeteris Paribus" clause*, Poznan Studies in the Philosophy of the Sciences and the Humanities 1, pp. 59–64.

PATRYAS, W., 1977, *Idealization and approximation*, Poznan Studies in the Philosophy of the Sciences and the Humanities 3, pp. 180–198.

PEARCE, D., 1982, *Logical properties of the structuralist concept of reduction*, Erkenntnis 18, pp. 307–333.

PEARCE, D. and RANTALA, V., 1983a, pp. *New foundations for metascience*, Synthese 56, pp. 1–26.

PEARCE, D. and RANTALA, V., 1983b, *Correspondence as an intertheory relation*, Studia Logica 42, pp. 363–371.

PEARCE, D. and RANTALA, V., 1983c, *Constructing general models of theory dynamics*, Studia Logica 42, pp. 347–362.

PITT, J.C., 1981, *Pictures, Images and Conceptual Change: An Analysis of Wilfrid Sellars' Philosophy of Science* (Reidel, Dordrecht).

POPPER, K., 1982, *Objective Knowledge: An Evolutionary Approach* (Oxford Univ. Press, Oxford), 2nd ed. 1979.

POST, H.R., 1971, *Correspondence, invariance and heuristics: in praise of conservative induction*, Studies in History and Philosophy of Science 2, pp. 213–255.

PRUGOVEČKI, E., 1979, *Stochastic phase spaces and master Liouville spaces in statistical mechanics*, Foundations of Physics 9, pp. 575–587.

PRZELECKI, M., 1976, *Fuzziness as multiplicity*, Erkenntnis 10, pp. 371–380.

PRZELECKI, M., 1978, *Some approach to inexact measurement*, Poznan Studies in the Philosophy of the Sciences and the Humanities 4, pp. 27–36.

PRZELECKI, M., SZANIAWSKI, K. and WOJCICKI, R., eds., 1976, *Formal Methods in the Methodology of Empirical Sciences* (Reidel, Dordrecht).

PUTNAM, H., 1981, *Reason, Truth, and History* (Cambridge Univ. Press, Cambridge).

RANTALA, V., 1979, *Correspondence and non-standard models: a case study*, in: NIINILUOTO and TUOMELA, 1979, pp. 366–378.

RICE, J.R., 1964, *The Approximation of Functions, Vol. 1: Linear Theory* (Addison-Wesley, Reading, MA).

RICE, J.R., 1969, *The Approximation of Functions, Vol. 2: Nonlinear and Multivariate Theory* (Addison-Wesley, Reading, MA).

ROSENKRANTZ, R., 1980, *Measuring truthlikeness*, Synthese 45, pp. 463–488.

RUDNER, R., 1966, *Philosophy of Social Science* (Prentice-Hall, Englewood Cliffs, NJ).

SCHAFFNER, K., 1967, *Approaches to Reduction*, Philosophy of Science 34, pp. 137–147.

SCHEIBE, E., 1973, *The approximative explanation and the development of physics*, in: P. Suppes, L. Henkin, A. Joja, and Gr.C. Moisil, eds., Logic, Methodology and Philosophy of Science IV (North-Holland, Amsterdam), pp. 931–942.

SCHEIBE, E., 1979, *On the structure of physical theories*, in: NIINILUOTO and TUOMELA, 1979, pp. 205–224.

SCHWARTZ, R.J., 1978, *Idealization and approximations in physics*, Philosophy of Science 45, pp. 595–603.

SCRIVEN, M., 1961, *The key property of physical laws — inaccuracy*, in: H. Feigl and G. Maxwell, eds., Current Issues in the Philosophy of Science (Holt, Rinehart, and Winston, New York), pp. 91–101.

SNEED, J.D., 1971, *The Logical Structure of Mathematical Physics* (Reidel, Dordrecht), 2nd ed. 1979.

SNEED, J.D., 1979, *Quantities as theoretical with respect to qualities*, Epistemologia 2, pp. 215–250.

SNEED, J.D., 1983, *Structuralism and scientific realism*, Erkenntnis 19, pp. 345–370.

STEGMÜLLER, W., 1979, *The Structuralist View of Theories* (Springer, Berlin).

SUCH, J., 1978, *Idealization and concretization in natural sciences*, Poznan Studies in the Philosophy of the Sciences and the Humanities 4, pp. 49–73.

SUPPE, F., 1974, *The search for philosophic understanding of scientific theories*, in: F. Suppe, ed., The Structure of Scientific Theories (Univ. of Illinois Press, Urbana), pp. 1–241.

SUPPE, F., 1976, *Theoretical laws*, in: PRZELECKI *et al.*, 1976, pp. 247–267.

SUPPES, P., 1962, *Models of data*, in: E. Nagel, P. Suppes and A. Tarski, eds., Logic, Methodology and Philosophy of Science: Proceedings of the 1960 International Congress (Stanford Univ. Press, Stanford), pp. 252–261.

SUPPES, P. and ZINNES, J.L., 1963, *Basic measurement theory*, in: R.D. Luce *et al.*, eds., Handbook of Mathematical Psychology, vol. 1 (Wiley, New York), pp. 1–76.

TUOMELA, R., 1979, *Scientific change and approximation*, in: NIINILUOTO and TUOMELA, 1979, pp. 265–297.

UCHII, S., 1977, *Induction and causality in a cellular space*, in: F. Suppe and P.D. Asquith, eds., *PSA 1976*, vol. 2 (Philosophy of Science Association, East Lansing), pp. 448–461.

WOJCICKI, R., 1974, *Set theoretic representations of empirical phenomena*, J. Philosophical Logic 3, pp. 337–343.

WOJCICKI, R., 1976, *Some problems of formal methodology*, in: PRZELECKI *et al.*, 1976, pp. 9–18.

WORRALL, J., 1982, *Scientific realism and scientific change*, The Philosophical Quarterly 32, pp. 201–231.

Barcan Marcus et al., eds., Logic, Methodology and Philosophy of Science VII

THE STRUCTURE OF EMPIRICAL SCIENCE: LOCAL AND GLOBAL

WOLFGANG BALZER*

Seminar für Philosophie, Logik und Wissenschaftstheorie, Univ. München, West Germany

C.-ULISES MOULINES

Inst. de Investigaciones Filosoficas, Univ. Nacional Autonoma de Mexico, Mexico

JOSEPH D. SNEED**

Dept. of Humanities and Social Sciences, Colorado School of Mines, Golden, CO 80401, U.S.A.

Introduction

We outline a method of describing the logical structure of related empirical theories employing a concept of intertheoretical link. The global structure of empirical science is represented as a net of linked theories. The "content" of such a net is the class of all structured arrays of individuals that are consistent with the theories and links in the net. This concept of "content" is used to formulate local empirical claims of theories in a net without formulating a global claim for the net. Characterizations of the distinction between theoretical and non-theoretical concepts, relative to a given theory, and the intended applications of a theory are provided. Our approach derives from [8] and later developments in [1], [2] and [8]. The fundamental concept of intertheoretical link is a generalization and clarification of the concept of "bridge laws" discussed in [4], [6] and [7].

Model elements

Empirical science may be represented as a net of linked theory elements. A theory element consists of some "concepts" K that are used to say

* This author's work was made possible by a fellowship at the Netherlands Institute for Advanced Studies.

** This author's work was supported by a grant from the Atlantic Richfield Foundation.

something about some array of things, the intended applications for the concepts, I. Thus a theory element is an ordered pair $T = \langle K, I(K)\rangle$. The "conceptual apparatus" K of the theory element consists of certain categories [5] of set-theoretic structures. In all categories $\|\chi\|$ associated with empirical theories it appears that the objects $|\chi|$ may be given by a "species of structures" in the sense of Bourbaki [3, p. 259]. For $x, y \in |\chi|$, $\chi(x, y)$ is the set of morphisms "from x to y" and the set of χ-isomorphisms is $I\chi$.[1]

The conceptual core K of a theory element consists of two categories: "potential models" and "models". Potential models are just the kinds of structures that one might claim to be models for a theory. They determine the formal properties of the theory element's conceptual apparatus without imposing any additional restrictions that correspond to empirical laws. We make this distinction more precise by defining a "model element". It is an ordered pair $K = \langle \|M_p\|, \|M\|\rangle$ in which $\|M\|$ is a full sub-category of $\|M_p\|$. We also require that the laws of T be "invariant under M_p-isomorphisms" in the sense that, for all $x, y \in |M_p|$, if $IM_p(x, y) \neq \Lambda$, $x \in |M|$ iff $y \in |M|$. Thus we consider theory elements $T = \langle K, I(K)\rangle$ where K is a model element and $I(K)$ is the range of intended applications of K. One might take the range of intended applications of K, as $I(K)$ — a sub-class of $|M_p|$ and provisionally formulate the empirical claim of T as $I(K) \subseteq |M|$.

Model element links

Intertheoretical links serve to carry information about the values of relations and functions from the applications of one theory to those of another or across different applications of the same theory. Here we shall consider only external links. Among other things, these links provide a kind of "empirical semantics" for a model element that make it more than "just a piece of mathematics". We begin by characterizing the purely formal properties of a binary intertheoretical link between model element cores K' and K. First, we take a step back and consider "relators" beween categories $\|\chi\|$ and $\|\psi\|$. A relator $\|R\|$ between $\|\chi\|$ and $\|\psi\|$ — a $\|\chi\|, \|\psi\|$ relator — is a sub-category of the category $\|\chi\| \times \|\psi\|$. A relator is a generalization of the usual category-theoretic concept of a functor and must satisfy analogous requirements.

---

[1] Consistent with the category-theoretic notation, we shall denote the class of all sets by "|SET|" and the set of all functions from X to Y by "SET(X, Y)".

A link between K' and K is a restriction of the potential models of both theories — a sub-set of $|M'_p| \times |M_p|$. But, a link may have associated with it some restriction of the morphisms of both $\|M'_p\|$ and $\|M_p\|$. The morphisms associated with the links will have to do with transformations of just those components of the structures in $|M'_p|$ and $|M_p|$ whose values are correlated by the link. This suggests that we might regard a binary link between the model element cores K' and K as an $\|M'_p\|, \|M_p\|$ relator $\|\lambda\|$ in which $|\lambda|$ characterized the mutual restriction of the potential models and

$$\lambda(\langle x', x\rangle, \langle y', y\rangle) \subseteq M'_p(x', y') \times M_p(x, y)$$

characterized the pairs of M'_p and M_p-morphisms associated with transformations of the linked components. Links should have certain properties that relators, in general, do not have. Generally, a $\|\chi\|, \|\psi\|$ relator need not contain any $\chi \times \psi$-isomorphisms at all since it is not a full sub-category of $\|\chi\| \times \|\psi\|$. We do not want all $M'_p \times M_p$-isomorphisms, but only those that are associated with the components in the potential models that are correlated by the link. The requirement (D1-A-2-a) below will suffice. If $\langle x', x\rangle$ and $\langle y', y\rangle$ are λ-linked, x' is M'_p-isomorphic to y' and x is M_p-isomorphic to y, then there is some λ-linked pair of isomorphisms "connecting" x' with x and y' with y. λ-isomorphic pairs will typically have many $M'_p \times M_p$-isomorphisms that are not λ-isomorphisms. These correspond to transformations of components of the potential models other than those that are correlated by λ. It also seems clear that, whenever x' and x are λ-linked and y' is empirically equivalent to x', there should be some y that is empirically equivalent to x that is λ-linked with y'. That is, a link $\|\lambda\|$ should be isomorphism-invariant in the sense of (D1-A-2-b).

To make these ideas precise, we must specify the corresponding components in K' and K that are linked. We will denote them by the places they occupy in the structures in the species of potential models using the following notation. If $|\Sigma|$ is a class of structures with m-components and $i_1, \ldots, i_n \in \{1, \ldots, m\}$ with $i_j \leq i_{j+1}$ and $X \subseteq |\Sigma|$ then "$|x: i_1, \ldots, i_n|$" denotes the class of all sequences of components appearing in the places $i_1, \ldots, i_n$ in some structure in the class X. We also need to be more explicit about what the components in the objects of the category of potential models are doing. We say K is a "k-l-n model element core" when the first k-components are the "base sets" of the structures, the next l are "auxiliary base sets" having to do with auxiliary mathematical structures like the real numbers and the last n components are "non-basic components" like relations and functions over the basic sets. (See [3, Ch. IV].) Thus when K' and K are respectively k'-l'-n' and k-l-n model element cores and $\langle i_1, \ldots, i_r\rangle$, $\langle j_1, \ldots, j_s\rangle$ are sequences of non-auxiliary "component positions" in members of $|M'_p|$ and $|M_p|$ respectively, we may define a $\langle i_1, \ldots, i_r\rangle$-$K'$, $\langle j_1, \ldots, j_s\rangle$-$K$ link to be a K', K link $\|\lambda\|$ in which the "values" of the components $\langle i_1, \ldots, i_r\rangle$ in the structures in $|M'_p|$ are correlated with the "values" of the components $\langle j_1, \ldots, j_s\rangle$ in the structures of $|M_p|$ and the values of no other components (D1–B–2). Further, for all uncorrelated components $\{a_1, \ldots, a_t\}$ and $\{b_1, \ldots, b_u\}$, all values that these components take in M'_p and M_p appear in the structures that are related by $|\lambda|$ (D1-B-3). We further require that the M'_p-morphisms linked by $\|\lambda\|$ with M_p-morphisms shall be only those that exist between structures in which the uncorrelated components $\{a_1, \ldots, a_t\}$ and $\{b_1, \ldots, b_u\}$ have the same values (D1–B–4). We summarize these ideas in the following definition.

(D1) For all categories $\|\lambda\|$, K, $K' \in |\mathrm{SET}|$, k, k', l, l', $n, n' \in \mathbb{N}^+$ if

(H$_1$) $K' = \langle \|M'_p\|, \|M'\|\rangle$ and $K = \langle \|M_p\|, \|M\|\rangle$ are respectively k'-l'-n' and k-l-n model element cores,

(H$_2$) $i_1, \ldots, i_r \in \{1, \ldots, k', k'+l'+1, \ldots, k'+l'+n'\}$, $i_h \leq i_{h+1}$ and $j_1, \ldots, j_s \in \{1, \ldots, k, k+l+1, \ldots, k+l+n\}$, $j_g \leq j_{g+1}$,

(H$_3$) $a_1, \ldots, a_t \in \{1, \ldots, k', k'+l'+1, \ldots, k'+l'+n'\}$, $a_h \leq a_{h+1}$ and $b_1, \ldots, b_u \in \{1, \ldots, k, k+l+1, \ldots, k+l+n\}$, $b_g \leq b_{g+1}$ so that:

$$\{a_1, \ldots, a_t\} \cap \{i_1, \ldots, i_r\} = \{b_1, \ldots, b_u\} \cap \{j_1, \ldots, j_s\} = \Lambda,$$

$$\{a_1, \ldots, a_t\} \cup \{i_1, \ldots, i_r\} = \{1, \ldots, k', k'+l'+1, \ldots, k'+l'+n'\},$$

$$\{b_1, \ldots, b_u\} \cap \{j_1, \ldots, j_s\} = \{1, \ldots, k, k+l+1, \ldots, k+l+n\},$$

then

(A) $\|\lambda\|$ is a K', K *link* iff

(1) $\|\lambda\|$ is an $\|M'_p\|, \|M_p\|$ relator,

(2) for all $\langle x', x\rangle, \langle y', y\rangle \in |M'_p| \times |M_p|$, $z' \in |M'_p|$:

(a) if $I[M'_p \times M_p](\langle x', x\rangle, \langle y', y\rangle) \neq \Lambda$ then

$$I[M'_p \times M_p](\langle x', x\rangle, \langle y', y\rangle) \cap \lambda(\langle x', x\rangle, \langle y', y\rangle) \neq \Lambda.$$

(b) if $\langle x', x\rangle \in |\lambda|$ and $IM'_p(x', z') \neq \Lambda$ then there is some $z \in |M_p|$ so that $IM_p(z', z) \neq \Lambda$ and $\langle z', z\rangle \in |\lambda|$.

(B) For all $X \subseteq |M'_p|$, $\|\lambda\|$ is a $\langle i_1, \ldots, i_r\rangle$-$K'$, $\langle j_1, \ldots, j_s\rangle$-$K$ *link in* X iff

(1) $\|\lambda\|$ is a K', K link,

(2) for all $h \in \{1, \ldots, r\}$, $g \in \{1, \ldots, s\}$,

(a) $|D_1(\lambda) \cap X : h| \neq |M'_p : h|$, (b) $|D_2(\lambda) : g| \neq |M'_p : g|$,

(3) $|D_1(\lambda): a_1, \ldots, a_t| \times |D_2(\lambda): b_1, \ldots, b_u|$

$$= |M'_p: a_1, \ldots, a_t| \times |M_p: b_1, \ldots, b_u|,$$

(4) for all $\langle x', x\rangle$, $\langle y', y\rangle \in |M'_p| \times |M_p|$ and $\langle \mu', \mu\rangle \in M'_p \times M_p(\langle x', x\rangle, \langle y', y\rangle)$, if $\langle \mu', \mu\rangle \in \lambda(\langle x', x\rangle, \langle y', y\rangle)$ then

$$|\{x'\}: a_1, \ldots, a_t| = |\{y'\}: a_1, \ldots, a_t|,$$

$$|\{x\}: b_1, \ldots, b_u| = |\{y\}: b_1, \ldots, b_u|.$$

Interpreting links

There are different kinds of intertheoretical links which function in different ways. Here we shall consider only "interpreting links". A K', K link is an interpreting link for K when models of K' serve as acceptable means of measuring or determining the values of components in potential models of K. More precisely, an interpreting $\langle i_1, \ldots, i_r\rangle$-$K'$, $\langle j_1, \ldots, j_s\rangle$-$K$ link is a link that allows us to infer something interesting about values of the components $\langle j_1, \ldots, j_s\rangle$ in at least some potential models of K from knowledge of the values of the components $\langle i_1, \ldots, i_r\rangle$ in models of K'.

The concept of an interpreting link is largely a pragmatic concept. Which links are used as interpreting links is a fact about the practice of empirical science, not a fact about the formal properties of links. Thus, we can not give a purely formal characterization of interpreting links. Nevertheless, we are able to give some formal necessary conditions for links to be used as interpreting links. First, it is clear that for interpreting $\langle i_1, \ldots, i_r\rangle$-$K'$, $\langle j_1, \ldots, j_s\rangle$-$K$ links, at least some models of K' must be linked ((D2-1) below). These models of K' are "acceptable" measuring devices or measuring situations for the components $\langle j_1, \ldots, j_s\rangle$. Members of $D_1(|\lambda|)$ outside this set correlate values of $\langle i_1, \ldots, i_r\rangle$ with classes of values for $\langle j_1, \ldots, j_s\rangle$, but these correlations are just "meaningless numbers" — readings from faulty instruments.

Consider the case of a K', K link in which the laws of K', together with the link, "entail" the laws of K. More precisely, $\|\lambda\|$ is such that for all $\langle x', x\rangle \in |\lambda|$, if $x' \in |M'|$ then $x \in |M|$. In this case, it is not plausible to regard K' as providing an interpretation of components in K. We may not think of models for K' providing acceptable methods of "measuring" values of components in K about which the laws of K say "something more". This suggests that links with this property should not count as interpreting links (D2-3-a).

Consider next a K', K link $\|\lambda\|$ in which the laws of K, together with the link, entail the laws of K'. That is for all $\langle x', x\rangle \in |\lambda|$, if $x \in |M|$ then $x' \in |M'|$. Were $\|\lambda\|$ to be regarded as a K-interpreting link, this would mean that all "data" that satisfied the laws of K had been obtained from acceptable measurements. There could be no "bad data" that just happened to satisfy the laws of K. We might attempt to formulate all empirical theories in a way so that this is true. In fact we do not appear to do this. The laws of isolated empirical theories are always formulated so that they entail nothing substantive about what counts as acceptable data for them (D2-3-b).

We summarize these ideas as follows.

(D2) For all categories $\|\lambda\|$, $K, K' \in |\mathrm{SET}|$, $k, k', l, l', n, n' \in \mathbb{N}^+$ so that $\|\lambda\|$ is a K', K link and $\|M_p'\| \neq \|M_p\|$, $\|\lambda\|$ is K-interpreting only if

(1) $D_1(|\lambda|) \cap |M'| \neq \Lambda$,

(2) there exist $\langle i_1, \ldots, i_r\rangle \in \{1, \ldots, k', k'+l'+1, \ldots, k'+l'+n'\}$, $i_n \leq i_{n+1}$, $\langle j_1, \ldots, j_s\rangle \in \{1, \ldots, k+l+1, \ldots, k+l+n\}$, $j_g \leq j_{g+1}$, so that $\|\lambda\|$ is a $\langle i_1, \ldots, i_r\rangle$-$K'$, $\langle j_1, \ldots, j_s\rangle$-$K$ link in $|M'|$,

(3) there exist $\langle x', x\rangle, \langle y', y\rangle \in |\lambda|$, so that:

(a) $x' \in |M'|$ and $x \notin |M|$, (b) $y \in |M|$ and $y' \notin |M'|$.

Model element nets

The logical structure of the whole of empirical science at any given time may be exhibited as a set of model elements together with the intertheoretical links *between* them. Here we restrict our attention to nets containing only binary links. Formally, we may think of a model element net N as an ordered pair $N = \langle |N|, L\rangle$ where $|N|$ is a set of model elements and L is a set of binary links linking members of $|N|$. The set L imposes a binary relational structure on $|N|$ in an obvious way and it is somewhat more convenient to discuss the properties of model element nets in terms of this relational structure. Let $L(K', K) \subseteq L$ be the set of all K', K links in L. We may think of $L(K', K)$ as containing just *one* link

$$\lambda[K', K] = \bigcap\{\|\lambda\| \in L(K, K)\}.$$

Consider the binary relation $L_r \subseteq |N| \times |N|$ that contains $\langle K', K\rangle$ just in case $L(K', K) \neq \Lambda$. That is, $L_r(K', K)$ just when there is some link between K' and K. Clearly, $\langle |N|, L_r\rangle$ is a binary relation structure.

We have just considered the most general properties of linked model element nets in which the links may be of any sort, including interpreting.

It is natural to expect that there are additional properties of nets that have to do with specific kinds of links. Here we consider only those special properties that have to do with interpreting links. If we think of interpreting links roughly as channels or paths for the transmission of information, then the relation between $L(K', K)$ and $L(K, K')$ is an important contingent fact about the logical structure of empirical science. Our conception of a K-interpreting K', K link is that it serves to transmit information *from* K' *to* K. This suggests that there should be a kind of asymmetry for interpreting links. Information should be conceived as flowing in only *one* direction between two adjacent linked model elements (D3–4).

We would like model element nets to do two things. First, we would like to use them to talk about the global structure of the whole of empirical science at any given time. Second, we would like to use them to explicitly exhibit fragments of this global structure. Considering the first task, it seems plausible to require that every model element in a net N representing the whole of empirical science have at least one interpreting link. There are neither uninterpreted nor self-interpreting formal structures in empirical science. Clearly, this entails either that $|N|$ is unbounded with respect to interpreting links or that there are closed "chains" of interpreting links. We do not find either "horn" of this dilemma obviously unacceptable. However, if we require that every model element have an interpreting link, we can only exhibit completely sub-nets of N that have closed interpreting chains. While it is not manifestly impossible that there are such chains in the structure of empirical science, there do not appear to be "small" ones. The "sub-nets" we want to use to illustrate "local" features of the net of empirical science will not be of this kind. In fact, if we require that every model element have an interpreting link, they will not be sub-nets at all because some "terminal" model elements will appear without interpreting links. On this point, we opt for simplicity and do not require that every model element have an interpreting link.

One technical point needs to be mentioned. Though we do not consider internal links here, we do want our concepts to be general enough that these additional features of model element nets can be added without much reformulation. Thus we define an "unconstrained model element net" to be one which contains the "vacuous internal link" $\|M_p\| \times \|M_p\|$ for all model elements in the net. "Constrained model element nets", which we do not consider, will be those that contain sub-links of this vacuous internal link. When we come to consider the "content" of model element nets our definition will entail that degenerate unconstrained nets consisting of only one model element will simply have $|M|$ as their content. We would not get this feature unless we explicitly required the presence of vacuous internal links.

We make these ideas more precise in the following necessary conditions without explicitly defining our special notation for links in a net introduced before. We give necessary conditions only since we restrict our attention to interpreting links.

(D3) For all $N \in |\mathrm{SET}|$, N is an *unconstrained model element net* only if there exist $|N|$ and $L \in |\mathrm{SET}|$ so that:

(1) $N = \langle |N|, L \rangle$,

(2) for all $K \in |N|$, there exist k, l, n so that K is a k-l-n model element,

(3) for all $\|\lambda\| \in L$, there exist $K', K \in |N|$ so that $\|\lambda\|$ is a K', K link.

(4) for all $K', K \in |N|$, if $\lambda[K', K]$ is a K-interpreting link then $L(K, K') = \Lambda$.

(5) for all $K \in |N|$, $\|M_{\mathrm{p}}\| \times \|M_{\mathrm{p}}\| \in L$.

Global content

The content of a model element net consists of structures that meet all the requirements this net imposes on "the way the world is". The model classes tell us what potential models are empirically possible in the absence of links. Links tell us what combinations of potential models are empirically possible. Together they tell us what combinations of models are possible. To make these ideas precise we think of the content of a model element net as a collection of binary relation structures consisting of ordered pairs of models.

First consider what the links tell us about empirically possible binary relational structures consisting of potential models. Consider the model element net N and the binary relation structure N_r associated with N. The links in N tell us that empirically possible sets of potential models must be linked together in the manner characterized by the binary relation structure N_r. Consider a set of ordered pairs of potential models σ_l consisting of at least one pair $\langle x^1, x^2 \rangle$ for each pair of linked model elements $\langle K^1, K^2 \rangle$ in L_r so that all pairs corresponding to $\langle K^1, K^2 \rangle$ are in the link $\lambda[K^1, K^2]$. Each link $\lambda[K^1, K^2]$ is "represented" in σ_l by at least one of its members ((D4–A–4–a) below). Further, we require that pairs of potential models "representing" the same pair of linked model elements do not "overlap". That is, in the notation of (D4-A-4-a), if $\kappa(\langle x^1, x^2 \rangle) = \kappa(\langle y^1, y^2 \rangle)$ then

$x^1 = y^1$ iff $x^2 = y^2$. This requirement is motivated by the idea that linked potential models will, in many cases, be the "same" objects described in the vocabulary of different theories. Similar intuitive considerations motivate a stronger requirement (D4-A-4-b).

It is easy to see what the laws of each model element add to the restrictions imposed by the links. Call the binary relation structures $C_l = \langle B(\sigma_l), \sigma_l \rangle$ we have just considered "potential model representations" of the net N. We may then simply add the requirement that the members of potential model representations of N be models of the model element they represent rather than just potential models. We will call such representations "model representations" of N and denote them by $C_p = \langle B(\sigma_p), \sigma_p \rangle$. For model element nets N, we may define

$$C_p[N] := \{C_p \mid C_p \text{ is a model representation of } N\}.$$

Intuitively, $C_p[N]$ is a plausible candidate to be called "the content of N". The sub-script "p" is used to indicate that this concept of "content" is a sub-class of potential models. It distinguishes this concept from the "non-theoretical content" considered below in Section 6. We may now see how the net operates to "narrow down" the content of each of its members. For $K \in |N|$, we may take $C_p[N](K)$ to be the set of all models of K that appear in some member of $C_p[N]$. That is

$$C_p[N](K) := \{x \mid x \in |M| \text{ and there is a } \langle B(\sigma_p), \sigma_p \rangle \in C_p[N] \text{ and } x \in B(\sigma_p)\}.$$

Alternatively, $C_p[N](K)$ is the class of all members of $|M|$ that are linked to some model of at least one of the model elements K' in $|N|$ that are linked with K.

Our discussion of the content of a net N has dealt only with features that may be described without mentioning specific kinds of links. We now consider what must be added content when some links are identified as interpreting links. Consider the net:

$$K^1 \xrightarrow{\lambda[K^1, K^2]} K^2 \xrightarrow{\lambda[K^2, K^3]} K^3$$

Suppose that $\lambda[K^1, K^2]$ is an interpreting link and $\lambda[K^2, K^3]$ is any kind of link, interpreting or otherwise. Focusing on the model element K^2, we may say that K^1, together with $\lambda[K^1, K^2]$, provides the empirical interpretation for K^2 and that K^3, together with $\lambda[K^2, K^3]$ "says something more" about the content of K^2. A link from K^2 like $\lambda[K^2, K^3]$ is only "interesting" when it links potential models that have been interpreted by interpreting links like $\lambda[K^1, K^2]$. In contrast, an interpreting link may be interesting even when the potential model it interprets is not linked "forward" by any other links. The link $\lambda[K^2, K^3]$ should be represented in the content by pairs containing potential models of K^2 that also appear in pairs representing $\lambda[K^1, K^2]$, while the interpreting link $\lambda[K^1, K^2]$ may be represented by pairs containing potential models of K^2 that do not appear in pairs representing $\lambda[K^2, K^3]$. This motivates (D4-A-4-c-i). Note that, when there is more than one interpreting link for K^2, (D4-A-4-c-i) requires that potential models in the content linked by non-K^2-interpreting links be linked to potential models in *all* the model elements that interpret K^2. This appears to be plausible in the case where the domains of all the interpreting links intersect in $|M_p^2|$. This is the typical case where one theory requires two or more other theories for its interpretation. There may be other cases where the same mathematical apparatus appears in two "theories" with different empirical interpretations. We do not consider these cases here.

It is not completely clear that we should not weaken the requirement to permit "partially interpreted" potential models of K^2 to be linked in the content of N by non-K^2-interpreting links. However, if we *do* choose to rule out partially interpreted potential models of K^2 from appearing in this way, it appears that we should also rule them out of the content altogether. That is, when there are several interpreting links for K^2 — say $\lambda[K^0, K^2]$ and $\lambda[K^1, K^2]$ — and $\langle x^0, x^2\rangle$ represents $\lambda[K^0, K^2]$, we should require that there be some $\langle x^1, x^2\rangle$ representing $\lambda[K^1, K^2]$. A little reflection suggests that we should actually strengthen this to allow $\lambda[K^1, K^3]$ to be any kind of link, interpreting or otherwise. Thus we require (D4-A-4-c-ii).

In general, we do not require that interpreting links "to" K^2 be connected to links "from" K^2 in the same way. Not all interpreted potential models need be linked "forward" by other links that may be present. We do however rquire that all interpreted potential models that have forward links exhibit these in σ_l (D4-A-4-d). We summarize these ideas in the following definition.

(D4) For all $N \in |\mathrm{SET}|$, if $N = \langle |N|, L\rangle$ is an unconstrained model element net then: for all $C_l \in |\mathrm{SET}|$,

(A) C_l is a *potential model representation for* N iff there exist $B(\sigma_l)$ and σ_l so that

(1) $C_l = \langle B(\sigma_l), \sigma_l\rangle$,

(2) $\sigma_l \subseteq \{\langle x^1, x^2\rangle \mid$ there exist $K^1, K^2 \in |N|$ and $\langle x^1, x^2\rangle \in |M_p^1| \times |M_p^2|\}$,

(3) $B(\sigma_l) = \{x \mid$ there is a $\langle y, z\rangle \in \sigma_l$ and $x = y$ or $x = z\}$,

(4) there is a $\kappa \in \mathrm{ONSET}(\sigma_l, L_r)$ so that for all $K^1, K^2, K^3 \in |N|$, $K^1 \neq K^2 \neq K^3$:

(a) for all $\langle x^1, x^2\rangle \in \sigma_l$, $\langle x^1, x^2\rangle \in \lambda[\pi_1(\kappa(\langle x^1, x^2\rangle)), \pi_2(\kappa(\langle x^1, x^2\rangle))]$

(b) if there exist $K^4, K^5 \in |N|$ so that $\langle K^1, K^4\rangle$, $\langle K^4, K^2\rangle$, $\langle K^1, K^5\rangle$, $\langle K^5, K^2\rangle \in L_r$ then:

(i) for all $\langle x^1, x^4\rangle, \langle x^4, x^2\rangle, \langle y^1, y^5\rangle, \langle y^5, y^2\rangle \in \sigma_l$, if, for $i, j \in \{1, 2, 4, 5\}$

$$\kappa(\langle x^i, x^j\rangle) = \kappa(\langle y^i, y^j\rangle) = \langle K^i, K^j\rangle$$

then $x^1 = y^1$ iff $x^2 = y^2$,

(ii) for all $\langle x^1, x^5\rangle, \langle x^1, x^4\rangle, \langle y^5, y^2\rangle, \langle y^4, y^2\rangle \in \sigma_l$, if, for $i, j \in \{1, 2, 4, 5\}$

$$\kappa(\langle x^i, x^j\rangle) = \kappa(\langle y^i, y^j\rangle) = \langle K^i, K^j\rangle$$

then $x^5 = y^5$ iff $x^4 = y^4$,

(c) if $\langle K^1, K^2\rangle$ and $\langle K^2, K^3\rangle \in L_r$, and $\lambda[K^1, K^2]$ is a K^2-interpreting link, then:

(i) for all $\langle x^2, x^3\rangle \in \sigma_l$ so that $\kappa(\langle x^2, x^3\rangle) = \langle K^2, K^3\rangle$, there is a $\langle x^1, x^2\rangle \in \sigma_l$ so that $\kappa(\langle x^1, x^2\rangle) = \langle K^1, K^2\rangle$,

(ii) for all $\langle x^1, x^2\rangle \in \sigma_l$ so that $\kappa(\langle x_1, x_2\rangle) = \langle K^1, K^2\rangle$, if $\lambda[K^2, K^3] \neq \Lambda$ then there is some $\langle x^2, x^3\rangle \in \sigma_l$ so that $\kappa(\langle x_2, x_3\rangle) = \langle K^2, K^3\rangle$,

(d) if $\langle K^1, K^3\rangle$ and $\langle K^2, K^3\rangle \in L_r$ and $\lambda[K^2, K^3]$ is an interpreting link then, for all $\langle x^1, x^3\rangle \in \sigma_l$ so that $\kappa(\langle x^1, x^3\rangle) = \langle K^1, K^3\rangle$ there is some $\langle x^2, x^3\rangle \in \sigma_l$ so that $\kappa(\langle x^2, x^3\rangle) = \langle K^2, K^3\rangle$.

(B) $C_l[N] := \{C_l \mid C_l$ is a potential model representation for $N\}$.

(C) For all $K \in |N|$, $C_l[N](K) := \{x \in |M_p| \mid$ there is a $\langle B(\sigma_l), \sigma_l\rangle \in C_l[N]$ and $x \in B(\sigma_l)\}$.

(D) for all $C_p \in |\mathrm{SET}|$, C_p is a model representation for N iff

(1) $C_p = \langle B(\sigma_p), \sigma_p\rangle$ is a potential model representation for N,

(2) for all $\kappa \in \mathrm{ONSET}(\sigma_p, L_r)$ satisfying (A-4), and all $\langle x, x'\rangle \in \sigma_p$, $x \in |M(\pi_1(\kappa(\langle x, x'\rangle))|$ and $x' \in |M(\pi_2(\kappa(\langle x, x'\rangle))|$.

(E) $C_p[N] := \{C_p \mid C_p$ is a model representation for $N\}$.

(F) For all $K \in |N|$, $C_p[N](K) := \{x \in |M_p| \mid$ there is a $\langle B(\sigma_p), \sigma_p\rangle \in C_p[N]$ and $x \in B(\sigma_p)\}$.

Non-theoretical structures

It is useful to distinguish those components in the potential models of K that are "interpretable" or "non-theoretical" in the net N from those that are "theoretical". The theoretical components in the potential models of K are those components in the potential model structures that are not affected by any of K's interpreting links. The non-theoretical components are those whose values are correlated in some way, by interpreting links for K, with values of components of potential models in other model element cores. We may make the distinction between theoretical and non-theoretical components in K precise in the following way.

(D5) For all $N \in |\mathrm{SET}|$, if N is a model element net and, for all $K = \langle \|M_p\|, \|M\| \rangle \in |N|$, if there exist k, l, n so that K is a k-l-n model element core, then, for all $i \in \{1, \ldots, k, k+l+1, \ldots, k+l+n\}$:

(A) $|M_p : i|$ is K-*non-theoretical* in N iff there is some $K' \in |N|$ so that:
  (1) there exist k', l', n' so that K' is a k'-l'-n' model element
  (2) $K' \neq K$,
  (3) there exist $j_1, \ldots, j_s \in \{1, \ldots, k', k'+l'+1, \ldots, k'+l'+n'\}$ so that $\lambda[K', K]$ is a $\langle j_1, \ldots, j_s \rangle$-$K'$, $\langle i \rangle$-K link in M',
  (4) $\lambda[K', K]$ is a K-interpreting link in N.

(B) $|M_p : i|$ is K-*theoretical* in N iff $|M_p : i|$ is not K-non-theoretical in N.

This definition of non-theoretical components is not quite adequate. Components in the potential models of K do not count as non-theoretical unless they are linked "singly" to K'. A $\langle j_1, \ldots, j_s \rangle$-$K'$, $\langle i_1, i_2 \rangle$-K link does not necessarily make i_1 K-non-theoretical. Of course, the same link *may* also be a $\langle j_1, \ldots, j_s \rangle$-$K'$, $\langle i_1 \rangle$-K link, but it need not be. If it just rules out pairs of values for $\langle i_1, i_2 \rangle$ while admitting all values for i_1, it is not. For example, the link between the pressure function (P) in classical hydrodynamics (CHD) and the energy (U) and volume (V) functions in simple equilibrium thermodynamics (SETH) provided by $P = -\mathrm{d}U/\mathrm{d}V$ makes neither U nor V SETH-non-theoretical because it only rules out $\langle U, V \rangle$ pairs, but not U-values or V-values. Thus, by our definition this link would produce no non-theoretical components. But, this link does play an essential role in interpreting SETH and somehow the interpreting information it provides should appear as restrictions on the values of some non-theoretical components. Intuitively, this link makes the defined SETH component "thermodynamic pressure", $\Pi := -\mathrm{d}U/\mathrm{d}V$, SETH-non-theoretical. Counting the defined component Π among the SETH-non-theoretical components would capture the intuition that this link is essential to the interpretation of SETH. Generalizing, one might think that our definition of non-theoretical components should be broadened to include the possibility that defined components are non-theoretical. But, doing this would mean that we could no longer uniquely define the non-theoretical structures associated with a model element in a net as we do below in (D6). Countenancing non-uniqueness here would considerably complicate the subsequent discussion. For this reason, we rest with the present, admittedly deficient, definition of non-theoretical components.

We may now characterize the objects of the category of non-theoretical structures or "partial potential models" $\|M_{\mathrm{pp}}[N](K)\|$ for model element K in net N.

(D6) For all $N \in |\mathrm{SET}|$, if N is a model element net and, for all $K = \langle \|M_{\mathrm{p}}\|, \|M\| \rangle \in |N|$, if there exist k_{p}, l_{p}, n_{p} so that K is a k_{p}-l_{p}-n_{p} model element, then $\|M_{\mathrm{pp}}[N](K)\|$ is the *category of partial potential models for K in N* only if there exist k_{pp}, l_{pp}, n_{pp}, $k_{\mathrm{pp}} \leq k_{\mathrm{p}}$, $l_{\mathrm{pp}} \leq l_{\mathrm{p}}$, $n_{\mathrm{pp}} \leq n_{\mathrm{p}}$ so that

(A) $\|M_{\mathrm{pp}}[N](K)\|$ is a category,

(B) $|M_{\mathrm{pp}}[N](K)|$ is a k_{pp}-l_{pp}-n_{pp} species of structures,

(C) for all $i_k \leq k_{\mathrm{pp}}$, $i_l \leq l_{\mathrm{pp}}$, $i_n \leq n_{\mathrm{pp}}$ there exist $j_k \leq k_{\mathrm{p}}$, $j_l \leq l_{\mathrm{p}}$, $j_n \leq n_{\mathrm{p}}$ so that, for all $x \in \{k, l, n\}$, $|M_{\mathrm{pp}}[N](K): i_x| = |M_{\mathrm{p}}: j_x|$,

(D) for all $i \in \{1, \ldots, k_{\mathrm{pp}}\} \cup \{k_{\mathrm{pp}} + l_{\mathrm{pp}} + 1, \ldots, k_{\mathrm{pp}} + l_{\mathrm{pp}} + n_{\mathrm{pp}}\}$, $|M_{\mathrm{pp}}[N](K): 1|$ is K-non-theoretical in N,

(E) there is no k'_{pp}, l'_{pp}, n'_{pp}; $k_{\mathrm{pp}} < k'_{\mathrm{pp}} \leq k_{\mathrm{p}}$, $l_{\mathrm{pp}} < l'_{\mathrm{pp}} \leq l_{\mathrm{p}}$, $n_{\mathrm{pp}} < n'_{\mathrm{pp}} \leq n_{\mathrm{p}}$ so that $M_{\mathrm{pp}}[N](K)'$ is a k'_{pp}-l'_{pp}-n'_{pp} species of structure satisfying (B) through (D) above and

$$|M_{\mathrm{pp}}[N](K)| = |M_{\mathrm{pp}}[N](K)': 1, \ldots, k_{\mathrm{pp}} + l_{\mathrm{pp}} + n_{\mathrm{pp}}|.$$

Fully characterizing the category $\|M_{\mathrm{pp}}[N](K)\|$ requires specifying its morphisms as well. We do not now have a fully satisfactory way to do this. See [9] for some idea of the problems with doing this.

The Ramsey functor — Ram — for K in N is just the "forgetful functor" from $\|M_{\mathrm{p}}\|$ to $\|M_{\mathrm{pp}}[N](K)\|$.

(D7) For all $N \in |\mathrm{SET}|$, if N is a model element net and, for all $K = \langle \|M_{\mathrm{p}}\|, \|M\| \rangle \in |N|$, if $\|M_{\mathrm{pp}}[N](K)\|$ is the category of partial potential models for K in N, then Ram is the *Ramsey functor for* $\|M_{\mathrm{p}}\|$ *in N* iff $\mathrm{Ram}: \|M_{\mathrm{p}}\| \to \|M_{\mathrm{pp}}[N](K)\|$ is a functor so that for all $x \in |M_{\mathrm{p}}|$,

$$\mathrm{Ram}_0(x) = |\{x\}: 1, \ldots, k_{\mathrm{pp}} + l_{\mathrm{pp}} + n_{\mathrm{pp}}|.$$

We may think of the laws of K as determining a sub-category of $\|M_{\mathrm{pp}}[N](K)\|$ whose objects are just those members of $|M_{\mathrm{pp}}[N](K)|$ that can be "filled out" with theoretical components in some way that satisfies the laws of K. We call this sub-category "the non-theoretical content of K"

and denote it by "$C_{pp}[N](K)$". Clearly, the non-theoretical content of K in N — $C_{pp}[N](K)$ — is just the Ramsey functor image of the model of K:

$$C_{pp}[N](K) := \overline{\mathrm{Ram}}(\|M\|).$$

Local empirical claims

How are we to regard the empirical claim of a single model element in a model element net? We have already suggested that the theory element $T = \langle K, I(K)\rangle$ claims roughly that $I(K) \subseteq |M|$. A more plausible rendition of the empirical claim of K is that $I(K) \subseteq C_{pp}[N](K)$. Here $I(K)$ is concieved as some sub-class of the non-theoretical structures of K — $|M_{pp}[N](K)|$. To say more about $I(K)$, let us think about what a single model element contributes to the content of a model element net. It seems natural to think of the intended applications of K as being provided by models of the model elements that are linked to K by interpreting links. But not all of these models will provide acceptable data for K. Some will be ruled out because they are not "interpreted" by model elements that are still "further back". Others may be ruled out by restrictions imposed by other model elements, besides K, that they interpret. Clearly, we do not want to include the restrictions imposed by K itself. Doing this would make K's claim trivially true. Further, it appears that we would not want to include restrictions imposed by model elements that K interprets. The reason is that the laws of K have an "indirect" effect on what these model elements rule out in the other model elements that are "behind" them. This suggests that we should think of the intended applications for K as being provided by the "net content" of interpreting model elements immediately "behind" K in the net N. But, the "net" whose content is relevant here is not N. Rather, it is N, less everything in N that is "before" K that K interprets.

We may make these ideas more precise in the following way. First consider the set of model elements that interpret K in N:

$$<K := \{K' \in |N| \mid \langle K', K\rangle \in L_r \text{ and } \lambda[K', K] \text{ is } K\text{-interpreting}\}$$

and the set of model elements that K interprets in N

$$>K := \{K' \in |N| \mid \langle K, K'\rangle \in L_r \text{ and } \lambda[K, K'] \text{ is } K'\text{-interpreting}\}.$$

A model element K in net N has "backward" and "forward" interpreting filters

$$N < K = \langle |N < K|, L < \rangle, \qquad N > K = \langle |N > K|, L > \rangle$$

associated with it, where

$$|N < K| = \{K' \in |N| \,|\, \langle K', K\rangle \in \mathbb{L}_r^i\},$$
$$|N > K| = \{K' \in |N| \,|\, \langle K, K'\rangle \in \mathbb{L}_r^i\}$$

are respectively all members of $|N|$ that are "backward" and "forward" linked to K by interpreting links in N, where "$\mathbb{L}_r^i(K', K)$" means that K' and K are linked by a "chain" of interpreting links. $L <$ and $L >$ are respectively all the links in L that are interpreting links between members of $|N < K|$ and $|N > K|$. Clearly, $N < K$ and $N > K$ are model element nets. Intuitively, the interpretation of everything in the net $N > K$ "presupposes" K. We may now delete from N every model element whose interpretation may "presuppose" K to obtain $N \sim N > K$. As we have rendered it in the preceding section, the claim of K is true of just things in the non-theoretical content of K — $C_{pp}[N](K)$. Formally, $\lambda[K', K]$ does not pair any members of $|M_p'|$ with members of $|M_{pp}[N](K)|$. But we may consider

$$\lambda_{pp}[K', K] := \{\langle x_p', x_{pp}\rangle \in |M_p'| \times |M_{pp}[N](K)| \,|\, \text{there is an } \langle x_p', x_p\rangle \in \lambda[K', K] \text{ so that } x_{pp} = \text{Ram}(x_p)\}.$$

Suppose $\lambda[K', K]$ is a $\langle j_1, \ldots, j_s\rangle$-$K'$, $\langle i_1, \ldots, i_r\rangle$-$K$ link. Then, for $K' \in\, < K$, consider every member of $|M_{pp}[N](K)|$ that is $\lambda_{pp}[K', K]$-linked with some member of the content of K' in $|N \sim N > K|$. That is

$$\overline{\lambda_{pp}[K', K] >}(C_p[N \sim N > K](K')).$$

Each of the interpretors of K in N may contribute in this way to specifying K's intended applications. So we obtain

$$\bigcap\{\overline{\lambda_{pp}[K', K] >}(C_p[N \sim N > K](K')) \,|\, K' \in\, < K\}.$$

We have, in effect, specified necessary conditions for the intended applications of K in net N. That is, we require at least that

$$I(K) \subseteq \bigcap\{\overline{\lambda_{pp}[K', K] >}(C_p[N \sim N > K](K')) \,|\, K' \in\, < K\}.$$

It may even be plausible to regard these conditions as sufficient as well. If we regard N as including all of empirical science, we might hope that it included enough to rule out all "empirically meaningless" structures. Were this so we could replace "$\subseteq$" above by "$=$". We need not commit ourselves on this question here. For the sake of notational convenience, we may let

$$I[N](K) := \bigcap\{\overline{\lambda_{pp}[K', K]} > (C_p[N \sim N > K](K')) \mid K' \in\, < K\}.$$

Intuitively, $I[N](K)$ is the intended applications of K as narrowly as they can be specified by N. Whether or not $I(K) = I[N](K)$ we leave open.

References

[1] BALZER, W. and SNEED, J.D., 1978, *Generalized net structures of empirical theories, I and II*, Studia Logica 36 (3) (1977), pp. 195–212; and Studia Logica 37 (2), pp. 168–194.

[2] BALZER, W. and MOULINES, C.-U., 1980, *On theoreticity*, Synthèse 44.

[3] BOURBAKI, N., 1968, *Elements of Mathematics: Theory of Sets* (Addison-Wesley, Reading, MA), Ch. IV.

[4] CARNAP, R., 1956, *The methodological character of theoretical concepts*, in: H. Feigl and M. Scriven, eds., The Foundations of Science and the Concepts of Psychology and Psychoanalysis (Univ. of Minnesota Press, Minneapolis), pp. 38–76.

[5] MACLANE, S., 1977, *Categories for the Working Mathematician* (Springer, New York).

[6] REICHENBACH, H., 1924. *Philosophie der Raumzeitlehre*, (F. Vieweg u. Sohn, Braunschweig), new edition: HANS REICHENBACH Gesammelte Werke, Band 2, A. Kamlah and M. Reichenbach eds. (Braunschweig–Wiesbaden, 1977).

[7] SCRIVEN, M., 1958, *Definitions, explanations and theories*, in: H. Feigl, M. Scriven, G. Maxwell, eds., Concepts, Theories, and the Mind-Body Problem (Univ. of Minnesota Press, Minneapolis), pp. 99–195.

[8] SNEED, J.D., 1979, *The Logical Structure of Mathematical Physics*, 2nd ed. (Reidel, Dordrecht).

[9] SNEED, J.D., 1979, *Theoretization and invariance principles*, in: I. Niiniluoto and R. Toumela, eds., The Logic and Epistemology of Scientific Change, Acta Philosophica Fennica 30 (North-Holland, Amsterdam), pp. 130–178.

Barcan Marcus et al., eds., Logic, Methodology and Philosophy of Science VII

AIM AND STRUCTURE OF SCIENTIFIC THEORIES*

BAS C. VAN FRAASSEN

Dept. of Philosophy, Princeton Univ., Princeton, NJ 08544, U.S.A.

Philosophy of science attempts to answer the question *What is science?* in just the sense in which philosophy of art, philosophy of religion, and the like answer the similar question about their subject. For better or for worse our tradition has focused on the scientific theory rather than on scientific activity itself (on the product, rather than on the aim, conditions, and process of production, to draw an analogy, which is already one that points in its terminology to the product as most salient feature). Yet all aspects of scientific activity must be illuminated if the whole is to become intelligible. Despite the announced subject of our symposium, therefore, I shall devote a preliminary section to the aim of science, and the proper form of epistemic or doxastic attitudes toward scientific theories, before entering upon the description of their structure.

1. The aim of science

The activity of constructing, testing, and refining of scientific theories — that is, the production of theories to be accepted within the scientific community and offered to the public — what is the aim of this activity?

I do not refer here either to the motives of individual scientists for participating, or the motives of the body civic for granting funds and otherwise supporting the activity. Nor do I ask for some theoretically postulated "fundamental project" which would explain this activity. It is

* The author wishes to thank the National Science Foundation and Princeton University for support for his research and sabbatical leave. The topics discussed in this paper will be further elaborated in a general "reply to critics" for the forthcoming volume *Scientific Realism versus Constructive Empiricism*, edited by C.A. Hooker and P. Churchland.

part of the straightforward description of any activity, communal or individual, large-scale or small, to describe the end that is pursued as one of its defining conditions. In the most general terms, the end pursued is success, and the question is what counts as success, what are the criteria of success in this particular case?

We cannot answer our particular questions here without some reflection on what sort of thing this product, the scientific theory, is. A scientific theory must be the sort of thing that we can accept or reject, and believe or disbelieve; accepting a theory implies the opinion that it is successful; science aims to give us acceptable theories. To put it more generally, a theory is an object for epistemic or at least doxastic attitudes. A typical object for such attitudes is a proposition, or a set of propositions, or more generally a body of putative information about what the world is like, what the facts are. If anyone wishes to be an instrumentalist, he has to deny the appearances which I have just described. An instrumentalist would have to say that the apparent expression of a doxastic attitude toward a theory is elliptical; "to believe theory *T*" he would have to construe as "to believe that theory *T* has certain qualities." I shall not follow that path. Let me state here at once, as a first assumption, that the theory itself is what is believed or disbelieved.

At this point we can see at once that there is a very simple possible answer to all our questions, the answer we call *scientific realism.* This philosophy says that a theory is the sort of thing which is either true or false; and that the criterion of success is truth. As corollaries we have that acceptance of a theory as successful is, or involves the belief that it is, true; and that the aim of science is to give us (literally) true theories about what the world is like.

The answer must of course be qualified in various ways to allow for our epistemic finitude and the consequent tentativeness of reasonable doxastic attitudes. Thus we add that although it cannot generally be *known* whether or not the criterion of success has been met, we may reasonably have a high degree of belief that it has been, or that it is met approximately (i.e., met exactly by one member of a set of "small variants" of the theory), and this imparts similar qualifications to acceptance in practice. And we add furthermore of course that empiricism precludes dogmatism, that is, *whatever* doxastic attitude we adopt, we stand ready to revise in face of further evidence. These are all qualifications of a sort that anyone must acknowledge, and should therefore really go without saying. They do not detract from the appealing and as it were pristine clarity of the scientific realist position.

I did not want to discuss the structure of theories before bringing this position into the open, and confronting it with alternatives. For it is very important, to my mind, to see that an analysis of theories — even one that is quite traditional with respect to what theories are — does not presuppose it. Let us keep assuming with the scientific realist, that theories are the sort of thing which can be true or false, that they say what the world is like. What they say may be true or false, but it is nevertheless literally meaningful information, in the neutral sense in which the truth value is "bracketed."

There are a number of reasons why I advocate an alternative to scientific realism. One point is that reasons for acceptance include many which, *ceteris paribus*, detract from the likelihood of truth. In constructing and evaluating theories, we follow our desires for information as well as our desire for truth. For belief, however, all but the desire for truth must be "ulterior motives." Since therefore there are reasons for acceptance which are not reasons for belief, I conclude that acceptance is not belief. It is to me an elementary logical point that a more informative theory can not be more likely to be true — and attempts to describe inductive or evidential support through features that require information (such as "inference to the best explanation") must either contradict themselves or equivocate.

It is still a long way from this point to a concrete alternative to scientific realism. Once we have driven the wedge between acceptance and belief, however, we can reconsider possible ways to make sense of science. Let me just end these preliminary remarks now by stating my own position, which I call *constructive empiricism*. It says that the aim of science is not truth as such but only *empirical adequacy*, that is, truth with respect to the observable phenomena. Acceptance of a theory involves as belief only that the theory is empirically adequate — but it involves more than belief.

While truth as such is therefore, according to me, irrelevant to success for theories, it is still a category that applies to scientific theories. Indeed, the *content* of a theory is what it says the world is like; and this is either true or false. The applicability of this notion of truth value remains here, as everywhere, the basis of all logical analysis. When we come to a specific theory, there is an immediate philosophical question, which concerns the content alone: *how could the world possibly be the way this theory says it is?*

This is for me the foundational question *par excellence*. And it is a question whose discussion presupposes no adherence to scientific realism, nor a choice between its alternatives. This is the area in philosophy of science where realists and anti-realists can meet and speak with perfect neutrality.

2. Theory structure — models and their logical space

In this section I shall present a view of theories which makes language largely irrelevant to the subject. Of course, to present a theory, we must present it in and by language. That is a trivial point, for any effective communication proceeds by language, except in those rare cases in which information can be conveyed by the immediate display of an object or happening. In addition, both because of our own history — the history of philosophy of science which became intensely language-oriented during the first half of this century — and because of its intrinsic importance, we cannot ignore the language of science. Hence I shall make it the subject of the last section.

In what is now called the received view, a theory was conceived of as an axiomatic theory. That means, as a set of sentences, defined as the class of logical consequences of a smaller set, the axioms of that theory. A distinction was drawn: since the class of axioms was normally taken to be effectively presentable, and hence syntactically describable, the theory could be thought of as in itself uninterpreted. The distinction is then that scientific theories have an associated interpretation, which links their terms with their intended domain. We all know the story of misery and pitfalls that followed. Only two varieties of this view of scientific theories as a special sort of interpreted theories, emerged as anywhere near tenable. The first variety insists on the formal character of the theory as such, and links it to the world by a partial interpretation. Of this variety the most appealing to me is still Reichenbach's, which said that the theoretical relations have *physical correlates*. Their partial characters stand out when we look at the paradigm example: light rays provide the physical correlate for straight lines. It will be immediately clear that not every line is the path of an actual light ray, so the language-world link is partial. The second variety, which came to maturity in Hempel's later writings, hinges for its success on treating the axioms as already stated in natural language. The interpretative principles have evolved into axioms among axioms. That is, the class of axioms may be divided into those which are purely theoretical, in which all nonlogical terms are ones specially introduced to write the theory, and those which are mixed, in which nontheoretical terms also appear. It will be readily appreciated that in both these developments, despite lip service to the contrary, the so-called problem of interpretation was left behind. We do not have the option of interpreting theoretical terms — we only have the choice of regarding them as either (a) terms we do not fully understand but know how to use in our reasoning, without detriment to the success of science, or (b) terms which are now part of natural language, and no less well understood than its other parts. The choice of the correct view about the meaning and understanding of newly introduced terms, makes no practical difference to philosophy of science, as far as one can tell. It is a good problem to pose to philosophers of language, and to leave them to it.

In any tragedy,[1] we suspect that some crucial mistake was made at the very beginning. The mistake, I think, was to confuse a theory with the formulation of a theory in a particular language. The first to turn the tide was Patrick Suppes, with his well-known slogan: the correct tool for philsophy of science is mathematics, *not* metamathematics. This happened in the fifties — bewitched by the wonders of logic and the theory of meaning, few wanted to listen. Suppes' idea was simple: *to present a theory, we define the class of its models directly*, without paying any attention to questions of axiomatizability, in any special language, however relevant or simple or logically interesting that might be. And if the theory as such, is to be identified with anything at all — if theories are to be reified — then a theory should be identified with its class of models.[2]

This procedure is in any case common in modern mathematics, where Suppes had found his inspiration. In a modern presentation of geometry we find not the axioms of Euclidean geometry, but the definition of a Euclidean space. Similarly Suppes and his collaborators sought to reformulate the foundations of Newtonian mechanics, by replacing Newton's axioms with the definition of a Newtonian mechanical system. This gives us, by example, a *format* for scientific theories. In Ronald Giere's recent encapsulation, a theory consists of (a) the *theoretical definition*, which defines a certain class of systems; (b) a *theoretical hypothesis*, which asserts that certain (sorts of) real systems belong to that class.

This is a step forward in the direction of less shallow analysis of the structure of a scientific theory. The first level of analysis addresses the notion of theory *überhaupt*, but we do not want to stop there. We can go still a bit further by making a division between relativistic and non-

---

[1] I use the word deliberately: it was a tragedy for philosophers of science to go off on these logico-linguistic tangles, which contributed nothing to the understanding of either science on logic or language. It is still unfortunately necessary to speak polemically about this, because so much philosophy of science is still couched in terminology based on a mistake.

[2] The impact of Suppes' innovation is lost if models are defined, as in many standard logic texts, to be partially linguistic entities, each yoked to a particular syntax. Here the models are mathematical structures, called models of a given theory only by virtue of belonging to the class defined to be the models of that theory. See Section 4.

relativistic theories. In the latter, the systems are physical entities developing in time. They have accordingly a space of possible states, which they take on and change during this development. This introduces the idea of cluster of models united by a common *state-space*; each has in addition a domain of objects plus a "history function" which assigns to each object a history, i.e., a trajectory in that space. As Ms. Lloyd will discuss in her paper on population genetics in this Congress, a real theory will have many such clusters of models, each with its state-space. So the presentation of the theory must proceed by describing a class of *state-space types.*

In the case of relativistic theories, early formulations can be described roughly as relativistically invariant descriptions of objects developing in time — say in their proper time, or in the universal time of a special cosmological model (e.g., Robertson–Noonan models). A more general approach, developed by Glymour and Michael Friedman, takes space-times themselves as the systems. Presentation of a space-time theory T may then proceed as follows: a (T-)*space-time* is a four-dimensional differential manifold M, with certain geometrical objects (defined on M) required to satisfy the *field equations* (of T), and a special class of curves (the possible trajectories of a certain class of physical particles) singled out by the *equations of motion* (of T).

Clearly we can further differentiate both sorts of theories in other general ways, for example with respect to the stochastic or deterministic character of imposed laws. (It must be noted however that except in such special cases as the flat space-time of special relativity — its curvature independent of the matter-energy distribution — there are serious conceptual obstacles to the introduction of indeterminism into the space-time picture.)

I do not wish to dwell on the details of foundational research in the sciences. But I want to point out that the point of view which I have been outlining — the *semantic view* as opposed to the received view — is much closer to practice there. The scientific literature on a theory makes it relatively easy to identify and isolate classes of structures to be included in the class of theoretical models. It is on the contrary usually quite hard to find laws which could be used as axioms for the theory as a whole. Apparent laws which frequently appear are often partial descriptions of special subclasses of models, their generalization being left vague and often shading off into logical vacuity. Let me give two examples. The first is from quantum mechanics: *Schrödinger's equation.* This is perhaps its best known and most pervasively employed law — but it cannot very well be an axiom of the theory since it holds only for conservative systems. If we look into the general case, we find that we can prove the equation to hold, for some constant Hamiltonian, under certain conditions — but this is a mathematical fact, hence empirically vacuous. The second is the Hardy–Weinberg law in population genetics. Again, it appears in any foundational discussion of the subject. But it could hardly be an axiom of the theory, since it holds only under certain special conditions. If we look into the general case, we find the logical fact that certain assumptions imply that it describes an equilibrium which can be reached in a single generation, and maintained. The assumptions are very special, and more complex variants of the law can be deduced for more realistic assumptions — in an open and indefinite sequence of sophistications.

What we have found, in this approach, is a way to describe relevant structures in ways that are also directly relevant, and seen to be relevant, to our subject matter. The scholastically logistical distinctions that the logical positivist tradition produced — observational and theoretical vocabulary, Craig reductions, Ramsey sentences, first-order axiomatizable theories, and also projectible predicates, reduction sentences, disposition terms, and all the unholy rest of it — had moved us *mille milles de toute habitation scientifique*, isolated in our own abstract dreams. Since Suppes' call to return to a nonlinguistic orientation, now about thirty years ago, we have slowly regained contact.

3. Theory structure — relation to the world

Above I mentioned Giere's elegant capsule formulation of the semantic view: a theory is presented by giving the definition of a certain kind (or kinds) of systems plus one or more hypotheses to the effect that certain real (kinds of) systems belong to the defined class(es). We speak then of the *theoretical definition* and the *theoretical hypotheses* which together constitute the given formulation (in, so to say, canonical form) of the theory. A "little" theory might for example define the class of Newtonian mechanical systems and assert that our solar system belongs to this class.

Truth and falsity offer no *special* perplexities in this context. The theory is true if those real systems in the world really do belong to the indicated defined classes. From a logical, or more generally semantic point of view, we may consider as implicitly given models of the world as a whole, which are as the theoretical hypotheses say it is. There is of course a very large class of models of the world as a whole, in which our solar system is a Newtonian mechanical system. In one such model, nothing except this solar system exists at all; in another the fixed stars also exist, and in a third, the solar system exists and dolphins are its only rational inhabitants. Now the world must be one way or another; so the theory is true if the real world itself is (or is isomorphic to) one of these models. This is equivalent to either of two familar sorts of formulations of the same point: the theory is true exactly if (a) one of the possible worlds allowed by the theory is the real world; or (b) all real things are the way the theory says they are.

But while the subject of truth yields no special conceptual difficulties in this context, I do not believe that it marks the relation to the world, which science pursues in its theories. This, as you will recall, is the point at issue between scientific realism and empiricism. But leaving the issue itself aside, I think that even scientific realists need to be acutely interested in a much closer, more empirical relation of theory to world. I call this relationship *empirical adequacy*.

The logical positivist tradition gave us a formulation of such a concept which was not only woefully inadequate but created a whole cluster of "artifact problems" (by this I mean problems which are artifacts of the philosophical approach and not inherent in its subject). In rough terms, the empirical content of a theory was identified with a set of sentences, the consequences of that theory in a certain "observational" vocabulary. In my own studies, I first came across formulations of more adequate concepts in the work of certain Polish writers (PRZELEWSKI 1969, WOJCICKI 1974), of DALLA CHIARA and TORALDO DI FRANCIA (1973 and 1977) and finally of course in PATRICK SUPPES' own writings on what he calls empirical algebras and data models (1967, 1969). While some of these formations were still more language-oriented than I liked, the similarity in their approach was clear: certain parts of the models were to be identified as *empirical substructures*, and these were the candidates for representation of the observable phenomena which science can confront within our experience.

At this point I perceived that the relationship thus explicated corresponds exactly to the one Reichenbach attempted to identify through this concept of coordinative definitions, once we abstract from the linguistic element. Thus in a space-time the geodesics are the candidates for the paths of light rays and particles in free fall. More generally, the identified spatio-temporal relations provide candidates for the relational structures constituted by actual genidentity and signal connections. These actual physical structures are to be embeddable in certain substructures of space-time, which allows however for many different possibilities, of which the actual is, so to say, some arbitrary fragment.

Thus we see that the empirical structures in the world are the parts which are at once *actual* and *observable*; and empirical adequacy consists in the embeddability of all these parts in some single model of the world allowed by the theory.

Patrick Suppes has very carefully investigated the construction of data models, and the empirical constraint they place on theoretical models. Thought of as concerned with exactly this topic, much apparently "a prioristic" theorizing on the foundations of physics takes on a new intelligibility (see my *PSA* 1980 and *Synthèse* 1982 papers). A reflection on the possible forms of structures definable from joint experimental outcomes yields constraints on the general form of the models of the theories "from below" which can then be narrowed down by the imposition of postulated general laws, symmetry constraints, and the like, "from above".

4. The language(s) of science

We arrive now finally at the subject which I have mostly tried to banish from our discussion: language. I must admit that I too was, to begin, overly impressed by certain successes of modern logic. Thus one reviewer (John Worrall) of my book *The Scientific Image*, was able to quote the remark in my first paper on the subject, that the interrelations between the syntactic and semantic characterizations of a theory "make implausible any claim of philosophical superiority for either approach". The interrelations referred to are of course those described by the generalized completeness proof. I have long since changed my mind about its significance, both theoretical and practical.

To begin with the theoretical point, when a theory is presented by defining the class of its models, that class of structures cannot generally be identified with an elementary class of models of any first-order language.[3] The reason is found in the limitative meta-theorems, which brought to light the dark side of completeness. To take only the most elementary example, if a scientist describes a class of models, the mathematical object he is most likely to include is the real number continuum. There is no elementary class of models of a denumerable first-order language each of which includes the real numbers. As soon as we go from mathematics to metamathematics, we reach a level of formalization where many mathematical distinctions cannot be captured — except of course by *fiat*, as

[3] This answers a question posed in another review, the one by Michael Friedman. Unfortunately Friedman assumed the contrary answer, and built part of his critique on that conjecture.

when we speak of "standard" or "intended" models. The moment we do so, we are using a method of description not accessible to the syntactic mode.

On the practical side we must mention the enormous distance between actual research on the foundations of science, and syntactically capturable axiomatics. While this disparity will not affect philosophical points which hinge only on what is possible "in principle", it may certainly affect the real possibility of understanding and clarification.

Given this initial appreciation of the situation, shall we address ourselves to language at all? The answer I think is *yes*, not on any general grounds, but for a number of specific reasons.

Before detailing those specific reasons, let us look for a moment at language and the study of language in a general way. Russell made familiar to us the idea of an underlying *ideal language*. This is the skeleton, natural language being the complete living organic body built on this skeleton, the flesh of course rather accidental, idiosyncratic, and molded by the local ecology. The skeleton, finally, is the language of logic; and for Russell's contemporaries the question was only whether *Principia Mathematica* needed to be augmented with some extra symbols to fully describe the skeleton.

Against this we must advance the conception of natural language as not being constituted by any one realization of any such logical skeleton. Logic has now provided us with a great many skeletons. Linguists have uncovered fragments of language in use for which no constructed logical skeleton yet provides any satisfactory model. Natural language consists in the resources we have for playing many different possible language games. Languages studied in logic texts are models, rather shallow models, of some of these specific language games, some of these fragments. To think that there must in principle exist a language in the sense of the objects described by logic, which is an adequate model for natural language taken as a whole, may be strictly analogous to the idea that there must exist a set which is the universe of set theory.

So if we now apply our logical methods in the philosophy of science we should, as elsewhere, set ourselves the task of modeling interesting fragments of language specially relevant to scientific discourse. These fragments may be large or small.

In my opinion, choosing the task of describing a language in which a given theory can be formulated, is a poor choice. The reason is that descriptions of structure in terms of satisfaction of sentences is, as far as I can see, generally less informative and less illuminating than direct mathematical (instead of metamathematical) description. It is the choice, explicit or implicit, to be formed in almost all linguistically oriented philosophical studies of science. It was the implicit choice behind, certainly, almost all logical positivist philosophy of science.

At the other extreme, we may choose a very small fragment, such as what I have called the fragment of *elementary statements*. Originally I characterized these as statements which attribute some value to a measurable physical magnitude. The syntactic form was therefore trivial — it is always something like "*m* has value *r*" — and therefore the semantic study alone has some significance. Under pressure of various problems in the foundations of quantum mechanics, I broadened my conception of elementary statements in two ways. First, I admitted as possibly logically distinct the attributions of ranges (or Borel sets) of values. Second, I admitted as possibly distinct the attributions of states of certain sorts from that of values to measurable magnitudes. (It should be added however that I soon found it much more advantagous to concentrate on the propositions expressible by elementary statements, rather than on the statements themselves. At that point there is not even a bow in the direction of syntactic description.)

There are points between these two extremes. I would point here especially to certain forms of natural discourse that are prevalent in the informal presentation of scientific theory, but which have a long history of philosophical perplexities. The main examples are causality and physical modality. From an empirical point of view, there are besides relations among actual matters of fact, only relations among words and ideas. Yet causal and modal locutions appear to introduce relations among possibilities, relations of the actual to the possible. Since irreducible probability is now a fact of life in physics, and probability is such a modality, there is no escaping this problem. Yet, if we wish to be empiricists, we have nowhere to turn besides thought and language for the locus of possibility. In other words, an empiricist position must entail that the philosophical exploration of modality, even where it occurs in science, is to be part of the theory of meaning.

In Sections 1 and 3 I already made clear one important point in the empiricist view of scientific models. They may, without detriment to their function, contain much structure which corresponds to no elements of reality. The part of the model which represents reality includes the representation of actual observable phenomena, and *perhaps* something more, but is explicitly allowed to be only a proper part of the whole model.

This gives us I think the required leeway for a program in the theory of meaning. If the link between language and reality is mediated by models, it may be a very incomplete link — without depriving the language of a complete semantic structure. The idea is that the interpretation of language is not simply an association of a real denotata with grammatical expressions. Instead the interpretation proceeds in two steps. First, certain expressions are assigned values in the family of models; the reference or denotation is gained indirectly because those model elements *may* correspond to elements of reality. The exploration of modal discourse may then draw largely on structure in the models which outstrips their representation of reality.

A graphic, if somewhat inaccurate way to put this would be: causal and modal discourse describes features of our models, not features of the world. The view of language presented here — that discourse is guided by models or pictures, and that the logic of discourse is constituted by this guidance — I recommend as a general empiricist approach for a theory of meaning without metaphysics.[4]

References

[1] DALLA CHIARA, M.L. and TORALDO DI FRANCIA, G., 1973, *A logical analysis of physical theories*, Rivista di Nuovo Cimento, Serie, 2, 3, pp. 1–20.

[2] DALLA CHIARA, M.L. and TORALDO DI FRANCIA, G., 1979, *Formal analysis of physical theories*, in: G. Toraldo di Francia, ed., Problems in the Foundations of Physics (North-Holland, Amsterdam).

[3] PRZELEWSKI, M., 1969, *The Logic of Empirical Theories* (Routledge and Kegan Paul, London).

[4] SUPPES, P., 1967, *What is a scientific theory?*, in: S. Morgenbesser, ed., Philosophy of Science Today (Basic Books, New York).

[5] SUPPES, P., 1969, *Studies in Methodology and Foundations of Science* (Reidel, Dordrecht).

[6] WOJCICKI, R., 1974, *Set theoretic representations of empirical phenomena*, J. Philosophical Logic 3, pp. 337–343.

---

[4] For the background of this paper I wish to refer the reader to pp. 221–30 of the Introduction to F. SUPPE, ed., *The Structure of Scientific Theories* (Univ. of Illinois Press, Urbana, 1974); Ch. 5 "Theories" of R. GIERE, *Understanding Scientific Reasoning* (Holt, Rinehart, Winston, New York, 1979); Ch. 3, Section 4 of my *An Introduction to the Philosophy of Time and Space* (Random House, New York, 1970); my book *Scientific Image* (Oxford Univ. Press, Oxford, 1980); and the following articles:

SUPPES, P., *What is a scientific theory?*, in: S. Morgenbesser, ed., Philosophy of Science Today (Basic Books, New York, 1967), pp. 55–67.

VAN FRAASSEN, B.C. *Theory construction and experiment: An empiricist view*, in: P. Asquith and R. Giere, eds., PSA 1980, Vol. 2 (Philosophy of Science Association, East Lansing, MI, 1981), pp. 663–678.

VAN FRAASSEN, B.C., *The charybdis of realism: Epistemological implications of Bell's inequality*, Synthèse 52 (1982), pp. 25–38.

7. Foundations of Probability and Induction

Barcan Marcus et al., eds., Logic, Methodology and Philosophy of Science VII

TOWARDS A UNIFIED CONCEPT OF PROBABILITY

HAIM GAIFMAN*

Inst. of Mathematics and Computer Science,
The Hebrew Univ. of Jerusalem, Israel

1. Introduction

Probability, like the ancient Roman God Janus, has more than one face. The main line of division is between the concepts of subjective and objective probability. The first interprets probability as a measure of certainty or of belief in the truth of certain statements, the second — as a factual concept which measures long-run frequencies. 'Degree of belief', 'credence' or 'plausibility' are among the terms used for the first concept, 'chance' 'objective chance' and 'propensity' — for the second. In addition to the main classification there are variants, or subcategories, within each class, as well as conceptions which share certain aspects of each. Whereas the mathematical apparatus is to a large extent agreed upon[1], there is great diversity and many conflicting views when it comes to its applications. In their zeal for clarity and neatness some philosophers have insisted on maintaining strictly separate terminologies and it has even been suggested to replace 'probability' in technical or philosophical contexts by a variety of other sharply defined terms. The suggestion has (fortunately) not taken root and the fact remains that the concept of probability owes much of its vitality and force to the diversity of

* This research was done with the support of the DFG for which the author is grateful. Part of this paper was presented in an invited address to the 7th International Congress of Logic, Methodology and Philosophy of Science, held in Salzburg, July 1983. The author wishes to thank Prof. Gideon Schwartz of the Hebrew University Statistics Department for several enlightening discussions during the work on this manuscript.

[1] There are of course several variants such as the setup based on conditional probabilities as a primitive notion, instead of the standard one-argument function. Other variants are designed to capture non-numerical concepts of probability: orderings, partial orderings and interval-valued functions; they belong all to the same family inasmuch as each can be refined to a numerical function satisfying Kolmogorov's axioms.

its aspects. The concept plays fundamental roles, theoretical and practical, in an extremely wide variety of domains ("almost everywhere" one is tempted to say) and has become an essential element in our picture of the world.

My aim in this paper is to argue for a conception according to which subjective and objective probabilities derive from one prototypal concept. They are its aspects, or the forms it takes in different contexts, under different circumstances. This is not to abolish the distinction between subjective and objective probability, but to recognize their common kernel and the fact that there is a wide spectrum between "the extreme objective" and "the extreme subjective" forms of probability. Since the debate has been going on several levels, from philosophy to statistics, a unified view of probability will have implications on these various levels. Philosophically the problem highlights central issues, for it is a nice illustration of the meeting and the merging of the epistemic into the ontological. It has clear implications concerning the various types of modality (where analysis along some of the lines of Section 2 can be clearly applied). What it implies for statistics is indicated to some extent in subsequent observations and in Section 3. But in this paper I am not concerned with drawing and developing the conclusions for either domain.

I start by assuming a probability function, interpreted as a degree of belief. This is not because I want to impose the subjectivist interpretation but because, as a starting point, it provides the more comprehensive system. In principle it allows us to consider probabilities defined over any collection of statements. The question posed is: What makes such probabilistic assignments objective? The answer is given by pointing out two basic aspects: *inner stability* and *success*. Inner stability is treated in Section 2, success is defined in Section 2 and is analyzed in more detail in Sections 3 and 4. The analysis on the whole is not intended as a reduction of the objective concept to the subjective one. Once a probability gets established as objective, it should be taken as seriously as any fact in the world. The property, or the propensity, of the coin to behave in a very particular way is not more "subjective" than other theoretical concepts of science. At the same time one realizes how this passage from the subjective to the objective is done; one also realizes its continuous nature and the wide and intricate class of varying possibilities. (This kind of passage is universal, but probability, by its very nature, is a particularly suitable example.)

The conclusions of the analysis are that it makes perfect sense to talk of unknown objective probabilities, even without postulating them in advance. Other conclusions concern the methodology of prior probabilities.

The presupposition of a prior probability is endorsed as a necessary principle, but this prior is regarded in a way that Bayesian (at least dyed-in-the-wool Bayesians) might not find palatable. It is evaluated by its success, sometimes it can be refuted by facts and replaced by another prior, not via conditioning. The theoretical arguments indicating why this cannot be avoided derive from considerations concerning the prior probability's complexity, they are given in Section 3.

The material concerning objective unknown probabilities, which formed part of the first draft, has been omitted to avoid excessive length; it is planned for a separate paper. Here I wish only to point out that by considering these objective probabilities some standard statistical practices of the classical school can be justified on rationality grounds. In particular, randomization in sampling can be regarded as a sound minimax, or risk-aversion, priciple. The very definition of randomness implies a high a priori probability that the sample is representative. (If some additional information reduces this probability then of course the sample should not be used.) The point is that randomization can be preferred to some deterministic procedure that has a higher a priori probability to yield a representative sample. The reason for the preference is that the higher value of the second procedure rests on the more subjective parts of the prior, i.e., on probabilistic subjective assignments that the agent himself regards as uncertain. The light in which the agent sees his own assignment of probabilities can be read from the prior itself, provided that it is defined over a sufficiently wide field of events. This idea underlies the notion of inner stability as defined in Section 2. (An explicit way of representing this higher level judgement would be to introduce higher order probabilities[2].) Indeed there is no denying that we judge some of our judgements, including probability judgements, to be more well founded, or certain, than others. Claims to objectiveness can be defined by means of those higher order judgements. Randomization may be preferred just as a risk-averse person may rationally prefer a gamble with lower expected gain and lower variance.

As against this, certain classical prescriptions are insufficient or incoherent. These are the well-known examples which do not enable one to incorporate relevant information in his decision making (e.g., information

---

[2] Perhaps it is the time to take a closer look at the possibility of a higher order probability calculus. Some suggestions have been recently made by Skyrms, Domator and others, but on the whole the subject is wide open.

indicating that the randomly chosen sample is not representative); or prescriptions which, given the same evidence, result in different decisions, depending on the intentions the statistician had when collecting his evidence. There is a long list of much discussed examples. A Bayesian-like approach appears to be the natural candidate for a comprehensive unified framework. But in order to fulfil this role successfully it should be enlarged by the inclusion of further distinctions on the theoretical level; in particular — distinctions reflecting second-order judgements. It is hoped that the present paper indicates such a direction.

The paper is intended mainly as a non-technical presentation of basic ideas. I found it necessary to include a minimum of technical details in Section 2.

I presuppose the usual setup of a probability function, P, which takes values in the interval $[0,1]$ and satisfies the Kolmogorov axioms. (Occasionally it is referred to as a probability distribution.) I do not enter into the question of σ-additiveness, which some philosophers (not myself) have found problematic; the points to be made do not depend on this assumption since the patterns are clearly indicated in finite, sufficiently large domains. The arguments of the function are the so-called *events* of probability theory. I find it convenient to use occasionally a mixed terminology and refer to them also as *statements*. The domain of the probability is a *field* (or Boolean algebra), i.e., it is closed under formation of unions, intersections and complements, or — in the parallel logical terminology — the class of statements is closed under disjunctions, conjunctions and negations. As customary in probability theory I use the Boolean notations '$\cup$', '$\cap$' and '$-$'.

Given the evidence e, the a priori probability function $P(\cdot)$ changes to the *conditional probability* $P(\cdot \mid e)$ defined as:

$$P(A \mid e) = P(A \cap e)/P(e).$$

This conditioning on e presupposes that $P(e) > 0$. The definition is sometimes extendible to cases where $P(e) = 0$, by passing to the limit (these are cases where e has the form $X = x$, with X a continuous random variable). Such extensions do not affect the general line of argument (cf. footnote 3).

The restriction $P(e) > 0$ can be removed if we take as primitive a two-argument conditional probability $C(\cdot, \cdot)$; but then we have to assume that $C(\cdot, e)$ is indeed defined. The usual setup of a one-argument function is much more convenient and, as will be indicated, the points to be made carry over to the two-argument setting.

2. The objective aspects of subjective probabilities: success and stability

Let us start with a trivial example. Assume that Adam assigns probability 0.6 to e. Neither the truth of e nor its falsity can prove or refute Adam's judgement. This is often interpreted as indicating the purely subjective, or logical, character of probabilistic assignments. The case would have been different had the assignment been 0 or 1. It has been sometimes argued that probability 0 does not necessarily signify that the event is ruled out as impossible. But if an event, e, of probability 0 is known to take place the prior probability function is no longer of any use, because we cannot derive the conditional probabilities.[3] The function has to be changed not by conditioning. It is as good as refuted. (Having a conditional two-argument prior $C(\cdot, \cdot)$, with e a possible value of the second argument, is equivalent at this point to having two one-argument priors: $C(\cdot, e_0)$ and $C(\cdot, e_0 \cap e)$, where e_0 represents the background knowledge. Since $C(e, e_0) = 0$, the first is refuted; but this time we have a ready substitute, provided in advance.)

The person behind the prior distribution may envisage the possibility of having to switch his function. But from the point of view which the function itself represents, the truth of a statement whose probability is 0 means the end of the game. Therefore assignments of extreme values 0 or 1 constitute factual claims. But then what is the status of assignments of values which are near the end-points? Do they not also constitute a sort of commitment? Suppose Moses assigns the same e probability 0.9 and then it is found that e is true. Although e's truth or falsity cannot "prove" either Adam's or Moses' assignment we feel that Moses was nearer to the truth. He was much more confident that e is true and would have been much more surprised had it turned out otherwise. His relative success is indicated by the smaller change in his probability value after conditioning on e. Both posterior probabilities of e, given e, are 1; so Adam's value for e has changed from 0.6 to 1 whereas Moses' value has changed from 0.9 to 1.

In general, consider a class **C** of statements and two prior probabilities P_1 and P_2. Let e be the accumulated evidence and assume that e decides each statement in **C**, i.e., implies its truth or its falsity. *Then, given e, P_1 is more successful than P_2 over* **C** *if, as a result of conditioning on e, the restriction of P_1 to* **C** *undergoes a smaller change than the restriction of P_2 to* **C**. In order

[3] Unless, of course, we have $P(A \mid X = x)$ where the conditional probability is defined as the limit of $P(A \mid x - \varepsilon \leq X \leq x + \varepsilon)$. But think of e in the present example as a statement of the form "John has come to the party", or "the coin will land on heads", or "$x - \varepsilon \leq X \leq x + \varepsilon$".

to measure change some distance function between probability distributions is needed. Smaller change means that the distance between $P_1(\cdot)$ and $P_1(\cdot \mid e)$, as functions over **C**, is smaller than the corresponding distance between $P_2(\cdot)$ and $P_2(\cdot \mid e)$.

I do not suggest that we fix, once for all, the same metric between probability distributions. The best way of defining the distance will depend on the context in which the evaluation of priors takes place. Some members of **C** may carry more weight than others. The scale of measurement may vary. When **C** is a rich field of events, we may have several aspects of "distance", so it becomes a vector; in that case being more successful is not decided by comparing two numbers only; there are several coordinates of success and we get a partial ordering. Yet, when all this has been taken into account there remain clear-cut cases when one probability has been more successful than another. Also certain basic patterns do not depend on the particulars of the distance function. These are the patterns which concern me here.

Note that, so far, we need distances only in cases when one member of a pair of distributions is a 0, 1-valued function over **C**; this is because e is assumed to decide each member of **C**; success is evaluated by considering statements which have been decided.

The survival value of a prior probability, i.e., its value as a guide to correct decisions, is directly related to its success over certain classes of statements. For example, if e has practical implications, Adam has to reckon more than Moses with the possibility that e is false. Moses' course of action is destined to yield higher benefits than Adam's if e is true, but he stands to lose more if e is false. Such will be also the relative success of their prior probabilities over $\{e, \bar{e}\}$.

This short account is meant to stress the fact that a prior probability may have a better or a worse accord with the actual world and that its refutation (by a true statement of probability 0) is only an extreme case of disaccord. Better accord means higher success over a wider class of statements and is directly related to its standing as an objective probability.

Of course, the most successful, as well as the most objective, probability is the function which assigns 1 to all true statements and 0 to all the rest. But such a probability is beyond human grasp. We can define it in the way I just did, but not in any way that will enable us to compute its values, even for the most elementary statements (such as "the coin will land on heads"). But we do have access to probabilities which are very successful over some rich classes of statements. We cannot decide with any confidence which side a fair coin will land on, or whether or not in a certain experiment an electron will be emitted within the next ten seconds. We have even good reasons to suppose that, in the second case, complete knowledge of the present state of the world is not going to help us. But we have probabilities which have been highly successful over classes of statements describing certain long-range phenomena. I shall later discuss the meaning of that success; here I only point out that the success of a prior probability depends on the prior as well as on the actual world. For example, the fact that certain, relatively simple, mathematical definitions yield probabilities which are successful for long-range phenomena means that some very strong regularities govern the world's behaviour.

Expected change

Let us now consider the prior probability's prediction concerning its own success, i.e., the change that will result from conditioning on the incoming evidence. Taking up again our elementary example, we see that Adam assigns probability 0.6 to a change from his present pair of values (0.6, 0.4) to $(1,0)$, and 0.4 to a change to $(0,1)$. Moses' probabilities are: 0.9 for a change from his present (0.9, 0.1) to $(1,0)$, and 0.1 for a change to $(0,1)$. Consequently, Adam forecasts an appreciable change for his present distribution, while Moses, admitting the possibility of a more radical change for *his* distribution, thinks it unlikely and is quite confident that the change will be small. Had Moses assigned probability 1 to e he would have been certain that there will be no change at all.

In general, put $p = P(e)$, then $1-p = P(\bar{e})$ and the change over $\{e, \bar{e}\}$ from $P(\cdot)$ to $P(\cdot \mid e)$ is the change from $(p, 1-p)$ to $(1,0)$. Take $1-p$ as a measure of this change. It is the simplest most intuitive measure: the absolute difference between e's a priori and a posteriori probabilities (it is also the difference between $\bar{e}$'s probabilities). Similarly, take p as the change from $(p, 1-p)$ to $(0,1)$. Consequently the prior P predicts with probability p a change to the amount of $1-p$ and with probability $1-p$ a change to the amount of p. Its *expected change* is the weighted average where each change is weighted according to its probability; it comes out as: $p(1-p)+(1-p)p$. This value is largest if $p = 0.5$ and decreases as p moves away from 0.5, becoming 0 at the end-points 0, 1. Adam's expected change is 0.48, that of Moses 0.18. Measuring probabilities on a percentage scale we could say that Adam expects 48% of change, Moses only 18%. (The percentage scale is feasible here because our change is a weighted average of probability differences, ranging from 0 to 1. It should not be applied if the scale is changed.) This difference in expected change reflects well their different attitudes concerning their own knowledge. Moses is much more certain, i.e,, thinks himself to be nearer to the truth than Adam thinks himself. Note that for $p = 0.5$ the expected change is the exact change that must take place when the truth concerning e is known. Similarly, Adam is guaranteed a change of 0.4 at the least.

The generalization to any finite field of events is straightforward. Let $\mathbb{F}$ be such a field and let $e_1, \ldots, e_n$ be its atoms; the e_i's are non-empty mutually exclusive events and every event in $\mathbb{F}$ is a union of e_i's. Let $P(e_i) = p_i$, then P is completely determined by $(p_1, \ldots, p_n)$. Assuming $P(e_1)$ to be non-zero, $P(\cdot \mid e_1)$ will have the corresponding form $(1, 0, \ldots, 0)$. Let us measure the distance between $(p_1, \ldots, p_n)$ and $(1, 0, \ldots, 0)$ by $1 - p_1$. It is the maximal difference in probabilities assigned by the two functions to any event in the field. (Either e_1 or $\bar{e}_1$ is an event for which the difference of probabilities is maximal.) In general let the distance function, d, be[4]:

$$d(P(\cdot), P(\cdot \mid e_i)) = 1 - p_i. \tag{1}$$

Then the expected change is:

$$\sum_i p_i (1 - p_i). \tag{2}$$

This value is equal to $1 - \sum_i p_i^2$ as well as to $\sum_{i \neq j} p_i p_j$. It is maximal if $p_i = 1/n$ for all i; it decreases as the distribution becomes less evenly spread and attains the value 0 at each of the end-points $(0, \ldots, 1, \ldots., 0)$. The same pattern will take place when we use any other reasonable distance function[5]. The exact meaning of being "evenly spread" is determined by the distance function but our general considerations will not depend on such finer shades[6].

A noteworthy expression is obtained if we change our scale to a logarithmic one and define the new distance to be $-\log(1 - x)$, where x is the distance just used. The previous maximal distance 1 (obtained for distributions which assign weight 1 to different atoms) now becomes infinite. Then the expected change is:

---

[4] This determines the distance only in cases where one of our functions is 0.1-valued. It can be extended to the general case in more than one way and, for the moment, I prefer not to consider specific functions in general.

[5] For example, the Euclidean distance between the vectors $(p_1, \ldots, p_n)$ and $(0, \ldots, 1, \ldots, 0)$. This however turns out, on later considerations, not to be a suitable choice, see footnote 8.

[6] In our particular case "evenly spread" has also the following meaning: If we change $(p_1, \ldots, p_n)$ to $(p'_1, \ldots, p'_n)$ so that $p'_i = p_i$ for all i except j and k, if $p'_j + p'_k = p_j + p_k$ and $|p'_j - p_k'| < |p_j - p_k|$, then $(p'_1, \ldots, p'_n)$ is more evenly spread than $(p_1, \ldots, p_n)$. The value of (2) increases with this change.

$$-\sum_i p_i \log p_i. \tag{3}$$

This is the probability's entropy. It has been introduced by Shanon as a measure of information: the amount of information given when in state of knowledge $(p_1, \ldots, p_n)$ one is told which of the e_i's is true. Again the value is maximal for $p_i = 1/n$, decreases to 0 as the distribution becomes less evenly spread and attains 0 at the end-points.

Setting aside technical details, the intuitive picture is clear: A high expected change means little assurance as to which of the (mutually exclusive) statements is true. It corresponds to evenly distributed weight. A low expected change signifies confidence of being successful, i.e., of being nearer to truth.

So far we have simplified the picture by not assuming any additional structure on the field of events besides that of the Boolean algebra. This amounts to treating our atomic events as anonymous points in an abstract set. But in fact the very definition of the events presupposes an additional structure which has to be taken into account when the number of atoms is very large and, in particular, when the field is infinite. Assume that our atomic events are of the form: $X = x_i$, where X is some magnitude ('random variable' in probabilistic parlance) and the x_i's are its possible values (whose number is assumed for the moment to be finite, but large). Assigning probability 1 to $X = 48/100$ one is still not far from the truth if $X = 46/100$ happens to be the case; certainly nearer than the one who assigns probability 1 to $X = 61/100$. To reflect the situation we should modify our distance function between probabilities by taking into account varying distances between the atomic events. This, in particular, is necessary if the x_i's are the possible outcomes of a measurement and we consider a possible future refinement which will yield a more accurate value. (In the limit we get an infinite field with a continuum of possible values.) Here is a possible modification: Let $w(e_i, e_j)$ be a number representing the relative distance between the atomic events e_i and e_j. Let $P(e_i) = p_i$. Our previous distance between $P(\cdot)$ and $P(\cdot \mid e_i)$ has been $1 - p_i$; it can be written as $\sum_{j \neq i} p_i$. Define the modified distance as[7]:

[7] There does not seem to be a straightforward generalization of (4) to a distance function between two arbitrary probability vectors $(p_1, \ldots, p_n)$, $(q_1, \ldots, q_n)$. The following appears however to offer some promise $\sum_{i,j} |p_i q_j - q_i p_j| \, w(e_i, e_j)$. If $(q_1, \ldots, q_n) = (0, \ldots, 1, \ldots, 0)$ we get (4). If $w(e_i, e_j) = 1$ for all $i \neq j$ we get an expression which has a clear interpretation as the expected gain in the following system of bets. Player I bets p_i on e_i and $1 - p_i$ on $\bar{e}_i$ and player II choses the bets according to $(q_1, \ldots, q_n)$.

$$\sum_j p_j w(e_j, e_i). \tag{4}$$

(We can omit '$j \neq i$' by setting $w(e_i, e_i) = 0$.) Our previous distance is thus the particular case where $w(e_i, e_j) = 1$ for all $i \neq j$ and $w(e_i, e_i) = 0$. The expected change becomes the double sum:

$$\sum_{i,j} p_i p_j w(e_i, e_j). \tag{5}$$

We can now pass directly to the case of a continuous magnitude: Write $w(X = x, X = y)$ as $\rho(x, y)$, then the expected change is

$$\sum_{x,y} \rho(x, y) P(X = x) P(X = y) \tag{6}$$

where the sum is interpreted as a double integral, obtained by the standard limit technique (ρ has to be measurable). If $\rho(x, y) = (x - y)^2$ then the expected change turns out to be twice the variance of X. Note that in general the values of X need not be real numbers; the construction makes sense with respect to any space which is provided with some distance function (the linear structure of the space is not needed).

The time-line induces an essential structure on fields whose events take place in time. This structure may influence our choice of distance function. With each t we get an associated field $\mathbb{F}_t$, containing possible events up to time t; change and expected change become dependent on t. I shall not pursue here this line of enquiry.

The importance of time for our subject is that it underlies the most common examples of evenly spread probabilities which are nonetheless regarded as objective. Someone who assigns probability 0.5 to "heads on the next toss" signifies thereby a total lack of assurance concerning that particular outcome. Yet if he thinks the coin to be fair he regards 0.5 as "the correct value" and judges other values to be objectively erroneous.

My main goal is to provide an analysis for such claims to objectivity, solely in terms of prior probabilities, using as a principal tool a more general version of the notion of expected change.

Inner stability

So far we have considered changes in some probability over a given field which are caused by conditioning on events from this same field. In the generalized definition a second field supplies the evidence:

Let $\mathbb{E}$ and $\mathbb{F}$ be two fields of events and let P be a prior probability defined over a field which includes both. *The expected change, under* $\mathbb{E}$, *of* P

over $\mathbb{F}$, *is the expected amount of change that P's restriction of* $\mathbb{F}$ *undergoes by conditioning on evidence from* $\mathbb{E}$.

In this context I shall refer to $\mathbb{E}$ as *the field of evidence* and to $\mathbb{F}$ as *the field of forecasts.* I use "forecasts" as a convenient term. Its temporal aspects should be ignored, for the definition applies to any pair of fields. Our previous concept of expected change over $\mathbb{F}$ turns out to be the special case where $\mathbb{E} = \mathbb{F}$.

Here is the formal definition of expected change, formulated first for a finite field of evidence $\mathbb{E}$. Let P_e be the conditional probability obtained by conditioning on e; let $P \mid \mathbb{F}$ and $P_e \mid \mathbb{F}$ be the restrictions of P and P_e to $\mathbb{F}$. Then

$$\mathrm{EC}_d(P, \mathbb{F}, \mathbb{E}) \stackrel{\text{def}}{=} \sum_e P(e)\, d(P_e \mid \mathbb{F}, P \mid \mathbb{F}) \tag{7}$$

where e ranges over all the atoms of $\mathbb{E}$ and d is the distance function between probability distributions. 'EC' stands for 'expected change'. In the sequel 'd' shall be omitted; occasionally I shall omit also 'P' and write $\mathrm{EC}(\mathbb{F}, \mathbb{E})$. The following simple example can clarify the meaning of expected change.

Let $\mathbb{F}$ be a finite field whose atomic events are written in the form '$Y = b$'. It is convenient to speak of b as the value of Y, but it need not be a number; 'Y' can refer, say to some object's colour, or to its shape and b may belong to a set of colours, or shapes. Similarly, write the atomic events of $\mathbb{E}$ in the form '$X = a$'. Then $\mathrm{EC}(\mathbb{F}, \mathbb{E})$ measures the a priori informativeness, or relevance, of the value of X for the value of Y. If one who is interested in the value of Y can buy information concerning X, then $\mathrm{EC}(\mathbb{F}, \mathbb{E})$ determines the worth of this information, evaluated according to his own prior probability. To conclude this example let us compute $\mathrm{EC}_d(\mathbb{F}, \mathbb{E})$ choosing as d the simplest distance function, namely the maximal difference of the probabilities of any event in the field: $d(P', P'') = \max(|P'(A) - P''(A)|)$, where A ranges over the common domain. I shall refer to it as the *total variation distance* (it is half of the total variation of $P' - P''$, as defined in measure theory). In our case the field is generated by $\mathbb{E}$ and $\mathbb{F}$ and its atoms are all the events $X = a \cap Y = b$. For d = total variation distance, $\mathrm{EC}(\mathbb{F}, \mathbb{E})$ comes out as

$$\tfrac{1}{2} \sum_{a,b} |P(X = a \cap Y = B) - P(X = a)P(Y = b)| \tag{8}$$

where a and b range (independently) over the values of X and Y. Note that for this distance $\mathrm{EC}(\mathbb{F}, \mathbb{E}) = \mathrm{EC}(\mathbb{E}, \mathbb{F})$. But, in general, the two can be different: $\mathbb{E}$ can be more informative for $\mathbb{F}$ than $\mathbb{F}$ for $\mathbb{E}$. Asymetries can take place when the distance function expresses additional structural features of the field besides that of a Boolean algebra.

Some natural requirements concerning distance functions suggest themselves at this point. (They are not essential to the main line of argument and the reader not interested in the details can skip this part.) Let P and Q be probabilities defined over some field which includes $\mathbb{F}$. Then:

$$\text{If } \mathbb{F}' \subseteq \mathbb{F} \text{ then } d(P \mid \mathbb{F}', Q \mid \mathbb{F}') \leq d(P \mid \mathbb{F}, Q \mid \mathbb{F}). \tag{I}$$

The intuition is obvious: d measures the amount of disagreement between P and Q; it does not increase when we restrict ourselves to a subfield. It is interesting to note that (I) rules out certain distances originating in other mathematical contexts which are unsuitable in the present one[8]. As a direct consequence of (I) we have:

$$\text{If } \mathbb{F}' \subseteq \mathbb{F} \text{ then } \mathrm{EC}(\mathbb{F}', \mathbb{E}) \leq \mathrm{EC}(\mathbb{F}, \mathbb{E}). \tag{I$'$}$$

A second requirement is a so-called covexity condition: Let P_1, P_2, Q be probabilities over F and let λ_1, λ_2 be non-negative numbers such that $\lambda_1 + \lambda_2 = 1$; then

$$d(\lambda_1 P_1 + \lambda_2 P_2, Q) \leq \lambda_1 d(P_1, Q) + \lambda_2 d(P_2, Q). \tag{II}$$

(Here $(\lambda_1 P_1 + \lambda_2 P_2)(e) \overset{\text{def}}{=} \lambda_1 P_1(e) + \lambda_2 P_2(e)$.) (II) is satisfied by all prima facie candidates for distance functions (including some which are ruled out by (I)). Its intuitive meaning, which may not be clear at first glance, is clarified by noting its connection with:

$$\text{If } \mathbb{E}' \subseteq \mathbb{E} \text{ then } \mathrm{EC}(\mathbb{F}, \mathbb{E}') \leq \mathrm{EC}(\mathbb{F}, \mathbb{E}). \tag{II$'$}$$

Indeed, the expected change under richer, or more refined, evidence should not decrease. (II) implies (II′); on the other hand (II′) implies (II) for the case in which Q dominates $\lambda_1 P_1 + \lambda_2 P_2$ (i.e., where, for all e, $(\lambda_1 P_1 + \lambda_2 P_2)(e) > 0$ implies $Q(e) > 0$); if we add some natural continuity condition on d, then (II′) implies (II). I shall not enter here into the proof. We can now define the expected change under an infinite field $\mathbb{E}$ of possible evidence as the supremum of the expected changes under all finite subfields:

$$\mathrm{EC}(\mathbb{F}, \mathbb{E}) \overset{\text{def}}{=} \sup\{\mathrm{EC}(\mathbb{F}, \mathbb{E}') : \mathbb{E}' \subseteq \mathbb{E},\ \mathbb{E}' \text{ finite}\}. \tag{9}$$

---

[8] For example, the Euclidean distance function, which yields for $(1,0,0)$ and $(1/3, 1/3, 1/3)$ a distance of $(6/9)^{1/2}$; passing to $(1,0)$ and $(1/3, 2/3)$ the value increases to $(8/9)^{1/2}$. It should not increase because the second distributions are induced over a subfield.

A natural stipulation on the distance function is that, for P and Q probabilities over $\mathbb{F}$, $d(P, Q) = \sup\{d(P \mid \mathbb{F}', Q \mid \mathbb{F}'): \mathbb{F}' \subseteq \mathbb{F}, \mathbb{F}' \text{ finite}\}$. Assuming it we have:

$$\mathrm{EC}(\mathbb{F}, \mathbb{E}) = \sup\{\mathrm{EC}(\mathbb{F}', \mathbb{E}'): \mathbb{F}' \subseteq \mathbb{F}, \mathbb{E}' \subseteq \mathbb{E}, \mathbb{F}', \mathbb{E}' \text{ finite}\}. \tag{10}$$

Another plausible condition[9] on d implies that if $\mathbb{F} \subseteq \mathbb{E}$ then $\mathrm{EC}(\mathbb{F}, \mathbb{E}) = \mathrm{EC}(\mathbb{F}, \mathbb{F})$. Its intuitive appeal is clear: $\mathbb{F}$ is the most informative field of evidence with respect to itself; any wider field yields no more information with respect to the statements in $\mathbb{F}$ than these statements themselves.

The concept of expected change can be further developed by considering the absolute information, or degree of refinement, of the evidence on which we condition. How much detail do we need in order to produce a certain amount of expected change? (The field of possible evidence $\mathbb{E}$ is to be represented as a union of an ascending chain of subfields each refining the previous ones.) But I shall not pursue this further.

Let us return to the coin toss. Let h = "heads on the next toss". Belief in the fairness of the coin (or, to be more precise, of the experimental setup) implies that the assignment $P(h) = 0.5$ is not liable to change by conditioning on all kinds of evidence, in particular on evidence of the past and the present. Thus, our probability is stable. To be sure, if the possibility of a biased coin is not ruled out completely, then strong evidence in this direction will make a big difference; for example, evidence that 26 of the last 30 tosses resulted in "heads". But from the prior's point of view such evidence is extremely unlikely. Hence the expected change of P over $\{h, \bar{h}\}$ under evidence of past and present events, is extremely small. The stronger the belief in the coin's fairness, the smaller the expected change; in the extreme case of a Bernoulli distribution the change under evidence of the past outcomes is always 0. *In declaring 0.5 to be the "correct probability of h" one declares that his total lack of assurance concerning the next outcome does not reflect ignorance of past and present events*; *this is the sense in which the probability is claimed to be objective.*

The claim is meaningful even if, extrapolating from classical physics, one believes in a deterministic world. From a deterministic point of view knowledge of the initial conditions makes a prediction of the outcome

---

[9] The condition is as follows: Let P and Q be probabilities over finite field $\mathbb{F}$. Let $\mathbb{F}'$ be obtained by splitting one atom, a, into k distinct atoms $a = a_1 \cup \cdots \cup a_k$ and let P' and Q' be defined by: $P'(b) = P(b)$, $Q'(b) = Q(b)$ for $b \in \mathbb{F}$, $P'(a_i) = (1/k)P(a_i)$, $Q'(a_i) = (1/k)Q(a_i)$, $i = 1, \ldots, k$. Then $d(P', Q') = d(P, Q)$. The total variation distance satisfies it. The "cross product distance" of footnote 7 satisfies it as well if the distances between atoms in the new field $\mathbb{F}'$ are defined in a certain natural way.

possible. But the outcome of a toss depends on many parameters and is sensitive to extremely small variations of magnitude. Hence the values of many parameters have to be known with very high precision; the amount of needed information is exorbitant. Conditioning on evidence which falls short of that enormous amount will not affect the prior assignment $P(h) = 0.5$. Thus the deterministic variant of a probability's stability over $\mathbb{F}$ is somewhat as follows:

If $\mathbb{E}$ is a field of statements of reasonable complexity, describing past and present possible events, then the expected change, under $\mathbb{E}$, of the probability distribution over $\mathbb{F}$ is very small.

Objective probability is therefore possible in a deterministic framework. (Note that the objective interpretation of epistemic modality can be constructed along these lines.)

The strongest claim to objectivity is made in the non-deterministic framework of quantum physics, where the expected change is 0, even under evidence which includes the world's history up to this moment spelled out in as much detail as is theoretically possible. But also in a deterministic framework the less extreme form of objective probability flourishes quite well on coins, dice and other lottery devices.

I shall use *inner stability* or, for short, *stability*, as a measure which increases as the expected change decreases. Thus P is more stable than P' (over $\mathbb{F}$, under $\mathbb{E}$) if it has smaller expected change. More generally, a probability is more stable to the extent that it has smaller expected change over larger fields of forecasts and under larger fields of possible evidence.

Brian SKYRMS [1977] has introduced 'resilience' to denote a related concept: The resilience of P for a statement A is

$$1 - \max(|P(A) - P(A \mid e)|)$$

where e ranges over all statements of some presupposed language which are consistent with A and with $\bar{A}$. Thus, resilience is defined for a single statement. But the main difference between resilience and inner stability is that resilience ignores the prior probabilities of the events e used in the conditioning. It uses maximal possible change, not the change expected by the prior. The stipulation that e should be consistent with A and $\bar{A}$ (which makes the definition useless in the case of a finite field with two atoms) is presumably intended to bar the trivial changes caused by conditioning on A or on $\bar{A}$. Nonetheless the undesired consequences of using maximal change remain. Probabilities expressing very high certainty may have very low resilience because there still exists evidence that will induce a big change. One may assign probability $1 - \varepsilon$ to a coin being fair but, as long as

$\varepsilon > 0$, the resilience (for the next outcome with respect to past outcomes) will not be more than 0.5; because a very long sequence of "heads" in the past may still move the probability of "heads" to the neighbourhood of 1. In the limit, when $\varepsilon = 0$, the resilience jumps from 0.5 to 1. In similar examples 0.5 can be replaced by some arbitrary small δ. Such difficulties do not arise if we consider expected change.

As noted above, the extreme conception of objective probability arises in quantum physics. At the other extreme we find very unstable probabilities, highly sensitive to all kinds of evidence, defined over fields which lack the homogeneous structure that is needed for describing long-range phenomena. Between these extremes there is a broad spectrum of varying subjective-objective degrees. 'Spectrum', to be sure, is a rough description, for we do not have a linear order. The various kinds, or aspects, of objectivity can be clarified by considering the probability's inner stability over various fields of evidence.

Consider the example of bills drawn randomly from a bundle of 100 bills, with "genuine" and "forged" as the two possible outcomes of each test. Let $\mathbb{F}$ consist of statements about the sequence of outcomes. Let 'X' stand for 'the number of genuine bills in the bundle', then:

$$P(A) = \sum_j P(A \mid X = j) P(X = j).$$

Now $P(\cdot \mid X = j)$ is extremely stable over $\mathbb{F}$ under a wide field, $\mathbb{E}'$, which contains, in addition to the statements '$X = i$', all sorts of evidence: the bills' origin, the way they have been arranged in the bundle, the identity of the man who brought them, etc. This is implied in the assumption of random drawings. Indeed, *randomization consists in creating fields of events over which certain conditional probabilities are extremely stable.* In our case the created field is $\mathbb{F}$. (The stability is necessary in order that we may draw from events in $\mathbb{F}$, e.g. the first 15 outcomes, reliable conclusions concerning the value of X. In order that the conclusions be "sharp", i.e., sufficiently informative, the different conditional probabilities $P(\cdot \mid X = j)$ should be sufficiently apart from each other).

Let $\mathbb{E}$ be the subfield of $\mathbb{E}'$ generated by the events $X = j, j = 0, \ldots, 100$. The extreme stability under $\mathbb{E}'$ of $P(\cdot \mid X = j)$ means that for any e in $\mathbb{E}'$ (such that $P(e \mid X = j) > 0$) $P(\cdot \mid X = j)$ and $P(\cdot \mid X = j \cap e)$ coincide over $\mathbb{F}$. This can be shown to imply:

$$\mathrm{EC}(\mathbb{F}, \mathbb{E}') = \mathrm{EC}(\mathbb{F}, \mathbb{E}).$$

Stated informally it means that, as far as probabilities of events in $\mathbb{F}$ are concerned, the only relevant information is information concerning the number of genuine bills — a conclusion which we knew already intuitively. If the sampling is without replacement then the maximal length of outcome sequences is 100 and each sequence of length 100 determines the number of genuine bills. Hence $\mathbb{E} \subset \mathbb{F}$. Moreover each atom of $\mathbb{E}$ is split into atoms of $\mathbb{F}$ of equal probability; because $P(\cdot \mid X = j)$ assigns equal probability to all maximal outcome sequences which are consistent with $X = j$. Assuming a certain plausible condition on the distance function (cf. footnote 9) it follows that:

$$\mathrm{EC}(\mathbb{F}, \mathbb{E}) = \mathrm{EC}(\mathbb{E}, \mathbb{E}).$$

If the sampling is with replacement then $\mathbb{F}$ is infinite and does not include $\mathbb{E}$. But also in this case the equality is derivable from certain plausible assumptions concerning the distance function.[10] Altogether we get:

$$\mathrm{EC}(P, \mathbb{F}, \mathbb{E}') = \mathrm{EC}(P, \mathbb{E}, \mathbb{E}).$$

This means that the instability of the probability over $\mathbb{F}$ under evidence from $\mathbb{E}'$ is given exactly by its instability as a probability over $\mathbb{E}$; in this context the subjective element is fully represented as lack of knowledge concerning the number of genuine bills.

But this is not the end of the story, for the probability over $\mathbb{E}$ might itself be stable or unstable under various kinds of evidence. Compare Adam, who knows that the bundle was picked by lottery from a collection of 100-bill bundles, with Eve who has no such knowledge. Assume moreover that Adam knows the various lottery proportions of bundles satisfying $X = i, i = 0, \ldots, 100$. Now the prior probabilities of Adam and Eve may coincide over $\mathbb{E}$; in which case they coincide also over $\mathbb{F}$. The difference between them is revealed when we note that over $\mathbb{E}$ Adam's prior is stable under E^*, where E^* contains evidence about events taking place before the lottery's outcome, or unrelated to it.

A situation which is similar in this respect arises if the sampling is without replacement, the value of X is known, but 50 tests have been already performed, whose outcome is known and we are forecasting the remaining tests. Evidence of the last 50 outcomes will change the probability over the field of forecasts. But, given such evidence, the probability is stable; moreover, on the evidence itself, it is stable under evidence concerning events up to the beginning of tests.

It should be emphasized that stability is a property of the prior

[10] Roughly speaking, it is a continuous version of (9), or a combination of (9) with a continuity condition,

distribution; hence it can no more than indicate the prior's own assessment of its objectivity. Whether this assessment is justified is another matter. As we saw, an unstable probability will as a rule undergo a certain amount of change, irrespective of the world's behaviour; starting with $P(e) = 0.6$, our final value, after conditioning on e or $\bar{e}$, will differ by at least 0.4. Thus, a claim to subjectivity is justified: a man who thinks himself ignorant is indeed ignorant (unless we introduce non-conscious knolwedge, which certainly cannot be expressed in his declared prior). But a claim to objectivity can be refuted by facts; in the extreme case — by the occurrence of an event of probability 0, in the less extreme case — by a very large change that the probability is forced to undergo on a certain field of events. *Real* objectivity obtains only to the extent that a prior probability, which claims to be objective, is successful. The question of success, or rather of failure, is taken up in the next sections.

3. Internal and external change

What happens when a coin thought to be fair yields 1111111111, i.e., ten successive "heads", on the first ten tosses? One would feel uncomfortable with a prior probability which treats the tosses as independent with probability 0.5 for "heads". Let b_p be the Bernoulli distribution with probability p for "*heads*", i.e., each sequence of i ones ("heads") and j zeroes ("tails") is assigned probability $p^i(1-p)^j$. Strictly speaking $b_{0.5}$ has not been refuted for it assigns positive probability to the outcome. But his argument is not very convincing and it becomes less convincing when the lucky streak of ones continues for the next twenty tosses. $b_{0.5}$ is a failure and we cast around for a better candidate. There is an alternative way of reading this story, according to which $b_{0.5}$ is not, or should not be the prior to start with. At best it is a simplified version used to save work. A full-fledged prior should not concentrate all the weight on $p = 0.5$, but should allow for the possibility of a biased coin by spreading the weight on the interval $[0, 1]$. Thus we get the well-known mixture of b_p's. After conditioning on the evidence the weight will be shifted in the direction of 1 giving prominence to b_p's with p in that neighbourhood, but reserving non-zero weight for the rest. Call a change in the prior which results from conditioning on some evidence *internal change*, call other changes *external*; in particular, an external change takes place when a prior rejected as a failure is replaced by another; it is with this kind of change that I am mostly concerned.

An internal change avoids the necessity of evaluating the prior's success; the probability, so to speak, adjusts itself. It also anwers the following sort of question that has puzzled many at first sight:

The distribution $b_{0.5}$ assigns to 1111111111 and 0110101000 the same probability: $(\frac{1}{2})^{10}$. Yet the first is very unexpected and inclines us to revise the distribution, whereas the second is quite satisfactory. Why the difference?[11]

The Bayesian answer is that when a mixture of Bernoulli distributions is conditioned on the evidence of n outcomes the factor which determines the shift in weight is the number of ones and there is a very big difference between ten out of ten or four out of ten. The parallel non-Bayesian answer is that $b_{0.5}$ (i.e., the hypothesis $p = 0.5$) is evaluated against other distributions of the form b_p (rival hypotheses); the factor which determines their relative success is, again, the number of ones. In Section 4, I shall outline a more general type of answer.

The convenience of a self-adjusting probability may have too high a price. Such a gadget is quite complicated and the more its capability of self-adjustment the higher its complexity. Quite often the probability appears to be out of reach. How, for example, should the weight be distributed over $[0, 1]$, i.e among the b_p's? Presumably one should analyze his beliefs concerning this particular coin; but little reflection will show the practical impossibility of arriving at a non-arbitrary assignment: Is 10^{-3} too high a probability for total bias (i.e., for $p = 0$ or $p = 1$)? How much should Adam, after searching his mind, assign to $p \leq 0.3$? One's general knowledge of coins will undoubtedly have its effect, but to what extent? Suppose the coin has been tested yesterday and its fairness confirmed to a high degree. The most plausible explanation to its present stange behaviour is that it is not the same coin; somebody has switched coins during last night. But only now does Adam realize that he has ignored this possibility in his prior assignment. What is then the a priori probability that he ought to have assigned to somebody switching the coin? These well-known difficulties point out the idealized nature of a prior distribution. In some contexts of decision making it is advisable to sort out one's beliefs in order to establish approximate probability values. But in other contexts the picture of somebody who tries to find his prior is misleading. One does not look

[11] The question arises naturally in view of the explanation, found often in books of statistics, that rejection takes place because, given the hypothesis, an event of very small probability has taken place.

into himself to uncover a hidden prior; one *posits* it as a theoretical construct in order to create a uniform and coherent framework.

The presupposition of some prior distribution is, I think, necessary on the following counts.

(i) It is a regulative principle and the only sufficiently comprehensive one for organizing diverse statistical practices and inference making. Paradoxes, puzzles and cases of contradictory advice which arise occasionally by following the rules of various methodologies are well known.[12] The best way of organizing the patchwork of different rules into a rational coherent system is by assuming a prior distribution, even when no computation of its values is intended.[13]

(ii) Some highly significant general patterns do not depend on the particulars of the prior but are shared by all distributions of a certain type. Noteworthy among these are patterns that emerge upon conditioning on large amounts of data, when the initial finer shades of the prior are swamped by the evidence.

(iii) There still remain those cases where significant parts of the prior are somewhow accessible and should be used. I do not refer here to undisputable probabilities deriving from objective chances, but to those cases of uncertainty where information of a frequentist nature has to be combined with non-quantified or semi-quantified information.

(iv) The concept of one event being more likely or less likely than another is very fundamental and it applies also in cases where quantitative measurement of belief appears arbitrary. Non-quantitative, as well as fuzzy (interval-valued) probabilities are concepts of long standing. They can be treated by considering families of probability distributions instead of single ones; namely, associate with a qualitative, or fuzzy probability the family of all precisely defined distributions which agree with it.

---

[12] See for example Savage's article in SAVAGE et al. [1962] and the relevant chapters in HACKING [1965], KYBURG [1974], SEIDENFELD [1979].

[13] Thus the mere assumption of some prior distribution which represents one's beliefs suffices to rule out "randomly chosen" samples which are suspected of being non-representative. This is not to say that evaluations based on prior probabilities are the most efficient tool for reasoning in cases of uncertainty. Deductive systems, or algorithms may prove better suited in those contexts for which they were designed. What I claim is that such systems should be consistent with a probabilistic approach, this being the only general way of ensuring coherence. In a recent article DOYLE [1983] presents such a system and observes that, in principle, conclusions reached through it can be also obtained probabilistically. The strength of the system is indicated by the fact that a simulation in the other direction does not work in general. Thus existence of a probabilistic simulation ensures the system's coherence.

Now an extreme subjectivist tends to regard the evolution of our belief-system, or large parts of it, as an evolution of some prior probability, i.e., as the inner changes that take place by conditioning on the accumutating data. De Finetti held such views[14] and Carnap was accused by Lakatos[15] of trying to reduce all analysis of scientific progress to computations of conditional probabilities. The subjectivist view is certainly not a good reproduction of our immediate experience. Acceptance and refutation form a familiar and common method: accept some initial assumptions as true and reject them when they are found unsatisfactory. By contrast, probabilistic conditioning is a complicated, often subtle, process, known in special contexts since Bayes and Laplace (whereas the first method is as old as Adam). The subjectivist view is thus proposed as a theoretical model whose merits are first of all philosophical. Its chief merit appears to be the uniform way of treating the evolution of beliefs within a conceptually well-defined economical system, while allowing for the use of sophisticated mathematics. (In de Finetti's case it ties up with his operationalist views and strong idealistic inclinations; but this does not concern us here.) Such a view may have less speculative consequences, by influencing the approach to statistical inference and decision theory. Let us now see how much "idealized" the prior of the extreme subjectivist has to be.

Suppose that the outcome of the first 14 tosses has been 10101010101010. Given this data all Bernoulli distributions, b_p, seem unsatisfactory. (Again, imagine a longer sequence of this type if necessary.) Distributions of the form b_p are now replaced by better candidates: Markov processes. Again, the subjectivist gives a different account: The symmetric probability which is a mixture of b_p's is not what the prior ought to be. It is better than $b_{0.5}$ but it is still an approximation. The prior should have a non-symmetric component, allowing for the possibility of Markov processes which are not b_p's. It is now obvious how the story can be repeated: We propose to the subjectivist another hypothetical outcome of the first n tosses (where n is sufficiently large) which indicates a kind of regularity that does not fit any Markov process. He then rectifies again his prior probability, by assigning

---

[14] "It is not a question of 'correcting' some opinions which have been 'refuted': it is simply a question of substituting for the initial evaluation of the probability the value of the probability which is conditioned on the occurrence of facts which have already been observed; this probability is a completely different thing from the other, and their values can very well not coincide without this non-coincidence having to be interpreted as the 'correction of refuted opinion'" (DE FINETTI [1964], p. 118).

[15] In LAKATOS [1966]. As phrased by him, his charge is unwarranted.

non-zero weight to probabilities that allow for this type of new regularity (say he now includes also 2-step Markov processes). And so on. The principle of the game is that at each stage we call attention to some new hypothesis whose a priori probability, at that stage, is 0 (i.e., it is 0 when the hypothesis is formulated for the infinite outcome sequence, while for finite outcomes it has approximations, or restrictions, whose probability tends to 0 as the number of tosses increases). Moreover, we point out hypothetical outcomes which back up this hypothesis. In the given example the simplest hypotheses suggested, respectively, by the first and second outcomes are: "all members are 1" and "1 and 0 always alternate". Actually, less extreme hypotheses would have done, e.g., "the relative frequency of 1's is $\geqslant 0.8$" and "the relative frequency of 2-blocs of the form 10 is $\geqslant 0.4$". We can also allow some lattitude in our hypothetical outcomes, e.g. 1110111101 is already sufficient to cast some doubt[16] on the prior $b_{0.5}$ and 10110101010110 will suggest a possibility of dependency.

As a rule the more we continue in this game the more involved the new hypothesis will get. Roughly speaking, the regularities ignored by the prior become more complex and difficult to spot; hence the number of outcomes which are needed to reveal them, or back them up, grows. But with sufficiently many outcomes any desired degree of support is obtainable. The crux of the matter is that the prior probability itself becomes highly complicated. Having started with a simple probability $b_{0.5}$, we had to replace it with a much more complicated mixture of b_p's and, in the next stage, we had to consider a mixture of Markov processes. This rule holds in general: *the prior must be essentially more complex than the class of hypotheses which it covers.*

A preciser formulation is obtained when we consider the languages which serve to state the hypotheses and to define the prior. The language contains no more than countably many expressions. Hence we are dealing with countably many hypotheses. There exists a prior which assigns positive weight to each consistent one[17]. Presupposing this prior the subjectivist can take care, in advance, of all eventual developments. But such a prior can be defined only by means of strictly more expressive

---

[16] If we presuppose Bernoulli distributions, $p = 0.5$ will be rejected, given such an outcome, at the significance level 0.055.

[17] Namely, fix a set of models Ω for the language such that every consistent statement, e, has a model in Ω. Consider the σ-field generated by the consistent statements, i.e. by the Mod(e)'s where Mod(e) = set of all models satisfying e. For each consistent e let P_e be a probability measure which assigns weight 1 to some point in Mod(e) and take a mixture, with positive weights, of all P_e's.

language. This informally stated conclusion sums up certain mathematical results of GAIFMAN and SNIR [1982]. They have considered languages which contain, in addition to the predicates describing the empirical events, a sufficiently rich mathematical apparatus (first-order arithmetic, or some version of it). An extremely wide class of hypotheses, including all the usual statistical ones, can be formulated in such a language[18]. It is important to emphasize that, when speaking of "defining a probability", one refers here *not* to the probability over all the statements of the language[19], but only over those finitely generated by the basic empirical events. In the present example we need an empirical predicate $H(\cdot)$, where '$H(n)$' asserts that the nth toss results in "heads"; to define a probability means to define it over the finite Boolean combinations of statements of this form; as is easily seen, it suffices to define it over all finite conjuntions of $H(n)$'s. (The definition of a real-valued function means a definition which determines for each member in its domain and each natural number, n, a rational value which constitutes an approximation with error $< 1/n$.) It is shown that for any probability definable at a certain level in our language, there is a consistent hypothesis of probability 0 and a sequence of outcomes to back it up — both definable at the same level. Here "levels" are determined as in the arithmetic hierarchy, according to alternating-quantifiers depth; but the same technique will establish the phenomenon for other natural hierarchies within the language, or for languages in general. (The first result in this direction is due to PUTNAM [1963].) *This means that any prior definable in the language can be refuted at the same level at which it is defined.* There are also consistent hypotheses that get value 0 under all probabilities definable in the language; but this time the back-up sequence is definable only in a strictly stronger language; and in this stronger language we can also define a probability which assigns the hypothesis a positive value.

As in GAIFMAN and SNIR [1982], let us use the term "*dogmatic*" to indicate an assignment of probability 0 to consistent hypotheses; in particular P_1 is dogmatic with respect to P_2 if, for some h, $P_2(h) > 0$ and $P_1(h) = 0$. A result of GAIFMAN and SNIR [1982] which is related to the first

[18] For example the hypothesis that there exists a Turing machine which produces the outcome sequence; or the hypothesis that the limiting frequency of any block x of zeroes and ones is $f(x)$, where f is any function definable in the language.

[19] As a rule, the definition of probability becomes more complicated when the statements over which it is defined increase in complexity. The results would have ben trivial if "to define a probability" meant to define it over all statements of the language; for such a definition would have yielded a truth-definition for the mathematical part of the language.

one is that for any probability P definable at some level, there is another probability at the same level such that P is dogmatic with respect to it. On the other hand, by going one level up (in the arithmetic hierarchy) we can define a probability which dominates all probabilities at the previous level (i.e., it assigns positive value to each hypothesis assigned positive value by some lower-level probability).

[The difficulties of defining a non-dogmatic prior persists, under a different form, if we use a two-place conditional probability: The problem of having $P(e)>0$ for consistent e's becomes the problem of having $C(\cdot, e)$ *defined* for these same e's. Though not noted in GAIFMAN and SNIR [1982], all the results carry over to conditional probabilities; e.g., to define a conditional probability in which the second argument ranges over the consistent statements in a given language, a second, essentially more powerful, language is needed, etc.]

The results furnish a sort of theoretical basis to what we know already from experience: *As a rule non-dogmatism must be bought at the price of increased complexity*. Possible regularities must be considered explicitly, for there is no way of constructing a simple probability that will take care of them automatically. Carnap who tried to construct probabilities along certain natural lines, "from the bottom up", ended with probabilities which assign 0 to simple universal generalizations.[20] In this case the problem is easily solved, because we are dealing with mixtures of Bernoulli distributions. But in general it is practically impossible to consider in advance the regularities that some phenomenon might exhibit. They can be viewed only from the point of a higher level language; but then this language itself can be used to define new regularities. Thus the practical impossibility is not merely due to our not being clever enough; it is grounded in the very nature of an organization that describes patterns involving a potential infinity of items by finite statements (or say, patterns involving numerous items — by much shorter descriptions).

Even as an idealized construct, the notion of a single comprehensive prior which underlies long-range evolutions of beliefs is thus seen to be problematic. It is even more doubtful when such evolutions involve changes in the language itself. A faithful account of both long-range and short-range evolutions should consider internal as well as external changes

[20] Such are the symmetric probabilities in CARNAP's λ-system [1952]. HINTIKKA [1966] proposed later a two-parameter system which had probabilities assigning positive values to universal hypotheses. In view of de Finetti's theorem there is, however, an obvious and universal method for solving such difficulties: assign positive weight to the desired components in the mixture.

of probabilities. We can represent many changes of beliefs *as if* they had been induced by conditioning on a postfactum reconstructed prior. I shall later argue that such backtracking has methodological values, but this does not mean that we can do without external change.

The admittance of factual refutations of prior subjective probabilities, as an irreducible element, has implications on several levels. Globally, the evolution of beliefs comes to be viewed as an open enterprise. The question of a built-in organization which determines in advance the possible world pictures, that human beings can develop, is set aside. As far as methodology is concerned external change introduces an essential element of trial and error.[21] External change is, of course, most commmon on the everyday level. What in the highly theoretical setting appears as the inaccessibility problem of non-dogmatic prior probabilities appears on more concrete levels as an efficiency question: the prior probability's efficiency as a guide to action. The advantage of a dogmatic but easy to evaluate prior is obvious. It becomes crucial when decisions have to be reached in limited time or with limited resources for evaluation. Rather than be bogged down in computations, or in sorting out the details of one's beliefs, it is better to use a gross simplified version. There is a wide spectrum of simplifications — from highly theoretical ones which are due to the very limits of the descriptive apparatus, down to simplifications which may involve inconsistencies of mistakes in elementary arithmetic. (The point about the usefulness of this last kind has been nicely put by HACKING [1967].) The implications of the prior's efficiency for survival are obvious and need not be elaborated. Dogmatism in itself can be either a necessary virtue or a disaster, depending on the environment. Note that while dogmatism limits the possibilities of learning from experience it can, within its narrower range, make the process of learning faster: one jumps sooner to conclusions and accepts generalizations on the force of fewer instances.

Retroactive probability adjustment

Consider now the case of a refuted prior. The refuting evidence will as a rule provide some insights concerning the desired change. These should be

[21] The methodology suggested here can be regarded as embodying both Carnap's and Popper's insights. Note however that Carnap's methodology can stand by itself at least on a highly speculative level, but that this is not true of Popper's approach. Popper's framework lacks sufficient structure. His advice to try to refute the most daring unrefuted hypothesis does not lead, by itself, to anything. There are infinitely many "most daring unrefuted hypotheses" which one can refute one after the other without advancing in the least.

combined with whatever is still valuable in the old prior, i.e., with all the beliefs that are still in force. Many probabilities need not be changed. For example, one does not expect the strange behaviour of some coin to affect one's probability for "rain tomorrow" or for "the next drawn spade will be the ace". We regard the evidence as irrelevant for these and other statements. In a probabilistic framework irrelevance is expressed as independence: $P(A \cap B) = P(A)P(B)$. But the unrelatedness of the event A and the change-causing evidence e is not expressible in this context as probabilistic independence. If P_0 and P_1 are, respectively, the old and the new probability distributions, then, evidently, $P_0(A \cap e) = P_0(A)P_0(e)$ and $P_1(A \cap e) = P_1(A)P_1(e)$ do not imply $P_1(A) = P_0(A)$ (note that the second equality is trivial, because $P_1(e) = 1$; if $P_0(e) = 0$ the first one is trivial as well, but in general $P_0(e)$ need not be 0, only sufficiently small). The irrelevance of e to A functions here as a belief which is outside the scope of what is expressed either by P_0 or by P_1. It is a bridge over which we pass from one probability to the other. A more general kind of irrelevance is conditional irrelevance: We want to have $P_1(A \mid e') = P_0(A \mid e')$ because we judge that when e' is given e is irrelevant to A. Irrelevance assumptions can be quite powerful in reducing the problem of the new probability, from the original field to a much smaller one. If $e_1, \ldots, e_n$ is a partition into mutually exclusive events and if, for each i, given e_i the cause of the probability change is irrelevant to A, then $P_1(A \mid e_i) = P_0(A \mid e_i)$ and by consequence:

$$P_1(A) = \sum_i P_0(A \mid e_i) P_1(e_i). \tag{11}$$

Hence the $P_1(e_i)$'s determine $P_1(A)$. Jeffery's rule of probability kinematics[22] has exactly this form; the difference is that Jeffery considers probability changes which are caused by perceptual experiences not formulated as statements in our language. Note that without the assumption of irrelevance there is no place for (11) and that this is also true if the change is caused by perceptual experience (relevance being, in that case, a relation between such experiences and the statements in our language). I shall not enter into questions concerning Jeffery's kinematics and the further ramifications of (11). My aim is only to touch upon the problem of prior-changing in as much as it is related to the previous observations about complexity; so let us consider cases where irrelevance does not obtain.

[22] See JEFFREY [1983], chapter 11, where further references to literature on the subject can be found.

Having refuted, for example, the prior $b_{0.5}$ by evidence of ten successive "heads" how is one going to reevaluate probabilities of events concerning succeeding tosses?

Evidently there are no clear-cut prescriptions; on a general level we can only consider a general heuristics. Now, the problem is due to the prior being too dogmatic. Had one started with a suitable mixture of Bernoulli distributions he would have been able to condition on the evidence. Why not rectify this oversight be resetting the prior to what "it ought to have been" and then use conditioning? Resetting the prior involves of course the difficulties, mentioned before, of arbitrariness, unclear intuitions concerning the weights in the mixture, etc. One cannot expect more than a rough estimate, but a rough estimate might do in the short range and, in the long range, conditioning on the accumulated evidence will yield a more precise pattern.

The point to be made here is that retroactive prior adjustment is a sound heuristics and is not open to charges of "cheating" or incoherence.[23] There is no cheating in as much as one does not pretend that the adjusted prior is the one he had at the beginning. There is no reason why, after admitting that his subjective probability embodied a mistaken belief, he should not be able to correct it. As a matter of fact the "oversight" could have been deliberate. He might have chosen $b_{0.5}$ in order to avoid the complications that he is faced with now, hoping at that time that this approximation would do. We have seen that any prior probability definable in some language can be regarded as a simplification of a more sophisticated and less dogmatic prior. In this respect there is no sharp dividing line between high-level and low-level simplifications. Also, as far as readjusting the prior is considered, it does not matter whether the simplification was deliberate or non-conscious. The methodology of readjustments after the fact is quite common in science. Faced with a complicated system, a physicist will construct a so-called first-approximation model by ignoring certain factors.

[23] SHAFER [1979] proposes a Bayesian argument and then comments on it as follows «"It is beyond the capacity of the human mind to do all that the Bayesian ideal requires. But since we want to be coherent, we try to come as close to the ideal as we can. In particular, we used Bayes' theorem and other rules of the probability calculus to reconstruct coherent probabilities." I do not understand this argument. I do not understand how an argument that assumes we can do something we cannot do can say anything about whether our actual thought is "coherent".» Shafer's mistake consists in using "can do" and "cannot do" in an absolute sense. That I cannot find the precise solution to some complicated equation does not mean that I cannot approximate it by stages, treating my present candidate as if it were the solution and using a better approximation later.

These factors are assumed to have only minor effects and their inclusion would cost too much in terms of complexity. If this assumption is shown to be false, or if higher precision is required, an improved model is constructed, using insights gained at the first stage; and so on. The system in question can even have a purely mathematical definition; say, it is a complicated set of equations. An approximate solution is obtained by neglecting certain nuisance factors and the checking consists in obtaining numerical solutions for test cases run on a computer.

The heuristics of retroactive prior adjustements advises one to ask himself the following question "Suppose I did not neglect the possibilities which the evidence now suggests and suppose that I had to assign probabilities accordingly, what would have been my probability distribution?" The merit of the question is that, by calling attention to previous beliefs, it suggests a more balanced view than the view formed under the impression of the evidence. It acts against the tendency to jump to conclusions and to ignore previous knowledge that underlied one's initial prior. For example, faced with ten successive "heads", one tends to concentrate all weight on a small neighbourhood of 1, i.e., to take for granted an almost total bias; resetting the prior and conditioning might lead to a less extreme distribution.

4. Randomness

Recall that the success of a probability P over a field $\mathbb{F}$ given e (where e decides all statements of $\mathbb{F}$) is determined by the closeness of $P(\cdot)$ and $P(\cdot \mid e)$ as functions over $\mathbb{F}$. Since, over $\mathbb{F}$, $P(\cdot \mid e)$ has only two values, 1 and 0, the possibility of high success requires that the values of P over $\mathbb{F}$ should, in general, be near the end-points. For example, if $\mathbb{F}$ has two atoms a, $\bar{a}$ then $P(a)=0.5$ means that both success and failure are excluded; there are no risks and no gains. The possibility of high success is tied with the possibility of high failure and requires extreme values. If the distance function between probabilities over $\mathbb{F}$ reflects an additional structure then the last observation has to be modified accordingly, but the general picture remains the same: The possibility of high success over $\mathbb{F}$ requires "sharp" predictions as far as events in $\mathbb{F}$ are concerned. For example, if $\mathbb{F}$ is provided with a distance between its atoms, P should assign a sufficiently high weight to some sufficiently small neighbourhood in the space of atoms; high success will take place if the atom which is implied by e (i.e., whose probability under $P(\cdot \mid e)$ is 1) is in this neighbourhood.

A statistical test of a so-called simple hypothesis consists in choosing a field $\mathbb{F}$ of the form $\{a, \bar{a}\}$ and in rejecting or provisionally accepting (for further tests) a given distribution according to its success over $\mathbb{F}$. The 0-hypothesis is rejected at level of significance α if the change over this field is $\geqslant 1-\alpha$; i.e., if it is a change from $(q, 1-q)$ to $(1,0)$ with $q \leqslant \alpha$. The choice of $\mathbb{F}$ is an intricate problem occupying a central place in statistical testing. It is required that high success of one probability means also high failure of rival candidates within the presupposed family (or, if there is some metric within the family, high failure of all rivals which are not too close) that is to say — the test should have sufficient separating power. All this and the problems concerning composite hypotheses within a non-Bayesian framework are outside the scope of this paper. Here I want only to point out one prominent feature: The field $\mathbb{F}$ over which success is evaluated is, usually, incomparably smaller than the field of all possible outcomes which decide the members of $\mathbb{F}$. A test of the form just described involves two atoms; so, if we apply 4 different tests in order to test the probability $b_{0.5}$ by the outcome of twenty tosses, we are actually evaluating success over a field of 2^4 atoms by using evidence from a field of 2^{20} atoms. (It seems that the general reason for this stems from the basic fact that we are trying to describe patterns involving a potential infinity of items by using finite statements.)

The picture becomes simpler when we pass to the σ-field and consider infinite sequences (say of ones and zeroes) as outcomes. As a rule there is a rich infinite class of hypotheses that are assigned probability 1. Success means that such a hypothesis is satisfied, failure that it is not. I do not want to enter here into philosophical questions concerning the status of hypothetical infinite outcomes. Here, as in other domains, infinity is an invaluable tool for brining out patterns in finite but very large domains. The force of the observations to be made does not depend on ontological assumptions concerning infinite sequences.

Passing to the σ-field we find that the factual commitment of a probability distribution is indicated clearly by those statements that have probability 1. In the case of Bernoulli distribution $b_{0.5}$ the satisfaction of such statements correspond to an intuitive notion of randomness. One of them asserts for example that any algorithm which predicts the next outcome from all previous ones will have a 0.5 limit frequency of successes. While this can be regarded as expressing a sort of lawlessness it constitutes itself an extremely binding law. The sequence is disorderly only if "order" is identified with the possibility of predicting the next outcome. Random sequences obey very strict rules, though rules of a different nature. Von Mises introduced random sequences as a primitive concept (under the term "Kollektiv"). But then it turned out that randomness is nothing but the satisfaction of a certain set of laws, i.e., of statements whose probability is 1. There is a long list of proposals for choosing these statements.[24] All the various proposals can be treated within a uniform comprehensive system by presupposing a sufficiently expressive language in which these statements are formulated. This approach has been developed in GAIFMAN and SNIR [1982]. Given a probability P, a sequence, or, more generally, a model for the language, is *Φ-random with respect to P* if it satisfies all statements in Φ whose probability is 1. Here Φ is some class of statements (or, to be precise, sentences) in the presupposed language. It was shown that all previous concepts of randomness, in particular the latest ones of MARTIN-LÖF [1966], and SCHNORR [1971], result from certain natural choices of Φ. The concept as defined in GAIFMAN and SNIR [1982] applies not only to Bernoulli distributions but to arbitrary probabilities and not only to sequences but to "worlds" in general (a world being any model for the language). I shall refer to statements assigned probability 1 as *randomness properties*. In general, the richer the class Φ the stronger the concept of randomness (this was shown in GAIFMAN and SNIR [1982]).

To sum up, the factual claim of a prior probability is that the world is random with respect to it, i.e., Φ-random, where Φ is some presupposed class of statements. Note, however, that the set of random worlds does not determine the probability uniquely. Any two probabilities which are non-dogmatic with respect to each other (i.e., which assign value 1 to the same statements) have the same random worlds. The randomness of the infinite sequence (or, in general, of the infinite model) is the most obvious factual claim that the probability makes. Finer claims must be expressed through finer analysis of what constitutes a probability's success, along lines indicated at the beginning of Section 2.

In general, statistical tests derive from randomness properties. They constitute backward-extrapolations from the infinite to the sufficiently large finite. On the other hand, only randomness properties of a special kind give rise to performable statistical tests. The property should have, so to speak, effective finite approximations. For example, if it consists in some relative frequency tending to λ (or being confined in the long range to some interval Δ) then we should have effective estimates of the prob-

[24] A very comprehensive survey of the notion is given in MARTIN-LÖF [1969].

abilities of ε-deviations from λ (or Δ) as a function of ε and of the number, n, of outcomes. Consequently the randomness properties that correspond to statistical tests can be defined by formulas belonging to a fixed level in the arithmetical hierarchy (namely, the Σ_2-level[25]).

As noted, a sequence can be Φ-random with respect to some probability P without being Φ'-random where Φ' is some other class. A particularly interesting case is that of the so-called pseudo-random sequences. These are generated by computers and are intended to play the role of sequences random with respect to a given probability, which can be $b_{0.5}$. Being generated by a computer means that it is completely defined by a (deterministic) algorithm; there is a program, or if you wish a mathematical definition, which determines completely the entire sequence. This seems to contradict strongly the very notion of randomness with respect to $b_{0.5}$. But in fact, the sequence is Φ-random for a certain rich class, Φ, of statements. Thus, if one tries to predict the next outcome by using various other algorithms, his frequency of hits will converge to 0.5 at the desired rate. The prior $b_{0.5}$ will pass the usual statistical tests and short of possessing the actual algorithm and the means to implement it, a human being will not be able to improve his rate of hits, no matter how many outcomes he has observed.[26] If Adam merely knew of the algorithm's existence his prior probability distribution would still be $b_{0.5}$. Adam's situation would be analogous to the believer in determinism of Section 2 who regards $b_{0.5}$ as an objective probability for fair tosses. Just as the inner stability of a probability is relative to certain fields of evidence and forecast, so its success is relative and depends on the presupposed class of randomness properties.

When we go back to finite outcomes two other features are added to this relativity. First, success is determined by the distance between $P(\cdot)$ and $P(\cdot \mid e)$ over $\mathbb{F}$; it is essentially continuous and no longer a yes-no matter of satisfying certain properties.[27] The second, more interesting, feature is that the field $\mathbb{F}$ used for evaluating the prior's success depends on the field of outcomes, in our example — on the length of outcome sequences. One does not perform too many statistical tests given a sequence of, say, twenty

[25] In addition, the probability should converge at an effectively calculable rate. For full details see GAIFMAN and SNIR [1982], Section 5.

[26] Unless he has practically unlimited computational resources. In principle, knowing the existence of some algorithm, he can check one algorithm after another, ruling out every algorithm for which he gets a counter-example.

[27] The yes-no character of statistical tests is obtained by imposing arbitrary cut-off points, like a significance level 0.01, for the purpsoes of decisions.

outcomes. In general, there is a certain implicitly presupposed order of simplicity. *A prior probability fails inasmuch as an event of very small probability has occurred which has a description sufficiently simple with respect to the amount of data.* For example the outcome 1111111111 undermines $b_{0.5}$ because, putting $g(n) = n$th outcome, the event described by: "For all $n \leqslant 10$, $g(n) = 1$" has occurred. Note that one of the aspects of simplicity is that the description is a bounded version of an unbounded simple universal hypothesis: "For all n $g(n) = 1$". The outcome 1001010111 can be described by "for all $n \leqslant 10$, $g(n) = 0$ if n is prime"; for a sequence of length 10 this is not simple enough, it would have been seriously considered if '10' were replaced, say by '30'.

Now, simplicity is a notoriously complex notion. What is simple depends on the conceptual apparatus used in the given context. By fixing in advance a restricted family of candidates (i.e., statistical hypotheses) the classical statistician has already decided what in the given context is to count as simple. A prior distribution, being a more complex and finer tool, yields more insight into the general problem, where the underlying class of hypotheses is wide and open. The role of the background becomes clear; for example the outcome 1001001001 will have very different effects depending on whether it is produced by a coin or by a black box. We do not expect programmable coins but for a black box a periodic behaviour is very reasonable.

If each hypothesis in our class specifies an algorithm (as a candidate for generating the sequence) then we can use Kolmogorov's complexity (i.e., the length of the program written as a word over some fixed alphabet) as a measure of the hypothesis' non-simplicity. Some natural considerations indicate that in this case the a priori probability should decrease as C^{-n}, where n is the complexity and C is some constant $\geqslant 2$. But it is much more difficult to find a guideline in the case of a class which includes different types of hypotheses, say, both statistical and algorithmic ones.

References

CARNAP, R., 1952, *The Continuum of Inductive Methods* (Univ. of Chicago Press, Chicago).

DOYLE, J., 1983, *Methodological simplicity in expert system: the case of judgement and reasoned assumptions*, Artificial Intelligence Magazine, pp. 39–43.

DE FINETTI, B., 1964, *Foresight, its logical laws, its subjective sources*, in: Kyburg and Smokler, eds., Studies in Subjective Probability (Wiley, New York), pp. 93–158.

GAIFMAN, H. and SNIR, M., 1982, *Probabilities over rich languages, testing and randomness*, J. Symbolic Logic 47, pp. 495–548.

HACKING, I., 1965, *Logic of Statistical Inference* (Cambridge Univ. Press, Cambridge).
HACKING, I., 1967, *Slightly more realistic personal probability*, Philosophy of Science 34, pp. 311–325.
HINTIKKA, J., 1966, *A two-dimensional continuum of inductive logic*, in: Hintikka and Suppes, eds., Aspects of Inductive Logic (North-Holland, Amsterdam), pp. 113–32.
JEFFERY, R., 1983, *The Logic of Decision*, 2nd ed. (Univ. of Chicago Press, Chicago).
KYBURG, H., 1974, *The Logical Foundations of Statistics* (Reidel, Dordrecht).
LAKATOS, I., 1966, *Changes in the problem of inductive logic*, in: Lakatos, ed., The Problem of Inductive Logic, pp. 315–417.
MARTIN-LÖF, P., 1966, *The definition of random sequences*, Information and Control 9, pp. 602–619.
MARTIN-LÖF, P., 1969, *The literature on von Mises Kollektiv revisited*, Theoria 35, pp. 12–37.
SAVAGE, L. et al., 1962. *The Foundation of Statistical Inference* (Wiley, New York).
SEIDENFELD, T., 1979, *Philosophical Problems of Statistical Inference* (Reidel, Dordrecht).
SHAFER, G., 1979, *Jeffrey's rule of conditioning*, Technical Report 131, Dept. of Statistics, Stanford.
SKYRMS, B., 1977, *Resilience and causal necessity*, J. Philosophy 74, pp. 704–13.
PUTNAM, H., 1963, Degree of confirmation and inductive logic, in: Schilp, ed., The Philosophy of Rudolf Carnap (Open Court), pp. 761–784.
SCHNORR, C., 1971, *Zufälligkeit and Wahrscheinlichkeit*, Lecture Notes in Math. 218 (Springer, Berlin).

Barcan Marcus et al., eds., Logic, Methodology and Philosophy of Science VII

A PROBABILISTIC APPROACH TO MORAL RESPONSIBILITY

FRANK JACKSON

Dept. of Philosophy, Monash Univ., Clayton, Victoria, Australia

The light that probabilistic considerations cast on confirmation theory and decision theory is familiar enough. In this paper I argue that they also cast light on our assessments of moral responsibility, most particularly of guilt or blameworthiness, both in general and in the notoriously tricky cases of negligence and culpable ignorance. I start in Section 1 by combining two very well-known ideas. One is well known in Ethics, the other in Decision Theory. Their (I trust, less well-known) combination gives our theory of one element in being morally guilty. In Section 2 I indicate how our theory gives the intuitively correct answers in two relatively simple cases. In Section 3 I digress with a remark on the principle of alternate possibilities. In Section 4 I discuss negligence, culpable ignorance, and recklessness, and in Section 5 cases of culpability without wrongdoing. Finally, in Section 6, I consider the Morgan Rape Case from the point of view of our theory.

1. The two familiar ideas

The familiar idea in Ethics is the distinction between subjective rightness and objective rightness[1] (though it's variously labelled by various authors). Suppose that my child suffers from chronic tonsillitis. Acting on the best advice, I agree to her receiving a tonsillectomy at a good hospital from a good surgeon. By incredible bad luck she dies under the anaesthetic. Did I do the right or the wrong thing? Obviously the correct response is not to settle on one or the other reply but to distinguish. I did the subjectively right thing. I desired good and chose a means very likely to achieve it. I

[1] See, e.g., BRANDT, R.B., *Ethical Theory* (Prentice-Hall, Englewood Cliffs, NJ, 1959), pp. 364–5.

would have failed in my duty had I done otherwise. Nevertheless bad, not good resulted. I did the subjectively right but objectively wrong thing.

The familiar idea in Decision Theory is that what you ought to do is maximize expected utility.[2] This can be spelt out in various ways, and the differences between them matter little for our purposes. I'll settle on the following way. Associated with you at a time is a value function and a (subjective) probability function. The value function, V, assigns positive numbers to possible states of affairs you desire, negative numbers to those you abhor, zero to those about which you are indifferent, in such a way as to measure how much you desire that the possible states actually obtain. The probability function, Pr, assigns numbers to possible states of affairs in the 0 to 1 interval (say) in such a way as to measure your degree of belief that the possible states actually obtain. If we use 'S_i' to range over the possible states of affairs that might obtain if action A were performed, the expected utility of A, $\mathrm{EU}(A), = \sum_i V(S_i) \times \Pr(S_i/A)$. Decision Theory enjoins you to maximize expected utility. Of the available actions, you should perform that having maximum expected utility.

The 'should' here, of course, is being judged by the lights of your own value function. What you are being enjoined to maximize is expected utility calculated from *your* V. But when we judged in the tonsillitis example that I did the right thing, we were judging not from my value function but from what we took to be the *right* value function. It was important that I was desiring what I *ought* desire. My desiring alone is not enough to make my act subjectively right from the moral point of view, I need also to be desiring what it is right to desire. The two ideas, therefore, can be put together by saying that the morally subjectively right thing to do is to maximize expected utility *as calculated from the right value function.* An agent's subjectively right action at t is the act available to him with maximum expected utility as calculated from *his* probability function and from the *right* value function, be it his or not; and an agent is to be held morally guilty or blameworthy or culpable to the extent that he fails to do what is subjectively right, as just defined. The subjective element comes from the fact that it is *his* probability function, the moral element from the fact that it is the *right* value function, which are involved in the calculation. If we call utility calculated in this way, *expected moral utility*, then the thesis of this paper is that one is morally guilty, culpable, blameworthy or whatever, to the extent that one fails to maximize expected moral utility.

(Of course *what* one is guilty of depends on what actually happens. You

---

[2] See, e.g., JEFFREY, R.C., *Logic of Decision* (Univ. of Chicago Press, New York, 1965).

are not guilty of murder if no-one actually dies, though you may be guilty of attempted murder or of seeking to murder someone.)

Notoriously, opinions vary markedly on what the right value function is. Pleasure utilitarians will say it is the function which ranks states of affairs according to the amount of pleasure in them, ideal utilitarians that it is the amount of ideal good that matters, deontologists that it is important to consider which and how many moral precepts are being followed in each state of affairs, and so on and so forth. We need take no stand on this hard question of what the right value function is. Our aim is to show how answers to the question of moral guilt follow from the answer — whatever it is — to this hard question. If we use 'RV' for the right value function, our thesis is that what is subjectively right is to maximize $\Sigma_i \mathrm{RV}(S_i) \times \Pr(S_i/A)$.

Of course, despite the hardness of this question about RV and the very diverse answers that have been given to it, there is considerable agreement about particular cases. For instance, everyone (almost) agrees that a state of affairs containing murder is worse *ceteribus paribus* than one not containing murder, that is, that RV(the former) < RV(the latter), while perhaps disagreeing profoundly about exactly why this is so. In the cases to be discussed below, I will stick to ones where there is wide agreement about the right value ranking.

Also it is perhaps unfortunate to talk of expected moral *utility*. The theory of moral guilt herein is not tied to consequentionalism. 'Maximize expected moral utility' means that you should maximize $\Sigma_i \mathrm{RV}(S_i) \times \Pr(S_i/A)$. This leaves it open, should you wish, to insist on, for instance, the distinction between the disvalue of *killing* ten people and that of doing something that *results in* the killing of ten people, and similarly to distinguish the moral value of states of affairs differing only in, for instance, whether the intentions of the agents were direct or oblique. Similarly, there need be nothing essentially forward-looking in the considerations relevant to the moral value of our states of affairs. An act may have the greatest moral utility because it would bring about a state of affairs of redressing a past wrong or one exemplifying courage. I am using 'utility' only because it is entrenched. Also, the S_i should be read as including A itself (or else replace 'S_i' by 'S_i & A' everywhere).

Finally there is an important class of cases about which I will be saying nothing at all. These are cases where your value function diverges from the right one because of something like post-hypnotic suggestion, mental illness, a brain tumour, the effects of drugs, and so on, and this fact leads you to fail to maximize expected moral utility. We often exempt from moral guilt in such cases despite the failure. Exactly why is important and controversial, but we will restrict the scope of this paper to actions that spring from "non-interfered-with" value functions, that is, to cases that are in this sense normal.

There would be three parts to a complete theory of moral responsibility. A part which specified how to weight up the moral value of any given state of affairs; that is, which specified RV. Another part which specified when, or to what degree, an agent is normal when acting, and so when, or to what degree, their action is fit for *possible* moral assessment. But these two parts are not enough by themselves to determine culpability in an agent for an action. You also need to know how the agent stands epistemologically with respect to the action. This paper aims, by combining the two familiar ideas explained above, to sketch an approach to this third, epistemological part of the theory of moral responsibility.

2. Two test cases

I now show that our theory gives the correct answer in two relatively simple cases. I describe them as test cases because it is intuitively obvious in these cases which is the right answer, and so the fact that our theory delivers it can serve as confirmation. In fact I take this feature of our theory to be one of its principal virtues. Many have proposed plausible and interesting lists of the conditions that exculpate. But they are lists. Typically, no general theory is provided *from which* the various conditions can be derived and so explained.

(i) *The tonsillitis case.* My choice was between agreeing to the operation and not agreeing to the operation. We can represent the calculation of my expected moral utility (EMU) thus

$$\begin{aligned}\text{EMU(agreeing)} = {} & \text{RV(cure)} \times \text{Pr(cure/agree)} \\ & + \text{RV(death)} \times \text{Pr(death/agree)},\end{aligned}$$

$$\begin{aligned}\text{EMU(not agreeing)} = {} & \text{RV(cure)} \times \text{Pr(cure/do not agree)} \\ & + \text{RV(death)} \times \text{Pr(death/do not agree)}.\end{aligned}$$

(I have simplified by omitting such facts as that tonsillitis operations often lead to partial cures, and that they are in themselves unpleasant. Thus such factors as 'RV(partial cure) × Pr(partial cure/. . .)' and 'RV(cure & operation)' have been left out.)

Now given that my Pr(cure/agree) was high and my Pr(death/agree) was very low in the case described, and in particular only marginally greater than my Pr(death/do not agree), then the high value of RV(cure) is sufficient — despite RV(death)'s very low value — to make EMU(agreeing) > EMU(not agreeing). Thus I am innocent. I did what had higher expected moral utility.

(ii) *Coercion.* An act that would be wrong in the absence of coercion can be right in the presence of coercion. A bank teller who hands over a thousand dollars to Bonnie because she asks nicely is blamed for so doing; a bank teller who hands over the thousand dollars because Bonnie has a gun at the manager's head is not.

Some have sought to explain this by saying that action under external constraint is not free, or at least not fully free. But our bank teller is not being mechancially or psychologically controlled. No-one has actually seized his arm, his brain is not being probed, he is not under the influence of a Svengali. He is acting within his abilities and he could have done otherwise (as we ordinarily reckon such matters).

Moreover constraint comes in degrees. Not all constraints are sufficient to justify or excuse all acts — sometimes you should act against the constraint and take the penalty — and some hold that there are acts no amount of constraint justifies. The 'degrees of freedom' approach provides no obvious way into such questions, for it provides no obvious way to balance degree of constraint against likelihood and magnitude of good and evil.

Suppose instead we calculate in both cases the EMU of handing over versus that of not handing over the money, incorporating the constraints in the states of affairs. That is, instead of treating the constraints as reducing the agent's freedom, we will view the agent's *knowledge* of the constraints as changing his opinions as to what is likely to follow his acting one way or the other. In the case where Bonnie asks nicely and has no gun, the calculation might look as follows.

EMU(hand over)
= RV(Bank loses \$1000 and someone who has asked nicely is happy)
× Pr(Bank loses \$1000 and someone who has asked nicely is happy/hand over)
+ RV(Bank doesn't lose \$1000 and someone who has asked nicely is unhappy)
× Pr(Bank doesn't lose \$1000 and someone who has asked nicely is unhappy/hand over).

EMU(do not hand over) is as above with 'do not hand over' substituted for 'hand over'. Given plausible and too-obvious-to-spell-out assumptions about RV and the teller's Pr, EMU(hand over) < EMU(do not hand over). Thus the teller would be rightly held morally responsible for handing the money over in this case, on our theory.

In practice, of course, the calculation would at least sometimes need to be complicated by consideration of factors like that money handed over need not be beyond eventual recovery and that robbers, even those who ask nicely and don't have guns, are liable to impose a penalty for non-compliance. But that's right; that's the sort of complication that does need on occasion to be taken into account in considering guilt. We would consider the question of the teller's guilt to be thrown into doubt if the teller's subjective probabilities of the money being recovered eventually and of a penalty for non-complicance were high enough. In our terms, in calculating the EMU of, say, handing the money over, we would need to distinguish among such possible states of affairs as: (The bank loses the money but gets it back later and the teller is not injured), (The bank loses the money and doesn't get it back and the teller is not injured), and (The bank doesn't lose the money at all and the teller is injured). And, without going into the fairly obvious details, the EMU of handing over the money might exceed that of not doing so as a consequence of RV(The bank loses the money but gets it back later and the teller is not injured) > RV(The bank doesn't lose the money at all and the teller is injured).

It is probably now clear how the calculation of EMU in the second case (where Bonnie holds a gun to the manager's head) shows that the existence of a constraint may alter assignment of guilt. Among the states of affairs that now need to be taken into account are such as (The bank doesn't lose $1000 and the manager is shot). The calculation of the teller's EMUs might look as follows.

EMU(hand over)
 = RV(Bank doesn't lose $1000 and the manager is shot)
 × Pr(Bank doesn't lose $1000 and the manager is shot/hand over)
 + RV(Bank does lose $1000 and the manager is safe)
 × Pr(Bank does lose $1000 and the manager is safe/hand over).

EMU(do not hand over)
 = RV(Bank doesn't lose $1000 and the manager is shot)
 × Pr(Bank doesn't lose $1000 and the manager is shot/do not hand over)
 + Pr(Bank does lose $1000 and the manager is safe/do not hand over).

In this case the assumptions implicit in the case lead to: EMU(hand over) > EMU(do not hand over). This is because RV(Bank doesn't lose $1000 and the manager is shot) < RV(Bank loses $1000 and the manager is safe), and the chance of the bank losing $1000 and the manager being safe is much greater if the teller hands over. Hence the teller should hand over, and no guilt attaches to him for doing so.

What the constraint, or more exactly the teller's knowledge of it, alters in the second case is the likely outcomes of his two courses of action, and so his relevant expected moral utilities. He is equally free in his choice, but what is likely to result is very different in the two cases.

3. Alternate possibilities

We can now see why the very appealing principle of alternate possibilities fails. This principle states that a necessary condition of being morally responsible is that a set of alternative possible courses of action be open, really open, to the agent. This fits ill with our approach; for it has been in terms of the probability the agent gives to various possible outcomes, be they in fact available to the agent or not. Expected moral utility is calculated from the agent's probability function — from what he holds to one degree of belief or another will happen — and not from what will actually happen.

This is not an objection, because the principle of alternate possibilities is any way false, as Harry Frankfurt has pointed out.[3]

Consider the following example. Our bank teller is presented with what he believes is a golden opportunity to defraud the bank without risk. He refrains from doing so. We praise him. Does it make any difference if, quite unknown to him, there is a security system which makes the refrained-from embezzlement quite impossible? Surely not. He still deserves praise despite not in fact having the choice of embezzling. To say otherwise is to say that a security officer who says "There is secret security system in these banks which makes embezzlement by tellers impossible. But they deserve the highest praise, the system has never been activated" is talking nonsense. Likewise, Fred doesn't evade blame for killing his uncle in order to inherit, even if, unknown to Fred, the evil demon had planned to make him kill his uncle, but was spared the trouble by Fred's own enthusiasm.

---

[3] Frankfurt, Harry G., *Alternate possibilities and moral responsibility*, J. Philosophy 66 (3) (1969), pp. 829–839; though his diagnosis of where things have gone wrong is different.

The explanation for the appeal of the false principle of alternate possibilities is that if in order to get a case where you yourself have feature *F* you need to suppose that *X* obtains, it is unclear straight off whether it is *X* itself or your supposition of *X* that is necessary to your having *F*. Now on our theory, every case where you judge yourself guilty must be one where you are supposing that you have a range of options (perhaps including doing nothing). Otherwise you couldn't have failed to maximize moral utility, because that is defined by the moral utility of what you actually do falling short of that of at least one *other* course of action that you take to be an option. It is only when you consider agents other than oneself, that it becomes clear that what matters is the agent's *opinion* about there being more than one option, not the fact of the matter.

4. Recklessness, negligence, and culpable ignorance

When bad results from reckless or negligent behaviour, we are held responsible, yet typically the bad is not intended. Why then are we held responsible? This is no ivory tower question. Writers on the law have felt there to be a serious philosophical problem (though maybe they wouldn't put it quite that way) about recklessness and, especially, negligence, because they couldn't find the requisite *mens rea*.[4] On our account, misdeeds exhibiting recklessness and negligence are of a piece with the other cases of culpable action; they too exhibit failure by an agent to maximize his or her expected moral utility. No special or *ad hoc* clauses to cover negligence and recklessness are called for.

I will impose the following, not uncommon, regimentation on our ordinary usage of the terms 'negligent' and 'reckless'. A negligent act is one where the possibility of harm is overlooked — culpably so, it being our task to explain this culpability; a reckless act is one where the possibility of harm is disregarded rather than overlooked. I leave my lawn mower on the grass verge. I'm aware that there is a ten percent chance that someone will trip over it and break his leg. That is not what I want to happen but nevertheless I can't be bothered bringing the mower inside. That is reckless behaviour. If, though, the chance of someone tripping over the mower simply doesn't occur to me but should have, we have a case of negligent behaviour.

---

[4] See, e.g., TURNER, J.W.C., *The Modern Approach to Criminal Law*, 1945, discussed in HART, H.L.A., *Punishment and Responsibility* (Oxford, 1968), see ch. VI.

It will, I hope, be obvious how to handle recklessness on our account. If my probability function at the time gives a significant, even if small, probability of harm resulting, this can make my expected moral utility of leaving the mower outside lower than that of bringing it in. Accordingly it can make my failure to bring it in culpable.

I say it *can* make my failure culpable, because obviously in special circumstances bringing the mower inside can itself have probable consequences that lower its expected moral utility more than the amount the probability of, say, someone's breaking a leg on the mower, lowers that of leaving it outside. Suppose, as I am going to bring the mower inside, my neighbour has a heart attack, and instead of bringing the mower in, I rush him to hospital. The probability, even if small, of dire consequences from delaying while I bring the mower inside, might well be sufficient to make the expected moral utility of bringing it inside lower than that of leaving it outside.

Negligence is trickier. Recklessness was comparatively easy to handle because the agent gives some probability to harm happening, and this probability multiplied by the disvalue of the harm lowers the expected moral utility enough to give an alternative action higher expected moral utility. But negligence, as we are defining it and as it has presented a problem in the literature, involves *overlooking* the possibility of harm. You give no probability to harm resulting simply because you overlook the possibility altogether. It may be pointed out that if your attention was drawn to the matter, you would give harm some probability of happening. True, but *as it stands* this enables us to explain only why *if* your attention had been drawn to the matter and you had done nothing, you would have been culpable; not why you actually are culpable.

I need to discuss culpable ignorance before I discuss negligence.

Choosing can be a multi-stage affair. I am deciding which horse to back in the Melbourne Cup. I do not then and there make my choice. I first read the sporting papers, ask the barber, read the weather forecast, and so on and so forth. Typically, it is at some stage in this process that I make my choice of horse — just where depends on how big the bet is going to be, what evidence turned up, what evidence is likely to turn up, my bank balance, and the like. At each stage I am faced with the choice of whether to get more evidence or rest on what I have and decide then and there. The process ends when I make this latter choice — to rest on what I have — and choose among the horses on its basis.

We can calculate expected utilities for all these choices. Our treatment of culpable ignorance is via a comparison of the expected moral utility of making one's decision on the basis of what one knows right then, with the expected moral utility of getting more evidence and then choosing on the basis of the more. One is culpably ignorant when, although the latter has higher expected moral utility, you nevertheless don't do it. Accordingly, our treatment follows the lines already laid out. Culpability arises from failure to maximize expected moral utility.

I will just outline the calculation, as the details are neither original nor controversial; and I will write 'EMU(x/y)' for the result of replacing the probabilities in 'EMU(x)' by their conditionalizations on y. Suppose, for simplicity, (a) that just A and B are the options I take to be open to me, (b) that the course of investigation under consideration will, if pursued, result either in determining for certain that E or that $\bar{E}$, and (c) that before the determination of whether E or $\bar{E}$, it is A that has the higher expected moral utility. We are concerned with the relationship between: I = EMU(act without further investigation), and II = EMU(act after investigation by doing what *then* has higher expected moral utility). The first, I, is, of course, EMU(A). To calculate the second, II, we start by distinguishing two cases; one where the investigation cannot change the fact that A is the right thing to do, the other where it can.

In the first case we have both EMU$(A/E) >$ EMU(B/E), and EMU$(A/\bar{E}) >$ EMU$(B/\bar{E})$; and so II $=$ Pr$(E) \times$ EMU$(A/E) +$ Pr$(\bar{E}) \times$ EMU$(A/\bar{E})$. It it easy to show that this $=$ EMU(A).

Thus, in the first case, there is no question of praise for carrying out or of blame for not carrying out, the further investigation: I and II are the same. This is intuitively the right result. Suppose, according to what I now believe, giving drug A is the right and giving drug B the wrong thing to do; and suppose a certain investigation cannot change this ranking no matter how it turns out. Clearly, there is no blame attached to giving drug A instead of drug B without further ado.

In the second case, where the investigation may change the fact that A is the right thing to do, it must either be the case that EMU$(B/E) >$ EMU(A/E), or the case the EMU$(B/\bar{E}) >$ EMU$(A/\bar{E})$, but it is easy to prove that both cannot obtain. That is, if the investigation can change the ranking, its turning out one way will change it, while its turning out the other will leave it unchanged. Suppose it is E turning out to be the case that will change it, so we have EMU$(B/E) >$ EMU(A/E), while EMU$(A/\bar{E}) >$ EMU$(B/\bar{E})$. Then II $=$ Pr$(E) \times$ EMU$(B/E) +$ Pr$(\bar{E}) \times$ EMU$(A/\bar{E})$. It is easy to prove that II must, in this second case, be greater than I.

What is true, therefore, is that — with a proviso to be mentioned in a moment — if getting more information can change what it is right to do, getting the information and then acting on its basis must have greater expected moral utility than acting straight away. You are, therefore, culpable if you fail to obain the additional information by the general principle of maximizing expected moral utility.

The proviso is that the game be worth the candle. Typically, enquiry involves effort, and so has a disvalue of its own. We need to balance the amount by which II exceeds I — that amount being (as, again, it is easy to show) $\Pr(E) \times [\mathrm{EMU}(B/E) - \mathrm{EMU}(A/E)]$ — against this disvalue. It follows from our theory, therefore, that the culpability of ignorance depends on four conditions: (i) its being the case that the new evidence may change the ranking between courses of action, (ii) how likely the evidence which would change the ranking is (the '$\Pr(E)$' factor), (iii) how big the possible change in expected moral utility would be (the '$\mathrm{EMU}(B/E) - \mathrm{EMU}(A/E)$' factor), and (iv) how much difficulty is involved in getting the information.

The interest, again, is in the obtaining of the result, not the result itself. The four conditions, or something more or less like them, are those anyone would come up with after a little thought, as being the crucial factors in assessing ignorance as morally culpable. The interest, and the confirmation, lies in our having a reasonably simple and unified theory — we did *not* need *ad hoc* clauses — from which the four conditions can be obtained.

We have seen a signal advantage of our treatment of culpable ignorance. It explains what it ought to explain. Also it effects a conceptual simplification of the problem of culpability in general. We extended the notion of subjective rightness and wrongness via decision theory into an account of guilt which used the probability function the agent actually has and the value function the agent ought to have. The phenomena of culpable ignorance threatened us with the need to add a quite new ingredient, for it suggested that the probability function the agent *ought to have* and not just the one the agent actually has mattered. We have seen, however, that a natural extension of our original terms is sufficient to account for the phenomena.

Negligence in general can now be treated via our treatment of culpable ignorance. Culpable ignorance is a species of negligence — negligent failure to get more information. Negligence in general arises when there is something you ought to do before acting — get more information, in which case it's culpable ignorance, think twice, or thrice, rack your brains, put yourself in the other person's shoes, or whatever — and you fail to do it.

Suppose I have my mower on the nature strip without thinking. It isn't that I give some significant chance to someone falling over it but can't, nevertheless, be bothered to bring it in. Rather, I simply don't think. But I should have thought. Before acting I should have given thought to what might happen. In the terms of our approach I had a choice. To think before deciding whether or not to leave the mower out, or to decide straight away. My choosing the second rather than the first reflects a failure to maximize expected moral utility, as can be shown by a calculation precisely parallel to that given for the case of culpable ignorance. And the result is essentially the same. The extent to which I am guilty of negligence depends on: (i) that the extra thought might have changed the ranking as between leaving the mower out and bringing it in, (ii) the chance of this possible change actually occurring, (iii) how big the change would be if it occurred, and (iv) the effort, or more generally the disvalue, involved in extra thought. Usually the disvalue involved in extra thought is minimal, consequently our earlier result that if you neglect this factor you must, when the first condition obtains, get an increase in expected moral utility, means you should nearly always do the extra thinking — as is intuitively plausible.

5. Culpability without wrongdoing

J.L. Mackie defends, with certain relatively minor reservations, what he calls the straight rule of moral responsibility: "an agent is responsible for all and only his intentional actions."[5] Thus in order to be culpable or blameworthy one must intentionally do something objectively wrong. One difficulty with this rule, as Mackie himself notes, is that some cases of culpable ignorance are cases of *unintentionally* doing what is objectively wrong, and yet, despite the lack of intentionality, are indeed cases of culpability. I have just been advertising the success our theory has with these cases.

I now want to advertise the success our theory has with yet more challenging cases for the straight rule of responsibility, namely, cases where you are culpable or blameworthy although you *intentionally* do what is objectively *right*. One can be open to blame for intentionally doing what is the objectively right thing to do.

Here is a simple case. I am a young, struggling dermatologist anxious to make my reputation with a spectacular cure. A prominent and worthy citizen comes to me with an irritating but in no way dangerous skin

[5] Mackie, J.L., *Ethics* (Penguin, 1977), see ch. 9.

condition. I prescribe a drug which I know has a 90% chance of effecting a complete cure but a 10% chance of killing my patient. The 90% chance comes off and my reputation is made.

I clearly did the wrong thing. I took a quite unacceptable risk with the life of another for essentially self-interested reasons. But I did not do the *objectively wrong* thing. Good resulted from what I did, as indeed was highly likely. The sense in which I did the wrong thing is that my action was morally blameworthy or culpable.

The difficulty for the straight rule is that I was culpable yet I *intentionally* did something *objectively good or right.* I intended to cure and I did cure. By contrast, our theory has no difficulty with this case. It is clear that I failed to maximise expected moral utility. When allowance is made for the very good value that attaches to life and the relatively small value that attaches to relieving minor skin complaints, it is clear that the EMU(not prescribing the drug) > EMU(prescribing the drug).

6. The Morgan rape case[6]

I have applied our theory to some cases about which we have reasonably clear intuitions, otherwise the cases would have been useless as test cases. However, the Law Lords were divided in the Morgan Rape Case, ruling by only three to two in favour of the rule that if an accused in fact believed that the woman was consenting, whether or not that belief was based on reasonable grounds, he could not be found guilty of rape. Moreover, although this ruling created a furore, it did so in large part because of fears about its import in legal practice, not because of objections to it as a piece of moral philosophy. It was understandably feared that it would be too easy for those accused of rape to convincingly pretend to have believed that the woman was consenting if the rationality of their belief was to be set aside as irrelevant.

Furthermore, two common objections to it as a piece of moral philosophy are weak. First, it is sometimes objected that the belief in consent must be *reasonable* in order to excuse. But the immediate relevance of this is hard to see. One who has irrational beliefs is to that extent deficient in mental powers, and typically such deficiencies *do* excuse. (We will see shortly though that rationality is relevant, but not as a

[6] For an interesting account of this case (to which I am indebted) see CURLEY, E.M., *Excusing rape*, Philosophy and Public Affairs 5 (4) (1976), pp. 325–60.

trait of the belief in question.) Secondly, it is sometimes objected that even if the defendants really believed in consent, it is clear from the transcript that they intended "to have intercourse willy-nilly, i.e. the intent to have intercourse *whether or not she consents*".[7]

But this seems dangerously close to holding someone morally guilty for what they would have done, not for what they actually did. Suppose I pick up and keep fifty dollars I see lying in the street, mistakenly but genuinely believing it to be mine. You may judge of me that I would still have picked it up and kept it even if I had realized that it was someone else's. In this case you will have a low opinion of my character, but I nevertheless I have not acted culpably.

The verdict our theory delivers on the controversial Morgan rule is that the rule is mistaken. (Thus, if you were convinced of that all along despite the controversy, what follows is further confirmation.) The crucial oversight in the ruling is that belief is not an all-or-nothing business, it comes in degrees; and when this is borne in mine, it is clear that the defendants failed to maximize expected moral utility. The defendants faced a choice between going ahead or stopping; and the relevant calculation will, in outline, look as follows.

$$\begin{aligned}\text{EMU(go ahead)} = {} & \text{RV(rape)} \times \text{Pr(rape/go ahead)} \\ & + \text{RV(intercourse between consenting adults)} \\ & \times \text{Pr(intercourse between consenting adults/go ahead)}.\end{aligned}$$

$$\begin{aligned}\text{EMU(stop)} = {} & \text{RV(dual disappointment)} \times \text{Pr(dual disappointment/stop)} \\ & + \text{RV(rape avoided)} \times \text{Pr(rape avoided/stop)}.\end{aligned}$$

This is, of course, a grossly oversimplified outline, but even it is enough to show their guilt. The circumstances of the case make it clear that their Pr(rape/go ahead) cannot have been sufficiently low. Even if we accept their claim that they believed in consent (the Courts didn't, which is why they were convicted despite the Law Lords' ruling), this only means that Pr(rape/go ahead) was low. It does not mean that it was sufficiently insignificant when multiplied by the *very* high disvalue of rape, to allow the expected moral utility of going ahead to be higher than that of stopping. They thus must have failed to maximize expected moral utility. This is so whether they believed in consent or not; though obviously the extent to which they fell short of maximizing expected moral utility will be lower if

---

[7] KENNY, A.J.P., *Freewill and Responsibility* (London, 1978), see p. 65.

they believed in consent, and so the degree of culpability will be lower — as is intuitively plausible.

I have, in effect, treated the Morgan case as one involving recklessness. It is impossible to believe that the defendents did not give at least a significant, if small, chance to rape resulting. But we can imagine a variant on the case involving negligence instead of recklessness; a case where they simply did not bother to make a relevant inquiry. In such a variant they will be guilty of culpable ignorance, for clearly the four conditions arrived at before will apply.

Reasonableness does come into the picture I have sketched, but as a trait of acting on a belief, not of believing itself. A belief can be a reasonable one in itself, but not a reasonable one to act on, having regard to the nature of the possible consequences. Suppose I have an annoying but not really harmful complaint and believe, reasonably, that a certain drug will probably cure it. If I also believe that there is a small but not insignificant chance that the drug will kill me, it would be unreasonable of me to take the drug, despite the reasonableness of believing in a cure. The question of reasonableness of belief is distinct from that of reasonableness of acting on belief. What is crucial in the Morgan case is that the defendants' *action* was unreasonable; and unreasonable independently of whether they believed in consent, and whether if they did, their belief was reasonable.

A similar conclusion obtains if we consider the variant on the Morgan case where the defendants are guilty of culpable ignorance, that is, of a species of negligence rather than recklessness. It is unreasonable to act even on reasonable belief if there is available evidence you haven't bothered to get and the stakes are high enough. Given my researches to date, it may be reasonable for me to nominate Hyperion as the winner; but if I plan to put my life savings on him, more research before acting might well be the rational course.

Barcan Marcus et al., eds., Logic, Methodology and Philosophy of Science VII
© Elsevier Science Publishers B.V. (1986) 367–385

PROBABILITY EXISTS (but just barely)!

ISAAC LEVI

Dept. of Philosophy, Columbia Univ., New York, NY 10027, USA

In the preface to his *Theory of Probability*, B. DE FINETTI declares the first article of his ontological creed to be:

PROBABILITY DOES NOT EXIST.

As de Finetti explains, his stochastic atheism is directed against objective probability in any of its guises:

> The abandonment of superstitious beliefs about the existence of Phlogiston, the Cosmic Ether, Absolute Space and Time, . . . , or Fairies and Witches, was an essential step along the road to scientific thinking. Probability, too, if regarded as something endowed with some kind of objective existence, is no less a misleading conception, an illusory attempt to exteriorize or materialize our true probabilistic beliefs. (DE FINETTI, 1975, v. 1, p. x)

Thus, de Finetti regards belief in objective probabilities as mere superstition deriving from a misguided reification of subjective attitudes — in particular "our true probabilistic beliefs".

Sympathy with de Finetti's opposition to facile reifications is easy. Glib reification is all the rage among advanced philosophical thinkers. The temptation to indulge in a fit of ontological housecleaning is nearly irresistible.

Even so, the temptation should be resisted. That all applications of the calculus of probability in the natural and social sciences are representations and evaluations of subjective attitudes is incredible. Quantum mechanics is not a description of opinion. Nor are statistical mechanics and genetics.

According to de Finetti, the study of possibility is covered by the logic of certainty which, so de Finetti suggests, has an objective domain (DE FINETTI, 1975, v. 1, p. 26). Nonetheless, de Finetti also insists quite explicitly that when an agent makes a judgement that a proposition is possibly true, that judgment depends on his state of information (p. 27). In that sense judgments of possiblity are as subjective as the agent's "state of information" is.

Thus, it will be true or false that "it is possibly true that *H* according to agent *X* at time *t*." It will be true or false that "it is possibly true that *H* according to agent *X*'s state of information at time *t*." In the same spirit, it is true or false that "it is probable to degree *r* that *H* according to *X* at time *t* or relative to *X*'s state of information at time *t*."

On the other hand, de Finetti would deny that when agent *X* at *t* assigns a probability *r* to *H*, his assignment of that probability is true or false, correct or incorrect. Probabilities are previsions and previsions, so de Finetti insists, are not predictions which are true or false (pp. 70–71).

Although de Finetti does not explicitly say this, it would appear to be the case according to de Finetti's doctrine that when an agent at *t* judges a hypothesis *H* to be possible, his assessment of possibility is no more truth value bearing than his assessment of probability is. Both the probability and the possibility assessments are expressions of states of judgment which are neither true nor false.

To think of the agent who judges that *H* is probable to degree *r* as taking for granted or assuming or being certain that it is true that *H* is probable to degree *r* would be to engage in an objectionable reification of subjective or credal probability. Likewise to think of the agent according to whom *H* is possible as assuming the truth of "it is possible that *H*" is to attribute to him a commitment to reified possibility.

As I understand him, de Finetti is opposed to both modes of reification.

Consequently, for de Finetti one cannot be certain or in suspense concerning the truth value of "it is possible that *H*". One cannot consider the possibility that it is possible that *H* and we cannot sensibly ask whether it is possible that it is probable to degree *r* that *H* is true.

Of course, not only are we forbidden to have possibilities of probabilities, but probabilities of possibilities or probabilities of probabilities.

In sum, de Finetti's opposition to reification of probability applies both to de dicto possibility and de dicto probability. Good sense may be made of subjective or credal probabilities de dicto; but efforts to transfer this good sense to objective de dicto probabilities are to be avoided.

Thus, de Finetti explicitly opposes the dreams of Harold Jeffreys and Rudolf Carnap who sought to construct criteria for rational probability judgment so powerful that all rational agents would be obliged to make the same credal probability judgments relative to the same evidence (DE FINETTI, 1975, v. 2, pp. 341–342). But he also stands opposed to the dream of Arthur Burks according to whom there is a real, causal or physical de dicto probability measure defined over a space of real, causal or physical de dicto possibilities (BURKS, 1977).

I side with de Finetti concerning these matters. Unlike many others who agree that de dicto possibility and probability are best understood in an epistemic way (for example, GÄRDENFORS and SAHLIN, 1983), I maintain (in agreement with de Finetti) that this position entails rejection of iterated de dicto possibilities and probabilities.

An important ramification of such a view is that counterfactual judgments of a de dicto form must also be considered to lack truth values. Hence, there can be no iterated subjunctive conditionals. There can be no appraisals of subjunctive conditionals with respect to possibility or probability. The probability of a conditional is not a conditional probability simply because there is no probability of a conditional.

De Finetti's opposition to objective probability extends beyond his entirely healthy antipathy to the reification of de dicto probability and possibility.

Probabilistic concepts are used in scientific applications to represent conditions under which objects and systems respond in various ways to experimentation. "The probability of a 6," so wrote von Mises, "is a physical property of a given die and is a property analogous to its mass, specific heat or electrical resistance" (VON MISES, 1957, p. 14).

Thus, stochastic attributions, like attributions of dispositions of dispositions, abilities and compulsions, are predications of properties to objects. We describe an objective feature of the coin a when we declare that the chance of a landing heads up on a toss is 0.5 and the chance of a landing tails on a toss is also 0.5.

Such predications formally involve a representation containing three constituents: (i) a description of an event-type — the experiment S; (ii) a system of event-type descriptions often themselves represented by sets of points belonging to a field generated by a sample space, and (iii) a probability measure defined over the field (or the predicates describing types of events represented by sets of points in the field). The sets of points in the field (i.e., the measurable ones) are intended to represent possible responses of various kinds by the object of which the stochastic predicate is alleged to be true conditional on the performance of an experiment of kind S.

There should be little dispute about this among those who have recognized legitimate applications of statistical probability in science. Controversy appears once we seek to clarify the interpretation of stochastic attributions to things.

Followers in the tradition of Venn and von Mises understand predications of stochastic attributes so that the probability measure over the field associated with the sample space represents limits of relative frequency for each of the kinds of responses represented in the field on infinitely many repetitions of the experiment of kind *S*. Sometimes the trials are required to be on the same system. Thus, tosses are tosses of the same coin. Sometimes the repetitions may be performed on different systems of which the same stochastic attribute is true (for example, in making measurements of the spins of "similarly prepared" particles or in dropping bottles and counting the number of fragments into which they break).

Frequentist interpretations of statistical predicates appear unsatisfactory when taken too strictly. Statistical predications may be true of objects even if no experiments of kind *S* (whatever it might be) ever take place on the system in question or, indeed, on any system. In any case, the number which will actually take place will be at most finite whereas an infinite sequence of repetitions is required.

These considerations suggest that frequentists will be driven either to a subjunctive account of the frequency interpretation or to a dispositional account.

According to the subjunctive account, the probability distribution over the sample space specifies the limits of relative frequency which would be attained were trials of kind *S* to be repeated ad infinitum. Such a view presupposes that we could provide an account of subjunctive conditionals as truth value bearing hypotheses. I wish to reject this alternative and, given his metaphysical prejudices, it would appear that de Finetti should do so as well.

Even if we waive this objection, there is another. Suppose an urn contains 100 balls 50 of which are black and 50 white. Let the trial on the urn with its contents consist of selecting a ball while blindfolded after having thoroughly mixed the contents, noting the color and then returning the ball to the urn. According to interpretations of subjunctive conditionals of either the Stalnacker or Lewis varieties, it is false that if trials of kind *S* were repeated an infinite number of times, the limit of relative frequency with which blacks would be selected would converge on 50%. A black ball could, for example, be drawn every time. In spite of this, it would be true that the objective probability of obtaining a black ball on a trial of kind *S* is equal to 0.5.

Dispositional analyses avoid the mysteries of realistically construed subjunctive conditional analyses by avoiding commitment to truth value bearing subjunctive conditionals. Each attribution of a statistical property is taken to be the predication of a disposition to some sort of frequency behavior in the infinite long run. The second of our two objections, however, returns to haunt us in a new guise. The urn does not have a sure-fire disposition to yield blacks 50% of the time in the limit on an infinite series of trials of kind *S*.

There are variants on subjunctive and dispositional analyses of objective probability in terms of long-run frequencies which appeal to some conception of a probability measure. Such proposals, however, cannot be used to explicate the notion of a probability measure over the sample space without circularity.

If these difficulties are not enough, there are two others: von Mises (1951) insisted that probability theory is a science like any other science. It is the science of collectives and, as a consequence, has as its domain all applications of probability in statistical mechanics, quantum mechanics, genetics, psychology, sociology and economics. While agreeing that there are applications of the calculus of probabilities in all these domains, I think we carve up the world in the wrong way if we see these applications as covered by a single science.

Furthermore, although it was von Mises who complained that the views of Kolmogorov and Cramer were too formal and mathematical (von Mises, 1951), the semantical investigations of von Mises and those who seek to interpret statistical probability in terms of elaborate modal structures have contributed nothing through this activity to the understanding of the connections between attributions of stochastic characteristics to objects and judgments about test behavior of such objects.

Kolmogorov (1933), Cramer (1945) and Braithwaite (1953) combined accounts of the formal structure of probability measures used in statistical applications with an account of how information about experimental behavior can give epistemic support and be supported by information about statistical probability. Semantical studies are ignored in favor of epistemological and methodological problems relevant to applications.

Refusal to give a semantical interpretation in advance of application is not denying truth values to statistical predicates. It reflects the view that no semantical interpretation of stochastic predicates other than trivial tarskian satisfaction conditions will be obtained without scientific inquiry into the subject matter to which the stochastic concepts are being applied.

An interpretation of the measures over phase space used in statistical mechanics relevant to the problems studied must perforce be quite different from the interpretations appropriate to the statistical concepts used in applications in genetics. These diverse interpretations will not automatically provide the basis for a stochastically unified science of the sort envisaged by von Mises.

Moreover, the problem of semantic interpretation is clearly separated from the important epistemological and methodological problems pertaining to the applications of conceptions of statistical probability.

None of this matters very much to de Finetti. His opposition to objective probability covers all variants of statistical probability from Venn and von Mises to Kolmogorov, Cramer and Braithwaite.

My outlook agrees in main outlines with the approaches of Kolmogorov, Cramer and Braithwaite (see LEVI, 1967 and 1980a). However, these authors tend to avoid consideration of subjective or credal probability in discussing links between attitudes toward statistical hypotheses and toward data. In my opinion, a dualist or, perhaps, pluralist view of probability is needed in order to understand these links adequately.

Thus, Cramer supposes that if one knows that all but one ball in an urn are black, that there are a million balls in the urn and that a ball is selected at random, it is practically certain that the ball selected is black (CRAMER, 1945, p. 149). Cramer does not explain what "practical certainty" amounts to . He does not explain what conclusion should be reached if half of the million balls in the urn were known to be black and the remainder white. However, one could readily extend his proposal by introducing a principle of direct inference which licenses the judgment that the credal probability is 0.999999 that ball is black in the first case and 0.5 in the second case. (See LEVI, 1967, pp. 205–208 for fuller discussion.)

So construed, Cramer's practical certainty principle is a variant of direct inference or "statistical deduction" as Peirce used to call it (LEVI, 1980b). Peirce appears officially to have been a monist about probability restricting his usage to statistical probability and not discussing credal or subjective probability officially except to dismiss it. The same is true of Reichenbach who represented direct inference as assigning "weights" to hypotheses about test behavior on the basis of assumptions concerning statistical probability interpreted as limits of relative frequency (REICHENBACH, 1938).

In one respect, the difference between such monism and my dualism is merely verbal. Some authors do not want to call subjective or epistemic probability "probability" at all even when they endow it with all the features of credal, subjective or epistemic probability.

There is, however, a deeper point associated with monism having little to do with the terminological issue. Von Mises, Reichenbach, Fisher, Cramer, Neyman and many others were rightly suspicious of the principle of insufficient reason in all its incarnations and would, I suspect, have remained so in spite of its revival in recent years. (For a crushing critique of the current revival, see SEIDENFELD, 1979.)

In the absence of an objectively grounded surrogate for insufficient reason, the only relatively context-independent principle for objectively justifying statements of credal probability was through direct inference from knowledge of statistical probability. Otherwise one was left with principles of coherence requiring that credal probability conform to the requirements of the calculus of probability and little else. The authors cited above took the position that the only circumstance under which one is justified in making numerically definite judgments of credal probability are when such judgments can be grounded on knowledge of statistical probability via direct inference.

Counter to the impression generally given, such a view presupposes that statistical probability and credal probability are quite different. One and the same coin *a* can have a 50% statistical probability of landing heads up on a toss and a 90% statistical probability of landing heads up on a toss by Morgenbesser. Both attributions of statistical probability are true of the coin at the same time.

Statistical probability is construed this way both according to the Venn–von Mises approach to interpreting statistical probability and the Kolmogorov–Cramer–Braithwaite approach.

Once this is understood, it should become apparent that credal probability cannot be equated with statistical probability. Suppose the agent knows that coin *a* was tossed by Morgenbesser and knows that the statistical probability of heads is 0.9 on a toss by Morgenbesser and 0.5 on a toss. His credal probability should not be equated with the statistical probability of heads on a toss — i.e., with 0.5. This is so even though that statistical probability is an objective feature of the coin.

Principles of direct inference are designed to specify the epistemic conditions under which credal probabilities are to be equated with one system of statistical probabilities rather than another — if they are to be equated with any at all. As the information varies, the credal probability for a given hypothesis may be equated with a different true statistical probability than it was before.

I call the outlook which says that the only conditions under which one is rationally entitled to assign numericallly definite credal probabilities to hypotheses are those where such credal probabilities are derivable from knowledge of statistical probability via direct inference "Objectivist necessitarian" (LEVI, 1980a). This view is objectivistic because it acknowledges the dictates of the calculus of probability and direct inference from knowledge of objective statistical probability and nothing else as context-independent principles of rational probability judgment. It is necessitarian because it prohibits ruling out any probability distribution as impermissible for use in assessing expected value unless it is ruled out according to the principles just mentioned. Hence, the only time that an agent is justified in assigning a numerically definite credal probability to a hypothesis is when direct inference from knowledge of chances together with the calculus of probabilities entail such numerical definiteness. This is, of course, just what Peirce, Neyman, Pearson, Fisher and Kyburg agree upon in opposition to writers like de Finetti.

As I understand it, objectivist necessitarianism motivated the great pioneers in modern statistics in the 1920's and 1930's and represents an important attitude among professional statisticians to this day. I do not entirely agree with such objectivist necessitarianism whether it is of the Neyman–Pearson or the Fisher–Kyburg variety. But advocates of this view are right in recognizing direct inference as a fundamental principle of probability judgment on a par with coherence and superior to principles of insufficient reason.

Even in contexts of inference from information about test behavior to judgments about statistical hypotheses, direct inference plays a critical role; for whether one favors a Neyman–Pearson, Fisher or Bayes approach, the conditional probability that a hypothesis E about the data will be true given a statistical hypothesis H plays a salient role in the analysis. Such conditional probabilities determine the likelihood functions for the various statistical hypotheses on the experimental data E. Determining these likelihood functions requires appeal to principles of direct inference.

My reason for taking time to sketch out roughly where I would locate myself among those who insist that probability does, after all, exist is to indicate how extensive, nonetheless, my agreement with de Finetti happens to be.

I entirely agree in dismissing objective probability measures for propositions or sentences. I agree that construals of statistical probability as limit of relative frequency in actual sequences or in hypothetical sequences are not to be taken seriously. I agree that statistical probability is not a disposition to exhibit frequency behavior of some kind on an infinite number of trials.

Nonetheless, I do want to retain the intelligibility of predications of statistical probability without the excess metaphsyical and conceptual baggage which the other versions of objective probability invoke.

But having agreed so extensively with de Finetti, why do I refuse to go all the way and join de Finetti in his subjectivist monism?

One important reason for refusing to do so is that statisticians (so it seems to be) have not been deceived in supposing that genuine problems of statistical estimation arise in the natural and social sciences. I do not know how to make sense out of problems of statistical estimation without a conception of objective statistical probability. Nor, for that matter, can one make much sense out of significance testing or hypothesis testing without an understanding of statistical probability. In problems of any of these kinds, something is unknown — something which involves ignorance of the value of a statistical probability or of a statistical probability distribution.

By denying the intelligibility of statistical probability, de Finetti denies the meaningfulness of statistical hypotheses. As a consequence, he threatens the intelligibility of the problematic of theories of statistical inference including Bayesian statistical inference.

De Finetti never intended to consign statisticians to the ranks of the unemployed. His important paper from the 1930's (DE FINETTI, 1937) took as its central task to show that even though objective probability is meaningless, those who think they are estimating unknown objective probabilities are indeed estimating something. They are deceived only in thinking that it is objective probability they are estimating. Chapters 11 and 12 of volume 2 of DE FINETTI's 1975 book are intended to furnish a short course in statistical inference.

De Finetti's idea is that one can save statistics from his stochastic atheism by showing how the role of statistical hypotheses can be played by other types of hypotheses which are not about statistical probability and are metaphysically acceptable.

Consider then de Finetti's own example of the bent coin (DE FINETTI, 1937.) The problem is to make an estimate of the unknown value of p — the statistical probability of the coin landing heads up on a single toss. According to de Finetti, talk of this objective statistical probability is nonsense. Even so, those who seek to make an estimate of the value of p are, indeed, estimating the value of something. According to de Finetti, they are estimating the value of the relative frequency M/N in a very large sequence of tosses (i.e., of trials of kind S) of the bent coin. The unknown value of M/N is a meaningful unknown in de Finetti's sense. One can form truth value bearing hypotheses specifying the value of M/N and assign credal probabilities to these hypotheses.

Under the assumption of exchangeability, de Finetti shows that the credal distribution over the possible values of M/N relative to data concerning the relative frequency r/S in some s-tuple of tosses belonging to the N-fold sequence (where N is very large in relation to s) will be approximated by the continuous posterior credal distribution over the values of p relative to the same data. The approximation becomes exact as N goes to infinity. If one is prepared (as de Finetti was not) to consider hypotheses as to the unknown limit of relative frequency and to use probability measures obeying countable additivity and these alone, the surrogates for hypotheses about p would have posterior distributions identical with the posterior distribution for p itself.

De Finetti does not propose a finite frequency interpretation of statistical probability. He seeks to dispense with objective statistical probabilities in all their important functionings in scientific inquiry and, in particular, in statistical estimation and to replace them in those functionings by hypotheses about finite frequencies. In this respect, de Finetti's view of statistical probability resembles that of the anti-Bayesian, Henry Kyburg, who also sought to dispense with objective statistical probability, retain an epistemic conception of probability and replace statistical probability in its function in inquiry with the conception of finite frequency (KYBURG, 1961, 1974).

Of course, Kyburg and de Finetti share little else in common and the marked differences in their views may have blinded commentators (including me for a long time) to this important similarity. (See LEVI, 1977, p. 9 for my first reference to this point.) As we shall see, the similarity is important because it reflects a difficulty which both positions must address and which is addressed differently by de Finetti and by Kyburg.

One difficulty facing de Finetti's approach concerns cases where there are no additional trials or very few additional trials beyond the s trials used to obtain data for estimation. Thus, when batches of mice are treated with drugs to detect how many contract cancer, the problem of estimating the unknown statistical probability of a mouse obtaining cancer from the treatment cannot be replaced by the problem of estimating the relative frequency with which mice in a large sample N of mice treated with the given dosage of the drug contract cancer. Even before the budgetary constraints imposed by Reagonomics were introduced, the available funds did not permit dosing such a large number of mice; and even if the funds had permitted this practice, the exercise would have been gratuitous. In cases like this, and they are typical cases, investigators seek to estimate the values of unknown parameters knowing all the while that there will not be a long N-fold extension of the s-fold sequence of experiments used to obtain data. From de Finetti's point of view, so it would seem, there is no unknown to estimate — unless a surrogate different from long-run frequency can be found for the unknown statistical probability.

It should be obvious that we cannot let the unknown hypotheses be conjectures as to the conditional credal probability distributions over values of M/N conditional on the number of trials being extended to a sequence of length N. An unknown conditional probability could be a hypothesis about the agent's state of credal probability; but in estimating the cancer rate, the FDA statistician is not estimating something about the state of his mind. And if the conditional probability has the relativity to an agent and time deleted, it cannot, according to de Finetti's principles, bear a truth value and, hence, cannot have a credal probability assigned to it.

A variant on the idea of treating conditional probability distributions as conjectures standing in for unknown statistical probability distributions relative to the agent's current state of knowledge is to consider hypotheses concerning unknown relative frequency distributions in N trials, relative not to the current state of knowledge but to a transformation of the current state of knowledge which presupposes that the N trials have indeed taken place.

This suggestion is no more acceptable than the previous one. De Finetti claims to have a replacement for hypotheses concerning unknown objective probabilities — i.e., unknown relative to the agent's current state of knowledge and not relative to some transformation of his current state of knowledge. A multitude of such transformations can be contemplated and one can specify what is and is not known relative to each of them. Presumably, however, what drives the problem of estimation is to ascertain what is unknown relative to the current state of knowledge. The suggestion under consideration cannot fulfill the demands of de Finetti's project. The trouble is that it changes the subject.

Instead of considering what is unknown relative to some hypothetical state of knowledge, one might construe the unknown hypotheses (according to the current state of knowledge) to be subjunctive conditional judgments concerning what the relative frequencies would be were a large number N of trials conducted.

To make this view work, the subjunctive conditionals would have to be construed as truth value bearing so that credal probabilities could be assigned to them. But, as I understand de Finetti's position, he should reject this solution. It is another expression of "superstitious beliefs about the existence of Phlogiston. The Cosmic Ether, Absolute Space and Time, ..., or Fairies and Witches."

Consider instead declaring that the hypothesis that the statistical probability of heads on a toss equals 0.5. is to be functionally replaced by the hypothesis that the bent coin has a surefire disposition to land heads up with a relative frequency M/N very near 0.5 in a large but finite number N of tosses of the coin. I have argued previously that this is not adequate as a specification of truth conditions for the statistical hypothesis. For de Finetti, who is not concerned with this question, it cannot be adequate as a functional replacement either. Such surefire dispositions are as metaphysically obnoxious as subjunctive conditionals bearing truth values.

Thus, de Finetti's claims for the metaphysial significance of his representation theorem do not seem sustainable. This is not because his representation theorem is false but because it fails to show that hypotheses about statistical probability can be replaced by hypotheses about relative frequencies in large numbers of trials.

Other considerations reinforce this conclusion.

Suppose that we face some unusual situation where the agent does know that a very large number N of future trials of kind S will be conducted. In that event, if the agent's credal probability judgments satisfy the requirements of exchangeability concerning those N trials, his credal probability judgments for the hypothesis that $M/N = p$ (for those values of p which are possible values of M/N is approximately equal to the credal probability that the true value for the unknown objective statistical probability is approximately equal to p. In this sense, the hypothesis about relative frequency becomes surrogate for the hypothesis about objective statistical probability.

Suppose that two investigators X and Y have the information that the bent coin will be tossed N times for large N. In addition they have obtained data concerning the first s tosses in which the coin has landed heads r times. They also know that in these s trials, the coin was tossed by Morgenbesser.

According to X, the information that the tosses are by Morgenbesser is stochastically irrelevant. Whatever the unknown statistical probability p of heads on a toss might be, that value is equal to the unknown statistical probability of heads on a toss by Morgenbesser. Hence, for him the entire sequence of N tosses meet exchangeability requirements and the hypothesis about objective probability seems replaceable about unknown long-run relative frequencies.

By way of contrast, according to Y the information that the first s tosses are by Morgenbesser is not stochastically irrelevant. As a result, the unknown statistical probability of heads on a toss, p, is not replaceable by hypotheses about the long-run relative frequency in the given N-fold sequence of tosses, some of which are by Morgenbesser and others not. In the typical case, Y's credal state for that sequence will fail to met the requirements for all exchangeability.

This result is an embarrassment for de Finetti's position. X and Y would normally be taken to be interested in the same problem — to wit, the estimation of the unknown value of p. But de Finetti does not have a way of specifying a problem intelligible to him which can be viewed as an estimation problem shared by X and Y which is also a surrogate for the problem of estimating the unknown value of p.

Of course, even if we admit that X and Y are concerned to estimate a statistical parameter, they differ as to the usefulness of the available data for the purpose of making judgments about the unknown parameter. This difference, however, is not to be confused with a difference in the problems they are trying to solve — which seems to be the way in which de Finetti would have to understand the situation.

Matters are still worse. The disagreement between X and Y over the stochastic relevance of the information that Morgenbesser tossed the coin the first s times is a dispute concerning statistical probabilities. If de Finetti were serious about functional replacement of statistical probability by long-run relative frequency, he would be obliged to see the disagreement as concerning the truth values of hypotheses about long-run relative frequencies.

That is not, however, de Fínetti's approach. For him, the dispute reduces to a difference in credal probability judgments. According to X, the process of N tosses is exchangeable. For Y, it is not. There is no disagreement concerning the truth value of any hypothesis. Both X and Y are certain of the same propositions and in suspense concerning the same propositions.

Turn now to the question of direct inference. X knows that the coin has been tossed N times and that the relative frequency of heads on these N tosses is, say, 0.8. To simplify, he is invited to consider the hypothesis that on the first of these tosses the coin landed heads up.

There is nothing in de Finetti's principles of rational probability judgment which mandates the rational agent to assign to that hypothesis the credal probability of 0.8 — even if all he knows about the first toss is that it is a toss in the N-fold sequence. De Finetti himself has been explicit about this in his own way (DE FINETTI, 1937, pp. 73–77).

Yet, if knowledge of the relative frequency 0.8 in the N tosses is to stand surrogate for knowledge of the statistical probability of obtaining heads on a toss, it should play the role which knowledge of statistical probability plays in direct inference. According to all competing accounts of direct inference, it is entirely noncontroversial that if X knew the objective statistical probability of heads on a toss to be 0.8 and that the coin had been tossed, he would be obligated to assign the credal probability of 0.8 to the hypothesis that the coin landed heads up.

To be sure, even de Finetti agrees that the agent X should assign 0.8 if, in addition to knowing the relative frequency in N trials to be 0.8, the agent assigns the same credal probability to heads up on each trial (DE FINETTI, 1937, p. 74). And the agent's credal state satisfies this requirement if he judges the sequence of N tosses to be exchangeable. But there is no obligation on the agent to make credal probability judgments in that way. Knowledge of the relative frequency in N trials is not a surrogate for knowledge of statistical probability in direct inference.

This lacuna might be filled by supplementing de Finetti's principle of coherence with an additional principle which mandates adopting a credal state meeting the requirements of exchangeability for an N-fold sequence given knowledge that M/N is some definite value.

The elaboration of such a principle will turn out, however, to breed additional problems for de Finetti's positivist perspective. In addition to knowing that the relative frequency in the N tosses of coin a is 0.8, the agent knows that the first toss is by Morgenebesser. We cannot mandate the judgment of exchangeability in that case.

It will then become critical to ascertain the long-run frequency of heads in tosses by Morgenbesser. If we know that this long-run relative frequency is also 0.8, we can, perhaps, rest content with mandating the judgment of exchangeability for the two long-run sequences we now consider. But if we do not judge them equal, perhaps because we know them to be different or perhaps because we do not know one way or the other, it becomes debatable what judgment should be made.

In short, the familiar problems generating controversies concerning statistical deduction or direct inference emerge to plague us. Moreover, they do so in a manner which requires us to posit more than one long-run sequence. It will become increasingly difficult to find cases where all the requisite long sequences exist. The pressure to move to hypothetical long-run sequences will become overwhelming.

It is interesting to contrast Kyburg's approach with de Finetti's at this point (KYBURG, 1961, 1974 and LEVI, 1977). Kyburg sought to treat hypotheses about relative frequencies as surrogates for hypotheses about objective statistical probability. Unlike de Finetti, he did not always require the "reference classes" to contain large finite numbers of elements or to be arranged in a sequence. In principle, the classes could be unit sets.

Also, in contrast to de Finetti, Kyburg proposed an account of direct inference from knowledge of relative frequency in appropriate reference classes. However, because Kyburg hewed to such an extensionalist understanding of the hypotheses which are surrogates for statistical hypotheses, he was driven to a criterion for selecting reference classes for direct inference which violates the Bayesian requirement of confirmational conditionalization — to which de Finetti is clearly committed. Hence, Kyburg's approach is not available to de Finetti (LEVI, 1977).

In any case, whether one takes Kyburg's extensionalist view or allows statistical probability realistically construed as a "major premise" of direct inference, the conception of direct inference he captures seems to represent to a remarkable degree the conception of direct inference intended by Fisher in his account of recognizable subsets and which he claimed to be the basis of his approach to fiducial probability (FISHER 1959, pp. 3–33 and p. 55).

Kyburg's view of direct inference deviates from de Finetti's approach in another important respect. Even though the sole restriction imposed by de Finetti on credal probability judgment is coherence — i.e., conformity to the calculus of probabilities — he insists that to be rational one should be committed to a numerically definite system of probability judgments regardless of whether there is any warrant for singling out one system of numbers rather than another.

Kyburg rightly sees (as Neyman, Pearson and Fisher saw before him) the arbitrariness in all this and insists that rational agents ought to refuse to assign numerically definite credal probabilities when there is no warrant for doing so.

De Finetti's failure to consider as rational indeterminacy in credal or subjective probability judgment has other unfortunate consequences as well. In statistical problems as typically formulated, the data obtained from experiment often determine a "likelihood function" for assessing the contribution of the data for the support or undermining of the statistical hypotheses under scrutiny. This likelihood function may be well defined even if the prior and posterior probability distributions over the statistical hypotheses (or over hypotheses concerning long-run relative frequencies in future experiments) are indeterminate. And, indeed, it is typically the case in scientific investigations that such priors and posteriors are indeterminate or are regarded to be such. Thus de Finetti must deny what is, for many, an obvious feature of scientific life.

I do not mean to suggest that likelihoods are always well defined. In poorly designed experiments, they often are not. But one of the aims of experimental design is to guarantee that likelihoods on the data can be assessed. It is not always as urgent to guarantee that prior and, hence, posterior probabilities be well defined.

Kyburg has a better appreciation of these points, in my opinion, than de Finetti in spite of his agreement with de Finetti on the question of stochastic atheism.

I have registered my own reservations concerning Kyburg's and Fisher's approach to direct inference elsewhere. (See especially, LEVI, 1980a, ch. 16.) In brief, I claim that the Kyburg–Fisher approach to direct inference violates confirmational conditionalization. I regard this as a serious objection to following the Kyburg–Fisher approach.

Although one can endorse the Kyburg–Fisher approach to direct inference without replacing statistical probability by frequency, if one does follows Kyburg's approach to eliminate statistical probability, one does need to endorse a reasonable facsimile of his view of direct inference and, hence, to violate confirmational conditionalization.

I conclude from this that an effort, far more important than de Finetti's, to replace statistical probability by frequency proves inadequate.

The considerations I have introduced contra de Finetti's stochastic atheism have been invoked from a point of view which shares with his the concern to save the good sense of the problematic of statistical estimation from the metaphysical fantasies generated by much contemporary modal and stochastic realism. I have been arguing that his effort to use his celebrated representation theorem and extensions thereof for this purpose necessitates his bringing in a whiff of that very realism he so stoutly deplores.

None of the arguments I have offered compel anyone to inhale this whiff of realism. John Stuart Mill and Rudolf Carnap were prepared to rest content with singular predictive inference. Anyone who shares their attitude is presumably ready to abandon the problematic of statistical estimation altogether and will not be touched by anything I have had to say. But if one takes such an extreme view (as I suspect de Finetti sometimes does), the de Finetti representation theorem is quite unnecessary; for one need not be concerned to find a surrogate for the problem of unknown statistical probabilities. As we have seen, on the other hand, positivists who seek to protect the problematic of statistics, as de Finetti sometimes wants to do, will find the de Finetti representation theorem useless for their purpose.

It may, perhaps, be said that the de Finetti representation theorem still plays a role from a positivistic perspective. While denying the meaningfulness of objective statistical probability, we may still proceed in making credal probability judgments about observable events concerning test behavior "as if" there were "hidden variables" specifying values of unknown objective statistical probability and our credal probability judgments were a kind of average of these unknown objective probabilities.

It is true that the de Finetti representation theorem allows us to indulge in such poetry — as Jan von Plato has illuminatingly pointed out (for example, in VON PLATO, 1982). But von Plato rightly observes that the possibility of speaking in this way, though permitted by the representation theorem, is not mandated by it. The representa-theorem is entirely neutral with respect to the direction of reduction just as are the analogous representation theorems concerning the relation between wave and matrix mechanics in quantum theory.

In any case, I am not under the impression that de Finetti thought the representation theorem was important in order for him to be able to indulge in fantasies concerning the "existence of Phlogiston, the Cosmic Ether, Absolute Space and Time, ..., or Fairies and Witches" without really meaning it. We have already seen that de Finetti's radical positivism was tempered by his concern to save the problematic of statistics. And he sought to do so by stripping away the fantasy — not by indulging in it. De Finetti did not want merely to talk as if there were statistical probabilities. He wanted to be able to talk quite literally and strictly about unknown magnitudes which are meaningful surrogates for statistical probabilities.

I have argued that de Finetti's effort to exploit his representation theorem to escape stochastic realism while preserving the problematic of statistical estimation utterly fails. If I am right about this, no amount of rhetoric about "as if" statistical probabilities will put the problematic of statistical estimation back together. We need more stochastic realism than that.

Retaining a whiff of realism is not embracing the various realist extravaganzas which have been presented for our mystification in recent years. I have argued that approaches along the lines of Kolmogorov–Cramer–Braithwaite are to be preferred because they seek to understand statistical probability by (a) articulating the formal requirements for a stochastic model and (b) offering an account of the epistemic or evidential connections between hypotheses about statistical probability and hypotheses about test behavior. This, I believe, is the core idea which drove HACKING's 1965 account of chance, my 1967 account of statistical probability and Donald GILLIES 1973 work. In spite of their rather substantial differences, these efforts share a minimalist approach to the realism of statistical probability specifying no more semantical structure than is required to grant statistical hypotheses truth values.

Such views do not rule out additional semantical elaboration. However they are in general suspicious of proposals for a unitary semantical structure for all statistical hypotheses. Instead, they expect additional structure should and sometimes does emerge from the special theoretical and experimental investigations of the special sciences (LEVI, 1967 and 1980a).

In this way, these views seek to identify that framework of ideas and methods which might be a minimum necessary to render the problematic of statistical theory intelligible (as de Finetti desired it to be) without begging questions concerning the many debatable issues about the foundations of statistics which currently exist. Conceptual contraptions like chance displays, single case chances, history to chance conditionals, and causal analogues of confirmation functions do not appear necessary for the rescue of statistics. We can safely consign these ideas to the domain of Fairies and Witches even if we refuse to abandon statistical probability. In this sense, probability exists — but just barely.

References

BRAITHWAITE, R.B., 1953, *Scientific Explanation* (Cambridge Univ. Press, Cambridge).

BRAITHWAITE, R.B., 1962, *On unknown probabilities*, Observation and Interpretation in Philosophy of Physics, ed. S. Körner, (Dover, New York), pp. 3–11.

BURKS, A.W., 1977, *Chance, Cause, Reason* (Univ of Chicago Press, Chicago).

CRAMER, H., 1945, *Mathematical Methods of Statistics* (Princeton Univ. Press, Princeton).

DE FINETTI, B., 1937, *Foresight: its logical laws, its subjective sources*, in: KYBURG and SMOKLER, 1980, pp. 53–118.

DE FINETTI, B., 1972, *Probability, Induction and Statistics* (Wiley, New York).

DE FINETTI, B., 1975, *Theory of Probability*, 2 vols. (Wiley, New York).

DE FINETTI, B., 1977, *Probability: beware of falsifications*, in: KYBURG and SMOKLER, 1980, pp. 195–224.

FISHER, R.A., 1959, *Statistical Methods and Statistical Inference* 2nd ed. (Hafner).

GÄRDENFORS, P. and SAHLIN, N.-E., 1982, *Unreliable probabilities, risk taking, and decision making*, Synthèse 53, pp. 361–386.

GILLIES, D.A., 1973, *An Objective Theory of Probability* (Methuen, London).

HACKING, I., 1965, *Logic of Statistical Inference* (Cambridge Univ. Press, Cambridge).

KOLMOGOROV, A., 1933, *Foundations of the Theory of Probability*, 2nd English edition (Chelsea, New York, 1956).

KYBURG, H.E., 1961, *Probability and the Logic of Rational Belief* (Wesleyan Univ. Press, Middledown).

KYBURG, H.E., 1974, *Logical Foundations of Statistical Inference* (Reidel, Dordrecht).

KYBURG, H.E. and SMOKLER, H.E., 1980, *Studies in Subjective Probability* 2nd edition (Krieger, Huntington).

LEVI, I., 1967, *Gambling with Truth* (A. Knopf, reissued in paperback without revision in 1973 by MIT Press).

LEVI, I., 1977, *Direct inference*, J. Philosophy 74, pp. 5–29.

LEVI, I., 1980a, *The Enterprise of Knowledge* (MIT Press, reissued in paper with minor revisions in 1983).

LEVI, I., 1980b, *Induction as self correcting according to Peirce*, Science, Belief and Behaviour, Essays in Honor of R.B. Braithwaite, ed. D.H. Mellor (Cambridge Univ. Press, Cambridge, pp. 127–140.

VON MISES, R., 1951, *Probability, Statistics and Truth*, 3rd rev. English edition (MacMillan, New York).

VON PLATO, J., 1982, *The significance of the ergodic decomposition of stationary measures for the interpretation of probability*, Synthèse 53, pp. 419–432.

REICHENBACH, H., 1938, *Experience and Prediction* (Univ. of Chicago Press, Chicago).

SEIDENFELD, T., 1979, *Why I am not an objective Bayesian*, Theory and Decision 11, pp. 413–440.

8. Foundations and Philosophy of the Physical Sciences

Barcan Marcus et al., eds., Logic, Methodology and Philosophy of Science VII

ON LIMIT RELATIONS BETWEEN, AND APPROXIMATIVE EXPLANATIONS OF, PHYSICAL THEORIES

JÜRGEN EHLERS

Max-Planck-Institut für Physik und Astrophysik, Institut für Astrophysik, Garching, F.R.G.

Thus in science, as distinct from theology, a critical comparison of the competing theories, of the competing frameworks, is always possible... In science (and only in science) can we say that we have made genuine progress: that we know more than we did before.

Karl POPPER[1]

1. Introduction

This paper is concerned with relations between two physical theories with overlapping domains of application one of which, T', is regarded as a limiting case of the other one, T; let this be indicated by

$$T \to T'. \tag{1}$$

Two examples will be considered. In the first one, T' is the Galilei-invariant, T the Lorentz-invariant theory of collisions between particles, and in the second one T' is the Newtonian, T the Einsteinian ("general-relativistic") theory of isolated systems of gravitationally interacting bodies.

I am not concerned with the histories of the theories in question or with the intuitive ideas, motivations, manner of presentation etc. of their originators, but with rational reconstructions of those theories, logical relations between them, and with their relations to experience. I maintain that in the cases to be considered below and presumably in other cases too, limit relations *can* be understood *rationally*, that in spite of alleged meaning changes and conceptual incommensurabilities the successor theory T does *explain* and *improve* its predecessor T', i.e. that there are reasons for asserting that there is progress in science.

I propose to reconstruct theory-pairs (T, T') which are candidates for a limit relation (1) according to the following plan. Firstly, one reformulates T and T' such that

A The mathematical entities used in T and T' (basic and auxiliary sets, relations, fields, constants, axioms) are members of specializations Σ, Σ' of one single "species of structures" $\hat{\Sigma}$ (see BOURBAKI[2] and, in the context of physics, LUDWIG[3], BALZER[4] and SCHEIBE[5]).

B The specialization Σ underlying the structure of T is obtained by assigning specific nonvanishing values to unspecified dimensional constants of $\hat{\Sigma}$ whereas Σ' is obtained by taking these constants to be zero.

C The correspondence rules relating mathematical entities to facts, or elementary descriptions of facts (LUDWIG[3], see also SCHEIBE[6]) are the same for T and T'.

These assertions concern the concepts, laws and interpretation rules of T and T'. The second step involves an investigation of the sets S, S' of *solutions* of the laws of T and T', respectively, which represent possible physical processes (possible worlds, or rather world-fragments) according to those theories. It consists of showing that

D S and S' contain subsets S_ε, S'_ε such that all elements of S'_ε are approximated by elements of S_ε and all those of S_ε are approximated by elements of S'_ε, ε being a positive numerical parameter which determines error bounds tending to zero with ε.

The purpose of D is to exhibit conditions under which T and T' are empirically equivalent in view of the fact that the results of observations and experiments are imprecise and both theories have to be applied to actual data by means of imprecision-sets[3,6,7]. The approximate imbedding of S'_ε into S_ε shows that all explanations by means of T' based on elements of S'_ε can be carried out equally well with T, using S_ε, and the inverse imbedding of S_ε into S'_ε demonstrates that under specified conditions the testable contents of T do not go beyond that of T'.

While the validity of D depends on the mathematical structures of T and T' only and the interpretation rules can enter only insofar as they may suggest approximation-relations (uniform structures), the subsequent three statements take into account empirical tests of the theories. They involve a judgement of empirical adequacy and cannot be based solely on logics and mathematics. They state:

E *Some* experiments or observed phenomena are represented satisfactorily (i.e. within the limits of observation inaccuracy) by elements of S, but not by elements of S', whereas

F *all* successful applications of T' have been based on elements of subsets S'_ε as mentioned in D, and

G S' contains a *fictitious subset* which has not been used in successful applications of T' and which describes situations deemed impossible according to T.

Steps A, B, C may not be necessary for a comparison of theories T and T' in a successor relation, but they will be carried out below for the two examples mentioned, and the relative simplicity of assertion D which, combined with F, is crucial for the approximative reduction of T' to T, presupposes A, B, C; moreover C requires A and B. I conjecture that the steps A–G can also be taken for classical and quantum mechanics. (See, e.g. refs. 8 and 9.)

A reformulation along these lines shows in which "harmless" sense meanings of terms change in a corresponding transition from T' to T, and that such changes do not exclude a rational comparison of the theories.

If D and F hold, T *explains* T' as an approximation. If in addition E is valid, it is justified to say that T *improves* T'. If G also holds — as it does in the examples mentioned — T represents progress relative to T' not only since it adds new successful applications, but also since it serves to remove a useless part from T'.

A relation between T and T' as formulated in B implies that the mathematical structure underlying T' is a *degenerate case*[10] of that of T, an *asymmetric* relation between the two theories. ("Degeneracy of structure" has a precise meaning in the examples and can probably be defined for more general, perhaps all pairs of species of structures occurring in relations (1).) This degeneracy relation can be used to understand, independently of history and of empirical tests, why T may improve T' whereas T' could not possibly improve T.

In the examples below the successor theory T contains the speed of light c as a fundamental constant while no such constant is in the structure of T'. This loss of a dimensional constant in the transition from the laws of T to those of T' implies that the solutions of the laws of T' admit scale transformations which are not permitted in S. This property of (1) can be related to the manner in which particular families of solutions of T and T' approximate each other, as will be seen below. I conjecture that both these asymmetries — degeneracy and additional scale invariance — are *typical* for limit relations. (For the pair classical/quantum mechanics, Planck's constant h "generates" these asymmetries[9].)

It seems to me that any comprehensive explication of a successor or limit relation between important theories of physics must contain logical and mathematical aspects such as occur in A, B, D; approximation concepts as in D, E, F; interpretation rules as in C; and judgements about empirical adequacy such as enter E, F and G; and it will involve "conservative" as well as "progressive", "deductive" and "factual" aspects.[6]

All steps A–G will be outlined for the simple example of collision theory (Section 2). In the more important, more difficult and controversial second case of Einstein's versus Newton's theory of spacetime and gravity steps A–D have been taken, and E and G are obviously valid. F has not been demonstrated because of mathematical difficulties, but the task of establishing it has been reduced to a problem in the theory of nonlinear partial differential equations; no conceptual obscurity remains. This will be indicated in Section 3.

The presentation in Sections 2 and 3 will be semi-formal. It seems to me desirable to study real-life scientific cases of limit relations (1) rather than artificial examples, even if the latter can be completely formalized while the former may be too complex for that. But I readily admit that many aspects mentioned here could, and should, be made more explicit and precise, particularly those which involve the relation between theories and observational data.

2. Collision theories[11]

In classical, i.e. non-quantal theories of collisions between particles such as billiard balls, atoms, nuclei etc., moving particles are represented by means of world lines in spacetime. Elastic and inelastic collisions are idealized, in the simplest description to which attention is restricted here, as events (spacetime points) at which some "incoming" particle world lines end and some "outgoing" world lines begin. Collision theories assign state variables to the "in" and "out" states and formulate laws which are to single out the "dynamically possible" collisions from the kinematically conceivable ones. We shall consider only the *conservation laws* which are the basic ones for these theories. Let T denote the "relativistic", i.e. Lorentz-invariant collision theory, and let T' denote the "non-relativistic", i.e. Galilei-invariant collision theory. I wish to show that T' is a limit theory of T in the sense that assertions A–G of the introduction hold.

A common *kinematical framework* for T and T' uses the following *concepts:*

A real, four-dimensional, differential manifold M; a two-contravariant, symmetric tensor field $h = (h^{ab})$ of dimension $(\text{length})^{-2}$ on M; a two-covariant, symmetric tensor field of dimension $(\text{time})^2$ on M; and a constant λ of dimension $(\text{speed})^{-2}$.

These objects are required to satisfy the following axioms:

ST_1 The index of inertia[12] of h is 3.

ST_2 The index of inertia of g is 1.

ST_3 $g_{ab}h^{bc} = -\lambda\delta_a^{\ c}$.

ST_4 $\lambda > 0$ (theory T), $\lambda = 0$ (theory T').

These axioms imply: If $\lambda > 0$, h is Lorentzian and $-\lambda^{-1}g = h^{-1}$; if $\lambda = 0$, there exists a covector (1-form) $t \neq 0$, unique up to sign, such that $g_{ab} = t_a t_b$ and $h^{ab}t_b = 0$. Thus if $\lambda > 0$, (M, h^{ab}) is a Lorentz manifold and if $\lambda = 0$, (M, h^{ab}, t_a) is a Galilei manifold. (See KÜNZLE[13] and the references given there.)

In both cases there exist, at each point x of M, orthonormal frames (E_a), i.e. bases of the tangent space M_x such that the components of h and g are given by

$$(h^{ab}) = \text{diag}(-\lambda, 1, 1, 1), \qquad (g_{ab}) = \text{diag}(1, -\lambda, -\lambda, -\lambda),$$

and that for $\lambda = 0$

$$(t_a) = (1, 0, 0, 0).$$

Orthonormal frames are related to each other by (homogeneous) Lorentz transformations if $\lambda > 0$, by (homogeneous) Galilei Transformations if $\lambda = 0$.

A vector $V = (V^a)$ is called timelike if $g_{ab}V^aV^b > 0$, spacelike if there exists a covector W_a such that $V^a = h^{ab}W_b$ and $h^{ab}W_aW_b > 0$. V is a timelike unit vector if $g_{ab}V^aV^b = 1$.

Two timelike vectors U, V are orthochronous if $g_{ab}U^aV^b > 0$. Orthochronality is an equivalence relation with two equivalence classes each of which defines a time-orientation on M_x, for each x. Let one such time orientation be chosen at each event x; it will be used below to distinguish, for collisions at x, "in" and "out" world lines.

A curve in M is said to be timelike (spacelike) if its tangent vector is timelike (spacelike).

The physical meaning of M, h, g is given by the following *interpretation rules:*

IR_1 The points of M represent *events*[14].

IR_2 Timelike curves in M represent *particle motions.* The *proper time* at a particle between two of its events x, y is the line integral

$$\int_x^y (g_{ab}\,\mathrm{d}x^a\,\mathrm{d}x^b)^{1/2}.$$

IR_3 The *proper length* of a spacelike curve $x^a(s)$ from x to z (representing an instantaneous configuration of a string, e.g.) is

$$\int_x^z (h^{ab}W_aW_b)^{1/2}\mathrm{d}s, \quad \text{where } \frac{\mathrm{d}x^a}{\mathrm{d}s} = h^{ab}W_b.$$

Next, we introduce the concepts and laws which refer to collisions. According to IR_2 the motion of a particle determines a unit timelike vector V, the normalized tangent of its world line, called its 4-velocity. If the particle participates in a collision at x, V is to be chosen furture-directed at x, according to the chosen time-orientation. (The choice of such an orientation is purely conventional for the purposes of the simple collision theory presented here, no "arrow of time" is needed.)

Later we shall use the following kinematical *lemma.* Let V, V' be orthochronous, timelike unit vectors. Then $V - V'$ is spacelike: $(V - V')^a = h^{ab}W_b$. The invariant $\beta_\lambda(V, V') = \frac{1}{2}h^{ab}W_aW_b$ is nonnegative and vanishes exactly if $V = V'$.

If V, V' are the 4-velocities of two particles meeting at x, $\beta_\lambda(V, V')$ has the following interpretation: If $\lambda = 0$, $\sqrt{2\beta_0} = v$ is the speed of V relative to V'; if $\lambda > 0$, $\gamma_\lambda(V, V') = 1 + \lambda\beta_\lambda(V, V') = (1 - \lambda v^2)^{-1/2}$ is the Lorentz factor of V relative to V'. For fixed V, V' (or v),

$$\lim_{\lambda\to 0} \beta_\lambda = \beta_0.$$

These assertions are easily verified by expressing them, e.g., in terms of orthonormal components.

We postulate the *collision laws:*

C_1 Between collisions the *state* of a particle is determined by its 4- velocity V (translation state), a *conserved mass*[15] $M > 0$, and an *internal energy* U.

C_2 At each collision the sum of the *4-momenta* $(M + \lambda U)V$ of the "in"-states equals that of the "out"-states. Moreover, for any timelike unit vector E the sum of the quantities $\gamma_\lambda(V, E)U + \beta_\lambda(V, E)M$ is conserved.

We have now in essence carried out steps A–C of the programme outlined in the introduction. Before proceeding to the remaining steps let us verify that C_2 is indeed equivalent to the well-known conservation laws of relativistic and non-relativistic collision theory, respectively.

Indeed, in terms of components with respect to an orthonormal frame at the collision event the five conserved quantities of C_2 are

$$(M + \lambda U)\gamma_\lambda, \qquad (M + \lambda U)\gamma_\lambda V^\mu, \qquad \gamma_\lambda U + \beta_\lambda M, \tag{1}$$

where $\boldsymbol{V} = V^\mu E_\mu$ is the 3-velocity with respect to the frame (E_a).

Applying the lemma and the remarks following it (with V, V' replaced by E, V) and writing

$$m = M + \lambda U \tag{2}$$

one obtains for the five quantities (1):

$$\text{if } \lambda > 0: \quad m\gamma_\lambda, m\gamma_\lambda v^\mu, \frac{m\gamma_\lambda - M}{\lambda}, \tag{3}$$

$$\text{if } \lambda = 0: \quad M, Mv^\mu, \tfrac{1}{2}Mv^2 + U. \tag{3'}$$

Conservation of the five quantities (3) is obviously equivalent to conservation of the 4-momentum mV and the 4-scalar M. Thus we recognize in (3) the relativistic conserved quantities 4-momentum and conserved mass, or energy $\lambda^{-1}m\gamma_\lambda$, 3-momentum $m\gamma_\lambda v^\mu$ and conserved mass M, provided $\lambda^{-1} = c^2 = (\text{fundamental speed})^2$. The relativistic proper mass $m = M + c^{-2}U$ consists of the conserved part and a binding energy contribution[15] $c^{-2}U$. Similarly we recognize in (3′) the Newtonian conserved quantities mass, momentum and energy, the latter consisting of a kinetic and an internal contribution.

The five quantities (3) transform according to a representation of the Lorentz group, the quantities (3′) according to one of the Galilei group. The latter (indecomposable[16]) representation arises from the former one (which is equivalent to the 4-vector $\oplus$ scalar repres.) by contraction[16]. This relationship and the kinematical formulae given before show how the laws of T' arise formally from those of T by the limit process $\lambda \to 0$.

Since 3-velocities are measurable, the values of the theoretical terms m, M and U can be determined provided sufficiently many collisions[17] can be carried out and one can identify internal states (particle types) in different collisions as being "the same"[18]; one particle has to be used to fix the mass

representation as a subrepresentation. (Similar "surprises" are pointed out in ref. 9.) But nevertheless T' can be justifiably called a limiting case of T, mathematically *and* physically.

All physical variables and laws have been treated here so that arbitrary units of length, time and mass may be used. In particular one may put $c = 1$. It can be recognized from this setup that the frequently used "device" $c \to \infty$ for the transition $T \to T'$ is just a misleading metaphor. There are only *two* theories, T and T'; in $T, \lambda = c^{-2}$, and in T', $\lambda = 0$. Approximations are governed by dimensionless parameters (see (4) and (5)). Nevertheless the mathematical fact that the laws of T have those of T' as their limits if $\lambda \to 0$ is one reason why some solutions of T are approximated by solutions of T'.

What has been said here will not surprise physicists; I have explicated it as a *model* for more complex limit relations (1).

3. Newton's and Einstein's theories of gravity[20]

> It would be simply false to say that the transition from Newton's theory of gravity to Einstein's is an irrational leap, and that the two are not rationally comparable. On the contrary, there are many points of comparison: It follows from Einstein's theory that Newton's theory is an excellent approximation (except for planets or comets moving in elliptic orbits with considerable eccentricities).
>
> Karl POPPER[1]

The aim of this section is to support and explicate the statement just quoted. I shall again proceed according to the plan outlined in the introduction and illustrated in the preceding section. Let T now denote Einstein's, T' Newton's theory of isolated systems of gravitationally interacting bodies.

As *common concepts* for both theories one can take:

A manifold M; two tensor fields h and g; a constant λ (as in Section 2); a symmetric linear connection on M (or, equivalently, the corresponding derivative operator D); a two-contravariant symmetric tensor field $T = (T^{ab})$ of dimension $(\text{time})^{-4}$ on M.

Let $R = (R^{a}{}_{bcd})$ denote the curvature tensor associated with the connection D. The tensors g and h will be used to lower and raise indices, but because of ST_3 (which we retain) these two operations are not inverses of each other. Therefore the original positions of the indices of a tensor will be indicated by dots as in

$$T^{a\cdot}{}_{\cdot b} = T^{ac} g_{cb}.$$

Then ST_3 can be rewritten as

$$g_{a\cdot}^{\ c} = h_{\cdot a}^{\cdot\ c} = -\lambda\delta_a^{\ c}.$$

The contracted curvature tensor is defined by $R_{ab} = R^c{}_{acb}$.

In this section, only length and time are used as independent physical dimensions. Mass is regarded as $(\text{length})^3 \times (\text{time})^{-2}$ which amounts to setting Newton's constant of gravity equal to one.

As *axioms* we take ST_1–ST_4 of Section 2 and the following ones:

ST_5 $D_a h^{bc} = 0, \qquad D_a g_{bc} = 0.$

ST_6 $R^{a\ c}_{\ b\cdot d} = R^{c\ a}_{\ d\cdot b}.$

ST_7 $R_{ab} = 8\pi(T^{\cdot\cdot}_{ab} - \frac{1}{2}g_{ab}T^{c\cdot}_{\ \cdot c}).$

ST_8 (M, g, h, D) is spatially asymptotically flat.

M_1 At all events where $T^{ab} \neq 0$, $T^{a\cdot}_{\ b}$ maps any timelike vector V^a into a timelike vector which is orthochronous to V^a.

M_2 $D_a T^{ab} = 0.$

M_3 The support of T is spatially compact.

The axioms ST_1–ST_8 determine the *structure of spacetime*, M_1–M_3 refer to *matter* and *mechanics.*

If $\lambda > 0$, some of the axioms are redundant. This is not so for $\lambda = 0$. In this respect Einstein's theory is simpler than Newton's; it requires fewer axioms.

The axioms ST_8 and M_3 impose *global* conditions on the *local structure* defined by the remaining axioms. M_3 asserts that the restriction of the matter tensor T to any spacelike hypersurface vanishes outside of some compact set. The meaning and importance of ST_8 are explained in an appendix.

With the preceding statements the first two steps, A and B, of the general plan of the introduction have been carried out for (T, T'). To obtain C, we retain the rules IR_1–IR_3 of the second section, with one change: "particle" in IR_1 and IR_2 is to be interpreted now *not* as referring to (the idealized model of) a body, but as referring to a "place" or "spot" marked on a body. This change is necessary since the model of a point-particle is incompatible with the field equations (ST_7) of general relativity. In Einstein's theory, bodies have to be represented as extended (and deformable). We add one more interpretation rule:

IR_4 The *mass density* at a place of a body or in a (e.g, radiation-) field, with respect to a world line ("observer") with 4-velocity U, is given by

$$\rho_U = T^{ab} U_a U_b. \tag{6}$$

(The meaning of "mass density" is taken for granted here. If one does not want to take it over from "ordinary, local physics", one has to infer its meaning from the way it enters observable relations which are meaningful in view of IR_1–IR_3. Roughly speaking, IR_4 says "matter is where ρ_U is positive". This could and should be analysed further.)

Having stated the fundamental mathematical and physical assumptions underlying both T and T', I should like to draw some conclusions and to make some explanatory comments.

M_3 says that we are concerned with systems considered as isolated, separated from the "rest of the world"; e.g. the solar system or a double star "by itself". ST_8 formalizes that the metric (g, h) and the connection D are "due to" the system, not to anything "outside".

M_1 implies that the mass density ρ_U (see (6)) is positive except in vacuo (defined by $T^{ab} = 0$) where it vanishes.

ST_7 is the Einstein–Hilbert field equation of gravity if $\lambda = c^{-2} > 0$, and if $\lambda = 0$ it is, combined with the other "ST-axioms", equivalent to Poisson's equation (see the references 20, in particular EHLERS (1981)). Since the density within bodies is positive, both these laws imply gravity to be attractive. M_2, the local law of motion, states local conservation of mass and Cauchy's equation of motion for $\lambda = 0$, and their relativistic analogs for $\lambda = c^{-2} > 0$.

In order to avoid complications we here consider only ordinary bodies, not black holes, as sources of gravitational fields although one can include the latter which turn out to be the closest analogs of point-particles in Einstein's theory (see EHLERS (1981)[20]).

The reformulation of Newton's and Einstein's theories outlined here shows, as in the example of Section 2, that the mathematical structure of T' is a degenerated special case of that of T. The spacetime structure underlying T is equivalent to a Lorentz bundle over M while that of T' amounts to a Galilei bundle. Such principal fibre bundles[21] are locally determined by their structure groups, and since the Galilei group is a contraction of the Lorentz group this degeneracy relation is inherited by the corresponding bundles.

Let us now turn to steps D–G, i.e. to the comparison of particular models or solutions of T and T'.

In the case of Einstein's theory, the axioms given above imply that bodies move with speeds less than $c = \lambda^{-1/2}$. (A theorem which establishes this and which makes use particularly of M_1 and M_2 has been given in ref. 22, sec. 4.3.) This statement is related to observations, e.g., to measurements of radar travel times or Doppler frequency shifts. In view of this, assertion G of Section 1 is easily established: a fictitious subset of S' is obtained by taking Newtonian solutions with bodies having relative velocities in excess of c.

It is also easy to argue in favour of assertion E: The "anomalous" advance of Mercury's perihelion has been explained by T, but not by T'; several other pertinent examples are now available[23].

As concerns statement D for the case under discussion, it has been possible to give examples of sets S_ε, S'_ε of solutions which are in the required relation of mutual approximation[20], but they are rather special and certainly not sufficiently general to justify assertion F (if one wants to have rigorous results and is not satisfied by "plausible approximations"). It is also possible to give fairly general criteria for one-parameter families of solutions of T to be in osculating approximation with one-parameter families of solutions of T', in close analogy to the proposition stated in Section 2 after equation (5). As in the simpler case of Section 2, use is made of similarity transformations which are possible in T', but not in T. (See DAUTCOURT (1964) and KÜNZLE (1976), quoted in ref. 20, and a forthcoming paper by myself.) However, it is not clear whether *every* family of Newtonian solutions which is obtained by rescaling one such solution according to

$$h^{ab} \to h^{ab}, \quad g_{ab} \to \alpha^{-2} g_{ab}, \quad D_a \to D_a, \quad T^{ab} \to \alpha^4 T^{ab} \tag{7}$$

$(0 < \alpha < \infty)$, can be so approximated.

To investigate the ***existence problem*** just indicated one has to study how the Einsteinian system of nonlinear partial differential equations formed by eqs. ST_7 and M_2 degenerates, for $\lambda \to 0$, into the analogous Newtonian system. It is well known[22] that this system, specialized to a perfect fluid source with an equation of state,

$$\begin{aligned} T^{ab} &= (\rho + \lambda p) U^a U^b + p h^{a,b}, \\ g_{ab} U^a U^b &= 1, \qquad p = f(\rho) \end{aligned} \tag{8}$$

in which p denotes the pressure, is equivalent to a system of *hyperbolic* (wave-like) *evolution equations* for the dynamical variables g_{ab}, ρ, U^a and a system of *elliptic constraints*, i.e. equations which restrict the initial data. I have shown that in suitable variables these evolution equations *partially degenerate*, for $\lambda \rightarrow 0$, into the hyperbolic subsystem of Euler's equations for the matter variables and an elliptic subsystem. These two systems together are equivalent to the laws of Newton's theory. Moreover, the constraint equations of T go over, for $\lambda \rightarrow 0$, into a system which is implied by the limits of the evolution equations. (Ehlers, to be published. See also, for a similar analysis, KÜNZLE and NESTER and SCHUTZ and FUTAMASE[24].) These mathematical results suggest the following procedure: Take an arbitrary solution of Newton's laws (as stated above). Its initial data satisfy the limit equations of the Einsteinian constraints. It appears that there always exists a one-parameter family, parametrized by λ, of solutions to the Einsteinian constraints which converges for $\lambda \rightarrow 0$ to the given set of Newtonian initial data. (D. Christodoulou, personal communication; to be published.) This one-parameter family of admissable initial data for T determines, at least locally in time, a corresponding family of solutions of the local laws of Einstein's theory. If it could be proven that (i) these solutions also obey the asymptotic condition ST_8 of T and (ii) the fields $(g(\lambda), h(\lambda), D(\lambda), R(\lambda))$ converge in a suitable sense for $\lambda \rightarrow 0$, then it would follow that any slow motion, weak-field solution of T' can be approximated by a solution of T, so that assertion F would be justified. Thus modulo these unsolved mathematical problems the relation between T and T' can be said to be understood, in the sense of the assertions given in the introduction.

In view of the examples presented in this and the preceding section and similar examples in the literature it seems to me that there is no reason to doubt the rationality of the succession of theories, at least in physics. (This does, of course, not exclude "irrational" intermediate steps in the actual course of history, but that is a different matter.) Whoever denies the possibility of meaningful comparisons of empirically successful theories in historical succession ought to present clearcut cases in which the preceding kind of reasoning, or a modification of it, definitely does not apply, or point out why explanations as advocated here are inadequate.

Acknowledgement

The considerations of this paper have been influenced in an essential way by papers of G. Ludwig and E. Scheibe to whom I am grateful also for discussions. Also, I am indebted to my "relativistic colleagues" at the Max-Planck-Institut for their criticisms, particularly to B.G. Schmidt.

Appendix

The purpose of this appendix is to explain the meaning of the axiom ST_8 of Section 3 and to indicate its importance for the theories T and T' and for the relation $T \rightarrow T'$ between them. (Details of proofs will be published elsewhere.)

It has been shown by A. Trautman[25] and reviewed in ref. 20 that the local axioms ST_1–ST_7, M_1 and M_2 with $\lambda = 0$ are *not* sufficient to characterize the local structure of Newton's theory; one needs in addition the restriction

$$R^{a\,b}_{\cdot\,\cdot cd} = 0 \tag{9}$$

on the curvature. It states that parallel transport of spacelike vectors is path-independent, or in physical terms that the rotation axes of freely falling, neighbouring gyroscopes do not exhibit relative rotations in the course of time. However, (9) cannot be taken as a *common* axiom for T and T' since for $\lambda > 0$ it would lead to flat spacetime without matter. However, in spite of this difficulty one can find a common formulation for both theories since in the case $\lambda = 0$ the law (9) can be deduced from a boundary condition at spatial infinity which is also meaningful in Einstein's theory and which expresses, in *both* theories, that the physical systems considered are idealized as being isolated. This can be achieved as follows:

In both theories the concept of a spacelike hypersurface is well defined; that is a hypersurface all tangent vectors of which are spacelike. Also, in both cases there exist, along such a hypersurface, timelike unit normal vector fields U, and by means of them one can define two tidal field tensors

$$E^a{}_b = R^a{}_{cdb} U^c U^d, \tag{10}$$

$$B_{ab} = \tfrac{1}{2}\eta_{acde} R^{c\,d}_{\cdot\,\cdot bf} U^e U^f. \tag{11}$$

η_{abcd} denotes the volume element[13] of M.

They measure spatial rates of change of the gravitational fields as "seen" by observers having 4-velocities U. In vacuo, E and B are gravitational analogs of the electric and magnetic components of Minkowski's field strength tensor of electrodynamics. Both E and B are spatial with respect to U in the sense that $E^a{}_b U^b = 0$, $U_a E^a{}_b = 0$, $B_{ab} U^b = 0$, $U^a B_{ab} = 0$. The restrictions of $E^a{}_b$ and B_{ab} to a spacelike hypersurface H may therefore be considered as "3-dimensional" tensors in H.

Any reasonable concept of isolation, or *asymptotic flatness* of spacetime, in both theories will contain the requirement that spacetime can be covered by non-intersecting spacelike hypersurfaces which resemble, at infinity,

Barcan Marcus et al., eds., Logic, Methodology and Philosophy of Science VII

GRAVITY AND SPATIAL GEOMETRY[1]

DAVID MALAMENT*

Dept. of Philosophy, Univ. of Chicago, Chicago, IL 60637, USA

Philosophers of science have written at great length about the geometric structure of physical space. But they have devoted their attention primarily to the question of the epistemic status of our attributions of geometric structure. They have debated whether our attributions are *a priori* truths, empirical discoveries, or, in a special sense, matters of stipulation or convention. It is the goal of this paper to explore a quite different issue — the role played by assumptions of spatial geometry *within physical theory*, specifically within Newtonian gravitational theory.

Standard formulations of Newtonian physics, of course, presuppose that space is Euclidean. But the question arises whether they must do so. After all, the geometric structure of physical space was a topic of intense interest in the 19th century long before Newtonian physics was abandoned. Think of Gauss, Riemann, Helmholtz, and Poincaré. It is probably most natural to assume, and perhaps these men *did* assume, that any hypotheses about spatial geometry function only as inessential auxiliary hypotheses within Newtonian physics — superimposed, as it were, on a core of basic underlying physical principles which themselves are neutral with respect to spatial geometry. Yet it turns out that there is an interesting sense in which this is just not so, a sense which is only revealed when one considers Newtonian gravitational theory from the vantage point of general relativity.

One can, and I think should, construe the former theory as a special limiting form of the latter in which relativistic effects become negligible.

---

[1] The following is extracted from a long, technical paper [3]. Proofs can be found there together with a good deal of supplemental material on spacetime structure in Newtonian physics. The results presented there draw on work of Künzle in [1] and [2].

* I am grateful to Jürgen Ehlers and Robert Geroch for comments on an earlier draft. Ehlers, in particular, saved me from making a number of seriously misleading statements.

That is, one can think of Newtonian gravitational theory as the so-called "classical limit" of general relativity. The big surprise, at least to me, however, is that when one *does* think about it this way one finds that the theory *must* posit that space is Euclidean. It's curious. The very limiting process which produces Newtonian physics and a well-defined, observer invariant spatial structure also generates strong constraints on spatial curvature. These constraints turn out to be *so* strong as to guarantee the Euclidean character of space. That, anyway, will be my principal claim today.

Claim. Insofar as it is the "classical limit" of general relitivity, Newtonian gravitational theory *must* posit that space is Euclidean.

A good bit of differential geometry will be required to make the claim precise. But the underlying idea is quite intuitive. It is absolutely fundamental to relativity theory that there is an upper bound to the speeds with which particles can travel (as measured by an observer). The existence of this upper bound is embodied in the null cones (or light cones) one finds in spacetime diagrams. In classical physics, however, there is no upper bound to particle speeds. The transition from general relativity to Newtonian physics is marked by this all important difference. The maximal particle speed goes to infinity. The transition can be conceived geometrically as a process in which the null cones at all spacetime points "flatten" and eventually become degenerate. In the limit the cones are all tangent to a family of hypersurfaces, each of which represents "space" at a given "time". The curious fact is this. If at every intermediate stage of the collapse process spacetime structure is in conformity with the dynamic constraints of general relativity (as embodied in Einstein's field equation), then the resulting induced hypersurfaces are necessarily flat, i.e. have vanishing Riemann curvature. One can think of it this way — the limiting process which effects the transition from general relativity to Newtonian gravitational theory "squeezes out" all spatial curvature!

The proposition which follows is intended to capture the collapsing light cone picture in a precise statement about relativistic spacetime models.

We take a relativistic spacetime model to be a triple (M, g_{ab}, T_{ab}) where M is a smooth, connected, four-dimensional manifold (representing the totality of all spacetime points); g_{ab} is a smooth Riemannian metric of Lorentz signature $(+1, -1, -1, -1)$ on M (which represents the metric of spacetime); T_{ab} is a smooth, symmetric field on M (which represents the mass-energy density present throughout spacetime); and where Einstein's equation

$$R_{ab} - \tfrac{1}{2} g_{ab} R = 8\pi T_{ab}$$

is satisfied. In the proposition we start with a one-parameter family of such models all sharing the same underlying manifold M:

$$(M, g_{ab}(\lambda), T_{ab}(\lambda)), \quad 0 < \lambda \leq 1.$$

Then we impose two constraints — one on the limiting behavior of the $g_{ab}(\lambda)$ as λ goes to 0, and one on that of the $T_{ab}(\lambda)$. The first guarantees that all null cones open up and become tangent to a family of hypersurfaces. The second guarantees that the limiting values of mass-energy density, momentum density, and material stress (as determined by any one observer) are all finite. Our conclusion is that as a result of the conditions imposed the limiting hypersurfaces have vanishing Riemann curvature.

To motivate the first constraint it will help to consider a special case which should look familiar. In Minkowski spacetime all curvature vanishes. One can find a global t, x, y, z coordinate system in which the metric g_{ab} and its inverse g^{ab} have coefficients

$$g_{ab} = \mathrm{diag}(+1, -1/c^2, -1/c^2, -1/c^2),$$
$$g^{ab} = \mathrm{diag}(+1, -c^2, -c^2, -c^2).$$

(That is, the coefficients of g_{ab} form a 4×4 matrix whose diagonal entries are $+1$, $-1/c^2$, $-1/c^2$, $-1/c^2$, and whose non-diagonal entries are all 0.) Now let us consider these as fields parametrized by c. The first has a limit as c goes to infinity. The other does too after it is suitably rescaled:

$$g_{ab}(c) \to \mathrm{diag}(+1, 0, 0, 0),$$
$$g^{ab}(c)/c^2 \to \mathrm{diag}(0, -1, -1, -1).$$

In a sense the limiting process has allowed us to recover separate temporal and spatial metrics. We have pulled apart a non-degenerate metric of signature $(+1, -1, -1, -1)$ to recover its degenerate positive and negative pieces.

This example is special in several respects. The null cones open symmetrically around the "time" axis at each point. The opening occurs uniformly across the manifold. (It is as if the cones were rigidly rigged to each other.) And background affine structure is kept fixed and flat throughout the process. These features cannot be retained when one considers arbitrary (curved) relativistic spacetime models. But the limit existence assertions *can* be generalized, and they turn out to be exactly what one needs.

Consider again our parametrized family of metrics. We are not going to regiment how their null cones open. We shall allow, intuitively, that the cones open at different rates at different points, that their axes wiggle as they open, and so forth. Our sole requirement is that, *somehow or other*, the cones do finally become tangent to a family of "constant-time" hypersurfaces, *and* that they do so in such a way that, after rescaling, a well-defined spatial metric is induced on the surfaces. Formally the requirement comes out this way. (Here and in what follows, all limits are taken as λ goes to 0.)

(1a) There exists a smooth, non-vanishing, closed field t_a on M such that $g_{ab}(\lambda)\to t_a t_b$.

(1b) There exists a smooth, non-vanishing field h^{ab} of signature $(0, +1, +1, +1)$ on M such that $\lambda g^{ab}(\lambda)\to -h^{ab}$.

Clearly the parameter λ corresponds to $1/c^2$.

Let's consider the first clause. I claim that it captures the intended collapsing null cone condition. Suppose t_a is as in (1a). Since it is closed, t_a must be locally exact. That is, at least locally it must be the gradient of some scalar field t on M. It is precisely the hypersurfaces of constant t value to which the cones of the $g_{ab}(\lambda)$ become tangent. [To see this let ∇_a be any derivative operator on M, and let η^a be any vector in the domain of t, tangent to the surface through that point. Then $t_a=\nabla_a t$ and $\eta^a\nabla_a t=0$. It follows that

$$g_{ab}(\lambda)\eta^a\eta^b\to t_a t_b\eta^a\eta^b=(\eta^a\nabla_a t)^2=0.$$

Thus, in the limit η^a becomes a null vector. The surfaces of constant t value are degenerate null cones!]

One can also easily verify that the scalar field t gives limiting values of elapsed proper time. [Suppose that $\gamma:[a,b]\to M$ is a timelike curve with respect to all the $g_{ab}(\lambda)$, and its image falls within the domain of t. The elapsed proper time between $\gamma(a)$ and $\gamma(b)$ along γ relative to $g_{ab}(\lambda)$ is given by

$$PT(\gamma,\lambda)=\int_a^b[g_{mn}(\lambda)\eta^m\eta^n]^{1/2}\mathrm{d}s$$

where η^a is the tangent field to γ. As λ goes to 0 we have

$$PT(\gamma,\lambda)\to\int_a^b(t_n\eta^n)\mathrm{d}s=\int_a^b(\eta^n\nabla_n t)\mathrm{d}s=t(\gamma(b))-t(\gamma(a)).$$

Thus the limiting value of proper time is independent of the choice of timelike curve connecting $\gamma(a)$ to $\gamma(b)$. It is given, simply, by the t-coordinate interval between the two points.]

It remains now to consider the constraint to be imposed on the mass-energy tensor fields $T_{ab}(\lambda)$. Suppose (M, g_{ab}, T_{ab}) is a relativistic spacetime model, and ξ^a is a unit timelike vector at some point of M representing an observer 0. 0 will decompose T_{ab} at the point into its temporal and spatial parts by contracting each index with $\xi^a\xi^m$ or $(\xi^a\xi^m - g^{am})$. (The latter is the "spatial metric" as determined by 0.) The components he determines have the following physical interpretation:[2]

$$T_{ab}\xi^a\xi^b = \text{mass-energy density relative to 0},$$

$$T_{ab}\xi^a(\xi^b\xi^n - g^{bn}) = \text{three-momentum density relative to 0},$$

$$T_{ab}(\xi^a\xi^m - g^{am})(\xi^b\xi^n - g^{bn}) = \text{three-dimensional stress tensor relative to 0}.$$

We shall require of the limiting process that it assign (finite) limiting values to these quantities as determined by some observer 0. The condition comes out as follows.

(2) There exists a smooth field T^{ab} on M such that $T^{ab}(\lambda) \to T^{ab}$.

Here $T^{ab}(\lambda) = T_{mn}(\lambda)g^{ma}(\lambda)g^{nb}(\lambda)$. [The condition is stronger than the requirement that the $T_{ab}(\lambda)$ have a finite limit. To see where it comes from, consider a family of coaligned vectors $\xi^a(\lambda)$, each of unit length with respect to $g_{ab}(\lambda)$. For each λ, perform the decomposition above. If $T_{ab}(\lambda)\xi^a(\lambda)\xi^b(\lambda)$, $T_{ab}(\lambda)\xi^a(\lambda)[\xi^b(\lambda)\xi^n(\lambda) - g^{bn}(\lambda)]$, and $T_{ab}(\lambda)[\xi^a(\lambda)\xi^m(\lambda) - g^{am}(\lambda)][\xi^b(\lambda)\xi^n(\lambda) - g^{bn}(\lambda)]$ are all to have finite limits, it follows that $T_{ab}(\lambda)g^{am}(\lambda)g^{bn}(\lambda)$ must have one too.] Now we can formulate the proposition.

PROPOSITION. *Suppose that for all* $\lambda \in (0,1]$, $(M, g_{ab}(\lambda), T_{ab}(\lambda))$ *is a relativistic spacetime model. Further suppose that conditions* (1) *and* (2) *above are satisfied. Finally suppose that S is any spacelike hypersurface in M as determined by* t_a *(i.e. any imbedded three-dimensional submanifold of M satisfying* $t_a\eta^a = 0$ *for all vectors* η^a *tangent to S). Then if* $\mathcal{R}^a{}_{bcd}(\lambda)$ *is the three-dimensional Riemann curvature tensor field on S induced by* $g_{ab}(\lambda)$, $\mathcal{R}^a{}_{bcd}(\lambda) \to 0$.

[2] See, e.g., MISNER, THORNE, and WHEELER [4], p. 131.

A proof is given in considerable detail in [3]. Here we simply indicate the structure of the argument. It proceeds in two stages. Suppose that for each λ, $\nabla_a(\lambda)$ is the unique derivative operator (or affine connection) on M compatible with $g_{ab}(\lambda)$. Further suppose that ρ is taken to be the scalar field $T^{ab}t_a t_b$. First one shows that there must exist a derivative operator ∇_a on M such that $\nabla_a(\lambda) \to \nabla_a$,[3] and such that the structure $(M, t_a, h^{ab}, \nabla_a, \rho)$ satisfies the conditions:

Compatibility	$\nabla_a t_b = 0 = \nabla_a h^{bc}$,
Orthogonality	$t_a h^{ab} = 0$,
Poisson's Equation	$R_{ab} = 4\pi\rho t_a t_b$,
Integrability	$R^{[a}{}_{(b}{}^{c]}{}_{d)} = 0$.

These conditions characterize a kind of generalized Newtonian spacetime structure introduced by Künzle in [1] and [2]. Thus the first stage of the argument is of interest in its own right. It makes precise one sense in which a generalized version of Newtonian gravitational theory is the "classical limit" of general relativity.[4] In particular it shows that Poisson's equation is a limiting form of Einstein's equation.

The second stage of the argument makes the connection with spatial geometry. It certainly need not be the case that the four-dimensional Riemann tensor field $R^a{}_{bcd}$ on M determined by ∇_a vanishes. But Poisson's equation (in the presence of the Compatibility and Orthogonality conditions) *does* imply that the three-dimensional Riemann field $\mathcal{R}^a{}_{bcd}$ induced on any spacelike hypersurface S does so. (The claim is that *space*, not *spacetime*, is necessarily flat in the "classical limit" of general relativity.) Once the dust clears, this second stage of the argument turns on a simple linear algebraic fact. In three dimensions (but not higher) the Ricci tensor field cannot vanish without the full Riemann tensor field doing so as well.

One has $\mathcal{R}^a{}_{bcd} = 0$; and $\mathcal{R}^a{}_{bcd}(\lambda) \to \mathcal{R}^a{}_{bcd}$ follows easily from $\nabla_a(\lambda) \to \nabla_a$. So the proposition follows.

Edmund Whittaker once said that "gravitation simply represents a continual effort of the universe to straighten itself out". I have tried to show that at least in the limiting Newtonian context that straightening process is so complete as to rule out any spatial curvature whatsoever.

---

[3] The condition $\nabla_a(\lambda) \to \nabla_a$ can be taken to mean that for any smooth vector field η^a on M, $\nabla_a(\lambda)\eta^b \to \nabla_a \eta^b$. See [3] for a detailed discussion of limit relations between tensor fields.

[4] The argument in [3] is a variant of that given by Künzle in [2].

References

[1] KÜNZLE, H., 1972, *Galilei and Lorentz structures on space-time: Comparison of the corresponding geometry and physics*, Ann. Inst. Henri Poincaré 17, p. 337.

[2] KÜNZLE, H., 1976, *Covariant Newtonian limit of Lorentz space-times*, General Relativity and Gravitation 7, p. 445.

[3] MALAMENT, D., *Newtonian gravity, limits, and the geometry of space*, forthcoming in: Pittsburgh Studies in the Philosophy of Science.

[4] MISNER, C., THORNE, K. and WHEELER, J., 1973, *Gravitation* (W. H. Freeman, San Francisco, CA).

Barcan Marcus et al., eds., Logic, Methodology and Philosophy of Science VII

CONCEPTUAL REFORM IN SCIENTIFIC REVOLUTIONS

ROBERTO TORRETTI

Univ. de Puerto Rico, Fac. de Humanidades, Rio Piedras, PR 00931, USA

Ἀλλὰ χρόνωι ζητοῦντες ἐφευρίσκουσιν ἄμεινον

I shall speak about an aspect of scientific revolutions which, though duly noted by Thomas S. Kuhn in his famous essay (Kuhn 1962, p. 88; cf. Kuhn 1964), has not, in my view, been fully appreciated by him, nor by his critics and successors. For reasons that have to do partly with my own limitations, but also with the matter at hand, I shall restrict my comments to and draw my examples from major revolutions in fundamental physics, by which I mean historical processes that have brought about a change in the very concepts in terms of which the phenomena of motion and the states of physical systems are described. The aspect of these processes that I wish to bring to your attention is the role played in them by direct argumentative criticism of the concepts that are being transformed or replaced. I believe that such a discursive or "dialectical" criticism of concepts has contributed significantly in several cases to precipitate the development of a new conceptual system from the generally accepted one, and has provided good reasons for giving up the latter. Whether conceptual criticism has played a comparable role in other branches of science and in lesser revolutions is an interesting question which I shall leave open.

It is evident that such radical changes as I wish to consider here, involving the basic ingredients of the physicist's rational reconstruction of nature, cannot occur incessantly, or else the daily labors of scientific inquiry would lack a clear direction. However, it is a remarkable — and not yet wholly assimilated — fact of contemporary history that no less than two — and I would rather say three — major revolutions of this kind took place in the first three decades of the 20th century. (I refer to the advent of Special Relativity in 1905, General Relativity in 1915, and Quantum Mechanics in 1925.) The proximity of such events makes it very hard for us to believe that the goal of physics is the accurate representation of a ready-made transcendent truth; unless we are also willing to endorse the sceptical conclusion that we can never tell how far we are from reaching that goal or how well we are progressing towards it. For if each revolutionary conceptual system of physics is liable to be swept away by the next one, we cannot even anticipate in what *terms* transcendent truth may be accurately represented. But the repetition of major scientific revolutions raises a difficult philosophical problem even for those of us who do not indulge in the fantasies of realism. Even if we are willing to appraise the advance of science purely from within, in the light of its own past and prospective development, it might seem impossible to draw a valid epistemic comparison between alternative conceptual systems and thus to ascertain the progress in knowledge brought about by a major scientific revolution. The reason for this seeming impossibility can be briefly stated as follows: Factual observation, which has hitherto been acknowledged as the court of last appeal for the settlement of scientific disputes, cannot be called upon to decide between two conceptual systems if these systems are involved in the very description of the facts observed; such, indeed, must be the case when the concepts in question include the basic categories of kinematics and other fundamental predicates of physical systems.

Immanuel Kant was probably the first philosopher who, to counter the onslaught of modern sensationism, uncompromisingly held that we need concepts and an intellectual framework even to have an experience. "Anschauungen ohne Begriffe sind blind" — he said — "sense awareness without concepts is blind". At any rate, we may remark, it is altogether dumb, for in order to be able to *say* what you are sensing you must sense it *as* something, i.e. you must, in Kant's words, subsume the particular intuition under a universal concept. Kant, however, did not run up against the difficulty I mentioned, because he believed that all concepts we might ever resort to for "spelling out sense appearances in order to read them as experience" must fall under a fixed set of "categories" of the human understanding, which moreover, in their application to the fixed "forms" of human sense awareness — namely, Euclidean space and Newtonian time — yield a set of "principles" — in effect, the quintessential assumptions of classical physics — to which Kant maintained we are invariably committed by the eternal nature of human reason. Thus, in Kant's view, conceptual change can never take place at the fundamental level at which we saw the aforementioned difficulty arise. The hard core of Newtonian kinematics and dynamics as expounded, say, in Kant's *Metaphysische Anfangsgründe der Naturwissenschaft*, was there to stay. The appropriateness of new scientific concepts and the validity of any new hypotheses involving them could always by judged in the light of experience ordered by the "categories" in accordance with their attending "principles". As we all know, not one of the Kantian principles has survived the revolutions of early 20th century physics. Contemporary science is in no wise committed to Euclidean geometry and Newtonian chronometry, to the conservation of massive matter and instantaneous distant interaction, to strict causal determinism and the continuity of intensive quantities. One could, indeed, still vindicate Kant's approach by giving up the specifics of his categorial framework while retaining its more general, as yet unquestioned features. But such an attempt must raise at least two doubts: Does not the Kantian framework, when purged of its Newtonian features, become too abstract to be of much use by itself — that is, without any adventitious complements and qualifications — in the constitution of experience? And, if it is still sufficiently rich to be useful, what assurance is there that it will not be swept aside by a forthcoming conceptual revolution?

Anyway, this is not the time to dwell on a possible revival of Kantianism. In the context of the present paper Kant was to be remembered only for having first realized the function of the basic conceptual structure of experience and having prepared us, by his analyses, to grasp the dramatic significance of its mutability. One may indeed conjecture that the epistemic implications of radical conceptual change in physics were not appreciated sooner, directly in the wake of Relativity and the Quantum, due to a general mistrust of the Kantian approach caused by the overpowering influence of logical empiricism. (This conjecture probably holds, at any rate, for the academic establishment in the United States.) Writers of that philosophical persuasion — in particular, Hans Reichenbach — often cited Relativity as material proof that science owed its cognitive content to observation alone, and that the non-empirical framework of scientific description, far from being the manifestation of unchanging Reason, was freely agreed upon as a matter of convenience. The alleged duality of observed facts and stipulated conventions showed up during the final, maturer stage of logical empiricism in the notorious distinction between observation and theoretical terms of a scientific vocabulary, which served, among other purposes, to trivialize and thus effectively to sidestep the issue of conceptual change in physics. If a term is observational it must be possible, "under suitable circumstances, to decide by means of direct observation whether [it] does or does not apply to a given situation" (HEMPEL 1965, p. 178). Theoretical terms, on the other hand, are those that do not meet this requirement. A theoretical term obtains its full physical meaning by "partial interpretation" in the observational vocabulary, i.e. by the stipulation that certain sentences in which it occurs are true if and only if certain other sentences in which none but observational terms occur are true. Theoretical terms were naturally supposed to include such expressions as *rest mass, proper time, spacetime curvature, state vector,* which have been the harbingers of conceptual change in 20th century physics. Although, as far as I can tell, nobody has claimed that observational words are fixed forever in form or meaning, it was understood that they remain undisturbed by even the most drastic changes in the theoretical vocabulary. Indeed, why should anyone wish to modify the scope of terms that are furnished, as they stand, with their own infallible decision criteria? The permanence of the observational vocabulary in times of scientific revolution would then ensure, through the partial interpretation of the theoretical words, the possibility of comparing the statements of successive theories among themselves and with the facts of observation.

It is now generally agreed that such a division of scientific language into observational and theoretical terms is untenable. There can be no decidable empirical predicates, no set of terms under which phenomena, merely by being watched, obligingly classify themselves. Moreover, the supposition that the peculiar vocabulary of a physical theory obtains its meaning by "partial interpretation" in terms of such ordinary words as would normally pass for "observational" clashes with one of the characteristic tendencies of modern physical science. From its beginnings in the 17th century, its practitioners have been wary of common sense notions and common sense judgments, and have admitted ordinary usage as a welcome auxiliary for the description of their field of study only under the condition that it should ultimately submit to the jurisdiction and corrective control of scientific discourse, couched in the accepted artificial terminology. No shared set of "observational terms" can therefore bridge the gap between different systems of fundamental physical concepts.

Thus it is understandable that the same authors — namely Paul K. Feyerabend and Norwood Russell Hanson — who first fought the distinction between observational and theoretical scientific terms, should also have been the first to claim that the several basic conceptual systems of physics were mutually incomparable — or "incommensurable", as it became fashionable to say. There is a sense in which they were doubtless right, for such systems, in order to do their job, must be somehow self-contained and autonomous, in the manner of a Kantian categorial framework. And yet the recent history of physics, in spite of the great changes it has gone through, does not exhibit such deep chasms as the word "incommensurable" suggests. If Relativity and Quantum Theory were wholly disconnected, in their conceptual set-up, from Classical Mechanics and Electrodynamics, why should physicists find it necessary to instruct their students in the latter in order that they gain access to the former? Note that it is primarily the *concepts* of the classical theories which must be mastered in order to make sense of their successors. In other words, the student is taught to analyse experimental situations in the manner of classical physics so that he may learn to see them in a different manner. The seemingly paradoxical mixture of continuity and discontinuity in the history of physics, the succession of independent, mutually exclusive intellectual systems that nevertheless coalesce to form a living unity, becomes comprehensible and even natural as soon as one considers that each conceptual revolution in modern physics has been carried out by men deeply at home in the manner of thinking they have eventually abandoned, that their innovation arose from their perplexities, that each new system, being born, so to speak, in the old and out of self-criticism by its supporters, does not only cancel but also preserves its predecessor, in a way that varies in each case and therefore merits careful study, but which anyhow explains the persistent use of the old mode of thought as a preparation for the new one. When internal criticism leads to the replacement of a conceptual system by another, the bond which is thereby established between them can also serve to join the second system, through the first, to the thought-patterns of everyday life, from which the successive intellectual frameworks of physics have become increasingly divorced. More significant perhaps from the perspective we have chosen is the fact that when a new mode of thought issues from conceptual reform the problem raised by its real or alleged incommensurability with its predecessor is automatically solved. For there can be no question of *choosing* between the old and the new if the very existence of the latter is predicated on a previous acknowledgement of the failings of the former. If the old becomes disqualified by the same exercise in self-criticism that finally gives rise to the new, a comparison between the rival systems is not really called forth — indeed, the birth of one of them is the other's death.

A neat example of theory dislogment through conceptual criticism can be found in the First Day of Galileo's *Dialogo sopra i due massimi sistemi del mondo.* As you well know, Aristotle's cosmology heavily depends on his doctrine about the natural motion of the elements. Being simple, elements must move simply, unless of course they are compelled by an external agent to move otherwise. Aristotle recognizes two kinds of simple local motion, corresponding to the two varieties of simple lines out of which all trajectories are compounded, namely the straight and the circular. Since the four known elements, earth, water, air and fire, move naturally in straight lines to and from a particular point, Aristotle concludes that there must exist a fifth element that naturally moves in circles about that same point (*De Caelo*, I, ii–iii; in particular, 268b11ff., 269a2ff., 270b27ff.). This element is the material of which the heavens are made and the said point is therefore the center of the world. This is the ground for Aristotle's separation of celestial and terrestrial physics, and indeed, as Galileo's spokesman Salviati says, it is "the cornerstone, basis and foundation of the entire structure of the Aristotelian universe" (GALILEO, EN, 7, 42). Now, even if we grant the premises, Aristotle's conclusion does not follow, for, as Galileo's Sagredo is quick to note, "if straight motion is simple with the simplicity of the straight line, and if simple motion is natural, then it remains so when made in any direction whatever; to wit, upward, downward, backward, forward, to the right, to the left; and if any other way can be imagined, provided only that it is straight, it will be suitable for some simple natural body." (EN, 7, 40.) Similarly, any circular motion is simple, no matter what the center about which it turns. "In the physical universe (*nell'università della natura*) there can be a thousand circular motions, and consequently a thousand centers", defining "a thousand motions upward and downward" (EN, 7, 40). Salviati goes even further: "Straight motion being by nature infinite (because a straight line is infinite and indeterminate), it is impossible that anything should have by nature the principle of moving in a straight line; or, in other words, toward a place where it is impossible to arrive, there being no finite end. For nature, as Aristotle well says himself, never undertakes to do that which cannot be done". (EN, 7, 43.) Thus, "the most that can be said for straight motion is that is is assigned by nature to its bodies (and their parts) whenever these are to be found outside their proper places, arranged badly, and are therefore in need of being restored to their natural state by the shortest path" (EN, 7, 56); but in a well-arranged world only circular motion, about multiple centers, is the proper natural local motion of natural bodies. Although the Copernican physics that Galileo was reaching for was eventually founded on the primacy of straight, not circular, motion, the Aristotelian physics and cosmology could not survive the internal criticisms voiced by Sagredo and Salviati at these and other places of the *Dialogo*. For, as the latter remarks, "whenever defects are seen in the foundations, it is reasonable to doubt everything else that is built upon them" (EN, 7, 42). No wonder, then, that the publication of Galileo's book in 1632 had such a devastating effect on Aristotelianism.

Perhaps the clearest instance of conceptual criticism leading to a scientific revolution is Einstein's modification of the classical concept of time in §1 of his paper "Zur Elektrodynamik bewegter Körper". To understand him properly we should bear in mind that the kinematics in which he was trained in the late 19th century was no longer that of Newton's *Principia*, supposedly based on the unapproachable notions of absolute time and space, but rather the revised critical version of it proposed by Carl Neumann in his inaugural lecture of 1869, "Ueber die Principien der Galilei–Newton'schen Theorie", and perfected in the 1880's by men like James Thomson and Ludwig Lange. Neumann and his followers developed the concept of an inertial frame of reference, which is Einstein's starting point. In fact, Lange's definition of an inertial frame — which, by the way, is equivalent to Thomson's — is much more approriate to Einstein's needs than the one that he himself, somewhat carelessly, gives. (As you will recall, Einstein characterizes his "ruhende System" as "ein Koordinatensystem ... in welchem die Newtonschen mechanischen Gleichungen gelten" [EINSTEIN 1905b, p. 892], a condition blatantly at odds with the subsequent development of his paper.) Lange defines an "inertial system" as a frame of reference in whose relative space three given free particles projected from a point in non-collinear directions move along straight lines. Following Neumann, Lange also defines an "inertial time scale", i.e. a time coordinate function adapted to such an inertial frame, as follows: A given free particle moving in the frame's space traverses equal distances in equal times (measured by the scale in question). Relatively to an inertial frame furnished with an inertial time scale, one can meaningfully assert the Principle of Inertia as an empirically testable law of nature: Any other free particle — besides those used as standards in the foregoing definitions — travels with constant velocity (unless it happens to be at rest in the frame). What apparently no one realized until Einstein made it obvious is that the Neumann–Lange definition of an inertial time scale is hopelessly ambiguous. If t is such a time coordinate function adapted to an inertial frame F, and x, y and z are Cartesian functions coordinates for the relative space of F, then any linear real-valued function $t' = at + bx + cy + dz + k$ is also an inertial time scale adapted to F. Einstein overcame this ambiguity with his famous definition of time by means of radar signals emitted from a source at rest in the chosen inertial frame. This yields a time coordinate function unique up to the choice of origin and unit: the Einstein time coordinate of the frame. Relatively to an inertial frame furnished with Einstein time one can meaningfully assert the Principle of the Constancy of the Velocity of Light as an empirically testable law of nature: Any light signal — besides those used as standards in the foregoing definition — travels with the same constant speed *in vacuo*, regardless of the state of motion of its source. Einstein's Principle of Relativity says that the laws of physics take the same form when referred to any kinematic system consisting of Einstein time and Cartesian space coordinates adapted to an arbitrary inertial frame. The joint assertion of the Relativity and the Constancy of Light Velocity Principles entails that any two such kinematic coordinate systems are related to each other by a homogeneous or inhomogeneous Lorentz transformation. Of the many well-known revolutionary implications of this result I need mention only one: two Einstein time coordinate functions adapted to inertial frames in relative motion with respect to each other do not determine the same universal time order of events. This alone spells the downfall of Newtonian physics.

The example I have just sketched suggests a few remarks of a more general nature. In the first place, let me recall that, even though the ambiguity of the Neumann–Lange definition of an inertial time scale may look like a major conceptual shortcoming, it was of no practical consequence before the advent of fast particles and high-precision optics shortly before Einstein. For, as Eddington showed some sixty years ago, under the assumptions of Special Relativity the Einstein time coordinate of an inertial frame virtually agrees with that defined by the fairly obvious method of very slow clock transport over that frame (EDDINGTON 1924, p. 15), and two such time coordinates adapted to two inertial frames will not differ significantly over short distances if the frames move past each other at a speed much less than that of light. This may help us understand why Einstein's criticism came when it did. Generally speaking, even if a conceptual system of physics has hidden or obvious defects, physicists will not normally criticize them out of a craze for intellectual perfection, but only when a conceptual improvement is required by the praxis of research. For it is concepts *in use*, i.e. insofar as they are involved in the design and interpretation of experiments, that form the living tissue of physical thought.

In the second place, it is worth noting that the ambiguity of Neumann–Lange inertial time can be corrected, without giving up the substance of Newtonian theory, by denying that the speed of signal propagation has an upper bound. If there is no such an upper bound, then, under the remaining assumptions of Special Relativity, the time defined by infinitely slow clock transport over an inertial frame will be the same for all such frames. Coordinate systems that consist of this time coordinate and Cartesian space coordinates adapted to different inertial frames are mutually related by so-called Galilei transformations, that preserve the form of the Newtonian laws. It is now common, therefore, to include in formal statements of Newtonian mechanics a postulate to the effect that there is no uppermost bound to signal velocities or that the symmetry group of nature is the Galilei group. Such postulates, indeed, did not occur to anyone before Einstein's work was published, and they somehow involve a reformulation of Newtonian mechanics within the relativistic mode of thought. As a matter of fact, such postulates are testable and thus provide a means of experimental comparison — subject, of course, to the categorial framework of Relativity — between Newtonian and relativistic laws. This illustrates a common effect of conceptual criticism, whereby the criticized theory is not immediately discarded, but corrected in a way that makes it "commensurable" with the theory that is meant to replace it. An even better illustration is provided by Elie Cartan's restatement of Newtonian gravitational theory as a theory of curved spacetime, in which the linear connection and hence the curvature depend on the distribution of matter, and freely falling test particles describe spacetime geodesics (CARTAN 1923; cf. HAVAS 1964). In this theory, inertia and responsiveness to gravity are one and the same *de iure*, and not just *de facto*, as in Newton's original formulation. This corrects the main conceptual defect that Einstein found in the latter (see below). And of course, when Newton's theory of gravity is thus expressed in the chronogeometrical idiom of General Relativity, who would dare to suggest that it is "incommensurable" with Einstein's theory?

In the third place, I must emphasize that I do not claim that Einstein actually achieved his conception of Special Relativity through the exercise in conceptual criticism that he prefixed to his first presentation of it. To establish a link between a given mode of thought and its successor conceptual criticism need not play a role in the actual genesis of the latter. It may just as well be put forward after conceptual change has occurred, as a reason for accepting it. Indeed, in order to recover the rational continuity of the scientific tradition it is sufficient that we, its heirs and current bearers, are able to find appropriate critical arguments that bridge the gaps of conceptual revolutions; it is not necessary that those arguments should really have been made at the time the revolutions took place.

Conceptual considerations have also guided Einstein's thought along the way from Special to General Relativity. As Einstein himself told the 85th Naturforscherversammlung in Vienna in 1913, the gravitational phenomena known at the time did not warrant a modification of the extraordinarily successful Newtonian theory of gravity. What made a change imperative, at least in Einstein's eyes, was the clash between Newton's theory and Special Relativity. To set his quest for a new theory of gravity upon a definite course, Einstein seized on a conceptual difficulty that afflicted Newton's theory from its inception, though nobody seems to have been worried by it until then. If we spell out the Newtonian gravitational force on a body using Newton's law of gravity on the one hand, and Newton's Second Law of Motion on the other, we obtain an equation in which the mass of the body occurs as a factor on both sides. This explains why, though the gravitational force on different bodies — as measured by a dynamometer — can vary greatly at a given location it exerts exactly the same accelerating effect on them all. What remains unexplained, however, is why the mass of a body possesses this twofold significance, as responsiveness to gravitational attraction or gravitational "charge", and as resistance to it or inertia. Indeed, if one reflects on how the Newtonian mass or "quantity of matter" of a falling body thus masks its own presence, by undoing on one side of the gravitational equation what it does on the other, one is reminded of the notorious Lorentz–Fitzgerald conjecture, according to which the motion of a solid body across the electromagnetic ether is masked by the effect of that very motion on the intermolecular forces that hold the body together. (This analogy may have inspired the curious association, in Einstein's 1913 Vienna lecture, of the Michelson–Morley attempt to measure the relative velocity of the earth and the ether, with Baron Eötvös' experiment confirming the equality of inertia and gravitational charge — EINSTEIN 1913, p. 1255.) The relativistic reform of received ideas about inertia made it seem probable that the twofold Newtonian mass concept would fall apart. Thus, Max Planck thought it very unlikely that thermic radiation in a void cavity surrounded by reflecting walls should have weight. But then — Planck concluded — as such thermic radiation "certainly possesses inertial mass ... the generally assumed identity of inertial and ponderable mass, confirmed hitherto by all experiments, is evidently destroyed" (PLANCK 1907, p. 544). Einstein, however, based his speculations on gravity on that very identity. Persuaded as he was that "science is fully justified in assigning ... a numerical equality only after this numerical equality is reduced to an equality of the real nature of the two concepts" at issue (EINSTEIN 1956, pp. 56f.), he set out to develop a theory of gravity in which the quantitative equation between ponderable and inertial mass was not just the idealized statement of an observed coincidence, as it had been for Newton, but flowed from their conceptual identity. For uniform gravitational fields the identity of gravity and inertia is assured by the Equivalence Principle that Einstein introduced in 1907. This principle extends to all physical laws the scope of Newton's Sixth Corollary to his Laws of Motion, in the same way as Einstein's

Relativity Principle of 1905 had extended the scope of Newton's Fifth Corollary. In its original formulation the Equivalence Principle postulated the perfect physical equivalence of a reference frame at rest in a uniform gravitational field of intensity $\mathbf{g}$, with a reference frame that moves relatively to an inertial frame with constant acceleration $-\mathbf{g}$. This in turn entailed that a frame falling freely in a uniform gravitational field behaves in all like an inertial frame. But of course in the real world gravitational fields are approximately uniform only within fairly short distances and durations. To establish the identity of inertia with gravity also when the latter reacts to the non-uniform fields of real life, Einstein resorted to the geometrical interpretation of Special Relativity proposed by Hermann Minkowski, for which, at first, he had shown little sympathy. Minkowski had proved that Special Relativity in effect treats the arena of physical becoming — or spacetime — as a 4-dimensional Riemannian manifold with flat indefinite metric, in which inertial particles describe geodesic worldlines. The Equivalence Principle implies then that a test particle falling freely in a uniform gravitational field also describes a geodesic of the flat Minkowski metric. Einstein's masterstroke was to postulate that *any* freely falling test particle follows a geodesic of a suitable metric characteristic — more exactly: constitutive — of the prevailing gravitational field. Such a metric is normally not flat, but if it is assumed that it has the same signature as the Minkowski metric, it follows at once that the latter approximates it tangentially at each spacetime point. This accounts for the local success of Special Relativity. The essential identity of gravity and inertia is now secured with full generality, for inertial motion is conceived simply as free fall at a great distance from gravitational sources or in the local limit while free fall is acknowledged as the genuine motion of matter left on its own (its "natural" motion, so to speak). The geometrical view of gravity also enabled Einstein to surmount what he eventually came to see as a serious conceptual difficulty in Special Relativity. In this theory, the spacetime metric is taken for granted as a physically ungrounded structure which nevertheless fixes the worldlines of inertial matter. The metric of General Relativity plays a similar role with respect to freely falling matter, but, as befits a gravitational field, it does not lack physical sources, but depends, through the field equations, on the spacetime distribution of matter.

Einstein's development of a geometrical theory of gravity made good use of Riemann's theory of manifolds, which provides on its own an excellent illustration of a very important type of conceptual criticism leading to conceptual reform. In his inaugural lecture of 1854, "Ueber die Hypothesen, welche der Geometrie zugrunde liegen", Riemann took the received form of physical geometry to task for relying on too narrow a concept of space. He showed that Euclidean 3-space, which classical physics had unquestioningly adopted as its basic framework for the description of phenomena, is only a very special case of a vast family of structures, now known as differentiable manifolds. Even the much more restricted subfamily of Riemannian manifolds, in which a notion of curve-length is defined by means of a symmetric non-singular covariant tensor field of rank 2, furnished the mathematical physicist with a far richer range of choices than he had ever dreamt of. Riemann's criticism of standard geometry did not by itself bring about a conceptual revolution in physics, but it paved the way for it, by enabling men like Minkowski and Einstein to think freely yet strictly of alternatives to the established framework of scientific thought. Similar instances of liberating generalization are not infrequent in the history of mathematics. They provide much of the soil from which innovative physics draws its nourishment. They also supply — as in the case I mentioned of Cartan's restatement of Newtonian gravitation theory — a background against which the old becomes commensurable with the new.

The development of Special and General Relativity from 19th century physics is probably unexcelled as an example of radical conceptual innovation issuing from the past through conceptual criticism. In the history of quantum theories conceptual links are less clear. The difference is due in part, no doubt, to the fact that, while the advent of Relativity was dominated, at both its stages, by the exceptionally lucid thinking of a single man, quantum physics was brought about by several scientists who were not equally anxious for intellectual clarity and coherence, and who in the best of cases would only agree with each other — as Pauli once said of Heisenberg and himself — "as much as this is at all possible for two independently thinking persons" (Wolfgang Pauli to Hendrik Kramers, 27 July 1925; quoted by MEHRA and RECHENBERG 1982, vol. 3, p. 322). But this very fact makes the history of quantum physics all the more interesting for a study of rationality in scientific change. For rationality, which is certainly not to be had as the outcome of an algorithm for the vindication of beliefs, exists, if at all, as a collective achievement of men and women, and therefore must rest on strivings often at cross-purposes with one another. (Though here, indeed, in contrast with other harsher collective enterprises of man, differences must be settled *dià lógou*, i.e. through argumentative discourse.) A chapter of just this manyfaced life of reason is what we find, for instance, in the history of (non-relativistic) Quantum Mechanics, both at its birth through the seemingly opposed yet unwittingly convergent efforts of Heisenberg and Schrödinger, and in the succession of its so-called interpretations. From the standpoint I have taken here one ought to try to see this chapter in its connection with those that preceded it in the evolution of physics. Indeed, I surmise that a clear grasp of that connection, based on a cogent "rational reconstruction" of the transition from the classical to the quantum-theoretic mode of thought, might enable us to attain at last a shared understanding of the quantum-mechanical concepts and conceptions and to dispel the uncertainties concerning their meaning and scope. As I noted earlier, such a reconstruction need not tell the story "wie es eigentlich gewesen", as it really happened. Nevertheless, it is not an easy task and may come across some insuperable difficulties. The necessary evidence has been gathered, critically ordered and elucidated by Max JAMMER (1966, 1974) and again, in greater detail, by Jagdish MEHRA and Helmut RECHENBERGER (1982). It shows that the said transition was far less perspicuous to the agents involved than Einstein's development of Relativity — which may help explain why the true purport of Quantum Mechanics has never ceased to be a disputed question.

There is not much more I can say on the subject at this time, but by recalling a few facts I may perhaps inject some blood into my rather abstract hints. In §1 of the paper in which he proposed the hypothesis of light quanta, EINSTEIN (1905a) proved that the black-body radiation formula entailed by classical electrodynamics and statistical mechanics — now known as the Rayleigh–Jeans Law — not only failed to agree with experiment, but was inherently absurd. According to that formula the energy density of radiation emitted by a black body within a small neighborhood $d\nu$ of a given frequency is proportional to ν^2, and therefore the energy density of black-body radiation at all frequencies exceeds every assignable quantity. This showed that classical physics stood in need of radical reform, but, as is often the case with such arguments from absurdity, it gave no hint as to what to do next. So Einstein turned for a lead to the black-body radiation law derived by PLANCK (1900) from dubious theoretical considerations but confirmed thereafter by all actual measurements. Since black-body radiation, as KIRCHHOFF (1860) had shown, does not depend on the nature of the radiating body, one was free to choose any working model of the latter. Planck assumed a black body consisting of a collection of harmonic oscillators vibrating at all conceivable frequencies. To derive his radiation law he postulated that the energy $U(\nu)$ of the oscillators vibrating at any particular frequency was not a continuous, infinitely divisible magnitude, but a discrete quantity, composed of an integral number of equal finite parts. Planck then proved from classical principles that such parts or "energy elements" $e(\nu)$ depend linearly on the frequency. Symbolically: $U(\nu) = ne(\nu) = nh\nu$, where n is an integer. The proportionality factor h, with the dimension of action (energy $\times$ time), is of course Planck's constant. Einstein argued that Planck's "determination of elementary quanta is, to a certain extent, independent of the theory of 'black-body radiation' constructed by him" (EINSTEIN 1905a, §2) and was able to conclude that "monochromatic radiation of low density behaves ... as if it consisted of mutually independent energy quanta of magnitude h" (ibid., §6). The hypothesis of energy quanta was in the next few years fruitfully applied to several phenomena, notably the vexing anomaly of specific heats at low temperatures (EINSTEIN 1907, DEBYE 1912), and, with the publication of Bohr's paper "On the constitution of atoms and molecules" (1913), it became the mainstay of a quick-paced research programme on atomic structure and spectral lines. But for all its astounding experimental successes, the quantum hypothesis remained throughout this period of the so-called Old Quantum Theory (until 1925/26) a fortunate yet gratuitous guess. The Old Quantum Theory conceived the atom as a classical mechanical system that can exist in a number of different stationary states, subject to the *quantum condition* that, for each generalized coordinate q and conjugate momentum p, the integral $\int p\, dq$ over any closed curve (in phase space) is equal to an integral multiple of h. As a consequence of this, a transition from one stationary state to another could only take place instantaneously, in a mysteriously discontinuous fashion, as the atom emitted or absorbed a fixed amount of energy, characteristic of the transition in question. Such behavior was of course utterly incompatible with the classical mechanical concepts and principles employed in the characterization of the stationary states. Since the quantum condition yielded good predictions but the theory gave no reason that might make it understandable, one clung to it fiercely as to a magic incantation. Thus, when young Heisenberg resorted to half-integral multiples of h to account for the "anomalous" Zeeman effect, Pauli objected that one would then "soon have to introduce quarters and eighths as well, until finally the whole quantum theory would crumble to dust" (quoted by MEHRA and RECHENBERGER 1982, vol. 3, p. 30). In the end, even Bohr acknowledged that mixing classical mechanics with the new quantum ideas was in fact a "swindle" though one which "even if it might be a crime from a logical point of view", could still "be fruitful in tracing the secrets of nature in many situations". (Bohr to Pauli, 11 December 1924.) As difficulties piled up and it turned out that Bohr's methods would not even provide a satisfactory model of the helium atom, it became increasingly clear that a wholly new scheme for the description and explanation of atomic processes had to be developed from first principles. Pauli expected that "not only the dynamical concept of force, but also the kinematic concept of motion of the classical theory [would] have to undergo profound modifications" (Pauli to Bohr, 12 December 1924). Heisenberg's revolutionary paper "Ueber quantentheoretischer Umdeutung kinematischer und mechanischer Beziehungen" (1925) answered to just such expectations. Born greeted it as "an attempt to account for the new facts not by a more or less artificial and forced adaptation of the old familiar concepts, but by creating a new, truly appropiate conceptual system" (BORN and JORDAN 1925, p. 858). Heisenberg began by recalling that the formal rules employed by the Old Quantum Theory for the calculation of observable quantities "may be seriously criticized on the ground that they contain, as essential ingredient, relations between quantities (such as, e.g., the position and period of revolution of the electron) which apparently cannot in principle be observed; so that those rules evidently lack a perspicuous physical foundation" (HEISENBERG 1925, p. 879). To overcome this objection to the theory's conceptual stock-in-trade Heisenberg proposed to build a new "quantum-theoretical mechanics, analogous to classical mechanics, in which only relations between observable quantities occur" (ibid.). The gist of Heisenberg's proposal — as elucidated and elaborated by BORN and JORDAN (1925) — lies in the substitution of a *matrix* of time-dependent complex-valued functions for each of the time-dependent real-valued functions (generalized coordinates and momenta) employed in the classical description of a mechanical system. From such matrices a "Hamiltonian matrix" characteristic of the system is constructed on the analogy of the classical Hamiltonian function, in accordance with the rules of matrix algebra and the new matrix differential calculus introduced by Born and Jordan. A quantum-mecanical system is then fully described by a suitable number of pairs of conjugate matrices Q_j, P_j, satisfying the commutation relations $Q_jP_k - P_kQ_j = (\mathrm{i}h/2\pi)\delta_{jk}$, $Q_jQ_k - Q_kQ_j = P_jP_k - P_kP_j = 0$ (where δ_{jk} stands for the unit matrix if $j = k$, and for the zero matrix if $j \neq k$), and the system's Hamiltonian matrix H. The evolution of the system is governed by the matrix equations $\mathrm{d}Q_j/\mathrm{d}t = \partial H/\partial P_j$, $\mathrm{d}P_j/\mathrm{d}t = -\partial H/\partial Q_j$, which are of course formally identical to the canonical equations of classical mechanics. The postulated commutation relations introduce Planck's constant at a point where some proportionality factor is inevitable, but no "magic numbers" show up in the principles of the theory. Although the rationale of Heisenberg's move cannot be made clear without further considerations (such as are persuasively spelled out in §1 of the mathematical appendix to HEISENBERG 1930), the foregoing sketch suffices to show how the new mechanics, while radically changing the kinematic concepts of the old, managed to preserve its dynamical laws. Hence, as P.A.M. DIRAC (1926, p. 642) was quick to note, "*all* the information supplied by the classical theory [could] be made use of in the new theory". An even stronger link with classical mechanics is apparent in the alternative response to the difficulties of the Old Quantum Theory, independently developed in 1926 by Erwin Schrödinger and subsequently known as Wave Mechanics. Schrödinger had a deep aversion to quantum discontinuity — "this damned quantum jumping", as he once called it in Bohr's seminar (quoted by JAMMER 1966, p. 324; see also SCHRÖDINGER 1952). He was proud to offer a new approach to atom mechanics, which replaced the usual quantum condition by a different requirement that made no mention of "whole numbers" — integers do eventually turn up in his theory, but "in the same natural way as in the case of the *node-numbers* of a vibrating string" (SCHRÖDINGER 1926a, p. 361). As is well known, Quantum Mechanics took its now familiar shape after Schrödinger proved that Heisenberg's scheme and his own, for all their overt discrepancy, shared the same underlying structure and were in effect mathematically equivalent. Schrödinger's proof was all the more surprising as — in his words — "the departure from classical mechanics in the two theories seems to occur in diametrically opposed directions". Thus, while Heisenberg's theory with its arrays of discrete quantities indexed by pairs of integers had been described by BORN and JORDAN (1925, p. 879) as a "true theory of a discontinuum", "wave mechanics shows just the reverse tendency"; it is a step from classical point-mechanics towards a *continuum-theory.* In place of a process described in terms of a finite number of dependent variables occurring in a finite number of total differential equations, we have a continuous *field-like* process in configuration space, which is governed by a single *partial* equation, derived from a [variational] principle of action". (SCHRÖDINGER 1926b, p. 734.) I cannot go further into this fascinating historical juncture. I do hope, however, that Schrödinger's choice of words will make clear to what an extent he saw himself as fulfilling, not destroying, the spirit of classical physics.

Summing up: My chief contention is that in order to perceive the rationality of radical conceptual changes in fundamental physics one must view them as episodes of an intellectual history, the history of physical thought. The intellectual nature of this history precludes thoughtless turnabouts: new modes of thought stem from the old through self-criticism prompted by its internal difficulties and inherent tendencies. This is not to deny that the basic concepts of physics are, as Einstein said, "freie Erfindungen des menschlichen Geistes," free inventions of the human mind (EINSTEIN 1934, p. 180), so that there is no way of calculating what they will be like in the future. Though unpredictable, they must be grounded, or else they would not be "of the mind" (nor, properly speaking, "free"). But this is just what we find in the sources, at least within the tradition of modern mathematical physics: the authors of conceptual innovations have always or nearly always sought to motivate them carefully and to exhibit their provenance from established notions, by drawing apposite analogies, setting up correspondences, and even by retaining for the new ideas suggestive old names. The view of physics as a form of intellectual life, whose major turning-points are catalyzed by critical reflection, is not favored by the tendency, endemic among philosophers of science, to treat physical theories — in the loose, ordinary sense of the word — as informal expressions of the logical entities that such philosophers call "scientific theories" in a contrived special sense. Whether a "theory" be conceived, in the strict philosophical acceptance, as a set of propositions closed under deducibility, or, after the current fashion, as a definite Bourbaki-like structure surrounded by a vaguely characterized host of applications, such a "theory" or "theory core" is not an open-ended enterprise of thought, but a fixed, finished ideal object, with no signs of origin or seeds of change, to which history — invention, criticism, reform — can only supervene as an external accident. "Theories", in either philosophical sense, can stand beside each other like Egyptian pyramids, can have their several features outwardly set into some kind of correspondence, but cannot proceed from one another inwardly. The structuralistic, "non-statement" model of theories is indeed more fitting than the other one, insofar as it somehow makes allowance for evolution within so-called "theory nets" through the exercise of genuine scientific thought in the development of applications. But it does not seem to be of much help for understanding genetic relations between successive modes of thought embodied in distinct theories (in the ordinary sense). It is by actually rethinking the great intellectual systems of the physical world, not by boiling them down to marrowless bones, that one may come to see reason in their history.

Note

In his Salzburg lecture, Joseph Sneed announced an innovation in the structuralist view of theories. He pictured the whole of empirical knowledge at any given time as a system of conceptually heterogeneous theories, held together by so-called *links*. The entire approach hinges on the adequacy and fruitfulness of this notion of an intertheoretic link, which I understand is further explicated in a joint paper by Balzer, Moulines and Sneed, to be included in this volume. I do not doubt that, if successful, their project should result in a momentous contribution to epistemology. I am afraid, however, that Sneed's new picture, which still is built exclusively from clear-cut, static, extensional set-theoretic predicates, is too oblivious of science's genesis and growth to be of much help in understanding its rational development.

References

BOHR, N., 1913, *On the constitution of atoms and molecules*, Philos. Mag. 26, pp. 1-25, 476–502, 857–875.

BORN, M. and JORDAN, P., 1925, *Zur Quantenmechanik*, Ztschr. f. Physik 34, pp. 858–888.

CARTAN, E., 1923, *Sur les variétés a connexion affine et la théorie de la relativité généralisée*, I, Ann. Ec. Norm. Sup. 40, pp. 325–412.

DEBYE, P., 1912, *Zur Theorie der spezifischen Wärme*, Ann. Physik (4) 39, pp. 789–839.

DIRAC, P.A.M., 1926, *The fundamental equations of quantum mechanics*, Proc. Roy. Soc. (London) A 109, pp. 642–653.

EDDINGTON, A.S., 1924, *The Mathematical Theory of Relativity*, 2nd ed. (Cambridge University Press, Cambridge).

EINSTEIN, A., 1905a, *Ueber einen die Erzeugung und Verwandlung des Lichtes betreffenden heuristischen Gesichtspunkt*, Ann. Physik (4) 17, pp. 132–148.

EINSTEIN, A., 1905b, *Zur Elektrodynamik bewegter Körper*, Ann. Physik (4) 17, pp. 891–921.

EINSTEIN, A., 1907, *Die Plancksche Theorie der Strahlung und die Theorie der spezifischen Wärme*, Ann. Physik (4) 22, pp. 180–190.

EINSTEIN, A., 1913, *Zum gegenwärtigen Stand des Gravitationsproblems*, Phys. Ztschr. 14, pp. 1249–1266.

EINSTEIN, A., 1934, *Mein Weltbild* (Querido Verlag, Amsterdam).

EINSTEIN, A., 1956, *The Meaning of Relativity*, 5th ed. (Princeton University Press, Princeton).

GALILEO GALILEI (EN), *Le Opere*, Nuova ristampa della Edizione Nazionale (Barbera, Firenze, 1964–66), 20 vols.

HAVAS, P., 1964, *Four-dimensional formulations of Newtonian mechanics and their relation to the special and the general theory of relativity*, Rev. Mod. Physics 36, pp. 938–965.

HEISENBERG, W., 1925, *Ueber quantentheoretische Umdeutung kinematischer und mechanischer Beziehungen*, Ztschr. f. Physik 33, pp. 879–893.

HEISENBERG, W., 1930, *The Physical Principles of Quantum Mechanics* (Chicago University Press, Chicago).

HEMPEL, C.G., 1965, *Aspects of Scientific Explanation and other essays in the philosophy of science*, The Free Press, New York.

JAMMER, M., 1966, *The Conceptual Development of Quantum Mechanics* (McGraw-Hill, New York).

JAMMER, M., 1974, *The Philosophy of Quantum Mechanics. The interpretations of quantum mechanics in historical perspective* (Wiley, New York).

KIRCHHOFF, G., 1860, *Ueber das Verhältnis zwischen dem Emissionsvermögen und dem Absorptionsvermögen der Körper für Wärme und Licht*, Ann. Physik (2) 109, pp. 275–301.

KUHN, T.S., 1962, *The Structure of Scientific Revolutions* (Chicago University Press, Chicago).

KUHN, T.S., 1964, *A function for thought experiments*, in: L'Aventure de la Science, Mélanges Alexandre Koyré (Hermann, Paris), vol. 2, pp. 307–334.

LANGE, L., 1885, *Ueber das Beharrungsgesetz*, K. Sächs. Ak. Wiss. Leipzig, Berichte Verh. math. phys. Kl. 37, pp. 333–351.

MEHRA, J. and RECHENBERGER, H., 1982, *The Historical Development of Quantum Theory* (Springer, New York), four volumes in five.

NEUMANN, C., 1870, *Ueber die Principien der Galilei–Newton'schen Theorie* (Teubner, Leipzig).

PLANCK, M., 1900, *Zur Theorie des Gesetzes der Energieverteilung im Normalspektrum*, Verh. d. Deutsch. Phys. Ges. (2) 2, pp. 237–245.

PLANCK, M., 1907, *Zur Dynamik bewegter Systeme*, Preuss. Ak. Wiss. Sitzungsber., pp. 542–570.

SCHROEDINGER, E., 1926a, *Quantisierung als Eigenwertproblem* (Erste Mitteilung), Ann. Physik (4) 79, pp. 361–376.

SCHROEDINGER, E., 1926b, *Ueber das Verhältnis der Heisenberg–Born–Jordanschen Quantenmechanik zu der meinen*, Ann. Physik 79, pp. 734–756.

SCHROEDINGER, E., 1952, *Are there quantum jumps*? Brit. J. Phil. Sci. 3, pp. 109–123, 233–242.

9. Foundations and Philosophy of Biology

Barcan Marcus et al., eds., Logic, Methodology and Philosophy of Science VII

PHILOSOPHY OF BIOLOGY 1983: PROBLEMS AND PROSPECTS

MARJORIE GRENE

Dept. of Philosophy, Univ. of California, Davis, CA 95616, USA

Philosophy of biology is a burgeoning field; the present review paper consists chiefly in an attempt to summarize recent achievements and major lines of ongoing research, work in which biologists, philosophers and historians of science are all, and often cooperatively, involved. It is my view that interest has shifted from very general problems to much more detailed questions, in more realistic contact with the work of the biological sciences than was formerly the case. Further — and, again, of course, in my view — it is chiefly in such contexts, that is, in close interaction with the developing problems and discoveries of other disciplines, that philosophical 'research' acquires content and significance. It therefore seems appropriate to inform an international audience of the present state of the art, as perceived by one of its practitioners. My perception, it should be admitted at the outset, is (like all perception) at the same time an interpretation. I believe — and I shall return to this point in closing — that in our field as in others an overabstract approach to the philosophy of science is at last giving way to a more promising historically *and* realistically oriented *Problemstellung*. The literature I stress is accordingly biased in the direction I find promising: the direction that, I have to say, *is* promising. Even that literature is, as I shall have occasion to remark, still in part under the ban of the older tradition; but that, too, will pass.

Two areas call for special mention. First, there are philosophical debates about evolutionary biology (*not* in relation to the creationism furore, which I am ignoring. It raises political and historical, but not philosophical issues: see Lewontin, 1983c). Second, there are problems connected with systematics. I shall deal with these before turning to some more general questions that have been, or are, of interest in this field.

I. Evolutionary biology

Within this area, two directions of recent and ongoing work may be distinguished: on the one hand, conceptual analysis of what is still the orthodox view, that is, neo-Darwinism or the synthetic theory, and on the other, current debates within evolutionary biology that have clear methodological and philosophical implications.

A. *Conceptual analysis of Darwinian theory*

The synthetic theory *is* the theory of natural selection. But what is the status of 'natural selection' as an explanatory principle? Both evolutionists and philosophers still debate this question. A clear, if schematic, answer to it is furnished in Robert Brandon's "Structural description of evolutionary theory" (BRANDON, 1981b). Natural selection, he argues, serves as an organizing principle for evolutionary biology, itself without empirical content, but dependent on three empirical propositions: (1) that "biological entities are chance set-ups with respect to reproduction", (2) that "some biological entities differ in their adaptedness to their common environment, this difference having its basis in differences in some traits of the entities", and (3) that "adaptedness values are to a degree heritable" (BRANDON, 1981b, pp. 437–438). Of the first two principles he writes: "When and where these presuppositions are satisfied the principle of natural selection is applicable to the relevant entities, that is, ... these differences in adaptedness will result in actual differences in fitness" (p. 437). The third principle is needed if natural selection is to result in evolutionary change. Not all writers follow Brandon's terminology or the major lines of his analysis (see e.g. M.B. WILLIAMS, 1970, 1973) but at the least it provides a good starting point for the discussion of the basic concepts and basic issues.

These basic issues may be summarized under four headings.

1. The theory of natural selection is grounded on a characteristic blending of three explanatory factors: chance, cause (of the standard, deterministic or stochastic-deterministic sort) and teleology. The question is, how these three factors relate to one another. The theory is characteristically a two-step theory: variation is random, i.e. occurring without relation to the needs of the entity to be selected; but, in relation to a given environment, inherited variations have causal efficacy in determining (probabilistically!) the make-up of future generations (WRIGHT, 1967; for the role of chance, see WIMSATT, 1980a; MAYR, 1982a, 1984). Teleology

comes in, thirdly, in the relative adaptedness (to use Brandon's terminology) of different traits, some of which, given a common environment, are more likely than others to issue in relative reproductive success for their bearers. Thus the overall structure of Darwinian theory seems by now relatively clear.

2. Work remains to be done, however, in clarifying — and in unifying usage with respect to — the central concepts of 'fitness' and 'adaptation'. 'Fitness' may be used to denote actual (relative) reproductive success, as by BRANDON (1981b), or to designate the propensity to such success (MILLS and BEATTY, 1979; cf. ROSENBERG, 1982; BURIAN, 1983). 'Adaptation' is even more equivocal. The term is used by ethologists and ecologists often without relation to evolution; at the other extreme it is sometimes defined only in terms of production by natural selection in evolutionary time (GOULD and VRBA, 1982). For the evolutionary use of the concept, Lewontin's discussion is a classic (LEWONTIN, 1978, complete in LEWONTIN, 1983b). The extensive literature on evolutionarily stable strategies and on optimization theory should also be consulted here (LEWONTIN, 1979a; MAYNARD SMITH, 1979; BEATTY, 1980: TUOMI *et al.*, 1983).

3. Analyses of the structure of neo-Darwinian theory also impinge on some broader questions about biological explanation that should be mentioned here. As in the Brandon analysis already cited, a number of writers see natural selection as a guiding principle, so that the theory serves as a 'hypertheory' (WASSERMANN, 1981) or a 'generic theory' (TUOMI, 1981; TUOMI and HAUKIOJA, 1979; cf. BRADIE and GROMKO, 1981). Writers such as M.B. Williams, working in the tradition of hypothetico-deductivism, disagree (see WILLIAMS, 1982). Such writers are concerned, as Professor Williams pointed out at the discussion in Salzburg, with the problems of prediction and falsification for natural selection theory — an unprofitable issue, in my view, not only because of the plausibility of the 'hypertheory' approach (which seems to respond much better to the actual work of biologists than does the alternative), but also because of the inadequacy of formal analysis as the primary tool of philosophy of science, and in particular the triviality of the 'prediction' question. (Of course scientists do make predictions on the ground of their hypotheses; but prediction is neither the aim of science nor a criterion for calling a discourse or procedure 'scientific'.) Much more promising in connection with the analysis of evolutionary theory, I believe, is reflection on the nature and role of *models* in biology (LEVINS, 1966, 1975; WIMSATT, 1980b). The papers presented by Richardson and Lloyd at the Congress provide useful starting points in this direction, as does Lloyd's recent paper on Darwin (RICHARD-

SON, 1983; LLOYD, 1983a, b). Both authors furnish excellent materials for the study of biological models; in Lloyd's case, unfortunately, her arguments are crippled by her reliance on the so-called semantic theory of theories, a singular red herring, which tries to side-step the so-called 'received' view — surely the *formerly* received view — without abandoning its presuppositions (cf. BEATTY, 1981). (There are more promising alternatives; see BROWN, 1979, POLANYI, 1969, GRENE, 1977 and III, G below.) Detailed work on particular biological models can prove illuminating, however, whatever its methodological presuppositions; see for example the beautiful piece by MITCHELL and WILLIAMS (1979) on ecological 'strategy models' in relation to Darwinian theory.

4. In addition, the Mitchell and Williams paper suggests a fourth point that should be included here: the question of the relation between evolutionary theory and ecology, which is by no means as simple as it might appear, and also (see B5 below) between evolutionary theory and the study of development. (For a historical perspective on the first, see KIMLER, 1983.) As evolutionists themselves are rethinking the relation of their discipline to population genetics (LEWONTIN, 1974; MAYR, 1982a, 1983), so its relation to other fields in biology appears more complex than it formerly seemed. Both historically and conceptually, the interactions of specialized biological perspectives with evolutionary theory merit further study.

B. *Current controversies in evolutionary theory*

Recent controversies within evolutionary theory also raise some methodological and ontological issues of interest to philosophers of science. Five such lines of discussion may be noted. Three of these seem to be family quarrels *within* an expanded Darwinian tradition.

1. The theory of punctuated equilibrium has been heralded as a challenge to Darwinian gradualism; so it is, but not to natural selection (ELDREDGE and GOULD, 1972; GOULD and ELDREDGE, 1977; GOULD, 1980; 1982a). Both phyletic gradualism and relatively rapid speciation after long periods of stasis are perfectly amenable to explanation by selection. The hypothesis of punctuated equilibrium was put forward by paleontologists chiefly as a defense of their discipline: no need to keep apologizing, as Darwin and his heirs had done, for 'gaps in the fossil record'. Maybe the record itself can be read in harmony with evolutionary, and even Darwinian, principles! Where and how often punctuated equilibirum occurs as against a gradual development of new forms seems to be an empirical question. (See for instance an exchange of letters in *Science:* SCHOPF,

HOFFMAN and GOULD, 1983; also STEBBINS and AYALA, 1981; STEBBINS, 1982). The sundering of micro- from macroevolution may perhaps raise a more fundamental challenge (STANLEY, 1979; GOULD, 1982b), although the more conservative synthetic theorists see this, too, as an empirical question (STEBBINS and AYALA, 1981; AYALA, 1985).

2. The critique of 'adaptationsim', again, has been put forward, notably by GOULD and LEWONTIN (1979; see also LEWONTIN, 1979b) in protest, not simply against natural selection as such, but against genic selectionism run rampant, that is, against a program that interprets every trait of every organism in isolation as an adaptation produced by genes for their (i.e. the genes') 'survival'. It seems to be the cryptoatomism underlying such hypotheses and the pseudoteleology irresponsibly used to support them ('just-so stories') that these critics object to. Ernst Mayr, however, who himself defends an adaptationist program, argues that it is the atomism and naive determinism, rather than the misplaced teleology, of such theories that Gould and Lewontin were objecting to (MAYR, 1983). Whichever interpretation is correct, their argument is strengthened by the fact that some of the responses to it, and to other evolutionary heresies, may in turn be found to exemplify the very excesses they deplored (CHARLESWORTH *et al.*, 1982, MICHOD, 1981). In close connection with the morass of meanings of 'adaptation', there is still work here for philosophers to do.

3. *The units of selection controversy.* Classically, Darwinism describes the multiplication of organisms slightly better suited than their conspecifics to leave descendants in a given environment. As many evolutionists have insisted, no matter how much genetics and biochemistry one knows, one must remember: it is *phenotypes* that are selected. Yet no organism goes on indefinitely; what appears to be multiplied, more or less, is not organisms, but *genes.* Not that genes go on forever, either; but they are the least, and ultimate, replicators. Indeed, from the point of view of what Gould calls the hardened synthesis (GOULD, 1983), evolution *was* differential gene frequencies. Against such a reductionist view, however, a theory was put forward by Wynne–Edwards in 1962 to the effect that in some cases it is not genes, but even phenotypes, that are selected, but *groups* (notably, for instance, in a bird's warning call that may endanger itself but help the group; WYNNE–EDWARDS, 1963). In a very influential book, G.C. Williams (WILLIAMS, 1966) argued that such "group selection" is probably an artefact and is in any case non-parsimonious and therefore to be avoided. Only individuals, and indeed, ultimately, only genes, the basic biological individuals, are selected. The theory of kin selection and inclusive fitness (HAMILTON, 1964, 1981), moreover, permitted Darwinian evolutionists to

assimilate phenomena like "altruism" (as they metaphorically and misleadingly call it) through a standard interpretation in terms of individual (genic) selection. With E.O. WILSON's *Sociobiology* (1975) and DAWKINS' *The Selfish Gene* (1976), the triumph of atomizing, genic selection seemed to many (and still seems to some) to be complete. At the same time, however, G.C. Williams had admitted one case of demic (= group) selection, the *t*-allele in the house mouse (LEWONTIN and DUNN, 1960). Further, LEWONTIN's paper on "Units of Selection" (1970) and experimental work by M. WADE (see review article, *Quart. Rev. Biol.* 1978) and more recently by D.S. Wilson and others, has led to a more careful weighing of the alternatives (WILSON, D.S., 1980; WILSON, D.S. and R.K. COLWELL, 1981). Recently SOBER and LEWONTIN (1982) have argued, conclusively, in my view, that if the original sweeping concept of group selection was probably an artefact, the cherished notion of *genic* selection is itself artefactual, and therefore fails as a guide to what goes on in evolutionary processes. (For a critique of some features of their argument, see ROSENBERG, 1983, but also their reply, SOBER and LEWONTIN, 1983. His criticism does not, I believe, defeat their case.) Remarks in a review by LEWONTIN (1982) on Campbell's contribution to MILKMAN (1982) also stress, beyond the genotype, the complex organization of the genome now acknowledged by geneticists (see also HUNKAPILLER *et al.*, 1982).

If genic selectionism is abandoned, what *are* the units of evolution? Starting with Lewontin's now classic 1970 paper (LEWONTIN, 1970), there has been a good deal of philosophical as well as scientific discussion of this issue. Unfortunately, I cannot go into this debate here, but may mention three approaches. SOBER (1981) defines group selection as acting ... "on a set of groups if, and only if, there is a *force* impinging on those groups which makes it the case that for each group, there is some property of the group which determines one component of the fitness of every member of the group", (SOBER, 1981, my ital.) Note the emphasis on causal force and on properties. (I shall return to these concepts below.) The former also emerges as central in the Brandon account from which I began this essay. There is also a problem here, as David Hull points out (personal communication), of establishing what (kinds of) properties are genuinely *group* properties. WIMSATT (1980b) raises instead the question of what he calls 'entification'; when does a group qualify as a *thing*, and so as a unit of selection? A third approach to this problem is used by M.B. Williams in her contribution to the Congress (M.B. WILLIAMS, 1983). The point here is that an expanded Darwinian theory — or perhaps an 'essentially' Darwinian theory? — can reasonably leave open the question, at what level natural selection occurs. (Brandon wants to distinguish further, not very effectively, I find, between 'units' and 'levels' of selection (BRANDON, 1982).) Perhaps the clearest and most authoritative statement of the outcome of this debate so far is Hull's "Units of Evolution" paper (HULL, 1981). From all these sources, it is clear that the question of the units of selection is posed within a Darwinian framework. (See the anthology on units of selection edited by BURIAN and BRANDON, 1984.)

Two other current controversies must be mentioned, finally, which *do* challenge the neo-Darwinian orthodoxy.

4. *The neutral mutation theory.* This view is described by its supporters (e.g. KIMURA, 1983b) as a theory of non-Darwinian evolution. It has been argued (FITCH, 1982) that 'non-Darwinian' does not mean 'anti-Darwinian' and that this view, too, can be assimilated to the synthetic theory. This seems questionable. If the major factor in evolution were the long term retention of mutations without any effect of natural selection, surely the synthetic theory would be mistaken (AYALA, 1974; but see KIMURA, 1983a, where the "neutral" view seems softened). Perhaps there is room for conceptual analysis of these issues, and certainly of their history.

5. *Endoetiological theories.* Finally, a literature proposing various *internal* mechanisms of evolution has been making its appearance, whether in analogy to structuralism (WEBSTER and GOODWIN, 1982) or to Prigoginian non-equilibrium thermodynamics (WILEY and BROOKS, 1982, 1983). Any version of Darwinism, of course, must insist that evolution is 'opportunistic', dependent basically on random mutation and on the demands of the environment. If the environment is stable, as a noted evolutionist remarks, nothing happens in evolution. To find, or to allege that one finds, an internal dynamic driving the process is a dramatically anti-Darwinian move. Whether 'theories' like those mentioned will prove effective may be questionable; they appear to this writer to rush rather too hastily to grand conclusions. Stuart Kauffman's work in a similar direction, however, exhibiting lines of self-organization in the patterns of ontogenesis, certainly offers a weighty addendum to the concept of selection as the motor of evolutionary change (KAUFFMAN, 1982, 1983, 1985). Along with work like Campbell's on the genome (CAMPBELL, 1982), it may serve to remind evolutionists that organization, the self-patterning of living entities, is to be taken seriously after all. Indeed, that epigenesis is the missing link in evolutionary theory has been widely acknowledged. (See e.g. RACHOOTIN and THOMSON, 1980; GOULD, 1982b.) Like other missing links, it may be difficult to fill in correctly, but work to come in this area is worth watching, especially (so far as I can tell) in connection with Kauffman's research.

II. Systematics

A. *The species concept*

A thesis proposed by GHISELIN (1974) and developed by Hull (HULL, 1976, 1978, 1980, 1981) has attracted and is still attracting much attention and controversy. (But see also anticipations in HENNIG, 1950 and WOODGER, 1952.) They propose that species taxa be considered, not as classes with members, but as individuals (wholes) with parts. Thus *Felis tigris* is simply all the tigers that have ever existed, do exist or will exist in ancestor-descendant relations; it is not characterized by any 'essential' properties or by 'similarities' between parts of the whole collection. The traditional evils of 'typology' and 'essentialism' are thus avoided, the anomaly of non-resemblance between members of the same species (as between males and females, or between various insect morphs) is evaded — and, besides, it is precisely lineage relations, and only such relations, that basically interest evolutionary theory. Thus if the proposal seems counter-commonsensical, says Hull, the 'intuitions' of common sense must be sacrificed to the demands of theory. Although the arguments of Hull's early papers was in part weak (or so it seemed to me), his "Individuality and Selection" (1980) is a most powerful statement, not only of this position on species, but of the relation between 'replicators' and the 'interactors' that also play a crucial role in the production of lineages (or 'evolvors') from sequences of replicators. Resistance to the 'species are individuals' view persists, however; the leader of the current opposition is perhaps Philip Kitcher, who is engaged in writing a book on the concept of species as sets (KITCHER, 1982b, 1984; cf. SOBER, 1982, 1984). The class interpretation is still defended also by some: e.g. R. CAPLAN (1981), KITTS (1983) and M. RUSE (in prep.). An odd aspect of this debate is that while the s-a-i-thesis was introduced as a necessary presupposition for Darwinian theory, the cladistic opposition to Darwinism (of which more in a moment) enbraces it just as vigorously, and even accuses the Darwinians of sticking to the class notion (WEBSTER and GOODWIN, 1982). The subsumption of the s-a-i-thesis under 'essentialist' theories ('essentialist', for once, in a nonpejorative sense) has recently been suggested by John Beatty (BEATTY, 1983).

B. *Foundations of taxonomy*

The past decades have also seen lively debates about the foundations of systematics and taxonomy. Three schools of taxonomists, representing (1) evolutionary taxonomy (see e.g. MAYR, 1982b), (2) numerical taxonomy or pheneticism (SNEATH and SOKAL, 1973) and (3) phylogenetic systematics or cladism (HENNIG, 1950, 1966), have put forward rival arguments for their respective positions (HULL, 1970). Pheneticism, which attempted to classify organisms by taking any and all characters into account (without weighting) seems (except perhaps in botany) to be receding; but the first and third continue, the third ever more vociferously, if not stridently. Cladism, founded on the work of HENNIG (1966), originally sought to establish phylogenetic relations through sorting out recently derived characters shared by sister groups (synapomorphies) from more primitive characters (plesiomorphies). Evolutionary taxonomy, in contrast, also considers evolutionary histories and lifestyles in judging what groups are more or less closely related. Moreover, it is willing to recognize cases of phyletic gradualism as well as splitting, as against the motto of the Hennigians: "only clades, no grades". Some cladists, shocked by such 'unscientific' procedures, have renounced the phylogenetic intent of the school's founder; as 'transformed cladists' they seek *only* to classify species by shared characters (and positive characters only, in contrast to pheneticists, who count negative characters as well) and want no part of phylogeny at all (NELSON and PLATNICK, 1981; PATTERSON, 1982). They seek only 'nature's hierarchy' — where 'hierarchy' seems to denote a network of species (not the relation of species to 'higher taxa' as in classical taxonomic hierarchies). This point is worth noting in connection with the question of the uses of 'hierarchy' in the biological literature; I shall return to it below. Whether the evolutionary and cladistic taxonomists should be at daggers drawn, as they often appear to be, is a question for further consideration. The senior evolutionary taxonomist has attempted a reasonable systhesis of the two approaches (MAYR, 1981); cladists, however, it seems, resist being reconciled with any one. The debate goes on. As a case in the politics, if not the philosophy, of science, it merits study. (For another moderate criticism of the cladists' claims, see HULL, 1979.)

A spinoff of cladism that may interest philosophers is a controversy about parsimony, in which FARRIS (1983) and SOBER (1983) take somewhat different positions.

III. General questions

A. *Reducibility*

The question, whether biology can be reduced to physics and chemistry, used to be *the* pressing problem in philosophy of biology. No longer. Even the staunchest reductivists have hedged their bets (SCHAFFNER, 1974). To all but a few archconservatives it is clear by now that a less monolithic approach to questions about relations among the sciences can prove more illuminating than the old unity-of-science view. 'Interfield theories' (DARDEN, 1983 and references) have been shown to provide complex, non-reductive relations among scientific disciplines or sub-disciplines. Other, historical, approaches also have shown up the spurious character of this 'issue'. (See G below; KITCHER, 1982a; LEVINS and LEWONTIN, 1980.)

B. *Teleology*

Again, this classic problem has lost interest. There was a spate of work on it about a decade ago, including some excellent analyses of the relation between 'functional' and 'teleological' statements or explanations (ACHINSTEIN, 1977; BRANDON, 1981a; HIRSCHMANN, 1973; MAYR, 1974; WIMSATT, 1972; L. WRIGHT, 1976). Now it seems that the place of teleological (or teleonomic) discourse in evolutionary theory has been tidily located (BRANDON, 1981b), and outside evolution functional discourse (see e.g. HORRIDGE, 1977), or even, in a limited fashion (as in ethology), teleological accounts, are routine and harmless.

C. *Hierarchy theory*

With the demise of reductionism (or near demise, see LEWONTIN, 1983a), on the other hand, the notion of hierarchical organization has become a central one — and from a number of directions. The study of interfield connections, for example, lends itself to this approach (DARDEN, 1983; BECHTEL, 1983). Problems of the origin of life (MAYNARD SMITH, 1982, part I: pp. 7–38) or of epigenesis (e.g. OSTER and ALBERCH, 1982) also raise the question of levels of organization. And a number of biologists are attempting to transform evolutionary theory and/or systematics through a resort to the concept of hierarchy (e.g. GOULD, 1982a; ARNOLD and FRISTRUP, 1982). There is a plethora of work in progress here that calls for philosophical clarification and analysis. Since this seems a fruitful field for further research, I shall present it in somewhat less cryptic fashion than most of my other points. (Not that D to G below lack importance; but D to F are so far just small clouds on the horizon, and G, while it could claim a paper, or book, of its own, harks back, in this context, to my introductory remarks.)

Let's start by taking, for example, a statement by S.J. Gould in *Science* in 1982:

> ··· I believe that the traditional Darwinian focus on individual bodies, and the attendant reductionist account of macroevolution, will be supplanted by a hierarchical approach recognizing legitimate Darwinian individuals at several levels of a structural hierarchy, including genes, bodies, demes, species, and clades. (GOULD, 1982a)

That sounds fine; but what is a "structural hierarchy", indeed, what is a hierarchy as such? A satisfactory answer is not so easy to find (GRENE, 1969). But several different meanings seem to be involved, in one way or another, in biological usage. (1) By now there is an extensive literature dealing with the question of levels of organization in biology. That is the angle from which I myself have been interested in the question of biological hierarchy. From macromolecules to cells to tissues to organs and organ systems to organisms (if not further), biological phenomena seem to get themselves organized in levels, such that lower levels impose limiting conditions, while in turn upper levels, by the very arrangement of their elements, constrain the activities of the lower levels. We may take as characteristic for the kind of constrained-constraining hierarchy involved here the notorious case of the genetic code: where organic bases arranged in a certain regular but highly improbable order constitute a message. The arrangement gives them a causal power they would not otherwise possess. Now this is a situation analogous to Aristotle's form/matter relation, if difficult to decribe exactly (see PATTEE, 1973). (2) This line of approach is complicated, however, by the fact that in taxonomy the concept "hierarchy" has quite a different use. In the taxonomic (or Linnaean) hierarchy, the lowest level is the species, then genus, then order, then class. But here, what is from an Aristotelian, form/matter point of view the lower level of organization (the genus) constitutes a higher level, while the highest level (the species) is lower. So, it seems clear, despite the smooth sequence of the Gould passage from which I started, taxonomists and hierarchy theorists cannot be using their central term in a univocal sense. Cladistic systematists, however, are at present loud in proclaiming the importance of "nature's hierarchy" and arguing that (especially Darwinian) evolutionary theory is confused, even unscientific, in its efforts to deal with this central fact of life. *Their* hierarchies are supposed to be presented unambiguously in what, partly following Hennig, they call "cladograms" (HENNIG, 1966; PATTERSON, 1982). But as now developed, a cladogram consists of units all of which are species; where has the "Linnaean hierarchy" gone, and what kind of hierarchy is it that remains? Thus what exactly (or at least less inexactly, or even less inconsistently) is meant by that term is still a thorny, and open, question. Work in progress by Hull, dealing chiefly with some recent cladistic arguments, and by a number of biologists, taking account both of PATTEE (1973) and of recent analysis of evolutionary theory (HULL, 1980, may help to clarify this situation (HULL, 1983; SALTHE, 1985; ELDREDGE and SALTHE, 1985; VRBA and ELDREDGE, 1984). S. Kauffman's work on non-standard components in evolutionary theory, already mentioned (KAUFFMAN, 1982, 1983, 1985), may also have a bearing on this difficult problem, or set of problems. But these ongoing analyses also complicate the matter. Vrba and Eldredge, for example, take notice of five different 'hierarchies' in nature: the genealogical hierarchy (genes, organisms, demes, species ...); the ecological hierarchy (enzymes, cells, organisms, populations, local ecosystems ...); the somatic hierarchy; the taxonomic hierarchies, and the hierarchy of homologies. Now for addicts of old-fashioned hierarchy theory (of a decade ago!), this presents a dizzying prospect. First, as I have already pointed out, the term "hierarchy" cannot be univocal in all these cases. Its senses need disentangling. Second, the new writers on hierarchy analyze their hierarchies in terms of three levels of organization: a level of focus related both to an upper and a lower level. The Pattee–Simon type hierarchies, to which I have been accustomed, are two-level affairs, and the switch from two to three needs assimilation, criticism or both. This is a growth point in the literature of biological methodology and evolutionary theory at which, in my view, philosophers and biologists could, and should, interact for the benefit of both.

D. *Causality* (*and explanation*)

In the context of scientific explanation, Humean and Kantian (let alone Millian) concepts of causality have proved inadequate (see e.g. BOYD, 1981, pp. 652–653). Philosophy needs the gap filled, and studies of biological causality may provide important assistance and instantiation both of problems and of possible solutions. The work of SOBER (1981) and BRANDON (1981b) already cited suggest an entering wedge; much more remains to be done, clearly in the field of evolutionary biology, perhaps in connection with other subdisciplines as well. Causality, of course, connects in turn with the question of explanation, which needs elucidation in more flexible and realistic terms than the older orthodoxy allowed. Again, some of the work already referred to above suggests directions for future research. Analysis of models, robustness, etc. also point in this direction. I can do no more than to suggest that this whole complex of problems demands a philosophical analysis foreshadowed but not yet actualized (so far as I know) in the literature available to date (that is, to September, 1983).

E. *Intensional discourse in biology*

As is clear from Sober's definition of a unit of selection, quoted above (I B 3), the causal force characteristic of evolutionary change is a force exerted by *traits* or *properties* of organisms (or of other units of selection). Reference to properties, far from being a neolithic vestige in science, as some have argued, plays a fundamental role in the discourse, experimental design, and explanatory power of evolutionary biology. Thus, however purely extensional species conceived as individuals may be — they are pointed to, not characterized — intensionality must find a home, and a centrally located home, in any adequate analysis of the methods and epistemic claims of biology. Species as classes used to admit this dimension, if only implicitly. In the newer approach, as in Sober's case, for example, intensionality is exiled from the species concept for the sake of evolutionary theory (SOBER, 1982); species are lineages, and you'd better not mix in anything else. But then Sober himself — again, for the sake of Darwinian theory — carefully reinstates properties in a crucial explanatory role. Maybe this is the right way to go; whether it is or not, the matter needs further exploration.

F. *Perception*

The point raised in I A 5: of the interconnections between evolutionary biology, ecology and theories of development, also bears on another problem of fundamental importance to particular methodological studies in philosophy of biology as well as to philosophy of science (and philosophy) in general. That is: the nature of perception and its differing role in different biological disciplines. (1) The traditional theory of perception (GREGORY, 1966) has sometimes been conceived in close conjunction with evolutionary theory: perceptions are hypotheses, and natural selection has selected those that would lead to survival and thus to differential reproductive success. J.J. Gibson's 'ecological theory of perception' (GIBSON, 1979; REED and JONES, 1982) offers an alternative and, in my view, more promising evolutionary interpretation of the way in which, through our perceptual systems, we learn to orient ourselves in our environment. (2) Given this more adequate approach to perception as such, it would be illuminating, I believe, to consider the varying roles of perception in various branches of biology: in paleontology, for instance, as against biochemistry, Mendelian genetics, field ethology — etc. etc. This would be work, not only on interfield *theories*, but on the interconnections, and separations, of fields through the necessary reliance of specialists on particular, carefully acquired and sustained sorts of observational skills. These are of course not rigidly divided from differences, and interconnections, in hypothesis-formation and theorizing: as we all know by now, all observation is theory-laden. But what I have occasionally referred to as 'the conceptual fabric of biology' could be usefully studied in this, as well as more obvious and abstract, dimensions. (For some reflections on perception and science in general, see COMPTON, 1983; GRENE, 1986.)

G. *The primacy of history*

Seventh, but perhaps most important of all, historico-philosophical case studies can shed new floodlights on the complexities of biological science in its methodological, epistemological and (inseparably from these) social dimensions. As I remarked at the outset, it is in large part the cooperation of biologists, philosophers and historians of biology that has already proven, and promises to prove in future, so fruitful in this context. Science, as James Franck is supposed to have remarked, is either something people do or it is nothing at all. Sciences are, in A.C. MacIntyre's terms, *practices* (a concept that has *nothing* to do with 'practical applications' of science, let alone with pragmatism, that quack medicine of philosophic pseudo-therapy). MacIntyre defines a practice as:

> any coherent and complex form of socially established cooperative human activity through which goods internal to that form of activity are realized in the course of trying to achieve those standards of excellence which are appropriate to, and partially definitive of, that form of activity, with the result that human powers to achieve excellence, and human conceptions of the ends and goods involved, are systematically extended (MACINTYRE, 1981, p. 175).

And in science, the chief good internal to the practice is the hope of achieving a correct understanding of how something in the real world really works. From this point of view, paths of discovery, internal values of science, changing (intellectual) goals within the sciences are all germane to the *philosophic* analysis of their ontological and epistemic claims. Thus the investigation, historical and philosophical, of many special growth points in science, past and present, can greatly assist in the study of general problems in philosophy of science as such, as well as in philosophy of biology in particular. Provine's study of the role of the concept of adaptation in the development of Sewall Wright's theory of evolution is a case in point (PROVINE, 1983, 1985). So is Kitcher's paper on "Genes" (KITCHER, 1982a). In this area, Richard Burian is working on a major project on the recent history of the concept of the gene, which is intended to bring up to date in terms of recent and contemporary history and experimental work the authoritative work of Carlson (see BURIAN, 1985; CARLSON, 1966). Both MAYR and PROVINE (1980) and MAYR (1982b) furnish storehouses on which such investigations can draw, but, once the primacy of history (and of perception!) comes to guide philosophical reflection about the sciences, there is an almost unlimited field here for exploration by imaginative philosophers, philosophically inclined biologists, and historians, one for all and all for each.

A concluding apology

It is to be regretted that, with the exception of three titles by Finnish authors and two (in part) by a South African, this review has dealt only with work from exclusively English-speaking countries. I believe my colleagues do not mean to be parochial; if there is convergent conceptual evolution of which they are unaware, they will certainly be delighted to hear of it. Addresses of some of those whose work in progress I have referred to are given at the close of the references. Others are available in the list of participants in the Congress. I am grateful to Prof. Mary Williams for her comments at Salzburg and to Mr. John Chiment for a critical reading of my penultimate draft.

References

This list is of course selective only. Where there are several treatments of the same topic by an individual, I have generally referred to the most recent one known to me, since it will include earlier references.

ACHINSTEIN, P., 1977, *Function statements*, Phil. Sci. 44, pp. 341–367.

ARNOLD, A.J. and FRISTRUP, K., 1982, *The theory of evolution by natural selection: a hierarchical expansion*, Paleobiol. 8, pp. 113–129.

AYALA, F.J., 1974, *Biological evolution: natural selection or random walk?* Am. Sci. 62, pp. 692–701.

AYALA, F.J., 1985, *Reduction in biology: a recent challenge*, in: Evolution at a Crossroads (Bradford Books, Cambridge, MA).

BEATTY, J., 1980, *Optimal-design models and the strategy of model building in evolutionary biology*, Phil. Sci. 74, pp. 532–561.

BEATTY, J., 1981, *What's wrong with the received view of evolutionary theory?* PSA 1980, 2, pp. 397–426.

BEATTY, J., 1983, Address at Williamstown, MA conference, on the species-are-individuals thesis and essentialism.

BECHTEL, W., 1983, *Building interlevel theories: the development of the Emden–Meyerhof pathway*, Abstrs. 7th Int. Cong. Logic, Meth. Phil. Sci. 4, pp. 277–280.

BOYD, R., 1981, *Scientific realism and naturalistic epistemology*, PSA 1980, 2, pp. 613–662.
BRADIE, M. and GROMKO, M., 1981, *The status of the principle of natural selection*, Nat. and Syst. 3, pp. 3–12.
BRANDON, R.N., 1981a, *Biological teleology: Questions and explanations*, Stud. Hist. Phil. Sci. 12, 2, pp. 91–105.
BRANDON, R.N., 1981b, *A structural description of evolutionary theory*, PSA 1980, 2, pp. 427–439.
BRANDON, R.N., 1982, *The levels of selection*, PSA 1982, 1, pp. 315–323.
BROWN, 1979, *Perception, Theory and Commitment* (University of Chicago Press, Chicago).
BURIAN, R., 1983, *Adaptation*, in: Dimensions of Darwinism, ed., M. Grene (Cambridge University Press, New York & Cambridge), pp. 287–314.
BURIAN, R., 1985, *On conceptual change in biology: the case of the gene*, in: Evolution at a Crossroads (Bradford Books, Cambridge, MA).
BURIAN, R. and BRANDON, R.N., eds., 1984, *Genes, organisims, populations: Controversies over the units of selection* (Bradford Books, Cambridge, MA).
CAMPBELL, D.H.,. 1982, *Autonomy in evolution*, in: MILKMAN, 1982, pp. 190–201.
CAPLAN, A., 1981, *Back to class: a note on the ontology of species*, Phil. Sci. 48, pp. 130–140.
CARLSON, E.A., 1966, *The Gene: A Critical History* (Saunders, Philadelphia).
CHARLESWORTH, B., LANDE, R. and SLATKIN, M., 1982, *A neo-Darwinian commentary on macroevolution*, Evol. 36, pp. 474–498.
COMPTON, J.J., 1983, *Natural science and being-in-the world*, Paper read at Pac. Div., Am. Phil. Assoc., Mar. 1983.
DARDEN, L., 1983, *Reasoning in theory construction, analogies, interfield connections and levels of organization*, Abstrs. 7th Int. Cong. Log. Meth. Phil. Sci. 4, pp. 288–291.
DAWKINS, R., 1976, *The Selfish Gene* (Oxford University Press, Oxford).
ELDREDGE, N. and GOULD, S.J., 1972, *Punctuated equilibria: an alternative to phyletic gradualism*, in: T.J.M. Schopf, ed., Models in Paleobiology (Freeman, Cooper & Co., San Francisco), pp. 82–115.
ELDREDGE, N. and SALTHE, S.N., 1985, *Hierarchy and evolution*, in: Oxford Surveys of Evolutionary Biology (Oxford Univ. Press, Oxford).
FARRIS, J.S., 1983, *The logical basis of phylogenetic analysis*, reprinted in: E. Sober, ed., Conceptual Issues in Evolutionary Biology (Bradford Books, Cambridge, MA), pp. 675–702.
FITCH, W., 1982, *The challenges to Darwinism since the last centennial and the impact of molecular studies*, Evolution 36, pp. 1133–1143.
GHISELIN, M., 1974, *A radical solution to the species problem*, Syst. Zool. 23, pp. 536–544.
GIBSON, J.J., 1979, *The Ecological Approach to Visual Perception* (Houghton Mifflin, New York).
GOULD, S.J., 1980, *Is a new and general theory of evolution emerging*? Paleobiol. 6, pp. 119–130.
GOULD, S.J., 1982a, *Darwinism and the expansion of evolutionary theory*, Science 216, pp. 380–387.
GOULD, S.J., 1982b, Introduction to reprint of K. Goldstein, *The Material Basis of Evolution* (Yale University Press, New Haven).
GOULD, S.J., 1983, *The hardening of the modern synthesis*, in: M. Grene, ed., Dimensions of Darwinism (Cambridge University Press, Cambridge and New York), pp. 71–99.
GOULD, S.J. and ELDREDGE, N., 1977, *Punctuated equilibria; the tempo and mode of evolution reconsidered*, Paleobiol. 3, pp. 115–151.
GOULD, S.J. and LEWONTIN, R.C., 1979, *The spandrels of San Marco and the Panglossian paradigm: a critique of the adaptationist programme*, Proc. Roy. Soc. London B 205, pp. 581–598.

GOULD, S.J. and VRBA, E., 1982, *Exaptation — a missing term in the science of form*, Paleobiol. 8, pp. 4–15.

GREGORY, R.L., 1966, *Eye and Brain* (McGraw-Hill, New York).

GRENE, M., 1969, *Hierarchy: one word, many concepts?* in: Hierarchical Structures, eds. L.L. Whyte, A.G. Wilson and D. Wilson (Elsevier, New York), pp. 56–58.

GRENE, M., 1977, *Philosophy of medicine: prolegomena to a philosophy of science*, PSA 1976, 2, pp. 77–93.

GRENE, M., 1986, *Perception and Interpretation in the Sciences*, to appear in the Bordet Lectures (Queen's Univ. Press, Belfast).

HAMILTON, W.D., 1964, *The genetical evolution of social behavior*, J. Th. Biol. 1, pp. 1–16, 17–52.

HAMILTON, W.D., 1981, *The evolution of cooperation*, Science 211, pp. 1390–1396.

HENNIG, W., 1950, *Grundzüge einer Theorie der phylogenetischen Systematik* (Deutscher Zentral Verlag, Berlin).

HENNIG, W., 1966, *Phylogenetic Systematics* (Univ. Illinois Press, Urbana).

HIRSCHMANN, D., 1973, *Function and explanation*, Ar. Soc. suppl. 47, pp. 19–38.

HORRIDGE, G.A., 1977, *Mechanistic teleology and explanation in neuroethology*, Bio. Sci. 27, pp. 725–732.

HULL, D.L., 1970, *Contemporary systematic philosophies*, Am. Rev. Ecol. Syst. 1, pp. 19–54.

HULL, D.L., 1976, *Are species really individuals?* Syst. Zool. 25, pp. 174–191.

HULL, D.L., 1978, *A matter of individuality*, Phil. Sci. 45, pp. 355–360.

HULL, D.L., 1979, *The limits of cladism*, Syst. Zool. 28, pp. 416–440.

HULL, D.L., 1980, *Individuality and selection*, Ann. Rev. Ecol. Syst. 11, pp. 311–332.

HULL, D.L., 1981, *Units of evolution: a metaphysical essay*, in: Philosophy of Evolution, eds. V.J. Jensen and R. Harré (Harvester Press, Brighton), pp. 23–44.

HULL, D.L., 1983, *Hierarchies and hierarchies*, manuscript.

HUNKAPILLER, T., HUANG H., HOOD, L. and CAMPBELL, J.H., *The impact of modern genetics on evolutionary theory*, in: MILKMAN, 1982, pp. 164–189.

KAUFFMAN, S., 1982, *Developmental constraints: internal facturs in evolution*, in: Brit. Soc. Devel. Biol. Symp. no. 6: Development and Evolution (Cambridge University Press, Cambridge), pp. 195–225.

KAUFFMAN, S., 1983, *Filling some epistemological gaps: new patterns of inference in evolutionary theory*, PSA 1982, 2.

KAUFFMAN, S., 1985, *Self-organization, adaptation and the limits of selection: a new pattern of inference in evolution and development*, in: Evolution at a Crossroads (Bradford Books, Cambridge, MA).

KIMLER, W.C., 1983, *Mimicry: views of naturalists and ecologists before the modern synthesis*, in: M. Grene, ed., Dimensions of Darwinism (Cambridge University Press, Cambridge and New York), pp. 97–127.

KIMURA, M., 1979, *The neutral theory of molecular evolution*, Sci. Amer. 240, pp. 98–126.

KIMURA, M., 1983a, Chapter in: M. Nei and R.K. Koehn, eds., Evolution of genes, and proteins (Sinaver, Sunderland, MA).

KIMURA, M., 1983b, *The Neutral Theory of Molecular Evolution* (Cambride University Press, Cambridge).

KITCHER, Ph., 1982a, *Genes*, Brit. J. Phil. Sci. 33, pp. 337–359.

KITCHER, Ph., 1982b, Paper on species delivered at East. Div., Am. Phil. Assoc., Dec. 1982. (Abstract: J. Phil. 79, pp. 721–722).

KITCHER, Ph., 1984, *Text of* 1982b, Phil. Sci. 51, June 1984.

KITTS, D., 1983, *Can baptism alone save a species?* Syst. Zool. 32, pp. 27–53.

LEVINS, R., 1966, *The strategy of model building in population biology*, Am. Sci. 54, pp. 421–431.

LEVINS, R., 1975, *Evolution in communities near equilibrium*, in: M.L. Cody and J.M.

Diamond, eds., Ecology and Evolution of Communities (Harvard Univ. Press, Cambridge, MA), pp. 16–50.
LEVINS, R. and LEWONTIN, R.C., 1980, *Dialectics and reductionism in ecology*, Synthese 43, pp. 47–78.
LEWONTIN, R.C., 1970, *Units of selection*, Ann. Rev. Ecol. Syst. 1, pp. 1, pp. 1–18.
LEWONTIN, R.C., 1974, *The Genetic Basis of Evolution* (Columbia Univ. Press, New York).
LEWONTIN, R.C., 1978, *Adaptation*, in: *Evolution* (Freeman, San Francisco), pp. 114–125.
LEWONTIN, R.C., 1979a, *Fitness, survival and optimality*, in: D.J. Horn *et al.*, eds., Analysis of Ecological Systems (Ohio State Univ. Press, Columbus, OH), pp. 387–405.
LEWONTIN, R.C., 1979b, *Sociobiology as an adaptationist program*, Behav. Sci. 24, pp. 5–14.
LEWONTIN, R.C., 1982, Review of "R. Milkman, ed., Perspectives on Evolution", Paleobiol. 8, pp. 309–313.
LEWONTIN, R.C., 1983a, *The corpse in the elevator*, N.Y. Rev. Bks., Jan. 20, 1983, pp. 34–37.
LEWONTIN, R.C., 1983b, *Adaptation* (longer version of 1978), reprinted in: E. Sober, ed., Conceptual Issues in Evolutionary Biology (Bradford Books, Cambridge, MA), pp. 235–251.
LEWONTIN, R.C., 1983c, *Darwin's revolution*, N.Y. Rev. Bks., June 16, 1983, pp. 21–27.
LEWONTIN, R.C. and DUNN, L.C., 1960, *The evolutionary dynamics of a polymorphism in the house mouse*, Gen. 45, pp. 705–722.
LLOYD, E.A., 1983a, *The nature of Darwin's support for the theory of natural selection*, Phil. Sci. 50, pp. 112–129.
LLOYD, E.A., 1983b, *Mathematical models in evolutionary theory and the semantic approach to theory structure*, in: Abstrs. 7th Int. Cong. Log. Meth. Phil. Sci. 5, pp. 277–280.
MACINTYRE, A.C., 1981, *After Virtue* (Notre Dame Univ. Press, Notre Dame, IN).
MAYNARD SMITH, J., 1979, *Optimization theory in evolution*, Ann. Rev. Ecol. Syst. 9, pp. 31–56.
MAYNARD SMITH, J. (ed.), 1982, *Evolution Now* (Freeman, San Francisco).
MAYR, E., 1974, *Teleological and teleonomic: a new analysis*, in: R.C. Cohen and M. Wartofsky, eds., Bost. Stud. Phil. Sci. 14 (Reidel, Dordrecht), pp. 91–117.
MAYR, E., 1981, *Biological classification: toward a synthesis of opposing methodologies*, Science 214, pp. 510–516.
MAYR, E., 1982a, *Adaptation and selection*, Biol. Zbl. 101, pp. 161–174.
MAYER, E., 1982b, *The Growth of Biological Thought* (Harvard Univ. Press, Cambridge, MA).
MAYR, E., 1983, *How to carry out the adaptationist program*, Am. Nat. 121, pp. 324–334.
MAYR, E., 1984, *How biology differs from the physical sciences*, in: Evolution at a Crossroads (Bradford Books, Cambridge, MA).
MAYR, E. and PROVINE, W.B., eds., 1980, *The Evolutionary Synthesis* (Harvard Univ. Press, Cambridge, MA).
MICHOD, R.E., 1981, *Positive heuristics in evolutionary biology*, Brit. J. Phil. Sci. 32, pp. 1–36.
MILKMAN, R., ed., 1982, *Perspectives on Evolution* (Sinauer, Sunderland, MA).
MILLS, S. and BEATTY, J., 1979, *The propensity interpretation of fitness*, Philosophy of Science 46, pp. 263–286.
MITCHELL, R.D. and WILLIAMS, M.B., 1979, *Darwinian analyses: the new natural history*, in: Analysis of Ecological Systems, eds. D.J. Horn, R.D. Mitchell and G.R. Stains (Ohio State Univ. Press, Columbus), pp. 23–50.
NELSON, G. and PLATNICK, N., 1981, *Systematics and biogeography*, in: Cladistics and Vicariance (Columbia Univ. Press, New York).
OSTER, G. and ALBERCH, P., 1982, *Evolution and bifurcation of developmental programs*, Evolution 36, pp. 444–459.
PATTEE, H.H. (ed.), 1973, *Hierarchy Theory* (Braziller, New York).
PATTERSON, C., 1982, *Cladistics*, in: MAYNARD SMITTH, ed., 1982, pp. 110–120.
POLANYI, M., 1969, *The logic of tacit inference*, reprinted from Phil. 44 (1966); address to Int.

Cong. Log. Meth. Phil. Sci., 1964; in: M. Grene, ed., Knowing and Being (Univ. of Chicago Press, Chicago), pp. 138–158.

PROVINE, W.B., 1983, *The development of Wright's theory of evolution: systematics, adaptation and drift*, in: M. Grene, ed., Dimensions of Darwinism (Cambridge Univ. Press, Cambridge and New York), pp. 43–70.

PROVINE, W.B., 1985, *Sewall Wright — Geneticist and Evolutionist* (Univ. of Chicago Press, Chicago).

RACHOOTIN, S.P. and THOMSON, K.S., 1980, *Epigenetics, paleontology and evolution*, in: G.G.E. Scudder and J.L. Reveal, eds., Evolution Today (Hunt Inst., Carnegie Mellon, Pittsburgh), pp. 181–193.

REED, E. and JONES, R., eds., 1982, *Reasons for Realism. Selected Essays of J.J. Gibson* (Erlbaum, Hillsdale, NJ).

RICHARDSON, R.C., 1982, *Grades of organization and the units of selection controversy*, PSA 1982, 1, pp. 324–340.

RICHARDSON, R.C., 1983, *The use of models in biological explanation*, Abstrs. 7th Int. Cong. Log. Meth. Phil. Sci. 4, pp. 333–336.

ROSEN, DONN, 1982, *Do current theories of evolution satisfy the basic requirements of explanation?* Syst. Zool. 31, pp. 76–85.

ROSENBERG, A., 1982, *Discussion: on the propensity definition of fitness*, Phil. Sci. 49, pp. 268–273.

ROSENBERG, A., 1983, *Discussion: coefficients, effects and genic selection*, Phil. Sci. 50, pp. 332–338.

RUSE, M., 1983, *Species: individuals, natural kinds or what?* manuscript.

SALTHE, S.N., 1985, *Evolving hierarchical systems: their structure and representation* (Columbia Univ. Press, New York).

SCHAFFNER, K.F., 1974, *The peripherality of reductionism in the development of molecular biology*, J. Hist. Biol. 7, pp. 111–139.

SCHOPF, W., HOFFMAN, A. and GOULD, S.J., 1983, *Letters*, Science 219, pp. 438–444.

SNEATH, P.H.A. and SOKAL, R.R., 1973, *Numerical Taxonomy* (Freeman, San Franciso).

SOBER, E., 1981, *Holism, individualism and the units of selection*, PSA 1980, 2, pp. 93–121.

SOBER, E., 1982, *Comment on Kitcher on species*, East. Div., Am. Phil. Assoc., Dec. 1982, manuscript.

SOBER, E., 1983, *Parsimony in systematics; philosophical issues*, Ann. Rev. Ecol. Syst. 14, in press.

SOBER, E. and LEWONTIN, R.C., 1982, *Artifact, cause and genic selection*, Phil. Sci. 49, pp. 157–180.

SOBER, E. and LEWONTIN, R.C., 1983, *Discussion: reply to Rosenberg on genetic selectionism*, Phil. Sci. 50, pp. 648–650.

STANLEY, S.M., 1979, *Macroevolution* (Freeman, San Francisco).

STEBBINS, G.L., 1982, *Perspectives in evolution*, Evolution 36, pp. 1109–1118.

STEBBINS, G.L. and AYALA, F.J., 1981, *Is a new evolutionary synthesis necessary?* Science 213, pp. 967–971.

TUOMI, J., 1981, *Structure and dynamics of Darwinian evolutionary theory*, Syst. Zool. 30, pp. 22–31.

TUOMI, J. and HAUKOIJA, E., 1979, *An analysis of natural selection in models of life-history theory*, Savonia 3, pp. 9–16.

TUOMI, J., SALO, J., HAUKIOJA, E., NIEMELA, P., HAKALA, T. and MANILLA, R., 1983, *The existential game of individual self-maintaining units: selection and defence tactics of trees*, Oikos 40, pp. 369–376.

VRBA, E. and ELDREDGE, N., 1984, *Individuals, hierarchies, selection and effects. Towards a more complete evolutionary theory*, Paleobiology 10.

Wade, M., 1978, *A critical review of the models of group selection*, Quart. Rev. Biol. 53, pp. 101–114.

Wassermann, G.D., 1981, *On the nature of the theory of evolution*, Phil. Sci. 48, pp. 416–437.

Webster, G. and Goodwin, B., 1982, *The origin of species: a structuralist approach*, J. Soc. Biol. Struc. 5, pp. 15–47.

Wiley, E. and Brooks, D., 1982, *Victims of history — a non-equilibrium approach to evolution*, Syst. Zool. 31, pp. 1–24.

Wiley, E. and Brooks, D., 1983, *Nonequilibrium thermodynamics and evolution: A response to Løvtrup*, Syst. Zool. 32, pp. 209–219.

Williams, G.C., 1966, *Adaptation and Natural Selection* (Princeton Univ. Press, Princeton).

Williams, M.B., 1970, *Deducing the consequences of selection: a mathematical model*, J. Theor. Biol. 29, pp. 343–385.

Williams, M.B., 1973, *The logical status of the theory of natural selection and other evolutionary controversies*, in: M. Bunge, ed., The Methodological Unity of Science (Reidel, Dordrecht), pp. 84–102.

Williams, M.B., 1982, *The importance of prediction tests in evolutionary biology*, Erkenntnis 17, pp. 1–15.

Williams, M.B., 1983, *The units of selection controversy: resolution by an axiomatization*, Abstrs. 7th Int. Cong. Log. Meth. Phil. Sci. 4, pp. 369–372.

Wilson, D.S., 1980, *The Natural Selection of Populations and Communities* (Benjamin/Cumming, Menlo Park, CA).

Wilson, D.S. and Colwell, R.K., 1981, *Evolution of sex ratio in structured demes*, Evolution 45, pp. 882–897.

Wilson, E.O., 1975, *Sociobiology* (Harvard Univ. Press, Cambridge, MA).

Wimsatt, W.C., 1972, *Teleology and the logical structure of function statements*, Stud. Hist. Phil. Sci. 3, pp. 1–80.

Wimsatt, W.C., 1980a, *Randomness and perceived randomness in evolutionary biology*, Synthese 43, pp. 287–329.

Wimsatt, W.C., 1980b, *Reductionistic research strategies and their biases in the units of selection controversy*, in: T. Nickles, ed., Scientific Discovery: Case Studies (Reidel, Dordrecht), pp. 213–259.

Woodger, J.H., 1952, *From biology to mathematics*, British J. Philos. Sci. 3, pp. 1–21.

Wright, L., 1976, *Teleological Explanations* (Univ. of California Press, Berkeley).

Wright, S., 1967, *Comments on preliminary working papers*, in: P.S. Moorhead and M.M. Kaplan, eds., Mathematical Challenges to the Neo-Darwinian Interpretation of Evolution (Wistar Inst., Philadelphia), pp. 117–120.

Wynne-Edwards, V.C., 1963, *Intergroup selection in the evolution of social systems*, Nat. 200, pp. 623–626.

Addresses of some authors of work-in-progress not present at the Salzburg Congress:

Prof. John Beatty, Dept. of Philosophy, Arizona State Univ., Tempo, AZ 85281.

Prof. Richard Burian, Dept. of Philosophy, Virginia Polytechnic Institute, Blacksburg, VA 24061.

Prof. John Compton, Dept. of Philosophy, Vanderbilt Univ., Nashville, TN 37235.

Dr. Niles Eldredge, American Museum of Natural History, Central Park W. at 79th Street, New York, NY 10024.

Prof. Stuart Kauffman, Dept. of Biochemistry and Biophysics, Univ. of Pennsylvania, Philadelphia, PA 19174.

Prof. Philip Kitcher, Dept. of Philosophy, Univ. of Minnesota, Minneapolis, MN 55455.

Prof. S.N. Salthe, Dept. of Biology, Brooklyn College, C.U.N.Y., Brooklyn, NY 11210.

Dr. Elizabeth Vrba, Transval Museum, P.O. Box 413, Pretoria 0001, South Africa.

Barcan Marcus et al., eds., Logic, Methodology and Philosophy of Science VII
© Elsevier Science Publishers B.V. (1986) 453–466

BIOLOGY AND VALUES: A FRESH LOOK

MICHAEL RUSE

Depts. of History and Philosophy, Univ. of Guelph, Ontario, Canada N1G 2W1

No one would deny that science and values impinge, sometimes happily, sometimes less so. My question is whether it is proper — whether it is *always* proper — to consider science and values as independent entities.

Ernest Nagel writes:

> There is a relatively clear distinction between factual and value judgments, and ... however difficult it may sometimes be to decide whether a given statement has a purely factual content, it is in principle possible to do so ... [I]t is possible to distinguish between, on the one hand, contributions to theoretical understanding (whose factual validity presumably does not depend on the social ideal to which a social scientist may subscribe), and on the other hand contributions to the dissemination or realization of some social ideal (which may not be accepted by all social scientists). (NAGEL 1961, pp. 488–9)

Most philosophers are not quite this blatant in their position; but, the message that science and values are things apart is there, nevertheless. It is true that philosophers do recognize things which are sometimes referred to as "epistemic values." Perhaps the best known of these is simplicity. It is allowed that the facts never completely determine choice of theories, and that in practice scientists are influenced by notions of elegance or beauty or "simplicity," so-called. Here, certainly, values enter into science (HEMPEL 1965).

But, while this is a concession — and an important one — its force is somewhat dampened by the fact that philosophers try desperately to devise formulae that will explain simplicity away. At least, they look for formulae that will take the personal element out of simplicity, reducing it to a rational non-value-impregnated method of choosing between rival hypotheses (SOBER 1975). And, in any case, such epistemic values as simplicity are rather anaemic compared to many of the value issues we encounter in real life. No philosopher will argue, for instance, that a scientist will choose one theory rather than another, because it has a more exalted view of women.

At least, no philosopher will argue that a *good* scientist will make this kind of choice, and that others should follow suit.

In recent years, thanks to detailed work on evolutionary biology, I have begun to worry about philosophers' stance on the non-role of values in science — a stance which I openly admit to having taken in the past, along with everyone else. Let me share with you some of those factors which have led me to doubt the value-neutrality of evolutionary biology. Then, I'll try to analyze some of the ways values can (and possibly should) enter into biology, and perhaps into the rest of science.

Modern evolutionary biology

I'll start with what I think is the best and most central of evolutionary thought. Then, I'll look for the values. But what is this best, most central thought? There are five areas I want to highlight.

First, there is the central mechanism of neo-Darwinian biology, *natural selection* or the survival of the fittest. Next, we have the *origin of life:* in which life might have been formed here on earth, naturally, from non-living components. Third is *ecology.* In the past two decades we have seen a merging of evolutionary population thinking with ecological population thinking, to form an extended and more powerful core to evolutionism.

Fourth, we have that area dealing with instinct and *behaviour.* Today, the study of behaviour from an evolutionary perspective — so-called "sociobiology" — is one of the brightest, albeit most controversial, stars in the evolutionary firmament. Finally, we have *paleontology.* The fossil record is key evidence for evolution.

I'll take up each of these five areas briefly, inquiring into values.

Natural selection

The major causal mechanism of change posited in modern evolutionary thought is that identified by Charles Darwin: natural selection. Not all organisms can survive and reproduce, and thus there is an ongoing winnowing or selection of favourable types (DARWIN 1859; AYALA and VALENTINE 1979).

Natural selection clearly bears evidence to its value-impregnated origins. It was introduced by Darwin to explain adaptations, which he and everyone else took to be indisputable evidence of a benevolent God's concern for the world: "irrefragable evidence of creative forethought" to quote the words of Darwin's contemporary, the anatomist Richard OWEN (1834).

Moreover, Darwin took over the whole linguistic and conceptual apparatus of the teleologists: "design," "purpose," "need," and so forth, and used the teleological aspect of organisms as a heuristic for solving problems. What point, what purpose, could a certain feature serve? And this way of thinking and speaking persists in evolutionary thought today. The organic world is seen as a testament to a good god, and biologists' work is a paean of praise (RUSE 1979a; 1981b).

Of course, I exaggerate. Late 20th century evolutionists are not natural theologians as was the early 19th century Archdeacon Paley, the most ardent exponent of design in nature. Indeed, most evolutionists, probably, would indignantly repudiate any suggestions that they or their work is "tainted" by religious belief. And, I have no reason to think they are concealing an underlying faith. But, still this does not deny that, at its heart, Darwinian evolutionary biology is riddled through and through — in language and in attitude — with a mode of thought which, when it entered biology, was highly value-impregnated. And that in itself is no small thing to note.

Moreover, I'm far from convinced that all the values have gone from the adaptationist way of thinking. I think there is still a fair amount of approval by biologists of the "success" of the "successful" in evolution. It may not be taken as evidence of a good god *per se*, but in itself is valued as a good thing. I'll return to this point later.

Apart from values coming through adaptation itself, much of the opposition to the supposed effects of natural selection is value-laden. One of the strongest attacks on selection and its resultant adaptation comes from biologists who are avowed Marxists, and who are quite candid about wanting to produce a Marxian-inspired evolutionary biology. The geneticist R.C. Lewontin and the paleontologist S.J. Gould have argued strenuously that pan-selectionism is on a par with Dr. Pangloss's arguments about everything being for the best, namely ridiculous. They feel that the assumption that selection is all, or near-all, powerful is just a dangerously misleading remnant of an out-dated theistic paradigm (GOULD and LEWONTIN 1979).

Moreover, it is claimed that such adaptationism is ultra-"reductionistic", breaking organisms into components, where the whole (the organism) is the sum of the parts (the separate, supposedly adaptive features). This is a cardinal dialectical-materialist sin. One should rather emphasize the integrative, holistic nature of organisms. Thus, Lewontin and Gould argue that organisms should be seen as wholes, where all the parts fit together, where the nature of parts may simply be dictated by the "engineering constraints" set by getting the organism working at all, and where there may therefore be no direct adaptive advantage to features. They cite the four-limbedness of vertebrates as a possible non-adaptive, yet crucial feature of organisms.

Obviously, values are influencing the Lewontin/Gould picture of organisms, no less than values influence a Marxist analysis of (say) religion, or of North American politics.

The origin of life

You might think that when we look at issues to do with the beginnings of life here on earth, we look at the one part of the evolutionary spectrum where values will not intrude. After all, life from non-life seems more of a physico-chemical issue than a biological issue, and physics is undoubtedly far from values. But, whatever the merits of this latter claim — and I don't find the value-neutrality of physics a truism — study soon shows that the origin-of-life question is as drenched with values, as any question could be (FARLEY 1977; GRAHAM 1972).

No one today believes that worms spring from mud, through a flash of lightening or some such thing. Rather, an inorganic "soup" made of elements occurring on the early globe probably produced some of the "building blocks" for life, particularly amino acids. That these can be produced naturally has been demonstrated experimentally. Then, bit by bit, these building blocks, could have joined to make larger, functioning molecules until finally one has fully working primitive life. Much is still unknown, but it is an unknown filled with hope not despair. (For details, see DICKERSON 1978.)

This scenario is one directly based on the dialectical materialist hypotheses of the Russian biochemist, A.I. Oparin (FARLEY 1977). He was quite open in his subscription to a Marxist-Leninist philosophy of nature, and consciously applied it to his work on the appearance of new life. This led to a two-pronged argument. On the one hand, Oparin severely criticized all attempts to show that life is nothing but physico-chemical processes, and could therefore simply come through chance rearrangement of molecules. This is an illegitimate mechanistic/reductionistic approach to life's origin (OPARIN 1938).

On the other hand, Oparin saw natural processes as developing through the force of circumstances, with their own momentum, with new properties (and laws even) emerging, as complexity grew.

> From the point of view of dialectical materialism life is material in nature, but it is not an inalienable property of all matter in general. On the contrary, it is inherent only in living beings and is not found in objects of the inorganic world. Life is a special, very complex and perfect form of motion of matter. It is not separated from the rest of the world by an unbridgeable gap, but arises in the process of the development of matter, at a definite stage of this development as a new, formerly absent quality. (OPARIN and FASENKOV 1961, p. 245; also OPARIN 1968)

I don't want to exaggerate. I don't want to say that Oparin sat down and "deduced" the origin of life from the basic principles of dialectical materialism, rather as Hegel "deduced" the number of planets from his philosophical principles. Much theoretical and empirical work went into Oparin's lifetime struggle with the origin-of-life question. I do argue, however, that dialectical materialism — a value-impregnated world view infused his approach and his results. And what we today work with is this legacy of Oparin. (See FARLEY 1968 for more on this point.)

Ecology

The very word "ecology" is value-laden — organic gardening, ugly sandals, and herbal everything. But, I'm not really concerned with the romantic yearnings of middle-class North Americans. Rather, my interest lies with the scientific study of populations, in their environments. It is this subject which has recently been integrated right into the core of the evolutionary synthesis (ROUGHGARDEN 1979).

I argue that values rule this part of biology. Let me give one example. One of the most celebrated and much discussed theories in modern ecology is the MacArthur/Wilson theory of island biogeography. This theory attempts to explain the numbers of species one finds on islands. The key assumption within the theory is that the number of species on an island (S) tends towards an equilibrium number. This equilibrium supposedly results from a balance between species coming into the island (immigrants), and species leaving the island (emigrants). The rates of immigration and emigration are functions of the island's area and position only. The equilibrium is, therefore, a dynamic balance, with equal numbers of species coming and going.

Now, what is the status of this central equilibrium assumption? The authors themselves defended it on the grounds that, at least, it lets them go beyond the purely descriptive. Biogeographers can make some predictions (MACARTHUR and WILSON 1967, pp. 20–1). This is fair enough. As a good traditional philosopher one will be loathe to commit the *faux pas* of asking about origins. Where or why MacArthur and Wilson got their equilibrium assumption is philosophically irrelevant.

But, matters are not quite this simple. There lurk uncomfortable questions about the equilibrium assumption: questions that a traditional philosophy of science — ignoring origins and considering only empirical evidence — quite fails to answer.

First, there is increasing evidence that the principle has a tenuous relationship with reality, to say the least (GILBERT 1980). The clear evidence for its truth is diminishingly small. Moreover, and more significantly from our perspective, it is clear that ecologists expound much effort defending it against attack. They do this, both by surrounding the assumption with other protective empirical claims, and by exploiting its fuzziness. What, for instance, counts as an immigrant? As two sympathetic critics write:

> The relationship between local and global populations of a species in an archipelago is problematic. If turnover is to be a significant concept ecologically, it should refer to real phenomena of population dynamics. But in a mosaic of small islands that are not effectively isolated, the dynamical patterns on single islands may only be understood if the global population breeding in the entire archipelago (or on the nearby mainland) is considered (HAILA et al. 1979). For example, the global population may consist of scattered pairs of birds breeding on several islands. As the breeding islands may be different in successive years, high turnover rates may be observed locally, even though the global population were stable. Ecological realism is thus needed in using the term turnover. (HAILA and JARVINEN 1982, 267)

Second, given the hypothesis' dubious empirical status, one cannot but note its close isomorphism to the traditional balance of nature picture. It's an exact exemplification. Input equals output. Island biogeography looks just like pre-Darwinian ecology expressed in mathematical symbols.

Third, there's good evidence that today's ecologists do in fact look upon the balance as a good thing. Moreover, this is the balance, as supported by the MacArthur/Wilson theory! Thus, Wilson himself see virtues in the evolutionary status quo. He thinks one of the greatest evils facing us today is the extinction of species, and the consequent upset of the present balance. This all has a very familiar ring to it. It is yet more reason for thinking that his equilibrium assumption is not the naive empirical hypothesis you might first take it to be. (Wilson has a book forthcoming bemoaning the loss of species, WILSON 1984. See also the pertinent recent articles on the controversy between ecologists over their models, LEWIN 1983.)

I won't labour the point. There are values at the heart of evolutionary ecology.

Sociobiology

In the past two decades, major strides have been made towards full evolutionary understanding of behaviour. Now, with its new name of "sociobiology," such understanding claims equal place with other members of the evolutionary family (WILSON 1975; MAYNARD SMITH 1978). Not only has sociobiology been a late developer, its adolescence has been usually traumatic. In particular, the coming of human sociobiology has brought major controversy, with staunch defenders and violent (sometimes physically violent) critics. Values abound!

Like DARWIN (1871), today's students of the evolution of behaviour want to apply their theories and findings directly to our species *Homo sapiens* (WILSON 1975; 1978; RUSE 1979b; 1981a; ALEXANDER 1979). And this is something which left-wing biologists find anathema. Such a view of humanity violates their most deeply held values (ALLEN et al. 1975; 1976; 1977).

They feel that a view of humans as products of genes, sifted by natural selection, is epistemologically inadequate because of its "reductionism." And, it is morally inadequate because of its "determinism." It fails to see that humans as humans are emergent beings. It condemns humans to their roles in life, because of their genes. This lays the way open for all kinds of capitalist, sexist, fascist, repressive practices. Thus, human sociobiology is a pernicious doctrine. What we must see rather is that humans are cultural beings. It is true that there are certain basic biological determinants (e.g. the need for food and for sleep). But, these are background to culture. And, it is through recognition of the primacy of culture that we must build our science of humankind.

That such a vision of humans is value-impregnated needs no argument. But what about human sociobiology? Is it value-impregnated? I believe it is, although not necessarily in the way claimed by critics.

Let me note one value assumption which runs right through the human sociobiological endeavour. This is a fundamental commitment to the unity and similarity of humankind. For Wilson, and for his fellows, there is the shared hypothesis that what motivates the Kalahari Bushman is precisely that which motivates the New York business executive.

> The building block of nearly all human societies is the nuclear family (REYNOLDS, 1968; LEIBOWITZ, 1968). The populace of an American industrial city, no less than a band of hunter-gatherers in the Australian desert, is organized around this unit. In both cases the family moves between regional communities, maintaining complex ties with primary kin by means of visits (or telephone calls and letters) and the exchange of gifts. During the day the women and children remain in the residential area while the men forage for game or its

symbolic equivalent in the form of barter and money. The males cooperate in bands to hunt or deal with neighboring groups. If not actually blood relations, they tend at least to act as "bands of brothers". Sexual bonds are carefully contracted in observance with tribal customs and are intended to be permanent. (WILSON 1975, p. 554)

There is an almost total lack of empirical evidence for the belief that we are all bound by common genes, which have been collected through common processes of selection. Hence, at this stage of the development of sociobiology, it is hard to interpret the sociobiological vision of humankind as other than a reflection of sociobiologists' own commitment to the worth and status of each and every member of our species. Human sociobiology is indeed thoroughly value-impregnated.

Paleontology

I come, finally, to the evolutionary study of the fossil record. In the past decade, this has been the subject of considerable controversy, as the orthodox Darwinian "phyletic gradualists" have battled with the saltationary "punctuated equilibrists" (RUSE 1982).

On the one side, we have those who argue that evolution is gradual, like a branching tree. Natural selection is taken to be the key agent of change, and adaptation is seen as important in the fossil record as it is in organisms living today (STEBBINS and AYALA 1981; MAYNARD SMITH 1981). On the other side, we have those who argue that evolution goes in jumps, occurring only when one group breaks from another (GOULD 1980; 1982; ELDREDGE and GOULD 1972; GOULD and ELDREDGE 1977). Causal factors other than selection are believed to be important in the evolutionary process.

One of the leading developers and supporters of the punctuated equilibria position is the paleontologist Stephen Jay Gould. Much of Gould's justification for his paleontological perspective comes from Marxist philosophy. In at least three ways, Gould defends his view of the fossil record by invoking his value-system. First, there is Gould's already discussed view that the adaptationist programme is a theistic throw-back, purveying an overly reductionistic picture of organisms (GOULD and LEWONTIN 1979). Gould himself prefers an integrated holistic view of organisms; a view of organisms where many features have no direct adaptive value, but are simply part of the "mechanics" of existence. The punctuated equilibria picture incorporates just such a conception of organisms.

Second, Gould criticizes Darwinian gradualism as being just an act of faith, reflecting Darwin's own 19th century liberal views about the virtues of gradual (as opposed to revolutionary) change. Gould, to the contrary, endorses a philosophy which leads him to expect rapid, abrupt breaks with the past. His view of the fossil record is therefore simply his own world picture made, if not flesh, then stone (GOULD 1980).

Third, there are the implications for humankind. According to the punctuated equilibria thesis, either you are in a species, or your are not. Specifically, either you are in *Homo sapiens*, or you are not. And, once you are in *Homo sapiens*, you look for a certain genetic uniformity. There is no question of significant evolutionary change between earliest and latest members. This, of course, is an attractive picture to Gould. If it is true, it means that it is simply bad science to look for significant biological differences between humans. All differences are a function of the environment — nationality, social status, upbringing, and the like.

In these various respects, therefore, Gould offers us a value-influenced picture of life's past history. Even in paleontology, values intrude.

Wherein lie the values of biology?

One final task remains in this essay. Briefly, let me sketch what should be a full study in its own right, namely an identification of the actual places wherein values enter evolutionary science. To begin, there is the simple act of *choosing a problem* as worthy of study (LAUDAN 1977). Surely the fact that virtually every biologist at some point or another turns to *Homo sapiens* tells us something? The implication is that there is something special about our own species.

Next, I think we get values in the whole manner in which the ideas and concepts of a theory are presented. What I have in mind here is a wide range of items which collectively come under what one might refer to as the "clothing," or perhaps even the "skin" of a theory. (I like biological metaphors!) I mean for instance the very *language* which is used to talk about a theory. WILSON (1975) for instance was criticized as a sexist for referring to *Homo sapiens* as "man." I mean also the *pictures* which are used to illustrate a point. Again Wilson was criticized because his pictures supposedly portrayed males more prominently than females (ALLEN et al. 1977). I think in this case the objection fails, but the general point is well taken.

Undoubtedly another item which must be included in the clothing of a theory are the *examples, metaphors, and analogies* which are used to support and clarify the theory. But, I think now you will have the general drift of what I have in mind. There may perhaps be more items operative here, but I am sure I have touched on some of the main ones. You will remember just above that I equivocated on whether to refer to these items as the "clothing" or the "skin" of a theory. This was not just a nice point of literary style. In a sense, both terms talk about something on the surface — not part of the real, central body — and this is certainly what I see as distinctive about the items I have listed above, vis-à-vis the theories in which they appear.

Wilson's pictures, for instance, are more "peripheral" than say his specific causal mechanisms. Without the mechanisms sociobiology as we know it today would not exist. One can and does publish books on sociobiology without pictures. However, there is a crucial difference between clothing and skin. Clothing one can take off, and although one might feel awkward about appearing in none at all, one can change clothes and still be the same person. Skin however may be on the surface, but is not something changed — at least, were one to change one's skin one would be a different person, even though one would not be quite so different as if one were to change one's sex (say).

Hopefully, the point I am trying to make is starting to become clear. Prima facie, it would seem that all the items I have listed could be changed and the theory itself would not be changed. One could for instance have different pictures. (This is the "clothing thesis.") But I am not quite so sure that matters are this simple (i.e. in respects I incline to the "skin thesis"). Influenced by Max BLACK's (1962) view of metaphor, I wonder if change of language, example, and so forth, would not run deeper. I will not argue the point here but simply leave my readers with the information that had Wilson's final picture, that illustrating modern *Homo sapiens*, been a nice rosy picture of Adolf Hitler, with lots of beaming, buxom German Mädchens in the background, then I for one would not have written a book sympathetic to sociobiology.

Continuing my list of ways in which values enter into science, I argue that the very *statements* of science itself reflect and endorse values. What I would argue is that many of claims of biology are supportive of particular value positions, and that without the scientific factual claims the positions would be unsupported. In this way I see even the factual claims as having value connotations. The sort of situation I have in mind is one which occurs in debate about the morality of vegetarianism. Both sides might agree that one ought not eat persons — their disagreement is over the personhood status of (say) cows. Here factual claims enter into the debate, as when for instance an unregenerate meat-eater like myself argues that biology shows that cows do not have the level of self-awareness of humans. (The debate might also involve factual claims about human bodily needs for certain proteins.)

You might think that although all I am saying is true, it is a little bit trivial. No one denies that scientific claims can be used to support moral positions — even the logical empiricist allows this much. But it is hardly to say that scientific claims are value-impregnated in any interesting sense. However, my position is a little stronger than this. Science constantly outstretches its reach, in the sense that given a number of basic facts (I am also enough of a logical empiricist to talk about "basic facts"), one can spin any number of explanatory hypotheses to account for them.

My claim is that science becomes value-impregnated because scientists want to argue to particular moral positions. Hence, needing a particular factual claim, he/she takes his/her limited range of facts and uses them to support this precise claim. And, all of this occurs notwithstanding the real possibility of the facts supporting a different claim, one which could indeed go against the scientist's moral position.

As I have argued about the equilibrium theory of island biogeography, the central premise specifying that input equals output (i.e. that species immigration numbers equal species emigration/extinction numbers) is value-impregnated. It is not something definitively decided by the facts. Values stand behind this premise. In the equilibrium theory case, it may perhaps be that the values are no longer held by biogeographers, although I've given reasons to doubt this. My point is that one cannot understand the claim's status without reference to values — whether these be held today or not.

As noted earlier, I am not now throwing everything to the wind, totally repudiating the claims of standard philosophy of science, and arguing that science is totally "subjective." I do not argue that a biologist can believe and argue precisely what he/she pleases. One still has all the checks on hypotheses that one had before. A claim supporting a value-position cannot violate the empirical evidence. It can still be judged by other criteria of good science, like the epistemic values of simplicity, consilience, and so forth. And it should be so judged. My point is simply that there is a place where a biologist's values, of all kinds, can and do enter into the hypotheses of his/her science.

The final place where I see values entering into biology is in what I like to follow the Kantians in calling "regulative principles" (KORNER 1966; RUSE 1980). By these I mean the standards and criteria to which theories must conform if they are to be judged "good" science, or indeed if they are to be judged "science" at all. I would say that many nineteenth-century evolutionists were showing their approval of the machine in particular and the Industrial Revolution in general in their endorsement and support of evolutionism, and in their insistence that a proper approach to biological origins must be mechanistic and conform to unbroken law.

In this century, I think Lewontin's Marxism functions as a regulative principle, inasmuch that for him the proper approach to science is one which stresses the integrative, holistic nature of the world. This belief is one which imposes itself, as it were, on what for him is constitutive of good science. "Marxism stresses the wholeness of things, both between organism and surroundings and within organism" (LEWONTIN and LEVINS 1976, p. 62). Oparin's Marxism earlier had fashioned as a regular principle for him.

At all levels therefore there is a place for values. They are there at the beginning, in the choice of problems. They stay to the end, as one decides what is and what is not an acceptable answer.

Conclusion

Let me make one final point. I welcome values. I do not think the worse of science because of them. Rather, I see science as a far more human activity than most philosophers would allow. The best science succeeds because of values, not despite them. Only someone fleeing from reality and from themselves would find this fact upsetting.

Bibliography

ALEXANDER, R.D., 1979, *Darwinism and Human Affairs* (Univ. of Washington Press, Seattle).

ALLEN, E. et al., 1975, *Letter to editor*, New York Review of Books 22, pp. 18, 43–44.

ALLEN, E., 1976, *Sociobiology: another biological determinism*, BioScience 26, pp. 182–186.

ALLEN, E., 1977, *Sociobiology: a new biological determinism*, in: Sociobiology Study Group of Boston, ed., Biology as a Social Weapon (Burgess, Minneapolis).

AYALA, F.J., and VALENTINE, J.W., 1979, *Evolving: The Theory and Processes of Organic Evolution* (Benjamin/Cummings, California).

BLACK, M., 1962, *Models and Metaphors* (Cornell Univ. Press, Ithaca, NY).

DARWIN, C., 1859, *On the Origin of Species* (John Murray, London).

DARWIN, C., 1871, *Descent of Man* (Murray, London).

DICKERSON, R.E., 1978, *Chemical evolution and the origin of life*, Scientific American, September, pp. 70–86.

ELDREDGE, N., and GOULD, S.J., 1972, *Punctuated equilibria: an alternative to phyletic gradualism*, in: T.J.M. Schopf, ed., Models in Paleobiology (Freeman/Cooper, San Francisco).

FARLEY, J., 1977, *The Spontaneous Generation Controversy: From Descartes to Oparin* (Johns Hopkins Press, Baltimore).

FEYERABEND, P., 1975, *Against Method* (New Left Books, London).

GILBERT, F.S., 1980, *The equilibrium theory of island biogeography: fact or fiction*? J. Biogeography 7, pp. 209–35.

GRAHAM, L., 1972, *Science and Philosophy in the Soviet Union* (Knopf, New York).

GOULD, S.J., 1980, *Is a new and general theory of evolution emerging*? Paleobiology 6, pp. 119–30.

GOULD, S.J., 1982a, *Darwinism and the expansion of evolutionary theory*, Science 216, pp. 380–7.

GOULD, S.J., 1982b, *Punctuated equlibrium — a different way of seeing*, in: J. Cherfas, ed., Darwin Up to Date (IPC Magazines, London), pp. 26–30.

GOULD, S.J., and ELDREDGE, N., 1977, *Punctuated equilibria: the tempo and mode of evolution reconsidered*, Paleobiology 3, pp. 115–51.

GOULD, S.J., and LEWONTIN, R., 1979, *The spandrels of San Marco and the Panglossian Paradiagm: a critique of the adaptationist programme*, Proc. Roy. Soc. London B 205, pp. 581–98.

HAILA, Y. and JARVINEN, O., 1982, *The role of theoretical concepts in understanding the ecological theatre: a case study on island biogeography*, in: E. Saarinen, ed., Conceptual Issues in Ecology (Reidel, Dordrecht), pp. 261–78.

HARDING, S. and HINTIKKA, M.B. 1983, *Discovering Reality: Feminist Perspectives on Epistemology, Metaphysics, Methodology, and Philosophy of Science* (Reidel, Dordrecht).

HEMPEL, C., 1966, *Philosophy of Natural Science* (Prentice-Hall, Englewood Cliffs).

KORNER, S., 1966, *Experience and Theory: An Essay in the Philosophy of Science* (Routledge and Kegan Paul, London).

KUHN, T.S., 1962, *The Structure of Scientific Revolutions* (Univ. of Chicago Press, Chicago).

LAUDAN, L., 1977, *Progress and its Problems: Towards a Theory of Scientific Growth* (Univ. of California Press, Berkeley).

LEWIN, R., 1983, *Santa Rosalia was a goat*, Science 221, pp. 636–9.

LEWONTIN, R. and LEVINS, R., 1976, *The problem of Lysenkoism*, in: H. and S. Rose, eds., The Radicalisation of Science (Macmillan, London), pp. 32–64.

MACARTHUR, R.H. and WILSON, E.O., 1967, *The Theory of Island Biogeography* (Princeton Univ. Press, Princeton).

MAYNARD SMITH, J., 1978, *The evolution of behavior*, Scientific American 239 (3), pp. 176–193.

MAYNARD SMITH, J., 1981, *Did Darwin get it right*?, London Review of Books 3 (11), pp. 10–11.

NAGEL, E., 1961, *The Structure of Science* (Routledge and Kegan Paul, London).

OPARIN, A., 1938, *The Origin of Life* (Macmillan, London).

OPARIN, A.I., 1968, *The Origin and Initial Development of Life* (Washington, DC).

OPARIN, A.I. and FESENKOV, V., 1961, *Life in the Universe* (Twayne, New York).

OWEN, R., 1834, *On the generation of the marsupial animals, with a description of the impregnated uterus of the kangaroo*, Phil. Trans., pp. 333–64.

ROUGHGARDEN, J., 1979, *Theory of Population Genetics and Evolutionary Ecology: An Introduction* (Macmillan, New York).

RUSE, M., 1979a, *The Darwinian Revolution: Science Red in Tooth and Claw* (Univ. of Chicago Press, Chicago).

RUSE, M., 1979b, *Sociobiology: Sense or Nonsense*? (Reidel, Dordrecht).

RUSE, M., 1980, *Philosophical aspects of the Darwinian revolution*, in: F. Wilson, ed., Pragmatism and Purpose (Univ. of Toronto Press, Toronto).

RUSE, M., 1981a, *Is Science Sexist? and Other Problems in the BioMedical Sciences* (Reidel, Dordrecht).

RUSE, M., 1981b, *Teleology redux*, in: J. Agassi, ed., Essays in Honour of Mario Bunge (Reidel, Dorecht).

RUSE, M., 1982, *Darwinism Defended: A Guide to the Evolution Controversies* (Addison-Wesley, Reading MA).

SOBER, E., 1975, *Simplicity* (Clarendon Press, Oxford).

STEBBINS, G.L. and AYALA, F.J., 1981, *Is a new evolutionary synthesis necessary*? Science 213, pp. 967–71.

WILSON, E.O., 1975, *Sociobiology: The New Synthesis* (Belknap, Cambridge, MA).

WILSON, E.O., 1978, *On Human Nature* (Harvard Univ. Press, Cambridge, MA).

WILSON, E.O., 1984, *Biophilia* (Harvard University Press, Cambridge, MA).

Barcan Marcus et al., eds., Logic, Methodology and Philosophy of Science VII

BIOLOGICAL COGNITION: ITS UNITY AND DIVERSITY

B.G. YUDIN

Academy of Sciences of the USSR, Moscow, USSR

It is customary to treat biology as a sphere of knowledge, placed between physico-chemical sciences on the one hand and social sciences on the other. However, it is not always realized and registered that "between" has two different meanings. First, such judgement has an ontological aspect since the subject matter of physico-chemical sciences is treated as more fundamental and universal than that of biology. Second, it also contains methodological substance since biological knowledge is regarded as being less developed, strict and substantiated, as well as having lesser explanatory and prognostic potentialities than physical and chemical knowledge. Quite often this judgement is, with corresponding modifications, extended to also cover the sphere of social knowledge.

The corollary of the first, ontological premise is a notion of biological knowledge being, in the final analysis, necessarily represented as a certain subarea of physico-chemical knowledge, defined by several specifying assumptions.[1] From the methodological premise it is inferred that a major way of biological cognition development lies in constructing such theories that, while possibly being also specifically biological in terms of language and substance, have, nevertheless, to originate from ideals and standards set by physics and chemistry.

These very reasons, naturally represented in the most general form, underlie the formulation of a basic methodological problem of biological cognition, i.e. a relation of biology to physical and chemical sciences, or, to be more specific, a possibility of reducing biology to physics and chemistry. This very problem will serve as a starting point while being considered

[1] This position in its detailed form is, for instance, contained in M. RUSE's, *The Philosophy of Biology*, 1973.

from another standpoint, different from the usual one. As a rule, its analysis explicitly, or more often implicitly, presupposes biology as representing something integral, one-dimensional and unidirectional.

It stands to reason that the inner differentiation of biological knowledge, being quite evident, has long drawn the attention of researchers, both biologists proper and other scholars, dealing with philosophical and methodological problems of biology. In particular, it found its reflection in the structural levels of the living matter concept, being actively developed at present. The range of problems, studied within the framework of this concept, covers not only the identification of discrete levels of existence and development of life, but also the analysis of relations between sections of biology, investigating each of these levels. This is only one of thc cxisting ways of considering the differentiation of biological knowledge.

This circumstance is, however, often disregarded when reducibility is brought into the picture. In such cases it is not only and even not so much the present state of biology that is usually meant but its possible (necessary) form to be assumed in the more or less distant future, i.e. the form of a sufficiently complete and, hence, integral science. As a result, both present biology and its future trends are, wittingly or unwittingly, assessed in terms of such a complete and integral form. This situation makes one treat the diversity of actually existing biological cognition as something temporary and transient, from which it is possible and necessary to abstract.

Meanwhile, a problem of reducibility appears to be interpreted quite differently if this implicitly postulated unity of biology is questioned. It is, of course, not a matter of totally negating the unity of biological cognition but of representing both this unity and inseparable diversity of biological cognition as two aspects, necessarily presupposing each other, of the dialectical opposition largely responsible for a pattern of biology development.

A natural basis of biological cognition unity is furnished by the unity of life and the living things as the sphere of reality facing biological cognition. The concepts of "life" and "the living things", paramount for biological cognition, cover an immense variety of objects, phenomena and processes, which are, nevertheless, regarded as having unity in some important respects. The nature of such unity is, in terms of its content, revealed in different ways at various stages of development of biological cognition and under various general biological concepts. Meanwhile, the very realization of this unity, preceding any specific biological study, is the most important premise, constitutive for biological cognition. To conceive something as living means conceiving it as being somewhat singled out from the order of physical and chemical objects. As compared to them, the object of biological cognition is given to us and perceived by us quite differently in some fundamental aspects.

It is not an exclusive prerogative of the biologist's professional thinking to register living things as singled out in a particular category of their own. On the contrary, biological cognition uses (naturally, specifying and developing it) the "living/non-living" opposition initially formed in mundane consciousness and rooted in the depths of centuries, in the sphere of man's active practical attitude towards the world. Categorizing, in terms of such opposition, is a first thought operation produced by man in coming across an as yet unknown object, with the results of categorizing also largely predetermining the man's attitude towards this object along with possible forms and methods of practical interaction with it.

Already for primitive man living things appeared both as a source of sustenance and as meeting other basic needs, while being at the same time a source of danger; as an object of hunting, collecting, tilling and cultivation; as something extremely vital to man's existence, also being very close to him, literally kindred (cf. totemism), and, finally, as a definition applicable to man himself. "Man has not yet opposed himself to the rest of nature: all objects and phenomena seemed to him to be "living". Such was the origin of primitive anthropomorphism and, as its consequence, the religious beliefs in the form of animism ... That very period of man's history had produced a notion of the "living" and "dead" things. All turns dead (man, animal, plant, stone, water, etc.) after "the soul leaving the body".[2] Such differentiation of the living and the non-living could proceed in various ways, though the very presence of such a borderline is most essential.[3] This circumstance has, in some way or other, found its reflection in varied systems of beliefs and different cults of fertility, animism, hylozoism, and, lastly, the biblical myth of creation where the creation of living things, immediately preceding the creation of man, occurs after the creation of inanimate nature.

The living, being fundamentally meaningful in terms of practical activity, becomes an object of religious, aesthetic and moral attitudes. A cognitive

---

[2] *A History of Biology from the Ancient Times up to the Early XX Century* (in Russian), edited by S.R. MIKULINSKY ("Nauka" Publ., Moscow, 1972), p. 16.

[3] "To primitive consciousness an opposition of the visible ("one's own", assimilated) and invisible worlds..., the living and the dead, unlike the corporeal and incorporeal, the animate and inanimate, ... was important. *Animism, A philosophic encyclopedic dictionary* ("Sovetskaya Entsiklopedia" Publ., Moscow, 1983), p. 25.

attitude to the living is also formed within such a highly value-oriented context. And what is more, a practical and value-oriented significance of the living also appears as a basis for initial identification of any possible regularity, order or rule to be found in the hard to grasp diversity of life phenomena. Hence, the man's biological notions are, from their inception, characterized not only by being referred to a particular type of object but also by specific and value-oriented association with these objects. Biological cognition represents a reflection of both objects, endowed with the property of life, and practice of man's interaction with the objects of such category.

A historical development of such practice proper entails a change in its value orientations and comprehension within the framework of biological cognition. In this respect biological cognition appears as a solution of a specific *task of culture*, i.e. establishing a common denominator for all the varied and often mutually contradictory notions of the living, originating from man's various practical attitudes towards objects endowed with life. It should be emphasized that such a task has its concrete solution at each stage of evolution of society, its interaction with living nature, and finally, development of biological cognition proper. It means that its solution cannot be delayed for the future, bringing refined and definitive formulations along with it. The answer of biological cognition to a request, proceeding from culture, may be necessary here and now, as representing, in essence, a self-determination of culture in one of its fundamental aspects.

Biological cognition, represented in such a way, evidently precedes any science in the true sense, being independent of it, while its unity and integrity is ensured (to a generally possible degree) by way of realizing a particular cultural function. Such a situation is, incidentally, radically changed with the advent of science and transformation of biological cognition into one of the spheres of sciences. A cognitive attitude towards the living becomes increasingly self-sufficient, with the impact of practical activity and value-oriented factors being more mediate. On the contrary, in the course of time the very results of scientific biological cognition start to generate an intense impact on practical activity, primarily in the spheres of agricultural production and medicine.

At the same time biological cognition experiences an ever increasing permanent impact on the part of the general context of scientific cognition, its standards and norms. Biological cognition becomes a component of a new whole, scientific cognition; but science, formed as a social institution and turned into one of the structure-forming elements of culture, has a cultural mission of its own, i.e. obtaining systematized, conclusive, substantiated and objectively true knowledge about the world, as well as providing explanations related to certain fragments of this world and complying with a certain historically changing totality of ideals and norms.[4] Such explanations represent, in terms of culture, replies, constructed by means of conceptual thinking and dealing with the meaningful queries, pertaining to world outlook, that are engendered, resolved or reproduced in the course of cultural advance, naturally also including science itself. While biology articulates an attitude to life as a value, providing a rational conceptual expression for it, science as a whole reveals and confirms the value of rational cognition per se and the rational approach to the world. Again at every given stage of scientific development its cultural task has a specific solution, however unsatisfactory this may be in retrospect.

Biology, being an extensive section of scientific cognition, does its bit in terms of solving this task. In this context, however, one may wonder whether a traditional cultural mission of biological cognition is retained in case of acquiring this new function. In terms of this study the same problem may be formulated in a different way — whether biology is qualitatively original enough to be treated as irreducible to physics and chemistry.

The answer here seems to be in the affirmative. Moreover, one problem of the specific cultural mission of biology manifests itself as being particularly urgent today. Mankind now has at its disposal — if not actually, then at least potentially — some novel and extremely powerful means of modifying life, including such means, developed by biology and related to manipulating living objects, as genetic engineering or various kinds of biotechnology. While not yet attaining the power of a demiurge, capable of recreating life, man has become capable of totally destroying it. It is quite clear, for instance, that current and very urgent problems of environmental protection are, in fact, nothing but the necessity of preserving the Earth's diversity of life. In this case a problem of preserving life in its diversity assumes, along with its utilitarian and practical meaning, also an apparent moral and aesthetic significance.

Turning of biological knowledge to man's concerns, his habitat, character and conditions of his vital activity, as noted by R.S. Karpinskaya "suggests a question of qualitatively specific nature of biological cognition, sovereignty of biology as a science, in terms of being an initial component

---

[4] *The Ideals and Norms of Scientific Research* (in Russian), (The V.I. Lenin Byelorussian State University, Minsk, 1981).

of resolving a whole range of problems related to methodology and world outlook"[5], stressing in this context a fundamental role of integral and aggregate knowledge of the essence of life and laws of its development.

In the present situation man has started to develop a quite particular perception of the uniqueness and value of life. Such realization is also largely due to biology, in particular the conception of the biosphere as a finely adjusted, though rather fragile organism, involving all the living things on the Earth. Naturally, the cultural mission of biology does not end here since biology provides a rational conceptual expression for such perception, and only scientific biological cognition, irrespective of existing assessments of its development and strictness, is, in the final analysis, capable of performing this function in modern culture.

A. Schweizer, developing the ethics of "a reverential attitude towards nature", has stressed the profound Weltanschauung meaning of the fact that man's existence presupposes as a necessary condition the preservation of life as such. However, this reverential stance, while being necessary, seems, due to its passive mood, to be insufficient. At present active efforts to preserve life on this planet are also needed. (It stands to reason that excessive and presumptuous action, taken without a serious consideration of one's own Weltanschauung and value orientations, may in this instance, as in any other case, become only counterproductive.) The charting of such a course of action, along with identifying and proving some technologically feasible impacts on living nature to be culturally unacceptable, also forms a part of the mission assigned by culture to modern biological cognition.

This analysis makes it possible to conclude that biological cognition is dealing with problems of culture and world outlook that go beyond the framework of problems dealt with by physical and chemical cognition. Is it possible, however, that biology, constructed exclusively by means of physics and chemistry, is potentially capable of solving these problems in a similar or even more effective way? Let us consider this possibility.

It was more than forty years ago that E. Schroedinger asked himself: "What is life in terms of physics?"[6] Such way of formulating the problem is, of course, necessary and, as time has shown, highly productive. Yet this is only one facet of a more general and fundamental question, namely "What is Life?". In terms of physics (and chemistry) one may find out why life *is possible* and in which forms; while the regularities, established by physics

---

[5] R.S. KARPINSKAYA, *Biology and World Outlook* (in Russian), ("Mysl" Publ., Moscow, 1980), p. 35.

[6] E. Schroedinger, *What is Life?* (Cambridge Univ. Press, London, 1944).

and chemistry, appear for biology as necessary (but not sufficient!) conditions for explaining the range of phenomena dealt with by it. When physical and chemical regularities are represented as limiting the existence of life or making it improbable, cognition is faced with rather complex problems of substance.

Such was, for instance, the case of the second principle of thermodynamics, being treated as applicable to biology. No lessening of free energy and no increase in entropy as a result of processes, occurring in the organism, has long caused confusion among the researchers, also somewhat contributing to animated vitalism since it was believed that a stable existence of organic structures runs counter to the laws of thermodynamics. A theory of open systems, clearing the way to a study of thermodynamic characteristics of biological objects, represented an important result in developing those problems.

On the whole, however, the laws of physics and chemistry permit no representation of the entire diversity of the living as being a necessary corollary of them. It is believed that in the future this task will finally be resolved; but there is as yet no valid reason to disregard an alternative, suspending a search along such lines of biological cognition that are distant from physics and chemistry. The development of biology is, of course, dependent upon physics (such dependence is even necessary for the self-determination of biology), while being far from following the lead of physics. Hence, the yardsticks, borrowed from physics, are often unfit for an adequate assessment of biological cognition; at any rate, each such borrowing requires a special justification for it.

Therefore, such pronouncements as "History teaches us that as time passes Biology tends towards physicalization and chemicalization", or assertions of biological development tending towards physical reduction,[7] seem to be hasty and lopsided. In fact, physicalization and chemicalization represent only *one of the trends* in the development of biological cognition.

In this context the development of cybernetics became meaningful for the self-determination of biology. Thanks to cybernetics a wide range of phenomena and processes, related to obtaining, storage and use of information to realize regulating and controlling interactions, was made available for study. Physical and chemical cognition paid no particular attention to these phenomena and processes, while soon after the advent of cybernetics they were found to play an essential part at most different

---

[7] J. Almog, *A quantum basis of heredity and mitosis?*, in: 6th Intern. Congress of Logic, Methodology and Philosophy of Science, Abstracts, Sections 8, 9 (Hannover, 1979), p. 213.

structural levels of the living — from a cell and intracellular structures to a population, biocenosis and the biosphere as a whole. At that time cybernetics was widely believed to be capable of solving almost all the cardinal problems of biology. At present, however, there is no particular need in proving that biological cognition is not confined to the limits of cybernetics either.

Generally speaking, the methodological peculiarities, inherent in biological cognition, reflect exactly its quantitative originality but not its limitations with respect to physical cognition. Hence, for instance, the elimination of functional explanation from biology through reducing it to a cause and effect pattern — recently quite a labour-consuming endeavour — is of little relevance to a real methodology, effectively applicable to biological research. It might be recalled that the two sharply different phases of sleep, "fast" and "slow", discovered several years ago, were first treated by researchers primarily in terms of functional meaning of each phase. Here we see that biological cognition differs from physical and chemical cognition by the very *type* of questions formulated by it and sought to be answered. The question, relating to the functions of this or that phenomenon or process with respect to a certain system, so essential for biology, becomes simply meaningless within the framework of physics and chemistry.

A study of the cognitive loading of biological notions, concepts and principles, registering and reflecting, in some way or other, the qualitative peculiarity of the organic world, seems to be much more promising in terms of the methodology of biological cognition. One may cite as an example the principle of survival of the fittest, along with the principles of corpuscular heredity and covariant reduplication, called specific biological general natural-historical principles by N.V. Timofeev-Resovsky and regarded by him as a foundation of future theoretical biology and a basis of experimental study and description of biological phenomena.[8] Such an approach apparently implies another direction of biological development quite different from physicalization.

A specific cognitive nature of biology seems to find its most general expression in the notion of biological reality, gaining an increasingly wide recognition in modern philosophic and methodological publications. It permits recognition not only of a particular nature of biological objects but also of specific means of cognition employed by biology. Biological reality

[8] N.V. Timofeev-Resovsky, *From the history of micro- and macroevolution relations*, in: Micro- and Macroevolution (in Russian), (Tartu, 1980), pp. 7–12.

is "a scientifically dissected reality of living nature, changing in the course of historical advance of cognition and largely dependent in this change upon the experimental means of impact on the object, as well as methods of its theoretical reproduction".[9] The notion of biological reality thus reflects a level of our knowledge about the living at a certain stage of biology development. At the same time, this notion underlies any general biological concept since its development is dependent upon the preceding initial idea, to be set in some way or other, about the diverse, varied and multifaceted phenomenon of life. The very complexity and multifaceted nature of this phenomenon is conducive to the possibility of parallel existence of a multitude of such sharply different initial ideas (this being another source of biological cognition diversity, responsible for competing research programmes), with each of those ideas subsequently proving to be somewhat incomplete, lopsided and hence potentially limited in terms of prospects for a future development of biological cognition. Nevertheless, the presence of such initial idea, such integral depiction of life, is vital for any general biological concept.

The conceptual means alone are obviously inadequate to express this initial notion, thus making it necessary for biological thinking to rely upon images and metaphors not to be discursively developed but rather simply "caught" by consciousness.[10] Each similar image cuts out of such phenomenon as life, inaccessible in its extensity to immediate perception and direct observation, certain cross-sections, which subsequently become objects of study.

A collation of Ch. Darwin's and L.S. Berg's evolutionary concepts is helpful in explaining this point. Berg noted in his "Nomogenesis": "For the process of the organic world evolution to be graphically depicted it has to be represented not as a growing tree, ever developing its new branches (here Berg opposes the metaphor, characteristic of Darwin — B. Yu), but in the form of pages, being turned over in the book where one page (form) follows after another: the turned over pages pass to history while the open ones continue to live for the time being".[11] Both metaphors produce vivid spatial images meant to cover and express the temporal dimensions of the

---

[9] I.T. FROLOV, *Life and Cognition* (in Russian), "Voprosy filosofii", No. 8 (1979), p. 25.

[10] For more detailed information on the metaphors and images as structural elements of scientific cognition see: B.G. YUDIN, *Explanation and understanding in scientific cognition* (in Russian), "Voprosy filosofii", No. 9 (1980).

[11] L.S. BERG, *Works on the Theory of Evolution* (in Russian), ("Nauka" Publ., Leningrad, 1977), p. 287.

entire evolutionary process, proceeding over periods impossible to perceive in ordinary terms.

Amplifying this collation it may, with a rough approximation, be said that the Darwin's notion of the living is more populational[12] and ecological, while that of Berg is more physiological and organismic, thus accounting for differences in emphasis. While Darwin regarded an individual organism primarily as something always accompanied by its like and having a limited access to necessary means of sustenance, Berg treated an individual organism as something characterized, first of all, by its internal organization.

The biological reality dealt with by Darwin was thus substantially different in several respects from the one dealt with by Berg, with each of them having his own characteristic viewpoint of life and its evolution. Unlike Darwin, Berg sought in particular to perceive the living in terms of physics and chemistry on the basis of thermodynamics. It is not a matter of this realm of knowledge being less developed in Darwin's time than Berg's. While Darwin largely assumed the approach of a naturalist, studying "natural history" in traditional sense, Berg tends, though more potentially than actually, to construct biology on the basis of experimental and analytical methods. It is manifest, for instance, in his suggestion of systematic relations to be established among the groups of organisms proceeding from the chemical composition of proteins contained in their cells.

Incidentally, it is noteworthy that Darwin, when making use of artificial selection as a hypothetical model for understanding natural selection, did in fact introduce a quasi-experimental procedure into the study of evolution. While artificiality of selection was regarded by traditional naturalists as an obstacle in the way of observing a phenomenon in its pure form, Darwin's awareness of artificially produced changes subordinated to regularities of naturally proceeding processes, has contributed not only to substantially extending the framework of observational biology but also constructing a concept much more in line with the standards of sciences than its precursors or contemporaries.

A collation of the Darwin's and Berg's concepts is also indicative of biological cognition being exposed in the course of its development to

---

[12] In this context see: Yu. V. CHAIKOVSKY, *The origin of Darwin's discovery* (in Russian), "Priroda", No. 6 (1982), p. 94.

impulses emanating from other spheres of scientific knowledge. This permits a more detailed treatment of the impact upon the development of biology on the part of general scientific context thus, in its turn, making it possible to represent the diversity of biological cognition from another viewpoint.

When speaking of modern biology it is possible to identify, as a first approximation, a number of vectors responsible for its development. Each of them is related to a more or less extensive area of biological research, involving a number of scientific disciplines; at the same time — and this is very important — more or less evidently claiming to represent biology in its entirety; each of them is characterized by a certain vision of biological reality and hence, a certain originality of methodological guidelines.

Firstly, there is a vector that orientates biological cognition towards physics and chemistry. It covers such sections of biology as biophysics and biochemistry, microbiology and molecular biology, cytology and embryology. In methodological terms, this sphere is characterized by a predominance of experimental and analytical methods of research. (It should be emphasized that a methodological description of research along these lines is, in this case and henceforth, in no way exhaustive. It is only a matter of identifying most representative methodological features necessary for a specific collation of this trend in research with others.) Here a leading structural level is the level of an individual cell. A substantive practical significance of research along these lines may be seen in various examples since an experimental pattern is often easily transformed into a pattern of production technology. Biotechnology and genetic engineering, microbiological industry and pharmacology represent only some practical applications taken to support this statement.

A second vector orientates biology towards systems-mathematical and cybernetic realms of knowledge. Disciplines developed along these lines include biocenology, various subsections of ecology, and the conception of the biosphere, thus proving the level of ecosystems to be central for this research. At the same time, the ideas of cybernetics, as mentioned above, are extensively applied to studying the processes of regulation and control at different structural levels, e.g. in researching the problems of coding, transfer and realization of heredity information, in treating the physiological mechanisms of regulation at the level of the organism, in analyzing the dynamics of evolution of both populations and ecosystems, etc.

Most characteristic problems of biological cognition, proceeding along these lines, include identifying a system through setting a certain pattern of system-forming linkages,[13] constructing a model that reflects informational, energetic and other interactions among the system's elements. As a rule, this model may be mathematically expressed, thus permitting a computerized simulation of the processes occurring in the system as well as its responses to various external impacts. In practical terms, such research is noted for its prospects for controlling the functioning of natural biosystems, making it possible to identify such impacts as ensure the maintenance of the system's homeostatic balance or its transformation into a required direction. Research, proceeding along these lines, may be of immediate practical use in terms of protecting the natural ecosystems against human and technological environmental impacts.

Another vector characterizing the development of modern biology, is orientated towards social cognition. In this context such disciplines as zoopsychology and ethology may be cited. Their orientation towards social cognition has nothing to do with the hasty and poorly substantiated extrapolations, attempting to provide an explanation for man's and society's life in terms of these disciplines. The point is that an ethologist seeks to *understand* the behaviour of the animal under study, identify the intent, motive and meaning of certain actions, as well as to somewhow *interpret* them. Such terms as "motive", "intent", and "meaning" are used in this case rather metaphorically than literally, but it is, however, important that they are borrowed from the sphere of social cognition. That is incidentally why explaining the relations in human society on the basis of ethological notions and conceptions proves to be methodologically incorrect. As to the anthropomorphism and sociomorphism characteristic of ethological studies and the conceptual apparatus of ethology, when controlled by methodological reflection, it seems to furnish no evidence that biological cognition proceeding along these lines, may be treated as underdeveloped. The models and analogies represent a vital and necessary means of cognitive activity in science, with both social relations and social cognition proper providing a possible source for them.

A particular role within the framework of the given direction of biological cognition is assigned to observation and description, made, as a rule, on the basis of field studies. It is noteworthy that ethological publications, both popular and strictly scholarly, often tend to treat the animals under study as "our lesser brethren" (Sergey Essenin). Moreover, recently this trend has been increasingly manifested in reflection of the

[13] See, for instance: K.M. KHAILOV, *Marine Ecological Metabolism* (in Russian), (Kiev, 1971).

ethologists, directed towards the basis of their researches. Apparently, the absence of such value orientation could make it simply impossible to conduct most studies, requiring a laborious and painstaking observation for months and even years.

The direction of the fourth and last (but, surely, not the least) vector may be exemplified by such sections of biology as systematics, morphology and evolutionism. Here an extensive use is made of observation and description, along with natural-historical explanation, the classical methods of biological study. In this case the value-oriented nuances just mentioned with respect to ethology, are not characteristic of observation and description. This vector seems to most clearly reflect a specific nature of biological cognition, thus occupying, in this respect, a prominent position. The basic structural levels taken up by researchers of this orientation are represented by the level of an individual, as well as species and other systematic ranks, with the level of population — where phenomena, constituting the substance of microevolutionary processes are registered — being of particular importance.

From the viewpoint of methodology, a relation between population, as an object of evolutionary study, and community, as an object of ethology, is noteworthy. Both these objects are, in terms of composition, often coincident with each other, while being approached in different ways. The population in an evolutionary study is, in essence, an integral system of species characterized by a common genetic pool. In its turn, the community, studied by ethologists, represents a system of species, having functional-complementary relations with each other. Each of them displays a definite pattern of behaviour and occupies a definite place in the hierarchic structure of community, etc.

The vectors represented here are in no way designed to reflect the motley and intricate picture of modern biological cognition in its entirety. It serves only as an illustration of the multitude of ways followed by biology in its development. No single direction can be properly regarded as major, with all the others being treated as collateral and irrelevant. Each of them covers a sufficiently wide range of its own cognitive problems, while the very diversity of them proves a wealth of empirical, theoretical and methodological instruments at the disposal of biology. In this context the development of modern biology appears as a multidimensional process, inevitably represented in a distorted way if account is taken of only one of its constituents.

A qualitatively specific nature of biological cognition in no way precludes, but on the contrary suggests, its active interaction not only with a single other sphere of scientific cognition but a whole gamut of them. The attempt to represent a trajectory, followed by the development of biological cognition as a whole, requires a construction of the resultant vector, situated within the field of differently directed forces.

At last, it has to be noted that each of these directions, apparently, interacts, to a lesser or greater degree, with all the rest of them, relying upon them, making use of their results and, in its turn, ensuring their advance. By and large, however, present relations among them can hardly be characterized as harmonious, meaning that the unity of biological cognition is internally controversial. It acts not only as an initial premise but as an ideal, having, incidentally, no predicted, direct and clearly marked way to attaining it.

10. Foundations and Philosophy of Psychology

Barcan Marcus et al., eds., Logic, Methodology and Philosophy of Science VII

THE EXPLICATION OF PSYCHOLOGICAL COMMON SENSE: IMPLICATIONS FOR THE SCIENCE OF PSYCHOLOGY

JAN SMEDSLUND

Inst. of Psychology, Univ. of Oslo, Blindern, Norway

In this paper a technical concept of common sense will be introduced. The various consequences of introducing this particular concept will be discussed in some detail. A view of psychology will be presented which is incompatible with the ones currently held by most researchers.

A concept of common sense

We will begin by immediately defining the concept: By *the common sense of culture C* will here be meant *the set of all implications taken for granted by all members of C.*

The term *implication* ordinarily designates the relation which holds between an ordered pair of propositions when the first cannot be true without the second also being true, i.e. when the truth of the first is a sufficient condition of the truth of the second. In other words, an implication states what *follows necessarily* from something. Stated in the possible worlds idiom, the necessity is expressed in the following definition: "a proposition *P* implies a proposition *Q* if and only if in each of all possible worlds if *P* is true then *Q* is also true" (BRADLEY and SWARTZ, 1979, p. 31).

However, the common sense of a culture includes other things than necessary relationships between actually formulated propositions. There are also implications where the antecedent is a nonverbal behavior (nodding, shaking one's head, smiling, frowning, pointing, etc.) or a nonbehavioral sign (traffic lights, arrows, maps, etc.), and where the consequents may also be nonverbal behaviors or other states of affairs. In order to include a wider range of referents, the term implication will, therefore, be used here in a more general sense than the one referring to the relationship between propositions only, while still retaining its formal properties. It will be taken to refer to the relation which holds between an ordered pair of any sort of states of affairs when the first cannot be the case without the second also being the case, i.e. when the first being the case is a sufficient condition of the second being the case. Paraphrasing Bradley and Swartz (see above) the new definition of implication, stated in the possible worlds idiom, will be as follows: *A state of affairs P implies a state of affairs Q if and only if, in each of all possible worlds, if P is the case then Q is also the case.*

If an implication, in the wide sense indicated above, is taken for granted by all members of a culture, then it is part of the common sense of that culture. To take $P \supset Q$ for granted implies that one acts in every way as if $P \supset Q$ were the case. It should be noted that there need be no awareness on the part of the person involved of taking $P \supset Q$ for granted. In fact, common sense is typically tacit, i.e. unreflected, and must be inferred from actual behavior. However, inferring from observed behavior to what is taken for granted, obviously, involves some uncertainty. A person may behave as if she is taking something for granted because she wants to deceive, or for reasons only accidentally related to the implication involved. The only way to increase diagnostic certainty is to keep varying the conditions and noting whether or not the person continues to behave consistently as if $P \supset Q$ were the case.

It should be noted that what is taken for granted in common sense as defined here is not the occurrence of any particular states of affairs, but only the relationship of implication between such states of affairs when they occur. If always when P is taken to be the case, Q is also taken to be the case, and if always when Q is taken not to be the case, P is also taken not to be the case, then the indications are that the implication $P \supset Q$ is being taken for granted.

The final term in the definition is member of a culture. What is a culture and who is a member? For the present purpose, the delimitation of a population sharing a culture is taken as a given. It is also assumed that a culture is characterized, at least partly, by the shared system of implications taken for granted by its members, i.e. precisely by its common sense. This circularity is, I think, unavoidable and benign. A population is distinguished according to certain social science criteria. The shared set of implications in this population is mapped and the outcome of this further contributes to a characterization of the culture and its boundaries. Individuals who differ sharply from the other members of the population in the implications they take for granted are, then, regarded as not being full members of the culture.

In conclusion, common sense means consensus. Given a population sharing a given culture, common sense refers to the set of implications taken for granted by *everyone* in that population. Actually everyone also takes for granted that everyone else takes this set of implications for granted. In other words, the tacit taking for granted is undifferentiated with respect to individuals and, hence, in a sense, is absolute.

How, then, can this realm of order, consisting of unformulated implications taken for granted, be studied scientifically?

A concept of explication of common sense

Since common sense is mainly implicit, and revealed only indirectly by what is taken for granted in action by everyone, it needs to be *explicated*, i.e. described in a valid manner. The process of putting into words something that hitherto has been unreflectively taken for granted (explication) cannot in itself be explicated. What we can do is to formulate a concept of what is a valid explication and, then, derive a method of validation from this. Here is the concept: *A proposition stating an implication is a valid explication of common sense to the extent that members of the culture involved agree that the proposition is correct and that its negation is incorrect.*

Several features of this definition merit some comment. First, the explication must take the form of a proposition stating an implication, since, by definition, this is the logical form of common sense.

Second, the criterion of degree of success of the explication is taken to be the amount of direct agreement among members of the culture. This presupposes that, presented with an explicit formulation, people have some access to their own unformulated intuitions (tacit assumptions) and may check the formulation against this. Since it is a presupposition for the successful use of language in general that one has access to one's unformulated intuitions, I think this is an admissible assumption. However, it does not preclude that, in particular instances, one may be mistaken. More specifically, a person may overlook certain possibilities and, hence, accept too narrow or too broad explications, or explications that are too narrow in some respects and too broad in other respects.

Two examples will serve to clarify this. The explication "*P* is angry if and only if *P* is frustrated" is too broad if respondents agree on the following: One cannot imagine a situation in which someone is angry yet not at all frustrated. On the other hand, one can imagine situations in which someone is very frustrated yet not at all angry. Therefore, the proper form of the explication should be "*P* is angry only if *P* is frustrated."

The explication "*P* is surprised only if *P* experiences something that she has expected not to occur" is too narrow if respondents agree on the following: One cannot imagine a situation in which someone is surprised yet has not experienced something that she had expected not to occur. However, one also cannot imagine a situation in which someone has experienced something that she had expected not to occur, yet does not feel any surprise. Therefore, the proper form of the explication should be "*P* is surprised if and only if *P* experiences something that she had expected not to occur."

Third, it should be noted that the definition of a successful explication of common sense does *not* refer to a criterion of correspondence between the explication and the actual behavior that it refers to. There is a very fundamental reason for this: Let us continue to use the example: "*P* is surprised if and only if *P* experiences something that she has expected not to occur." The criterion of the degree of correctness of this explication is, by definition, the amount of agreement among members of the culture involved that it is correct and that its negation is incorrect. The alternative criterion of correspondence with actual behavior would amount to observing the extent to which people are, in fact, surprised when they experience something they had expected not to occur, and the extent to which when people are surprised they have experienced something they had expected not to occur. The reason why this criterion is unacceptable is that it cannot lead to a falsification. It simply does not make sense to say that *P* is surprised, yet *P* has not experienced anything that *P* had expected not to occur. Similarly, it does not make sense to say that *P* has experienced something that *P* had expected not to happen, yet *P* is not at all surprised. Such descriptions are not acceptable and always require additional explanations that restore the validity of the original explication. Hence, it appears that the successfulness of an explication of common sense does not depend on its conformity with observations, but on consensus about its correctness.

Fourth, the terms "correct" and "incorrect" have quite specific meanings here. They refer to judgments to the effect that the implicational relationship does or does not in fact hold up, i.e. that *P* actually implies *Q* or that *Q* actually follows from *P*. Another way of expressing this is to say that the implication exists because of the commonly accepted meanings of *P* and *Q*. Common sense is agreement about what follows from what.

Given a delimitation of the subject matter to be investigated (psychological common sense) and the goal of research (the successful explication of psychological common sense), the next step is to discuss how the goal may be achieved.

Method

It has already been mentioned that the process of arriving at an explication of common sense cannot in itself be explicated. However, once a formulation has been achieved, it needs to be evaluated and, if necessary, to be improved. In order to perform the evaluation we need to rely on adequate methods.

It follows from the definition of a successful explication of common sense that an evaluation must involve a study of the degree of *consensus* among members of a culture as to the correctness of the explication. An optimal study of this sort of consensus would have to include several features:

Each person studied should have to make her judgments independently of others. The degree of consensus arrived at should not depend on social pressure or influence of any kind, but should simply reflect the person's being a member of a given culture.

Since expressions such as "imply" and "follow from" which refer directly to the implicative relationship may be difficult and ambiguous, each person could instead be asked only the following four elementary types of questions: Is it conceivable that (could it possibly be the case that) (1) *P* and *Q*, (2) *P* and not-*Q*, (3) not-*P* and *Q*, (4) not-*P* and not-*Q*?

However, because abstract symbols can be difficult and confusing and do not belong to the vernacular, the questions should only be asked in concrete form. Example: Is it conceivable that (could it possibly be the case that (1) Jane has passed an examination that she had definitely expected to fail *and* she is very surprised (yes or no), (2) Jane has passed an examination that she had definitely expected to fail *and* she is not at all surprised (yes or no), (3) Jane has passed an examination that she had definitely expected to pass and she is very surprised (yes or no), (4) Jane has passed an examination that she had definitely expected to pass *and* she is not at all suprised (yes or no).

There remains one serious difficulty with the proposed sort of procedure. Since questions such as "is it conceivable that?" or "could it possibly be the case that?" are often taken as a challenge to the inventiveness of a person, they are frequently responded to in ways which appear to defeat the purpose of the investigation. Examples: In response to question (2) a person may answer: "Yes, Jane may be so happy that she has no time to be surprised", or "Yes, Jane is so depressed because her boyfriend has left her, that she doesn't respond to events at all." In response to question (3) a person may answer: "Yes, Jane has had so many surprises these last days and nothing has gone as she expected. Hence, she was actually surprised that something *did* come out as expected." The conclusion from such answers should not be to give up the project of studying common sense. Rather, one should take appropriate measures to exclude considerations going beyond the scope of the given questions. There is one feature common to all the three deviating answers mentioned, namely that they introduce some *additional* factor (happiness, depression, other recent experiences). These can be excluded by a rule stating that "no other circumstances intervene". This rule may be repeatedly emphasized in the instructions. It sometimes appears to work well, but it also serves to make the original questions "is it conceivable that/could it possibly be the case that" even more difficult and confusing. This is so because they directly challenge the person to try to invent some circumstances under which a proposition would or would not be true while, at the same time, all such circumstances are to be excluded.

A way out this dilemma that I have found useful is to introduce the concept of *acceptable explanation.* An acceptable explanation is one that is intelligible (meaningful, makes sense) as it is formulated without any additional information or assumptions. An unacceptable explanation is one that is not intelligible (not meaningful, does not make sense) as it is formulated, but requires additional information or assumptions in order to become intelligible (meaningful, make sense). Since P and Q in an implication $P \supset Q$ are an ordered pair, it is actually close to ordinary language to talk about explanation; P explains Q, Q occurs because of P. But this means that the four standard questions mentioned earlier may be reformulated as follows: is this an acceptable explanation (yes or no): (1) Jane is very surprised *because* she has passed an examination that she had definitely expected to fail, (2) Jane is not at all surprised *because* she has passed an examination that she had definitely expected to fail, (3) Jane is very surprised *because* she has passed an examination that she had definitely expected to pass, (4) Jane is not at all surprised *because* Jane has passed an examination that she had definitely expected to pass. When people are asked about which of these four explanations that are acceptable and which are unacceptable, the outcome is a very high degree of consensus to the effect that the two states of affairs referred to (surprise and the experience of something expected not to occur) mutually imply each other.

I have now described briefly the sort of method that may be used to evaluate the successfulness of an attempted explication of common sense. In a recent study (SMEDSLUND 1982b) involving 36 formulations of common sense concerning behavior modification, judgments of the acceptability of explanations turned out to yield a high degree of consensus: Explanations consistent with the proposed formulations were judged to be acceptable by 92% of the participants and explanations inconsistent with the proposed formulations were judged to be unacceptable by 96% of the participants. The reader is referred to this study for further details on methodology. The outcome shows that, even in a pilot study which had methodological shortcomings, it is possible to formulate highly successful explications of psychological common sense. In another study, less methodologically developed (SMEDSLUND 1982a), some progress was reported in exploring the common sense involved in interpersonal relations in psychological treatment.

The metatheoretical status of valid explications of common sense

Let us suppose that it is indeed generally possible to formulate valid explications of common sense, i.e. explications that yield approximate consensus within a given culture. What sort of propositions are these?

First, they are regarded as necessarily correct by all members of the culture. Since the psychological researchers are also members of the culture, they too will regard them as necessarily correct. Technically, they may also be given this status by introducing consensually acceptable definitions of the terms. In the example used, surprise may be defined as "the feeling that accompanies the experiencing of something that has been expected not to occur." Inserting this in the original formulation we get: "*P* gets the feeling that accompanies the experiencing of something that has been expected not to occur if and only if *P* experiences something that she has expected not to occur." It is possible, in this way, to prove all other valid explications of common sense too. For examples, including technical proofs of 36 relatively complex formulations, see SMEDSLUND (1978).

We may, then, conclude with respect to the modal status of successful explications of psychological common sense that they are *noncontingently true.*

Second, the propositions involved do not require any particular new experiences, but rely on the already existing intuitions of all members of a culture. They have the status of *a priori* for the persons participating in the explication process. Cf. BRADLEY and SWARTZ's definition: "*P* is knowable a priori" $=_{def}$ "It is humanly possible to known *P other than* experientially" (1979, p. 150). The implication follows from what the words mean, and what the words mean is known to the members of the culture. It is another matter, of no relevance here, that this knowledge was once acquired through experience in the socialization of each individual.

The conclusion is that the epistemic status of the explications of common sense is *a priori.*

Third, the propositions involved appear to be *normative* rather than descriptive. More specifically, they may be characterized as expressing *obligations* (rules that *must* be followed). There are three main arguments for this view: First, an obligation can be transgressed, whereas a factual necessity cannot. One can *say* "I am very surprised although I have not experienced anything that I had expected not to happen" but this is a wrong (forbidden) way of speaking. Second, a transgression does not invalidate an obligation in the way a deviation invalidates a description. Even though a transgression occurs, the obligation remains valid. Finally, there are usually direct and indirect *sanctions* after a transgression aimed at restoring normal usage of language.

These matters are complicated and need further clarification. Meanwhile, one may conclude, tentatively, that successful explications of psychological common sense are *normative* and have the form of *obligations.* They describe implications that exist given the proper (obligatory) meanings of words.

Summarizing the preceding it may, then, be said that valid explications of psychological common sense are *noncontingent, a priori* and *normative.*

The utility of common sense psychology

Let us assume that it is generally possible to explicate psychological common sense in a valid manner, and also that the presented account of the metatheoretical status of the outcome is correct. What, then, can one do with such a system of propositions?

To begin with, it should be noted that common sense explications are, and must be, phrased in the ordinary language of the culture involved. This language is used by people to *describe, explain* and *predict* their own and each others behavior. It is suggested here that common sense psychology formulates the explanatory and descriptive structure of that language. The implications formulated ($P \supset Q$) can yield explanations, "Q becauseof P" and predictions "Given P one may expect Q", as well as retroactive inferences "given not-Q one may infer not-P". Furthermore, common sense formulations may be a help in solving practical problems, including those of professional psychology: "In order to bring about Q, try to bring about P!"

Applied to the concrete example used above, we get: "She was surprised *because* something happened that she had expected would not happen", "if P experiences something that she has expected not to happen, then one may predict that she will be surprised", "in order to surprise P, try to introduce something that she expects not to happen!" With more complicated examples than the one given, many useful and nontrivial explanations, predictions and practical procedures may, hopefully, be generated.

Another, equally useful, aspect of explications of common sense has to do with the failure of prediction. If you expose a person to an event that she definitely expects not to occur, and she shows no signs of surprise whatsoever, there are only two possible inferences one can make. Either she is hiding here surprise successfully, or she is really not surprised. In the latter case she either cannot 'after all' have been certain that the event would not occur, or other concerns may have masked or drowned out the surprise. It remains true that failure of prediction means that the sought-for conditions have not been established. Hence, common sense formulations are useful in the evaluation of the success of practical procedures.

In summary, common sense psychology appears to be practically useful by virtue of being an explication and systematization of the predictive potentialities embedded in ordinary language. (For examples of practically relevant analyses see SMEDSLUND 1978, 1980, 1981, 1982a, 1982b.)

Weaknesses of traditional psychology

It is immediately apparent that common sense psychology and traditional psychology are diametrical opposites when seen from a metatheoretical point of view. Common sense psychology is *noncontingent, a priori,* and *normative,* whereas traditional psychology is taken to be *contingent, empirical* and *descriptive.* They will be regarded here as incompatible and competing total views. Having presented briefly the case for common sense psychology, I will now turn to a consideration of certain features of the traditional view of psychology.

The main thrust of my argument will be that traditional psychology has some very serious intrinsic weaknesses which cannot be eliminated, and which will, therefore, eventually lead to its demise. It will be replaced by a psychology consisting of explications of common sense. The weaknesses of traditional psychology may be summarized under three main headings:

(1) Psychologists have generally taken for granted that their theoretical propositions are contingent and, hence, falsifiable. However, there has been no routine checking of the actual modal status of theoretical statements, and there is a corresponding scarcity of stringent conceptual definitions and derivations. When checks of modal status *are* done (SMEDSLUND 1984b), it turns out that many psychological propositions of the traditional kind are actually noncontingent and equivalent to more or less successful explications of common sense. From this it follows that the empirical research allegedly supporting such propositions must be characterized as *pseudoempirical*, i.e. as senseless attempts to test noncontingent propositions by empirical methods.

It is asserted here that a very considerable proportion of traditional psychological theories, even though masquerading as contingent, really consists of more or less successful explications of common sense, and, hence, is noncontingent. Psychologists have been guided by their sense of what is plausible. However, because of their false metatheory, they have not generally realized that the plausibility of a theoretical formulation may not stem from data and general experience, but from their own implicit common sense.

(2) It is traditionally assumed that psychological knowledge stems from experience, particularly in the form of research data. This view is protected and maintained by two fundamentally erroneous general assumptions:

The first assumption is that if a formulation appears to arise in connection with, or appears to be consistent with, some experience, then it must be empirical. It is not generally recognized that the achievement of a priori knowledge may also be facilitated by reflections upon experience. The reason for this lack of recognition is, I believe, that psychologists tend to regard a priori knowledge as belonging to the domain of philosophy or even as "unscientific" and, in both cases, as uninteresting. In order to clarify this issue, let us consider the definitions of "empirical" and "a priori": "*P* is knowable empirically" $=_{\text{def}}$ "It is humanly possible to know *P* *only* experientially" (BRADLEY and SWARTZ 1979, p. 150) and "*P* is knowable a priori" $=_{\text{def}}$ "It is humanly possible to know *P other than* experientially" (BRADLEY and SWARTZ 1979, p. 150). It follows that only by considering carefully the *possibilities* involved, can one determine the epistemic status of a proposition. This is almost never done in psychological research reports, and, consequently, an uncritical empiricist bias is perpetuated.

The second erroneous assumption is that experience is generally conducive to the acquisition of knowledge. This means that one is seldom considering seriously whether or not a certain set of experiences actually provides sufficient information to make possible the formulation of a given proposition. Psychologists "have come to have what can only be called a perverse conception of the nature of experience" (BREHMER 1980, p. 224). Briefly, this consists in assuming that truth is generally manifest and accessible in experience. However, it is actually difficult and frequently almost impossible to learn from experience because "experience often gives us very little information to learn from" (BREHMER 1980, p. 240). It follows that only by considering carefully the *content* of the available experience, can one determine to what extent a given proposition could possibly be based on it. Again, such considerations are almost never found in psychological research reports. This also permits an undisturbed perpetuation of the general idea that everything can be accounted for by experience.

The preceding points open up the possibility that traditional type psychological propositions may frequently not be what they pretend to be, namely contingent and empirical, but, on the contrary, may be noncontingent and a priori explications of common sense. However, the most serious difficulty of traditional psychology lies in its ignorance of the role of culture, and, hence, of the normative aspect of its subject matter.

(3) There are two factual conditions concerning the role of culture which have not been taken seriously in the traditional approach. When they *are* taken seriously, the traditional point of view will necessarily break down.

First, it is impossible to do psychological research unless the researcher participates in the culture of the persons studied (see SMEDSLUND 1984a). To participate in the culture means to share that which is taken for granted by the members of the culture (the meanings of words, acts, situations, i.e. what follows from these). Only then can the researcher *communicate* with the subjects, grasp the meaning of their behavior and of the situation, etc.

Second, the persons studied describe, explain, and predict the behavior of others and their own behavior in terms of the ordinary language of the culture and according to the common sense psychology embedded in this. Accordingly, the research psychologist has the task of describing, explaining and predicting the behavior of persons who already share a system of describing, explaining and predicting, which guides their behavior. In other words, scientific psychology must not only presuppose common sense psychology, but also account for it.

Common sense psychology expresses what follows from what in matters psychological in a given culture. The scientific researcher must account for that, but in terms of propositions which do not contradict those of common sense psychology. Because to contradict common sense is to break the rules about what words mean and, therefore, to become unable to predict and explain the behavior of ordinary people as well as to become nonsensical to them. Said in yet another way, the behavior of people in a culture is structured by the common sense psychology of that culture and any valid psychological account must conform with this. But this means that scientific and common sense descriptions, explanations and predictions become indistinguishable and rely on propositions which are noncontingent, a priori and normative. This means that the traditional approach to psychology with its inadequate metatheory will disappear.

The last stand of empirical psychology

The deeply engrained tradition of empiricism in psychology will not die easily. However, in the future it can only survive under very much harsher conditions than today. Its formulations must routinely be shown to be contingent *and* empirical. This requires explicit definitions of the terms involved, as well as detailed analyses of the sort of data that permit given formulations. Furthermore, they must not contradict psychological common sense in the given culture.

Let us use the example of surprise to illustrate what this means. An empirical psychologist may want to study the conditions of surprise. The relation between presence/absence of surprise and presence/absence of experience of something that is not expected lies within common sense and cannot be studied empirically. The researcher may want to study the relationship in more detail than this. It may turn out that it is common sense that the relation between amount of surprise and amount of unexpectedness must be a *direct* and *monotonic* one. On the other hand, the *shape* of the function may arguably go beyond the explicable and shared intuitions of members of the culture. People may share the notion "the more unexpected the event, the stronger the surprise", but nothing more refined. It may then be a task for an empirical psychologist to establish whether or not people actually are able to or can be brought to predict reliably with a ratio scale of surprise and a ratio scale of unexpectedness, whether or not any generalizable functions emerge, etc. It is not the place here to discuss the formidable problems encountered by such research, notably with respect to such matters as ***relevance***, ***representativeness*** and ***replicability***. (See e.g. GERGEN (1973, 1976) and SMEDSLUND (1979).)

Since empirical psychology is restricted to problems where common sense is demonstrably silent (absent), it clearly runs the risk of being of little relevance for real life problems, artifactual and hard to replicate. The prospects for the empirical approach are brightest in biological psychology where common sense is generally silent, whereas the prospects are dimmest in social psychology where common sense psychology regulates everything.

Conclusion

Although the form of this paper has been polemic with respect to traditional empirical psychology, let me now retreat a little and simply summarize what has been tentatively established.

A concept of common sense and a concept of a valid explication of common sense have been elaborated. A methodology of how to establish explications of common sense has been described and some successful implementations have been mentioned. The metatheoretical characteristics of explications of common sense have been elaborated and contrasted with traditional psychological propositions.

Given these analyses, it has been shown that traditional empirical psychology is highly vulnerable to attack. It can survive only when purged of noncontingent and a priori features and by being restricted to areas where psychological common sense is silent, notably in biological domains.

Meanwhile, the central and fundamental role of psychological common sense has been established. Hopefully, this will lead to more rapid advances in psychological theory, but also in practical psychology which has suffered greatly from the empiricist bias. Notably this has occurred through the view that professional competence is somehow achieved through great quantities of relevant experience, and that scientific knowledge about e.g. treatment is achieved through the accumulation of more data. The alternative view defended here is that advances in professional competence and in theoretical knowledge about treatment are the result of an increasingly penetrating analysis and exploitation of psychological common sense.

References

BRADLEY, R. and SWARTZ, N., 1979, *Possible Worlds. An Introduction to Logic and its Philosophy* (Basil Blackwell, Oxford).

BREHMER, B., 1980, *In one word: not from experience*, Acta Psychologica 45, pp. 223–241.

GERGEN, K.J., 1973, *Social psychology as history*, J. Personality and Social Psychology 36, pp. 309–320.

GERGEN, K.J., 1976, *Social psychology, science and history*, Personality and Social Psychology Bull. 2, pp. 373–383.

SMEDSLUND, J., 1978, *Bandura's theory of self-efficacy: a set of common sense theorems*, Scandinavian J. Psychology 19, pp. 1–14.

SMEDSLUND, J., 1979, *Between the analytic and the arbitrary: a case study of psychological research*, Scandinavian J. Psychology 20, pp. 129–140.

SMEDSLUND, J., 1980, *Analyzing the primary code: from empiricism to apriorism*, in: The social foundations of language and thought. Essays in honor of J.S. Bruner, ed. D. Olson (Norton New York).

SMEDSLUND, J., 1981, *The logic of psychological treatment*, Scandinavian J. Psychology 22, pp. 65–77.

SMEDSLUND, J., 1982a, *Seven common sense rules of psychological treatment*, J. Norwegian Psychological Assoc. 19, pp. 441–449.

SMEDSLUND, J., 1982b, *Revising explications of common sense through dialogue: Thirty-six psychological theorems*, Scandinavian J. Psychology 23, pp. 299–305.

SMEDSLUND, J., 1984a, *The invisible obvious: culture in psychology*, in: Lagerspetz, K.M.J. and Niemi, P., eds., Psychology in the 1990's (Elsevier, Amsterdam), pp. 443–452.

SMEDSLUND, J., 1984b, *What is necessarily true in psychology?*, Ann. Theoret. Psychology 2, pp. 241–272.

Barcan Marcus et al., eds., Logic, Methodology and Philosophy of Science VII
© Elsevier Science Publishers B.V. (1986) 495–502

RESEARCH STRATEGY IN PSYCHOPHYSIOLOGY

EUGEN N. SOKOLOV

Moscow State Univ., Marx Avenue 18, Moscow, USSR

This paper deals with the foundation of the research strategy in psychophysiology based on the "man–neuron-model" principle. Data obtained in psychophysical experiments in man and neuronal mechanisms studied in animals are integrated in the framework of a model constructed from neuron-like elements. The output of the model as a whole simulates the macrolevel and the responses of neuron-like elements simulate the microlevel of information processing. The constructed model representing a working hypothesis is used for quantitative predictions in planning of experiments.

The "man–neuron-model" research strategy can be demonstrated in the area of psychophysiology of colour vision. The integration of data from colour psychophysics and colour-coding neurons is achieved in a model of a colour analyser compiled from neuron-like elements.

The perceived colours are located on a sphere in the four-dimensional space with the euclidian distances between the points representing colours equal to the subjective difference between the colours. The red–green, blue–yellow, white–black and gray neurons being linear combinations of cone responses represent orthogonal coordinates of the sphere. The colour-selective detectors represent colours on the surface of the sphere.

In his *Philosophical Notebooks*, V.I. Lenin has formulated the concept of cognition as a process which being not strictly linear is represented by a spiral of scientific approximation. The cognitive spiral is a sequence of experiments, theory and its practical application which, proving the correctness of theory, stimulates in turn new concepts and new experiments.

This paper deals with the application of the cognitive spiral concept to the formulation of the research strategy in psychophysiology.

Psychophysiology is the science which studies the physiology of the psychic functions and the brain–body behaviour interrelationships of the living organism in conjunction with the environment.

The most remarkable step in recent development of psychophysiology is single-unit recording in a conscious man. It turned out that the responses of some neurons are directly connected with psychic phenomena. Now the core substance of psychophysiology is a study of neuronal mechanisms of psychic processes and states.

The research strategy in psychophysiology is based on the "man–neuron-model" principle. The first step of the research consists in the evaluation of stimulus–response characteristics at the psychophysical level. The most powerful method in the area of psychophysics is multi-dimensional scaling. A person perceiving signals can estimate in numbers their subjective differences. The matrix obtained from these estimates obeys the axioms of metric space. This suggests a calculation of coordinates representing the signals in an n-dimensional perceptual space. The euclidian distances between the points representing the signals in an n-dimensional space closely fit with the subjective differences between appropriate signals. The fitness of the n-dimensional representation of signals is evaluated by a coefficient of correlation between the subjective differences and the euclidian distances.

The points representing the signals are not randomly distributed in the n-dimensional space, but are located on the surface of a sphere.

The second step of the research deals with single-unit responses obtained with the same set of stimuli which were used the in psychophysical experiments. Two main types of afferent neurons are found. The predetector neurons with gradual responses fit the orthogonal coordinates of the n-dimensional perceptual space. The selective detector neurons selectively tuned with respect to particular stimulus characteristics fit the local patches of the spherical surface.

The third step of the research is the integration of psychophysical and neurophysiological data achieved by constructing a model from neuron-like elements. The neuron-like element performs a linear summation weighing the inputs. Each selective detector having many inputs is characterized by a vector of synaptic connections transmitting a vector of excitations from gradual predetectors. The connection vectors of a given set of detectors are of a constant length. The response of a selective detector is a scalar product of the excitation vector and the connection vector. The excitation vector acting on the set of selective detectors produces in each detector dependent on its connection vector a response of particular magnitude. The maximum response is generated in a detector having a connection vector collinear with the given excitation vector.

The change of a stimulus results in a modification of an excitation vector generated by predetectors and accordingly a transition of the excitation maximum with respect to the set of detectors. The selective detectors characterized by connection vectors of a constant length are located on a sphere. Thus, the stimulus change is coded by the change of the position of the excitation maximum on the sphere composed from selective detectors. The differences between signals are modeled by euclidian distances between the points representing the signals. Thus, the model constructed from neuron-like elements as a whole simulates the responses at the psychophysical level. The neuron-like elements of the model generate responses analogous with the reactions of the nerve cells participating in the function under investigation. The computerized version of the model is a most effective research tool.

Two functions of the model should be separated in the psychophysiological research. The integrative function of the model serves to summarize the psychophysical and neurophysiological data into an uncontradictory system. As soon as the model is constructed it performs the function of a working hypothesis generating predictions concerning the outcomes of the planned experiments. The advantage of the working hypothesis in the form of a computerized model consists in its predictive power. The quantitative predictions are obtained from the model presenting it with signals extending the range of signals used in preceding experiments. The computerized model as a working hypothesis can simulate the outcomes of the psychophysical and neurophysiological experiments. The responses of the model as a whole are taken into consideration by analogy with the psychophysical experiments. The characteristics of single neuron-like elements are obtained from the computer by analogy with the neurophysiological experiments. The predictions derived from the model are tested at the next stage of psychophysical and neurophysiological experiments. If the results of these experiments fit the characteristics predicted from the model, the working hypothesis is tested further. If, however, the results of the experiments do not correspond with characteristics derived from the model, the model is modified in accordance with the total bulk of experimental data. The modifications of the model can refer to the reconstruction of connections between the neuron-like elements, to the introduction of new elements into the system and to changes of properties of single elements. The modified model integrating the results of an extended set of experiments is functioning as a working hypothesis at the next stage of experimentation. Thus, the model plays in the psychophysical research sequentially two roles: that of an accumulator of data and that of a predictor of the experimental results. These two functions of the model interchangeably incorporated into experiments are building up a spiral of a cognitive process. The validity of the models obtained in psychophysiological research are checked by their practical application for computerized signal recognition and robot technology.

The "man–neuron-model" formula for research strategy in psychophysiology can be illustrated in the area of colour vision.

The differences between colours can be perceived and numerically estimated by the subject in psychophysical experiments. The matrix of the subjective differences between colours obeys the axioms of metric space. This allows the representation of colours as points in an n-dimensional space using a multi-dimensional scale. The evaluation of the results obtained from psychophysical experiments with aperture and pigment colours shows that colours of equal luminosity are located on a surface of a sphere in a three-dimensional space. The euclidian distances between the points representing these colours closely correlate with subjective differences between the perceived colours. The coordinate axes of the orthogonal colour space are represented by red–green, blue–yellow and an achromatic system. The polar coordinates of the sphere correspond with subjective aspects of colour perception. The horizontal angle fits the hue and the vertical angle fits the saturation.

The white colour is located on the pole and the monochromatic lights which produce with purple colours a continuous curve are positioned above the equator plane.

The colours of different luminosity are positioned on a surface of a sphere in a four-dimensional space with red–green, blue–yellow, white–black and grey orthogonal axis. Three polar coordinates of a four-dimensional sphere correspond to hue, saturation and luminosity, highly correlating with hue, saturation and the value of Mansell colour body. Thus, the achromatic lights of various luminosity represented by two-dimensional vectors with black–white and grey components are located on a semicircle.

The experiments at the neuronal level demonstrate several stages in the processing of colour information. The cones absorbing light in short- middle- and long-wave range generate a three-dimensional vector in a non-orthogonal space. The responses of red–green, blue–yellow and achromatic photopic horizontal cells compose the components of an excitation vector in an orthogonal coordinate system. The length of the excitation vector in which three types of the horizontal cells are participating corresponds to the subjective luminosity. It means that colours of equal luminosity are located on the surface of a sphere in the three-dimensional space. At the level of bipolar cells in the retina, the three-dimensional vector with the luminosity-dependent length is transformed into a four-dimensional vector of a constant length. This is achieved by a formation of seven colour-coding channels: red+ green−; red− green+; yellow+ blue−; yellow− blue+; black+ white−; black− white+ and grey. Because of the opponent characteristics of these bipolar cells (except grey) only four channels are activated by any particular colour generating a four-dimensional excitation vector of a constant length.

The upper layers of the geniculate body and some area of the visual cortex contain colour detectors selectively tuned to different colours.

The model of a local colour analyser integrating psychophsyical and neurophysiological data is constructed from neuron-like elements simulating colour-coding cells. The light generates in the cone-analogues a three-dimensional vector. The responses of the photopic horizontal cell-analogues being a linear combination of cone inputs represent orthogonal coordinates for a colour-coding system. This three-dimensional vector with the length representing the luminosity is transformed by analogues of bipolar cells into a four-dimensional excitation vector of a constant length. The four-dimensional vector activates in parallel manner a set of selective colour detector-analogues. Each selective colour detector-analogue is characterized by a four-dimensional vector simulating synaptic connections converging on this neuron-like element. The response of such a selective colour detector-analogue is a scalar product of a given excitation vector and a connection vector. The connection vectors in a given set of selective colour detector-analogues are assumed to be of a constant length. It means that the maximum scalar product will be obtained when the excitation vector will be collinear with the connection vector. In other words, the excitation maximum will be generated at a particular selective colour detector-analogue having the vector of synaptic connections collinear with the given excitation vector. The change of the spectrum at the input of the model results in a modification of the four-dimensional excitation vector and a transition of the excitation maximum from one selective colour detector-analogue to the other.

The model as a whole simulates by euclidian distances between the points in the four-dimensional space the subjective differences between the colours in man. At the same time the responses of neuron-like elements of the model simulate the responses of appropriate neurons participating in the colour-coding process.

This model representing a local colour analyser does not simulate the influence of a colour background observed in psychophysical experiments in the form of a simultaneous colour contrast. The next stage of the cognitive spiral in the psychophysiology of colour vision consists in such a transformation of the model which preserving all other effects would demonstrate a simultaneous colour contrast. This modification of the model is achieved by combination of several local colour analysers into a system with inhibitory connections between identical colour predectectors constituting the components of the four-dimensional vector.

Thus, the extension of the test conditions stimulates the perfection of the model. Such an iterative procedure of experiments and model perfection compiles a spiral of cognition.

The integration of psychophysical and neurophysiological data into a framework of a model compiled from neuron-like elements opens new perspectives for interpretations. Some examples are to be mentioned.

1. The interpretation of the coordinates obtained by multi-dimensional scaling

The basic problem confronting the multi-dimensional scaling is an appropriate choice of the dimensions. Because of random errors present during estimations of subjective differences between the signals the number of dimensions obtained by multi-dimensional scaling surpass the number of real dimensions. Two types of mistakes are admitted: either some relevant dimensions are excluded or some non-relevant dimensions are incorporated.

The adequate choice of the coordinates is based on their neuronal functions as the responses of the predetector neurons. The number of dimensions of the perceptual space should equal the number of independent neuronal channels generating an excitation vector at the predetector level.

The other problem which the multi-dimensional scaling is facing refers to the orientation of the axis of the perceptual space. The multi-dimensional scaling is lacking such data. Because the orthogonal coordinate system is defined by responses of the predetector neurons the axis of the perceptual space should be oriented in such a way that the coordinates of the signals obtained by multi-dimensional scaling correspond with the responses of predectector neurons evoked by the same set of signals.

2. The interpretation of subjective differences between signals

The difference between two signals on the sphere is characterized by a spherical angle dividing the excitation vectors produced by these signals. The subjective difference between them is measured by a euclidian distance between the ends of the excitation vectors. In other words, the subjective difference is a euclidian distance between selective feature-detectors representing signals on the surface of the sphere.

3. The interpretation of the threshold

The spherical model of perception predicts that the just noticeable differences between signals are characterized by a spherical angle seperating two neighbouring feature-detectors located on the surface of the n-dimensional sphere. Two different signals generating excitation vectors which produce response maxima on the same feature-detector are not discriminated.

4. The re-interpretation of Weber–Fechner's and Steven's laws

The intensity of the signal in the spherical model of perception is coded at the level of predetectors by a two-dimensional vector and at the level of selective detectors by neurons selectively tuned to different degrees of intensity. The spherical model for perception of intensity is reduced to a semicircle. The differential threshold in the spherical model of intensity perception is characterized by the angle between neighbouring intensity selective feature-detectors. The introduction of an angle as a measure of the threshold is a re-interpretation of Weber's law in the framework of the spherical model.

The integration results in a linear function between logarithm of the stimulus intensity and the angle separating the background excitation vector and the vector representing particular stimulus intensity. This result was also obtained from multi-dimensional scaling. The functional relationship between the logarithm of stimulus intensity and the angle characterizing the position of the stimulus on the sphere is a re-interpretation of Fechner's law.

The angle between the excitation vectors representing stimuli of different intensity does not correlate directly with subjective differences which are euclidian distances. The euclidian distances are sine functions of the semi-angles between vectors. This non-linear effect of subjective scaling with respect to logarithm of stimulus intensity is the re-interpretation of Steven's law.

The re-interpretated versions of Weber–Fechner's and Steven's laws fit the spherical model of perception which resulted from the "man–neuron-model" strategy in psychophysiology.

Barcan Marcus et al., eds., Logic, Methodology and Philosophy of Science VII
© Elsevier Science Publishers B.V. (1986) 503–520

THE FRAMING OF DECISIONS AND THE EVALUATION OF PROSPECTS

AMOS TVERSKY and DANIEL KAHNEMAN

Dept. of Psychology, Stanford Univ., Stanford, CA, U.S.A.

The modern theory of individual decision making under risk, as formulated by BERNOULLI (1738), axiomatized by VON NEUMANN and MORGENSTERN (1947) and generalized by SAVAGE (1954), has emerged from a logical analysis of games of chance, not from a psychological analysis of choice behavior. The theory has been developed primarily as a normative model that describes the behavior of an idealized rational person, not as a descriptive model that explains the behavior of real people. As one noted economist put it, the theory "has a much better claim to being called a logic of choice than a psychology of value" (SCHUMPETER, 1954, p. 1058).

However, the tension between logical and psychological considerations and the interaction between normative and descriptive arguments have characterized decision theory from its early days. Bernoulli and Cramer introduced concave utility functions in order to explain for money to rationalize risk aversion and reconcile individual differences in risk bearing with the concept of mathematical expectations. Similarly, the modern theory of personal probability, developed by RAMSEY (1931), DE FINETTI (1937) and SAVAGE (1954), can also be viewed as an attempt to generalize decision theory so as to permit individuals to assign different probabilities to the same event — if they do not have the same information or if they hold different beliefs. Hence, the normative analysis of value and belief has been extended to accommodate psychological considerations.

On the other hand, it has been widely believed that an adequate normative theory of choice must also provide an acceptable descriptive model because (i) people are generally effective in pursuing their goals and (ii) more effective individuals, organizations and modes of action are more likely to survive than the less effective ones. Indeed, the expected utility model has been extensively used to explain personal, economic and political decisions. These applications have been based on the assumption that the axioms of rational choice (e.g., transitivity, substitution) represent an acceptable idealization of human behavior and that the expected utility model, which follows from these axioms, provides a reasonable approximation of individual decision making under risk or uncertainty.

This position has been challenged by two lines of evidence. The first, initiated by the French economist Maurice ALLAIS (1953), indicates that the axioms of independence and substitution, which underlie the expected utility model, are consistently violated in a predictable manner. The second line of evidence came from psychological experiments showing that the preference order between prospects depend critically on the manner in which they are represented or framed. This work challenges not only the axioms of expected utility theory but the more fundamental principle that preferences are independent of the manner in which the choices are described. In this article we illustrate the effects of framing, outline a descriptive theory of choice constructed to account for them and discuss some of its implications. The present paper follows closely TVERSKY and KAHNEMAN (1981); see also KAHNEMAN and TVERSKY (1979).

The options among which one must choose are defined by their possible outcomes and the probabilities (or contingencies) with which they occur. The outcomes and the contingencies associated with a particular choice can be described or framed in different ways. The frame that a decision maker adopts is controlled partly by the formulation of the problem, and partly by the decision maker's norms, habits and personal characteristics. Alternative frames for a decision problem may be compared to alternative perspectives on the same visual scene. Veridical perception requires that the perceived relative height of two neighboring mountains, say, should not reverse with changes of vantage point. Similarly, rational choice requires that the preference between options should not reverse with changes of frame. Because of imperfections of human perception and decision, however, changes of perspective often reverse the relative apparent size of objects and the relative desirability of options.

Systematic reversals of preference, induced by variations in the framing of contingencies or outcomes, have been observed in a variety of problems and in different groups of respondents. Here we present selected illustrations of preference reversals with data obtained from students at Stanford University and at the University of British Columbia, who answered brief questionnaires in a classroom setting. The total number of respondents for each problem is denoted by N, and the percentage who chose each option is indicated in parentheses.

The effects of variations in framing is illustrated in Problems 1 and 2.

Problem 1 ($N = 152$). Imagine that the U.S. is preparing for the outbreak of an unusual Asian disease, which is expected to kill 600 people. Two alternative programs to combat the disease have been proposed. Assume that the exact scientific estimate of the consequences of the programs are as follows:

If Program A is adopted, 200 people will be saved. (72%)

If Program B is adopted, there is 1/3 probability that 600 people will be saved, and 2/3 probability that no people will be saved. (28%)

Which of the two programs would you favor?

The majority choice in this problem is risk averse: the prospect of certainty saving 200 lives is more attractive than a risky prospect of equal expected value, i.e., one chance in three to save 600 lives.

A second group of respondents received the cover story of Problem 1 and a different formulation of the alternative program as follows.

Problem 2 ($N = 155$). If Program C is adopted 400 people will die. (22%)

If Program D is adopted there is 1/3 probability that nobody will die, and 2/3 probability that 600 people will die. (78%)

Which of the two programs would you favor?

The majority choice in Problem 2 is risk seeking: the prospect of 400 people certainly dying is less acceptable than two chances in three of a loss of 600 lives. The preferences in Problems 1 and 2 illustrate a common pattern: choices involving gains are often risk averse while choices involving losses are often risk seeking. However, it is easy to see that Problems 1 and 2 are in fact identical. The only difference between them is that the outcomes are described in Problem 2 by the number of lives lost, and in Problem 1 by the number of lives saved relative to an anticipated loss of 600 lives. The change in the description of the outcomes, from lives saved to lives lost, is accompanied by a pronounced shift from risk aversion to risk seeking. We have observed this reversal in several groups of respondents, including university faculty and physicians. The inconsistent responses to Problems 1 and 2 arise from the conjunction of a framing effect with contradictory attitudes toward risks involving gains and losses.

In order to explain such findings we have developed a descriptive model of choice, called prospect theory (KAHNEMAN and TVERSKY, 1979). Prospect theory distinguishes two phases in the choice process: an initial phase in which outcomes and contingencies are framed, and a subsequent phase of evaluation. For simplicity, we restrict the formal treatment of the theory to choices involving stated numerical probabilities and quantitative outcomes, such as money, time or number of lives. Consider a prospect that yields outcome x with probability p, outcome y with probability q, and the status quo with probability $1-p-q$. According to prospect theory, there are values $v(\cdot)$ associated with outcomes, and decision weights $\pi(\cdot)$ associated with probabilities, such that the overall value of the prospect equals $\pi(p)v(x)+\pi(q)v(y)$. A slightly different equation should be applied if all outcomes of a prospect are on the same side of the zero point.[1]

In prospect theory, outcomes are expressed as positive or negative deviations (gains or losses) from a neutral reference outcome, which is assigned a value of zero. Although subjective values differ among individuals and attributes, we propose that the value function is commonly S-shaped, concave above the reference point and convex below it, as illustrated in Fig. 1. For example, the difference in subjective value between gains of \$10 and \$20 is greater than the subjective difference between gains of \$110 and \$120. The same relation between value-differences holds for the corresponding losses. Another property of the value function is that the response to losses is more extreme than the response to gains. The displeasure of losing a sum of money is generally greater than the pleasure associated with winning the same amount, as is reflected in people's reluctance to accept fair bets on a toss of a coin. Several studies of decision making and judgment[2] have confirmed these properties of the value function.

The second major departure of prospect theory from the expected utility model involves the treatment of probabilities. In expected utility theory, the utility of an uncertain outcome is weighted by its probability, while in prospect theory the value of an uncertain outcome is multiplied by a decision weight $\pi(p)$, which is a monotonic function of p but is not a probabability. The weighting function π has the following properties. First, impossible events are discarded, i.e., $\pi(0)=0$, and the scale is normalized

[1] If $p+q=1$ and either $x>y>0$ or $x<y<0$, the equation in the text is replaced by $v(y)+\pi(p)\,((v(x)-v(y))$, so that decision weights are not applied to sure outcomes.

[2] P. FISHBURN and G. KOCHENBERGER, Decision Sciences 10 (1979), pp. 503–518; S.A. ERAKER and H.C. SOX, Medical Decision Making 1 (1981). In the last study, several hundred clinic patients made hypothetical choices between drug therapies for severe headaches, hypertension and chest pain. Most patients were risk averse when the outcomes were decribed as positive (e.g., reduced pain, or increased life expectancy), and risk taking when the outcomes were described as negative (e.g., increased pain, or reduced life expectancy). No significant differences were found between patients who actually suffered from the ailments described and patients who did not.

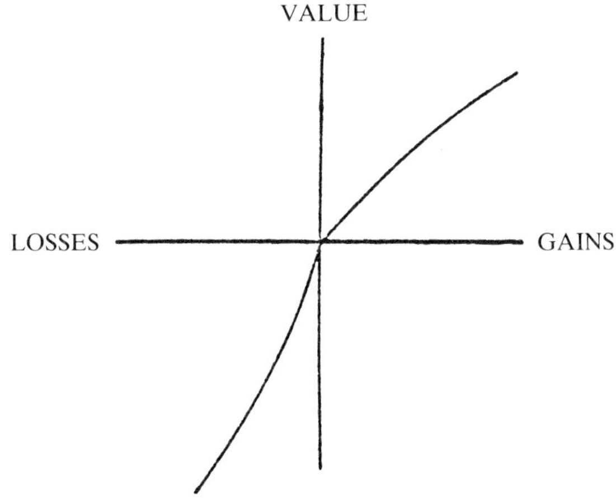

Fig. 1. A hypothetical value function.

so that $\pi(1)=1$, but the function is not well behaved near the endpoints. Second, for low probabilities $\pi(p)>p$, but $\pi(p)+\pi(1-p)<1$. Thus low probabilities are overweighted, moderate and high probabilities are underweighted, and the latter effect is more pronounced than the former. Third, $\pi(pq)/\pi(p)<\pi(pqr)/\pi(pr)$ for all $0<p, q, r\leqslant 1$. That is, for any fixed probability ratio q, the ratio of decision weights is closer to unity when the probabilities are low than when they are high, e.g., $\pi(0.1)/\pi(0.2)>\pi(0.4)/\pi(0.8)$. A hypothetical weighting function which satisfies these properties is shown in Fig. 2. The major qualitative properties of decision weights can be extended to cases in which the probabilities of outcomes are subjectively assessed rather than explicitly given. In these situations, however, decision weights may also be affected by other characteristics of an event, such as ambiguity or vagueness.

Prospect theory, and the scales illustrated in Figs. 1 and 2, should be viewed as an approximate, incomplete and simplified description of the evaluation of risky prospects. Although the properties of v and π summarize a common pattern of choice, they are not universal: the preferences of some individuals are not well described by an S-shaped value function and a consistent set of decision weights. The simultaneous measurement of values and decision weights involves serious experimental and statistical difficulties.

If π and v were linear throughout, the preference order between options would be independent of the framing of acts, outcomes, or contingencies.

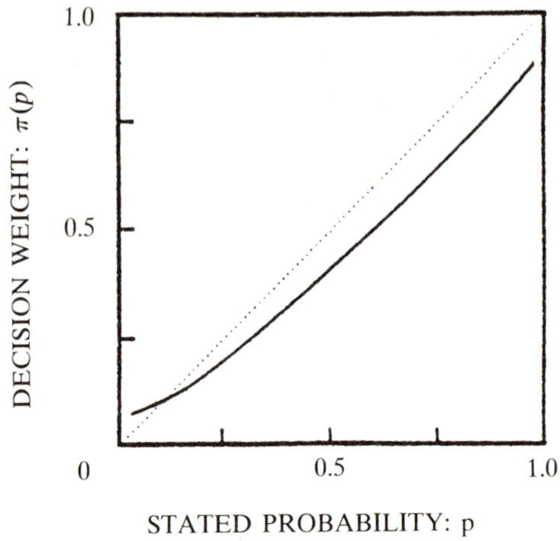

Fig. 2. A hypothetical weighting function.

Because of the characteristic non-linearities of π and v, however, different frames can lead to different choices. We first illustrate how the interaction of the cognitive rules of framing with the psychophysical principles of evaluation produce predictable violations of the dominance principle. The next two sections demonstrate reversals of preferences induced, respectively, by the framing of probabilities and outcomes.

Problem 3 ($N = 150$). Imagine that you face the following pair of concurrent decisions. First examine both decisions, then indicate the options you prefer.

Decision (i): Choose between:
A. a sure gain of $240 (84%),
B. 25% chance to gain $1000, and
75% chance to gain nothing (16%).

Decision (ii): Choose between:
C. a sure loss of $750 (13%),
D. 75% chance to lose $1000, and
25% chance to lose nothing (87%).

The modal choice in Decision (i) is risk averse: a riskless prospect is preferred to a risky prospect of equal or greater expected value. In contrast, the majority choice in Decision (ii) is risk seeking: a risky prospect is preferred to a riskless prospect of equal expected value. This pattern of risk aversion in choices involving gains and risk seeking in choices involving losses is attributable to the properties of v and π. Because the value function is S-shaped, the value associated with a gain of \$240 is greater than 24% of the value associated with a gain of \$1000, and the (negative) value associated with a loss of \$750 is smaller than 75% of the value associated with a loss of \$1000. Thus, the shape of the value function contributes to risk aversion in Decision (i) and to risk seeking in Decision (ii). Moreover, the underweighting of moderate and high probabilities contributes to the relative attractiveness of the sure gain in (i) and to the relative aversiveness of the sure loss in (ii). The same analysis also applies to Problems 1 and 2.

Because (i) and (ii) were presented together, the participants in fact had to choose one prospect from the set: A&C, B&C, A&D, B&D. The most common pattern (A&D) was chosen by 73% of respondents, while the least popular pattern (B&C) was chosen by only 3% of respondents. However, the combination of B&C actually dominates the combination of A&D, as is readily seen in Problem 4.

Problem 4 ($N = 86$). Choose between:
A&D. 25% chance to win \$240, and
75% chance to lose \$760 (0%).
B&C. 25% chance to win \$250, and
75% chance to lose \$750 (100%).

When the prospects were combined and the relation of dominance was made transparent, all respondents chose the dominant option. The popularity of the dominated alternative in Problem 3 implies that this problem was framed as a pair of separate choices. The respondents apparently failed to entertain the possibility that the conjunction of two seemingly reasonable choices could lead to an untenable result.

The violations of dominance observed in Problem 3 do not disappear in the presence of monetary incentives. A different group of respondents who answered a modified version of Problem 3, with real payoffs, produced the same pattern of choices[3]. Other studies have also reported that

---

[3] A new group of respondents ($N = 126$) was presented with a modified version of Problem 3, in which the outcomes were expressed as points. The participants were informed that the gambles would actually be played by tossing a pair of fair coins, that one participant in ten would be selected at random to play the gambes of his or her choice, that the payoffs would be proportional to the total number of points accumulated, and that it was possible to win as

violations of the rules of rational choice, originally observed in hypothetical equations, were not eliminated by payoffs.

We suspect that many concurrent decisions in the real world are framed independently, and that the preference order would often be reversed if the decisions were combined. The respondents in Problem 3 failed to combine options, although the integration was relatively simple and was encouraged by instructions. The complexity of practical problems of concurrent decisions, such as portfolio selection, would not allow people to integrate options without computational aids, even if they were inclined to do so.

The framing of contingencies

The following triple of problems illustrates the framing of contingencies. Each problem was presented to a different group of respondents, who were informed that one participant in ten, preselected at random, would actually play the prospect of his or her choice. Chance events were realized, in the respondents' presence, by drawing a single ball from a bag containing a known proportion of balls of the winning color, and the winners were paid immediately.

Problem 5 ($N = 77$). Which of the following options do you prefer?
A. a sure win of $30 (78%),
B. 80% chance to win $45 (22%).

Problem 6 ($N = 85$). Consider the following two-stage game. In the first stage, there is a 75% chance to end the game without winning anything, and a 25% chance to move into the second stage. If you reach the second stage you have a choice between:
C. a sure win of $30 (74%),
D. 80% chance to win $45 (26%).
Your choice must be made before the game starts, i.e., before the outcome of the first stage is known. Please indicate the option you prefer.

---

much as $26. To ensure a positive return for the entire set, a third decision was included, between an even chance to win 800 or 1600 points and an even chance to win 1000 or 1400 points. These payoff conditions, which produced considerable involvement, did not alter the pattern of preferences observed in the hypothetical problem: 67% of respondents chose prospect A, and 86% chose prospect D. The dominated combination of A & D was chosen by 60% of respondents, and only 6% favored the dominant combination of B & C.

Problem 7 ($N = 81$). Which of the following options do you prefer?
E. 25% chance to win $30 (42%),
F. 20% chance to win $45 (58%).

Let us examine the structure of these problems.[4] First, note that Problems 6 and 7 are identical in terms of probabilities and outcomes, because prospect C offers a 0.25 chance to win $30, while prospect D offers a probability of $0.25 \times 0.80 = 0.20$ to win $45. Consistency therefore requires that the same choice be made in Problems 6 and 7. Second, note that Problem 6 differs from Problem 5 only by the introduction of a preliminary stage. If the second stage of the game is reached, then Problem 6 reduces to Problem 5; if the game ends at the first stage, the decision does not affect the outcome. Hence, there seems to be no reason to make a different choice in Problems 5 and 6. By this logical analysis, Problem 6 is equivalent to Problem 7 on the one hand, and to Problem 5 on the other. The participants, however, responded similarly to Problems 5 and 6, but differently to Problem 7. This pattern of responses exhibits two phenomena of choice: the certainty effect, and the pseudo-certainty effect.

The contrast between Problems 5 and 7 illustrates a phenomenon discovered by ALLAIS (1953), which we have labeled the certainty effect: a reduction of the probability of an outcome by a constant factor has more impact when the outcome was initially certain than when it was merely probable. Prospect theory attributes this effect to the properties of π. It is easy to verify, by applying the equation of prospect theory to Problems 5 and 7, that people for whom the value ratio $v(30)/v(45)$ lies between the weight ratios $\pi(0.20)/\pi(0.25)$ and $\pi(0.80)/\pi(1.0)$ will prefer A to B and F to E, contrary to expected utility theory. Note that prospect theory does not predict a reversal of preference for every individual in Problems 5 and 7. It only requires that an individual who is indifferent between A and B prefer F over E. For group data, the theory predicts the observed directional shift of preference between the two problems.

[4] Another group of respondents ($N = 205$) were presented with all three problems, in different orders, without monetary payoffs. The joint frequency distribution of choices in Problems 5, 6 and 7 was as follows: ACE: 22, ACF: 65, ADE: 4, ADF: 20, BCE: 7, BCF: 18, BDE: 17, BDF: 52. These data confirm in a within-subject design the analysis of conditional evaluation proposed in the text. More than 75% of respondents made compatible choices (AC or BD) in Problems 5 and 6, while less than half of the respondents made compatible choices in Problems 6 and 7 (CE or DF), or in Problems 5 and 7 (AE or BF). The elimination of payoffs in these questions reduced risk aversion but did not substantially alter the effects of certainty and pseudo-certainty.

The first stage of Problem 6 yields the same outcome (no gain) for both acts. Consequently, we propose, people evaluate the options conditionally, as if the second stage had been reached. In this framing, of course, Problem 6 reduces to Problem 5. More generally, we suggest that a decision problem is evaluated conditionally when (i) there is a state in which all acts yield the same outcome, e.g., failing to reach the second stage of the game in Problem 6; (ii) the stated probabilities of other outcomes are conditional on the non-occurrence of this state.

The striking discrepancy between the responses to Problems 6 and 7, which are identical in outcomes and probabilites, could be described as a pseudo-certainty effect. The prospect yielding $30 is relatively more attractive in Problem 6 than in Problem 7, as if it had the advantage of certainty. The sense of certainty associated with option C is illusory, however, since the gain is in fact contingent on reaching the second stage of the game.

We have observed the certainty effect in several sets of problems, with outcomes ranging from vacation trips to the loss of human lives. In the negative domain, certainty exaggerates the aversiveness of losses that are certain relative to losses that are merely probable. In a question dealing with the response to an epidemic, for example, most respondents found "a sure loss of 75 lives" more aversive than "80% chance to lose 100 lives", but preferred "10% chance to lose 75 lives" over "8% chance to lose 100 lives", contrary to expected utility theory.

We also obtained the pseudo-certainty effect in several studies, where the description of the decision problems favored conditional evaluation. Pseudo-certainty can be induced either by sequential formulation, as in Problem 6, or by the introduction of causal contingencies. In another version of the epidemic problem, for instance, respondents were told that risk to life existed only in the event (probability 0.10) of the disease being carried by a particular virus. Two alternative programs were said to yield "a sure loss of 75 lives" or "80% chance to lose 100 lives" if the critical virus was involved, and no loss of life in the event (probability 0.90) of the disease being carried by another virus. In effect, the respondents were asked to choose between 10% chance to lose 75 lives and 8% chance to lose 100 lives, but their preferences were the same as when the choice was between a sure loss of 75 lives and 80% chance to lose 100 lives. A conditional framing was evidently adopted, in which the contingency of the non-critical virus was eliminated, giving rise to a pseudo-certainty effect. Note that the certainty effect reveals attitudes toward risk that are inconsistent with the axioms of rational choice, while the pseudo-certainty effect violates the more fundamental requirement that preferences should be independent of problem description.

Many significant decisions concern actions which reduce or eliminate the probability of a hazard, at some cost. The shape of π in the range of low probabilities suggests that a protective action which reduces the probability of a harm from 1% to zero, say, will be valued more highly than an action which reduces the probability of the same harm from 2% to 1%. Indeed, probabilistic insurance, which halves the probability of hazard, is judged to be worth less than half the price of regular insurance, which eliminates the risk altogether (KAHNEMAN and TVERSKY, 1979).

It is often possible to frame protective action in either conditional or unconditional form. For example, an insurance policy that covers fire but not flood could be evaluated either as full protection against the specific risk of fire or as reduction in the overall probability of property loss. The preceding analysis suggests that insurance should appear more attractive when it is presented as the elimination of risk than when it is described as a reduction of risk. Relevant evidence was obtained by SLOVIC, FISCHHOFF and LICHTENSTEIN (1982). These investigators found that a hypothetical vaccine which reduces the probability of contracting a disease from 0.20 to 0.10 is less attractive if it is described as effective in half the cases than if it is presented as fully effective against one of two (exclusive and equiprobable) virus strains, which produce identical symptoms. In accord with the present analysis of pseudo-certainty, the respondents valued full protection against an identified virus more than probabilistic protection against the disease.

The preceding discussion highlights the sharp contrast between lay responses to the reduction and the elimination of risk. Because no form of protective action can cover all risks to human welfare, all insurance is essentially probabilistic: it reduces but does not eliminate risk. The probabilistic nature of insurance is commonly masked by formulations which emphasize the completeness of protection against identified harms, but the sense of security that such formulations provide is an illusion of conditional framing. It appears that insurance is bought as protection against worry, not only against risk, and that worry can be manipulated by the labeling of outcomes and by the framing of contingencies. It is not easy to determine whether people value the elimination of risk too much, or the reduction of risk too little. The contrasting attitudes to the two forms of protective action, however, are difficult to justify on normative grounds.

The framing of outcomes

Outcomes are commonly perceived as positive or negative in relation to a reference outcome which is judged neutral. Variations of the reference point can therefore determine whether a given outcome is evaluated as a gain or as a loss. Because the value function is generally concave for gains, convex for losses, and steeper for losses than for gains, shifts of reference can change the value difference between outcomes and thereby reverse the preference order between options. Problems 1 and 2 illustrated a preference reversal, induced by a shift of reference that transformed gains into losses.

For another example consider a person who has spent an afternoon at the race track, has already lost $140 and is considering a $10 bet on a 15 : 1 longshot in the last race. This decision can be framed in two ways, which correspond to two natural reference points. If the status quo is the reference point, the outcomes of the bet are framed as a gain of $140 and a loss of $10. On the other hand, it may be more natural to view the present state as a loss of $140, for the betting day, and accordingly frame the last bet as a chance to return to the reference point or to increase the loss to $150. Prospect theory implies that the latter frame will produce more risk seeking than the former. Hence people who do not adjust their reference point as they lose are expected to take bets that they would normally find unacceptable. This analysis is supported by the observation that bets on longshots are most popular on the last race of the day (McGlothlin 1956).

Because the value function is steeper for losses than for gains, a difference between options will loom larger when it is framed as a disadvantage of one option, rather than as an advantage of the other option. An interesting example of such an effect in a riskless context has been noted by Thaler (1980). In a debate on a proposal to pass to the consumer some of the costs associated with the processing of credit-card purchases, representatives of the credit-card industry requested that the price difference be labeled a cash discount rather than a credit-card surcharge. The two labels induce different reference points, by implicitly designating as normal reference the higher or the lower of the two prices. Because losses loom larger than gains, consumers are less willing to accept a surcharge than to forego a discount. A similar effect has been observed in experimental studies of insurance: the proportion of respondents who preferred a sure loss to a larger probable loss was significantly greater when the former was called an insurance premium (Slovic et al., 1982).

These observations highlight the lability of reference outcomes, as well as their role in decision making. In the examples discussed so far, the neutral reference point was identified by the labeling of outcomes. A diversity of factors determine the reference outcome in everyday life. The reference outcome is usually a state to which one has adapted; it is sometimes set by social norms and expectations; it sometimes corresponds to a level of aspiration, which may or may not be realistic.

We have dealt so far with elementary outcomes, such as gains or losses in a single attribute. In many situations, however, an action gives rise to a compound outcome, which joins a series of changes in a single attribute, e.g., a sequence of monetary gains and losses, or a set of concurrent changes in several attributes. To describe the framing and evaluation of compound outcomes, we use the notion of a psychological account, defined as an outcome frame which specifies (i) the set of elementary outcomes that are evaluated jointly and the manner in which they are combined; (ii) a reference outcome that is considered neutral or normal. In the account that is set up for the purchase of a car, for example, the cost of the purchase is not treated as a loss, nor is the car viewed as a gift. Rather, the transaction as a whole is evaluated as positive, negative or neutral, depending on such factors as the performance of the car and the price of similar cars in the market. A closely related treatment has been offered by THALER (1980).

We propose that people generally evaluate acts in terms of a minimal acount, which includes only the direct consequences of the act. The minimal account associated with the decision to accept a gamble, for example, includes the money won or lost in that gamble, and excludes other assets or the outcome of previous gambles. People commonly adopt minimal accounts because this mode of framing (i) simplifies evaluation and reduces cognitive strain; (ii) reflects the intuition that consequences should be causally linked to acts; (iii) matches the properties of hedonic experience, which is more sensitive to desirable and undesirable changes than to steady states.

There are situations, however, in which the outcomes of an act affect the balance in an account that was previously set up by a related act. In these cases, the decision at hand may be evaluated in terms of a more inclusive account, as in the case of the bettor who views the last race in the context of earlier losses. More generally, a sunk-cost effect arises when a decision is referred to an existing account in which the current balance is negative. Because of the non-linearities of the evaluation process, the minimal account and a more inclusive one often lead to different choices.

Problems 8 and 9 illustrate another class of situations in which an existing account affects a decision:

Problem 8 ($N = 183$). Imagine that you have decided to see a play where admission is $10 per ticket. As you enter the theater you discover that you have lost a $10 bill.

Would you still pay $10 for a ticket for the play?

Yes (88%), No (12%).

Problem 9 ($N = 200$). Imagine that you have decided to see a play and paid the admission price of $10 per ticket. As you enter the theater you discover that you have lost the ticket. The seat was not marked and the ticket cannot be recovered.

Would you pay $10 for another ticket?

Yes (46%), No (54%).

The marked difference between the responses to Problems 8 and 9 is an effect of psychological accounting. We propose that the purchase of a new ticket in Problem 9 is entered in the account that was set up by the purchase of the original ticket. In terms of this account, the expense required to see the show is $20, a cost which many of our respondents apparently found excessive. In Problem 8, on the other hand, the loss of $10 is not linked specifically to the ticket purchase and its effect on the decision is accordingly slight.

The following problem, based on examples by SAVAGE (1954) and THALER (1980) further illustrates the effect of embedding an option in different accounts. Two versions of this problem were presented to different groups of subjects. One group ($N = 93$) where given the values that appear in parentheses, and the other group ($N = 88$) were given the values shown in brackets.

Problem 10. Imagine that you are about to purchase a jacket for ($125) [$15], and a calculator for ($15) [$125]. The calculator salesman informs you that the calculator you wish to buy is on sale for ($10) [$120] at the other branch of the store, located 20 minutes drive away. Would you make the trip to the other store?

The responses to the two versions of Problem 10 were markedly different: 68% of the respondents were willing to make an extra trip to save $5 on a $15 calculator, while only 29% were willing to exert the same amount of effort when the calculator's price was $125. Evidently, the respondents do not frame Problem 10 in the minimal account, which involves only a benefit of $5 and a cost of some inconvenience. Instead, they evaluate the potential saving in a more inclusive account, which includes the purchase of the calculator, but not of the jacket. By the curvature of v, a discount of $5 has a greater impact when the calculator's price is low than when it is high.

A closely related observation has been reported by PRATT, WISE and ZECKHAUSER (1979) who found that the variability of the prices at which a given product is sold by different stores is roughly proportional to the mean price of that product. The same pattern was observed for both frequently and infrequently purchased items. Overall, a ratio of 2 : 1 in the mean price of two products is associated with a ratio of 1.86 : 1 in the standard deviation of the respective quoted prices. If the effort that consumers exert to save each dollar on a purchase, e.g., by a phone call, were independent of price, the dispersion of quoted prices should be about the same for all products. In contrast, the data of PRATT et al. (1979) are consistent with the hypothesis that consumers hardly exert more effort to save $15 on a $150 purchase than to save $5 on a $50 purchase. Many readers will recognize the temporary devaluation of money which facilitates extra spending and reduces the significance of small discounts in the context of a large expenditure, such as buying a house or car. This paradoxical variation in the value of money is incompatible with the standard analysis of consumer behavior.

Discussion

This paper has presented a series of demonstrations in which seemingly inconsequential changes in the formulation of choice problems caused significant shifts of preference. The inconsistencies were traced to the interaction of two sets of factors: variation in the framing of acts, contingencies and outcomes, and the characteristic non-linearities of values and decision weights. The demonstrated effects are large and systematic, although by no means universal. They occur when the outcomes concern the loss of human lives as well as in choices about money; they are not restricted to hypothetical questions and are not eliminated by monetary incentives.

The dependence of preferences on frames can be compared to the dependence of perceptual appearance on perspective. Imagine that you notice, while traveling in a mountain range, that the apparent relative height of mountain peaks varies with your vantage point. You will conclude that some impressions of relative height must be erroneous, even when you have no access to the correct answer. Similarly, one may discover that the relative attractiveness of options varies when the same decision problem is framed in different ways. Such a discovery will normally lead the decision maker to reconsider the original preferences, even when there is no simple way to resolve the conflicting preferences. The susceptibility to perspective effects is of special concern in the domain of decision making because of the absence of objective standards such as the true height of mountains.

The metaphor of changing perspective can be applied to other phenomena of choice, in addition to the framing effects with which this paper was concerned.

The problem of self-control is naturally construed in these terms. The story of Ulysses' request to be bound to the mast of the ship in anticipation of the irresistible temptation of the Sirens' call is often used as a paradigm case. In this example of precommitment, an action taken in the present renders inoperative an anticipated future preference. An unusual feature of the problem of intertemporal conflict is that the agent who views a problem from a particular temporal perspective is also aware of the conflicting views that future perspectives will offer. In most other situations, decision makers are not normally aware of the potential effects of different decision frames on their preferences.

The perspective metaphor highlights the following aspects of the psychology of choice. Individuals who face a decision problem and have a definite preference (i) might have a different preference in a different framing of the same problem; (ii) are normally unaware of alternative frames and of their potential effects on the relative attractiveness of options; (iii) would wish their preferences to be frame-independent; but (iv) are often uncertain how to resolve detected inconsistencies. In some cases (e.g., Problems 3 and 4, and perhaps Problems 8 and 9) the advantage of one frame becomes evident once the competing frames are compared, but in other cases (e.g., Problems 1 and 2, and Problems 6 and 7) it is not obvious which preferences should be abandoned.

These observations do not imply that preference reversals, or other errors of choice and judgment are necessarily irrational. Like other intellectual limitations, discussed by SIMON (1955) under the heading of bounded rationality, the practice of acting on the most readily available frame can sometimes be justified by the mental effort required to explore alternative frames and avoid potential inconsistencies. However, we pro-

Barcan Marcus et al., eds., Logic, Methodology and Philosophy of Science VII

THE SOCIAL CONSTRUCTION OF MIND

ROM HARRÉ

Sub-Faculty of Philosophy, Univ. of Oxford, Oxford, England

For some fifty years the idea has been abroad that mind is a product whose genesis in individuals in closely bound up with the social and historical conditions within which those individuals grow up and are constrained to act. (The ultimate origins of the idea, of course, go back to Marx and Hegel.) In different but closely connected ways, MEAD (1934), VYGOTSKY (1967) and WITTGENSTEIN (1980) have contributed refinements and clarifications to the thesis that the mentality of individuals is a product of their circumambient social orders. Can these various insights be brought together into a systematic acount of mind sufficiently clear and simple to serve as the basis of research programmes in psychology, attractive enough to displace the current confusions? In this paper I shall try to show how, by bringing out the distinctive metaphysics implicit in the social construction point of view, and by setting out a clear alternative system of conceptual controls to the Cartesianism that animates most of recent psychology, the main projects of such a programme can be delineated.

A. Metaphysics

I take the metaphysics of a science to be the open set of background assumptions that controls the total practice of a community of scientists. Our analysis of this will bring out ontological as well as methodological issues.

Traditional psychology has been based on the same metaphysical scheme as that of the physical sciences. People are taken to be internally complex things, located in a physical space-time, the space-time of Euclid and Newton. Whether the internal complexity of people–things is describable wholly in terms of the concepts of the physical sciences or whether some residual mentalistic concepts are needed, is an issue contained within this general framework. (For instance, contemporary cognitive psychology is set up within a Newtonian metaphysics.) One effect of the unexamined acceptance of this metaphysics is to direct research concentration on the individual internal causal mechanisms, be they phsyiological or cognitive, Humean or generative. Social constraints on thinking, feeling, intending and acting appear as addenda which would be reduced by the methodology of individualistic experimentation to the status of *ceteris parabus* conditions.

From the social constructionist point of view this metaphysical scheme is radically inappropriate (as its Cartesian sibling the idealism that underwrites phenomenology). An important manifestation of mindedness is speech, perhaps the most important of the human social practices in which mindedness appears. Let us begin with an analysis of a conversation among (not 'between') a community of speakers. By a 'conversation' I mean the playing out of a Wittgensteinean language-game, that is the using of speech within and integral to some activity which is a recognisable practice within a form of life. For instance, the instructions, orders, questions, classifications, reprimands, etc., integral to the communal practice of building a house, christening a child, and so on. The animating thought is that — in many ways (though not in all) — it does not matter *where* in Newtonian/Euclidean space-time, an insult, an order to fire etc. is delivered, but *who* delivers it and to whom. In important ways it is the fact that I insult you rather than that the insult occurs in the dining room rather than in the car that matters. Only certain people can bring about nuclear annihilation by speaking and their speech counts whether they are in Camp David or the White House. Plainly, indexicality is going to be a major clue to the metaphysical constraints I want to bring out.

Let us try out the idea of the people participating in the conversation as an array of locations for the speech-acts that are their conversational contributions and compare this with the Newtonian scheme.

The physical world is manifested in an array of locations, which are the places and moments of a spatial temporal grid. These places (and moments) are picked out by indexicals (typified by 'here', and 'now') whose sense cannot be reduced to geographical and calendrical references. The world is constituted by causal interactions (events) between internally complex things variously located in the spatio-temporal array. The question of whether things can finally be eliminated in favour of colocations of causal interactions is open.

The social-psychological world is manifested in a grid of locations, which are the array of people of a social order. People are picked out by indexicals (typified by 'I', 'you', etc.) whose sense cannot be reduced to proper name referents. (If I say "Rom Harré insults you" our subsequent relationship would depend upon whether you know that I am Rom Harré.) The world is constituted by the speech-acts of the community in which a conversation comes into being.

Assigning priority to this second scheme for the purposes of delineating the project of human psychology has the immediate effect of suggesting that human interactions will, amongst other constraints, be conducted within the conventions of conversations. For instance, for certain sorts of conversations intelligibility, sincerity, turn-taking etc., but for others different conventions will be in force, e.g. telling exemplary anecdotes, witty exchanges, shouting people down, etc.)

Several comments are in order. I have developed the parallel between Newtonian ontology and a conversational community only with respect to space as the set of possible locations for things. The temporal aspects of speech-acts in an array of people–places are a topic in their own right (for instance, is there anything corresponding to absolute time?) of some complexity which I shall not address in this paper. (See my *Personal Being*, Ch. 2.)

As the Newtonian ontology developed in the hands of Boscovich, Faraday and others 'things' were placed by 'powers'. The material beings which existed in space and time were conceived of as the nodes of a causal structure created by productive or efficacious relations between spatio-temporally located states (potentials). A parallel can be drawn to this throughout psychological space. The conversation is also a relational structure. A speech-act proper consists in utterances at one (some) place (that is person(s)) and uptake at another (others). Efficacy is achieved not through causality but through meaning. The reciprocal pair 'intention–understanding' whose linkage is semantic, corresponds to 'cause–effect' whose linkage is generative. (Searle calls this 'an intentional structure'.)

The people-array, the set of locations that corresponds to space differs in an important way from the Euclidean flat space of the Newtonian scheme. It is not isotropic, that is it has a structure by which one region differs from another. The people-array of a psychological space is structured by differential *rights* to speak, and, for example, some people in some social orders can always interject (commanding officers in armies) while others may need prior speech-acts from their superiors (permissions) to make a remark (say during a court-martial). In most societies certain topics are reserved to particular persons. One cannot, for example, reprimand someone else's child. In terms of the metaphor the space of conversational interchanges is 'curved'. The curvatures of physical space are associated with gravitational fields. The structured psychological 'space' is associated with moral orders since the structure of a person-array follows from the existence of differential rights to speak. This is a matter of enormous significance for the practice of scientific psychology.

Empty places are people without the right to speak. A stenographer secretary at a board meeting is a location in psychological space, a possible place for the utterance of a speech-act, but without the appropriate right that place remains empty during that episode. The principle that children should be seen but not heard creates a few empty places in some person-arrays.

Let us call this the 'primary structure'. What of its status? I have offered it as an ontology, a basic system of categories intended to pick out what is real, rather than as a myth or image for controlling a methodology. Taken as an ontology the question of its relation to the ontology of the physical sciences has to be addressed. It is no part of my purpose to deny that all individual psychological processes (at least as token-events) are materially grounded in physiology. I owe the impetus to take up this question at least in outline, to some critical comments by John Searle. The alternative strategy is to insist upon a physicalist ontology from the beginning, so that the scheme of people-arrays and conversational interchanges would simply be a way of picking out certain structures from the causal network of spatio-temporally located physical powers. But this cavalier preempting of issues can be objected to on several grounds.

(1) Since the conversation and physiology of the speakers are given as categorially diverse phenomena, the question of the nature of their relationship should be treated as contingent. Room must be left by the conceptual systems for it to turn out to be the case that some features of the conversation do not have physiological correlates (even token-correlates) in which they can be materially grounded. In short, as in the physical science the way explanations are developed by grounding dispositions in deeper levels of micro- or macro-structure, ought to be controlled by the aim of establishing a common ontology (cf. ARONSON, 1984) through the empirical testing of *a posteriori* hypotheses about such groundings, which may turn out to be implausible.

Thus it may happen that type–type correlations, by which psychologically defined types (such as 'feelings of moral outrage') are related to physiologically defined types (such and such neurophysiological states) are rarely achieved. Even in the case of pain, the common ontology of '*c*-fibres' etc. which enables 'hurts' to be given alternative individuating criteria probably of universal application in either a phenomenological or a physiological conceptual scheme, is far from adequate to give an account of the use of the conceptual cluster around 'pain', given the diversity of language-games and deeply different forms of life within which pain has some place or other. Compare the conventional displays by injured Italian soccer players and the stoicism of Chinese peasants being operated on under acupuncture, with the *couvarde.* These examples illustrate a philosophical point about criteria of identity and individuation of the speech-acts and other semiotic displays which different cultures indulge in, in relation to physical injury — these criteria not being determined by the physiological accompaniments of the injury. But may not the writhing of the footballer and the stiff upper lip of the colonial administrator ontologically be the product of the information-processing workings of individual brains? This brings us to the second objection.

(2) The maintenance of an ontological independence for the conversation (as a public and collective intentional structure) may be required by the possibility that there are structural properties of the conversation that are not features of the intentional structures of the contributions (projected or actually delivered) of individual speakers, taken singly. This qualification is required because a preemptive ontology of physiological categories in physical space-time must be individualistic, since it is only in individuals that the token-identities of intentional entities with physiological states etc. can be realised. There is any amount of evidence for the existence of such properties. One line comes from business studies on the workings of committees as natural decision makers. A conversation may be produced which has cognitive properties of rationality etc., independent of the intentions and cognitive operations of individual members. But more convincing still is the work of Pearce and Cronen on the analysis and reconstruction of conversation between intimates which has shown that the stability of a relationship may depend on properties of the conversation unintended by either of the participants in preparing their contributions and mutually opaque to each of the speakers.

Even though everything that takes place in a conversation is and must be materially grounded criteria of individuation and identity may not be found in the material properties of speech as noises in the air, and the physiology of utterance and uptake. Speech acts may differ not by what they are in themselves, but by the nature of their neighbourhood in a conversation. It is this fact which forces us to experiment with a dual ontology. But as an interpersonal, public and collective ontology it is as different as could be from idealism.

Taking for the moment the independence of the conversational ontology as a working hypothesis, there are some methodological consequences to be noted. Considered as locations for speech-acts people are simple elements of the system. This is plainly an artificial assumption. But it permits the formulation of a range of questions whose possibility is concealed by the traditional metaphysics, in which people as things are to be treated as indefinitely internally complex. Our ontology admits of the possibility of a culture in which *all* matters that we take to be psychological, for instance rational (or irrational) thought, the labelling of feelings as emotions etc., are carried out publicly and collectively through interpersonal conversational exchanges. Such people would be only physiologically complex. Their utterances would be thoughtless and spontaneous, but in accordance with social conventions, though not consciously intended as such. I do not suppose that there is any such tribe, but it is a possibility within the conversationalist ontology and provides the necessary foil for the question of how psychologically complex with respect to what kinds of episodes and topics different categories of real people are. There may be enormous cultural diversity.

How much we take to be researchable as cognitive or other attributes of individuals is, or an aspect of which is, to be found in properties of conversations? A simple example is the discovery that delinquency is not the mark of an individual falling under a psychological type 'delinquent' but a social type status arrived at by an essentially conversational practice of formal labelling. I will take up the example of rationality in the third part of this paper, where I will try to show that it has some features in common with delinquency. Some psychological concepts seem to have a dual role. On the one hand they pick out properties of conversational display from collective imperatives: on the other they identify psychological states of individuals, not always specifically related to one another. Intentions are a striking example. Certain speech-acts such as "I am going..." are conventionally taken to express intentions (but need not manifest a reciprocal 'inner' state of intending) which can become part of a public record by which the others may maintain the resolution of the speaker even if he or she has forgotten the occasion of the speech-act by such reproaches as "Daddy (Mr. President) you *said* you would...".

Philosophers have tended to treat memory as if the deep issue was the problematic grounds for verisimilitude of private recollections of events whose retreat into the past necessarily makes them unavailable for current checking. But the psychology of memory as a real human faculty cannot properly be confined to that sort of vocabulary. The social constructionist point of view helps to draw attention to another aspect. Memories are those reminiscences (as conversational contributions prefaced by "I remember (etc.) ...") which have been authenticated by those who have memorial rights. Typically in a family the mother exercises memorial rights and reminiscences become recollections once they have her imprimature (cf. the work of M. Kreckel on performative speech in family circles). Again, we run across a matter of social structure and moral order (differential rights) in one aspect of a phenomenon which has traditionally been studied by experimenting with the recollective capacities of isolated individuals. The point is not to deny that such capacities are of interest, but to emphasize that the language games of remembering cannot be explained by reference to those capacities alone.

These points have come up through exploring some of the consequences of taking individuals as simple locations in people-arrays at which speech-acts can occur. But people are psychologically complex. Our scheme must now be elaborated to make room for hypotheses about various ways that the psychology of individuals is, or could be, thought of as socially constructed.

B. A conceptual 'space' for psychological concepts

The conceptual scheme upon which both subjectivist (phenomenalist) and objectivist (behaviourist) psychologies have been constructed is a linear 'space' based on a polar opposition between the 'inner-subjective' and the 'outer-objective', generating as its typical problem the Cartesian paradox of 'other minds'. (For a recent reatment of the project of opposing that opposition, see D. Rubinstein (1981).)

To help define the projects of a social constructivist psychology I propose a two-dimensional 'space' based on the following oppositions.

(1) A dimension of display, having public display at one pole and private display at the other. For many psychological purposes it does not matter how something is concealed — behind one's face or behind one's hand (cf. Wittgenstein on locking up one's diary). Public display depends on such matters as reciprocal understanding, which itself depends on actor and interactor, subscribing to local social conventions.

(2) A dimension of realisation, having matters which are individually

realised at one pole and those which are collectively realised at the other. I have already pointed out the interesting case of remembering which has both an individual and a collective realisation.

Other dimensions could be added, for instance, active/passive; but the 'two-space' is enough for my purposes in this paper. Diagrammatically we have the following picture

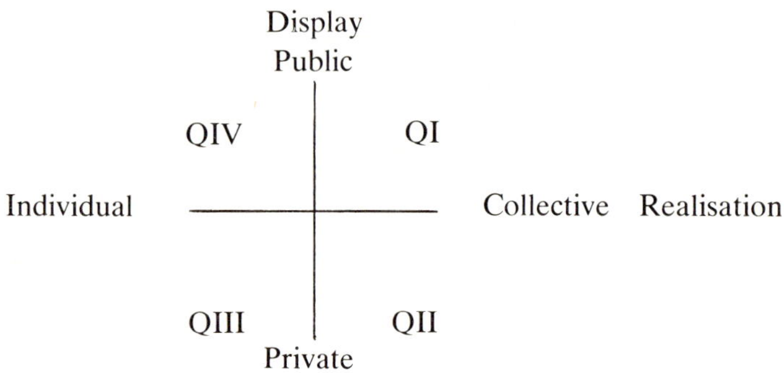

I label the quadrants QI–QIV. QI I shall refer to as the 'social' quadrant and QIII the 'personal'. I take Marx and Veblen to be typical philosophers of QI and Freud of QIII.

There are two uses to which such a scheme can be put.

(i) To control conceptual analysis and through it, research design. Knowledge, though not a popular topic of research among psychologists, has an interesting duality when analysed within this framework. Philosophers, working within a Cartesian metaphysics, have analysed knowledge in terms of the individualistic concept of belief, coming up with such definitions as 'true or certified belief'. In so far as there is a psychology of belief (cf. DENNETT, 1982) the psychology of knowledge would be part of it. But the knowledge picked out by this concept is typically that which is assessed by a written examination, the individual person's store of these beliefs. However, a glance at the variety of language-games in which 'to know' appears shows that there is a public-collective practice in which performatives of trust such as "I know...", "You can take it from me..." etc. are used to create a shared corpus which can be used as a resource by actors in various tasks and contexts. Sociologists have shown that such a corpus is socially structured by a complex moral order of rights to assert and rights to display, which characterises different categories of persons. For example, there is lay and professional knowledge, common and esoteric knowledge, men's and women's knowledge (cf. SCHUTZ and LUCKMANN 1973).

(ii) This scheme can also be used to represent that process which I have called, following Mead and Coulter, the social construction of mind. As a thesis of psychological development it was already fully formulated by Lev VYGOTSKY (1967). According to the Vygotskean point of view, human psychological development proceeds from performances in Quadrant I through Quadrant II to III, the process Vygotsky called 'appropriation'. In Vygotsky's view the main instrument for the social construction of mind was speech, the acquisition of which shapes the inchoate flux of feeling and cognition into that we recognise as mind, to which I would add the Wittgensteinean insight that one must always bear in mind that speech occurs in social/practical episodes, the so-called 'language games'. This conceptual 'two-space' allows for the formulation of the Vygotskean–Wittgensteinean thesis in a rather precise way: that mind is a personal appropriation from social, that is public-collective practices.

But what sort of practices which would be classified in my Quadrant I would be appropriately picked on as possible progenitors of the mentality of individuals? If this mentality is a secondary psychological structure from what features of the primary structure of public conversations is it appropriated? The Wittgensteinean notions of language-game and form of life are germane but both need spelling out if this conceptual scheme is to sustain any interesting research programme. By 'language-game' I take Wittgenstein to have meant any social practice (including building, counting, christening, and so on) in which speech played a significant part. (We might be wise to widen this to include other semiotic systems than speech, such as gestures, road-signs, heraldic devices, etc.) By 'form of life', a notion central to but vague in Wittgenstein's thought, one can usefully mean roughly what anthropologists mean by a culture, encompassing every kind of language-game, from agricultural practices and beliefs in a cosmic realm, but, within which, for reasons which will emerge, I wish to emphasize the moral order of a society, particularly its distributions of rights and duties, and its modes and criteria of moral evaluation of persons, their actions and their feelings.

What is it that language-games offer for appropriation by an individual (in ways I will not discuss in this paper) as forms to shape generalized mentation into a personal mind? I can briefly touch only on two.

(a) Public-collective conversation includes ways of speaking (and writing) that can serve as grammatical models for organising concepts of the mind. For instance, I believe a strong case can be made, both philosophically and anthropologically, for the view that the grammar of personal address in use in the practices of Quadrant I, the public-collective realm, serves as a model for the organisation of individual thought around the private 'centre' of transcendental self and so for the structure of consciousness; and that the form of the latter varies with the way the former is employed in this or that culture (cf. the relevant studies in HEELAS and LOCK, 1981).

(b) But the public-collective conversation also includes exemplary conversational episodes from which models of self-address can be taken. A.J.P. KENNY (1979) has proposed a theory of that which was traditionally called 'the will' in terms of privately appropriated models of ways of command to be found in public-collective episodes for the exercise of authority. Personal agency is an individual power whose scope and degree depends on public-collective practices.

These are the briefest sketches of Vygotskean suggestions which could be fleshed out into substantial research programmes. I cite them in this context to make clear by example part of what is meant by 'the social construction of mind'. Our capacity to formulate such hypotheses is facilitated by adopting the conceptual scheme I have called the 'two-space', in preference to the Cartesian opposition of inner and outer.

C. Two kinds of social construction

I have sketched several examples to illustrate aspects of the social constructivist approach to the understanding of human mentation. Thought as a public-collective phenomenon manifested in a wide variety of language-games, is both logically and aetiologically primary. By Vygotskean appropriation secondary and derivative processes of thinking become available in very different degrees for individuals. I use the term 'Vygotskean appropriation' to stand in for the imperfectly known ways that individuals come to be microcosms of the social linguistic world around them. Important among these ways is the relationship developmental psychologists call 'psychological symbiosis'. In such a relationship the dominant partner creates and later supplements by the use of certain ways of speaking and acting, the deficits in personal thought capacity of thc subordinate partner. Further pursuit of this topic would carry us away from philosophy of mind to theoretical issues in developmental pschyology. I turn now to illustrate some of the many ways in which properties of mind taken individualistically in the Cartesian framework must be seen socially constructively once that framework is abandoned.

(1) The case of rationality can be used to illustrate the social constructiv-

ist principle that conventions of discourse can engender individual mental habits. In this case social constructionism appears as a causal theory by which the existence and properties of an individual mental characteristic is explained by reference to a public-collective practice. The theory can be illustrated by two recent pieces of research.

Social historians have shown how rapidly accepted beliefs about the rationality of women have changed. These changes can be researched by studying changing presumptions in the laws dealing with various rights to management (for instance of property) and to competence in public representation, coupled with the studies of portrayal of women in plays and novels. In the mid-nineteenth century women are portrayed as enthusiastic, swayed by impulse and generally irrational in their approach to problems. (See for instance the contrast in the proposals for helping Phineas Finn in his trial for murder offered by Mr. Low and the Duchess of Omnium. The conversation between them is a marvellous exemplar of the point (see A. TROLLOPE, *Phineas Redux*, London 1874, Ch. 54).) Can it be that in one short century the cognitive capacity of half the human race has changed so drastically that we now demand the same standards of cognitive skill from both men and women? No account of rationality in terms of information processing and brain physiology can possibly be plausible.

A clue to the resolution to this puzzle can be found in some recent studies of the rationality of scientists. The Cartesian picture requires an inner cognitive 'engine' which processes information according to some logical algorithm, producing those epitomes of rational discourse, the printed scientific papers. Microsociologists of science have shown that this picture is quite misleading. Knorr–Cetina's study of scientific papers shows how logical properties are inserted into a discourse in the course of a complex series of social interactions by which community conventions are enforced. Assertions must be provided with reasons which, if they are to count as such in a scientific community, must be demonstrably logical in their interrelationships. The question of whether the convention of discourse can be grounded epistemologically is another matter. Realists think it can, conventionalists think it cannot. The production of rational papers by scientists does not depend on, nor does it prove, the existence of rationality as a property of the mentation of individual scientists. One supposes that certain sorts of mental habits will be fostered in a community with so strong a convention as to the form of public discourse. (Compare here the recent suggestion by Mühlhäusler in a seminar paper (Oxford 1983) that syntax is a socially motivated conventional modification of speech inessential to its cognitive properties.)

Can one identify a conversational convention that would serve as the basis of a complementary explanation of the alleged irrationality of women? Confusingly, ethnomethodologists have pointed out two features of conversations, both of which they have called ‘accountability’. In the sense germane to this paper a speech-act is accountable if the speaker or someone on his or her behalf, could be required to give reasons for what has been said — either as to the very saying of it, or more particularly as to its content. (The other sense of ‘accountable’ refers to the question of whether what has been said falls or does not fall within the expected repertoire of sayings of a person of that sort.) It is clearly a social convention whether the speech-acts of a conversation are accountable in the first sense, that is whether routinely people are expected to give or to be able to give reasons for what they have said. Women's rationality according to the social constructivist point of view, is to be understood in terms of changing social conventions of conversation, in particular, whether or not the speech-acts of women are accountable. The historical evidence suggests that the conventions of accountability have changed and that in mid-nineteenth century women's speech was not, in general, accountable. Women were not called upon nor expected routinely to provide reasons for their conversational contributions. In these social circumstances the habit of preparing for the request for an account could hardly have been expected to take root. Women's ‘intuition’ is the mythic cognitive faculty that Cartesianism calls into being ‘behind’ non-accountable speech, and is no more really an inbuilt property of feminine psychology than would a corresponding ‘rationality’ be for our contemporary women colleagues. If women have reacted to the changing social conventions governing their speech by acquiring the habit of preparing for demands for accounts, as I suppose those women who take part as of right in accountable conversations no doubt have, so they have acquired ‘rationality’.

Once again a feature of processes to be found in my Quadrant III, the realm of the private and individual, has been shown to derive from the grammatical models and typical episode structures of the conversation which fills the public-collective realm, Quadrant I. But why have the conventions of women's speech changed? The socio-economic upheavals of the First World War must have had some important role to play in altering economic and status relationships. As I have hinted, further speculations along these lines ought to be controlled by reference to the sociologies of Marx and Veblen.

K.V. Wilkes has pointed out to me by way of criticism that this treatment of rationality as a discourse-convention engendering a habit of thought, ties rationality rather tightly to language. But what about the rationality of practices? Even in the heyday of women's intuition, wouldn't there have been many expectations of rationality of women's practices which would require an explanation in terms of features of individual cognitive processing? While I would agree that the cognitive processes behind practices including the use of manual skills is a matter of great psychological and philosophical interest, it sows the seeds of confusion to lump where we should split. Rationality has to do paradigmatically, I think, with the giving, the having and the asking for reasons. In this sense, rationality is conversation-bound and clearly related to conventions in speech and writing. Practical skills routinely involve other assessments and criteria of excellence in performance, for instance, efficient, neat, skillful, effective, useful, and their opposites. To speak of the rationality of cooking, child-rearing and other practices often assigned to women in traditional roles and societies can only make sense if we have in mind a discourse about these practices in which questions like 'Why do you prick an egg before boiling it?' call for answers that are related in some principled way to the question. "That's what mother did" or "It's always done that way" are not answers that can satisfy a rationality-convention.

(2) Another way that the social (public-collective) enters into the personal (private-individual) is through the essential involvement of a social component in a concept which is used by an individual in relation to (in interpreting) his or her experience. This case is quite different from (1) in that the social context enters into an individual's assessment of certain features of that context. Thus, as I hope to show, in interpreting a feeling as an emotion, an individual must take account both of the phenomenal quality of the feeling and of the context in which the feeling occurs — and in many cases, the context is social. The cognitive processing can be individual (though sometimes it is collective) but draws on both intra- and inter-personal information.

In the case of many emotions an individual's assessment of certain social relations in the context determines, in part, how a feeling is interpreted as an emotion. This can be illustrated with a development and elaboration of Sabini's and Silver's theory of envy and jealousy (J. SABINI and M. SILVER, 1982).

B is envious of *A* if the following conditions obtain: (1) *A* has *x*: (2) *A* has a right to *x*; (3) *B* wants *x*; (4) *B* knows or believes (1), (2), and (3).

B is jealous of *A* if the following conditions obtain: (1) *A* has *x*; (2) *A* and *B* have an equal right to *x*; (3) *B* wants *x*; (4) *B* knows or believes (1), (2) and (3).

Differentially under these conditions *B* will interpret the nasty feeling when he thinks of (sees etc.) *A* with *x* as envy or jealousy — or more probably others will make the judgement for him.

These conditions are not to be read as contingent causal accompaniments of the emotions, but analytical components of the concepts of envy and jealousy. (Rules for the use of the word 'envious' and 'jealous' in a public discourse or a private discourse derived from it). As such they cannot be disputed by citing 'data' from psychological experiments. They can be upset only by counter-arguments drawn from philosophical (linguistic) analysis.

These conditions reveal themselves in characteristic strategy (which we might use as Wittgensteinean 'criteria' for assessing the emotional state of someone with whose history and actions we are acquainted).

The envious man or woman has a number of possible moves. He, or she, (*B*) may denigrate *A*, challenge *A*'s right to *x*, or even denigrate *x*. Thus, if one of one's colleagues in knighted, one might say he wasn't such a good philosopher, or that he didn't really deserve a knighthood, or that academic knighthoods are two a penny, etc., The jealous individual either tries to deprive *A* of *x* when appropriate, or to get *x* for himself, preferably in such a way as to deny it to *A*.

This is a tentative analysis and no doubt can be further refined by exploring marginal cases, sibling language-games, the extent to which the conditions are severally necessary and/or jointly sufficient for the attribution of the emotional state. The key point for my argument is that in interpreting his feeling as a green emotion, *B* is taking into account the social relations that obtain between himself and *A* with respect to the local moral order, that is with respect to the rights each has to some good, status, or advantage. Emotions, or at least some of them, not only involve a characteristic feeling (James–Langer); a characteristic display (Darwin–Leventhal); a hypothesis as to the cause of the feeling (Schochter–Singer), but also a moral assessment dependent on social context. An emotion of this sort is not socially caused, but is socially constructed. It is *B*'s beliefs about rights rather than a public assessment of these rights that is the efficacious element.

A very similar point has been made about anger, since Aristotle's classical analysis. To interpret an aggressive and irritated response to something that happens (is done by another etc.) as anger, the suffer *must* take the interference to which he or she has been subject as some sort of transgression. This is not a contingent feature of certain happenings to be studied by accumulating experimental data, but a deep grammatical rule governing the language-games of 'anger'. And there are different rules for 'cross', 'irritated', 'mad', 'furious', 'offended', 'righteously indignant', 'vengeful', and so on across the spectrum of the red emotions.

Heelas has suggested that there may be tribes without emotions. By that he means that a people may have quite other ways of managing, interpreting and classifying the feelings (passions in the literal sense) which are caused by social and physical situations (bereavements, physical dangers etc.) than ours. If those ways lack the causal and normal component which was so prominent in our psychology (and the Champoy do indeed seem to make no use of them in the deep grammar of the language-games by which they handle their passions), then it would be grossly misleading to say that their psychology includes emotions. An important consequence would be that there would be no place in the practices of that culture for the moral assessment of those feelings. For as it is right to feel pity and hope, it is wrong to feel anger and despair.

D. Summary

Social constructionism is not a doctrine about how experience in general is possible. It does purport to provide the conceptual basis for explanations of the way experience is organised into individual minds. With classifiable and communicable attributes, it opens up the possibility that as societies and their language-games differ so the minds whose structures are modelled on features of their socio-linguistic environments may differ too. It makes of psychology a science like biology that recognises a diversity of organisms and of 'solutions' to the maintenance of life in different environments; unlike physics, which seeks for underlying universals in the diversity of particular phenomena. This should not be surprising since after all human beings are organisms.

References

Aronson, J., 1983, *A Theory of Explanation* (Macmillan, London).
Dennett, D., 1979, *Brainstorms* (Harvester Press, Hassocks).
Harré, R., 1983, *Personal Being* (Blackwell, Oxford).
Heelas, P. and Lock, A., 1981, *Indigenous Psychologies* (Academic Press, London).

KENNY, A.J.P., 1979, *Aristotle's Theory of the Will* (Yale Univ. Press, New Haven).
MEAD, G.H., 1934, *Mind, Self and Society* (Chicago Univ. Press, Chicago).
PEARCE, W.B. and CRONEN, V.E., 1980, *Communication, Action and Meaning* (Praeger, New York).
RUBINSTEIN, D., 1981, *Marx and Wittgenstein* (Routledge and Keegan Paul, London).
SABINI, J. and SILVER, M., 1982, *Moralities of Everyday Life* (Oxford Univ. Press, New York).
SCHUTZ, A. and LUCKMANN, T., 1973, *Structures of the Life World* (North Western Univ. Press, Evanston).
SEARLE, J., 1983, *Intentionality* (Cambridge Univ. Press, Cambridge).
VYGOTSKY, L., 1967, *Thought and Language* (MIT Press, Cambridge, MA).
WITTGENSTEIN, L., 1980, *Remarks on the Philosophy of Psychology*, (Blackwell, Oxford).

Barcan Marcus et al., eds., Logic, Methodology and Philosophy of Science VII

THE CONCEPT OF ROLE AND HUMAN BEHAVIOUR

LARS HERTZBERG

Dept. of Philosophy, Åbo Academy, Åbo, Finland

Prologue

1. Many social scientists nowadays think of the concept of role as providing a key to understanding the ways in which a person's behaviour and identity are related to the culture to which he belongs. And accordingly, they hold that roles provide the most important of the links between sociology and social psychology.

Role theory does, indeed, seem to constitute an important advance over associationist or behaviourist accounts of the socializing process. It seems better able to do justice to familiar facts like these:

> When an individual becomes a participant in the shared forms of conduct of his community, the process through which this happens is to a much larger extent a matter of his taking over whole patterns of activity, patterns of living, from others, than of specific habits or ways of responding being implanted in him one by one.
>
> Much of this process takes place, not through organized instruction, but through the learner's spontaneously coming to identify with certain individuals in his environment.
>
> Every society provides a variety of models with whom it is possible to identify, and this variety is largely bound up with the distribution of tasks among members of the society.
>
> What is shaped in this process is not simply behaviour, but features of personality as well: motives, ways of thinking and feeling.
>
> The existence of such recurrent patterns of behaviour and of personality tends to guide our understanding of others as well as the ways in which we appraise our own actions.

On the other hand, the use made of the concept of role is surrounded by problems. There are divergent views on the definition of the concept, on its field of application. Roles and some of the problems connected with them have been central in recent philosophical writings on the social sciences.[1]

[1] See R. HARRÉ and P.F. SECORD, *The Explanation of Social Behaviour* (Oxford, 1972), as well as Martin HOLLIS, *Models of Man* (Cambridge, 1977). Also, Jeff COULTER, *The Social*

These problems, I believe, are symptomatic of a situation that is not uncommon in the sciences of man. Social scientists coin new concepts to refer to various phenomena of human life, or they take over current concepts and provide them with technical definitions. However, the sense of these terms will ultimately depend on their ability to attach themselves to pre-scientific concepts and to the forms of thought to which these belong. (Indeed, I would argue, this is a condition for their having an application to human conduct as described in our common language.) In practice, these pre-scientific forms of thought, like underwater currents, will very largely guide the scientists' use and understanding of the scientific terms, sometimes in a way that is quite independent of, or runs counter to, the technical definitions.

This in itself is not a matter for criticism. Where role theory is concerned, however, my contention is that it is guided by divergent, indeed incongruent forms of thought. Scientists seem to be pulled now along one current, now along another, and, not realizing this, will occasionally be led to arguing at cross-purposes, to worrying about illusory problems, to lumping together quite disparate phenomena, etc.[2] The situation seems to call for conceptual "depth psychology". The problems involved are quite complex; this paper is a very modest attempt at addressing a few of them.[3]

Above all, the use of the concept of role by social scientists (and connected concepts, such as role-internalization, role-expectation, etc.) seems to be drawn in two different directions. On the one hand, roles will be regarded as assignments, that is, as something that a person can be given by others, and that he may or may not accept. On the other hand, roles will be regarded as affectations, that is, as ways in which persons present themselves, and which others may or may not credit. The issue concerning the relation between roles and behaviour will be altogether different,

---

Construction of Mind (London, 1979), esp. pp. 109–115. (I find much of what Coulter has to say congenial.) For a sociologist's discussion about some of these problems, see Ralf DAHRENDORF, *Homo Sociologicus* (Opladen, 1964), Chapter VII. Having surveyed a variety of definitions of the role concept, Dahrendorf concludes: "So verwirrend die Fülle der Definitionen scheinen mag, so deutlich ist andererseits in den meisten von ihnen ein gemeinsamer Kern, der jenseits terminologischer Unterschiede liegt..." (p. 66). What I shall try to argue in this paper is that the appearance of such a "common core" is an illusion.

[2] For a case in point, see Peter L. BERGER, *Invitation to Sociology* (Garden City, New York, 1963).

[3] My aim is not to map the common uses of the *word* "role"; my concern is rather with the common forms of thought and expression that seem to have a bearing on the way the word has come to be used by social scientists.

depending on which of these notions one has in mind. My paper will consist of two parts, each devoted to one of the two notions.

The two notions converge in the theatrical context which is the historical source of the role concept. Being cast as Ophelia means being given an assignment in the context of a theatrical performance. In order to carry out her assignment, the actress must affect to be a certain sort of person: to be in love with Hamlet and be driven to madness by his indifference, etc. However, a theatrical production will also include a number of role-assignments that do not involve having to affect anything: e.g. the roles of prompter, curtain-raiser, lighting director, etc. (Of course, the prompter may affect to be, say, absent-minded, but this then is no part of his assignment.)

Roles as assignments

2. For a role to be assigned to a person, we might say, is for his conduct to hold a certain place within a system of human activities; it means that he is committed to contributing to the system in some more or less specific way, or that the system is dependent on his contributing in certain ways if it is to work well or to work at all. (The dependence may be logical or practical.) Words like "job", "task", "function" are often used in talking about different kinds of assigned roles. A role commitment may be incurred through an explicit decision, or tacitly, through one's simply starting to carry out a task. It may also be due to one's holding a particular position which is bound up with a function or carries with it certain obligations.

The variety of cases falling under this notion is very large. Many of the roles that are commonly assigned to people are of a more or less standard kind, such as the roles of bishop, witness at a trial, baby-sitter or cockswain in a rowing team. On the other hand, roles may take shape momentarily and without design, as in the distribution of tasks among a number of passers-by trying to rescue a child from a river.

As the examples show, even standard roles vary considerably with respect to the scope and definition of the tasks involved, as well as with respect to their permanence and importance in the role-bearer's life. A role may be instituted for the sake of some specific need that is perceived to exist; on the other hand, tasks may gradually develop or combine around holders of various positions, or be developed or combined by those who hold them. What belongs to a role may be fixed or fluid; it may be a matter of general agreement or controversy; some roles are mainly determined by the concerns of those who are served by them or dependent on them (nurses, telephone operators), in the case of other roles, standards are primarily set and maintained by the holders themselves (as in the case of creative professions). And so on.

3. Assigned roles are typically spoken of in the context of rules, decisions, promises or agreements of which the persons involved or the members of the community are aware, and which they generally consider binding. It is possible to distinguish between a person's *having* a role, and his *fulfilling* the role, for one may have a role that one never fulfills or even tries to fulfill: one may refuse the tasks assigned, neglect them or forget about them. But the presumption will be that the rules or decisions are such as the persons involved comply with; otherwise they could not be said to be in force. (Sometimes, an observer may ascribe roles to the members of a group or community simply on the basis of actual behaviour. Such a use of the concept of role is not illegitimate, even though in this case the agents themselves may not be aware of the existence of the roles — consider, say, leadership in informal groups. But this seems to be a secondary use of the notion: here, it seems, there could be no talk — other than metaphorically — of role-obligations or commitments. And here the distinction between having and fulfilling a role cannot be as clearly maintained as in the other sort of case.)

4. An important qualification must be made at this point: many of the regulations or regularities that are connected with membership in a category of persons do not constitute roles. The reason for this seems to be that we reserve this term for those cases in which the relevant behaviour is regarded as contributing to the performance of some task or the fulfillment of some need of the community. (The needs may be as perceived by the members of the community, or by the observer.) In our culture, for instance, there are various legal and non-legal norms relating to the behaviour of children; also, there are ways of behaving that are typical of children of various ages; yet to be a child is not, in itself, to have any specific role, for there is not, in our culture, any particular contribution that children, *as* children, are required to make.

Sociologists often, maybe intentionally, ignore this restriction. Thus, being male or female, Jew or Gentile, an aristocrat, an immigrant, a music lover or card-player will sometimes be mentioned, alongside professional positions, parenthood, or the like, as role-connected categories, even though in most cases there is no clear connection with any *specific* social contribution.[4] The issue is not just verbal. Take the role of immigrant: what is it supposed to consist in? Legal obligations of immigrants? Occupations (or forms of conduct?) typical of (or suitable for?) immigrants? Occupations (or conduct) held to be typical (or suitable)? Norms of conduct that immigrants tend to subscribe to? Given the right sort of context, I suppose the word "role" could be used to refer to any *one* of these aspects, but to think that one could somehow create a unitary concept by lumping them all together as "the role of immigrant" would be a bad confusion.

Related to this is another feature of many sociological discussions of the topic: the failure to distinguish, in the case of role-connected categories, between those aspects of a person's behaviour that belong to his role, and those that simply tend to go together with it. Though it may be the case, say, that doctors tend to live in certain kinds of neighbourhoods, to drive certain kinds of car, tell certain kinds of jokes, etc., it should be obvious that, if doing these things is said to belong to the doctor's role, this is a very different sense of "role" from that in which the word refers to his professional function.

To be sure it may sometimes be said that a physician is "obliged" to drive a Cadillac or to live in a serene neighbourhood. Such remarks bring attention to what is, in itself, an important fact of social life: the holders of certain positions — especially if these carry with them some degree of social esteem — are often under pressure of various sorts to project a certain kind of personality, or to adopt a certain life-style. Such demands (imposed by members of a profession or from outside) will sometimes arise from a serious concern with the dignity of the profession; however, they often shade over by degrees into a philistine preoccupation with status and privilege. It belongs to the nature of the case that disentangling the frivolous from the serious is difficult here; and self-deception will undoubtedly play an important part in these connnections. One should also remember that, whatever the source of these demands, even if an individual has the courage to ignore them, doing so may interfere with his ability to carry out his real obligations. (Consider, for instance, the difficulties faced by Dr. Lydgate in George ELIOT's *Middlemarch.*)

Now it might be argued that when sociologists ignore distinctions like these they do so on purpose, having found that they are not essential to their scientific aims. And of course this in itself is not fallacious. Yet in

---

[4] Consider, in this context, Hollis's worrying about whether there are such roles as those of drug-pusher, television personality, cyclist or Englishman (*op. cit.*, p. 79).

ignoring distinctions of our language, one incurs the risk of lumping together matters that, though apparently similar, are in fact profoundly different. In using an undifferentiated notion of role, there is some danger that one may come to ignore, for instance, the great variety of phenomena that come under the heading of social pressure, or the large variety of factors (tangible sanctions, subtler forms of pressure, spontaneous identification, etc.) that may guide a person's conduct towards conformity with a standard. The most important case of this, as I see it, is the fact that sociologists by and large tend to bypass the question which ones among the demands imposed by the community on the holders of a role can be considered legitimate and which are spurious. To be sure, this is a question on which one sometimes cannot take a stand without committing oneself on controversial matters of social life, and so it might be thought to lie outside the professional competence of social scientists. Against this it could be argued that the use of the concepts of role and role-obligations requires such commitments, and that, by refusing to commit himself, the sociologist will contribute to creating an atmosphere in which, all social requirements being regarded as equally valid, none will be regarded as *really* valid.

5. Sometimes this tendency towards assimilation is carried even further. Consider, for instance, the following passage by Theodore Sarbin:

> Role expectations may be viewed as *actions* or as *qualities* expected of the occupant of a position. If viewed as actions, role expectations are codified as in a job description; the occupant is expected, for instance, to call the roll, open the windows, secure the doors. If viewed as qualities, role expectations are codified in adjectival terms; for example, the occupant is expected to be warm, friendly, outgoing, sincere, and cautious.[5]

It should be clear that personality traits do not constitute role-assignments in the sense that the performance of certain tasks does. If personality traits are important for an assignment, it is because they are necessary or desirable if one is to succeed in one's tasks. One may fail as a nurse or teacher if one is unfriendly or cold, but one fails *in one's tasks, not in being* a certain kind of person. To speak of an obligation to be warm, or of neglecting to be warm, etc. (as opposed to acting warmly), would be senseless.

---

[5] Theodore R. SARBIN, *Role: psychological aspects*, in: D.L. Sills, ed., International Encyclopedia of the Social Sciences (New York, 1968), Vol. 13, p. 546.

6. The passage by Sarbin illustrates another basic unclarity in many writings on role theory. In the literature, the word "expectation" is often used to refer to that which determines the content of roles. However, this is of course an ambiguous word, carrying, on the one hand, a sense of "belief", and, on the other hand, a sense of "requirement" (or "obligation").[6] The use of this ambiguous term may have been in part responsible for the blurring of the role concept in sociology; under its shield, it seems, a continuous shift has been taking place between regarding roles as *typical* conduct (and characteristics), and regarding them as *required* conduct. This said, it must be admitted that, in human communities, expectations as beliefs and expectations as requirements are often interdependent in complex ways: the ambiguity of this term is no mere accident. This interdependence, however, is a matter for inquiry; an undifferentiated use of the term only serves to obscure it.

7. What, then, is the relation between role-assignments and personality? Can the formation of personality, as some social scientists have argued, be explained through the internalization of role-assignments?[7] In approaching this issue we should take note of the fact that the word "internalization", too, has been used in various ways.

Sometimes, the concept of role-internalization is being used to mean, roughly, that one carries out one's assignments voluntarily rather than because one is pressured, persuaded or enticed into doing so. This use of the term, however, is apt to mislead, since it seems to imply that this can only come about by a process through which one has gradually come to be different from what one originally was; and hence that no one ever naturally accepts an obligation.[8] Of course many roles are undertaken

---

[6] This ambiguity is pointed out, for instance, by Erving GOFFMAN in *Encounters* (Harmondsworth, 1972), p. 81n.

[7] "[The] significance of role theory could be summarized by saying that, in a sociological perspective, identity is socially bestowed, socially sustained and socially transformed" (BERGER, *op. cit.*, p. 98).

[8] This seems to be the way in which the term is taken, for instance, by Georg Henrik von Wright:

"It is essential to the well-functioning of the institutionalized behaviour patterns in a society that agents should, *on the whole*, conform to them and that the reasons for conformity should, *on the whole*, *not* be the impact of normative pressure, but simply acceptance of the rule. When these conditions are satisfied, we say that the rules are *internalized* with the members of the society. ···

Internalization, however, is also loss of freedom of a kind. That the institutionalized behaviour pattern is internalized means that our action in agreement with this pattern is *externally* determined." (G.H. von WRIGHT, *Freedom and Determination* (Amsterdam, 1980), p. 47.)

simply for the sake of sanctions and rewards, or in response to inducements of other, maybe subtler, kinds. And frequently what one does at first because of inducements may then come to be done voluntarily, say, because one comes to value or to like one's tasks, or out of sheer habit. But there seems to be no reason for assuming that this is what happens always, or normally, when people come to accept social tasks: this would only seem plausible to someone who takes for granted Hobbes's view that man is by nature unsocial. Looking after one's children; helping one's fellow beings; applying oneself to all kinds of tasks and problems; trying to be of some use — these are things we naturally do, things that we see people doing all around us. It needs no special process to explain this. In very many cases, people take on tasks, not *in spite* of what they are, but *because* of what they are. This sense of internalization, then, does not seem to promise an explanation of how personality is formed.

At times, again, what is meant by role-internalization is that having, and fulfilling, a role may come to have an effect on one's life in general: on one's conception of oneself, one's attitudes and responses, one's outlook. (The role performance is in this case being regarded as cause rather than effect.) Thus in being trained for a profession a person will be given new concepts, a new perspective. In this connection, of course, what was said earlier about the demands and attitudes attendant on the status of certain roles is pertinent. Another matter which has caught the attention of many writers on occupational roles is the extent to which the performance of certain functions, e.g. presiding at court, conducting a funeral, waiting at table, has a ritual aspect, demanding that one should project certain feelings, or the absence of feelings.[9] It is hardly surprising if circumstances like these come to influence a role-holder's personality. But we should not forget that, when this does happen, it is just a side-effect; it is no part of what we mean by fulfilling a role that the role-holder should be changed by it. One can become an excellent judge, become deeply committed to one's duties, etc., and yet one's personality need not change at all, need take on nothing of the judicial stereotype.

Obviously, too, it is a far cry from saying that someone's personality has been influenced by his role, to saying that it is moulded by it, or to saying that his personality is determined by (or is even identical with) the totality of his roles. Saying the latter hardly makes sense: it would be meaningless to say that a person's sense of humour, his liking or not liking children, his

---

[9] For a discussion of the waiter's role, made notorious by Sartre, see D.Z. PHILLIPS, *Bad faith and Sartre's waiter*, Philosophy 56 (1981), pp. 23–31.

being lazy or honest, calm or gullible could be fully explained by his holding such and such roles. To be sure, there are cases in which a person's professional role will colour his whole life, where he seems to have no real interests or concerns apart from those dictated by his position. But such cases are exceptional rather than normal. And in such a case we should not say that this person's professional self *was* his real self, but rather we should think that he lacked the ability to be his real self.

The idea, then, that the formation of personality could be regarded as an effect of a person's having various social roles does not seem acceptable.

8. It might be asked why it should ever have seemed tempting to embrace this idea. One reason might be a tendency to emphasize occupations in which the relation between role and personality is particularly close. With respect to someone who works as a scholar, for instance, there are other issues that may be raised concerning his relationship to his role, beside the question of whether he fulfills his professional duties: there is also, as we might put it, the issue of whether or not he is a scholar at heart. This involves more than his having his heart in his work: one may set one's heart at becoming a scholar for reasons external to one's work, say, in order to follow a family tradition, or because one does not have a clear understanding of the real nature of the work. Being a scholar at heart, on the other hand, involves being at one with one's work, living a life in which the demands of one's work become one's *own* demands.[10] This means aiming beyond one's duties towards one's institution, and beyond the approval of one's peers.

Now it could, I think, be said that, if one is not a scholar at heart, there is a sense in which one is not completely fulfilling the role of scholar. There are many roles concerning which the corresponding thing cannot be said, indeed, many roles with respect to which the whole issue could not be raised. (On the other hand, there are several kinds of role where the issue does arise, though in other ways: say, those of mother or teacher, of clergyman or doctor, of artist or politician, etc.) The reason for this is not, I believe, that if one is not a scholar at heart one will not be able to master one's tasks; I doubt if this is true. It is rather connected with the fact that the standards of scholarly work are set only through the ways in which scholars actually commit themselves in their work. Their commitments show what it is to find meaning in a life of scholarship. And this can only be

---

[10] For an illuminating discussion of this point, or one closely related to it, see Peter WINCH, *Human Nature* in his book "Ethics and Action", esp. pp. 84 ff.

shown by someone for whom that life has meaning in itself; that is, by someone who is a scholar at heart.

As was already said, the fact that certain roles involve issues of personality in this way makes them special cases. But it is plausible to think that these are the cases that present themselves most readily to social scientists, and hence come to be regarded as paradigms of occupational roles. Perhaps this explains some of the attraction of the idea that types of personality in general correspond to, or are formed by, social roles.

An additional circumstance which may have obscured the issue of roles and personality, I believe, is a failure to take note of the distinction between role-assignments and role-affectations. I shall now proceed to speak about the latter.

Roles as affectations

9. For an example of what I mean by a role-affectation we might consider the short story "The Hitchhiking Game" by the Czech writer Milan Kundera.[11] In this story, a young man and his girl friend are riding in a car on their way to spend their holidays together. The girl, who is rather childish, shy and inexperienced, on an impulse starts playing the role of a vulgar, seductive woman of the world. The young man gets into the spirit of the game, and as the game continues, neither has the power to break it off. Inevitably, in the man's perspective, appearance and reality begin to merge:

> It irritated the young man more and more how *well able* the girl was to become the lascivious miss. If she was able to do it so well, he thought, it meant that she really *was* like that. After all, no alien soul had entered into her from somewhere in space. What she was acting now was she herself; perhaps it was that part of her being which had formerly been locked up and which the pretext of the game had let out of its cage. Perhaps the girl supposed that by means of the game she was *disowning* herself, but wasn't it the other way around? Wasn't she becoming herself only through the game? Wasn't she freeing herself through the game? No, opposite him was not sitting a strange woman in his girl's body; it was his girl, herself, no one else.

As the story ends, the girl tries to turn everything back to where it began, but it is too late: the man's regard and affection for her have now completely vanished. The girl sobs:

---

[11] In Milan KUNDERA, *Laughable Loves* (Harmondsworth, 1974).

"I am me, I am me..."
The young man was silent, he didn't move, and he was aware of the sad emptiness of the girl's assertion, in which the unknown was defined in terms of the same unknown quantity.

It should be clear that the way in which the concept of role enters here is quite different from that spoken of before. For one thing, there is no question here of the girl fulfilling a demand or an obligation. She simply starts acting a part because she takes it into her head to do so. (There should be less temptation to confuse the two role notions for speakers of English than of other European languages, for English has two words, "role" and "part", where the other languages have one, the latter, it seems, being the word commonly used in speaking about role-affectations.)[12]

The role the girl plays could not be spoken of as a role she "has" or "holds"; it is not something she has been entrusted with, or has taken upon herself, say, in order to relieve someone else. There is no suggestion that, by playing her role, she is performing some specific function in the social context where she does so; her part, unlike an assigned role, is not logically dependent on her conduct being needed or called for in any sense. (This is not to deny that a person may occasionally be *required* to act a part, or that one's acting a part may occasionally serve a function.) The particular role she plays could not be described as belonging to any system of roles that exists independently of her; when a society is described by sociologists as a system of interlocking roles, what is intended are not roles in this sense.

In saying that this girl is acting a part (assuming that that is, indeed, the correct reading of the story), what we are emphasizing is primarily the fact that she is not being her real self; that there is something artificial or affected about her conduct. She is affecting a character other than her own. This sort of remark we occasionally make concerning people we know. Sometimes we find it hard to make up our minds whether someone is being his real self, or whether he is making himself out to be something he is not. Or we may find ourselves puzzled by the conduct of someone we thought we knew well, and then discover that he is putting on a performance. One may act a part with deliberate intent to deceive, or may do it simply in jest (as, apparently, at first in the story by Kundera). In the latter case, one's acting may be more or less stylized, making use of stereotyped forms of expression. The most common, and most intractable, cases of play-acting, however, are evidently those that fall between the two extremes, cases in which the performer's intentions are more indeterminate. The following

[12] I wish to thank Craig Dilworth, who straightened me out (at least tried to) on some points of English usage in this connection.

phenomenon, I believe, is not unfamiliar: when faced with certain people, or when placed in a certain type of situation, a person may, more or less automatically, fall into a particular part. He may be wholly unaware of this though he may discover it on reflection. This need not involve any element of deception: perhaps at bottom he realizes that nobody is taken in by his performance, and he may even know that people would accept him more readily if he were himself, yet for all that he may not be able to give up his acting.

There is often a note of reprehension in saying that someone is acting a part, but this is not a necessary feature. One may act a part without compromising one's integrity; indeed, one may sometimes have to do so in order to avoid compromising it. One may act courageous in order to encourage others; one's succeeding in this might be a mark of real courage (although, in this case, the courage one had would not be the courage one showed). We may see that a boy is playing his father and find something touching about this; this sort of thing we consider quite normal, and it could indeed be said to be a necessary part of becoming an adult: by acting a part, in certain cases, one may gradually become what one acts. (But we should beware of generalizing this into an idea that play-acting is always necessary, or always sufficient, for coming to be a certain kind of person.)

10. For a clearer understanding of what is involved in the concept of role-affectation, we must take note of a distinction that can be made between two types of description of what someone is doing or experiencing. Consider the following lists of expressions:

(I) casting a vote,
greeting someone,
serving a net ball,
making a fire,
looking out a window,
cooking a steak;

(II) being delighted with a gift,
speaking sincerely to someone,
getting absorbed in a task,
worrying about a test result,
recalling a friend with fondness,
wincing with pain.

Suppose that I wanted to contest the claim that one of these expressions could be correctly applied to a person. Where the first set of expressions is concerned, I could, it seems, sustain such a denial only by pointing to something about the actual performance in the context that was missing or had gone amiss: the ball did not actually touch the net; the fire had not really got started, or it was not his doing; the ballot used had not been properly validated; from where he stood, the glare prevented his actually seeing anything through the window; the oven was not working. For my denial to be acceptable, there would have to be a general understanding that, if the condition in question was not fulfilled, the performance could not be said to have taken place. (This does not mean that in such cases it would always have been possible to specify in advance what an acceptable case would have to be like.)

In the case of expressions of the other sort, this is not so. I can meaningfully doubt whether an expression of this type can be correctly applied to someone even if I recognize that nothing about his performance may be of such a nature that anyone who saw it would have to agree that the expression did not apply. Someone receives a gift and seems to be delighted with it. But a person who knows him well may realize that he is only putting on a performance. Still he may not be able to give grounds that others would find convincing, or specify what it was about the other's conduct that gave him away, apart from saying, perhaps, that, *for him*, the expression with which he received the gift was "not right".

This seems to me to constitute a conceptual difference between these two types of descriptions. It is, one might suggest, this difference between the two types of description that has traditionally led philosophers to say that descriptions of the latter type presuppose the fulfillment of certain inner, or mental, conditions. These, it has been thought, are only accessible to the agent himself, and hence, in the case of these descriptions, as opposed to the others, the crucial part of the performance is one that is hidden from observers. The view of the mind expressed in this account is notoriously problematic; there is little need, however, to go into that debate here. I simply wish to say that I agree with those who reject the notion of the metaphysical inaccessibility of the mind, but that, even so, an important truth is expressed in the idea that descriptions of the latter type require the fulfillment of certain "inner" conditions. What needs to be realized is simply that this is a conceptual rather than a metaphysical truth.[13]

The point made in saying that descriptions of the latter type presuppose

[13] These points are discussed in Lars HERTZBERG, *The indeterminacy of the mental*, Proceedings of the Aristotelian Society, Supplementary Volume LVII (1983), pp. 91–109.

the fulfillment of certain inner conditions could also be made by saying that applying them requires our taking a stand on some such thing as the sincerity or naturalness of the agent's performance, whereas no corresponding question arises with respect to descriptions of the former sort. I wish to leave the matter at that for the present, though fully aware that concepts like those of sincerity or naturalness, etc., stand in need of clarification themselves. I shall for now simply appeal to the understanding we all have of the common use of these concepts.

This distinction, I would like to suggest, can be used to throw light on the concept of role-affectation in the following way: the question of whether someone is or is not playing a role, in this sense, is always ultimately a question concerning the applicability of descriptions of the latter type. In other words, to say that someone is affecting a role is to say that he is giving a misleading representation of his "inner" life; that he is, in fact, doing artificially something that can only be said really to be done where it is done naturally. Role-play, in other words, is never external conduct regarded by itself, but only in relation to the light it seems to throw on what a person is, what he feels or how he experiences things.

If this is correct, it helps to make even clearer the distinction between role-assignments and role-affectations. The behaviour characteristically required of someone with an assigned role is behaviour of the first type. The tasks assigned to members of a society are typically of a sort to which inner conditions do not matter. Hence in the case of most role-assignments, performing them does not involve role-affectation. It is true that there are some roles that require that one should at least appear to have, or to lack, certain feelings or attitudes (judges, waiters, etc.), but since it may be natural for the role-holder to show, or not to show, those feelings, no role-affectation need be involved.

When a person is said to perform an assigned role, there is no suggestion of artificiality or insincerity. On the contrary, in the case of many assigned roles, performing them implies that one actually holds them, that is, that one actually belongs to the category of persons to which the role pertains. Whoever does the cockswain's job in a rowing competition *is* the cockswain, at least as long as that competition lasts. This is not true in all cases, of course: taking on the role of mother to a child does not make someone the child's mother. But it does not follow that in that case one is affecting to be the mother. (One may, on the other hand, affect to be *maternal*, but so may the real mother. A woman may, to some extent, fulfill the role of mother without being maternal, and she may fail to fulfill the role even though she is maternal.)

11. The concept of role-affectation has close connections with that of pretending. One may carry on a pretence as a means of giving off a certain character, and one may affect a character as part of a pretence.[14] Pretence, like play-acting, may be deliberate or unreflected, stylized or realistic, it may be carried out with or without serious intent at deception. There are, however, some important differences between the concept of pretence and role-affectation.

(i) While an imputation of role-affectation concerns the non-fulfillment of the "inner" conditions of conduct, pretence may concern conduct of both kinds. (A person may pretend to be casting a vote, provided he does not actually carry out the requisite performance, nor believe he does. He cannot affect to be casting a vote, though he may accompany his pretence, for instance, with affecting to be deeply concerned with the outcome of the election.)

(ii) Even where pretence involves the "inner" sphere, pretence and play-acting do not coincide.

If someone receives a gift and pretends to be delighted with it, this need not be a case of role-play. His reason for pretending may be the fact that he does not want to disappoint the giver. His conduct will be role-play if his *reason* for engaging in it is the fact that (in his view) such conduct is the conduct of a certain kind of person. This entails that in some sense he must be taken to act from an understanding of what it is to be a person of the kind he is affecting to be, whereas a genuine display of delight (or a simple pretence) need not proceed from an understanding of the kind of person one is: one simply responds (or pretends to).

(In saying that affectation proceeds from an understanding, I am not implying that it must be consciously engaged in; even when a person's role-play proceeds from self-deception, his understanding of things will in a sense determine the ways in which it is possible for him to deceive himself.)

It might be suggested that I should have said: role-play is conduct that someone engages in *for the sake* of creating the impression that he is a certain kind of person. While this may often be the case, there are two reasons why I did not say this: for one thing (as was already suggested) one need have no *further aims* in view in engaging in role-play; for another

---

[14] For an instance of the latter, consider the episode in *Bekenntnisse des Hochstaplers Felix Krull* by Thomas MANN (Frankfurt am Main, 1954, pp. 115 ff.), in which the protagonist, to evade conscription, feigns an attack of illness, and, in order to make the pretence more convincing, affects to be ashamed of his illness and anxious to conceal it. Mann's novel, as a whole, provides interesting illustrations of the phenomena of pretence and role-affectation.

thing, we should, perhaps, allow for cases in which there is no element of *presentation* involved, as when someone plays a part by himself.

(iii) Role-play can on occasion be carried out by *not* pretending. Thus, in the Kundera story, it could be said that it was natural for the girl to pretend not to feel any sexual desire for the young man. Showing her desire, then, was part of her role-play, by which she suggested that she was the kind of person for whom showing it would be natural.

(iv) Where pretending concerns the "inner" sphere, pretending to be something implies that one either is not, or believes one is not, that which one pretends to be. This is not true for play-acting. When I receive a gift I may be surprised and delighted, and yet the way I show my surprise and delight may be an affectation. This only means that I am not showing my feelings in the way that is natural for me, as the kind of person I really am, but as if I were a kind of person that I am not.

12. In the previous sections I have made use of the distinction between conduct being natural and artificial for a person, between a person being and not being his real self. If my remarks about role-affectation are to be understood, I must be able to clarify the nature of this distinction. The issues enountered here are complex; I have space only for a few tentative remarks.

It will occasionally be asserted, as a matter of conventional wisdom, that people are *always* playing parts, that the idea of a distinction between a person's roles and his real self is an illusion. Similar views are sometimes adopted by social scientists, as, for instance, by Erving Goffman when he writes:

> At one extreme, one finds that the performer can be fully taken in by his own act; he can be sincerely convinced that the impression of reality he stages is the real reality. · · · At the other extreme, we find that the performer may not be taken in at all by his own routine. · · · When the individual has no belief in his own act and no ultimate concern with the beliefs of his audience, we may call him cynical, reserving the term 'sincere' for individuals who believe in the impression fostered by their own performance.[15]

Goffman is evidently leaving no room for the possibility that a person's behaviour might be natural, that it might not spring from a concern with impressions in the first place. To this extent, we might say, his view is itself

---

[15] Erving GOFFMAN, *The Presentation of Self in Everyday Life* (Harmondsworth, 1971), p. 28. The same view is expressed by BERGER (*op. cit.*, p. 109). For a short, perceptive commentary on Goffman's views, see Alasdair MACINTYRE, *The self as work of art*, New Statesman, 28 March, 1969.

cynical. (However, Goffman does not seem to have held to this view consistently.)

The attraction of the cynical view may come, in part, from its seeming to be the *sober* view of human affairs; the cynic presumes to stand outside the bustle of life, never to be carried along and never to be disappointed by the actions of others. Such, we might think, is the stance the scientist should take. In addition, the difficulty we may often have of deciding whether someone is being his real self or not, the impossibility of identifying criteria in human behaviour for distinguishing the artificial from the natural, make it appear that no scientific import can be given to the distinction. (Actually, we must distinguish between two alternative attitudes here: on the one hand, the refusal to commit oneself as to whether a person's conduct is natural or affected, and, on the other hand, committing oneself to the view that behaviour is always affected.)

A further reason for wanting to reject the distinction between nature and affectation, it might be thought, is this: we find that people are not consistent even with respect to the sorts of attitudes and responses that we should be the most inclined to regard as expressive of their natures. A person's anger, his sense of humour, his fears, his ways of responding to works of art, may to some extent depend on the social context, on his understanding of those present and his relations to them. Connected with this is the fact that such responses, too, are partly shaped by learning, in a process in which the models provided by others play an important part.

Considerations like these seem to have weighed with R. Harré and P.F. Secord, as when they write:

> One and the same biological individual may present a variety of internally consistent personae in different social situations, no one of which is more authentically *him* than the others.[16]

This is obviously their basis for citing with apparent approval the view that

> if the affectation is taken away then nothing man-like remains. The Stoics' aim of stripping away appearance leaves nothing but a stone god. In short the task of the social psychologists is to identify the parts, not to look for a common identity behind the parts.[17]

Yet I wish to contend that the view that human behaviour is always role-play is unintelligible, and that the distinction between a person's roles

[16] HARRÉ and SECORD, *op. cit.*, p. 208.

[17] *Ibid.*, p. 210.

and his real self is one that we cannot avoid making, whether in our relations with others or in doing social science.

The difficulties that I have mentioned here, which seem to lie in the way of upholding this distinction, are, I believe, only apparent, stemming from a misunderstanding of the nature of the distinction itself. The error, I want to argue, lies in supposing that it is an empirical matter whether or not there is such a distinction to be made, and how it should be applied.

I shall try to explain why I believe this to be an error. Let us, to begin with, consider how we go about applying this distinction. Suppose we meet someone and his conduct strikes us as affected in some way. If we get more closely acquainted with him, our initial impression may either be borne out or belied. However, it would be a mistake to think about this situation in terms of the forming and testing of a hypothesis. For the bearing that this man's later conduct has on our first impression is not independent of the way in which *it* strikes us; in particular, of what we find natural or affected about it. The distinction between the natural and the affected belongs to the way in which we *regard* him; it determines, for one thing, the ways in which we will describe his conduct.

It would perhaps be illuminating to think of what we do in applying this distinction, and the changes that our application of it may undergo, by analogy with the notion of "seeing as" and the change of aspects discussed by Wittgenstein.[18] There are, in particular, two points that this comparison serves to bring out. For one thing, the way in which we see an ambiguous picture is an immediate response to the picture rather than an inference. And furthermore, it is normally only to the extent that we are capable of seeing a picture in a specific way that we are able to see details of the picture as providing evidence for or against the correctness of this way of seeing it.

As I get to know someone, it might be said, I gradually develop a knack for seeing *him* in the various things he says and does. What I learn in this way is not necessarily something that I should be capable of putting into words. On the whole, developing such a knack need not have anything to do with coming to identify a single unifying principle in all he does. If this is so, Harré and Secord are in error when they argue from the impossibility of identifying a single authentic self to the non-existence of a real self apart from all affectations.[19]

---

[18] Ludwing WITTGENSTEIN, *Philosophical Investigations* (Oxford, 1958), p. IIxi.

[19] We should not that the word "self" has a variety of uses. A person may apologize for something he did when he was tired or in a rage, for instance, by saying that he was "not

What I regard as belonging to a person's real self, we might say, constitutes what he *is* for me: the significance that his actions and words may come to have for me. It determines the kinds of feelings that I can have for him; and on the other hand, my feelings for someone may shape the way in which I regard him. This mutual dependence is brought out in a passage in the story by Kundera, in which the young man muses over the change that his girl friend appears to have undergone:

This was all the worse because he worshipped rather than loved her. It had always seemed to him that her inward nature was *real* only within the bounds of fidelity and purity, and that beyond these bounds it simply didn't exist. Beyond these bounds she would cease to be herself, as water ceases to be water beyond the boiling point. When he now saw her crossing this horrifying boundary with nonchalant elegance, he was filled with anger.

This point, I believe, could also be expressed in roughly the following way: it is only to the extent that we see something about a person's conduct as expressive of his real self that we are able to regard him as a person.[20] For this reason, I want to say, the distinction between what belongs to the real self and what is an affectation is one that will enter into all human relations. It is also, I would argue, necessarily involved in the study of social life. In giving an account of life in a social group, the scientist must commit himself concerning the distinction between what is real and what is affectation in the lives of the people involved. Otherwise, what he is putting forward will not be an account of a *life*. (This does not mean that he must look for features that all human life-forms have in common. It would be a misunderstanding to think that a person's real self, what is natural for him, cannot be a reflection of the particular culture to which he belongs.)

It is true that the commitments that are called for here may be controversial, and that there cannot, in matters like these, be any method

himself". The point of such a remark is the abrogation of responsibility; it has no direct connection with the distinction discussed here. Indeed, the cases of which it will be said that a person was not himself will typically be ones in which there will be little doubt that his action expressed his real self.

We also speak about a person's inmost self, his true self, etc. It would take us too far to go into the differences between these concepts here. It seems to me that Anthony MANSER, in *Problems with the self* (Proceedings of the Aristotelian Society 84 (1983–84), pp. 1–13), overlooks the variety of self-concepts. His point about a unitary self as another social role, something that a person may strive to develop, is perhaps concerned with a different concept than that discussed here, with the self as an ideal rather than a reality.

[20] For a discussion of matters pertinent in this connection, see D.W. HAMLYN, *Person-perception and our understanding of others*, in his book "Perception, Learning and the Self" (Oxford, 1982).

which guarantees the correctness of the judgements arrived at. But this can be no objection to the point that such commitments must be made. One could only suppose that it was one if one thought that scientific inquiry must concern itself only with matters in which conclusive truth can be attained. But this would surely be a misunderstanding of the nature of science.[21]

[21] I wish to thank Heikki Kannisto for useful comments on an earlier draft of this paper.

Barcan Marcus et al., eds., Logic, Methodology and Philosophy of Science VII

ISSUES IN THE ONTOLOGY OF CULTURE[1]

DAN SPERBER

School of Social Science, Inst. for Advanced Studies,
Princeton, NJ, USA

What kind of things are cultural things? Are they psychological things? Are they of an irreducible nature? What should be the relationship between a science of culture and other sciences, psychology in particular? These are the issues I want to touch on today. They have been much discussed in the past, by Durkheim, Boas, Kroeber, Radcliffe-Brown, Sapir, Leslie White, Geertz, and Sahlins among many others. We would agree, I suppose, that the arguments they used were not always as strong as the convictions they expressed.

Our present aim, it seems to me, should be to raise the level of argument and to achieve a better grasp of the issues, rather than to arrive at some final conclusion. Our understanding of cultural phenomena is far too limited to warrant a definite acceptance or rejection of psychological reductionism, or an elaborate real definition of culture.

The notion of the reduction of one *theory* to another is fairly well understood and is illustrated by famous cases such as the reduction of thermodynamics to statistical mechanics (see NAGEL 1961, Chapter 11). Loosely speaking, a theory B can be reduced to a theory A if all the generalisations of B can be re-formulated in the vocabulary of A, and if all these re-formulated generalisations of B can be shown to follow from the generalisations of A.

The notion of the reduction of one field of inquiry to another, such as the reduction of cultural anthropology to psychology, is much vaguer, and particularly so when either of the fields is not characterised by a well-established theory, or by a well-established theoretical programme. In such cases, assertions to the effect that one field can, or cannot, be reduced to

[1] My thanks to Scott Atran, Martin Hollis and Jerry Katz for useful comments on erlier drafts.

the other are generally based on a priori convictions rather than on specific arguments. Some people believe in the Unity of Science, others believe in Emergent Evolution. I am an agnostic in these matters. I am not too concerned about the ultimate reducibility of a full-fledged culture theory to a full-fledged psychological theory. Besides, as exemplified by recent work in the philosophy of biology (Darden and Maull 1977, Darden 1978), relationships between fields are too varied and subtle to be analysed solely in terms of reduction or non-reduction.

What I would like to know is how to go about developing a theoretical understanding of things cultural: in particular, should we bother with psychology, or ignore it? This is why I am interested in the ontological question: are cultural things psychological things? For, clearly, if cultural things are in no way psychological things, then psychology is irrelevant to their study; and if they are psychological things, then... well, then the issue is somewhat more complicated, but there is some hope that psychology might be relevant to the study of culture.

This is not to say that one could take psychological theories and somehow "apply" them to cultural data. Rather, it could be the case that at least some anthropological hypotheses have definite psychological implications, and that at least some psychological hypotheses have definite anthropological implications. If so, the need for mutual consistency would be a welcome source of constraint on theorising in both fields, and developments in each field might be suggestive of developments in the other. My estimate is that, given the present state of the arts, anthropology has more to receive and less to give than psychology, but this need not always be so.

I shall borrow from Jerry Fodor (1974) a nice way of distinguishing a more general ontological issue from the more particular issue of reductionism, or, if you prefer, a way of distinguishing two ontological questions which could be raised regarding culture: are *types* of cultural things *types* of psychological things? And here, of course, we are asking about the types that a science of culture and a psychology does, or would recognise, types about which there are, or there might be interesting generalisations. Or: are *tokens* of cultural things *tokens* of psychological things? The point of Fodor's distinction being that it is possible to have token-identity without having type-identity; we could have for instance some interesting generalisation about a class of psychological events each of which could be described as a physical event, while the class itself could not be characterised in the physical terms available to us.

I take it that if we answered "yes" to the type-identity question, it would be but a short step, or no step at all, to psychological reductionism. However, I shall argue, we are in no position to answer the type-identity question. The reason for this is simple, even though it will take a bit of elaboration: we don't know what types of cultural things there are (and, arguably, we don't know too well either what types of psychological things there are), hence discussions of type-identity are premature.

On the other hand, a case could be made for token-identity between cultural things and psychological things. Take for instance my reading this paper in front of you: this is clearly something cultural. At the same time it is a complex of psychological and more particularly psycholinguistic events. Now, a description of it in psychological terms might be cumbersome to the point of utter irrelevance,[2] but this is not to say that it would be incomplete. Possibly, from a full psychological description, a proper anthropological description (whatever that might be) could be reconstructed. If we believed that such a situation generally obtained, we would be token-psychologicalists without having to be psychological reductionists.

What would being token-psychologicalists buy us? Two things: first, we could avoid having to make the assumption that cultural things belong to an independent ontological level, without having to commit ourselves to strict reductionism (and, of course, by a similar reasoning, we could be token-physicalists with respect to psychological things — see FODOR 1981, Chapters 5 and 6 — and hence consider every token cultural thing to be a token physical thing).

Second, there would be an initial plausibility to the view that some psychological generalisations might be of relevance to the study of culture. Of course, this might turn out not to be the case at all; types of psychological things might be entirely unrelated to types of cultural things, in the same way as, I suppose most of us would want to claim, types of physical things are unrelated to types of cultural things, in spite of a possible token-identity between them.

There is, however, a difference between token-psychologicalism and token-physicalism with respect to cultural things. Token-identity of cultural things to physical things, while easy enough to accept speculatively, is hard to imagine in a direct fashion. If we tried to work out cases, say a physical description of my reading this paper to you which would refer strictly to the same event as a description of it in anthropological terms, we would have to imagine an intermediary psychological description. We

[2] See PUTNAM 1975, pp. 295–298 for a detailed discussion of a similar example.

would have to do so in order, for instance, to be able to select the right neurological events, those involved in my speaking and in your listening, and to leave out other neurological events, say those taking place in some of you who, instead of listening, are thinking: "if only I had not had so much Schnaps last night!" The latter neurological events, even though they might take place in this room, do not properly belong to this specific cultural event (they do belong to the wider cultural event of our Salzburg Congress, I suppose).

Token-identity of cultural things to psychological things involves no intermediary level. This suggests — though, I repeat, it does not guarantee — that there might be some degree of correspondence or even of overlap between psychological and anthropological typologies, and hence some degree of mutual relevance between the two fields.

Are we, though, in a position to maintain that token cultural things *are* token psychological things? Not even that, I am afraid. There are strong grounds to hold that environmental factors such as population density, seasons, climate, and also man-made devices from everyday artifacts to irrigation systems or telephone networks are to be taken into account in a description of cultural things, and, surely, these are neither type- nor token-identical to psychological things. However, even something weaker than strict token-psychologicalism, namely, the assumption that every token cultural thing is a complex of token psychological and environmental things — all of which are physical things — is enough for our purpose: it allows us to consider that cultural things have no ontological independence, while keeping an open mind about reductionism; it gives us reasons to hope that psychology — and ecology — might be of relevance to the study of culture.

These then are the two ontological points I want to develop here: we don't know what types of cultural things there are, and we have good grounds to believe that token cultural things are a mixture of token psychological and environmental things. In order to establish these points, I have to turn now to the analysis of anthropological concepts.

Cultural things, anthropological terms, and types of family resemblance

Do we know what types of cultural things there are? "But, of course, we do!" most of my fellow anthropologists would answer: we don't know all of the types, we don't know them too well, but we know that there are clans, and lineages, and marriages, and kinship systems, and agricultural techniques, and myths, and rituals, and sacrifices, and political systems, and legal codes, and scholarly institutions, etc. Now, these cultural types are not, and do not correspond to, psychological types. There are good grounds therefore, to oppose psychological reductionism in the study of culture, rather than be agnostic about it, and good grounds to treat culture as autonomous (and we can leave it to philosophers to decide whether this is to be understood ontologically or methodologically).

This view has been most cogently developed by David Kaplan:

> Anthropology has formulated concepts, theoretical entities, laws (or if one prefers, generalizations) and theories which do not form any part of the theoretical apparatus of psychology and cannot be reduced to it. This is the logical basis for treating culture as an autonomous sphere of phenomena, explainable in terms of itself. It is wholly beside the point to maintain that anthropologists cannot proceed that way, for the brute fact of the matter is that in their empirical research this is the way they do most often proceed. (KAPLAN 1965, p. 973)

Kaplan's argument rests on an evaluation of anthropology's achievements. This evaluation can be challenged in two ways, one which I shall mention but not pursue, since I believe it to be unfair and unproductive, and another one, which I have developed elsewhere (SPERBER 1982) and of which I shall try to show that it has some interesting ontological implications.

The fact is that there is very little agreement among anthropologists about anything, beyond rejecting a few old-fashioned theories, e.g. meteorological interpretations of religious symbolism, and defending the profession against external attacks: no single concept is shared by all practitioners, no theoretical entity is universally acknowledged, no theory is generally accepted. In such conditions, it could be argued, nothing can be inferred about the autonomy of culture from the state of the art. I won't pursue this argument because I am convinced that anthropologists, without arriving at any kind of theoretical consensus, have, somehow, developed a genuine common competence in the study of socio-cultural phenomena. An evaluation of anthropology's achievements which does not include an explication of this competence is incomplete, and therefore insufficient to refute Kaplan's contention.

What I want to argue, rather, is that what looks like "concepts, theoretical entities, laws and theories" of anthropology are really intellectual tools of another kind; they are interpretive tools. From their existence and usefulness it is impossible to draw ontological conclusions (or what Kaplan sees as "methodological" conclusions).

It is not just that anthropologists don't *share* theoretical concepts; it is that they don't *have* theoretical concepts of their own. What they do have is a collection of technical terms. They are technical in the sense that they are terms of the trade rather than ordinary language terms (or they are ordinary language terms used in a non-ordinary way). They are not theoretical, though, in that their origin, development, meaning and use are largely independent of the development or content of any genuine theory.

Throughout the history of anthropology, a number of these technical terms have been critically analysed, for instance "taboo" by Franz STEINER (1956) and Mary DOUGLAS (1966), "totemism" by GOLDENWEISER (1910) and LÉVI-STRAUSS (1966), "patri-" and "matri-linearity" by LEACH (1961), "belief" by NEEDHAM (1992), and, of course, "culture" by a great many anthropologists (see KROEBER and KLUCKHOHN 1952, GAMST and NORBECK 1976). The vagueness or the arbitrariness of these terms has been repeatedly pointed out. Yet, in spite of this critical work, there are no signs that anthropologists are converging on a set of progressively better defined and better motivated notions. If anything, there is more divergence and no greater conceptual precision today than there was half a century ago.

Edmund LEACH (1961) and Rodney NEEDHAM (1971, 1972, 1975) have convincingly argued that this vagueness of anthropological terms is not accidental, that it has to do with the way these terms have been developed and with the kinds of things they are being used to refer to, so that, if we want proper theoretical terms in anthropology, we should construct altogether new ones.

Rodney Needham has further argued that anthropological technical terms are best understood as "family resemblance" or "polythetic" terms, that is as terms referring to things among which resemblances exist, but which don't fall under a single definition. More technically, a polythetic term is characterised by a set of features such that none of them is necessary, but that any large enough subset of them is sufficient for something to fall under the term. A polythetic term need not be fully polythetic: all its referents may share one or even several features, but as long as these necessary features are not jointly sufficient, the term is still polythetic.

Actually, it is dubious that fully polythetic terms (i.e. terms without any necessary feature) are ever used. All the members of a useful polythetic class normally belong to the same domain, which determines at least one common feature. All the members of the class of "games", to take Wittgenstein's famous example of family resemblance, share the feature of being activities.[3] Or, when Needham writes: "the members of a class of

[3] An example I owe to Jerrold Katz (personal communication).

social facts may share no feature in common" (1981, p. 3), he does not mean to deny, presumably, that they share the feature of being social facts. If actual polythetic terms are only partly polythetic then their use commits one not only to the existence of a resemblance but also to the presence of at least one definite feature in the object referred to.

Now, I want to argue that anthropological terms do indeed have some kind of family resemblance organisation, but that it may be a different kind of family resemblance than the one Wittgenstein and Needham had in mind. They had in mind a resemblance between the things described by the same term. For instance, every thing described as a game resembles some other things described as games. Let us call this a "descriptive resemblance". I shall suggest, however, that anthropological technical terms are not used descriptively, but interpretively. They are not directly used to describe; they are used to translate or render native terms or notions (or notions that the anthropologist attributes to the natives). The resemblance involved is an "interpretive resemblance" between the particular notions interpreted and the notion generally conveyed by the interpretive term. As a consequence, all the notions that can be properly interpreted by means of the same term will exhibit a typical family resemblance pattern: i.e. two such notions need not directly resemble one another, but there will be at least one further notion (the notion conveyed by the interpretive term) that they both resemble.

In the case of descriptive resemblance, there is normally no resemblance between the term itself and the things it is used to describe. In the case of interpretive resemblance, a term with some notional content (more about that later) is used to interpret other notions (expressed or not by a term); it is the resemblance in content between the interpretive term and the term or notion interpreted that makes the interpretive use possible.

The view that anthropology is an interpretive science is a well-known one and has been brilliantly defended by Clifford GEERTZ (1973). This is not, however, the view of anthropology I am defending. I agree that anthropologists studying individual cultures are mainly involved in an interpretive task, i.e. in representing native representations by means of translations, paraphrases, summaries, and syntheses understandable to their readers. On the other hand, I see the task of theoretical anthropology not as an interpretive, but as a descriptive and explanatory one, i.e. as similar to the theoretical task of the natural sciences. Furthermore, I am, I believe, on my own in arguing that the technical vocabulary of anthropology is — as a matter of fact, not of necessity — itself interpretive and not descriptive or properly theoretical. It is precisely because of its ontological implications that this claim is likely to be resisted by those who otherwise see anthropology as a fully interpretive activity.

An example: "marriage"

From now on, I shall briefly recapitulate and then pursue the argument with reference to an example. Take "marriage". Now, here is a true technical term of anthropology, and as good a type of cultural thing as you will ever get. But how good a *type* is it? Do all marriages fall under a single definition, or do we have reasons otherwise to believe that they share some yet unanalysed common essence?

Let us look first at a couple of characterisations of marriage that have been proposed. The *Notes and Queries* (1951) suggested: "Marriage is a union between a man and a woman such that children born to the woman are recognized legitimate offspring of both partners". Here, you don't have to look for exotic counter-examples. In most Western societies, the distinction between legitimate and illegitimate offsprings is becoming abolished. Children born in or out of wedlock may enjoy the same rights. The only sense in which some children may still be called "illegitimate" is precisely that they are born out of wedlock. But this, of course, makes a definition of marriage in terms of the legitimacy of offspring quite circular.

Or consider Lévi-Strauss's claim: "If there are many types of marriage to be observed in human societies ... the striking fact is that everywhere a distinction exists between marriage, i.e. *a legal, group-sanctioned bond between a man and a woman*, and the type of permanent or temporary union resulting either from violence or consent alone" (LÉVI-STRAUSS 1956, p. 268; italics added). In the very same paper, Lévi-Strauss gives a counter-example to his own characterisation. He argues that many "so-called polygamous societies ... make a strong difference between the "first" wife who is the only true one, endowed with the full rights attached to the marital status while the other ones are sometimes little more than official concubines" (ibid., p. 267). Now, there may well be a group-sanctioned bond between a man and his "official" concubine. Therefore if Lévi-Strauss wants to distinguish this bond from true marriage, then his characterisation of marriage fails.

Such failures to properly define "marriage" are not accidental. Edmund Leach has argued that "marriage is ... 'a bundle of rights'; hence all universal definitions of marriage are vain" (1961, p. 105). The point being here that the rights bundled together vary from society to society. Leach lists ten kinds of rights, from: "to establish the legal father of a woman's children", to: "to establish a socially significant 'relationship of affinity' between the husband and his wife's brothers." He shows that that there is not a single one of these rights which is present in all cases of marriage.

Developing Leach's argument, Needham concludes that "marriage" "is an odd-job word: very handy in all sorts of descriptive sentences, but worse than misleading in comparison and of no real use at all in analysis" (NEEDHAM 1971, p. 8). There are two ways in which "marriage" can be said to do odd jobs: it does a few different jobs for all anthropologists, and also, and I believe more importantly, it does a different job for each anthropologist in his own field.

Imagine an anthropologist who goes to study the Ebelo. She might, in principle, wonder whether the Ebelo have at all the institution of marriage, but it would be surprising if she did. It is generally taken for granted in the profession that marriage is universal. She would not however expect to find something which would fall under a well-established definition of marriage, since there is no such definition. What she expects to find is some native institution which she may call "marriage" with as much justification as other anthropologists in their use of the term.

The problem she faces is not whether the Ebelo have marriage, but, as P.G. Rivière puts it, "which of the forms of relationship between the sexes is ... to be regarded as the marital one" (RIVIÈRE 1971, p. 65). The logic is one of a party game, really: "if one of these forms of relationship were a form of marriage, which one would it be?" It would take a very odd society, or a very uncooperative anthropologist, for the question to remain without an answer. It is not surprising then that marriage should be found in every society. This is made possible, however, precisely by the fact that "marriage", whatever it does otherwise, does not denote a precise type of cultural thing.

But how does our anthropologist go about identifying which Ebelo form of relationship is "the marital one"? Does she *look* at relationships? No, relationships are not the kind of things you can look at. What she does, mostly, is to get Ebelo people to describe in their own terms the types of relationships they entertain among themselves. She then decides which of the native notions, and, possibly which of the native terms, is best rendered by "marriage". "Marriage" in English designates simultaneously a status, a change of status, and a jural relationship. In other languages, these three notions may not come under a single term. In such cases, "marriage" would be used to render a cluster of native categories.

Our anthropologist comes to the conclusion that "marriage" corresponds to the Ebelo term *kwiss*. She then goes on to explain what she understands the Ebelo to believe, namely that marriage, i.e. *kwiss*, is a bond between a man and a woman blessed by ancestral spirits. Note in passing that "bond", "blessed", "ancestral spirit" are also used interpretively in this characterisation of the meaning of "marriage"/*kwiss*, i.e. they are not used to describe things, but to render further Ebelo notions.

Now a new case of marriage, the Ebelo case, has been added to the anthropological stock. It has been added on the basis of a resemblance. So had all previous cases, even the first one. "Marriage" became a technical term of anthropology when an anthropologist — or was it an historian? — decided that some exotic notion was best rendered by the ordinary language word "marriage". From then on, "marriage" began to swell and loose its contours as more and more different notions were interpreted by means of it. The notional content of "marriage", in anthropological writings, became a loose synthesis or compound of the sundry particular notions the term served to interpret. The point to stress is that, for a new notion to be rendered by "marriage", it need not fall under the general notion conveyed by the term, it need merely resemble it. That is why, also, the fuzziness of anthropological terms is no obstacle to their use: fuzziness is no hindrance — if anything it is a help — to the establishement of resemblances. Other terms such as "taboo" or "totem", became technical when an anthropologist decided to borrow a native word rather than translate it, and the family of interpretive resemblances was then built around this first exotic notion.

That the anthropological notion of marriage should be a family resemblance notion is thus no accident. It is a result of the very way in which it has been and is being developed. Resemblance — and not the possession of definite features — determines where "marriage" is to be applied. There is no reason to expect the development of anthropology to reverse this state of affairs. Actually, the better anthropologists come to know a greater variety of cases, the looser becomes the resemblance between instances of "marriage".

Is, however, the resemblance involved in determining the applicability of "marriage" one among the things called marriage, or one between the notion interpreted and the notion (or notions) generally conveyed by the term used to interpret it? Is it, in other words, a descriptive or an interpretive resemblance? If the account I have sketched of how anthropologists go about identifying new cases of marriage is correct, then, clearly, the resemblance involved is an interpretive one.

Ontological implications

The two types of family resemblance, the descriptive and the interpretive one, have different ontological implications. If you take "marriage" to be based on descriptive resemblance, you should envisage that the term is only partly polythetic. Surely, all marriages are relationships; plausibly, they are all jural relationships. So, when you describe something as a marriage, this may well commit you to the existence of jural relationships as a fundamental *type* of cultural things. Now, unless you are prepared to argue that jural relationships are also a proper type of psychological things, using the polythetic notion of marriage so understood does not allow to keep an open mind about psychological reductionism: it should set you against it.

Not so with interpretive resemblance. Imagine that our anthropologist reports that Peter and Mary, two Ebelo individuals, are married. Is she, in so doing, *stating* that there is a bond between Peter and Mary that has been blessed by ancestral spirits? Presumably not, if only because it would commit her to the existence of ancestral spirits. She is reporting, rather (in the free indirect style — see SPERBER 1982), what the Ebelo people involved believe about Peter and Mary. She is interpreting Ebelo ideas. What does such an interpretation commit her to, ontologically speaking? It commits her to the existence of certain Ebelo people, and to the existence of certain representations in the minds of these people. Does it commit her to the existence of a thing, or a state of affairs (which could be called a marriage), and which would be distinct from the fact that certain Ebelo people hold the view that Peter and Mary are *kwissed*? I don't see how. Our anthropologist might *want* to further commit herself in that way, but her report would give us no ground to follow her in such a commitment.

If "marriage" is an interpretive term, used to render a variety of native notions (or notions attributed to the natives, or notions synthesised from several native notions), then every anthropological account of a case of "marriage" is, when properly analysed, an account of a set of psychological facts. More specifically, to say that two people are married is to say that representations to the effect that these people are *kwissed* (or whatever native term is rendered by "marriage") are properly distributed in the population. What constitutes a proper distribution is determined by the native notion of *kwiss* itself. For instance, it may be part of the native notion of a *kwiss* that once a priest and the spouses hold it that the latter are *kwissed*, then they are. To say that two people are undergoing a marriage ceremony is to say that some physical interaction is taking place between people such that will cause the proper representation to be properly distributed.

But what about "marriage" in theoretical or comparative work? Doesn't it, there, correspond to a general concept? Well, if you believe it does, say *which* concept. I am not claiming that it would be impossible to define a general concept which could reasonably be expressed by "marriage". I am suggesting that there is no obvious reason why you would want to define a concept meeting this particular condition, and that anthropologists, notwithstanding the appearances, have never truly bothered. They have found it useful to abstract from interpretive ethnographic reports in order to arrive at general interpretive models. These models are not *true* of anything; what they do is help the reader get a synthetic view of ethnographic knowledge. They also serve as sketches of possible interpretations for further ethnographic work. So, "marriage" in these general anthropological writings is both a loose topic-indicating word, and an interpretive term considered not in relation to any one of its particular uses, but in relation to several of its actual or potential uses.

What is true of "marriage" is true of the technical vocabulary of anthropology in general. "Tribe", "caste", "clan", "slavery", "chiefship", "state", "war", "ritual", "religion", "magic", "witchcraft", "possession", "myth", "tales", etc. are interpretive terms[4]. There is a family resemblance, but an interpretive one, between all the notions each of these terms serves to render. When these terms are used to report specific instances of events or states of affairs, they help the reader get an idea of the way in which the people concerned perceive the situation ("seeing things from the native's point of view", as the phrase goes). What do these interpretive reports tell us about the *nature* of whatever is taking place? Well, they tell us that there are some psychological and some physical things going on, and that's it.

A few general terms used in anthropology are not interpretive in that sense, but nor do they suggest the existence of a distinct ontological level of culture. Some are straightforwardly psychological, such as "color classification". Others are straightforwardly ecological, such as "dam". What differentiates these psychological or ecological terms used in anthropology from the proprietary vocabulary of the field, is that they apply quite independently of the "point of view" of the subjects concerned. People can

---

[4] See DETIENNE 1981 for a relevant discussion of "myth" and SPERBER 1982, Chapter 1, for one of "sacrifice".

have a color classification without being aware of the existence of such things as classifications, and we can agree that beavers build dams without attributing to them any cultural point of view. On the other hand, a marriage cannot take place without some people entertaining the idea that a marriage (or a *kwiss*, or something of the sort) is taking place. Moreover, it is unclear what other necessary conditions there are for something to be a marriage beyond its being represented as such in the appropriate minds.

Conclusion

I have tried to make three points:

(1) The technical vocabulary of anthropology is neither observational nor theoretical, it is an interpretive vocabulary. Moreover, each anthropological term serves to interpret a great variety of native notions which share among themselves a mere family resemblance. Terms the use of which is determined by interpretive resemblance carry neither ontological nor typological implications.

(2) We just don't know, therefore, what *types* of cultural things a science of culture would recognise. We don't know whether these types would be reducible to psychological types. There is no a priori reason, either, to assume that these types would correspond, even approximately, to the technical terms of current anthropology, since these terms are not even aimed at identifying such types.

(3) Interpretive accounts of particular cultural phenomena, an Ebelo marriage, or the reading of a paper at a philosophical congress for instance, allow one kind of description: culture phenomena are mental representations being distributed, over time and space, in a human population as a result of physical interactions and cognitive processes.

I want to suggest, in conclusion, that this token-identity of cultural events to distributions of ideas is what we have to start from if we are interested in a scientific understanding of culture. What does that imply?

A distribution of ideas would not be likely to fall under any type recognised in the cognitive psychology of the individual organism, nor is there any reason why it should. The study of the distribution of ideas stands to the study of individual cognitive phenomena the way epidemiology stands to the study of individual pathology. Epidemiology and individual pathology use different data, concepts and method; epidemiology takes into account environmental variables of various ontological status; but, whether it is approached from the point of view of individual pathology or from that of epidemiology, the ontology of diseases is basically the same, and the two fields are highly mutually relevant. Epidemiology does not "reduce" to individual pathology, but they both work — individual pathology exclusively, and epidemiology essentially — within the same ontological level of biological facts.

Similarly, I am not arguing for a reduction of cultural anthropology to individual psychology. I am suggesting rather that the scientific study of culture might take the form of an epidemiology of ideas (a notion that has been toyed with by a few anthropologists, social psychologists and biologists at various times, but never been properly developed). Like the ontology of disease epidemiology, the ontology of an epidemiology of ideas would be somewhat messy; it would take into account a variety of environmental variables; but its basic subject-matter would be of course psychological. Hence a relationship of mutual relevance with individual organism psychology, and in particular with cognitive psychology, could be expected. I am aware, though, that there is no direct path going from the type of ontological clarification we have been concerned with, to what is truly essential and will ultimately decide the issue, namely the development of a scientific understanding of culture.

References

DARDEN, L. and MAULL, N., 1977, *Interfield theories*, Philosophy of Science 44, pp. 43–64.

DARDEN, L., 1978, *Discoveries and the emergence of new fields in science*, PSA, 1978, 1, ed. P. Asquith and I. Hacking (Philosophy of Science Association, East Lansing, MI), pp. 149–160.

DETIENNE, M., 1981, *L'Invention de la Mythologie* (Gallimard, Paris).

DOUGLAS, M., 1966, *Purity and Danger: An Analysis of the Concepts of Pollution and Taboo* (Routledge and Kegan Paul, London).

FODOR, J., 1974, *Special sciences*, Synthese 28, pp. 77–115. Reprinted in FODOR 1981.

FODOR, J., 1981, *Representations: Philosophical Essays on the Foundations of Cognitive Science* (The MIT Press, Cambridge, MA).

GAMST, F. and NORBECK, E., 1976, *Ideas of Cultures* (Holt, Rinehart and Winston, New York).

GEERTZ, C. 1973, *The Interpretation of Cultures* (Basic Books, New York).

GOLDENWEISER, A., 1910, *Totemism, an analytical study*, J. Amer. Folklore 23, pp. 178–298.

KAPLAN, D., 1965, *The superorganic: science or metaphysics?*, Amer. Anthropologist 67(4), pp. 958–976.

KROEBER, A.L. and KLUCKHOHN, C., 1952, *Culture: a critical review of concepts and definitions*, Papers of the Peabody Museum of American Archeology and Ethnology 47(1), pp. 1–223.

LEACH, E., 1961, *Rethinking Anthropology* (The Athlone Press, London).

LÉVI-STRAUSS, C., 1956, *The family*, in: Man, Culture and Society, ed. H.L. Shapiro (Oxford Univ. Press, Oxford), pp. 261–285.

LÉVI-STRAUSS, C., 1966, *Totemism* (Beacon Press, Boston).
NAGEL, E., 1961, *The Structure of Science* (Harcourt, Brace and World, New York).
NEEDHAM, R., ed., 1971, *Rethinking Kinship and Marriage* (Tavistock, London).
NEEDHAM, R., 1972, *Belief, Language and Experience* (Blackwell, Oxford).
NEEDHAM, R., 1975, *Polythetic classification*, Man 10, pp. 349–369.
NEEDHAM, R., 1981, *Circumstantial Deliveries* (Univ. of California Press, Berkeley and Los Angeles).
PUTNAM, H., 1965, *Mind, Language and Reality: Philosophical Papers*, Vol. 2 (Cambridge Univ. Press, Cambridge).
RIVIERE, P.B., 1971, *Marriage: a reassessment*, in: Rethinking Kinship and Marriage, ed. R. Needham (Tavistock, London).
Royal Anthropological Institute, 1951, *Notes and Queries in Anthropology*, 6th ed. (London).
SPERBER, D., 1982, *Le Savoir des Anthropologues* (Hermann, Paris).
STEINER, F., 1956, *Taboo* (Cohen and West, London).

12. Foundations and Philosophy of Linguistics

Barcan Marcus et al., eds., Logic, Methodology and Philosophy of Science VII
© Elsevier Science Publishers B.V. (1986) 573–595

NATURAL LANGUAGE METAPHYSICS

EMMON BACH

Dept. of Linguistics, Univ. of Massachusetts, Amherst, MA 01002, USA

0. Introduction

Metaphysics I take to be the study of how things are. It deals with questions like these:

What is there?

What kinds of things are there and how are they related?

Weighty questions, indeed, but no concern of mine as a linguist trying to understand natural language. Nevertheless, anyone who deals with the semantics of natural language is driven to ask questions that mimic those just given:

What do people talk as if there is?

What kinds of things and relations among them does one need in order to exhibit the structure of meanings that natural languages seem to have?

Questions of this latter sort lead us into *natural language metaphysics.* In this paper, I want to show how we are driven to such questions when we try to give a serious account of the semantics of natural language and I want to say something about possible answers.

Linguistics, like any other field of inquiry, lives off of puzzles. Why can we say this and not that? If I say a certain sentence, does that commit me to the truth of certain other sentences? Why does no language do this and every language do that? Why do languages that put the verb at the end of the clause overwhelmingly tend to use postpositions rather than prepositions? Why are Dutch *weten*, German *wissen* ('know') etymologically related to Latin *video* ('see') (cf. Greek *oida* 'know', present in meaning but perfect in form, thus historically 'I have seen')? Linguists, like other seekers after understanding, usually follow the maxim: Divide and conquer! That is, we try to understand complex phenomena like those just alluded to by setting up various components in our explanatory theories, by idealizing, and trying to do justice to the complexity of the phenomena by appealing to the interactions among the various subsystems. This sort of strategy has led to many satisfying results and interesting new questions. To the extent that we are successful we think we are finding out something about the nature of language and the users of language.

The general framework for descriptions of natural language that I start from draws upon two traditions: that of generative theory as developed in the last several decades under the leadership of Noam Chomsky and others; that of model-theoretic semantics as inspired especially by Richard Montague. I like to look at the relationship between these two lines of research as encapsulated in two theses:

I. Chomsky's thesis: natural languages can be described as formal systems.

II. Montague's thesis: natural languages can be described as *interpreted* formal systems.

Note that in this way of looking at matters, Thesis II embodies or presupposes Thesis I. (DAVIDSON (1967) must be mentioned along with Montague as one of the first to propose that the methods of interpretation developed by Tarski, Carnap, and others, could after all be applied to the study of natural language semantics.) I interpret 'interpreted' here in the sense of providing models of various sorts that contain non-linguistic objects (in general) that are assigned to linguistic expressions as their "semantic values" (LEWIS, 1972). Thus, I understand 'semantics' in the sense of a theory of the relationship between language and something that is not language.

Now even at this most general level we run into some fundamental problems. If we ask what these non-linguistic objects are that we assume for our model structures, there are two quite different kinds of answers, at least, that have been proposed or presupposed or defended: one tradition, probably the most prominent one in the philosophical tradition, has it that they are real objects and relationships in the world (as well as, perhaps, their analogues in other possible worlds); the other, which seems most prominent in the tradition of generative theory, says that they are mental objects: concepts, representations, or the like. Thus, in his latest major book, Chomsky draws a distinction between 'real semantic interpretations' and properly linguistic or psychological semantic interpretations, presumably of the second sort (1982, p. 324).

Are these two answers genuine alternatives or is there some way to reconcile them? Chomsky's few remarks (*ibid.*) seem to suggest that they are not incompatible. I quote:

> Note that this step in the process of interpretation [the positing of a domain D of individals in "mental space" EB] ... should be considered to be in effect an extension of syntax, the construction of another level of mental representation beyond LF ['Logical Form'], a level at which arguments at LF are paired with entities of mental representation, this further level then entering into "real semantic interpretation."

Elsewhere (1975), CHOMSKY has suggested that the step from LF to semantic representation (SR) takes us beyond linguistics in that it requires us to bring in systems of 'knowledge and belief.' On the other hand, in a recent book, Jerrold KATZ (1981) makes a plea for what he calls a Platonist conception of linguistics in which the objects that we study are purely abstract, and linguistics is something like a specialized branch of mathematics (cf. Thomason's similar interpretation of Montague's position in his introduction to MONTAGUE, 1974, and the reaction in CHOMSKY, 1980, pp. 29 f). Finally, there are plenty of passages in Chomsky's writings which display the attitude that 'real semantics' has very little or nothing to do with genuine linguistic questions.

Now, it is a frequent move in linguistics, as in other fields, to mark off some domain of questions and — tentatively — leave other questions to be dealt with in some other, perhaps not yet invented discipline or theory. Thus, in the early days of generative grammar it was quite usual to argue that various facts were semantic and not to be represented in a linguistic grammar (i.e. syntax and phonology). Once 'real semantics' began to play an important role in linguistics other wastebaskets were appealed to: 'pragmatics' was one, 'real world knowledge' was another. But each such decision must be backed up at some point by providing or pointing to a genuine theory of some sort about what is left out. If something is claimed to be outside of semantics because it is a matter of pragmatics (in one of its several senses), then we had better be able to work out a pragmatic account or at least give plausibility arguments for why we think a pragmatic theory would provide us with such an account. The present paper is concerned with asking about just such a program or theory with respect to certain phenomena and puzzles that seem to go beyond pure semantics as usually conceived. The first point that I will try to argue for is this: No semantics without metaphysics!

We can come at the sort of questions I want to raise here from a somewhat different angle. Many writers contend that it is possible and desirable to draw a sharp line between what we might call 'constructional' (or 'structural') semantics and lexical semantics. Let me quote a sentence from Richmond Thomason (MONTAGUE, 1974, p. 48):

> *But we should not expect a semantic theory to furnish an account of how any two expressions belonging to the same syntactic category differ in meaning.*

And in a footnote to that sentence:

> The sentence is italicized because I believe that failure to appreciate this point, and to distinguish lexicography from semantic theory, is a persistent and harmful source of misunderstanding in matters of semantic methodology. The distinction is perhaps the most significant for linguists of the strategies of logical semantics.

The idea here is something like this: in constructing a syntax and semantics for a fragment of a natural language like English we start for the most part with unanalyzed chunks of meaning for individual lexical items like *fish* or *walk* or *kiss*. All that we need to know about such items is what *kind* of meanings they have, for example the *fish* and *walk* denote properties or sets of individuals, *kiss* a two-place relation between individuals, and so on. We offer analyses for certain 'logical' words: *the*, *a*, *every*, *be* (cf. MONTAGUE, 1974, p. 261); other words perhaps receive no direct interpretation at all: their semantic effects are exhibited only in conjunction with the rules that introduce them: thus *and*, *or* as well as grammatical morphemes like past tense markers and the like (so-called syncategorematic items). Now we build up the meanings of complex expressions by stating explicit semantic rules for each of the constructional rules of our syntax, requiring that the resultant semantic value be a function of the semantic value of the component parts. Thus, for a subject-predicate rule that gets us sentence like *John walks* we need to state a general rule that will license the particular theorem: *John walks* is true iff the individual denoted by *John* is in the set of walkers (at the time and world of the evaluation, say). Thomason's point is that this general rule should in no way depend on differences among the meanings of *walk* and *jump* or *run* or *shout* or *exult*. This kind of distinction is reminiscent of the traditional distinction between logical and non-logical constants. And it is subject to similar suspicions, as I will try to show. So the second main point I want to make is this: No constructional semantics without lexical semantics!

There are at least two parts of the enterprise of doing the semantics of natural languages where metaphysical questions rear their (ugly or beautiful?) heads: in making decisions about the general structure and content of our models and their relation to the things in our syntaxes, and at points where it seems that we have to 'go inside' the meanings of particular lexical items in order to state compositional rules of the semantics. Let's consider some examples of each of these areas in turn. At the end I will return to the more general question: What are these things that we talk about in our models for the meanings of natural language expressions?

1. General characteristics of model structures

By a *model structure* I mean the collection of various kinds of things and possible relationships among them that are used to give an interpretation for a language. Thus, a standard semantics for a first order theory has a very simple model structure: a set of individuals, a set of two truth values (and a set of assignments of values to variables in a Tarski style quantification theory — I will henceforth omit mention of assignments of values to variables, supposing it to be a constant feature of all the model structures I want to talk about). A model structure is then a candidate for the range of a an interpretation function for a language.

Let me use a quick review of the model structure assumed in Montague's best known paper on English as a way of getting into the kinds of worries that I want to consider here (that is, 'The proper treatment of quantification in ordinary English,' henceforth 'PTQ' = Paper 8 in MONTAGUE, 1974). It makes use of the following sets:

A: the set of possible individuals,
I: the set of possible worlds,
J: the set of times, with simple ordering $\leqslant$ on $J \times J$,
2 $(=\{0, 1\})$: the set of truth values.

In PTQ, (disambiguated) English is interpreted indirectly via a translation function that takes English expressions into expressions in a typed intensional logic (IL) which is in turn interpreted by an assignment of elements constructed out of the above sets to the expressions of the logic. (Given the two-step functional relationship between disambiguated English, IL, and the model, we know that there is a direct function from disambiguated English to the model structure. This means that for Montague's purposes, IL is merely a convenient way of exhibiting the structure of the interpretation and is theoretically dispensable.) What kinds of things are made out of these ingredients? The answer is given by a recursive definition that starts with the sets A, 2, and the set of *indices* $I \times J$ (worlds cross times) and allows us to construct all total functions from sets of possible denotations to sets of possible denotations and in addition from indices (worlds plus times) to sets of denotations (these are the *senses* or *intensions* that make the logic intensional). This gets complicated very quickly: the denotation of the

English word *John*, for example (given Montague's theory of interpreting noun phrases as generalized quantifiers), is a function from functions from indices to (functions from (functions from (indices to individuals) to truth values)) to truth values (if I got that all straight!) You can imagine (you probably can't) what kind of denotation the preposition *in* has in a sentence like *John kissed Mary in the garden*!

So what answer to the query *What is there?* do we get from PTQ? (Or to the question *What does the speaker of PTQ-ese talk as if there is?*) The answer is: a whole lot! Given that the intensional logical allows variables of all types, if to be is to be a possible value of a variable available for quantification, the PTQ gives us an infinite collection of different kinds of beasts to put into our ontological zoo. Yet all of this is built up in a way that is set-theoretically quite respectable and as far as the primitive elements are concerned quite simple. How well does this apparatus perform when we take it to be a candidate for providing insights into the meanings of natural language expressions? Well, there are problems. And these problems seem to be of two kinds: the system seems to give us too much and too little.

Before getting into the details of these embarrassments of riches and poverty, let me give some quick illustrations (cf. BACH, 1981) of the sort of metaphysical questions that arise just out of the model structure of PTQ itself quite apart from the hierarchy of functions we've just looked at. You will note that the set of times (with its ordering relation) is an independent ingredient in the model structure, that is, it is outside of possible worlds, hence, we can always get a definite answer to questions about the temporal relations of happenings in different possible worlds. Example: *If Mary had left on the space flight yesterday she would now be eating breakfast.* Now, I'm not at all sure that we want to make this sort of a claim even for one world, say *our* world, as part of our semantics. Moreover, even the relatively minimal assumptions about the ordering relation (it is transitive, antisymmetric, reflexive) commit us to certain views about such questions as this: Is time travel possible? Two well-known native speakers of English, Peter GEACH (1965) and David LEWIS (1976), appear to have or have had diametrically opposed views on this matter. (It was worries of this sort that first got me to thinking about English ethnometaphysics.) And physicists make good money thinking seriously about time reversal. Arthur Prior once characterized the job of tense-logicians as being 'lawyers of time' who do up briefs for their clients. Well, we are the clients and we have to think seriously about just what sort of temporal systems we want to adopt in our model structures. A few more such worries follow.

What are the members of J? Montague doesn't tell us. Early tense logic *assumed* that they were something like dimensionless moments or instants (and Montague uses this kind of talk in his paper, 'On the nature of certain philosophical entities,' Paper 5 in MONTAGUE, 1974). There seems to be a growing consensus that intervals are better or at least necessary in addition. Should we take instants as basic and *construct* intervals as convex sets of instants (v. BENTHEM, 1982)? Should we take intervals as basic and construct instants (*ibid.*; KAMP, 1980) in the manner of Wiener, Russell, or Whitehead? Should we say more or less than Montague does about the structure of time? Is it discrete? I would argue that the set of times must be at least countably infinite (cf. BACH, 1981). Then, if time is discrete and homogeneous, as has been argued by some linguists, there cannot both be a first and a last moment of time. Are questions about the Big Bang and the Final Whimper *linguistic* questions? Is time dense, continuous, Dedekindian? Some physicists want to say that time comes in little smallest granules called chronons (one estimate: 10^{-43} seconds). Before anything happened, how long did it last? Can time pass if nothing happens? Native speakers of Standard Average European languages (Whorf's phrase: SAE) differ on this point too (Archimedes and Aristotle, Newton and Leibniz, for example (cf. NEWTON-SMITH, 1980)). WHORF (1936) claims that we speakers of SAE differ on many of these points from speakers of Hopi, we being Newtonian absolutists, they being relativistic. But Einstein was no Hopi! I'll get back to some of these questions below, but let's now return to the main thread.

Montague's semantics makes a nice start toward solving some puzzles. Consider for example his reconstruction of the notion of a property as a function from possible worlds to sets. This analysis allows us to distinguish between the property of being human and the property of being a two-legged rational (?) animal. *Maybe* in this world the two properties pick out the same set of entities. but surely not in every possible world (for example, very likely *not* in this possible world).

Now, it seems that Montague's reconstructions of things like intensional meanings (Fregean senses) don't go far enough along the highroad of intensionality. It's been known for a long time that belief-contexts and the like make for insuperable difficulties for Montague-style propositions, properties, relations-in-intension and so on. Here's a different kind of argument (due to Gennaro Chierchia, cf. CHIERCHIA, 1984) based on rather mundane linguistics facts. Consider English phrases like these:

(1) sold,

(2) bought,

(3) sold by Mary,

(4) bought by Mary.

Everything we believe about linguistic methodology and English syntax urges us to say that phrases like (3) and (4) are built up out of phrases like (1) and (2) by the addition of an agent phrase in *by* to a passive (or passivized) verb phrase or verb. Now it can be plausibly argued that in every possible world in which there is buying and selling the set of things that are bought will be coextensive with the set of things that are sold. But on Montague's analysis this means that the property of being sold is identical to the property of being bought. Ergo, there is no way to get the function that makes the meaning of (3) and (4) to give us a different result when we combine *by Mary* with the two passive phases in (1) and (2). Hence, Montague grammarians have been forced, within the confines of PTQ semantics, to posit two completely unrelated passive rules for cases like (1) and (2) as against (3) and (4) (THOMASON, 1974; BACH, 1981; cf. COOPER, 1979, for a dissenting view which tries to turn this vice into a virtue). Here, I think we have not been paying enough attention to what I like to call Montague's advice: Take natural languages seriously! Perhaps they are trying to tell you something. In this case, as Chierchia very convincingly argues, English is trying to tell us that we need to have in our models properties as entities of some sort that can be distinguished even if they pick out the same sets in all possible worlds. That is, PTQ isn't intensional enough; natural language is *very* intensional.

That was an example showing how PTQ model structures aren't rich enough to do what we need. An opposite sort of example showing that the type structure and the assignment of types to syntactic categories is doing too much, that is, forcing distinctions that make life ugly for the semanticist, is the following: consider the sentence *Mary loves everything!* There is no obvious way in which we can give a meaning to this sentence in PTQ semantics that will allow us to conclude that Mary loves, for example, dancing, Chinese cabbage, the Pythagorean theorem, and Montague semantics (to say nothing of generalized quantifiers, propositions, and Sam). Or consider the sentence: *It's boring to be boring.* We can construct a higher order predicate applicable to (say) properties but on the one hand it can't have anything to do with the predicate in *John is boring* and on the other we can construct sentences *ad libitum* that keep pushing us toward higher and higher order functions: *It's boring for it to be boring to be boring* ... (cf. PARSONS, 1979; CHIERCHIA, 1982; TURNER, 1983). We can see this problem from a slightly different angle. PTQ is a quite rich fragment but is deliberately restricted at its base: there are only singular NP's and sets of individual concepts at the bottom of the interpretation. As various workers have extended the coverage of the fragments (e.g. BENNETT, 1974, for plurals; DELACRUZ, 1976, for propositional predicates and noun-phrases) it has seemed to be necessary to introduce new sets of syntactic categories and rules for these extensions. But now it becomes a complete accident that the kinds of constructions that English uses for expressions involving these various kinds of new categories resemble the constructions that are used for the simpler categories. Put more plainly, once we learn how to construct sentences about people, cabbages, kings, and pigs, we don't expect to have to learn a whole new syntax to be able to talk about first principles, purposes, propositions, groups, and so on.

Montague's theory requires a functional mapping from syntactic categories to logical types. In some cases this seems good. As we look at new languages we find that the predictions about the existence and behavior of categories like those of sentence and noun-phrase hold good: sentences correspond to truth values (or maybe propositions), noun-phrases to generalized quantifiers, verb-phrases to sets or properties. We are beginning to gain some insight into differences that are not reflected in PTQ semantics among other 'parts of spech' like intransitive verbs, adjectives, and nouns (GUPTA, 1980; CHIERCHIA, 1984). What we don't expect to find are syntactic differences within these broad general syntactic classes (well, maybe some, e.g. mass versus count nouns, etc., but see below) that depend on the kinds of things, actions, qualities denoted by the various lexical items. But the attempt to reconstruct some of these semantic differences on the basis of the type theory inevitably leads to an explosion of syntactic categories.

One move that was made early on by CRESSWELL (1973) arises from reflecting on what sorts of things (and what *sorts* of things) we want to include in our domain of possible individuals (PTQ's A, Cresswell's D_1). Montague is already fairly liberal: there are four individual constants in his fragment that denote (rigidly) the individuals Bill, John, Mary and ninety, and the sets of things countenanced by PTQ include fish, unicorns, men, women, prices and temperatures. Cresswell is not only quite explicit about letting us put into the domain anything whatsoever that we want to talk about but offers us the option of pumping various higher order things like propositions and properties back down into the domain.

Now, once we start putting new and unusual things into our domains, it begins to look as if we might want to do some sorting (THOMASON, 1972; WALDO, 1979). Let me mention some of the sorts that have been introduced into our domain in the last five years or so.

I've already mentioned properties as primitive elements. Here I'm most familiar with the work of Chierchia (already mentioned) who is building on the work of Nino Cocchiarella. This move is also made in the situation semantics of BARWISE and PERRY (1981) and in the work of BEALER (1982). Some other but more recently (re-)discovered species represented in various zoos around the world are kinds and stages (G. CARLSON, 1977), plural individuals and quantities of stuff (LINK, 1983), events, processes, states, situations. Of course most if not all of these animals had been reported to exist at one time or other long since. What is recent is getting them into the model-structures of a certain family of related ways of doing semantics. As these new things come in it is reasonable to want to put a little more structure into our domains. In the next part of this paper I want to take up this kind of question in the area of events and situations. But before doing so, let me mention one more move that seems to be common across several theories: that is 'going partial.'

Theories have been going partial in two (not unrelated) ways. One way is using partial functions in the semantics, the other is in providing for partial worlds, so to speak (both already in CRESSWELL, 1973).

People usually seem to think of worlds as pretty big things, as entire ways in which things — everything — could be. Now for lots of purposes it seems as if it would be nice to have something intermediate in 'size' between individual things and entire worlds in the usual sense. In one way or another, various workers have used the idea of something like a partial world or part of a world in a crucial way. I will mention two theories that I know a little bit about (there are no doubt many more I don't know about): CRESSWELL'S (1973) metaphysics of propositions and categorial languages and BARWISE and PARRY'S situation semantics (1980).

In keeping with his laudable decision to say something definite about the metaphysics of his possible world semantics, Cresswell offers the following analysis of possible worlds:

> We are to suppose that we are given a set *B* of 'basic particular situations.' The idea is that any subset *w* of *B* determines a world. The elements of *B* which are members of *w* might be thought of as the 'atomic facts' of world *w*. (1973, p. 38)

Although Cresswell writes 'determines' here, elsewhere (e.g. p. 42) the power set of *B* just *is* the set of possible worlds. That he is really thinking of possibly very small portions of a world as being themselves worlds is clear from the discussion of individuals in a later chapter. For Cresswell, an individual is a function from a world to a subpart of that world (it is thus something more like Montague's individual concept). He writes:

Strictly, since a subset of a world can itself be a world a basic individual ρ is a function from possible worlds into possible worlds provided that for any world w, $\rho(w)$ is a subset of w. (p. 94)

Cresswell calls the value of the function applied to a world w the "manifestation" of ρ in w.

As its name implies, situation semantics takes seriously the idea of a situation, something like a limited portion of the world (note the definite article). (This discussion is couched in terms of the first presentations of the theory, e.g. in BARWISE and PERRY, 1981. I am not sure to what extent what I say is consonant with the later versions (1983) of the theory.) As I mentioned, the theory accepts properties and relations as irreducible elements of the model structure. Corresponding to situations are situation types, partial functions from ingredients of situations to truth values, the ingredients being ordered n-tuples of ($n-1$ place) relations and individuals. The World Type is a total function of the same sort. There are many points of interest in the theory for people interested in linguistic semantics, but this is not the place to explore the theory. The point I want to make here is that, just as in Cresswell's metaphysics of propositions, we have world-like things that are possibly smaller than worlds, can stand in a part-whole relation to each other and are of the same logical type as worlds. I want to take over much of these theories in the following discussion. For concreteness and because it represents a less radical departure from familiar model-structures I will take Cresswell's proposals as a base.

2. Eventology

I now want to look in some detail at a topic that we might call (rather barbarously) *eventology*. Under this heading I wish to consider two questions. The first is this:

(1) Do we want or need to include something like events in our model structure?

I associate the insistence on the importance of this question as well as a vigorous defense of the answer 'Yes!" above all with writings of Donald Davidson (recently made conveniently available together with further commentary in DAVIDSON, 1980).

The second question, brought into prominence in modern times especially by Anthony KENNY (1963) and Zeno VENDLER (1967), has to do with a classification of what I have called 'eventualities' (BACH, 1981), that is things that go variously under the names of 'events, processes, states' (my favored terminology), 'activities, accomplishments, achievements' (VENDLER), 'performances' (KENNY). (There is a vast literature, both linguistic and philosophical, on such matters. For a comprehensive discussion from a linguistic point of view and within the same general framework of assumptions as my own, see DOWTY, 1979.) So the second question is this:

(2) What kinds of eventualities are there and what are their properties? Whatever one thinks about the philosophical, that is, metaphysical answers to these questions, I think there is ample evidence that we want to answer "Yes" to question (1) and provide an answer to question (2) if we want to do natural language semantics.

I think it is reasonable to assume that the proper place in our theories to try to come to terms with the second question is in our semantics rather than (or rather than only in) the syntax. I don't want to deny that the classification can have syntactic correlates, in fact, it is clear that it often does if we look around at the languages of the world. But the distinctions feel semantic. So I want to claim that they are not *just* syntactic. One reason for supposing that this is correct is that the distinctions turn up in language after language as overt or covert categories, but with wildly differing syntactic and morphological reflexes. Thus, our theories of Universal Grammar should provide a place for talk about the classification but it seems hopeless to build such a theory on the basis of pure linguistic form. Another, perhaps more compelling, argument is that the classification plays another role in licensing certain inferences, as we shall see; indeed, this was one of the important species of arguments for establishing the classification in the first place. In general, it is not the case that purely syntactic distinctions license (non-syntactic) inferences, cf. grammatical gender.

Moreover, I am going to start from the position that we want to reflect the classification somehow in our model-structures. The question is how?

Perhaps it would be well to inject a brief reminder of just what it is I am talking about. Consider these eight sentences:

1. Bill loves Mary.
2. Mary finds a unicorn.
3. Bertha builds a cabin.
4. John runs.
5. Bill is loving Mary.
6. Mary is finding a unicorn.
7. Bertha is building a cabin.

8. John is running.

The contrasts in acceptability and 'specialness' of interpretation when we compare English sentences in the simple present and present progressive establishes the distinction between states (1) and non-states on the other (2, 3, 4). Differences in the interpretations of progressives help to establish the differences on the one hand between processes (4) and events (2, 3) and further between momentary or punctual events (Vendler's achievements: 2) and protracted events (3) (I am less secure about the linguistic necessity for the last distinction than I am about the others).

I am not going to argue extensively here for the necessity of including eventualities as entities in our models (cf. DAVIDSON, 1980, passim for events). Let me limit myself to one argument that has the nice property of helping to establish the irreducibility of events. Consider the kind of patterns of inference given by KENNY (1963) as part of the justification for the difference between activities or processes and performances (a species of events).

9. Mary is building a cabin. Therefore, Mary has not built a cabin. (Mary is *V*-ing. Therefore, Mary has not *V*-ed.)

Now, never mind that this inference is clearly not valid, there is a genuinely correct intuition that Kenny is trying to get at and one that we constantly use in our everyday decisions. (Cf. *Bill is dying. Therefore, Bill has not died.* This inference requires additional premises.) Suppose the mechanic at my garage tells me: We are replacing your carburetor. Then I will correctly infer that the car is not yet ready to pick up and will ask some reasonable question like: When do you think you will be done? What is the basis of this inference? Well, the most direct way to say something about (9) is this: If Mary is building a cabin then that cabin-building event is not yet over. This way of talking makes direct reference to something like Davidson's particular ephemeral events.

This example is also nice in that it helps to establish that we need to have different *kinds* of eventualities. Kenny offers the following as a diagnostic for processes (activities):

10. Mary is running. Therefore, Mary has run.

Again there are problems about whether this is a genuine semantic entailment (consider the very first instant of Mary's running — you have to say the first sentence VERY FAST). But once again I think we must admit that there is a real insight here and the sharp contrast with (9) helps establish the necessity to separate processes from events.

Now, if you will grant me the conclusions just drawn, I think you will be able to see why I have some doubts about the advice of Thomason to stay clear of the meanings of individual lexical iterms. In (9) and (10) we have instances of a quite general constructional rule of English: formation of progressives. But as the examples show, the truth conditions for the resultant expressions might very well depend on the meanings of the individual verbs.

Now the first obvious stab at getting at these distinctions in our model-structures is just to incorporate eventualities into our domain E and say something about their properties (cumulative reference, indivisibility, etc.). I don't claim that it is impossible to construct them out of things otherwise needed, just that all of the attempts to do so that I know about don't seem to work. For example, MONTAGUE in NCPE (1974, p. 6) gave an analysis of events and other kinds of entities on the basis of a model-structure including just worlds, times, and individuals, and there have been many attempts to follow out his ideas. For example, instantaneous events were reconstructed as properties of moments of time, protracted events as properties of intervals, considered as sets of moments (that was just hinted at). But the Kenny intuition about examples like (9) can't be captured in this way. Further (gratia Terry Parsons) what are we to say about two events of the same species that occur at the very same instant? So we might try to add places and think of events as properties of space-time locations. Well that works for some but not others. Suppose Mary suddenly realized that she had forgotten to turn off the power drill. Where did that happen? (Cf. DAVIDSON, 1969: 'The individuation of events' in DAVIDSON, 1980.) Now although there may be real metaphysical doubts about whether the Great Pyramid and the Battle of Waterloo are at bottom entities that are fundamentally different in kind (cf. WHITEHEAD, 1920), natural language seems to advise us to treat them as different, so we will probably want to sort E into at least two main kinds of things: eventualities and objects, with further distinctions in each subdomain. This was the move I adopted a number of years ago in some work on English tense, aspect, and temporal adverbials. At the time it seemed wildly innovative, today it doesn't seem so adventurous. Davidson, again, was my main inspiration. (This work, a very rough draft of some chapters of which was circulated to some extent under the title 'Topics in English Metaphysics,' will probably never see the light of day. Two papers that grew out of that work are BACH, 1980, 1981. The present paper draws in part from a larger work in progress with the same working title. Some of the ideas were presented in a joint seminar with Terry Parsons on tense and aspect at U. Mass., Spring 1978. The basic subcategorization adopted was the one hinted at already: states, processes, events (punctual or instantaneous) and protracted. In this work all the types of eventualities were treated on a par.)

There have been persistent attempts to take one or another type as basic and to derive the others. Probably the best known is that of VON WRIGHT'S logic of change (1963), in which some kinds of events are analyzed as changes of states. Von Wright is himself quite careful not to claim that all events and processes can be analyzed in this way. DAVIDSON (1967) gave good reasons for doubting that this would work for all events: for many events the only state that comes about as a result of the event is the trivial one of the event's having taken place: walking around the block, playing a game of chess. There is apparently a strong tendency to think that states are somehow basic, a sort of filmstrip view of reality which I do not share. If anything, quite the opposite seems to be true. It took about two millennia to come up with a satisfactory way of coping with Zeno's questions about what it could possibly mean to be in a state of motion at an instant or how you could possibly add together dimensionless instants to get changes (you can't).

One immediate bonus of making this move is that we now have all the ingredients to construct a theory of time on the basis of simple relations among events along the lines of Wiener, Whitehead, Russell and others (cf. WHITROW, 1980; KAMP, 1980). The most natural and immediate kind of temporal system that we get out of the primitive relations of precedence and overlap is an interval system but it is possible to define instants with all of the right properties (cf. *ibid.*: instants are proper filters on the set of events, or, if you wish, the intervals associated with them). In this way it is possible to think about possible histories as sets of eventualities and certain specified temporal relations among them.

One interesting metaphysical question is this: is time independent of things happening 'in' it? An ancient question which I don't think should be answered in a semantics for English (cf. WHITROW, 1980; NEWTON-SMITH, 1980). It seems most parsimonious not to assume an independent time-series. As I indicated before, it seems downright wrong to insist that everything that happens in a possible history, let alone separate possible histories, be mappable onto a single time line. If we take sets of events as basic then time can remain nicely imminent. And we need not insist on total connectedness (KAMP, 1980, argues for this freedom on the basis of two applications: the indeterminacy of temporal relations among happenings involving vague predicates of change; the construction of narrative structures where only some of the relations of precedence and overlap are specified).

One way to get at the essential differences among states, processes, and events is to perform Gedanken-experiments in which we imagine various possible histories (Bach, 1981; M. Moneglia in unpublished work). Imagine a history in which nothing happens. Clearly, such a history could contain no events or processes, but only states. This is not to say that there are no states that presuppose a dynamic changing universe, there clearly are (cf. being in orbit). But states *per se* do not require change. So let us say that events and processes have the property of temporality, states do not. This observation leads me to the following speculation: states have a different ontological status than events and processes. The latter are the primary ingredients of possible histories (together with the individuals and stuff involved in the events and processes). States have a more derivative and abstract status. Perhaps it is only states that can be profitably thought of as properties of moments — that is, instants — of time. Another one of my native informants was quite insistent on this point. Whitehead (1920) shows how to construct instants out of families of events and then argues that the idea of nature at an instant is something that we need for our theories but does not have the same immediate reality as the processes and events that make up the process-event that we call Nature (putting things this way is very un-Whiteheadian to be sure!).

It is interesting that the sentences that Davidson uses in his arguments for events are all about genuine flesh and blood events: butterings of toast, explosions of boilers, raisings of arms, kickings of Shem and Sean and the like. There are problems lurking in some of the Davidsonian paraphrases: "There was an *x* such that *x* was an explosion and *x* was of the boiler" for

The boiler exploded.

(The problems have to do with some of the conjuncts like '*x* was of the boiler.') But stative sentences resist this kind of paraphrase even more:

The satellite was in orbit:

"There was an *x* such that *x* was a being in orbit and *x* was of the satellite."

John loved Mary:

"There was an *x* such that *x* was a loving and *x* was of Mary and *x* was by John"

Sally was in New York:

"There was an *x* such that *x* was a being and *x* was in New York and *x* was of/by Sally".

Or maybe *x* was a being in New York?

Now what is the difference between events and processes? Events are bounded: they have a beginning and an end and maybe a middle. If there are instantaneous events, their beginnings and their endpoints are identical and they have no middles. Processes need not be bounded. So let us say that events have the property of boundedness. Now a number of familiar properties of events and processes follow. Events are countable, processes as such are not. Events in general cannot be subdivided into subevents of the same kind. Processes may be. (Not ad libitum!) Processes have the cumulative reference property: running + running = running. Events don't. It also follows that events are the primary hooks on which we hang our temporal structures. I think this is the *reason* why events and states act the way they do in narrative structures (KAMP, 1981; HINRICHS, 1981); events move the story line forward, states don't.

Now, all of this reminds us strongly of important distinctions in the realm of things and stuff, a point that has not gone unnoticed in the literature (ALLEN, 1966; L. CARLSON, 1981; HOEPELMAN and ROHRER, 1980; MOURELATOS, 1978): mass, count, plural and so on. Processes are to events as stufl is to things. There seems to be something very basic about this articulation of the world and/or our experience of it. (It is fun to consider the analogies in the realm of sound systems and the most basic distinctive features of phonemes: ontology recapitulates phonology!)

Godehard LINK (1983) has proposed an analysis of the nominal domain that goes like this: The domain of ordinary individuals is extended to include 'plural individuals' with the extended domain making up a Boolean algebra which is complete with an individual-join and part-whole relationship. In addition, the domain includes a special set of atoms, 'quantities of matter,' with its own algebraic structure (a join semilattice with a material part-whole relation). The subdomain of quantities of matter is systematically related to the big domain by a homomorphism. Barbara Partee and I have explored the consequences of carrying Link's ideas over to the domain of eventualities along the lines of the proportion mentioned above: events are linked to 'quantities of process' in much the same way that things are linked to quantities of matter in Link's construction (cf. BACH, forthcoming, for details).

Let me make a parenthetical remark about the formal properties of the kind of distinctions we are drawing here. I have said that events and processes are temporal and that it is not the case that states are temporal and further that events are bounded but processes need not be. In each case we have attributed a certain property to classes of eventualities but it is important to be clear about the claims as they apply to the other classes of eventualities in each case. States *may* be temporal and processes *may* be bounded. The point is that they don't have to be. In linguistic jargon, it is natural to think of these properties as something like features which may have unspecified values. Semantically speaking, we may think of them as abbreviations of or references to something like meaning postulates, that is, restrictions on the class of admissible models used to interpret natural languages.

2. TNT: The Nicest Theory

Let me now say a little bit about how we might build some of these distinctions into our model-structures for interpreting English. (I am freely stealing from all sorts of people here, and hope only that I haven't forgotten to mention any of the stealees.)

Let us follow CRESSWELL (1973) in taking as basic a set B of basic particular situations and letting the set H of possible histories simply be the power set of B. In order to think clearly about what we are doing we might adopt Cresswell's suggestion of taking each member of B to be a set of space-time points, but in no way do I (nor does Cresswell) want to be restricted to such a physicalist interpretation.

Along with the inherent relations of the Boolean structure H, we must have two further relations: strict temporal precedence ($<$) and overlap (o, as in KAMP, 1980). (I will have nothing to say about modality here, but assume that further relations such as accessibility can be specified to hold among worlds or histories.)

Now let us superimpose on this picture the additional elements of our model: individuals and properties (using this word in a general way to include relations of various adicity). Again following and extending Cresswell, let us say that individuals and properties determine functions from possible histories to parts (subsets) of those histories and let us call the values for these functions for a history h the *manifestations* of the individual or the property in h. I say 'determine' rather than 'are' functions because I want to allow individuals and properties to be different even if they are manifested identically in every possible history. Thus the property of being bought and the property of being sold can be different, and the Morning Star can be different from the Evening Star. We thus need to assume that there is a function — call it EXT — that takes us from individuals and relations to functions from histories to their parts. Notice that at this point there is no difference between individuals and properties (this would please Whitehead, I think). EXT(JOE) and EXT(KISSING) both pick out subparts of histories: the first, all those space-time points (in our physicalist illustration) which are manifestations of Joe; the second, all of which are kissings.

We now want to notice some differences between these different sorts of entities and others. One important difference between an individual like Joe Blow and a property like Kissing is that Kissings require a kisser and a kissee but Joe Blow manifestations do not require a 'Joe-er' or a 'Joe-ee.' And in respect to the number of such extra things required Kissings differ from Laughings and Rainings. So, following CHIERCHIA, 1984, let us say that there is a function AD defined on the set of properties into the natural numbers. This function tells us how many argument places each property requires, from 0 on up, with the 0-place properties to be thought of as propositions.

Let us notice some other differences among kinds of things as far as properties of their manifestations go. The value of EXT(Joe Blow) in any history will give us either temporally limited sections of a space-time worm, or an entire space-time worm, depending on the size of the history. Moreover, any manifestation of Joe Blow (let me henceforth drop the EXT part when there's no danger of confusion) will exhibit a space-time continuity that we may not find for other kinds of things, for example, presidents of the United States, Mr. America's, and so on. Following Greg Carlson, Chierchia, and others, let's allow our set of individuals to include things like Kinds and Types of Matter. Now the manifestations of Dogs will be lots of continuous and non-overlapping space-time regions and we will notice that given any one of these we can find an individual like Fido, such that that continuous portion of space-time is the value of EXT(Fido) in that history. (PTQ: 'meaning postulate' (2), p. 263 in MONTAGUE, 1974.) Given an individual like Mud we will again find lots of space-time regions as its manifestation in any history (all the portions of mud in the history). Now it is a fact about the interpretation of English that there will be important differences in the interpretation of sentences using predicates like *being a dog* and *being mud* (the indivisibility and cumulative reference properties we notice above). Let me stress that the denotation of "mud" (used as a term phrase) is *not* some scattered individual (all the mud quantities in the history). The latter is the value for the function EXT(Mud) given the history as argument. I assume that the general structure of the models we use for different languages will be the same, although this is just a guess and needs empirical verification. but the mappings from expressions to the model-structures can vary a great deal. Thus the meaning of the Japanese word *inu* is something like the union of the meanings of the English words *dog* and *dogs.*

Note that manifestations are themselves worlds or histories. Thus we have a built-in basis for a characterization of different kinds of predicates, namely those that take worlds as arguments and those that take other kinds of things (individuals, properties, propositions) as arguments. (I follow CHIERCHIA, 1984, in assuming a one–one relation between properties and their corresponding predicates.) It is tempting to identify this distinction with the stative-nonstative distinction. Thus the semantic value of stative sentences would be exactly propositions in the classical sense, functions from worlds to truth values or equivalently sets of possible worlds. This would solve a longstanding mystery: what is it that unifies the interpretation of such diverse sentences as the following (all stative by the usual tests):

1. Two plus two equals four.
2. Mary is intelligent.
3. Dogs are mammals.
4. Oscar was drunk.
5. Sally was running.
6. Phillip has left.

Moreover, we can explain the differences between (1)–(3) and (4)–(6) on the basis of the "size" of the worlds or histories that they pick out.

Finally, it seems that the approach outlined here, programmatic as it is, offers a properly mysterious status to manifestations, stages, bare happenings, and stuff. They are, in this setup, completely dependent on the linguistic (and conceptual?) functions which pick them out. Thus, the question of what the ultimate stuff of the world is remains comfortably open in our semantic theories: it can be atoms, wavicles, pure mass-energy, pure spirit, or air, fire, earth, and water.

4. What are we talking about?

I've now said a little (but perhaps more than enough) about some of the kinds of things we seem to need in our ontology for English and a little bit (not near enough) about how we might get them into a semantics for English. It would be immoral of me as a linguist (I'm stealing a phrase from Montague) to make claims one way or the other about whether or not these sorts of things correspond to real things in the world, perceptual or conceptual categories that are independent of language, or nothing at all.

But it is impossible for me as a human and puzzler about the world and our place in it to refrain from thinking about these larger questions.

Let me first say that the kinds of distinctions in the realm of things and of events I've illustrated here really do seem to be very basic to Language with a capital L. I have yet to see a language that does not show some reflex of the state-event-process distinction or the thing-happening distinction. I've worked quite a bit on languages that have been claimed not to make a distinction between nouns and verbs (Wakashan languages like Nootka, Kwakiutl, Xa'isla — the language of Kitamaat Village, near Kitimat, B.C.). Well, the distinction is there after all, even though it doesn't come out quite like it does in English. I once foolishly wrote a paper (Bach, 1968) in which I tried to argue that English was basically like Nootka or standard predicate calculus. I couldn't have been wronger. I now think there is more truth than madness in the old idea that nouns are names for persons, places, and things, verbs names for actions and qualities, adjectives for qualities. But the kinds of "semantices" accessible to most of us before Montague were simply not rich enough to give a good fit to natural language meanings. I have only touched on a couple of the big areas that seem at one and the same time to be basic to the semantics of natural language and also basic to our picture of what the world is like: time and the structure of happenings, things and the stuff that constitutes them. Others are space and locational relations, causation and human responsibility. You can see why I sometimes wonder whether what I am doing is linguistics or philosophy: a philosopher once said to me when I was expounding on 'natural language metaphysics' "But Emmon, that *is* metaphysics!" I've puzzled for a long time about what the difference is between certain kinds of philosophy and certain kinds of linguistics and finally decided that the main difference lies in whether you're embarrassed about not knowing about a paper in *Linguistic Inquiry* or the *Journal of Philosophy*.

But then, after all, the academic divisions we make our livings in don't at bottom necessarily reflect the way the world is. In the final judgment we'll all say we were just trying to put it all together by taking little bits and pieces because that's all we could do. Is there a natural language metaphysics? How could there not be? One of our main resources for coming to understand the world is, after all, language, a sort of tool box for doing whatever it is we want to do. Do the fundamental distinctions that are reflected in the overt and covert categories of natural language correspond in any way to the structure of the world? How could they not? But this is where linguistics stops. And so will I.

Acknowledgements

Most of the work in this paper was done at the Max Planck Institute for Psycholinguistics, Nijmegen. I wish to thank the Institute and its staff for support. For discussion of the issues raised I am grateful to many people, especially Terry Parsons, Gennaro Chierchia, and to Barbara H. Partee, in addition for comments on my paper. None of these people should be held responsible for any remaining errors of style or substance.

Bibliography

ALLEN, R.L., 1966, *The Verb System of Present-Day American English* (Mouton, The Hague).

BACH, E., 1968, *Nouns and nounphrases*, in: Universals in Linguistics Theory, eds. E. Bach and R.T. Harms (Holt, Rinehart, and Winston, New York), pp. 90–122.

BACH, E., 1980, *Tenses and aspects as functions on verb-phrases*, in: C. Rohrer, ed., Time, Tense, and Quantifiers (Niemeyer, Tuebingen), pp. 19–37.

BACH, E., 1981, *On time, tense, and aspect: an essay in English metaphysics*, in: P. Cole, ed., Radical Pragmatics (Academic Press, New York), pp. 63–81.

BACH, E., forthcoming, *The algebra of events*, to appear in: Linguistics and Philosophy.

BARWISE, J. and PERRY, J., 1980, *The situation underground*, Stanford Working Papers in Semantics 1.

BARWISE, J. and PERRY, J., 1983, *Situations and Attitudes* (MIT Press, Cambridge, MA).

BEALER, G., 1982, *Quality and Concept* (Clarendon Press, Oxford).

BENNETT, M., 1974, *Some Extensions of a Montague Fragment of English*, Ph.D. Dissertation, UCLA.

VAN BENTHEM, J.F.A.K., 1982, *The Logic of Time* (Reidel, Dordrecht).

CARLSON, G., 1977, *Reference to Kinds in English*, Ph.D. Dissertation, University of Massachusetts/Amherst.

CARLSON, L., 1981, *Aspects and quantification*, Syntax and Semantics 14, pp. 31–64.

CHIERCHIA, G., 1982, *Nominalizations and Montague grammar: A semantics without types for natural language*, Linguistics and Philosophy 5, pp. 303–354.

CHIERCHIA, G., 1984, *Topics in the Syntax and Semantics of Infinitives and Gerunds*, Ph.D. Dissertation, University of Massachusetts/Amherst.

CHOMSKY, N., 1975. *Reflections on Language* (New York).

CHOMSKY, N., 1980, *Rules and Representations* (Columbia, New York).

CHOMSKY, N., 1982, *Lectures on Government and Binding* (Foris, Dordrecht).

COOPER, R., 1979, *Bach's passive, polysynthetic languages, temporal adverbs, and free deletion*, in: E. Engdahl and M.J. Stein, eds., Studies Presented to Emmon Bach by his Students (G.L.S.A., Amherst, MA).

CRESSWELL, M.J., 1973, *Logics and Languages* (Methuen, London).

DAVIDSON, D., 1967, *Truth and meaning*, Synthese 17, pp. 304–323.

DAVIDSON, D., 1980, *Essays on Actions and Events* (Clarendon Press, Oxford).

DELACRUZ, E.B., 1976, *Factives and proposition level constructions in Montague grammar*, in: B. Partee, ed., Montague Grammar (Academic Press, New York).

DOWTY, D.R., 1979, *Word Meaning and Montague Grammar* (Reidel, Dordrecht).

GEACH, P., 1965, *Some problems about time*, in: P. Geach, Logic Matters (Univ. of California Press, Berkeley and Los Angeles, 1972), pp. 302–318.

GUPTA, A., 1980, *The Logic of Common Nouns* (Yale Univ. Press, New Haven).
HINRICHS, E., 1981, *Temporale Anaphora im Englischen*, Magisterarbeit, Universität Tübingen, Tübingen.
HOEPELMAN, J. and ROHRER, C., 1980, *On the mass count distinction and the French imparfait and passé simple*, in: C. Rohrer, ed., Time, Tense, and Quantifiers (Niemeyer, Tuebingen), pp. 629–645.
KAMP, H., 1980, *Some remarks on the logic of change, Part I*, in: C. Rohrer, ed., Time, Tense and Quantifiers (Niemeyer, Tuebingen).
KAMP, H., 1981, *Evénements, représentations, discursives et référence temporelle*, Langages 64, pp. 39–64.
KATZ, J.J., 1981, *Language and Other Abstract Objects* (Rowman and Littlefield, Totowa, NJ).
KENNY, A., 1963, *Action, Emotion, and Will* (Humanities, New York).
LEWIS, D., 1972, *General semantics*, in: D. Davidson and G. Harman, eds., Semantics of Natural Language (Reidel, Dordrecht), pp. 169–218.
LEWIS, D., 1976, *The paradoxes of time travel*, Amer. Philosophical Quart. 13, pp. 145–152.
LINK, G., 1983, *The logical analysis of plurals and mass terms*, in: R. Bäuerle, Ch. Schwarze, and A. von Stechow, eds., Meaning, Use, and Interpretation of Language (de Gruyter, Berlin), pp. 302–323.
MONTAGUE, R., 1974, *Formal Philosophy*, R. Thomason, ed. (Yale Univ. Press, New Haven).
MOURELATOS, A.P.D., 1978, *Events, processes, and states*, Linguistics and Philosophy 2, pp. 415–434.
NEWTON-SMITH, W.H., 1980, *The Structure of Time* (Routledge and Kegan Paul, London).
PARSONS, T., 1979, *Type theory and ordinary language*, in: S. Davis and M. Mithun, eds., Linguistics, Philosophy, and Montague Grammar. (Univ. of Texas Press, Austin and London).
THOMASON, R., 1972, *A semantic theory of sortal incorrectness*, J. Philosophical Logic 1, pp. 209–258.
TURNER, R., 1983, *Montague semantics, nominalizations, and Scott's domains*, Linguistics and Philosophy 6, pp. 259–288.
VENDLER, Z., 1957, *Verbs and times*, Philosophical Review 56, pp. 143–60.
WALDO, J., 1979, *A PTQ semantics for sortal incorrectness*, in: S. Davis and M. Mithun, eds., Linguistics, Philosophy, and Montague Grammar (Univ. of Texas Press, Austin and London).
WHITEHEAD, A.N., 1920, *Concept of Nature* (Cambridge Univ. Press, Cambridge).
WHITROW, G.J., 1980, *The Natural Philosophy of Time* (Clarendon, Oxford).
WHORF, B.L., 1950, 1936, *An American Indian model of the universe*, IJAL 16, pp. 67–72. Reprinted in J.B. Carroll, ed., Language, Thought, and Reality: Selected Writings of Benjamin Lee Whorf (MIT Press, Cambridge, MA).
VON WRIGHT, G.H., 1963, *Norm and Action* (Routledge and Kegan Paul, London).

Barcan Marcus et al., eds., Logic, Methodology and Philosophy of Science VII

SEMANTICS AND THE COMPUTATIONAL METAPHOR

L. JONATHAN COHEN

The Queen's College, Oxford Univ., Oxford, England

I

Thirty or forty years ago, when computers were a novelty and we seemed to know much less about them than about human thinking, it was natural for the question 'Can computers think?' to be among those in the forefront of discussion. This question soon dissolved very fruitfully, under the pressure of enquiry, into a vast number of more specific issues in the science of computer hardware and software. So that we have eventually come to have rather more extensive and accurate knowledge — knowledge that is still growing rapidly — about what computers can do and how they can do it than we have ever had about the actual mechanisms of human thinking. For the last decade or two, therefore, the guiding question in the science of thought has tended to point in the opposite direction. The question raised has been not so much 'Can computers think?' but rather 'Do humans compute?'

The computational hypothesis, which has come to dominate cognitive psychology, assumes an affirmative answer to this question, in the sense that it expects the computational analogy to be more successful than any other in generating a variety of theories that are not only testable, but also worthwhile testing, about how particular mental processes operate. Memory, visual imagery, concept formation, problem solving, speech comprehension, etc., are treated as fields of research in which experiments may be used to test theories that such-and-such a combination of iteration, recursion, chunking, horizontal searching, vertical searching, geometrical coding, linguistic coding or other mode of information-processing is at work. Commonly the researcher first constructs or sketches or surveys a suitably wide range of computer-programmes (implementable on a suitably wide range of computer-architectures) that, when compared with the mental process in question, would provide analogous outputs for analogous inputs. He then devises experiments on the performance of human subjects in order to determine which of these computer-simulations is closest to the kind of explanatory mechanism required by the results of the experiments. For example, differences in reaction-time may be used as prima facie evidence about relative complexity in explanatory mechanisms or subjects' verbal protocols may be taken as prima facie evidence for the structure of the computational routines that they prefer.

But an important new scientific idea often provokes a tangle of philosophical accretions. Implications are imputed that go well beyond the original, experimentally oriented scope of the idea. Notoriousy evolution, relativity and quantum mechanics, for example, have in turn all suffered in this way. And within the last few years the computational hypothesis, irrespective of any particular experimental outcomes, has been claimed to carry some rather specific implications for the semantics of natural language. The purpose of the present paper is to criticise that claim. So the paper is occupied solely with the implications of the computational hypothesis, and not at all with its merits as a strategy for scientific research, which I take to be an issue for experimental psychology to determine.

The argument to be criticised runs roughly as follows (e.g. FODOR, 1975, 1980 and 1981; cf. also PYLYSHYN, 1980, and HAUGELAND, 1981; but I am criticising the argument, not the people, whose current — July, 1983 — views on the subject are unknown to me). Considered action, concept learning and perceptual integration are familiar achievements of infra-human organisms and pre-verbal children. So, if these cognitive achievements are to be explained computationally, their authors must be attributed representational systems that are not natural languages. At the same time it is argued that humans cannot learn a language whose terms express types of meaning not expressed by the elements of some representational system that they are already able to use. Hence the innate non-conventional system of representation which has in any case to be postulated in order to account for non-verbal and pre-verbal thought must be supposed to have a vocabulary that suffices to express anything expressible in any natural language.[1] And the computational programs of tacit human reasoning must

[1] This thesis is not discussed in the present paper. It seems to assume that there is no other way to acquire new concepts than by defining them in terms of old ones. But that is a procedure which fits such paradigms as 'bachelor' or 'brother' better than it fits the names of animal species, say, or raw materials. Where the acquisition of natural language vocabulary is "organised around stereo-types, exemplars, images or what have you" (FODOR, 1975, p. 153), the role of Mentalese is more controversial. I hope to explore these issues in a forthcoming paper "Innateness and the computational metaphor".

be supposed capable of operating on representations constructed from this private vocabulary. But if those mental operations are indeed computational, then, it is is said, they have access only to the formal properties of such verbal or non-verbal representations of the world as the senses induce: they have no access to the semantic properties of such private representations. They operate on the representations themselves, not on relationships between such representations and the real world. So to characterise a mental state for psychological purposes one should never have to specify or imply the truth-value, for example, of any representation that it embraces, nor should one have to accept or reject the existence of anything to which such a representation refers. Thus belief is a mental state, but to ascribe knowledge is to go beyond the domain of purely psychological characterisation, because what is known has to be true. Moreover it is not only this familiar principle of intentionality that is alleged to sit well with a formalist account of human computations. The computational importance of the *form* of representation is also said to explain the supposed existence of referential opacity — resistance to the substitutivity of identically referring terms — within a person's own internal representations of what he believes, wants, etc. Oedipus's courtship behaviour, for example, is said to depend on which form of desription he (literally) has in mind for the object of his affection, viz. 'Jocasta' or 'my mother'. But we still have to ask: to what do the basic, non-verbal representations owe their representational status? In many cases this status cannot be wholly accounted for by some resemblance between what represents and what is represented, since pictures on their own may fail to identify what they represent at a sufficiently determinate level of specificity: is the picture of a tree, a deciduous tree, an oak in winter, or an oak, etc.? Nor can this representational status be accounted for by computational rules if semantic properties are irrelevant to the computational process. So a non-verbal type of mental representation must owe at least part of its status as representation to some other mode of connexion with what it represents — and this can only be the causal linkage that normally exists between a fact of the kind that it represents and assent to a token of its type. In the end, therefore, the meaning of a sentence in English, say, or French is tied to the causal susceptibilities of the internal representation in Mentalese which interprets it. The semantics of natural language would ideally be founded on a systematic catalogue — or perhaps better on a recursive characterisation — of those susceptibilities. They provide the gold backing, as it were, for the paper currency of semantic markers.

All this, then, is a superstructure of argument that some philosophers have erected on the foundation of the computational hypothesis. At its heart lies the doctrine that so far as mental operations are computational they have access only to the formal properties of internal representations.[2] Once that formalist gloss is accepted, the causal semantics is a scarcely avoidable consequence. And the formalist doctrine, I shall try to show, is totally without foundation. It has no warranty either in the computational hypothesis itself, as a methodology of psychological investigation, or in any as yet acquired empirical evidence.

II

Of course, many other philosophers have discussed the problem of internal representations already. My reason for joining them is just that the formalist issue, despite its centrality, does not seem to me to have yet been brought into sufficiently clear focus. For example STICH (1980) argued that mental states cannot be classified by their form rather than their content without serious paradoxes arising about lookalike worlds on remote planets. If Hither-Fodor believes that Jimmy Carter is from Georgia, he would, as Stich points out, entertain a representation that is of just the some *formal* type as Yon-Fodor (who has a corresponding belief about Yon-Carter and Yon-Georgia), despite the fact that the two beliefs are actually about different men and different places of origin. To avoid such a possibility, argues Stich, we should need to classify mental representations by their contents rather than their forms. But, from the fact that Stich's type of example provides a reason not to classify *another* person's internal representations by form alone, it does not follow that this is how each person should be supposed by psychologists to treat *his own* internal representations. And that is what is at issue here, since we are concerned with the question whether or not a person's mental operations have access only to the formal properties of such representations.

---

[2] In FODOR's latest book (1983) he still insists (p. 39): "If cognitive procedures are *computational* systems, they have access to such information solely in virtue of the *form* of the representations in which it is couched. Computational processes are, by definition, *syntactic*" (Fodor's italics). And this insistence on the *definitionally* syntactic character of computational processes is all the more remarkable for occurring only two pages after the assertion (p. 37): "I don't think that theoretical terms usually have definitions". Although little is said in his (1983) about the supposed language of thought it is clear (p. 9) that Fodor regards his current interest in faculty psychology as being quite consistent with the Neocartesian position that he defended in his (1975).

Nevertheless there is a familiar logical consideration that at first sight stands patently opposed to the formalist doctrine. We familiarly distinguish (due mainly to the seminal work of TARSKI, 1936, and CARNAP, 1948) between a syntactical and a semantical characterisation of logical relations. But apparently the formalist doctrine allows only the former to be a basis for our mental computations.

It will help here to review briefly the prototypical operation on purely formal properties that was invented by Aristotle. His theory of syllogistic reasoning exploited, in effect, the possibility of stating rules that permit or forbid certain transformations of specified sequences of symbols so as to treat those rules as criteria for the correctness or incorrectness, respectively, of corresponding inferences. Similarly it is possible to obtain a criterion of provability from given axioms in terms of conformity to rules for transforming the corresponding well-formed formulas in a specified symbolism. Standardly such a symbolism is a listed set of types of visually recognisable objects, such as two-dimensional letter-types, numeral-types, punctuation-types, etc. Tangible objects would serve as well, at least for those logicians who have well-developed tactile discrimination. Other sensory modalities too could be used. But they are less convenient to invoke here, because smells, tastes and sounds cannot be so easily arranged as shapes in checkable and recheckable sequences of tokens. So what is crucial to the idea of logical formalisation is just that a specified procedure of sensory verification stands guarantor for intuitive intellectual operations such as inference or proof. This procedure, stipulated in a syntactic meta-language, pays no attention to any meanings or truth-conditions or translational equivalences that may be independently assigned to well-formed formulas of the object-language.

In the logical context, therefore, formalisation consists in a deliberate and self-conscious substitution of manipulative procedures dependent on the perception of forms (in general, on the visual recognition of shapes) for intellectual procedures apparently dependent on an intuitive grasp of semantic properties such as the meanings of the words 'some', 'any', 'if', 'not', etc. According to the computational hypothesis, however, those intellectual procedures are themselves to be understood by analogy with the operation of appropriate computer programmes. How then can these programmes be construed as operating solely on formal, *non*-semantic properties? The formalist doctrine seems to assert the paradoxical claim that the very same intellectual operations that logicians, over more than two thousand years, have sought to replace by formalised proxies are already intrinsically formal. Nor do exponents of the formalist doctrine offer any explanation why those operations have appeared, and still appear, to so many as dependent on intuitions of semantic properties.

Perhaps, however, formalists who consider the matter will be inclined to object that the apparent dependence of logical operations on semantic properties of the object-language is an artifact of linguistic processing. Meaning seems to enter into the situation, they may say, only when we formulate our inferences in natural language, because we then have to connect up pivotal elements in our derivational structure with particular linguistic morphemes, such as 'some', 'any', 'if' or 'not'. The apparent meanings of these morphemes are thus explained away as being parasitic on an underlying system of operations on purely formal properties (compare BELNAP, 1962), and so the formalist doctrine is not refuted by the fact that intuitive judgements of inferential accuracy appear to depend on knowledge about the meanings of key words.

But such a defence of formalism makes two crucial assumptions.

First, it assumes that the functions of logical particles in natural language can be satisfactorily characterised in purely syntactic terms. And this assumption is not only not implicit in the computational hypothesis: it is also out of accordance with the familiar experience of glossing a natural language statement like "Either you go or I shall call the police" with its explicitly truth-functional counterpart like "At least one of the two propositions "you go" and "I shall call the police" is true."

Secondly, an assumption is made that pre-verbal or sub-verbal reasoning follows a syntactical rather than a semantical pattern. And this assumption too is by no means implicit in the computational hypothesis. Admittedly some psychologists have proposed accounts of deductive reasoning that construe it as operating in accordance with syntactical procedures (in the sense of CARNAP, 1937 and 1948). Thus RIPS (1984) has constructed a model ANDS which works rather like a person struggling systematically to solve text-book exercises in propositional calculus natural deduction by searching diligently through what conclusions can be derived from relevant premisses, and what premisses can serve for the derivation of relevant conclusions, where the legitimacy of derivations is governed solely by rules for permissible transformations. But semantical accounts are also current. For example, JOHNSON-LAIRD (1981 and 1982) has constructed a model of deductive inference that relies, not on conformity with stipulated rules of derivation, but on knowledge of the truth conditions of propositional connectives for inferences involving relations between propositions and on an interpretational decision procedure for inferences involving quantification over monadic predicates. Both Rips and Johnson-Laird claim that their respective computational programmes are consistent with the kinds of error that actually occur in human reasoning. And no crucial experiments have yet been proposed that might help to determine whether tacit human reasoning operates on a syntactical or a semantic basis, or sometimes on one and sometimes on the other. Perhaps one day we shall have such experiments. But this issue, whatever its eventual outcome, is at any rate best taken as an open, empirical issue. It is an issue of the kind that the computational hypothesis indicates a general strategy for resolving — by means of computer-simulation and experiment. So philosophers risk obstructing the path of enquiry if they try to anticipate such a resolution by reading some partisan implication into the computational hypothesis itself. Indeed, one inherent disadvantage of the formalist doctrine is that it appears to improverish computational methodology by thus closing it off from certain otherwise respectable avenues of psychological research involving semantic considerations.

It may well be objected, however, that the formalist doctrine should not be construed as excluding the possibility that human reasoning operates in terms of what logicians call semantical properties. For example, a mechanism that executes a truth-table decision procedure on given sentences of a particular language, in order to determine whether or not they are truth-functional tautologies, need not be attributed any knowledge of their meanings: it may be described instead as just manipulating symbols according to predetermined rules. So the formal properties on which the mind operates may be taken to include those within the domain of what CARNAP (1948) once called 'pure semantics' (and many people now call 'formal semantics') as well as those within the domain of logical syntax. What the formalist doctrine does exclude, it will be claimed, is the possibility that human computational operations may sometimes be concerned with those meanings that a person cannot specify without making some reference to features of his own environment. But *such* knowledge is certainly not needed for executing the procedures of pure semantics.

All right. Let us grant then that the formalist doctrine, in an appropriately restricted version, is not to be ruled out *ab initio* on this score, any more than by Stich's argument. But does the doctrine have any positive merit?

III

The first point that needs to be made in any assessment of the arguments advanced in favour of the formalist doctrine, is that the doctrine obtains no support from the principle of intentionality — the principle that when you ascribe someone a certain mental attitude towards a specified proposition you do not imply either the truth of the proposition or even the existence of anything to which it refers. The appearance of support springs from a failure to distinguish adequately between two ways in which it is possible to have regard for the semantic properties of an expression-token. In one sense (call it *de dicto*) a person or a programme would have regard for semantic properties if he or it had regard to the type-meaning of the expression or to any component of this meaning, and in quite another sense (call it *de re*) if he or it had regard to the actual truth-value, or to the actual, real-world referent, of the expression-token. It is only in the latter, *de re* sense that the principle of intentionality precludes the psychological description of mental states from being concerned with semantic properties: such a concern would then introduce an admixture of alien, environmental matters into what should be a purely psychological description. But presumably no-one would hold that knowledge of the actual truth-value of an expression-token, or knowledge of its actual, real-world referent, is a necessary prerequisite for exploring its inferential or computational liaisons. So the view to which the formalist doctrine poses an interesting and important challenge is the *de dicto* one according to which human computational processes do at least sometimes have regard for the meanings of internal representations or of components within them. And against this view the principle of intentionality exerts no leverage.

Nor does the formalist doctrine obtain any support from the existence of referential opacity within a person's own internal representations of what he believes, wants, etc. For no such referential opacity exists (*pace* FODOR, 1980, p. 66 and 1981, p. 158). More specifically, we can see that Oedipus' behaviour must depend, not on which description he has in mind for the object of his affection, but on what he believes about this object. For, even though he had the name 'Jocasta' in mind for the object of his affection, because he liked the sound of the word, he might have believed also (would that he had!) that Jocasta was his mother and he would then have been in a position to infer, without any difficulty from resistance to substitution, that the object of his affection was his mother. Thus the concept of referential opacity has no part to play in characterising how a rationally coherent thinker should regard the representations embraced within his own current mental states. In a statement made by Teiresias, whether publicly or privately, the sentence "Oedipus wants to marry Jocasta" may contain a referentially opaque occurrence of 'Jocasta', since whatever belief

Teiresias has about Jocasta's identity, he may wish not to impute that belief to Oedipus. But in a statement made privately by Oedipus to himself the sentence "I want to marry Jocasta" ought to contain a referentially transparent occurrence of 'Jocasta',[3] since Oedipus would not be a rationally coherent thinker if he had said anything thereby to exclude himself from exploiting, in his *own* reasoning, whatever beliefs he may hold about Jocasta's identity. A representation towards which a rationally coherent thinker is privately ascribed a certain current mental attitude can be characterised as referentially opaque only if we are out to articulate the logical constraints that operate when the person using that ascription of it as a premiss for his reasoning is not also the person to whom it is ascribed.

In other words an ascription of belief normally generates referential opacity because any speaker or hearer of the ascription might well accept some identity-statements to which the believer does not himself subscribe, and it would therefore risk turning a true ascription into a false one if any substitution on the basis of these statements were carried out within the formulation of the belief. But in the special case when the speaker, the hearer, and the believer are all just one and the same person at one and the same moment no such risk arises. So, if you currently adopt a certain mental attitude towards a a particular representation, you would be irrational if you do not, at least in your private reasoning, treat it as referentially transparent. Consequently any computational processing to which your self-ascription of this attitude is subject is not necessarily anchored to a particular form of reference at any point: to the extent that there is such an anchoring, your thoughts lack rational coherence. It is not the form of your self-ascription that will normally determine its computational liaisons at any one moment, but rather the total interlocking system of your mental attitudes at that moment.

IV

So far I have been examining the computational hypothesis only from the side of what it is a hypothesis about, not from the side of the model or

---

[3] Perhaps it will be objected that even Oedipus might have distinguished in his own mind between "I want to marry Jocasta" and "I want to marry the king's widow", even though he knew that Jocasta was the king's widow. But the distinction that is relevant to Oedipus' state of mind here is between one set of reasons for wanting to marry Jocasta and another, not between one form of description for this desire and another.

analogy that it proposes. In enquiring whether the formalist doctrine gives a correct interpretation of that hypothesis I have discussed mental features such as belief, deduction, intentionality and referential transparency, not computational features such as programming languages, compilers and circuity. And the formalist doctrine has not come off well so far. But supporters of the doctrine might claim that whatever genuine support it has lies rather in the nature of the model than in that of the modeled. If the model is indeed essentially formalist, then any difficulties about mental features, such as the referential transparency of mental representations within a rationally coherent pattern of reasoning, are relevant rather to the very truth or applicability of the computational hypothesis, in its formalist interpretation, than to the question whether that interpretation correctly elucidates what the hypothesis implies.

Now the formalist doctrine does acquire some *prima facie* support from the fact that in the writing of artificial intelligence programmes considerable use is often made of a certain ambivalent category of expression-type. An expression-type *E* of this kind is characterised by having both of two mutually distinct sets of properties. On the one hand there are the properties that *E* has in virtue of its belonging to a computer-programming language where *E* has no meaning or function other than as a name of itself or as part of an expression that names it by inclusion within the scope of quotation-marks or a quote-function. On the other hand there are the semantic properties that the same expression-type *E* has in virtue of its also belonging to the ordinary vocabulary of a natural language. Thus in running SHORTLIFFE'S (1976) interactive programme MYCIN, which is written in Interlisp, the clinician may input 'NO' in response to the machine's output "ANY OTHER SIGNIFICANT EARLIER CULTURES FROM WHICH PATHOGENS WERE ISOLATED?" But this output has no reference in Interlisp to events in the outside world. It merely serves to invite a response-input of a kind that the computer has been programmed to register and process whenever this output occurs. In English the expression asks whether there were any other significant earlier cultures from which pathogens were isolated. But in Interlisp it has no such meaning, because expressions like CULTURES and PATHOGENS have only formal properties, one might say. They are like self-naming expressions of a syntactic meta-language masquerading as expressions of the relevant object-language. And if computer programmes had only expressions of this kind as their inputs or outputs they would admittedly be operating only on formal properties, not on semantic ones. Whenever they were put forward as models of actual human reasoning, the modeling — we might call it 'simulated parroting' — would always be a kind of formalisation. Indeed the claim has often been advanced that a computer understands a natural language just so far as it can thus formally participate in dialogue in that language (TURING, 1950; RAPHAEL, 1968; BOBROW, 1968). And, on the other side, it is this kind of claim that SEARLE (1980) seems to have in mind when he argues that computer programmes can*not* serve as adequate models of human understanding.

But it is a mistake to suppose that the computational hypothesis directs us only towards such formal simulations. In fact it makes available a distinction evident in familiar high-level programming languages, which is closely analogous to the difference between operating on forms and operating on meanings.

For example, the Basic formula PRINT *X*, *Y*, *Z* orders a procedure that does operate solely on the formal properties of whatever '*X*', '*Y*' and '*Z*' stand in for, though not on the formal properties of '*X*', '*Y*' and '*Z*'. Basic instructions such as READ or RESTORE are like PRINT in this respect. To anyone running through a programme in his head they would appear as requiring him to mention, rather than to use, the expressions concerned. But the formula PRINT $X + Y$ involves both formal and non-formal properties: the numbers denoted by '*X*' and '*Y*' are to be added and the numerical expression denoting that sum is to be printed. And the formula IF $X > 7$ THEN GO LINE 30 instructs execution of the instruction on line 30 if the value of *X* is greater than 7. So the condition on which this instruction hinges concerns the size of the number denoted in the context by the expression '*X*'. The semantic property of denoting that number is crucial here. Nor is it possible to understand the condition as a purely formal one by thinking of it as depending in some way on the shapes of *X* and '7', since the very same symbol-type '7' has also to be used for the operation of counting as in the setting up of arrays by DIM statements: compare RUSSELL'S (1919, p. 10) view that we cannot give a purely formal characterisation of arithmetical procedure because 'we want our numbers to be such as can be used for counting common objects'. The expression '7' may be said to designate the number 7 on such occasions in the sense that when the computer takes that expression as input the computer's ensuing behaviour depends on what the expression is said to designate. This sense of 'designate' is defined by NEWELL (1981) in highly general terms, i.e. for any symbol system whatever, and when applied to the particular case of numerals in a high-level programming language it helps to sustain the analogy between what is being said here about numerical expressions in such a language and what Russell said, à propos of Peano's axiom-system, about numerical expressions in ordinary use. Moreover, by using numerals to designate a set of positions in a two- or three-dimensional matrix it is possible to use the programming language to represent a geometrical structure and by programming the machine to change these positions in real time it is possible to represent movement. The peripheral graphic displays that can then be generated are a way of exploiting the sensitivity of the programme to the semantic properties of expressions that are part of the data or input. These expressions may therefore be said to 'represent' the relevant geometrical structure. How else should we describe the relationship between the numerical evaluation of the matrix, on the one side, and, on the other, the structure-types of which tokens appear on the display-screen or print-out?

FODOR (1980) is certainly right to insist that WINOGRAD's (1972) programme does not set up a real robot, but only instructs a computer to dream or imagine, as it were, that it's a real robot: the memory states of the machine are so arranged that its available data are whatever natural-language descriptions they would be *if* there were objects for the machine to perceive. But even a dream can be described in meaningful language or pictured intelligibly, as can the life of a fictional character. And what Fodor fails to take account of is that Winograd includes a set of display routines in his system of programmes, which thus contrives to simulate each state of the machine's hypothetical world on a graphic screen. It follows that within Winograd's system of programmes the data are assigned some semantic content by being linked to this graphic display. Specifically, in order to achieve the linkage, the system has to construct a three-dimensional matrix and thus operate on the semantic or non-formal properties of numerical expressions in the programming language that are integrally linked within the system to the state of the data.

We can thus distinguish between simulated parroting, as already defined, and what we can call 'simulated understanding'. So far as a computer is programmed to simulate uncomprehending participation in dialogue within some section of natural language, for example, it may be said to display simulated parroting of that section of natural language. A person taught to perform the same kind of simulation as the computer would certainly be aware of needing a good memory about which speech-sounds to utter in response to which, but might not be aware of needing any other intellectual skill. On the other hand, so far as a computer programme operates on the semantic properties of expressions in its programming language, it may be said to display simulated understanding, since it enables a machine to execute procedures (counting, calculating, mapping one pattern on to another, etc.) which substitute for other conscious skills than textually cued memory of text. Simulated parroting deals only with forms, simulated understanding only with meanings, though in practice elements of both tend to co-exist, often in an intricately inter-woven pattern, in most artificial intelligence programs. A machine simulates parroting insofar as it behaves as if it is miming the outputs of the relevant type of person each time it is confronted with a particular category of input. It simulates understanding insofar as it behaves as if it works things out for itself. In doing so it has to provide an artificial means of executing specified procedures that humans know how to execute in their own minds, and it thus behaves as if it understands the specification.

Admittedly there is a rather strict constraint on the range of meaningful specifications expressable in a high-level programming language. Nothing can be said in such a language that is not reducible eventually to procedures executable by the logic gates, circuitry and peripheral devices of the digital computers on which it is to be implemented.[4] So we cannot yet designate a flower or the parts of a rifle in one of these languages: we can only designate expression-types isomorphic with those that name such things in natural language (insofar as we can register control of a circuit that will print out tokens of the required expression-type in an exercise of simulated parroting). But there is no *a priori* constraint on what could count as a peripheral device. So the variety of procedures executable in the implementation of a program could in principle be substantially extended if a robot's sensory transducers were linked to the computer's input and its output was linked to the robot's behaviour as well as to the display screen. Simulated understanding could thus be extended to respond to the specification of procedures for investigating and describing the computer's environment and for deliberating and deciding about actions in that environment: it would then not respond just to the specification of procedures for counting, calculating, etc. And in the course of achieving such an extension an expression in the programming language could come to designate, relative to the peripherally enhanced computer, a particular

[4] Of course, if we were interested in artificial intelligence programmes only for their possible use as props, aids, short-cuts, etc in the execution of human tasks — i.e. as extensions of human performance in medicine, administration, business, etc. — it would be reasonable to adopt a conception of semantics for them that allowed all their expressions to have much the same meanings as these expressions would have in natural language. But it would be viciously circular to apply such a conception of semantics when the programmes are being cited in accordance with the computational hypothesis and they therefore function as models for the explanation of semantic processing, etc. in the human mind.

environmental entity or event of any desired kind, as NEWELL (1981, pp. 58–59) points out — at least in the sense that, when the computer takes the expression as input, the computer's behaviour will depend on what that entity or event is.

V

A different defence of the formalist doctrine may now be put forward. We are told (FODOR, 1980) that the notion of formality must remain 'intuitive and metaphoric', because formal properties are essentially contrasted with semantic ones and there are said to be difficulties about giving a complete list of semantic properties. But it is clear that the computational hypothesis cannot be construed as restricting mental operations to a concern with formal properties in any sense of 'formal' that would restrict computational models of human understanding to programmes for simulated parroting and exclude them from concern with programmes that provide also for simulated understanding. There must somehow be a relevant sense of 'formal' that applies even when both types of programme are under consideration. This must be a sense in which *every* programme for a computer's operations on given inputs has regard only for formal, i.e. non-semantic, properties of those inputs. Whatever a computer does, when it responds to an input, it must be construed as showing no evidence, in what it does, of comprehending meanings. It must presumably be described instead as responding in a predetermined pattern to the symbolic shapes that spell out its input, whether this input reaches it via the keyboard of an on-line consol or in any other way.

Now if such a construal is possible it raises an important question about the nature of the model that the computational hypothesis is supposed to afford. Every explanatory or heuristic model in natural science has both positive and negative analogies with that which it models. If it had no positive analogy, it would be irrelevant, and if it had no negative analogy it would be just another instance of the puzzling process that needs instead to be modeled. But the precise extent of the positive analogy, as against the negative one, is often open to discussion. And in the case of the computational hypothesis it is worth while raising this issue with reference to the programming language that is implemented in the articulation of any model generated by the hypothesis. More specifically, since formula-tokens of an appropriate programming language normally play an integral role in programming a computer and in supplying programmes with inputs, do such tokens of software formula-types belong to the model's positive analogy? If they do, then it is easy enough to see a sense in which tacit mental processes may be said to be operations on formal properties, since the analogous computer operations may be said to latch on to (or be triggered by) the formal or non-semantic, properties of these programming-language utterance-tokens.

Admittedly the people who design or use programming languages have to attach meanings to their formulas. When a programme-writer tries out a short programme, executing it consciously in his imagination, he exploits these meanings. But we need no more suppose that a computer is responsive to the meanings of formulas tapped out on its keyboard than that an electric kettle understands the meaning of the word 'off' when the switch is depressed which bears that label. In the case of the kettle it suffices for the off-switch to be the one that is depressed. Similarly in the case of the computer it suffices for the '7' and '4' levers, say, to be depressed in succession. The comuter does not need actually to understand the meaning of the numerical expression '74', let alone the semantic force of utterances in which that expression plays a part. The computer does sufficiently well if it behaves as if it understands. It may thus be said to operate on the formal, non-semantic properties of its input. And so far as the human mind works analogously the human mind too must operate on the formal properties of appropriate pieces of brainwriting (as DENNETT, 1979, pp. 39–50, calls it).

Such a conclusion would not be at odds with the distinction already drawn between simulated parroting and simulated understanding. That distinction depends essentially on an analysis of programming-language formula-types, since it differentiates between those computer operations that correspond to the mention, and those that correspond to the use, of expressions in the construction of such formula-types. It is a distinction that remains valid even when programmes are executed on paper or consciously — in a programme-writer's imagination — rather than in a suitable machine. It is thus a distinction that runs quite across the distinction between what a machine responds to and what such a programme-writer does. On the other hand the present distinction between formal and non-formal properties is more or less parallel to the difference between what a machine responds to and what a programme-writer does. It certainly makes every machine computation an operation on formal properties; and as a consequence it makes the computational hypothesis imply that every tacit mental process which is analogous to a machine computation will also be an operation on formal properties.

But the formalist doctrine is successfully propounded in this sense only if programming-language formula-tokens do indeed belong to the positive analogy. If instead they have no counterparts in the mind and should therefore be taken to belong to the negative analogy, the formalist doctrine gets no purchase on the situation. So let us examine how far, and for what reasons, these formula-tokens should be regarded as part of the positive analogy.

They certainly have an obvious role in that analogy where the mental process under investigation involves the use of actual or sub-vocal speech. But this is not important in the present connection, since the formalist doctrine is fundamentally about pre-verbal and sub-verbal thought. And there it is not at all so obvious that programming-language formula-tokens have a part to play in the positive analogy. Perhaps a person is programmed, as it were, with the help of natural-language sentence-tokens, when he is taught a foreign language. But the intellectual skills that gradually mature in the neonate are like programmes that, in default of positive evidence to the contrary, do not need to be written in any language. So long as they come to be there, and are stored available for execution when an appropriate occasion arises, they may well be hard-wired: they do not need to exist at any time in a quasi-written form as well. Of course, the exprimental psychologist is encouraged (by the computational hypothesis) to write a programme in some suitable high-level programming language that will match the infant's apparent programme as closely as possible. But the sequence of programming-language formulas that the psychologist writes down forms part of his explanatory theory. By stating that the infant behaves as if programmed in this way, he describes what kind of a mechanism he imputes to the child's mind. And the same is true for the study of tacit mental mechanisms at any later stage of human life also. The psychologist does not need also to impute to his subjects a counterpart of his own written programme.

Indeed, the great virtue of information-processing programmes as explanatory mechanisms is the catholicity with which they admit of embodiment. They can be embodied in many different kinds of hardware (in computers of different architecture or different manufacture); in written-out memoranda (sometimes a list of formulas, sometimes a flow-diagram) for consciously obedient execution by human readers (whether orally, manipulatively or in imagination); and also, according to the computational hypothesis, in our neuronal networks. Why then should we have to suppose that, whenever they are embodied in our neuronal networks, they must always be embodied in *both* of two quite different ways — one a counterpart of the way in which they are embodied in a computer and the other a counterpart of the way in which they are embodied in written-out memoranda for consciously obedient human execution? Just as an artificial computer may be hardwired, so too the brain may be.

Precisely the same point can be made about the representations that may constitute the input or output of pre-verbal or sub-verbal programmes in the human mind. Such inputs or outputs are in any case to be conceived as being immanent in the functioning of a person's neuronal network, as in the functioning of a factory-made computer. An internal representation may thus be stored, retrieved or operated upon. But we do not need to postulate the occurrence also of some mental counterpart of the programming-language utterance-token, or of the print-out or graphic display, which occurs peripherally when an analogous representation is processed by a computer. Such an occurrence triggers, or is triggered by, the information-processing operations of the computer. But in the human being the analogous information-processing operations may be supposed (when not triggered by other information-processing operations) to be triggered directly by the sensory transducers and also to trigger directly any effector nerves that are to come into play.

The issue at stake here may be further clarified by bearing in mind a distinction that is familiar in other fields within the philosophy of science. We often need to distinguish between an instrumentalist and a realist interpretation of certain expressions in scientific theories. Are these expressions just convenient instruments for use in constructing informative descriptions of the activity of familiar entities, or do they denote the existence of relatively unfamiliar — and normally unobservable — entities? Thus in the philosophy of meachnics BERKELEY (1721) was a pioneer of instrumentalism as LOCKE (1706) had been of realism. And in the present issue the expressions open to either an instrumentalist or a realist disambiguation are those constructed in accordance with the computational hypothesis and ascribing internal representations to a person in the metaphorical terms of formulas in a high-level programming language. Are these ascriptions just a convenient way to characterise the information-processing aspect of the person's neuronal system at a particular moment, or do they assert the real existence of tacit sentence-tokens in Mentalese — bits of physically salient brainwriting — every time the person's neuronal computer receives the corresponding input or delivers the corresponding output? Is the psychologist claiming merely that a person's neuronal computer behaves *as if* it has received an appropriate software input or is he claiming that it has *actually* received this in some form?

Both the instrumentalist and the realist mode of interpretation allow the possibility of conscious representations in natural language or mental imagery. Both modes of interpretation are compatible with all sorts of differing theories about the richness of our innate conceptual apparatus. Indeed, there does not seem to be any possible *psychological* evidence — even of the kind cited by FODOR (1981, pp. 28–29) — that is explicable on the realist assumption and not on the instrumentalist one, *since every computationally relevant feature of the alleged sentence-tokens in natural language or Mentalese has in any case to be got into, or out of, the neuronal hardware.* Admittedly the realist interpretation carries over more positive analogy from any computational model when it imports this extra element into what is postulated by psychological explanations. But it does so at the cost of greater ontological extravagance. So in default of any relevant *neurological* evidence, considerations of ontological economy favour rejection of the realist interpretation here.[5] The computational hypothesis makes bold enough assumptions already about the working of the brain as an information-processing system, without our saddling it with yet further, and speculative, assumptions about the existence of tacit programming-language counterparts. By dissociating the computational hypothesis from

[5] MATURANA (1978) points to many biological phenomena that can usefully be characterised for an observer in terms of a software representation, such as a set of generative rules, but can nevertheless be understood solely as the activity of a structure-determined system, with no mechanism constituting a software representation. The current controversy about visual imagery (see the debate between PYLYSHYN, 1981, and KOSSLYN, 1981) draws attention to another such issue. Those who doubt the reality of Mentalese sentence-tokens will also be inclined to doubt the reality not only of Kosslyn's visual buffer but also of any propositional alternative to it. For such an instrumentalist the choice here is just between the neuronal process characterisable by a software representation of a picture and the neuronal process characterisable by a software representation of a proposition (with appropriate time differences between the two processes), however difficult it may be (see ANDERSON, 1978) to resolve that issue on the basis of behavioural data. Here too it is conceivable that physiological evidence might support the realist point of view. But in default of such evidence the instrumentalist has a stronger case: it is quite unnecessarily extravagant to suppose actual pictures in the head as well as the necessary neuronal mechanism, or actual list structures as well as the necessary neuronal mechanism — let alone to suppose actual sounds in the head whenever we have auditory imagery! On the instrumentalist approach, moreover, what DENNETT (1979, p. 122) calls Hume's problem does not arise: if there are no pictures in a person's head, there is no problem about how these may be inspected.

these further assumptions we make it clear that when phenomenologists or behaviourists criticise the thesis of brain-writing they are not thereby saying anything to weaken the computational hypothesis itself.

Nor is acceptance of a relational analysis of statements about tacit beliefs compatible only with a realist, and not with an intrumentalist, interpretation of the computational hypothesis in respect of questions about how such belief occurs. It is tempting to suppose that, if the statement "George tacitly believes that rain is falling" is true if and only if a certain kind of relationship obtains between George and a representation of rain's falling, then that representation itself must somehow exist in George's mind like a formula-token of an internal programming language. That kind of representation must really exist, we may suppose, in order for there to be something to which George is appropriately related.

Certainly that is how the realist interpretation seems to be supported by the relational analysis. But on the instrumentalist interpretation there is no internal formula-token. So the relation in question has just to be construed as being between George and a saying-type, instead of between George and a saying-token: a token of that saying-type would occur in George's mind if and only if his belief were not a tacit one. There is then no need to suppose that a relational analysis of belief-statements supports the mental reality of formula-tokens in an internal programming language. (Indeed, even if the pre-historic culture that long ago originated our concept of belief had somehow implicitly developed a realist interpretation of the computational hypothesis, that would be no reason why we should now adopt the same interpretation.)

It follows that the formalist doctrine is on rather weak ground even if it is propounded in the sense currently under consideration. Since the mental reality of tacit programming-language counterparts is an unnecessary and extravagant assumption, there is no reason to suppose that mental representations have any formal properties analogous to those software ones on which computers or their programmes may be said to operate. So there is no reason here to suppose that mental processes operate *characteristically* on the formal properties of internal representations.[6]

[6] Nothing is said in the present paper about the so-called mind-body problem, because the hard core of that problem, posed by the distinctive quality of conscious experience, does not appear to be addressable by computationalist psychology.

VI

There is one last kick that might be left in the formalist horse. Even if we suppose utterance-tokens of programme software, or of input or output formulas, to belong to the negative analogy, there is still a sense, it might be claimed, in which formal properties have an indispensable part to play in the positive analogy. When a computer is programmed, or a programme is executed, the form or structure of the hardware is altered: certain logic gates are opened and others shut in an appropriate sequence or sequences. So the state of the hardware is to be regarded — the formalist might claim — as being itself a translation or encoding of the written programme. Indeed the computer might then be said to show its understanding of programme-tokens or input-tokens by virtue of its ability to produce such translations. And, though the computer's user may think in terms of the meanings of these tokens if he executes the programme in his conscious imagination, the machine has regard only for the state of its own physical structure. In other words, the machine operates on formal properties, not semantic ones. Analogously the formalist's claim might be that the human brain-state undergoes a physical transformation whenever it registers a representation of anything, and it is this transformation that may be supposed to trigger the execution of any appropriate programmes. Mentalese is not a programming language that stands to the brain's computing circuitry as software to hardware. Instead its tokens, according to the formalist, are identical with relevantly transformed states of people's brains: such a transformation, and no more, is what the registering of a sub-verbal representation consists in. As FODOR himself once (1981, p. 174) puts it, "there is good reason for treating some neurological states as linguistic tokens". So tacit mental computations are on this view just as much concerned with formal properties, and just as little with semantic ones, as are the operations of programmed artefacts.

This version of the formalist position has the virtue of not multiplying entities beyond necessity. Its extravagance is conceptual rather than ontological. The trouble now is that a metaphor is being used to explicate a metaphor. The original metaphor to be explicated was the computational conception of the mind.[7] But now the computationally relevant properties

---

[7] To speak of the computational conception of the mind as a metaphor is not to deny the possibility that both mind and computer may be usefully seen as falling under some more general and non-metaphorical concept such as NEWELL's (1981) concept of a physical symbol system.

of computer hardware are being viewed as linguistic forms that have the same meanings as relevant programming-language formulas. Instead of treating the input and output of programming-language tokens in the usual manner as a two-way flow of information between human user on the one side and computer hardware on the other, we are now being asked to treat the states of this hardware as being themselves the sentences of a language, albeit a language that has no communicative role. And this highly strained and metaphorical conception of how a computer functions puts a serious restriction on the extent to which the computational hypothesis can be regarded as supplying a directly apposite model of the mind. Apparently one has first to conceive the functioning of a computer by reference to the model constituted by the functioning of a language and then conceive mental processes on the model of corresponding computational ones. Instead of a suggestion how familiar but puzzling mental events such as language-processing may be explained in terms that belong to a different and better understood category of description — the terminology of computing — we are offered an interpretation of the computational hypothesis that seeks in effect to explain one kind of language in terms of another. It looked like being a very promising feature of the computational analogy that it could lead us to see how information that is familiarly processed in lingusitic form can also be processed by a machine. But our great expectations are doomed to disappointment if computers are just languages in disguise. Indeed it now becomes more natural to supose that the human mind is being regarded as a model for understanding how a computer works, rather than the other way about. FODOR (1981, p. 203) explicitly recognizes this to be a consequence of his own view, but does not explain how he reconciles such a conception with the aims and methods of cognitive psychology.

Moreover, even if one were prepared to stomach this blurring of the model's direction in order to be able to talk about formality in the required sense, the computational hypothesis could still not be construed as implying that all mental processes are formal in that sense. The reason is that even if we now think of the machine's execution of a programme as essentially an operation on formal properties, a person's conscious execution of the programme in his own imagination may be just as essentially an operation on semantic properties, if anything ever is. Such a person has to have learned the functions of the various expressions in the software version of the programme, quite as much as anyone who understands instructions about how to calculate square roots, say, must know the meanings of the sentences that formulate those instructions or that list the numbers to be so treated. Moreover the computerised and the consciously humanised versions of the programme must be precisely similar to one another in structure. Precisely the same sequence of iterations, recursions, chunkings, horizontal searches, vertical searches, etc. must occur in each. In so far as a model is to be provided for a particular kind of tacit explanatory mechanism in the mind, it does not matter which version of the programme is invoked. Hence the computational hypothesis is certainly not to be taken as implying that all mental processes are formal in the sense under examination. But neither does it imply that they are all semantical.

It should be noted too that, as soon as brain-states are treated as formulas of an internal language, a deep and important philosophical problem seems to open up (FODOR, 1981, p. 203). Specifically, such a language, like any other, ought to have a semantics as well as a syntax: so what would constitute that semantics? This awesomely blown-up problem is totally deflated, however, if psychological theories are constructed in accordance with the interpretation of the computational model that I am proposing. We do not then treat brain-states as representational formulas of a real internal language but suppose instead that, by referring as our model to the linguistic (software) formulation of an appropriate programme that can be run on factory-made computers, we have a tool for describing the programmed brain-processes that are responsible for certain experimentally detectable results. As we extend the range of simulated understanding that is operative in such computational models, we may hope thereby to extend the range of behaviour for which such an explanation is available. So internal representations do exist, just so far as the brain-processes exist that are describable with the help of representational formulas in a program that models the activity of such processes. But no internal *language* needs to be postulated, and *a fortiori* we do not need to by puzzled by the problem of what would count as a semantics for such a language.

Perhaps someone will be tempted to object: surely any move towards extending the range of simulated understanding that is operative in a computational model is *ipso facto* a move towards building up a semantics for the language of neuronal machine states? Not at all. In such a model simulated understanding is the appropriate machine response to certain types of programme. A machine which exhibits this kind of response is not a language, nor are its states sentences, any more than a human being who understands written instructions is himself a language. In applying the computational hypothesis to tacit mental processes we should treat simulated understanding as a model for a certain type of intelligence or conceptual competence — e.g. the ability to take account of numbers or shapes in a sensory input — rather than for a certain type of semantics.

VII

The upshot of all this is that the formalist doctrine is quite unfounded. There is no relevant sense in which the computational hypothesis implies that all mental processes are formal. The modern concept of computation is a highly fruitful new idea, linked closely with the development of new technology, and so perhaps we should not be altogether surprised that it turns out to resist capture within the framework of old distinctions, drawn from other fields, like the distinction between form and content or between syntax and semantics. If we insist on applying those rather trite descriptive frameworks, what we find is that either both terms of the antithesis may be said to be applicable or neither. On the one hand, the computational hypothesis should be capable of generating both syntactic and semantic accounts of deductive reasoning, and it should also be able to exploit models that exhibit both simulated parroting and simulated understanding. On the other hand, if the instrumentalist interpretation of the computational hypothesis is correct and there is no internal language or 'brain-writing', there are then no relevant sentences to have either form or meaning in the sense in which these may be contrasted with one another, and so tacit mental computations cannot properly be said either to operate on forms or to operate on meanings.

What follows for the psychology of meaning is that the computational hypothesis does not after all place any narrow *a priori* constraints on the nature of meaning in natural language. Rather it permits the construction of a wide variety of computational models for language-learning, speech-production, speech-comprehension, etc. in accordance with differing psychological theories about how those activities take place and, in particular, about how they are connected with non-verbal thinking. Such models will exhibit varying proportions of simulated parroting and simulated understanding, and will not be restricted either to a syntactic or to a semantic conception of the structure of human reasoning. An this is just as it should be. If a philosophical gloss on the computational hypothesis imposes restrictions on the range of programmes imputable to the human mind, that gloss is obstructing possible paths of psychological enquiry and is best discounted, so as to leave experiment as the actual arbiter between alternative theories.[8]

References

ANDERSON, J.R., 1978, *Arguments concerning representations for mental imagery*, Philosophical Review 85, pp. 249–277.

BELNAP, N.D., 1962, *Tonk, plank and plink*, Analysis 22, pp. 130–134.

BERKELEY, G., 1721, *De Motu*, in: A.C. Luce and T.E. Jessop, eds., The Works of George Berkeley Bishop of Cloyne, Vol. 4 (Nelson, London, 1951), (first published in 1721).

BOBROW, D.G., 1968, *Natural language input for a computer problem-solving system*, in: M. Minsky, ed., Semantic Information Processing (MIT Press, Cambridge, MA), pp. 135–213.

CARNAP, R., 1937, *The Logical Syntax of Language* (Routledge and Kegan Paul, London).

CARNAP, R., 1948, *Introduction to Semiotics* (Harvard Univ. Press, Cambridge).

DENNETT, D.C., 1979, *Brainstorms: Philosophical Essays in Mind and Psychology* (Harvester Press, Hassocks).

FODOR, J.A., 1975, *The Language of Thought* (Crowell, New York).

FODOR, J.A., 1980, *Methodological solipsism considered as a research strategy in cognitive psychology*, The Behavioral and Brain Sciences 3, pp. 63–109.

FODOR, J.A., 1981, *Representations: Philosophical Essays on the Foundations of Cognitive Science* (Harvester, Brighton).

FODOR, J.A., 1983, *The Modularity of Mind: An Essay on Faculty Psychology* (MIT Press, Cambridge, MA).

HAUGELAND, J., 1981, *Semantic engines: an introduction to mind design*, in: J. HAUGELAND, ed., Mind Design: Philosophy, Psychology, Artificial Intelligence (MIT Press, Cambridge, MA), pp. 1–34.

JOHNSON-LAIRD, P.N., 1981, *Mental models in cognitive science*, in: D.A. Norman, ed., *Perspectives on Cognitive Science* (Ablex, Norwood), pp. 147–191.

JOHNSON-LAIRD, P.N., 1982, *Propositional representations, procedural semantics, and mental models*, in: J. Mehler, E.C.T. Walker and M. Garrett, eds., Perspective on Mental Representation: Experimental and Theoretical Studies of Cognitive Processes and Capacities (Erlbaum, Hillsdale), pp. 111–131.

KOSSLYN, S.M., 1981, *The medium and the message in mental imagery: a theory*, Psychological Review 88, pp. 46–66.

LOCKE, J., 1706, *An Essay Concerning Human Understanding*, 5th ed. (Awnsham, Churchill and Manship, London).

MATURANA, H.R., 1978, *The biology of language: the epistemology of reality*, in: G.A. Miller and E. Lenneberg, eds., Psychology and Biology of Language and Thought: Essays in Honor of Eric Lenneberg (Academic Press, New York), pp. 27–63.

---

[8] I am grateful for some very helpful comments on earlier drafts of this paper by Jonathan Adler, Gillian Cohen, William Ewald, Yorick Wilks and discussants at a Greater Philadelphia Philosophy Consortium Conference on Language and Semiotics (October 23, 1982), a Columbia University Philosophy Department Seminar (October 26, 1982), an Oxford University Philosophy Subfaculty Graduate Class (April 27, 1983) and an Essex University Cognitive Science Seminar (May 5, 1983). The paper was written during tenure of a British Academy Readership in the Humanities.

NEWELL, A., 1981, *Physical symbol systems*, in: D.A. Norman, ed., Perspectives on Cognitive Science (Ablex, Norwood), pp. 37–85.

PYLYSHYN, Z.W., 1980 *Computation and cognition: issues in the foundations of cognitive science*, The Behavioral and Brain Sciences 3, pp. 111–169.

PYLYSHYN, Z.W., 1981, *The imagery debate: analogous media versus tacit knowledge*, Psychological Review 88, pp. 16–45.

RAPHAEL, B., 1968, *SIR: Semantic Information Retrieval*, in: M. Minsky, ed., Semantic Information Processing (MIT Press, Cambridge, MA), pp. 33–134.

RIPS, L.J., 1984, *Reasoning as a central intellective ability*, in: R.J. Sternberg, ed., Advances in the Study of Human Intelligence, Vol. 2 (Erlbaum, Hillsdale), pp. 105–147.

RUSSELL, B., 1919, *Introduction to Mathematical Philosophy* (Allen and Unwin, London).

SEARLE, J.R., 1980, *Minds, brains and programs*, The Behavioral and Brain Sciences 3, pp. 417–457.

SHORTLIFFE, E.H., 1976, *Computer-Based Medical Consultations: MYCIN* (Elsevier, New York).

STICH, S.P., 1980, *Paying the price for methodological solipsism*, The Behavioral and Brain Sciences 3, pp. 97–98.

TARSKI, A., 1936, *Der Wahrheitsbegriff in den formalisierten Sprachen*, Studia Philosophica 1, pp. 261–405.

TURING, A.M., 1950, *Computing machinery and intelligence*, Mind 59, pp. 433–460. Reprinted in E.A. Feigenbaum and J. Feldman, eds., Computers and Thought (McGraw-Hill, San Francisco, 1963), pp. 11–38.

WINOGRAD, T., 1972, *Understanding Natural Language* (Academic Press, New York).

13. History of Logic, Methodology and Philosophy of Science

Barcan Marcus et al., eds., Logic, Methodology and Philosophy of Science VII

METAPHYSICAL AND INTERNAL REALISM: THE RELATIONS BETWEEN ONTOLOGY AND METHODOLOGY IN KANT'S PHILOSOPHY OF SCIENCE

GERD BUCHDAHL

Dept. of History and Philosophy of Science, Univ. of Cambridge, Cambridge, England

I

I want to discuss the issue of Metaphysical *versus* Internal Realism, as recently formulated by Putnam, in terms of Kant's, or at least: of a Kantian account of the philosophy of scientific theory construction. Putnam's espousal of Internal Realism in his later philosophy, involving as it does a type of transcendental approach, makes this particularly topical; so does Putnam's description of the world of the metaphysical realist as "noumenal".[1] If we take the metaphysical realist as operating with some kind of correspondence theory of truth, this can be shown to lead almost inevitably to a deeply skepticist position: All scientific theories are corrigible; therefore we can never know whether any theory of research programme, however successful and progressive, will correspond to the way things really are. — Obviously, the metaphysical realist position turns on the interpretation of this 'really are', of the way the world is, independently of any theorizing about it. Following Putnam, we may then describe the realist as converting the notion of the world as it really is into that of a world which is 'theory-independent'; with the consequence that our epistemic ignorance as to whether we have arrived at a finally successful theory is converted into a sort of constitutive incapacity, to the effect that it would not even *make* sense to claim that there could be a *final theory* about such a world; the reason being that any such assumption still operates with the notion of a world that is relative to some theory or other; whilst the idea of a 'world, simpliciter' — again Putnam's term — which our final theory is supposed to be 'about', would have to be theory-independent.

[1] Hilary PUTNAM, *Meaning and the Moral Sciences* (London, 1978), pp. 125–135.

Now the obvious way of avoiding the skepticist conclusions flowing from such a position — and it is a way Putnam espouses also — is to proscribe the notion of a theory-independent world; a move made the more convincing if we interpret the notion of a theory-independent world as being equivalent to that of the world in a 'noumenal' sense; as something possessing the status of a Kantian 'thing is itself'; with the contrasting 'theory-relative world' bearing a corresponding resemblance to Kant's world as "appearance"; being conditional upon a set of presuppositions with transcendental status; thereby making the concept of 'world' 'internal' to these conditions — hence Putnam's term "internal realism", corresponding to Kant's own "empirical realism", defined here as situated within the boundaries of a "transcendental-idealist" framework. Putnam's development of the theme on these lines (though somewhat sketchy) is attractive, not the least because it holds out the possibility of re-establishing some historical continuity with the richness and subtlety of Kant's own scheme. — However, tracing historical roots and continuities demands — in line with good hermeneutical doctrine — also a fresh interpretation of the older philosophical — here: Kantian — scene itself; both the old and the new thereby acquiring a fresh significance. So I want here to develop the theme of internal realism — still somewhat sketchy in Putnam's own presentation — in terms of a fresh analysis of Kant's own position, by developing its structual intricacies in ways not hitherto appreciated very clearly.

To this end, I want to develop the ramifications of Putnam's internal realist position by refining the notion of 'theory', in his expression 'theory-relative world', in terms both of the structure lines of Kant's own philosophy of science, as well as of his general philosophy; an incidental aim being to introduce some order into our understanding of the overall Kantian scheme itself — clarity concerning which is still somewhat missing from most expositions of Kant.

II

First, some terminological matters. Kant's chief model for a theory was of course Newton's *Principia*; I shall say that such a theory provides a 'phenomenology' of the world, using this term in the literal OED sense of 'giving an account of the phenomena', both descriptive and explanatory, in terms of hypotheses, fundamental concepts and principles.

As to 'theory': we shall assume — with Kant and most philosophers of science — that this is controlled by a certain 'methodological structure', or 'methodology' for short [label 'M'], defining the constraints on the formulation of hypotheses. Now Kant's own methodological scheme is itself exceedingly subtle, and anticipates much later, and indeed very recent, views about scientific methodology. He defines three constraints, or methodological components, which are postulated as determining, respectively, what he calls the "[inductive] probability", the "possibility", and finally, the "unity", or systemicity, of a set of hypotheses that make up some given theory.[2]

The first of these components — let us call it the 'probative component' [PC] (see Fig. 1) since it chiefly concerns the various proof structures of science — is envisaged by Kant still in terms of a purely inductive logic; but we may usefully extend this so as to include also later hypothetico-deductive or corroborational procedures; or again, confirmation theory, or even Bayesian logic; all concerned with determining the 'evidential support' for some suggested hypothesis.

A second component which, again for obvious reasons, we may term the 'Systemic Component' [SC], is concerned with the systematic articulation of the individual hypotheses composing a theory, on lines which since Kant's time have been explained in terms of, say, the notion of the 'consilience of inductions'; more recently, as the dynamics of a Lakatosian 'research programme', involving intertheoretical relations, as well as the provision of additional background information, etc. But further, as is well known, and as Kant already made prominent, the process of systematization traditionally involves recourse to certain additional maxims or principles, such as those of economy, simplicity, continuity, homogeneity; it also invokes certain judgments of preference concerning the choice of explanatory theories as such; something whose real significance and importance for the nature of scientific explanation has only become clear since the writings of Duhem, Mach, Campbell and their later followers. For instance, the centre of gravity of a proper explanation may be placed in causal theories of a mechanistic type, or alternatively, of a teleological kind; again, different scientists express differential preferences for macro- as against micro- (or deep-structure) theory types; and so on.

---

[2] Immanuel KANT, *Logic* (trsl. R. Hartman and W. Schwarz, Indianapolis, 1974), Introduction, sect. X, pp. 92–93. Cf. also *Critique of Pure Reason* (trsl. Norman Kemp Smith, London, 1953), A770/B798. [A and B refer to 1st and 2nd ed. of CpR, respectively.]

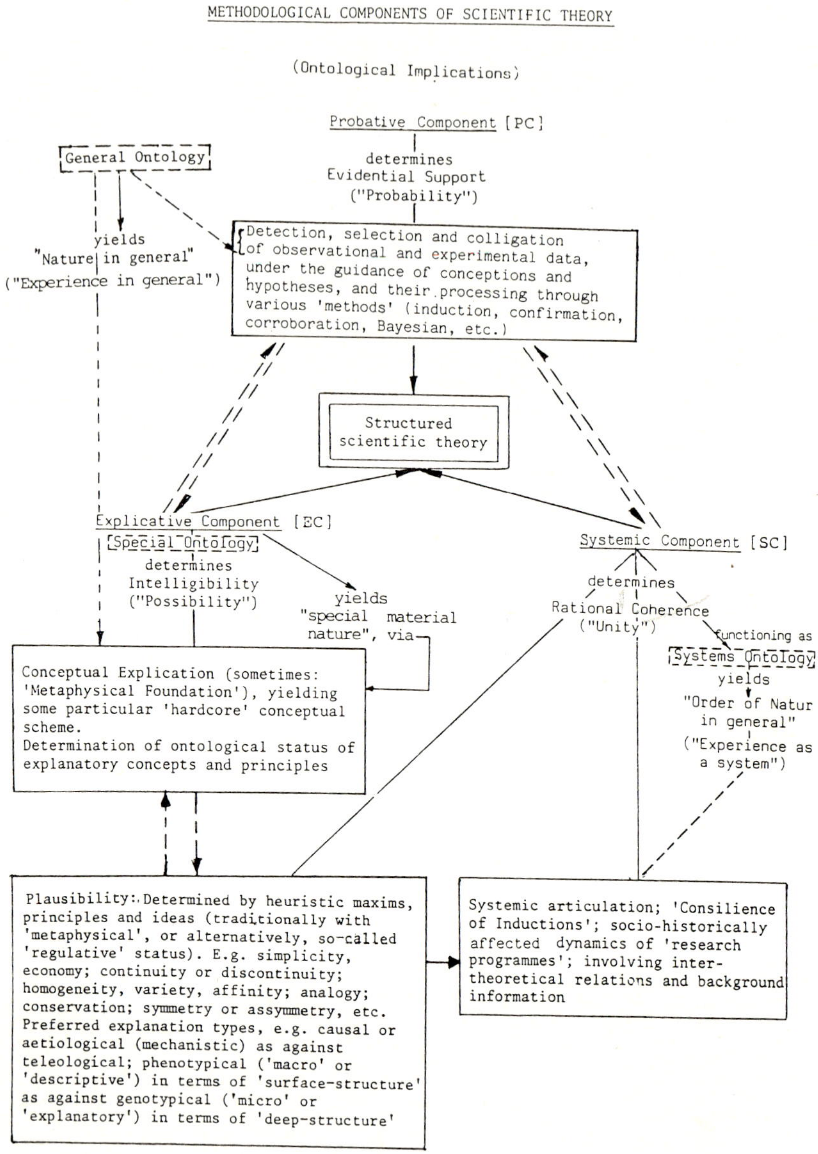

Fig. 1

Kant's third methodological component — I will call it the 'Explicative Component' [EC], echoing William Whewell's notion of the 'explication of conceptions' — is more difficult to describe. It is really an anticipation of more recent ideas on scientific explanation, such as, e.g., Lakatos's idea of the 'metaphysical hardcore' of some given research programme; though the general notion figures, as I say, prominently already in Whewell.[3] Kant himself develops it as (what he calls) a "Special Metaphysics of Nature", to distinguish it from the "General Metaphysics of Nature",[4] which corresponds to the material of the first half of the *Critique of Pure Reason*, i.e. the account of space, time and the principles of the categories. EC boils down to an analysis of the fundamental concepts and principles of a theory — in Kant's Newtonian paradigm this is space, matter, motion, force, and some of the basic laws of the *Principia*, those of the conservation of mass, inertia and momentum. Kant's objective of demonstrating the "real possibility" of these concepts and principles is then met by showing them to be integrable with the basic structure of experience; i.e. both with the *empirical* semantics of Newtonian language, and with the categorical framework which (in Kant's view) is an essential element of what he calls "nature in general", or "experience in general". (A modern comparison would be the demand that certain notions in recent quantum physics, such as a-causal discontinuities, time reversal, etc., should not only be capable of a purely mathematical treatment, and confirmable by observation, but can also be shown to 'make general sense'.) In this way Kant himself sought to show that gravitational attraction, or again, momentum transfer, can be demonstrated to make sense within the general structure of experience.[5]

The three components of Kant's triadic scheme [M.3] are not separable but must be taken to exert a mutual and combined influence on one another, though it is interesting to note that historically each of these components has at times been singled out as the sole, or at least primary, determinant for the acceptance of hypotheses. Thus PC mirrors the Baconian, inductivist, and more generally, empiricist approach to science; EC echoes rationalist treatments such as those to be found in Descartes' account of the foundations of dynamics; whilst philosophies of science such as those of Leibniz, or Berkeley, and of many modern writers put the

[3] For LAKATOS's 'metaphysical hardcore' notion, cf. his *Falsification and the Methodology of Scientific Research Programmes*, in: I. Lakatos and A. Musgrave, eds., Criticism and the Growth of Knowledge (Cambridge, 1970), pp. 133–134. For WHEWELL, cf. *Works* (ed. G. Buchdahl and L. Laudan, vol. vi, The Philosophy of the Inductive Sciences), Part Two, Book xi, Ch. II, *Of the explication of conceptions*, pp. 5–25.

[4] KANT, *Metaphysical Foundations of Natural Science* (trsl. J. Ellington, Indianapolis, 1970), with Ak. ed. pagination referring to vol. 4, in margin. [All page references to this edn.], p. 470.

[5] See for instance, ch. 2, p. 508; ch. 3, p. 550.

primary emphasis on coherence and harmony, i.e. SC. Finally, Kant's own claim that all three components of M.3, acting jointly, are relevant for a proper assessment of theories anticipates very recent philosophical developments in the philosophy of science; witness Kuhn,[6] and Lakatos (already referred to); there is no need to stress the similarities at length.

III

So far we have explicated Putnam's notion of a theory-relative world as one whose phenomenology is conditional on the assumption of some methodological M.3 structure that governs scientific theorizing as such. However, this clearly does not as yet dispose of the metaphysical realist assumption of a theory-independent world. Or rather, supposing we reject this assumption, then we need something to replace it, such that whilst we retain the notion of physical and theoretical reality, we interpret this 'internally', i.e. in such a way that it has significance in principle only when standing in an *essentially* conditional relation to M.3. (In other words, we shall widen Putnam's notion of 'theory'.)

Putting it differently: the old metaphysical realist question: 'how can we ever know whether a theory coincides with the way the world really is?' needs reinterpretation in such a way that our understanding of the phrase 'the way the world really is' will make it possible to give *at least* in principle — and that is after all what is in question here — an affirmative answer.

Interestingly, Kant himself, in connection with the problem of the possibility of any cognitive access to the systemicity or "order" of nature as such, at first sight formulates this in a way that echoes the metaphysical realist's skeptical locution (just cited), by asking how we can ever be "certain" that nature will conform to our chosen methodological framework, i.e. how we can be certain that such a methodological scheme possesses "objective validity"?[7] However, we soon discover that Kant's way of dealing with this question at once moves in an internal realist direction. Very roughly, his answer is that not only the actual account, the phenomenology yielded by a theory, is a function of the methodology, but so, likewise, is the very "possibility" of the order or systemicity of nature as such — that is, as far as concerns the systems case (SC).

---

[6] Th.S. KUHN, *The Structure of Scientific Revolutions* (Chicago, 1962).

[7] A651/B679; for "objective validity", cf. A669/B697; A664/B692.

Here we need some further terminology. Any account of the *real possibility in principle* of anything I shall term its 'ontology'. This expression I am using here in the sense in which it occurs, for instance, in the *Critique of Judgment*, where Kant refers to the categories, viewed as conditions of the possibility of "nature in general" ("experience in general") as "ontological predicates".[8] Now of course, ontology, so defined, can be construed in different ways. Classically, we meet with at least two modes which may be labelled 'metaphysical' and 'transcendental'. The former posits principles postulated as accounting for the possibility of *things*; such as we meet in Leibniz's theological version, as well as in the pre-critical Kant; the latter confines its account to the possibility of any *cognitive grasp* (Kant's "Erkenntnis") of things; of what brings "things in general" within a cognitive horizon.[9]

In line with this, we may now say that to Kant's three components of M.3 there must thus correspond, respectively, three distinct ontologies which we will label 'General Ontology' (Kant calls it "General Metaphysics"), 'Special Ontology' (Kant's "Special Metaphysics"), and finally, 'Systems Ontology'; Kant has no term for this. Interestingly, it turns out that the relations between these ontologies and the corresponding methodological components display considerable variations in our three cases, with corresponding differences also in their validational status, as we shall see. Let us consider them in turn.

The clearest case, and one that receives Kant's special attention, occurs in connection with SC. As defined, this determines, to start with, the systematic aspect of the construction of some scientific theory or other, yielding a corresponding phenomenology; i.e. a theoretical account of the phenomena. Such an account is an account of (what Kant calls) a "unity of nature" — so far to be understood in a purely phenomenological sense; Kant's rather 'instrumentalist' contention being that no 'object' ever corresponds to such unities; they are not "given" as objects but have only the status of something for ever "projected", as something "set as a task".[10] It follows however further that ipso facto any corresponding ontological account will not be able to, nor will it have to, generate the possibility of such unities qua 'objects' either. For which reason Kant

---

[8] KANT, *Critique of Judgment*, Introduction, sect. 5 (trsl. J.H. Bernard, New York, 1951), p. 17.

[9] For the "horizon of human reason", cf. A760/B788.

[10] A647/B675; cf. A498/B526.

describes systems ontology as having only a "regulative" and not a "constitutive" status.[11] But more of this in a moment.

Using now a Putnamian locution, we might thus say that the 'unity' or 'system' expressed by some theoretical construction is nothing in abstraction from theory, in the sense of SC. But now: Kant's contentions go further since as far as we have gone it might still be the case that there is some systemic order of nature, whether or not our employment of SC happens to have discovered it. This supposition of course leads us straight back to the metaphysical realist's skeptical complaint; which shows that the supposition just expressed must stem from that quarter, and must therefore require an internal realist Kantian estoppel. Now remember that Kant had himself formulated a question which at first sight seemed to lead straight to the metaphysical realist camp; asking how we can be certain that our methodological principles will lead us to theories that conform to nature; or more strictly, conform to nature's order and unity? But Kant's response to this question soon reveals that nothing of the sort is intended; on the contrary, he goes on to produce an answer in terms of what we might call a 'postulate of ontological relativity'. The notion of the 'order of nature' (its unity or systemicity) is now interpreted in such a way that its ontology is grounded, not independently, subsisting, as it were, 'in itself', or alternatively — using the theological version of this — in Divine Reason, but in human reason; interpreted here of course as the operations of the Systems Component (SC) itself; the latter now functioning *simultaneously* as a systems ontology, i.e. as yielding 'the order of nature in general'. As Kant expresses this: the methodological principles comprising SC must simultaneously be viewed as possessing a "transcendental" function also, by viewing them as necessary conditions of the possibility of any cognitive significance of the idea of the order of nature as such (cf. A650/B678; A654/B682).

Now what makes this response interesting is the fact that it elegantly incorporates the basic outlines of Kant's transcendental approach in general. The "idea of the unity of nature", he tells us, cannot be "realized"

[11] This denial of constitutive import is somewhat confusedly connected by Kant — and has certainly confused most commentators — with his contention that we cannot escape this conclusion by producing an ontological account of the unity as an 'object in itself'. Since phenomenologically speaking there is no object, the only alternative, he says, might be thought, to obtain such an object by viewing it as grounded in a divine substratum; equivalent to the fiction or analogy of the unity of nature as a "reality in itself" (A679/B707). Kant, as will be shown presently, of course rejects this possibility.

apart from the operations of SC, and its accompanying transcendental function (cf. A677/B705). Furthermore, apart from SC, that idea 'reduces' to something with the status merely of "a transcendental object" (A698/B726); an expression which is always used by Kant to denote the state of something (e.g. objects, systems of objects, etc.) still awaiting a 'realization', in the sense just used; i.e. the realization of the idea of unity as something grounding the process of theory construction.

Evidently the basic postulate here is, that apart from some 'realization procedure' or other [RP for short] — which for Kant always involves the epistemological functions of sensibility, understanding and reason — to repeat: apart from RP the ontological status of whatever it is that is in question, is 'reduced' to something whose ontology is as yet undefined. I have shown elsewhere that this 'reduction-realization process' [RRP] lies indeed at the basis of the whole Kantian metaphysic; i.e. the claim that apart from a realization everything subsists in a state of reduction, and is not something just given or found; RRP thus expressing Kant's basic postulate of ontological relativity.[12] Kantian relativity, of course, insists on confining the realization account to the operations of *human* reason — in the present case, its "theoretical employment", via SC. It leaves no opening for a realization of the idea of the unity of nature, moving forward from its status as "transcendental object", via the operations of Divine Reason, which for Kant would be regarding the idea of the unity of nature precisely as something existing "in itself" (cf. e.g. A678/B706f). For Kant this is merely a useful analogy or fiction, in line with which we may picture the systematization process as having achieved completion (cf. A678/B706; A698/B726).

The concept of the transcendental object therefore expresses the fact that prior to realization, the notion of a 'world' is completely undetermined, and as yet to be determined. The world, apart from the realization process is, as Kant puts it in the *Opus Postumum*, a mere "*determinabile*"[13]; one of the characterizations of the term "thing in itself" (as in this passage); the very opposite of the other use of the term where it denotes a realization that has come about without the use of the human

---

[12] For RRP, see also my *Reduction-realization: a key to the structure of Kant's thought*, in: J.N. Mohanty and R.W. Shahan, eds., Essays on Kant's Critique of Pure Reason (Norman, 1982), pp. 39–98; and my *Neo-transcendental approaches towards scientific theory appraisal*, in: D.H. Mellor, ed., Science, Belief and Behaviour (Cambridge, 1980), pp. 1–22.

[13] Kant, *Schriften* 22 (Berlin, 1938), pp. 418–422.

cognitive faculties, i.e. in a pre-critical theological, or post-critical moral context. Lack of clarity on this double use of the term 'thing in itself' lies probably at the root of many of the most pervasive critical responses to and misunderstandings of Kant. — Anyway, since under reduction there is no real world — no 'frozen universe' — with which to compare (per impossibile) the world that is realized in the context of the various theorifications of science, the skepticism of the metaphysical realist is banished.

But to return to our systems case as defined by Kant: We have seen that systems methodology has the simultaneous function of serving as a systems ontology: this is its "transcendental deduction" (A671/B699); something which, in accordance with Kant's usual formula, bestows upon the methodology a kind of "objective validity" (A669/B697; A664/B692). Needless to say, the 'validity' in question here is transcendental, and not — what we may call — heuristic. Whether our methodological concepts prove successful depends upon the details of the working out of the scientific research programme. Similarly, though the methodological principles are said to possess "synthetic a priori" status (A663/B691), this applies only to their transcendental, and not their methodological function. Still less, of course, does Kant claim (as is so often stated) that the theory resulting from the employment of the methodological principles is itself 'a priori certain', a view reinforced by Kant's formally a priori approach to the fundamental laws of Newtonian mechanics, yet to be considered. In all this, the subtle interplay between the two functions of SC has confused people into believing Kant to be holding what would plainly be an absurd position — not an uncommon response in the vast literature on this philosopher! Instead of which, all he claims is that the unity of nature as such is not something 'existing in itself' but something that comes to life simply and solely via actual acts of systemic construction, under the constraints of a certain methodology. In line with a similar position to be found in Wittgenstein, the unity here in question simply "shows itself" — compare *Tractatus* 6.36, referring there to 'the law of causality' instead of Kant's 'order of nature': "If there were a law of causality, it might run: 'There are laws of nature'. But that cannot of course be said: it shows itself".

Now the transcendental deduction of the idea of unity, when compared with the better-known deduction of the categories, is clearly of a relatively 'weaker' kind, yielding therefore also a weaker 'validation' of the relevant principles, for three reasons: (1) As already mentioned, the "unity of nature" lacks the status of an 'object', but yields only rules or principles for theoretical research, as Kant puts it at A665/B693. 'Systematic experience' is less basic, and more controversial, than "experience in general". Hence this deduction yields only "subjective necessity".[14] (2) The rules and principles composing SC are less deeply entrenched in logic than are the categories; they do no more than incorporate the scientific wisdom of traditional sciences, as Kant frequently emphasizes himself.[15] (3) The individual principles of SC are not demonstrated singly, unlike the principles of the categories; being simply introduced as a collective batch, drawn, as just mentioned, from the details of the history of scientific method.

IV

Mention of the categories brings us to the Probative Component [PC] of Kant's methodology, and to the 'General Ontology' associated therewith — though, as we shall find, far more loosely than was the case for SC. The methodological principles comprising PC determine solely the evidential strength of actual scientific theories; that is, they concern — with a proviso to be mentioned later — their phenomenological aspect only. Consequently, such success as is achieved in the course of theory construction bestows only *heuristic* validity upon the method, since — unlike the case of SC — PC does not function transcendentally. In particular, PC-methodology does not simultaneously generate General Ontology, the latter involving a different set of principles; i.e. those dealt with in the sections on the Aesthetic and Analytic of CpR; above all, the principles of the categories.

Thus, whilst Systems Ontology runs parallel with the operations of Systems Methodology, General Ontology is confined to generating the possibility of "nature in general", as Kant terms it[16]; which here means: the possibility of that evidential, observational basis which forms the starting point for the operations of PC; and on which are built the subsequent processes of induction, confirmation, etc.

To explain: The observational basis we may imagine to be expressed basically via statements concerning a plurality of objects and their states, as

---

[14] KANT, *First Introduction to the Critique of Judgment* (trsl. James Haden, Indianapolis, 1965), p. 15; Ak. ed., 20, p. 209 Cf. A666/B694.

[15] Cf. A654/B680; *Crit. Judg.*, op. cit., note 8, Intor., sect. v, p. 18; *Met. Found.*, op. cit., note 4, p. 472.

[16] Cf. *Met. Found*, op. cit., note 4, p. 469.

well as the changes which such objects undergo. General Ontology thus involves, a.o., the elucidation of the conditions of the possibility of cognitive access to such objects and their changing states; conditions whose validational force is supposed to be proved in the Second and Third Analogies of Experience.

Thus the Second Analogy seeks to demonstrate that the possibility of any cognitive grasp of objective change in principle — that is, note well, *causal or otherwise*, presupppposes that our individual perceptions should be linked by a relation which Kant defines as the relation of causality. General Ontology, so far, does not therefore extend to, nor provide a foundation for, the operations of empirical logic, e.g. inductive generalizations, involving functional or causal connections, inferred from the just mentioned contingent sequences and coexistences.

Contrast this with a theory of induction like that of J.St. Mill, for instance. In that theory, induction involves a reference to a universal law of causation, independently proved in some fashion or other by Mill. By contrast, the justificative force of such a law can not — under the Kantian account — derive from the General Ontology, the validational strength of whose principles extends solely to the grounding of the possibility of contingent existence and change in general, whether causal or otherwise, as just noted. On the contrary, if anything, causation, insofar as it is involved in the actual processes of nature, is rather a function of SC; flowing from the methodological maxims of SC into the PC-component. Furthermore, the contribution from SC should not of course be regarded as serving as a kind of 'foundation' (on the lines of Mill's proof); not 'metaphysical' (as it is for the latter), since this is excluded by Kant in principle; and not transcendental, since as we have seen, transcendental validation for the case of SC extends only to the idea of the unity of nature *in general.*

What, however, of the principle of causality, as a component of General Ontology? Now I have insisted that there it exhausts its strength in grounding the possibility of any change as such; so it cannot simultaneously serve as a foundation for *causal* change in particular. The true situation is rather this: The causal principle, which Kant has shown to be an intrinsic component of experience in general, is then extracted from that context, and freshly applied in a purely methodological context, including that of SC, whose successful operation is a purely heuristic, and not a transcendental matter (ignoring for the moment its bearing on Systems Ontology). Indeed, Kant states all this perfectly clearly when he writes:

The unity of reason [involved in the operations of SC] is ... not the unity of a possible experience [generated by General Ontology]. That everyting which happens has a cause, is not a principle known and prescribed by reason. That principle makes possible [only] the unity of experience. (A307/B363)

And the same point is made even more clearly in the *Critique of Judgment*, which states that (in our terminology) SC can only "ascribe" causality to nature, whereas, as part of General Ontology, that principle "prescribes" causality to nature in general.[17] Apart from this, Kant has no 'justification' of causal reasoning in science; its centre of gravity is located within the operations of SC; very much in the spirit of Quine's later dictum that far from theory depending on induction, it is induction that depends on the achievements of scientific theory.[18]

It is thus not by deduction from the principle of the Second Analogy, but more by virtue of good philosophical sense that Kant employs a principle, shown previously to be an essential element of "experience in general", also methodologically in the context of "experience as a system".[19] Thus, whereas the principles of General Ontology possess intrinsic validational strength, of a transcendental type, with respect to "nature in general", their methodological function within PC and SC gives them no more than heuristic validity. If this represents Kant's views more or less accurately, it explains also his attitude towards observation and induction in the context of scientific research; an attitude perhaps appreciated better nowadays (in the age of Kuhn, Feyerabend and Lakatos). He remarks that *modus ponens* (hypothetico-deductive logic) is "for the most part of little importance" in science (A792/B820); instead, it is "systematic unity" which must serve as the "sufficient criterion of empirical truth" (A651/B679).

V

Mention of modern views brings us, finally, to the Explicative Component [EC], equivalent to what Kant labels the "Special Metaphysics of Science", dealt with in his *Metaphysical Foundations of Natural Science* [MF]. Whereas in the case of PC, General Ontology subsists separately from the methodological aspect of scientific theorizing, and whilst Systems Ontology runs, in ways we have explained, 'in parallel', yet separate from

[17] *Crit. Judg.*, op. cit., note 8, Introd., sect. vi, p. 23.
[18] W.V. Quine, *Ontological Relativity and other Essays* (New York, 1969), p. 135.
[19] Kant, *First Intro. to the Crit. Judg.*, op. cit., note 14, p. 15; Ak. ed., 20, p. 209.

systems methodology, in the case of EC the phenomenological and the ontological aspects in some sense coincide; ontology is conflated into methodology, since it is the explicit task of EC to demonstrate only the *possibility* of such specific scientific concepts and laws as make up the 'metaphysical hardcore' of some given scientific theory — in Kant's example, of Newtonian mechanics and dynamics. It is important always to remember that this is Kant's specific objective; in the Preface to MF he remarks that "to cognize anything a priori is to cognize it from its mere possibility" (470) — and of course, *vice versa*. Since the operations of EC, as we shall presently see, proceed in some sense a priori, and certainly purely conceptually, it is important to realize that its results concern solely the possibility (in a stretched sense: the merely ontological aspect) of the matter. Thus Kant's purported a priori "proofs" of some of Newton's laws of motion are not intended by him to compete with Newton's empirical — indeed, as Newton claimed: inductive — demonstrations of these laws, in their phenomenological respect, but have as their sole objective the demonstration of the "possibility" of these laws.

What this means, we shall indicate briefly in what follows. But first, let us be clear on the general outlines of Kant's approach. EC, and its associated Special Ontology, differs from SC, in that it does not employ a separate and independent set of methodological principles, with a merely *associated* transcendental import (in respect of the idea of unity in general), but borrows any such principles as it requires from General Ontology (thus somewhat similar to the case of Kant's borrowing of the causal principle with respect to the operations of PC); "applying" — as he explicitly terms it at MF.470 — the principles of General Ontology to certain "empirical concepts" — in his example, those of matter, motion and force — in order to demonstrate their possibility. Any validational strength accruing to these principles in the context of EC is thus indeed of a transcendental kind, but this in rather a weaker and more Pickwickean sense of yielding the possibility (and realizability) of the 'foundational' laws of phenomenological physics.

I turn now to some details: It is the task of EC to generate the metaphysical hardcore of Newtonian mechanics. In effect, this involves the combined use of certain empirical as well as transcendental notions, the former obtained from an "analysis" (472) of the empirical concept of matter, the latter — as just noted — from the principles of General Ontology. Not surprisingly, the semantic analysis of matter turns out to be (in Kant's treatment) heavily indebted to the Newtonian 'scientific exemplar'. Thus, Kant defines matter, a.o., as that which fills space; a fact which we experience by virtue of a force of resistance or repulsion — all of which is purely empirical semantics. For instance, he explicitly writes that we know that matter fills space because when another material body seeks to intrude, its motion is gradually diminished, and — note the precise wording! — "the cause of motion is *called* moving force" (497; my italics). Again, "to fill a space *means*, to resist everything movable that strives by its motion to press into a certain space" (496; my italics); all appeals to linguistic usage.

But how does all this connect with the transcendental side of General Ontology? Well, in the case of force, Kant relates this to the transcendental principle of the Anticipations of Perception, and the associated categories of quality, i.e. reality, negation and limitation, which are correlated with "sensation" or "feeling", by means of which we come to know "the size and shape of an extended thing" (510; cf. 523). Thus in ways we cannot follow here in detail, Kant correlates the three categories with the balance which he claims to exist between the repulsive and attractive force which in this way defines the existence of matter: hence — rather loosely — establishing a connection between the categories and the empirical semantics of matter; the whole argument being intended to 'make sense' of the attraction of matter at a distance.

As a second example, let us take inertia (Newton's First Law of Motion). Here Kant invokes again — a third time over![20] — the principle of causation, previously shown to be (as explained) an essential ingredient of *all* change. To prove the law of inertia, Kant now has to import again certain semantic interpretations of the terms 'change' and 'cause'. The former is interpreted, on good Cartesian-Newtonian lines, as change of motion, or indeed, change of velocity; the latter (cause) is interpreted as externally acting force. Like the rest of the analysis, this last interpretation is by no means obvious, and simply embodies one of the basic assumptions of seventeenth-century mechanistic physics (the mechanical world picture). Kant evidently is quite clear on this, for he notes explicitly that to interpret force as *other* than merely 'external action', and instead to allow that action might also be internal; such an interpretation "would be hylozoism" and thus spell "the death of all natural philosophy" (544); i.e. the paradigm of modern physics.

Conceptual explication thus turns out to be a method of showing that some given basic laws or concepts of science express certain fundamental

[20] *Met. Found.*, op. cit., note 4 ch. 3, p. 104.

meanings, inherent in the substance of the branch of science under investigation; and furthermore, that these meanings can be integrated within the basic ontological features of all experience in general. It is thus evident that Kant's conceptual explication must not be viewed as a deductive process, but instead as a fairly loose kind of analytical exploration which seeks to integrate the basic features of science within a more general semantic and transcendental structure; Kant's special account being both an anticipation and — more interestingly — a very definition of the modern idea of the formation of the 'metaphysical hardcore' of a science, involved in the foundations, not only of physics, but also — and perhaps more so — of the 'softer' sciences of biology, geology, etc. Indeed, whilst at the outset I talked with Newton as though the foundational laws like Newton's laws of motion had a phenomenological, in the sense of inductive, foundation, modern views have tended to the conclusion that such foundational principles have a rather thin empirical basis; being rarely testable directly by observational means, unlike the modern ordinary second-order hypotheses of science; giving additional sense to my contention that in the case of EC the phenomenological and ontological aspects coincide.

But to return to EC and its associated Special Ontology: we see that this involves again a somewhat 'weaker' argument structure than the one encountered in General Ontology. Firstly, it employs categorial principles (like that of causation), whose *transcendental* validity is confined to the context of General Ontology, but subsequently applying them in a fresh and more 'problematic' context. Thus, whilst the causal principle of the Second Analogy is (as noted) validated by the fact that it makes *experience of change in general possible*, the *possibility of inertial motion* is clearly a more controversial matter; after all, Newtonian physics is a highly falsifiable subject, as its subsequent history has demonstrated. Quite generally, repulsive and attractive force, inertia, momentum conservation, etc., are ultimately all part of the overall structure of an empirical and hence contingent scientific research programme, and being 'hypothetical', belong to the changeable corpus of physical knowledge. Like the 'metaphysical hardcore' of Imre Lakatos, they are ultimately drawn from a conspectus of the accumulated body of scientific empirical knowledge, where it is only subsequently, and usually temporally, 'hardened' into a foundational complex. Thus whilst Special Ontology employs principles (the categories) that are stronger than the methodological principles of Systems Ontology, it brings them to bear on an "empirical concept" (matter) whose meaning is derived from Newtonian styles of thinking. In the case of General

Ontology, we can hardly imagine living without the experimental notions of change and coexistence, whereas one might be prepared to get along without "the possibility" of the principles of Newtonian theory, such as the laws of motion, conservation of mass, etc., and replace them by something else — as has indeed been the case for some of these conceptions.

However, this 'looseness of fit' which surrounds the relations between the phenomenological and the ontological aspects of PC, EC and SC, and indeed, between these three components themselves, has the advantage of not bringing the central themes of Kantian transcendentalism crashing down everytime there is a change in the paradigms of natural science. Lack of appreciation of this 'looseness' has led most scholars — we need only think of the case of Reichenbach — to the opposite conclusion. Against this, Kant's transcendental approach, in its application to both the ontology and the phenomenology of theory construction, turns out to be something much more informal, much more subtly and messily articulated, more tentative and general, than the usual more formal elucidations of the so-called 'transcendental argument' would lead us to expect. Evidently it is not so much a matter here of any formal deductions, or of the demonstration of the uniqueness and necessity of this or that 'a priori' condition; still less — as we have already noted — should Kant be saddled with the absurdity of having wanted to prove the 'a priori certainty' of, for instance, Newtonian mechanics; after all, he did say, in CpR, that "in natural science ... there is endless conjecture, and certainty is [here] not to be counted upon" (A480/B508). Rather, his aim is to explore the contextual structure of the language of physics, of science, and of life. Above all, it is clear that the transcendental approach with respect to "special" and "systems ontology" is a far more tentative, historicist and pragmatic affair than the conventional spellbound preoccupation with the a priori side of transcendentalism has usually led people to expect; and even the a priorism of the pure transcendental principles, in this context, turns out to have a very different significance and a different type of entrenchment from what most commentators have usually led their readers to believe.

At any rate, once we grasp the complexities of the Kantian version of Putnam's 'theory-relative' ontology (with which we began), incorporating as it does a necessary reference, firstly, to scientific theory proper; secondly, to a triadically formulated methodology; and finally, to a threefold ontology associated therewith, we come to appreciate that it is difficult to conceive the possibility of understanding the overall structure of Kant's own thought without keeping in mind the structurelines outlined in this paper.

VI

In conclusion, I want to add a philosophical postscript, in order to provide a slightly more generalized perspective of the Putnam–Kant internal realist position, explained in what has gone before. We have claimed that Kant's transcendental approach may be viewed as a sort of realization procedure [RP] which takes the object, or the aggregate of objects ('the world') through a number of stages, in a quasi-dynamic fashion. At the initial stage such a world has what we may term 'zero-ontology', i.e. no account so far being given, or capable of being given, of the possibility of such a world — or as in Kant's case: of the cognizability of the world. Let us say that at this stage, Kant's transcendental framework is as yet not 'engaged', is still 'idling' (t_0-stage). The world then subsists only in a transcendental — or t_0 — sense, as "transcendental object", or "transcendental thing" (A682/B710): T_0 for short. — Then, at a subsequent stage, where certain epistemological and/or methodological functions (viewed as transcendental conditions) become 'engaged', via the mediating activity of sensibility, imagination, understanding and reason, we obtain a 'realization' (as explained in the earlier parts of this paper), yielding a 'positive ontology'; T_0 being brought now to a state of "appearing" (in Kant's technical sense of the term), as an empirical object or system of such objects (T_a; T_a^{θ}).

Now the T_0-stage we may imagine as corresponding to Putnam's "world, simpliciter", a world whose ontological significance he rejects. But such a rejection procedure needs to be made explicit; we need a postulate explicitly denying a positive ontology to the unrealized object.[21] For otherwise it is not obvious why there might not after all be intelligible world states, with a positive ontology, subsisting apart from some explicit realization procedure, e.g. on Kantian lines. After all, common sense usually assumes this, and it is a position enshrined also in pre-critical rationalist philosophy whose theological formulation unconsciously assumes a positive ontology grounded in the Deity.

The older logical positivist school already had such a postulate, stating that in the absence of a verification or verificability procedure, a proposi-

---

[21] As noted earlier, T_0 thus represents a new version of the notion of 'the world in itself'; the alternative version being a *realization* of T_0, albeit counterfactually, in a theory-independent way: the world-in-itself as, say, a divinely grounded noumenon. Kant is perfectly clear that there are these *two* senses; but his commentators, and indeed, Putnam himself, have almost invariably conflated the two versions.

tion is strictly meaningless, has zero semantic value. But we require something more general, relating realization to an ontological context in general. Now such a generalization may be defined in terms of the Husserlian notion of 'reduction' — a notion which I have already invoked informally in what has preceded. In Husserl, we first meet with the explicit statement that apart from some RP, any world *is to be considered* as 'reduced' to a state of "epistemological", and thus ontological "nullity".[22] Putting all this still more generally, in order to emphasize the philosophical method behind this position, we must define it more positively and explicitly through the following postulate: No realization process can get off the ground unless the world has previously been subjected to a reduction. For otherwise, some world or state of things might still be supposed to subsist prior to some given RP, with whose results it might then compete; the position of metaphysical realism again.

Such a reduction postulate is really a generalized version of Kant's idea of the philosophical equivalent of the Copernican Revolution. Whilst previous to Kant, ontologies are imagined to be pre-given, for Kant they must be viewed as the result of a subjective, humanly grounded, RP. But as just noted, such a position requires the postulate of a prior universal reduction. Only then can RP generate a realization with Kant's claimed a priori status, e.g. of the principles of the categories, or of the scientific methodology; in line with his famous slogan that "we can have a priori knowledge only of that in things which we place ourselves in them" (Bxviii). For this to be effective, we have to supplement RP by reduction, RRP for short; representing a generalization of the transcendental approach, and a representation of what Quine labelled "Ontological Relativity" (though no doubt not with the same meaning). Such an approach is not of course confined to Kant; it is assumed also in such varied philosophical positions as those of Husserl, Wittgenstein, Heidegger, Habermas, and of course Putnam.

RRP is really the philosophical equivalent of the subjectivist trend that has become predominant in recent philosophy of science, in Kuhn, Feyerabend, Lakatos, etc. But it seems important to make explicit the general philosophical assumptions lying behind these schemes, and their historical roots, as generalized here in the formulas of RRP.

---

[22] Edmund HUSSERL, *The Idea of Phenomenology* (trsl. W.P. Alston and C. Nakhnikian, The Hague, 1964), p. 31.

Barcan Marcus et al., eds., Logic, Methodology and Philosophy of Science VII

CONCEPTUAL EVOLUTION AND THE EYE OF THE OCTOPUS

DAVID L. HULL

Dept. of Philosophy, Northwestern Univ., Evanston, IL 60201, USA

The major reason for evolutionary analyses of conceptual change being so unsatisfactory is that they are modeled on an inadequate understanding of biological evolution. Genes are not in the least like beads on a string, an entire meadow can function as a single organism in the evolutionary process, and species are not classes of similar organisms, regardless of the dictates of common sense and ordinary usage. To make matters worse, even when such biological concepts as genes, organisms and species are adequately understood, they are still not general enough to function in a truly general theory of evolution (HULL 1980). In order for evolution to occur by variation and selection, certain entities (replicators) must pass on their structure largely intact. In biological evolution replication occurs primarily at the level of the genetic material. Certain entities must also interact with their environments in such a way that this replication is differential. Entities at a wide variety of levels, including genes, chromosomes, cells, organisms, kinship groups, and possibly even populations and entire species can perform this function. As a result of these two processes, other entities (lineages) change indefinitely through time — they evolve. For our purposes the most important feature of the preceding process is that at all levels it is spatiotemporally localized. Evolution is a local, not global, phenomenon. Both the entities that function in selection processes and the entities that evolve as a result are spatiotemporally localized entities. They are individuated in terms of location, internal cohesiveness and continuity through time. They are historical entities. The sort of identity that is fundamental is genidentity.

Neither biologists nor philosophers of biology are unanimous in their acceptance of the preceding characterization of the evolutionary process. In general, opponents consider the things that evolve (species) as spatiotemporally unrestricted classes such that sufficiently similar organisms, living at different times and widely separated places, can be interpreted as belonging to the same species (KITTS and KITTS 1979, CAPLAN 1980, 1981, KITCHER 1984). Defenders maintain that such a conception of species may have some warrant, but whatever the use these classes may have in other contexts, they cannot evolve (GHISELIN 1974, HULL 1976, 1978a, WILEY 1981, SOBER 1984). In this paper, I can present only the briefest sketch of the reasons for treating both the entities that function in selection processes and the entities that evolve as a consequence of them as historical entities. My main purpose is to extend this line of reasoning to conceptual evolution, especially the sorts of conceptual change that occur in science. Concepts are traditionally treated as classes of similar (or identical) tokens. For example, in this paper thus far, five tokens of the term-type "evolution" have appeared. Clearly, there are many contexts in which concepts can and should be viewed as tokens of the same type, but the evolutionary context is not one of them. If concepts are to evolve, then they can no more be construed as similar tokens of the same type than can species be interpreted as classes of similar organisms. Both must be construed as lineages. Just as not all eyes are "eyes," not every set of axioms with the same assertive content can count as the same theory. Not all instances of Darwin's theory count as "Darwin's theory."

In this paper I begin with a brief justification of treating biological phenomena the way I do, using the eye of the octopus as an example. I then proceed to present parallel arguments for treating concepts in the same way. Objections are likely to arise at three junctures. First, certain readers may disagree with my understanding of the evolutionary process; e.g., it is not necessarily localized in space and time. Second, they might object to any attempt to treat conceptual evolution as analogous to biological evolution, regardless of the nature of the biological analogy. Finally, they might be willing to accept my analysis of biological evolution and the attempt to present an analogous treatment of conceptual evolution but disagree with my particular attempt. I have dealt with the first two sorts of objections elsewhere (HULL 1981, 1982) and cannot go over this same ground here. In this paper I must confine myself to objections of the third sort.

When is an eye not an eye?

The structure/function distinction runs throughout biology. Should anything that functions as an eye count as an eye, or must eyes exhibit a certain structure, or both? When De Voe (1981, p. 433) states that eyes are "sensory organs that direct onto thin layers of neural tissues — retinas — whatever of the external world can be conveyed by emitted, refracted, scattered, or reflected quanta of light," he is defining eyes in terms of a combination of structural and functional criteria. On this definition, many organs that organisms use to see don't count as eyes, including some vertebrate eyes. I am not about to address the tangle of problems surrounding the relative primacy of functional and structural criteria in defining classes of organs in biology. Instead I propose to add yet another dimension to the problem — descent. Evolutionary biologists are unwilling to consider two organs the "same" organ, no matter how similar they are in structure and/or function, if they lack the requisite ancestor-descendant connections (Wiley 1981). The eye of the octopus is the usual example.

Because nearly all organisms living here on Earth use exactly the same genetic code and those that do not depart in only minor ways, biologists conclude that all terrestrial life had a single origin. Thus, if present-day cephalopods and vertebrates are traced far enough back in time, they have common ancestors. When organs in present-day organisms that are structurally and/or functionally the same are traced back in this tree of life, they do not always converge on a common ancestral organ. They are not evolutionarily the "same." They are not evolutionary homologies. Conversely, when organs in present-day organisms that are structurally and/or functionally quite different are traced back in time, sometimes they rapidly converge on the same organ in a common ancestor. Similarity to one side, they are instances of the same organ. For example, no matter how similar the eyes of the octopus are to vertebrate eyes, they are not homologous. Conversely, no matter how different the spiracle in sharks may seem to the eustachian tube in human beings, they are evolutionary homologies.

Similarities and differences in structure are certainly part of the evidence that biologists use to decide which structures are or are not evolutionary homologies, but any degree of similarity in structure and/or function that excludes cephalopod eyes is likely to exclude certain vertebrate eyes as well. If one looks closely enough, systematic differences between vertebrate eyes and the eye of the octopus can be discerned, but the only reason to look that closely is that differences in overall structure (*Bauplan*), embryonic development, and fossil record imply that these two organs are evolutionarily different. They have different histories. They emerged in different lineages.

The distinction between similar organs and evolutionary homologies is so commonplace, even among non-evolutionary biologists, that it is liable to arouse little resistance. Schoolchildren are taught that the wings of birds, bats, flying fish and insects are not all "wings" in the same sense. The "forearms" of birds and bats are homologous as "forearms" but not as wings. At no level of analysis are the "wings" of insects and flying fish homologous to each other or to the "forearms" of bats and birds. Evolutionary biologists begin to meet resistance when they argue that two gene-tokens with exactly the same structure do not necessarily belong to the same gene-type. Population geneticists distinguish between genes with similar structures but only the most distant ancestral connections (independent genes) and those that are identical by descent. Ordinary molecular biologists do not. In their terminology, "homologous" genes need not be evolutionary homologies. CUU is CUU, and that is that.

One common justification for distinguishing evolutionary homologies from analogies and convergences is that failure to do so produces mistakes in historical reconstructions. If one wants to reconstruct phylogeny accurately, similarity due to common descent must be distinguished from similarity due to common functional and/or structural requirements, responses to common selection pressures, etc. The scope of the problem can be seen by comparing the organisms that inhabit the Mohave and Sahara Deserts (CODY 1974). Although the niches in these two deserts are very similar and the organisms that occupy these niches are comparably similar, the organisms living in these widely separated deserts are just as distantly related phylogenetically. A niche occupied in the Mohave Desert by a lizard might be occupied in the Sahara Desert by a snake. Which sort of similarity should take precedence? From the point of view of accurate historical reconstruction, discerning evolutionary homologies is essential. The story for the evolutionary process is not so straightforward.

The process-product distinction is common in philosophy. Selection, competition, and mutation are all part of the evolutionary process. The product is phylogeny. Because the term "evolution" is frequently used for both, they are easy to confuse. When a scientist entitles a paper "The Evolution of the House Mouse," he might be referring to its phylogenetic development, the processes through which it arose, or both. As it turns out, knowledge of the details of the evolutionary process are not much help in reconstructing phylogeny. Regardless of the resolution of the various controversies currently exercising evolutionary biologists, paleontologists will continue to reconstruct phylogeny as they have in the past. One exception is the controversy over the tempo of evolution. The more saltative the evolutionary process turns out to be, the less likely paleontologists are to find fossils that are intermediate between related species.

But evolutionary biologists want to do more than reconstruct phylogeny. They want to understand the evolutionary process. To do so, they have to discern regularities in this process and the entities that function in these regularities. For example, if speciation usually occurs by the isolation of peripheral populations, then those species with longer and more convoluted peripheries are likely to speciate much more frequently than other species. On this view, species with extensive, convoluted peripheries is a biological "natural kind," and particular species exemplifying these characteristics count as instances of this natural kind. Other possibilities for evolutionary natural kinds are the continua between monotypic and polytypic species, eurytopic and stenotopic species, r-selection and K-selection, sexual reproduction, asexual reproduction, and various combinations thereof.

It might seem that paleontologists and evolutionary biologists are engaged in incompatible, possibly incommensurable activities. Paleontologists must identify evolutionary homologies in the face of the deceptive similarities introduced by constraints imposed by the evolutionary process, while evolutionary biologists are constantly frustrated by the historical constraints imposed on the evolutionary process by common ancestry. What counts as the message for one is noise for the other. As long as one views biological species and the evolutionary natural kinds listed above equally as kinds, then the two activities do seem at cross-purposes. However, once one distinguishes between genuine kinds (such as polytypic species) and instances of these kinds (e.g., *Homo sapiens* as a polytypic species), the two perspectives become totally compatible. The natural kinds of evolutionary theory are spatiotemporally unrestricted in the traditional sense; e.g., sexual reproduction can occur anywhere in the universe at any time just so long as the conditions are right. Particular instances of sexual reproduction require physical contact. There is no fertilization at a distance. At least propagules must come into contact. It is the distinction between kinds and instances of kinds that KITTS (1983) ignores in his otherwise perfectly cogent criticisms of treating species as individuals and their names as proper.

Another example might help. Haplodiploidy is a particular mode of reproduction and sex determination. One sex has a particular complement of chromosomes; the other has double this complement. One is haploid; the other diploid. Haplodiploidy is especially interesting because it is one way for sociality to evolve. However, it has arisen numerous times in the course of evolution. As it is used by evolutionary biologists, haplodiploidy is a natural kind, not an evolutionary homology. It can function in evolutionary theory because it is in no way spatiotemporally localized or restricted. This same distinction applies to the eye example. Eyes as any structure capable of seeing might well function as a natural kind in biological theory; eyes as evolutionary homologies cannot. One can explain vertebrate eyes as functional eyes by reference to the relevant process theories but not their peculiar status as *vertebrate* eyes. To explain this, one must make recourse to the contingencies of their place and conditions of origin. As long as different names are given to natural kinds and evolutionary homologies, little confusion is likely to arise, but in biology the two are frequently given the same or very similar names. Not all carnivores belong to Carnivora, and not all organisms that belong to Carnivora are carnivorous. "Carnivorous" and "functional eyes" might function in laws of nature; Carnivora as a monophyletic taxon and vertebrate eyes as an evolutionary homology cannot.

As simple as the preceding distinction is, it is sufficient to resolve the apparent conflict between the historical and process perspectives in biology. But of greater importance is the fact that built into the processes of biological evolution are spatiotemporal requirements. In crucial instances, the formulae of population genetics range over entities that are not just similar in structure but *identical by descent*, in particular the entire apparatus of kin selection. In this context, the coefficient of relationship concerns the likelihood that one allele rather than another gets passed on. To be sure, any two organisms picked at random from a population are likely to have similar alleles at the vast majority of their loci, but kin selection requires that this similarity in structure be acquired by immediate descent. When an organism passes on fifty percent of its alleles in sexual reproduction, these descendant alleles will be similar to the ancestor alleles that gave rise to them, but the converse is not true. Many alleles with the same structure, when traced back in time, do not converge on the same ancestral allele.

At the lower levels in the evolutionary process, spatiotemporal localization is essential. No one is likely to object to evolutionary biologists treating genes (introns, etc. notwithstanding) or particular organisms (tillers and tussocks notwithstanding) as historical entities. Replication is itself clearly a spatiotemporally localized process. Replicators also produce more inclusive entities that intercede for them with increasingly more inclusive environments. Particular instances of the interplay between replication and environmental interaction are somewhat less obviously spatiotemporally localized. Resistance begins to start in earnest when one insists that the results of these processes — lineages — are themselves spatiotemporally localized, but anyone who insists on treating lineages as types of similar tokens is going to make hash of the evolutionary process. However, once the immediate effects of selection have left their mark, nothing about selection processes entails that these same decisions be made with respect to more macro-phenomena. The insistence of evolutionary biologists on recognizing evolutionary homologies (such as the vertebrate eye) and monophyletic taxa (like Vertebrata) is an extrapolation from decisions that are absolutely necessary at the microlevel to higher level phenomena. The entities that evolve must be treated as lineages. Arguments to the effect that higher taxa must be treated in the same way are not so conclusive. Not all biologists are willing to treat higher taxa strictly as clades. Some biologists prefer to intercalate "paraphyletic" groups into a phylogenetic classification. The primacy of the lineage perspective can be seen, however, both in the fact that phylogenetic relationships set limits to the introduction of paraphyletic taxa and in the need to justify departures from strict monophyly.[1]

In summary, evolutionary natural kinds are absolutely necessary if we are ever to have an adequate theory of biological evolution. Neither evolutionary homologies nor monophyletic taxa are candidates for these natural kinds because they are inherently spatiotemporally restricted. There can be no "laws of the vertebrate eye." Vertebrata might well be an *instance* of a natural kind, but it itself cannot *be* one. Neither single lineages nor monophyletic chunks of the phylogenetic tree are candidates for spatiotemporally unrestricted classes. If one finds this distinction difficult to maintain in the context of biological evolution, wait until the same distinctions are introduced into conceptual evolution.

When is Darwin's theory not Darwin's theory?

One need not be a Platonist to see a role in language for concepts modeled along the traditional type-token analysis. The type-token distinction in language parallels the individual-class distinction at the level of

---

[1] According to HENNIG (1966) and his followers, monophyletic taxa are those that include all and only the descendants of a particular ancestral species. Evolutionary systematists prefer to term such taxa "holophyletic" and limit the term "monophyly" to those taxa descended from a single ancestral species without requiring that all the descendants of a particular ancestral species be included in the same higher taxon. The result is taxa that HENNIG (1966) terms "paraphyletic".

things. Just as individuals are instances of classes, term-tokens are instances of term-types. In language, proper names denote individuals (or particulars of any sort) while general terms denote classes (as well as such things as properties, processes, etc.). But the type-token distinction applies equally to all terms. Numerous instances (tokens) of a term-type can exist for any term-type, regardless of whether it is proper or more general. Several tokens of the term-type "Darwin" have already appeared in this paper, even though Charles Darwin is a paradigm individual and his name a paradigm proper name.

A minimal requirement for something's being an individual and the same individual through time is genidentity — continuous development through time and internal cohesiveness at any one time. Some authors maintain that this minimal requirement is also sufficient. An individual can change any or all of its properties and still remain numerically the same individual in the process. Other authors are willing to allow individuals to change as much as they please just so long as they do not lose their "essential" characteristics — the characteristics that make them the kind of thing that they are. If a comet captured by a star can become a planet while remaining the same individual, then neither comethoood nor planethood is part of its essence. Similarly, if an organism can change from one sex to another during the course of ontogenetic development (as many do) without becoming numerically a new individual, then it follows that sexuality is not part of an organism's essence. No one is essentially male or essentially female.

A common objection to the more demanding criteria for individuality is that there are no essences. If essences do not exist, then retention of essence can hardly be a requirement for individuality. Although I think that the notion of essential characteristics has been repeatedly abused through the centuries, I also happen to believe that we cannot do without it. Even so, requiring that an individual be considered a new individual when it changes its essence leads to some very peculiar results. Genidentity does not always covary with retention of particular characteristics, no matter what characteristics are chosen. For example, organisms are paradigm individuals, and yet there is nothing about an organism that cannot change during the course of its ontogenic development, including material substance, overall structure and even genetic make-up. On the stronger requirement for individuality, either an entity that is changing continuously through time must be divided sequentially into a series of distinct individuals, or else one must conclude that organisms do not belong to any natural kinds whatsoever. As inconvenient as it may be, organisms that have mated successfully with members of one species can and do proceed to mate successfully with members of other species. On the stronger requirement, it follows that an organism becomes numerically a new individual when it ceases to be part of one genealogical nexus and begins to participate in another.

As a result of the preceding considerations, in this paper I have adopted the weaker, minimal requirement for individuality — genidentity. Thus, a proper name denotes the individual it does, and that is that. For the purposes of naming, nothing more is required of lineages (historical entities of all sorts) than genidentity. Gargantua was named "Gargantua", *Gorilla gorilla* was named "*Gorilla gorilla*", and Darwin's theory was named "Darwin's theory." Each of these historical entities may change very little during the course of their development, or they may change extensively. It does not matter. All that matters is that these changes be continuous and that the entity remain sufficiently cohesive in the process (given the appropriate scale). Before pursuing the implications of this perspective for conceptual historical entities, a few words must be said about general terms. For the sake of simplicity, I will limit myself just to classes of the sort that are at least candidates for the status of natural kinds.

On the traditional type-token analysis, numerous different term-tokens belong to the same term-type if and only if they "mean the same thing" and/or denote the same entities. For example, on this view, every instance of the term "evolution" is a token of the same type if it refers to the same sort of process. In natural languages, applying the type-token analysis is difficult because of the extensive ambiguity, redundancy and vagueness so characteristic of natural languages, not to mention changes in meaning. For example, in the 19th century, the term-type "evolution" meant orderly, cyclical, "programmed" change of the sort that occurs in embryological development. Because in 1859 DARWIN thought that the transmutation of species had none of these characteristics, he carefully avoided using the term "evolution" in his *Origin of Species.* However, a pre-*Origin* evolutionist, Herbert SPENCER (1852), thought that embryological development and evolutionary change were basically the same sort of process and, hence, used "evolution" for both. His usage caught on. Today most evolutionary biologists agree with Darwin about the nature of the evolutionary process but retain Spencer's term. To make matters worse, "evolution" is also used to mean change of any sort, so that we hear about the "evolution" of galaxies and tail fins in Cadillacs.

In the face of such linguistic facts of life, one can understand why philosophers have toyed with the idea of replacing natural languages with languages that are carefully and consciously constructed to avoid the near chaos of natural languages. In the early years, the task was simplified even further by the widely held assumption that the primary function of language is description. WITTGENSTEIN'S (1921) *Tractatus* is a good case in point. In this highly influential work, Wittgenstein took for granted that the primary task of language is accurate, unambiguous description of the world in which we live. Each atomic sentence refers to one and only one atomic fact. On this perspective, the traditional type-token analysis of concepts appears more than adequate, but even philosophers who favor a more wholistic, less atomic view of language do not always feel the need to question traditional type-token conceptual identity. To be sure, the inter-connections between words in a single language make translating from one language to another more difficult, but they do not raise special problems for a single language mirroring the world. Real difficulties surfaced when philosophers such as WITTGENSTEIN (1953) decided that the fundamental function of language is communication, introducing a social dimension to language. A speaker can still describe the world but only in the context of community-based language games. The reverberations of this change in perspective are still making themselves felt throughout philosophy. One impact has been on the old type-token notion of conceptual identity.

Because of continuing difficulties with the notion of "meaning," contemporary analytic philosophers have attempted to deal with general terms without recourse to it. One such suggestion looks initially like the one I propose in this paper, KRIPKE'S (1972) extremely influential notion of rigid designation. Proper names have commonly been viewed as designating their referents rigidly. KRIPKE (1972) suggested that at least some kind terms might be treated profitably in much the same way. On his analysis, the reference of a speaker's term-token is fixed by means of an initiating event, a sort of "baptism," and then transmitted in a link-on-link reference-preserving chain. It might seem that, on Kripke's view, "sameness of reference" is an historical notion because these reference-preserving chains look very much like historical entities. However, appearances are deceiving. Because reference is preserved in Kripke's chains, all term-tokens that belong to a single chain refer to the same thing in the old similarity sense. For example, if an early token referred to gold, all subsequent tokens must also refer to gold. However, if Kripke's reference-preserving chains are literally link-on-link, the opposite need not be the case. Term-tokens that "referred to the same thing" in the traditional sense might not be part of the chain that emanated from the initial intitiating event for that term-type. Several independent link-on-link chains, each with its own initiating event, are possible. If they all refer to the same thing, I suspect that Kripke would want them to be considered term-tokens of the same type. I fail to see how Kripke's sets of link-on-link chains that refer to the same thing are any advance over sets of independent tokens that all refer to the same thing, whether or not they form link-on-link reference-preserving chains.

A whole generation of philosophers have found Kripke's suggestion about rigid designation intriguing, but as always, discontent has set in, albeit not for the reason I have just presented in the preceding paragraph. For instance, KITCHER (1978, p. 182) has set out a theory of reference in which link-on-link reference chains can change their referents, and even when they do not, the ways in which the reference of term-tokens is fixed can change. An example of the first sort of change is the term-type "planet." Initially the extension of "planet" did not include Earth. After the acceptance of Copernicus's theory, it did. As an example of the way in which the mode of reference for a term-type can change, KITCHER (1982) notes that early on in the history of genetics, tokens of the term-type "gene" were applied only when Mendelian ratios were discerned. Later, the *cis-trans* test was added. KITCHER (1982) proposes to handle both sorts of changes by means of his notion of *reference potential.*

KITCHER'S (1982, p. 345) basic notion is the reference potential of a term-type for an individual speaker, the "class of events which, given the speech dispositions of the speaker, can initiate productions of the type." A derivative notion for KITCHER (1982, p. 340) is the reference potential for a community, a "compendium of the ways in which the referents of tokens of the term are fixed for members of the community." Although Kitcher (1982, p. 339) rejects "mysterious intensional entities," he does acknowledge a role for the intentions of particular language users, which he summarizes in three maxims — conformity, naturalism, and clarity. As Kitcher sees it, in most cases a language user intends to conform to the usage of his community, although in crucial cases he may not. For example, for Darwin as well as for most of his contemporaries, the term "species" referred to such things as dogs, horses and people, but Darwin's contemporaries were firmly convinced that species are immutable. If one putative species is actually descended from another putative species, then automatically they become one species. If Darwin was right, and all present-day species arose from one or a few original species, then only one or a few species actually exist, appearances notwithstanding.

The preceding example also helps to illustrate Kitcher's second maxim.

Sometimes people in general and scientists in particular intend to refer to natural kinds, even though the identifying descriptions that they are using may be abandoned by later workers. On occasion, however, one might decide to stick with the description even if it means admitting that one had not been referring to a natural kind (KITCHER 1982, p. 344). For example, most of Darwin's contemporaries defined the term "species" so that particular species have the traditional characteristics of natural kinds. Species at the time were thought to be, in the appropriate sense, eternal, immutable and discrete. On Darwin's view, particular species have none of these characteristics. *Canis familiaris* is in no sense eternal. At one time it did not exist. At some future date, it is sure to go extinct. All species are the contingent effects of natural forces and not part of the framework of the universe. Nor are species immutable. The only way that species arise is through evolution, and for Darwin, evolution was gradual. The boundaries between species are anything but discrete. Perhaps Darwin and his contemporaries intended for the names of particular species to refer to natural kinds, but it is difficult to see how anyone can reject every traditional characteristic of natural kinds for species and still claim that species are natural kinds (but see KITCHER 1984). Of course, one possibility is that the term-type "natural kind" was in the midst of changing its reference. As far as I can tell, it was not. In fact, it still has not.

KITCHER's (1982) third maxim is that, on occasion, scientists intend to refer to that which they can specify. Philosophers have not had much patience with the emphasis that scientists place on operational definitions. After all, they are not "definitions." But Kitcher sees an important role for the tests which scientists devise to help them to apply their terms. For example, if a *cis-trans* test turns out in a particular way, a geneticist is likely to produce a token of the term-type "gene." Although there is much more to a theoretically significant term-type than the methods scientists use at any one time to apply tokens of it, I do not think that it seriously misrepresents the behavior of scientists to claim that they do intend a referent in such circumstances to be whatever was causally linked to the production of this token or satisfied a particular description.

The chief problem for Kitcher, as I see it, is how his notion of reference potential can actually handle changes in reference as well as changes in mode of fixing reference. If Kitcher's linguistic communities were genuine social groups, then they could be used to integrate "diverse initiating events" so that they can refer to the same entities as scientists renew and extend the connections between their terms and the world. But KITCHER (1982, p. 346) defines linguistic communities in terms of agreement on initiating events. "With respect to a particular expression type, two speakers belong to the same linguistic community if they are disposed to count exactly the same events as initiating events for production of tokens of the type." As a result, Kitcher is forced to conclude that scientists who belong to the same research team might well belong to different "linguistic communities" with respect to certain terms and the same community with respect to others. Under such circumstances I fail to see how the "dispositions of the community to use tokens of a particular term to refer to a particular range of ways may vary through time" (KITCHER 1982, p. 340) or how terms can have "heterogeneous reference potentials" (KITCHER 1982, p. 345). If the dispositions of a community to use tokens of a term vary through time, then that is not the same community. To the extent that people accept diverse initiating events for a term, they do not belong to the same community.

Kitcher is having much the same problem as KUHN (1970, p. 176) confronted when he attempted to clarify his notion of paradigm by noting that a "paradigm is what the members of a scientific community share, *and*, conversely, a scientific community consists of men who share a paradigm." Kuhn is not unaware of the apparent circularity in the preceding quotation and proposes to avoid it by delimiting his scientific communities sociologically in terms of their professional relations. However, when scientific communities are delimited in this way, an inconvenient fact arises: members of the same scientific community are not always in agreement even over fundamentals. As strange as it may seem, scientists can cooperate with each other even when they are not in total agreement. I think that Kitcher is on the right track in his analysis of reference, but like Kuhn, I think that he must ground reference potential in sociologically delimited communities. In the rest of this paper, I propose to treat Kitcher's linguistic communities as genuine social groups. This means that not everyone who belongs to a particular linguistic community need agree totally about the initiating events for a particular term. Instead of a person belonging to as many linguistic communities as there are terms in his vocabulary, he is likely to belong to only a very few. There are only so many hours in a day. The time necessary to participate in a social group imposes an upper limit to how many social groups one can belong to.

On my reformulation of Kitcher's theory of reference, actual descent from an actual initiating event is *necessary* for two or more term-tokens to be tokens of the same term-type in the evolutionary sense of "term-type." Conceptual tokens must be organized first and formost into link-on-link chains — lineages. Conceptual lineages can merge, but *merger* is partially independent of both agreement and similarity in references. For example, in 1831 Patrick MATTHEW published his version of the principle of natural selection, DARWIN formulated his version in 1838, while A.R. WALLACE did not stumble upon natural selection until 1855. From the perspective of similarity in assertive content, all three of these conceptual tokens are tokens of the same type. Put differently, "natural selection" was initiated three times. Wallace's initiation was not totally independent of Darwin's initiation because both men had read Lyell and Malthus, Wallace had read Darwin's early work, and both men had exchanged letters on the species problem prior to Wallace's bolt from the blue. As far as I can tell, neither Darwin nor Wallace had read MATTHEW's (1831) *Naval Timbre and Arboriculture* prior to their own initiating events; see EISELEY's (1979, p. 72) contrary claim in his continuing anti-Darwin vendetta. Even though the initiating events for Darwin and Wallance were at least partially independent of each other, they belong in the same conceptual lineage because of the interplay between Darwin and Wallace after Wallace sent Darwin a copy of his species paper. Although the two men were never in total agreement about the nature and impact of natural selection, they developed their theories in consort. Matthew's initiating event remained totally outside this conceptual lineage.

From the point of view of natural selection as a conceptual lineage, Matthew's clear statement does not count as the "same" term-type as that of Darwin and Wallace. In conceptual evolution as elsewhere, similarity does not universally covary with identity by descent. In general, this means that truly unappreciated precursors do not count. They are of antiquarian interest only. "But that is not fair!" Whoever said that natural processes, including conceptual evolution, were fair? Fairness at a local level in science at least does matter. By and large scientists are given credit by their colleagues for that work that these colleagues find useful. Citations serve three main functions in science: to support the work of the scientist who is publishing the citation, to deflect blame if the work that is being cited turns out to be mistaken, and to give credit where credit is due. Scientists attempt to accrue to themselves as much credit as possible, but they also need support. They cannot gain support without giving credit. The success of this system of use and credit depends on the frequency of blatant stealing not becoming too high. Fairness to one's graduate students, fellow team members, and even distant colleagues is relevant to the ongoing process of science. Fairness to scientists who have long been dead is not (HULL 1978b).

To the extent that setting the record straight is actually operative, attempts to be fair to scientists in the past is a "misfiring" of an otherwise adaptive mechanism. In point of fact, reference to unappreciated precursors in the ongoing process of science serves quite different functions. For example, one way to counter present-day critics is to rail against the close-minded bigotry of the opponents of earlier scientists who proposed similar views. For evolution, Lamarck is always useful because he suffered so pitifully at Cuvier's hands. That Lamarck's notion of "evolution" was significantly different from that of Darwin and Wallace is conveniently neglected. Rarely are the views of unappreciated precursors all that similar to the later views that initiated the search for precursors. I do not intend to denigrate the usefulness of precursors as patron saints in the ongoing process of science, but misplaced romanticism to one side, they do not belong in conceptual lineages on which they had no influence.

The conceptual lineages initiated by Darwin and Wallace actually merged in the middle years of the 19th century. They need not have. If Wallace had become incensed by the treatment of his paper by Darwin and his friends or if Darwin had behaved much more territorially than he did, the two men might have become enemies and developed independent conceptual lineages right from the start. Scientists cooperate and compete with each other, but the character of this cooperation differs markedly within and between communities. Wallace became a (peripheral) member of the Darwinians. Other scientists who were in larger agreement with Darwin about the nature of the evolutionary process became adamant anti-Darwinians. A good example is St. George Jackson Mivart. When his efforts to work his way into the Darwinians failed, he came out as a vehement and very effective critic. In general, no simple relation exists between how much scientists agree with each other and how much they cooperate in forming scientific communities. Significant variation of opinion existed within the Darwinians. Conceptual evolution is impossible without it. But disagreement within a group is treated very differently from disagreement between groups; (for a more detailed discussion of Darwinism as an historical entity, see HULL 1985).

All of the preceding concerns descent as a necessary requirement for conceptual identity in science. I agree with GOULD (1977) that in the study of science, there is a point to distinguishing between "eternal metaphors" and conceptual lineages, but in the ongoing process of science, lineages are fundamental. Perhaps there is some role in science for everyone who happens on the "same" idea, but such a role has yet to be suggested. The only term-tokens that function in science are replicates. To count as replicate, they must be passed on. Periodically scientists also test their ideas by confronting them with the non-linguistic world. The chief vehicles for both processes (communication and description) are scientists. To the individual scientist, personal credit is the mechanism which makes the machine work. Scientists are also organized into cooperating groups of scientists primarily by means of the use that they make of each other's work. In order to have their research program assessed as being progressive, they must have achievements credited to members of their program (LAKATOS 1970). The interplay between replication and interaction with the world to make replication differential operates in conceptual as well as biological evolution.

As if treating actual replication as a necessary condition for conceptual identity were not problematic enough, treating it as sufficient is even more strongly counter-intuitive. Darwinism as a complex conceptual lineage has undergone many transformations in its long history. Darwin thought that evolution is gradual, undirected, possibly in some obscure sense progressive, and largely under the control of natural selection. I have yet to find a fellow-Darwinian who agreed with Darwin on every one of these tenets. The Darwinism that rushed in like a flood was saltative, directed, progressive, and natural selection was usually thought of as only of minor importance. Later, under the influence of August Weismann, Lamarckism was expelled from Darwinism and natural selection ruled supreme. Prior to the *Origin* Darwin thought that geographic isolation is necessary for speciation, but by 1859, he had convinced himself that it is not. Later, when Moritz WAGNER (1872) urged the importance of geographic isolation, Darwin responded hostilely. Later still, with the work of Ernst MAYR (1942), the importance of geographic isolation became one of the pillars of Darwinism. When neutralists first argued that many traits become widely distributed even though they had no adaptive significance, they claimed to be either anti-Darwinian or at very least non-Darwinian (KING and JUKES 1969), and the Darwinians agreed (AYALA 1976). Now, the selectionist and neutralist views are "competing hypotheses within the framework of the synthetic theory of evolution" (STEBBINS and AYALA 1981, p. 967). In the early years of the new synthesis, Richard GOLDSCHMIDT's (1940) saltative view of evolution was anathema to the Darwinians. Now Goldschmidt is the patron saint of those authors who want to take the "final step in the modern synthesis" (RAFF and KAUFMAN 1983, p. 24). Darwin was wrong in thinking that gradualness was part of the essence of Darwinism (GOULD 1982).

In the preceding paragraph, I have only sketched the wide variety of views that have gone under the name of "Darwinism." The issue is whether

Darwinism as an historical entity should be treated as an historical entity individuated by the minimal requirements of genidentity or whether stronger criteria should be added as well. Must Darwinism retain its "essence" in order to remain a single conceptual lineage, or can it evolve indefinitely just so long as a certain degree of internal coherence is retained and the development continuous? Most participants in the dispute over the essence of Darwinism opt for the first alternative. Darwinism definitely has an essence, but no two of these disputants can agree on the precise character of this essence. None too surprisingly, every Darwinian thinks that *his* preferred tenets are essential to Darwinism while those of his opponents are only accidental. Prior to ELDREGE and GOULD's (1972) pushing a somewhat saltative view of speciation (saltative not at the organismic level, or the level of higher taxa, but at the populational level), no one seems to have doubted that gradualism was of the essence of Darwinism. GOULD (1982) is now convinced that it is not. Darwinians have also exaggerated the extent and effect of selection as well. Nevertheless, Gould maintains that his version of evolutionary theory is carrying on in the Darwinian tradition. STEBBINS and AYALA (1981) agree. Punctualism is also embraced by the all-encompassing arms of the synthetic theory.

As historians have shown, Darwin changed his mind on numerous issues prior to the appearance of Wallace's paper forcing him to go public. Because of the largely private character of these early versions of his theory and all the attention that was paid to the *Origin of Species*, Darwin's views in 1859 have some priority in determining the essence of Darwinism, but Darwin continued to change his mind about evolution after the *Origins*, and his views as expressed in the *Origin* did not gain wide acceptance until long after his death. "Well, in spite of disagreements about every other aspect of the evolutionary process, at least every Darwinian agrees that species evolve. Any theory that does not include at least this much cannot be considered Darwinian." But according to the model of evolution suggested by ELDREDGE and GOULD (1972), species do not evolve. They come into existence rather abruptly and change very little thereafter until they go extinct. Instead of evolving themselves, species are the elements in species lineages. Species lineages are the things that evolve in a step-by-step fashion. Although there is considerable point to evolutionary biologists defining themselves as Darwinian, non-Darwinian, or anti-Darwinian, what Darwin actually said does not play much of a role in this activity. Any attempt to impose some essence on the Darwinian lineage is sure to elicit howls of indignation no matter which tenets are chosen as essential. I have no strong preferences in the matter just so long as the Darwinian lineage is recognized initially on the minimal requirements of genidentity and subsequent subdivisions do not obscure these connections.

The traits that characterize organisms change through time. Bones that were originally part of the jaws of ancient reptiles became modified through the years into the middle ear ossicles in mammals. Are they the "same" bones or "different" bones? I do not see that it matters. They are stages in a transformation series. I see no reason not to use different terms to refer to different stages in a transformation series just so long as these same terms are not used either for the entire transformation series or for characters not in the appropriate transformation series. Early structures that gave rise to present-day vertebrate eyes belong in the same transformation series with present-day vertebrate eyes even if they are not called "eyes." Similarly, one is only asking for confusion if one refers to both the entire eye transformation series and one stage in this series as "eyes." Finally, all hope is lost if cephalopod eyes are also termed "eyes." The notion of a transformation series is as fundamental to conceptual evolution as it is in biological evolution. Some biologists are willing to subdivide a gradually evolving lineage into successive chronospecies even in the absence of speciation events if the change is great enough (SIMPSON 1961). Others are not (WILEY 1981). I do not see that it matters much just so long as the criteria one uses are made explicit and names are assigned unambiguously. I see no reason not to subdivide the Darwinian lineage into Darwin's Darwinism, late 19th century Darwinism, neo-Darwinian Darwinism, the new synthesis Darwinism, etc., etc. if one is careful not to assume that these stages in conceptual evolution share some common features — because they may not.

The preceding discussion has concerned complex conceptual lineages. Because they contain so many elements, disagreements about each of the elements ramify. Although some of this complexity is decreased when one focuses on less inclusive conceptual entities, it is not eliminated. Reference potential, on my view, is determined by genuine social groups. No matter how narrowly these groups are defined, there is likely to be some disagreement not only about the referents of particular tokens of a term-type but even about proper modes of fixing reference. At times, reference may be in transition and usage highly ambiguous. That is the price one pays for a genuinely evolutionary analysis of conceptual change. One cannot insist on greater clarity than actually exists. To repeat, within-group variation is one of the chief mechanisms for evolutionary change. Defining groups to eliminate it purchases ease of expression at the cost of obscuring the processes that are producing the change. "But might not one postulate minimal conceptual atoms such that each denotes uniquely?" I think not, but even if such minimal units were plausible, within-group disagreement is still possible.

Conclusion and a consequence

By now the reader may understand the particular theory of conceptual identity that I am explicating but fail to see what difference it makes. The ramifications of this shift in perspective are numerous and fundamental. I can discuss only one here. If "Darwin's theory of evolution" is taken to refer to a set of tenets, regardless of their occasion of enunciation, then Darwinism in this sense is at least a candidate for a natural kind. One might expect it to function in a putative "law" of conceptual development. One should expect the same factors to be operative in its genesis and/or acceptance whenever a token of it arises. After all, on this view, the world is the way it is whenever the conditions are right, and Darwin's theory is Darwin's theory whenever tokens of the same term-types are expressed. On this view then, if one is an internalist, one should expect the same reasons, arguments and evidence to be operative in the genesis or at least acceptance of Darwin's theory whenever this theory arises. If the progress so appearent in the fossil record counted in favor of Darwin's theory in England, it should also count in its favor in Germany, France, and America. If not, then variations must be explained away. If one is an externalist, then one should expect the same socioeconomic factors to contribute to the genesis and/or acceptance of Darwin's theory around the world. If the competitive, individualistic character of British society led Darwin to formulate a competitive, individualistic theory of biological evolution, then only scientists living in such countries should formulate Darwin's theory. Parallel observations hold for acceptance as well. People in competitive, individualistic societies should be more prone to accept Darwin's theory.

However, historians who have studied both the genesis and acceptance of Darwin's theory have found no such correlations (GLICK 1974). Internalists found lots of relevant reasons, arguments and evidence, but they varied from country to country. Externalists found an even greater panoply of external factors, but they too varied irregularly from country to country. Externalists at least, are not deterred by these findings. They are not committed to the principle of same cause/same effect. The operative factors in scientists' formulating and/or accepting the views that they do are primarily such things as class allegiance, socioeconomic conditions, etc., but they "respond to local social and cultural conditions" (SHAPIN 1979, p. 144). In one country the rise of the mercantile middle class might contribute to the acceptance of Darwinism; in another it might contribute to its rejection. In one country conservatives might be attracted to Darwin's theory; in another liberals might find it attractive. But, no matter what these external factors turn out to be, *they* are what caused the acceptance or rejection of the theory.

As much as this view begs to be abused, as much as it looks like a game that anyone can play and no one can lose, I think that it is appropriate. It would be inappropriate for conceptual systems as natural kinds, but it is totally appropriate for conceptual systems as lineages. The only conditions that are relevant to them are *local* conditions. The failure to see that the same observations hold for internalist explanations as well has caused considerable confusion in traditional histories. In Great Britain, "idealism" was just getting a toehold when Darwin published his *Origin*, while special creation was the most recent widely accepted view. Hence, Darwin argued against special creation and ignored the idealists. According to Darwin, neither sort of explanation of species is "scientific," but he emphasized those features of his theory that made it superior to the views of, say, Adam Sedgwick, but avoided confronting the boundary between his conception of science and the "idealism" preferred by Richard Owen. Darwinism was presented with just the opposite state of affairs in Germany, where idealistic views of science reigned and special creation had never been all that popular. In Great Britain, the Darwinians played down the anti-religious implications of evolution; French Darwinians emphasized them. In Germany the apparently progressive nature of the fossil record was an important datum in favor of evolution; in Great Britain its effect was mixed because Lyell and Huxley had both argued against it. Perhaps all these intellectual variations should not have existed and should not have influenced the reception of Darwin's theory, but they did. Perhaps there is something properly termed Darwin's theory such that all possible evidence counts eternally for it or against it, but such theories have yet to play any role in science.

On traditional analyses of "explanation," one cannot explain a particular *qua* that particular but only as an *instance* of some universal. Thus, on the view I am advocating, Darwin's theory *qua* Darwin's theory cannot be explained either. It can be an instance of a natural kind, but it itself is not a natural kind. Hence, anyone who proposes to explain the genesis or acceptance of Darwin's theory must find an appropriate natural kind that it exemplifies. Such natural kinds have been extremely difficult to uncover in biological evolution; comparable natural kinds are likely to be even more difficult to discern in conceptual evolution, but at least the task is clear. The lesson that biological evolution has to teach us is that the tokens that are selected must be organized into lineages and that these lineages are not natural kinds. They are not the place to look for evolutionary regularities. The appropriate locus of evolutionary explanations is *kinds* of lineages.

One might explain the prevalence of one sort of reproduction over another in certain circumstances, the ratio of predators to prey, even the evolution of perceptual organs, but there can be no law of the vertebrate eye *qua* vertebrate eye. Similarly, one might explain the occurrence of highly innovative conceptual shifts, the rise of interfield theories, and so on, but not the evolution of Darwin's theory *qua* Darwin's theory. In the absence of relevant conceptual natural kinds and laws of conceptual change, "new knowledge can neither be predicted in advance nor explained after the fact by historians" (KOERTGE 1981, p. 19). Or to put the point differently, the only sorts of explanations possible are highly particularistic statements of the operative particular circumstances. COCK (1983, pp. 39, 57) apologizes for presenting such a "motley" and "miscellaneous collection" of reasons for William Bateson's reluctance to accept the chromosome theory, but these are the only sorts of explanations that one can present for such particulars in the absence of anything that might count as "laws" of conceptual change.

In conclusion, I cannot pretend to have made a strong case for the plausibility of a theory of conceptual evolution. However, I have shown two things. I have shown the lengths to which a proponent of such a theory must go if he hopes to be successful. I have also pointed out a systematic ambiguity in our conceptualization of conceptual change that has frustrated intelligent discussion of it — the distinction between a term-type as a set of similar tokens and conceptual lineages as causal sequences of term-tokens. The fact that a consistent treatment of notions such as "evolution" and "Darwin's theory of evolution" from either perpective seems counter-intuitive suggests that in our ordinary way of thinking, we vacillate between the two perspectives. By and large, organisms that belong to the same species are *locally* similar to each other. By and large, term-tokens that belong to the same term-lineages are *locally* similar to each other. The mistake is to extrapolate from local to global similarity.[2]

---

[2] Appreciation is owed to Philip Kitcher for reading and commenting on an early version of this paper.

References

AYALA, F.J., ed., 1976, *Molecular Evolution* (Sinauer, Sunderland, MA).

CAPLAN, A., 1980, *Have species become déclassé?* PSA 1980, eds. P.D. Asquith and R.N. Giere (Philosophy of Science Association, Ann Arbor, MI).

CAPLAN, A., 1981, *Back to class: a note on the ontology of species*, Philosophy of Science 48, pp. 130–140.

COCK, A.G., 1983, *William Bateson's rejection and eventual acceptance of chromosome theory*, Annals of Science, 40, pp. 19–60.

CODY, M.L., 1974, *Optimization in ecology*, Science 183, pp. 1156–1184.

DARWIN, C., 1859, *On the Origin of Species*, a facsimile of the first edition (Harvard Univ. Press, Cambridge, MA).

DE VOE, R.D., 1981, *Review of comparative physiology and evolution of vision in invertebrates*, in: I.H. Autrum, ed. (Springer, New York).

EISELEY, L., 1979, *Darwin and the Mysterious Mr. X* (Harcourt Brace Janovich, New York).

ELDREDGE, N. and GOULD, S.J., 1972, *Punctuated equilibra: an alternative to phyletic gradualism*, in: Models in Paleobiology, ed. T.J.M. Schopf (Freeman, Cooper and Co., San Francisco), pp. 82–115.

GHISELIN, M., 1974, *A radical solution to the species problem*, Systematic Zoology, 23, pp. 536–544.

GLICK, T.G., ed., 1974, *The Comparative Reception of Darwinism* (Univ. of Texas Press, Austin).

GOLDSCHMIDT, R., 1940, *The Material Basis of Evolution* (Yale Univ. Press, New Haven, CT).

GOULD, S.J., 1977, *Eternal metaphors of paleontology*, in: Patterns of Evolution as Illustrated by the Fossil Record, ed. A. Hallam (Elsevier, New York), pp. 1–26.

GOULD, S.J., 1982, *Darwinism and the expansion of evolutionary theory*, Science 216, pp. 380–387.

HENNIG, W., 1966, *Phylogenetic Systematics* (Univ. of Illinois Press, Chicago).

HULL, D.L., 1976, *Are species really individuals?* Systematic Zoology 25, pp. 174–191.

HULL, D.L., 1978a, *A matter of individuality*, Philosophy of Science 45, pp. 335–360.

HULL, D.L., 1978b, *Altruism in science: a sociobiological model of cooperative behaviour among scientists*, Animal Behaviour 26, pp. 685–697.

HULL, D.L., 1980, *Individuality and selection*, Annual Review of Ecology and Systematics 11, pp. 311–332.

HULL, D.L., 1981, *Kitts and Kitts and Caplan on species*, Philosophy of Science 48, pp. 141–152.

HULL, D.L., 1982, *The naked meme*, in: Learning, Development, and Culture, ed. H.C. Plotkin (Wiley, New York), pp. 273–327.

HULL, D.L., 1985, *Darwinism as an historical entity*, in: The Darwinian Heritage, ed. D. Kohn (Nova Pacifica, Wellington, New Zealand).

KING, J.L. and JUKES, T.H., 1969, *Non-Darwinian evolution*, Science 164, pp. 788–798.

KITCHER, P., 1978, *Theories, theorists and theoretical change*, Philosophical Review 87, pp. 519–547.

KITCHER, P., 1982, *Genes*, British J. Philosophy of Science 33, pp. 337–359.

KITCHER, P., 1984, *Species*, Philosophy of Science 51, pp. 308–333.

KITTS, D.B., 1983, *Can baptism alone save a species?* Systematic Zoology 32, pp. 27–33.

KITTS, D.B. and KITTS, D.J., 1979, *Biological species as natural kinds*, Philosophy of Science 46, pp. 613–622.

KOERTGE, N., 1981, *Explaining scientific discovery*, PSA 1982, eds. P.D. Asquith and T. Nickles (Philosophy of Science Association, Ann Arbor, MI), pp. 14–28.

KRIPKE, S.A., 1972, *Naming and necessity*, in: Semantics and Natural Language, eds. D. Davidson and G. Harman (D. Reidel, Dordrecht, Holland), pp. 253–355.

KUHN, T., 1970, *The Structure of Scientific Revolutions*, 2nd ed. (Univ. of Chicago Press, Chicago).

LAKATOS, I., 1970, *Falsification and the methodology of scientific research programmes*, in: Criticism and the Growth of Knowledge, eds. I. Lakatos and A. Musgrave (Cambridge, Univ. Press, Cambridge), pp. 91–196.

MATTHEW, P., 1831, *Naval Timbre and Arboriculture* (Longman and Company, London).

MAYR, E., 1942, *Systematics and the Origin of Species* (Columbia Univ. Press, New York).

RAFF, R. and KAUFMAN, T.C., 1983, *Embryos, Genes, and Evolution* (Macmillan, New York).

SHAPIN, S., 1975, *Phrenological knowledge and the social structure of early 19th-century Edinburgh*, Annals of Science 32, pp. 219–243.

SIMPSON, G.G., 1961, *Principles of Animal Taxonomy* (Columbia Univ. Press, New York).

SOBER, E., 1984, *Discussion: Sets, species, and evolution: Comment on Philip Kitcher's "Species"*, Philosophy of Science 51, pp. 334–341.

SPENCER, H., 1852, *The development hypothesis, The Leader*; reprinted in: Essay Scientific, Political, and Speculative (Appleton, New York), I, pp. 1–7.

STEBBINS, G.L. and AYALA, F.J., 1981, *Is a new evolutionary synthesis necessary?* Science 213, pp. 967–971.

WAGNER, M., 1872, *The Darwinian Theory and the Law of the Migration of Organisms* (Edward Stanford, London).

WILEY, E.O., 1981, *Phylogenetics* (Wiley, New York).

WITTGENSTEIN, L., 1921, *Tractatus Logico-Philosophicus* (Routledge and Kegan Paul, London).

WITTGENSTEIN, L., 1953, *Philosophical Investigations* (Macmillan, New York).

Barcan Marcus et al., eds., Logic, Methodology and Philosophy of Science VII

HISTORICAL SOURCES OF POPPER'S LOGIC OF SCIENCE

VADIM N. SADOVSKY

Inst. for Systems Studies, 9, Prospect 60-letija Octyabrya, Moscow, 117312, USSR

First, a brief comment on why Popper's logic of science or the logic of growth of scientific knowledge has been chosen as the topic of this paper. Prof. Popper is our contemporary and we have an opportunity of seeing him actively participating in this Congress. There are several reasons for the choice of Popper's logic and methodological ideas for historical analysis.

The first one is quite personal; for the last several years I was heavily engaged in the analysis of Popper's papers, editing a volume of his selected works for publication in Russian (cf. POPPER, 1983b). Other reasons are of more substantive nature. I believe Popper's logic and methodology of science is one of the most prominent in the Western philosophy of this century. In some respects — integrity, systematic character, consistency, popularity, etc. — it is probably superior to the contribution of other prominent Western philosophers like RUSSELL, WITTGENSTEIN, CARNAP and of the so-called historical school headed by KUHN. (It is worth noting that the Congress proceedings abound with papers for and against POPPER and only a few mention other well-known Western logicians and methodologists of science). Of course, I do not claim that Popper's methodology is true, nevertheless it is, undoubtedly, a considerable achievement of the 20th century. Hence, those living in the closing period of this century should cast a retrospective glance and try to assess its contribution to logic and methodology of science. Naturally, we shall first turn to Popper's works.

This approach proves even more valid since the study of history and of historical development of any phenomenon, including that of logical and methodological theories, is the most efficient method for understanding them. Another point is that Popper's Logic and Methodology of Science has a good 60-year history, the time of its "internal history", so to speak, which might be a clue to the perception of Popper's ideas in their integrity and unity.

That is the basic reason for my writing the paper on "Historical sources of Popper's logic of science". I think nobody doubts that as a proponent of dialectical and materialistic philosophy I will present the Marxist view on Popper's ideas and their historical evolution. I am fully aware that this view may be argued by both Popper himself and other philosophers adhering to different basic philosophical attitudes. This is not only possible but even necessary. Indeed, a constructive argument, a critical discussion, "battles with words rather than with swords" as Popper put it (POPPER, 1980, p. 396) — that is what all philosophers, Marxists and non-Marxists, from the West and from the East, should strive for.[1]

Prior to presenting the paper I would like first to stress that Popper, as is mentioned in his intellectual autobiography *Unended Quest*, received no standard philosophical education (he graduated from Vienna University in 1928 as a teacher of mathematics and physics) and, as he confessed himself, he barely passed an examination on philosophy by Schlick (POPPER, 1976, p. 78). Nevertheless, through self-education he acquired thorough knowledge in philosophy and in the history of philosophy and from his very first works on he has been a philosopher rather than, say, a logician, psychologist or naturalist. Accordingly, his views, in contrast to those of logical positivists dominant in the West during the 1920's–1950's, have throughout their evolution been so to speak "philosophically laden"; they had deep roots in philosophy and were largely philosophical in nature. It is noteworthy that Popper has never shared the neopositivist dogma on the meaninglessness of First Philosophy — Metaphysics. In all of his works he sought to substantitate the real existence of deep philosophical problems.

As regards Popper's views as a whole, we can, of course with a certain measure of conventionality, distinguish three major components thereof:

— philosophical concept;
— logical and methodological views;
— sociological and political ideas.

---

[1] This paper was presented on the opening day of the Salzburg Congress, July 11, 1983, Prof. Popper was still missing from the meeting. Later on, however, he became acquainted with the theses of the paper (SADOVSKY, 1983a) and was kind enough to comment on it first orally while in Salzburg and then in the letters to the author of August 16 and November 9, 1983 (POPPER, 1983c). I think the letters are of great interest as in them Prof. Popper named those philosophers and scholars who, he believes, had exerted the strongest influence on him. I was very grateful to Prof. Popper for the information and took it into account in writing the final version of this paper.

I think Popper's sociological and socio-political views are not organically linked with the logic of growth of scientific knowledge.[2] This component of his views is most vulnerable and, I would say, the most poorly developed part thereof. As a consequence, Popper's sociological and political ideas will not be considered in this paper. As for the two remaining components of his doctrines — philosophy and logic of science — I will treat them within the frameworks of:

— "external history", i.e. the historical sources and intellectual background of their evolution;

— "internal history", i.e. their evolution in Poppper's works.

To the best of my knowledge, Popper has never given a systematic description of the historical roots of his conceptions, though some statements to this end are scattered across his publications. Thus, in substantiating non-inductivism he usually refers to Hume; in presenting the theory of three worlds he stresses its relationship with the ideas of Plato, Hegel, Bolzano and especially Frege; in his theory of truth Popper resorts to Xenophanes and Tarski; his epistemological evolutionism is based on the theories of Darwin and Spencer, and so on. Inasmuch as Popper generally describes his basic philosophical position as realistic, it follows that his conception was to some or other extent influenced by different schools of philosophical realism. Among these he especially distinguishes Einstein's ideas. The writings of recent years note the similarity of some of Popper's views with the ideas of Bacon, Nietzsche and others. Hence, judging primarily from Popper's own statements, one obtains a rather vague picture: practically each of Popper's ideas has its predecessor and it is hardly possible to make out the core.

Our analysis, therefore, should be much more comprehensive than a mere factual record of Popper's own admissions as to who influenced his views and how. The general analytical procedure is as follows. Its key components are:

[2] There are different opinions in this respect. Thus, T. Burke, for example, writes: "There is no difficulty in showing how the characteristic features of his (Popper's) theory of knowledge determine, in essential parts, characteristic features of his political and social thought" (BURKE, 1983, p. 9). To my mind, this is a recording of, so to say, external determination which not only does not represent the problem but rather obscures the very crux of it. Popper's models of scientific knowledge, as I will ilustrate hereafter (see also SADOVSKY, 1983b), are relatively independent of their philosophical interpretations, particularly of Popper's interpretations. To a much greater extent this holds true for the models relative to Popper's socio-political views which, undoubtedly, represent some subjective preferences of his. Hence, we are safe in asserting that there is no organic linkage between Popper's scoiology and his logic of scientific knowledge.

(a) the regular reconstruction of Popper's philosophical and logical ideas in their integrity and identification of their central points.

Without such theoretical analysis we risk overlooking certain historical sources of the views concerned or at least we will lack a conceptual framework for their evaluation;

(b) Popper's own factual evidence regarding some or other influences on his views;

(c) the existing reconstructions of the historical sources of Popper's ideas, i.e. works of his critics and commentators both in the East and in the West;

(d) the general picture evolved through a unique theoretical interaction between the preceding points, naturally within the framework of the basic philosophical views of the analyst — in this case it is the Marxist philosophy.

We shall start with the first point. Despite the well-known evolution of Popper's views — from problems of demarcation and induction to the logic of scientific discovery, the theory of three worlds, epistemological and even cosmological evolutionism — many of his theories are quite consistent forming a rather explicit system, though not presented as such by Popper himself. We shall now single out the principal points of the system. Naturally, we may just reconstruct Popper's ideas. The key elements of the resulting construct are:

— realism, non-instrumentalism;

— development of logical theory of scientific (empirical) method;

— non-psychologism, anti-naturalism and non-essentialism;

— normativism based on conventionalism; scientific method as a system of methodological rules;

— non-fundamentalism, priority of conjectures and hypotheses with respect to experience and observation;

— non-inductivism, extreme deductivism;

— falsificationism as a method of assigning demarcation criterion;

— strong emphasis on problems and problem solving[3];

— criticism, critical rationalism, transition from dogmatic to critical thinking;

— fallibilism as the assumption of basic inaccuracy of human knowledge;

— skepticism as regards the feasibility of establishing the truth;

— theory of verisimilitude as approximation to the truth;

---

[3] This point was included on Prof. POPPER's advice (1983c).

— dualism of facts and norms;
— emergentism, non-reductionism;
— indeterminism;
— epistemological evolutionism attributing the growth of scientific knowledge to transition (evolution) from some to other more profound problems;
— the theory of three worlds as a means of constructing a foundation for the objective growth of scientific knowledge;
— cosmological evolutionism as an attempt to substantiate epistemological evolutionism.

We shall now focus on some points of the list for two purposes: to identify the historical sources of the points; and to assess the real significance of each one of them for the logic and methodology of science.

As for the general estimate of Popper's philosophical and logical views, it is as is well known mostly negative. Thus, J.H. Hattiangadi, a Canadian philosopher, said at the Fifth International Congress on Logic, Methodology and Philosophy of Science that "Popper's theory of science, though an inspiring and fertile methodology, has run aground on some rocks. A new approach is necessary" (HATTIANGADI, 1975, p. V-49). Similar opinion was voiced recently, if I understand them correctly, by COHEN (1980) concerning the theory of three worlds, by RUSE (1977) relative to some serious errors in Popper's philosophy of biology, and by HAACK (1979) in connection with Popper's epistemology without a knowing subject and by many other Western philosophers. The respective Marxist estimates of different aspects of Popper's logical and methodological views were in many instances given even earlier (KHABAROVA, 1968; GRYAZNOV, 1982; EVSEVICHEV, NALIOTOV, 1974; KUZINA, 1978; METLOV, 1979; NARSKY, 1981; OISERMAN, 1982; PANIN, 1981; RAKITOV, 1977; SADOVSKY, 1979; SEROV, 1975; YULINA, 1979). Few Western philosophers are familiar with them, therefore I shall focus on them.

Of course, the deficiency of Popper's philosophical, logical, and methodological ideas at large does not in the least imply the falsity of any of his statements, every idea or formulated concept. Besides, according to Marxist approach to historical and philosophical studies, the evolution of Popper's critical rationalism should not be confined to the judgements of the philosophy of dialectical materialism but embrace the interrelationship of Popper's philosophy with those philosophical views amid which and in the struggle against which it evolved and developed. Without due account of the second relation we shall not be able to correctly assess the first one.

Popper's logical and methodological ideas are, undoubtedly, the most interesting part of his views. His logic of science provides a bright illustration of the major trends, widespread in the West in the 20th century, in the philosophical theory of scientific knowledge, epistemology, theory of scientific method on the basis of anti-psychologism, wide application of the methods and tools of modern formal logic, etc. That is to say, its historical framework is made up of the entire range of preceding attempts to formulate a theory of science by some accurate scientific methods.

As compared with logical empiricism, linguistic analysis, instrumentalism, conventionalism and the historical school in methodology of science, Popper's theory of scientific method has a number of obvious advantages. Popper aimed at a logical theory of scientific method, especially in his early studies, which he referred to as empirical. This theory, he believes, thereby associating himself with the critique of psychologism in logic at the turn of the 19th century, differs from the psychological, empirical description of research activities as it is largely built with the tools of mathematical logic. This abstraction is quite acceptable since its fruitfulness is illustrated, in particular, by the research into the logic of science intensively pursued in recent years in different countries, including Marxist philsophers.

According to Popper, the theory of scientific method must be philosophical, epistemological rather than empirical since its specifically philosophical methods of construction are not confined to generalization of results of empirical, positive knowledge. Accordingly, Popper is right in criticizing positivism for its naturalism in interpretation of the essence of epistemological problems.

Thus, Popper turns out to be a more refined, shrewd philosopher of empirical trend challenging the naive empiricism of positivists in general and logical positivists in particular.

Popper is equally steadfast in opposing another extreme in perception of the theory of scientific method, notably the nature of apriorism. In many of his works Popper comes out against the implicit apriorism of instrumentalists, in particular, of Mach, Wittgenstein and Schlick, who treat the theories only as a tool for prediction, and hence as of no real epistemic significance. In this case Popper deliberately takes a reasonable and flexible stand as compared with some trends in modern philosophy. However, his views are deeply permeated by implicit conventional apriorism when he arrives at the conclusion that methodology must represent a set of methodological rules, i.e. conventions, and that his criterion of scientific knowledge demarcation from a non-scientific one is a certain convention. The idea of translating philosophical concepts into methodological rules was first advanced by T. Gomperz and this was noted by Popper himself in his *Logic of Scientific Discovery* (POPPER, 1959, p. 62). Of course, Popper is far from dogmatically accepting any methodological conventions. He points out that methodological rules can be justified only by their fruitfulness which we judge from their conclusions. Besides, such conventions lend themselves readily to criticism, as a result of which we may drop them. Yet, despite all his reservations designed to shield him from apriorism, his theory of scientific method fails to break away from it. Indeed, since his methodology of science is not based on the theory of reflection, stating that empirical, theoretical and methodological knowledge as a whole constitute images and/or copies of different facets of reality, he has no other way out except for recognizing either vulgar empiricism (Popper rejects it) or apriorism. So, Popper's methodology provides no clue to perceiving the role of conventions in scientific knowledge.

Another essential shortcoming of Popper's theory of scientific method is that he actually reduces epistemology, i.e. theory of cognition, to the logical theory of scientific knowledge. Popper himself feels the identification to be incorrect, and in his later works, in particular, in the theory of three worlds and in the conception of epistemological evolutionism, he considerably extends the purposes of epistemology and makes an attempt to produce its ontological substantiation. We shall see thereafter the extent to which he succeded in it.

Popper's non-inductivism is the most essential element of his theory. Popper believes it is he who deserves the merit for solving an important philosophical problem of induction. It seems, however, that his real contribution to this problem is not as significant. It was Hume who provided the proof of the logical untenability of inductive conclusions, and Popper agrees with him completely. In the middle of the 19th century Liebig and later Duhem expressed their negative attitude toward induction from the viewpoint of naturalists (POPPER, 1959, p. 30). In this respect Popper proceeds from their ideas. He holds that we should not adopt Hume's psychological interpretation of inductive assertions about causality (POPPER, 1963, p. 42–46) and claims, stemming from the logical and psychological criticism of induction, that there is no place for induction either in scientific activities or in logic and methodology of science.

In his non-inductive stand Popper, like with many other problems of his methodological concept, proceeds from real aspects of scientific knowledge. However, instead of determining their relationships and limitations, he makes an absolute of them. The fact that the problem of induction is not conducive to solution by purely logical methods makes Popper ignore induction altogether which runs counter to the real practice of scientific knowledge. Being quite correct in interrelating empirical and theoretical knowledge, Popper arrives at an unacceptable conclusion that experience never precedes theory which actually implies the inexplicability of the rational origin of theoretical knowledge. Still when Popper tries to do that he lets induction in his concept "through the back door" in the form of methodological rules (cf. KUZINA, 1978, p. 82–87). In other words, Popper's attempt to divide the inseparably bound induction and deduction could not succeed in general, hence many contemporary logicians and science methodologists oppose Popper's non-inductivism, and with reason.

There is yet another key principle of Popper's theory closely related to his non-inductivism, namely the principle of falsification. Accordingly, Popper's entire methodological conception is often referred to as falsificationism. Historically, the idea of falsification, as was noted by Popper himself, was first suggested by Dubislav and Frank (POPPER, 1959, p. 41). Popper treats this principle as general methodological rather than logical, hence he runs into some difficulties. The fact that a theory is disproved, given the truth of a singular statement contradicting it, was long common knowledge at least since the rise of modern natural science. As to Popper's principle of falsification, it goes well beyond the limits of this absolutely true contention and, in fact, means firstly the recognition of such a mechanism of falsification as the sole method of refuting scientific theories implying that in case a theory is refuted it must be immediately discarded and, secondly, which is more essential, rating the principle as a criterion of demarcation: only those theories may be regarded as scientific that can in principle be refuted, i.e. which are capable of being proved false. Given this global interpretation, the principle of falsification comes into conflict with reality: the scientific community has often to retain the refuted (in Popper's sense) theories until new more valid ones are developed; the very principle of falsification is not conducive to falsification, etc. Its main fault, however, is a distorted notion about the relationship between absolute and relative truths: Popper centers only on the relative truth of knowledge and, absolutizing the element of its relative falsity he is, I believe, unable to resolve the contradictions present in his theory of truth and the theory of verisimilitude.

Because of lack of space I will not analyze the faults of Popper's theory of verisimilitude, which may be historically traced back, in particular, to Peirce. Following the well-known papers of TICHY, HARRIS and MILLER of 1974, we may not use Popper's definition of this notion in our estimations of the degree of verisimilitude (POPPER, 1963, pp. 233–234). Altogether the notion as such is undoubtedly fruitful. Indicative of it is, in particular, the study of the problem of verisimilitude on the basis of the theory of distributive normal forms by Hintikka (cf. HINTIKKA, 1973; NIINILUOTO, 1978; SADOVSKY, 1979).

Using this notion of Popper's and his idea of falsification as an example, I would like to illustrate an important element of the Marxist approach to logical and methodological research. Within the framework of logic and methodology of science one must distinguish between (a) models of scientific knowledge, including the one of growth of science, and (b) philosophical interpretations of these models.

Examples of the former are provided by deductive models of science, a model of reduction, models of verification and falsification, Popper's definition of verisimilitude and the like. I think we have every reason to speak about the relative independence of the models from philosophical concepts within the frameworks of which they are used and developed (SADOVSKY, 1983b).

Behind the relative independence is the basically different nature of the compared theoretical constructs. In fact, on the one hand, the point is about the elements of essentially philosophical knowledge in the case of philosophical conceptions of scientific knowledge, and about special scientific knowledge, mostly logical and mathematical — in the case of scientific knowledge models. As a rule such models are not deduced from philosophical concepts of scientific knowledge and the area of their application does not coincide with that of the corresponding philosophical concepts. Indeed, though scientific realism and instrumentalism are almost extremities in the philosophical perspective, the proponents of these two philosophical trends actually use identical logical and methodological models (hypothetical-deductive model of scientific theory, deductive model of explanation, methods of inductive evaluation of hypotheses, etc.). The thesis concerning relative independence means that on the whole models are independent of the respective philosophical notions but the extent of independence of each model can differ and such models may serve as a tool for realizing some or other philosophical attitudes. Hence, each one of the models requires special analysis. At the same time, this thesis means that, so to speak, the "destiny" of these two most important components of the methodology of science is different. Philosophical conceptions continuously clash with one another and successively replace each other in the course of historical development. The central issue as regards the models is determination of the area and conditions of their application; as soon as this is known, the models may co-exist or replace the inadequate ones with more sophisticated models. The aforesaid means, in particular, that models may well commute from one philosophical conception to others, something which was frequently the case during the development of the methodology of science.

Making use of the introduced distinction, we may speak about the possibility of non-Popperian life of his models, say the models of falsification, verisimilitude and the like, naturally provided the general philosophical interpretation thereof by Popper is abandoned.

Let us finally touch upon some of Popper's philosophical concepts proper and, of course, in the historical dimension in which we are interested.

Popper usually refers to his philosophy as "critical rationalism", "metaphysical realism" and the like (POPPER, 1974, p. 963). In developing it, Popper wanted to avoid: firstly, naive empiricism, secondly, abstract speculations in the spirit of the German classical idealistic philosophy, and, thirdly, different forms of irrationalism.

Popper presents his critical rationalism, based on non-inductivism and falsificationism, as a theory of scientific rationality, that is a system of standards and norms of the rational growth of scientific knowledge. Its main object is scientific knowledge, its principal constraint is non-inductivism, and its functioning mechanism are the principles of falsification and criticism. Accordingly, Popper's critical rationalism turns out to be a system of methodological rules, i.e. a self-sustaining system. Hence, as Lakatos correctly noted: "the rules of the game, the methodology stand on their own feet; but these feet dangle in the air without philosophical support" (LAKATOS, 1974, p. 253). In other words, having refuted inductivism Popper, during the early period of his activities, attempted to develop a philosophy of his own — critical rationalism — doing his best to minimize the role of truly philosophical problems therein which could not but lead to emergence of basically unsolvable problems.

This situation was realized both by Popper himself and his disciples. Consequently, Popper tried to supplement critical rationalism with the "theory of three worlds", i.e. to complement epistemology with ontological discourse.

In postulating the existence of world 3, Popper tries to solve one of the cardinal philosophical problems of determining the objective nature of human knowledge. He does not accept the Marxist solution of this problem according to which the objectiveness and the true nature of human knowledge are proved and substantiated by practical activities of social man; instead he chooses the trend, in his solution of the problem, which was strongly influenced by the theory of ideas or forms by Plato and the absolute spirit of Hegel. Popper says that he "regards world 3 as being essentially the product of human mind. We [Popper adds] create the objects of world 3" (POPPER, 1976, p. 186). It follows from this important statement that in postulating the world of objective knowledge Popper rejects its objective idealistic interpretation. At the same time, his world 3 contains considerable contradictions.

In fact, Popper's world 3 included not only true but also false theories since otherwise no process of the growth of scientific knowledge could occur. In these circumstances, as was recently shown by COHEN, taking into account that world 3 also includes all inferences of scientific theories irrespective of whether they have been discovered by people or not, Popper's concept runs into considerable logical difficulties: indeed, any inference can be made from a contradictory theory, hence world 3 is simultaneously a collection of contradictory statements and irregular gigantic aggregate of everything objectively thinkable (COHEN, 1980). In both cases world 3 cannot perform the functions that Popper assigns to it.

Popper's concept in question is also marked by profound philosophical shortcomings relating primarily to the unsolvability of the ratio between subjective and objective consciousness therein. So far as, according to Popper, world 3 is autonomous, theories there have "their ideal existence even before they become the property of individual consciousness", and "the task of objective spirit is reduced to provoking realization of ideal inferences out of the spiritual material available in culture" (YULINA, 1979, p. 103). Such an approach, which ignores the interrelation between the subjective and the objective in consciousness, i.e. between world 2 and 3, I believe, is manifestly incapable of rationally explaining the origin of culture.

Popper's concept of three worlds has yet another essential deficiency: he interprets the changes in world 3 as those governed by the laws of the Darwin theory of the growth of knowledge. Accordingly, he is unable to answer the following question: what are the species and what are the individuals of epistemological evolutionism? If theories are species and individuals are the notions about these theories held by separate scholars, then, contrary to Popper's opinion, the growth of knowledge occurs in worlds 1 and 2 rather than in world 3. If the theories are individuals of species then it is impossible to show the corresponding species. Hence, it turns out that Popper's epistemological Darwinism is at best a metaphor rather than genuine science and philosophy (cf. COHEN, 1980).

The aforesaid holds true also for a broader context in which Popper has been developing epistemological evolutionism in his recent works, i.e. for the context of cosmological and metaphysical evolutionism. According to this concept the mechanism of adaptation is basically the same at all major levels of adaptation in the world — "genetic adaptation, adaptation behavior and scientific discovery" (POPPER, 1975, pp. 73, 75–76).

Experts point out that the biological foundation of Popper's views is full of factual and theoretical mistakes (cf. RUSE, 1977). This is not the main thing, however. More essential for evaluation of Popper's ideas is that he (in fact, not for the first time) failed to offer a consistent philosophical stand corresponding to the modern level of scientific knowledge.

Thus, Popper's cosmological evolutionism, like his earlier philosophical conceptions — critical rationalism and theory of three worlds — turns out to be one-sided hyperbolization of some real facets of the cognition process.

In considering Popper's philosophical views I find it important to focus on the implicit Hegelianism which undoubtedly can be found in his conceptions. This was analyzed in detail in Soviet philosophical literature, for example, by Khabarova back in the 1960's (KHABAROVA, 1969). It is also the central topic of the paper presented to the Congress by Yulina (YULINA, 1983). It is quite characteristic in this respect that in *Objective Knowledge* Popper treats his procedure of the growth of scientific knowledge as an "improvement and rationalization of Hegel's dialectical scheme" (POPPER, 1972, p. 297). Besides, Popper's very conception of theoretical ladenness of observation is largely contained in his conception of criticism of naive empiricism. I think that in judging the historical sources of Popper's ideas we should not understimate Hegel's philosophy, and through it numerous Marxian theses. As to Popper himself, he repeatedly stressed in his works the influence Marx had exerted on him. He was quite definite to this end in his letter to me of August 16, 1983; "The two philosophers who influenced me most strongly and for the longest period of time were Marx and Kant, I should say" (POPPER, 1983c).

In conclusion I would like to stress that my main questions — What lines in the development of philosophy does Popper accept and which ones have influenced him most strongly? — can be answered as follows. Popper operates within the following lines in the development of philosophy:

— logical theory of science in general and of scientific methods in particular (Leibniz, Frege, Russell, Wittgenstein, Carnap, Tarski);

— critical and dialectical modes of reasoning (Plato, Kant, Hegel, Marx);

— epistemological skepticism (Xenophanes, Hume, Kant) limited only by the objective progress of scientific knowledge (Marx, Peirce, Collingwood);

— different theories of objective knowledge (Plato, Hegel, Bolzano, Frege, Marx);

— philosophical conceptions of evolutionism (Darwin, Spencer).

Such are in my opinion the basic objective relations between Popper's philosophy and the history of development of philosophical ideas. The aforementioned philosophers and scholars have objectively exerted the greatest influence on the evolution of Popper's ideas in spite of the fact that subjectively Popper himself, more often than not, strongly opposes some of them (as is the case with Marxism and logical positivism). Scientific analysis, however, should be based on the objective state of affairs rather than on subjective likes and dislikes.

References

ACKERMANN, R.J., 1976, *The Philosophy of Karl Popper* (Amherst).

BURKE, T.E., 1983, *The Philosophy of Popper* (Manchester Univ. Press, Manchester).

COHEN, L.J., 1980, *Some comments on third world epistemology*, British J. Philosophy of Science 31 (2), pp. 175–180.

EVSEVICHEV, V.I. and NALIOTOV, I.Z., 1974, *The conception of the "Third World" in K. Popper's epistemology*, Voprosy Filosofii 10, in Russian.

GRYAZNOV, B.S., 1982, *K. Popper's philosophy of science*, in: Logic, Rationality and Creativity (Nauka Publ., Moscow), pp. 143–166, in Russian.

GRUNBAUM, A., 1976a, *Is the method of bold conjectures and attempted refutations justifiably the method of science?*, British J. Philosophy of Science 27 (2), pp. 105–136.

GRUNBAUM, A., 1976b, *Is falsifiability the touchstone of scientific rationality? Karl Popper versus inductivism*, in: R.S. Cohen, P.K. Feyerabend and M.W. Wartofsky, eds., Essays in Memory of Imre Lakatos. Boston Studies in the Philosophy of Science, vol. 39 (D. Reidel, Dordrecht, Boston), pp. 213–252.

HAACK, S., 1979, *Epistemology with a known subject*, Review of Metaphysics 33 (2), pp. 309–335.

HARRIS, J.H., 1974, *Popper's definitions of verisimilitude*, British J. Philosophy of Science 25 (2), pp. 160–166.

HATTIANGADI, J.N., 1975, *After verisimilitude*, in: 5th International Congress of Logic, Methodology and Philosophy of Science, London (Ontario), 1975, Contributed Papers (London, Canada), pp. V-49-50.

HINTIKKA, K.J., 1973, *Logic, Language-Games and Information* (Oxford Univ. Press, Oxford).

KHABAROVA, T.M., 1968, *K. Popper's conception as a turning point in the development of positivism*, in: Modern Idealistic Epistemology (Mysl Publ., Moscow), pp. 296–324, in Russian.

KUZINA, YE. B., 1978, *Anti-inductivism in Karl Popper's epistemology*, Filosofskiye Nauki 3, pp. 80–90, in Russian.

LAKATOS, I., 1974, *Popper on demarcation and induction*, in: Schilpp, A., ed., The Philosophy of Karl Popper, Part I (Open Court, La Salle), pp. 241–273.

LAKATOS, I., 1978, *The methodology of scientific research programmes*, in: Philosophical Papers, vol. I, ed. J. Worral and G. Currie (Cambridge Univ. Press, Cambridge).

METLOV, V.I., 1979, *Evolutionary approach in K. Popper's epistemology (critical analysis)*, Voprosy Filosofii 2, pp. 75–85, in Russian.

MILLER, D., 1974, *Popper's qualitative theory of verisimilitude*, British J. Philosophy of Science 25 (2), pp. 166–177.

NARSKY, I.S., 1981, *Essence and consequences of K. Popper's methodology and epistemology*, in: Critical Rationalism, Philosophy and Politics (Mysl Publ., Moscow), pp. 66–120, in Russian.

NIINILUOTO, I., 1978, *Truthlikeness: comments on recent discussion*, Synthese 38 (2), pp. 281–329.

OISERMAN, T.I., 1982, *Some problems of scientific and philosophical theory of truth*, Voprosy Filosofii 7, pp. 70–84, in Russian.

PANIN, A.V., 1981, *Dialectical Materialism and Postpositivism*, (Moscow State Univ. Press, Moscow), pp. 5–135, in Russian.

POPPER, K.R., 1959 (1980), *The Logic of Scientific Discovery* (Hutchinson, London).

POPPER, K.R., 1963 (1972), *Conjectures and Refutations. The Growth of Scientific Knowledge* (Routlege and Kegan Paul, London and Henley).

POPPER, K.R., 1972 (1979), *Objective Knowledge. An Evolutionary Approach* (Clarendon Press, Oxford).

POPPER, K.R., 1974. *Replies to my critics*, in: Schilpp, A., ed., The Philosophy of Karl Popper, Part II (Open Court, La Salle), pp. 961–1197.

POPPER, K.R., 1975, *The rationality of scientific revolutions.* in: Harré, R., ed., Problems of Scientific Revolutions: Progress and Obstacles to Progress in the Sciences (Clarendon Press, Oxford), pp. 72–101.

POPPER, K.R., 1976, *Unended Quest. An Intellectual Autobiography* (Open Court, La Salle).

POPPER, K.R., 1980, *Facts, standards and truth: a further criticism of relativism*, in: Popper, K.R., The Open Society and Its Enemies, vol. II, Addendum (Routlege and Kegan Paul, London), pp. 369–396.

POPPER, K.R., 1983a, *Realism and the Aim of Science*, From the: Post-Script to the Logic of Scientific Discovery, ed., W.W. Bartley, III (Rowman and Littlefield, Totowa, NJ).

POPPER, K.R., 1983b, *Logika i Rost Nauchnogo Znaniya* (Logic and the Growth of Scientific Knowledge), Selected Works, Translation into Russian, ed. Sadovsky, V.N. (Progress Publ., Moscow).

POPPER, K.R., 1983c, Letters to Prof. V. Sadovsky dated August 16 and November 9, 1983.

RAKITOV, A.I., 1977, *Philosophical Problems of Science. Systems Approach* (Mysl Publ., Moscow), in Russian.

RUSE, M., 1977, *Karl Popper's philosophy of biology*, Philosophy of Science 44 (4), pp. 638–661.

SADOVSKY, V.N., 1979, *Verisimilitude of scientific theories: logical and methodological analysis*, Voprosy Filosofii 9, pp. 97–110, in Russian.

SADOVSKY, V.N., 1983a, *Historical sources of Popper's theory of scientific growth*, in: 7th International Congress of Logic, Methodology and Philosophy of Science. Salzburg, July 11th–16th, 1983, Additional Abstracts (Salzburg, June 1983), pp. 1–4.

SADOVSKY, V.N., 1983b, *The models of scientific knowledge and their philosophical interpretations*, Voprosy Filosofii 6, pp. 38–48, in Russian.

SEROV, YU.N., 1975, *K. Popper's conception of conjectural knowledge*, in: Positivism and Science (Nauka Publ., Moscow), in Russian.

TICHÝ, P., 1974, *On Popper's definitions of verisimilitude*, British J. Philosophy of Science 25 (2), pp. 155–160.

YULINA, N.S., 1979, *K. Popper's emergentist realism against reductionist materialism*, Voprosy Filosofii 6, pp. 96–107, in Russian.

YULINA, N.S., 1983, *On Popper's implicit Hegelianism*, in: 7th International Congress of Logic, Methodology and Philosophy of Science, Salzburg, July 11th–16th, 1983, vol. 3, Abstracts of Section 6 (Salzburg), pp. 285–288.

14. Fundamental Principles of the Ethics of Science

Barcan Marcus et al., eds., Logic, Methodology and Philosophy of Science VII

THE ETHICS OF CLINICAL EXPERIMENTATION ON HUMAN CHILDREN

RICHARD M. HARE

Corpus Christi College, Univ. of Oxford, Oxford, England

Introduction

This paper is the result of work done for a working group on the subject under the chairmanship of Professor Gordon Dunstan, sponsored by the Society for the Study of Medical Ethics of London, funded by the Leverhulme Foundation, and hosted by the Ciba Foundation. I am grateful to all these and to my colleagues of the working group; but the opinions are my own. In the introduction to our report you will find something about the history of the issue since it was raised in an acute form by the crimes of the Nazis, and about the various declarations and reports that there have been on it. These are for the most part both vague and inconsistent with one another, and lack any fully reasoned ground for their affirmations. So the picture is still far from clear. I would make an exception of the recent U.S. report on the subject, which is clear and sensible enough, though it still lacks the philosophical discussion, which is needed to make its conclusions secure, and which it is my purpose in this paper to begin to supply. The problem is posed by the fact that medical research into diseases affecting children cannot progress without experimentation, but this may involve physical and other interventions which have been thought to contravene the children's rights. My paper uses this question to illustrate the methodology of medical ethics. I shall argue for a combination of utilitarianism and deontology, thus showing, I hope, that so far from being irreconcilable opponents, these positions are each, when properly formulated, correct accounts of our moral thinking at different levels; once the levels are distinguished, the conflict between them disappears. The distinction between levels of moral thinking (which I did not invent; it goes back to Plato and Socrates) is explained more fully in my recent book *Moral Thinking.*[1]

The utilitarian approach

In much of the literature, a lot is said about calculations of the risks and the benefits attached to particular research projects. The idea is that if the expectation of benefit (more exactly, the product of the amount of the benefit and its probability) exceeds the expectation of harm, then the research is justified. The benefits and harms may be to the individual patient, or to the public through the advance of medical science in general or of knowledge of the patient's disorder in particular; and many people think that it makes a difference whether it is the patient or somebody else who benefits or is harmed. This kind of calculation is often called risk-benefit or cost-benefit analysis. It is used in many fields, but is, as we shall see, far from being universally accepted as a safe way of making decisions having a moral component. But it has a natural appeal to our benevolent feelings: what could be more obvious, it might be said, than that we ought, when engaging in any activity involving risk to people, to balance the risks against the likelihood of benefit, not incurring any risk that is not justified by the benefits to be expected?

This is in accord with a simple version (too simple to be acceptable) of the philosophical position known as utilitarianism. Not all risk-benefit analyses, however, are sanctioned by utilitarianism; they have in addition to be impartial: that is, risks and benefits to one individual must be weighed equally with risks and benefits to another. This impartiality secures the doctrine against one possible objection but lays it open to another. On the one hand, indictments of it for supporting the atrocities of, for example, the Vietnam War, though common, are unjust; for the cost-benefit analysis made by strategists in that war were not impartially directed to the good of the Americans and Vietnamese alike, but to an American victory. But on the other hand it is often alleged against utilitarians that they bid us ignore certain duties that we have to particular people (for example a doctor to his patient); he ought not, according to this too simple version of utilitarianism, to give any weight to the fact that a certain subject is *his* patient, but

[1] R.M. HARE, *Moral Thinking: Its Levels, Method and Point* (Oxford Univ. Press, Oxford, 1981).

should treat benefits and harms to him on a par with those to anybody else. Taken strictly, this would make the distinction between what are called 'therapeutic' and 'non-therapeutic' research, on which so much weight has been placed in the literature, of much less relevance than has been claimed; for the fact that benefits are expected for this patient, the subject of the research, would have no more and no less importance than equal expected benefits to somebody else in the future who would be cured as a result of the research. Yet Paul Ramsey, for example, has held that non-therapeutic research on children (i.e. interventions which could not help cure the patient, but only help advance knowledge) ought to be completely ruled out as an infringement of the subject's rights, given that a child cannot give informed consent to such interventions.[2]

I shall ask in a moment whether such criticisms of utilitarianism can be sustained, or whether it can be modified to escape them. They are only one example of a kind of criticism of utilitarianism which has been extremely common and has seemed conclusive to many. This proceeds by bringing utilitarianism into conflict with common moral convictions, adducing cases in which a utilitarian, seeking to maximize expected utility, would have not merely to condone, but to prescribe actions which most people would agree to be manifestly wrong. Most such criticisms rely on appeals to the *rights* of people not to have certain things done to them. This notion has now to be examined.

Rights and deontology

Appeals to rights are commonly made by a school of philosophers which is generally thought of as at the opposite pole to utilitarianism, namely the so-called deontological school. Often philosophers of this school claim Immanuel Kant as their father; but this is questionable, because a more careful examination of the doctrines of Kant and of utilitarians like Jeremy Bentham and John Stuart Mill would reveal that they have much in common, as indeed Mill recognized.[3] The deontologists insist that there are certain duties, including duties to respect rights of other people, which are incumbent on us regardless of what good or bad consequences are foreseen.

---

[2] Paul Ramsey, *The Patient as Person* (Yale Univ. Press, New Haven, CT, 1970), p. 17.

[3] For refs. see my *Moral Thinking*, p. 4.

In our present question, it is obvious that appeals to rights are going to play an important part. One of the most important of the rights appealed to is the right not to be subjected to physical interventions without one's informed consent. Another, the right to privacy, is held to preclude even behavioural observations or the passing on to researchers of medical details about patients. Another is the right to fair and equal treatment, such as is said to be denied if in a controlled experiment some subjects are given a therapy but others are given placebos. There has seemed to be a tension between utilitarian thinking, with its risk-benefit analyses and its stress on the consequences of actions and on impartiality between the interests of all, and deontological thinking, with its appeals to rights irrespective of consequences, and to duties to particular people which are not owed to others. It is therefore hardly surprising that in most of the literature, and in the discussion of particular projects, these two kinds of consideration are both invoked, often without any clear idea of what the role of each should be in our decision-making, or of what to do when they conflict, as they often do.

Pluralism and the 'trumps' theory

One possible solution is that advocated by a position best described as pluralism, though often it is known, misleadingly, as intuitionism. The term 'intuitionism' is best reserved for any position, whether monistic or pluralistic, which regards intuitions or moral convictions as the ultimate court of appeal in moral thinking, the basic data from which all reasoning has to start, or against which all proposed moral 'theories' have to be tested. Pluralism (or more accurately pluralistic intuitionism) is the view that there can be many such intuitions, yielding different and possibly conflicting principles, some of them about rights. The pluralist is in a difficulty when it comes to resolving such conflicts; usually he will recommend a further appeal to intuition to decide the conflict in the individual case. Since it is difficult individual cases, in which different principles, all of them very important, conflict, that create most of the problems in this field, it is not likely that such a pluralistic solution is going to be of much help to us; just when we need help, we shall be told to consult our intuitions, and shall find them utterly uncertain.

If we are unsatisfied with a pluralistic solution, is there any other way of reconciling the claims of utility and rights? A suggestion has been made that we should operate as utilitarians in the main, but should treat certain entrenched rights as side-constraints on our utilitarian calculations. That is to say, if to do the best for all considered impartially would involve infringing one of these rights, the right is to prevail. Rights, as it has been put, are to be 'trumps'.[4] This, it is suggested, is why we are not to commit murder however great the balance of benefits over costs. Although in that particular case the suggestion is plausible, it does not overcome all the difficulties. It does not tell us how to decide *which* are the rights which ought to be so entrenched; and it does not allow us to say, as we may often want to say, that some rights can be overridden when *great* disasters will ensue if they are respected (as when we avert plagues by imposing quarantine rules which are undoubtedly infringements of liberty), but that lesser benefits will not justify their infringement. This difficulty is especially pressing in our present field; for often an experiment which everybody would think justified if the benefits to medical science were enormous would be thought unjustified if they were only small. On the 'trumps' theory there could be no difference between these two cases.

Levels of moral thinking

A more sophisticated suggestion relies on the separation of moral thinking into at least two levels. At the practical level at which most of us operate for most of the time, what the deontologists and pluralists and intuitionists say is largely correct. We do have a plurality of intuitive principles, many of them concerned with people's rights; and we do treat these principles as having great authority (religious thinkers like Joseph Butler say that they are the precepts of God revealed to our consciences); we react instantly and with strong repugnance to proposed breaches of them by ourselves or by others; for practical purposes we treat many of them as sacrosanct. And it is good that we think like this, because if we allowed ourselves to carry out elaborate risk-benefit analyses on particular occasions, we should nearly always get them wrong, either from a human inability to predict the consequences of our actions, or from an equally human tendency to deceive ourselves about the balance of harms and benefits likely to result. A researcher may convince himself that he will revolutionize the treatement of cancer by carrying out some questionable

[4] See R.M. DWORKIN, *Taking Rights Seriously* (Harvard Univ. Press, Cambridge, MA, 1977), p. xv.

experiment, when his more judicious and less involved colleagues could tell him that the risks were much greater than he thinks and the likelihood of a major breakthrough very small.

However, though this seems a good account of most of our moral thinking at the practical level, it may be suggested that it is not adequate by itself. This is because intuitive thinking is, as we have seen, not self-sustaining. We need a way of deciding what principles we should entrench, what intuitions we should cultivate, and what to do when they conflict. When people are training to be doctors, or becoming researchers, what ought their seniors to say to them? And how are they to know whether their seniors are giving the best advice, when so much in the field of medicine is changing that different principles may be appropriate to new conditions? To cope with this problem, a higher level of thinking is needed, by which we can criticize the principles, and adjudicate between them in particular conflicts. This higher level of thinking, it is suggested, is utilitarian. We should have, and teach, and cultivate in ourselves those intuitions and those intuitive principles whose general acceptance in the profession and outside it will do the best, all in all, for those affected, considered impartially. And the advocates of this view will often go on to say that nearly all the traditional principles and rights could be justified, in general, by that kind of thinking. It *is* for the best, for example, that doctors should acknowledge special duties to their own patients, including the duty to respect their privacy and bodily integrity, unless they consent to invasions of them. If people, and doctors in particular, were not brought up to respect these principles, much more harm than good would result.

It will however sometimes happen that sound general principles conflict with one another in particular cases. Our present topic provides obvious examples. Most people would agree that research leading to the advancement of medical knowledge ought not to be hampered. And most people would agree that patients and other subjects have a right not to be experimented on without their consent. To say that most people would agree is not to prove anything; these intuitions might be like the intuitions that everybody had a short time ago that husbands had a right to the obedience of their wives. However, deontologists will try, with greater or less success, to derive these convictions from more general principles which they think are self-evident. Those, like myself, who are not content with this sort of argument are likely to seek a utilitarian justification for the acceptance of the principles. They can be justified by the obvious utility of accepting them. We do not need much imagination to see the harm that would come of allowing researchers (or for that matter surgeons seeking organs for transplants) to shanghai people against their will and cut them up; confidence in the medical profession would be impaired, to put it mildly. And yet the benefits of improved medical knowledge are immense.

Those who are charged with approving research projects involving human subjects, like the hospital ethical committees we have in Britain, or the institutional review boards they have in the U.S., are thus in a dilemma. If they bless a project, patients' rights will be infringed; but if they do not, important and useful research may be inhibited. Such dilemmas will confront us even when the proposed research subjects are adults; with children, as we shall see, there are further complications.

To some extent these problems can be coped with in advance by adopting general guidelines which attempt to hold a balance between the principles. The working group for which I originally wrote this paper will be proposing some guidelines of this sort; since we have not finished our discussions I cannot tell you exactly what they will be, but I will give you a general idea. We shall assume, I am sure, that the requirement of consent for experimentation should be treated as absolute in the case of adult subjects. I have given the reason for this already, namely the harm that would come from any weakening of the principle. And in extending the principle to children, we are likely to propose that proxy consent by parents or guardians should be allowable, subject to certain safeguards and to severe restrictions as to the extent of the risk to which children may be thus submitted. In determining the precise degree of allowable risk, those of different philosophical persuasions will as before proceed differently. Someone who thinks that our common convictions need no justification will appeal to general consensus, among those who have reflected on the problem, that that is the limit of acceptable risk. They may say, for example, that to take small blood samples by pricking children's fingers, in order to determine the amount of lead in their blood in the course of research into the problem of atmospheric lead pollution, is all right if the children themselves do not object, and that this is obvious, but that taking a renal biopsy from a child who has nothing wrong with its kidneys but is having an appendectomy is obviously not all right. But if we cannot find any such consensus, or do not regard mere consensus as enough (and this would be my own position) we shall seek deeper foundations for the same practical prescriptions by pointing out the evil consequences that would result if the limits were put higher, or lower, than we propose. If they are put lower, then research which could be of value will be stopped in order to preserve rights of children whose preservation, *in those cases*, would make a negligible difference to the children's welfare or safety. I should myself put the lead-pollution-research example just mentioned in this category. On the other hand, if the limits were put higher, then there would be the danger that potentially quite harmful experiments would be allowed just because researchers with gleams in their eyes had prevailed on parents who did not fully understand what was being proposed, or upon gullible or perfunctory research ethical committees.

But even the best general guidelines cannot absolve us from the careful scrutiny of individual cases. This is because the guidelines, being general, have to be to some degree vague. For example, we may use the expressions 'negligible risk' and 'minimal risk', as many reports on this subject do, and we may try to give some idea in statistical terms of what these expressions mean; but obviously individual cases do not carry on their faces a numerical quantification of the risk involved; we have to judge this in the light of common sense and experience, and, if we have it, of our scientific knowledge. Likewise, the probability that useful results will arise from the research, and how useful, has to be assessed by similar means.

Such an approach helps with a crucial problem already mentioned, that of 'proxy consent'. If the reason for insisting on consent is to protect the interests of the subjects and preserve them from harm, while at the same time allowing experiments to be carried out which do not seriously infringe these interests; and if we do need sometimes to obtain data about children in order to advance our knowledge for the benefit of other children, we have to ask how these often conflicting aims can best be achieved in cases where a child (for example a neonate) cannot be asked for consent. Once we see the *object* of acknowledging the right in the case of adults, we can better see what is to be done to achieve the same object in the case of children. Since the object in both cases is to protect the interest of the subject, a common procedure is to ask someone who may be presumed to have the child's interest at heart to give or withhold consent on the child's behalf, protecting his interest in the same way as an adult would his own; but not withholding consent unreasonably, because it was also our purpose to allow valuable research to be done when this did not seriously affect the interest of the subject.

It is generally agreed that such consent should not be at the discretion of anybody directly involved in the research, because he may allow his own ambitions to tip the scales, and so give consent when this is not for the best. Often parents are cast in the role of proxy consenter; because they are almost universally regarded as the trustees of their children's interests, it is natural to make them responsible for their protection in this area too. But in some cases there can be an objection to this, if the interests of parent and child diverge (suppose, for example, that an indigent parent were offered a substantial sum of money to hire out his child as a research subject). For this reason it has sometimes been suggested that a special officer should be appointed to protect the child, with a status somewhat like that of a guardian *ad litem*. And in certain cases appeal might even be made to the courts. In other cases the scrutiny of an institutional review board, containing lay members, is thought sufficient.

The problem can be solved once we look at the *purpose* of according rights; and this crucial problem about consent illustrates very well the procedure we should follow elsewhere too: of looking always to the good or ill that comes from the acceptance of certain guidelines and restrictions. If a procedure or a principle is established in this way as sound, it should be safeguarded in professional practice and if necessary by law; but because the law, as Aristotle said, 'speaks in general terms',[5] there is a need for wise particular decisions and for sufficient latitude to make them, in the light of the purposes for which the laws and principles and procedures were devised.

---

[5] Aristotle, *Nicomachean Ethics* 1137b, 14–25.

Barcan Marcus et al., eds., Logic, Methodology and Philosophy of Science VII

EXPERIMENTATION ON CHILDREN: WIDENING THE CONTEXT

Comments on R.M. Hare's paper, "The Ethics of Clinical Experimentation on Human Children"

KNUT E. TRANØY

Dept. of Philosophy, Univ. of Oslo, P.O. Box 1024, Blindern, Oslo 3, Norway

1. My main problem with Professor Hare's paper is that I agree with practically all of it. It would be boring and unproductive if I were to state my agreement point by point. The remaining divergencies would not be significant or interesting enough to justify such a procedure. What I shall therefore try to do is to supplement Professor Hare's presentation by widening the context.

I shall do this, first, by bringing into focus the distinction between two main areas of medical ethics: clinical medical ethics (the ethics of clinical medical practice), and the ethics of biomedical research, a distinction which is presupposed in Professor Hare's paper. Contemporary clinical medicine can be described as a science-based practice, i.e. a practice which presupposes theory gained by scientific research. But as far as medical ethics is concerned, it is rather the other way round. Biomedical research ethics presupposes the ethics of clinical medical practice which, in turn, has its foundations in our common and shared morality.

One conclusion to be drawn from this is that although, in some sense, medical ethics is specific and different from "ordinary ethics" — the commonly shared morality prevailing in a given society — neither branch of medical ethics is autonomous or independent of general morality. The issue of clinical experimentation on children is well suited to illustrate this. To these points I shall return later. Before I go on, however, there is one general comment I want to make.

Professor Hare's paper is held in a general and abstract vein in the sense that it contains few references to case material. The same will have to be said about my own contribution. I wish it had been easier than in fact it is to avoid this philosophical onesidedness. When dealing with problems of medical ethics in cooperation with physicians it is part of a natural division of labour that the physicians deliver the case materials. They supply the essential facts from which ethical reflection takes off. We have no physician, no pediatrician, on the panel.

There is, however one basic physiological or biomedical fact about children which is so fundamental that I think it should be stated even if by a philosopher. This is the fact that children *are not adults in miniature.* If children were simply small-scale adults, the problem set for this symposium would hardly have arisen at all. For then results obtained by research on consenting and informed adults would have been convertible to suit children simply by a change of scale. But this is not so.

Therefore there is no substitute for research on children. So the choice we face is that between conducting experiments on children — e.g. when trying out new drugs —, or simply foregoing the knowledge which only controlled experiments on children can give.

There is another basic "moral" fact about children which a philosopher may well venture to state without the support of medical authority. This is the exceptionally powerful *moral appeal* which children possess, and quite especially when they are struck by illness and suffering. The relevance of this for our topic is too clear to be in need of further comment. This does mean that there are strong *moral* objections against voluntarily foregoing knowledge which could be used to prevent or cure illness and suffering among children. So, although informed consent from subjects of experimentation is now the order of the day for adults, it is not necessarily so in the case of children.

Now let me return to the more philosophical lines of my argument. I mentioned the distinction between two types of medical ethics — clinical ethics and research ethics. A variety of classical problems in moral philosophy reappear in both branches of medical ethics. One is the problem of deontological ethics versus utilitarianism, and, as part and parcel of this, the issue of intuitionism in ethics.

It is, then, only natural that these issues receive a great deal of attention in Professor Hare's paper, and I shall follow the lead. But in so doing, let me also make the following statement. I happen to believe that general *theoretical* ethics has something to learn from *practical*, or applied ethics. Perhaps in ethics the theoretical importance of practice is greater than the practical importance of theory.

2. Then let me elaborate on the suggestion that in ethics the theoretical importance of practice may be greater than the practical importance of theory.

By moral theory I now mean academic moral philosophy. For me, here and now, the outstanding example of practical ethics is medical ethics. It does *not* seem to me to be obvious that moral theory should have much to contribute to practical ethics. What I seem to have experienced, however, is that practical ethics may have much of importance to contribute to theoretical ethics. The experience I refer to is cooperation with medical people, clinicians and researchers, in the field of medical ethics. The benefit which I, as a moral theorist, have derived from this is of the nature of a reality test. It is impossible to be in doubt about the very real importance of many of the problems which confront us in medical ethics. Small children struck by leukemia or cancer are apposite examples. With regard to leukemia in children, great progress has been made in recent years. With regard to cancer much is still unknown. Is it ethically defensible at all *not* to try to supply the lacking knowledge?

I think it is very good for a moral philosopher to have the experience of something which is beyond doubt important, real and of moral relevance, and not only to the moral philosopher, but to others as well.

The situation I thus describe is not altogether unlike some encounters between working scientists and philosophers of science. Scientists can be in doubt about the reality contact and the importance of philosophy of science. Philosophers of science cannot possibly doubt the reality and importance of science.

Before I go on, let me now try to forestall the misunderstanding that I should consider moral philosophers useless in medical ethics. In fact, it is when cooperating with medical people that I have felt — for the first time in my philosophical life — that I have been of reasonable use to others. But it is by no means self-evident precisely wherein this usefulness of moral philosophers consists in such a joint interdisciplinary enterprise as medical ethics is. I shall try to supply some examples as I go along.

3. I agree with Professor Hare that the distinction between deontological and utilitarian moral theory and argumentation is of importance in medical ethics, and certainly when it comes to experimentation on children. My own experience also tells me that it is one of the fairly few technicalities of the moral theorist which people in the medical profession find useful and intelligible. But, to me, as a moral philosopher, the distinction itself has become less clear than it used to be.

In academic moral philosophy the matter has sometimes been discussed as though one had to choose one or the other side of the distinction. Either you are a deontologist, or you are a utilitarian (or, possibly, a "consequentialist" of some other sort). Alternatively, the question may be given the form of which is the more basic of the two.

Now, part of the trouble here may be that there are many different ways of spelling out the distinction more concretely. I do not think we need get involved in that aspect of the matter here. I shall assume, however, that the deontological alternative involves such things as *norms:* rules, rights and obligations, while the consequentialist alternative involves *values*, of one or more kinds, as well as judgements of good and bad, especially as applied to the consequences of various actions open to us.

However, I now find it difficult or impossible to make good sense of the idea that we have to choose between the two. I think that for me, this is an effect of my work in medical ethics. It is difficult to imagine a medical ethics which did not need *both:* basic, normative rules or principles *and* a systematic consideration and evaluation of consequences. This is one point where Hare and I agree completely.

It is, indeed, tempting to say that medical ethics must contain a considerable diversity of values which are not always compatible. What I have in mind are the familiar dilemmas: having to choose between prolonging life and alleviating pain; or between offending the integrity of a patient and furthering his/her health; or between the health and welfare of *this* single individual and the health and welfare of many other present and future individuals. The latter dilemma is obviously involved in experimentation on children.

On the other hand medical ethics can supply very good examples of normative principles which seem to be in some sense basic and self-sufficient. It may suffice to mention the notions of human worth and dignity and the respect such moral qualities command (in our culture, today). But there are others, as well, and it may be one of the functions of the moral philosopher to elicit and articulate such normative principles. Let me mention 3 examples — where I feel responsible for the formulations but not for the principles.

(1) "Those who have to live with the consequences of a decision, should also have the right to make it (or at least to influence it heavily)". We know that this principle has been relevant in the case of abortion. But its relevance for our issue is that it lends support to the argument that parents should be allowed to give proxy consent for their children.

(2) "No one else should make morally relevant decisions for adult and informed persons". This is one of the basic props of the case against

paternalism. Its relevance — if any — for our case is again to strengthen the position of the parents as decision makers on behalf of their children.

(3) "Adult and normal persons have the right to decide about the *access* to and the *use* of information about themselves — and information about those for whom they go proxy".

The relevance of this for my paper is twofold. In the case of children who can not themselves usefully receive relevant medical information, nor give free consent, it would in general be their parents to whom these rights are transferred. And, secondly, if I have a general right to deny others access to information about me, then this can be shown to be a normative principle which to a certain extent seems quite independent of consequentialist considerations.

The kind of conclusion I am now tempted to draw for moral philosophy is this: of course we must have both, although they are not reducible to each other or, rather, because they are not so reducible. (It does seem to me that they are not so reducible. In any case the disagreement over an alleged reducibility is so great that it would be damaging to the general usefulness of a reducibility principle — a fact about moral philosophers which is not unimportant when they enter into cooperation with the medical professions.) This — underlining the need for deontological as well as utilitarian considerations — may be an indirect way of saying that we should think of medical ethics as a *normative system*. By that I mean a finite and ordered set of norms and values. Sometimes we can increase the amount of order in such a system by using carefully chosen sub-sets of norms to reduce the conflict and tension between the more or less incompatible values of the system. The Declaration of Helsinki seems to me to be an excellent example of this in the area of biomedical research ethics.

A consideration of this very declaration seems to invite another conclusion as well. Obviously the growth of knowledge — increasing the truth content — is an important value in science. But — with a reference to Wednesday's symposium (Niiniluoto, Sneed, van Fraassen) on the structure of theories — truth *tout court* could not possibly be the only value, or *the* aim, of science. It must be a question of maximizing or optimizing fruitful, or interesting, or desirable, or informative truth. Contemporary biomedical research ethics, as expressed in the Declaration of Helsinki, does subordinate the growth of knowledge to other values, for instance therapeutic utility combined with the protection of personal integrity, and in so doing it is in accord with this principle. Obviously, moreover, this is a very important aspect in experimentation on children.

By choice to adopt one rule rather than another is a kind of legislation. Since it is always possible, and perhaps even reasonable and desirable, to evaluate the consequences of adopting alternative moral rules and principles (in Professor Hare's words: "looking always to the good or ill that comes from the acceptance of certain procedures and principles"), this seems to be at least an argument in favor of *rule utilitarianism* in medical ethics, and perhaps even in favor of consequentialism — of some kind.

4. Two things impress me particularly in connection with the issue we are discussing (experimentation on children). One is how our moral reactions tend to *change* as our experience deepens and our knowledge increases. A striking illustration of this seems to me to be provided by the history of biomedical research ethics over the last 50 years — from the 1930's through the war and the subsequent Nuremberg trials, and through the two versions of the Declaration of Helsinki (1964 and 1975).

I think we have to take it for a fact that our moral reactions *will* change in consequence of changing circumstances, and sometimes surprisingly fast and much. Modern medical ethics may be an example of this. Moreover, I think that some such changes are desirable, but that not all of them are. So it seems desirable that we should be able to *control* such changes to some extent.

I am not now thinking of a "research police" or a new system of courts. But I am thinking of at least three different instantiations of the idea of such control: one is the IRBs (Institutional Review Boards) of the United States in particular. The other is the advisory ethical committees for biomedical research, set up under the Declaration of Helsinki. The third is the less formal but public forums for consideration and discussion of such issues. The present symposium may well serve to illustrate the third category.

The situations for which we need research ethical rules are usually very complex, both factually and morally. It is just not possible to encompass such situations — for instance, the issue of clinical experimentation on children — in one moral swoop. To clarify one's view of such matters requires a different kind of *process:* a deliberate cultivation and development of moral experience.

"Reflective equilibrium" is a term which has now gained a certain currency in moral philosophy, thanks to John Rawls. (There is no explicit reference to this in Hare's paper, but perhaps he is not so far from it in what he says bout pluralistic intuitionism.)

The notion of a reflective equilibrium can be made sense of in two ways: as applied to the individual, and as applied to a social group. Rawls himself is perhaps mainly thinking of the individual. What I have in mind is the application of this kind of notion to a social context.

In the first place I want here to recall two of the points made above. One is the dependence relations between biomedical research ethics, clinical medical ethics, and our common and shared "ordinary" morality. The other is the idea of regarding medical ethics as a normative system — a finite and ordered set of norms and values — which is subject to controlled change.

The emergence and the history of biomedical research ethics can perhaps best be understood as the product of something which might be called a socially dynamic reflective equilibrium (or: a dynamic social consensus formation.) Characteristic principles of contemporary medical ethics, many of them cited or referred to in Professor Hare's paper and in mine, appear to have emerged in consequence of a process of this kind. In its early stages the process was restricted to the medical profession and its organizations. Gradually it took the form of an interaction between the medical profession and other more or less well defined groups of experts: lawyers, theologians, philosophers, and more recently sociologists, psychologists and others.

Two milestones along this road are the Nuremberg trials (1946–1948) and the Declaration of Helsinki (1964 and 1975). These events deserve to be called milestones not because they established certain medical ethical "results", but for other reasons. They were both new departures; they were breaks with the past negotiated under considerable dissent and uncertainty. And they have proved to be powerful incentives to a continued questioning and public discussion.

A major feature of this development is precisely that it has taken place in public. In the course of the last couple of decenniums this has brought medical ethical issues of all kinds into the public arena in an unprecedented manner. The contrast is striking between the present situation and that prevailing until not so long ago — when medical ethics was a legitimate concern for physicians only. And, on the whole, it seems difficult today to find good arguments to deplore this development, in spite of the often obviously disturbing effects of unrelenting and full publicity.

In conclusion let me add that I think the admission of problems of biomedical research ethics to the agenda of these international congresses is to be welcomed as a desirable contribution to the kind of process I have tried to describe.

Section 14 of this congress on (i.a.) methodology is devoted to the fundamental principles of the ethics of science. What *our* topic may suggest is that some such fundamental principles are presupposed by, or in, *any* set of sound and acceptable methodological rules. If *truth* can not be a value, or an aim, of science unless qualified by some adjective — as fruitful, or interesting, or informative, or relevant truth — that in itself is an argument for such a view. This is particularly easy to see in medicine and biomedical research, as summarized in the statement that the fundamental principles of medical research ethics presuppose the ethics of clinical medical practice —, which in turn presupposes our shared and ordinary general morality.

Barcan Marcus et al., eds., Logic, Methodology and Philosophy of Science VII

SCIENTIFIC AND ETHICAL RATIONALITY

JEAN LADRIÈRE

Inst. Supérieur de Philosophie, Univ. de Louvain, Louvain-la-Neuve, Belgium

The aim of this symposium is to examine if there is a rationality in ethics as there is evidently one in science, to compare, if this is the case, those two types of rationality and to investigate their relationships.

In order to meet the question, we should, at best, make appeal first to a concept of rationality by reference to which we could appreciate ethics on the one hand and science on the other. We could then try to determine the kinds of rationality proper to ethics and to science, respectively, by adding adequate specifications to the general concept of rationality. The difficulty is that we are not able to invoke an a priori concept of rationality; if we have recourse to the concept of rationality, it is in order to make explicit the specific character of certain proceedings. We must construct this concept by leaning on the historical experience which we have of some remarkable procedures which can be distinguished in a relative clear way from other procedures present in our experience.

Scientific activity is for us the most reliable paradigm of rationality. The question which must be raised is to know if there exists no rationality outside science or if this character which we call rationality in the domain of science is only a specification of a more general character which is found also, under other specifications, in other domains. This question is central for our subject because if only science is rational, than either it must be shown that ethics can be based on science or it must be declared that ethics is irrational.

Inside science itself, the paradigm of what we call rationality is without doubt the domain of mathematics. Two remarkable features appear here: the objects which are studied are defined in an exact manner, and what is said about those objects can be demonstrated. The essential element is, to be sure, demonstration. Now to demonstrate consists in giving the reasons for the validity of what is stated. Mathematical knowledge gives us so a model of a knowing supported by reasons. But this does not go without problems. Strictly speaking, a demonstration shows that it is possible to obtain a proposition A from a set Σ of propositions by application of certain rules. We know that the systems of rules have a relative character. What is thus established is only this: that, given such or such systems of rules, if the propositions of Σ are valid, the proposition A is equally valid. There is there a moment when appeal is to be made to intuition: when it comes to be recognized that the passage from a proposition A to another proposition A' which derives immediately from it by virtue of a rule R, is actually a concretization of the scheme in which the rule is expressed. The rationality of demonstration consists then, it seems, in the decomposition of a complex process in a sequence of elementary processes such that the intuition necessary for the realization of an elementary process be reduced to a minimum. But what is the value of the rules, and what is the value of the premises? It is possible to go back to more and more general systems of rules and of premises but finally, as the historical experience has shown, are met foundational problems for which the demonstrative procedure as such is of no more use. It remains then either to accept as unsuperable a situation where there is only a relative validity, or to have recourse to arguments of plausibility, for example by accepting as valid a constructive procedure of justification.

In the case of the empirical sciences, the notion of demonstration is no more useful as such. But the very idea which is at the basis of this notion can be retaken: we have to give reason, that is to say to propose justifications. This can be done in two ways. Either by advocating principles which can be themselves justified a priori, that is to say independently from empirical experience, or by showing that the available empirical data give a sufficient support, be it provisory, to the propositions submitted to a critical examination. The first way is illustrated by the Kantian enterprise of justification of Newtonian physics. The principles of classical mechanics are built by simple application to the empirical datum of motion of the principles which constitute the very condition of possibility of an empirical science in general. In such a way that the agreement with experience does not have to be established afterwards, but is warranted in advance by the very nature of the said principles, whose validity is established finally under the seal of necessity. Science such as we know it today has clearly followed the second way. This leaves open the question of the possible role of a priori principles in the construction of theories. But even for a neo-Kantian type of epistemology, the control of empirical experience remains necessary.

We could summarize all this by saying that rational knowledge is characterized by a recourse to criteria of justification. The current discussions bear on the precise manner in which such criteria must be formulated. But we must notice that, in any case, the criteria do not have an absolute character and therefore are not definitive. And this for two reasons. On the one hand, they are evolving with science itself. At measure where we extend the domain of knowledge, even inside a given discipline, we are led to introduce a greater number of parameters, and, correlatively, what is demanded for the acceptance of an experimental result becomes more and more complicated. And on the other hand, as an experimental result is never a simple registration of data but an interpretation, an appeal to the sole perceptual evidence is not sufficient. Some measure of appreciation must necessarily intervene and finally the criterion to which we must have recourse is the agreement of the experts.

However justification is not the only element which characterizes rational knowledge. Two other characters occur in a very evident manner: the ascent towards the foundations and the search for integration. According to an indication given already by Aristotle and still valid, rational knowledge is a knowledge according to principles. And with this idea of principle appears the idea of universality. The virtue of the principle is that it precontains in a certain way what it is founding. It gives reason of what is asserted in the sense that it gives a foundation. And the foundation is that without which the thing cannot exist. Again, the ideal case is the one where the link with the foundation is secured by a deduction. Demonstrative justification is actually what relates a proposition to the principles which found it. But there is more in the idea of foundation than in the idea of demonstration. A demonstration goes back to some premises. But a premise is not necessarily a foundation. What is proper to a foundation is to be in a certain sense ultimate, that is to say to be such as not having to be at its turn justified. The idea of foundation is reconcilable with the idea of justification only if it connotes the possibility of a self-justification. What is authentically a principle can play its role of foundation only on the condition of giving by itself the warrant of its own validity.

To tell the truth, it is difficult to admit that science ever gives real foundations in this sense. But it is true that it tries to establish its results on more and more profound bases, and this as well in the case of the purely formal sciences (where the known objects and their properties are brought back to more and more general structures) as in the case of the empirical sciences (where theories with a more and more extended explanatory power are constructed). What is considered at a given moment as "foundation" is thus such only in a relative sense. But what gives sense to this progression towards more and more deep levels of understanding, is well, it seems, the search for principles, in the most radical sense of the term. This search is besides marked by a strange circularity. The virtue of the principle is to give reason. When we have to do with an empirical science, we can say that a provisionally adequate theory gives reason of the phenomena which it covers. But the principles themselves must be justified. And if this is not possible a priori, they can be justified only by the demonstration of their elucidating power with respect to the particular propositions which they are supposed to found: it is thus finally those particular propositions which make possible the justification of those principles. The idea of justification plays thus in the two directions: justification of the principles by their consequences, justification of the consequences by their connection with the principles.

The principles have, to be sure, a role of unification. But after all, each discipline could have its proper principles, and the different systems of principles could be independent of each other. There is however, it seems, in the project of rational knowledge, this idea that the principles must form a unified system. The extreme form of this idea is that a unique principle must exist, from which all the others depend. But other forms of unity are conceivable. What seems to give support to this search of unity is the conviction that reality is one and that it belongs precisely to reason to accord itself with reality, thus to discover what makes the supposed unity of reality. This wish for unity is even probably what is really constitutive of reason. We do not have the certainty that the real is one, we do not have the certainty that we are able to think the real under the form of unity, we have at most very limited experiences, as impressive as they are, of unification, in restricted domains of our knowledge. But, as reasonable beings, we are animated by the project, which is perhaps fantastic, of a unifying reconstruction of our experience. Of course, this unity which is to be made is only a form. And if the project of unity is constitutive of reason, reason itself is only a form. But it is a form which is not simply an envelope, a simple configuration or a simple model. It is a form which is an idea, it is to say an exigency of realization. We can see in the history of knowledge how this form of unification finds its effectivity, how it is really operating in actual research. We are meeting here reason as work, it is to say as production of itself, more exactly as production of its own content. And we know that this production is not a simple accumulation: from what is acquired at a given moment a new domain of constitution opens itself and this domain, in its turn, by creating new objects, make appear ulterior possibilities of constitution.

We can find analogous characters in ethics. The term "ethics" is understood here not in the sense of denoting the domain of action governed by intrinsic norms, but in order to designate those very norms which action gives to itself and which are expressed in prescriptive propositions. Such propositions can correspond partly to ethical intuition. But they are able to be submitted to a process of rationalization. And this under the three forms of justification, of search for principles and of unification. Justification here can appear under two versions: connection with more general prescriptions, confrontation with experience. The first form of justification is important because it is associated with the second character of rationality, the search for principles. But it is thoroughly insufficient. A norm is illuminating for action only in the measure where it takes account of the real conditions of action, that is to say of the situations. And the situations are extremely variable, very complex and partially unforeseeable. A remarkable feature of the situations created by the techno-science is precisely their complexity and their novelty. We can, to be sure, recognize that a particular norm, relatively not too distant from the effective situations, is in conformity with some general norms. But a particular norm is not simply a particular case, deducible from some general statement; it adds to the general indications some specific indications, taken precisely from the analysis of some particular situations.

But how is it possible to justify with respect to experience? A classical answer is this: by the appreciation of the consequences. Past experience gives the possibility, in certain cases, to establish that a line of conduct, dictated by such or such a norm, produces such or such consequences. And in other cases, the knowledge which experience gives about human actions, enriched by the knowledge which science gives of the course of the world, gives the posssibility of predicting, with a sufficient probability, that a line of conduct, dictated by such or such a norm, will result in such and such consequences. It is then possible to compare the norms analogously to the comparison of hypotheses. It is after all the scheme of the theory of decision, which gives us a relatively precise model of a calculus of utilities. To be sure, such a type of reasoning cannot be used in an effective manner unless utilities can be attached to the foreseeable consequences and this in such a way that they become comparable, at least ordinally. As it is difficult to determine intersubjectively utilities, the concept of utility can be replaced usefully by the concept of preference. We shall have then to determine under what conditions a system of preferences is coherent and under what conditions different systems of preferences are compatible. This can be satisfactory for a so-called rational theory of decision, which concerns the adaptation of the means to given ends, not for an understanding of the ethical decision, which concerns finally the ends themselves. How are we going to fix utilities? How are we going to justify preferences? We could say: by judging their conformity with accepted norms. But we are falling then apparently in a circle: we had to afford precisely the justification of norms.

However, the analysis of the consequences enables us, at least, to verify that we are not introducing contradictions in our conduct, when the norm wherewith we are judging the consequences is the same as the norm which inspires our action. And it enables us to verify that we are not introducing a contradiction in our system of norms, when the norm wherewith we are judging the consequences is different from the norm which inspires our action. What is important is that a norm must be effectively capable of illuminating at least partially the situation and to bring a contribution to the weighting of the motivations for the decision. We could speak from this point of view of an ethical utility. But what does mean "to illuminate the situation"? It is to make perceptible, at least partially, its ethical import. Do we not meet here a kind of limit to rationality, where we must make appeal to intuition? Let us remark that there is always, between the norms and the effective action, intervention of a judgment which must precisely appreciate on the one hand the relevance of the norms with respect to the situation, and on the other hand the significance of the situation with respect to the norms. But this role of the judgment is analogous to the role of the interpretation in the cognitive practice.

But is there a rational approach of the ethical import of a situation? Here exactly intervene the principles and the search for unity. An important aspect of the rationality of ethics is that it consists not simply in connecting conducts with some prescriptions but in connecting the prescriptions themselves with a system of principles. After all, it is always possible to transform any proposition, describing a conduct, in a proposition prescribing this conduct. But this does not mean anything. What is really meaningful is to show that it is possible to inscribe the prescriptive propositions in a system. And there is a system only where there is a principle of systematicity. We find again here the idea of principle. But this time in the realm of prescriptivity, no more in the order of descriptivity. The term "validity" can be used in the two cases, but in different senses. On the one hand, we have to do with a cognitive validity, on the other hand with a prescriptive validity. In the first case, the principles have as function to justify the cognitive validity as such, that is to say to make appear what confers to a given proposition its validity character. In the second case, the principles have as function to justify the prescriptive validity as such, that is to say to make appear what confers to a given prescriptive proposition its prescriptive character. Thus the principle is not only a kind of norm of supreme generality, precontaining all the particular norms. It is a founding norm, or a norming norm, which gives the whole system of norms precisely its normative force.

Now, just as in the case of knowledge, the search for principles is animated by an exigency of unity. Here appears a third character of rationality, probably the most fundamental: reason is a unifying power. But we must understand how this aspect of rationality contains the two others: it is by this that it is the most fundamental. The coming back to the principles, in the order of action, is the search of what gives reason finally of every norm, that is to say of that which precontains the instauration of an order of normativity. It is thus the search of an ultimate justification of every regulation of action. And such a justification, if it is ultimate, must be able to give reason of itself, or again to be for itself its own foundation. The ascent to the principles, key of all justification, is the search of a position capable of showing that it does not ask any external justification but that it justifies itself by itself, inasmuch as it is a source of normativity. And the only manner for such a position to show this is to expose itself in its founding structure, that is to say to make apparent its own constitution as instauration of a normative order. Or, in more concrete terms, as containing, in its internal organization, the datum of an irreducible responsibility with respect to itself. The exigency of unity contains effectively the two other moments: it is the exigency of an ultimate foundation of justification, under the form of an ultimate principle of normativity. But the unifying principle, in order to correspond effectively to this exigency, must be able to give account at the same time of the normativity as such and of the very content to which the system of norms must give expression.

Reason is precisely such a unifying principle. If the ethical order refers finally to such a principle, it is rational in its very foundation. Reason is a unifying principle by this that it is self-foundation. This means that it is in itself exigency of unity and that it gives itself its own foundation by exposing itself, or by developing itself as this exigency of unity. The unity which is at stake here is not the simply formal unity of a system of propositions, it is the substantial unity of a complex totality which is given to itself as having to construct itself. In order to indicate that we have to do here with a concrete unity, we could use the term of integrity. In order to designate the complex totality whose unity is at stake, we could use the term of existence, in the sense where it denotes human concrete reality as the process of its own emergence (from the natural situations). Reason is existence itself, that is to say the concrete human reality, inasmuch as it is capable of reflecting itself and of understanding itself, and thus of giving itself its own foundation in the explicitation of its internal constitution.

What reflection makes thus appear is that existence has a double aspect: on the one hand, it is given to itself, as inscribed in a situation from which it emerges but which it does not master, a situation which is made of a superposition of cosmical, biological and historico-cultural determinations, and on the other hand, it is gifted with the power of constructing itself, of transforming the situation in a personal work, in conferring to itself a sense. This power is action. Action itself is sustained by a dynamism which is not limited by any determination but opens existence on the indetermination of the totality. In its effectivity, that is to say in the decision, action projects this dynamism in concrete figures, which are the stages through which existence constructs itself. The movement which bears action is always exceeding what this one realizes effectively. This tension between the moving inspiration which attunes so to say by advance action to the "telos" of the totality, and its effectivity, induces in it an exigency, which is really constitutive: the exigency of having to reappropriate in itself, in the concrete endeavour of its effective will, all the amplitude of the instituting force which calls it in advance of itself. It is this exigency which founds normativity. It is the wish, or the desire of existence of reconciliating itself with itself, it is to say to assume all the conditions of its own construction in the resolved will of its own integrity. The founding principle is thus unifying in the strongest sense: it poses itself as the exigency of a real unity which has to be constructed, as an appeal to a real integration of existence, in the two senses of a unification at the same time integrating and integral.

Of course, we have to do here only to a form. But, as for theoretical reason, it is a form which is an exigency of realization. And the encounter with the situation is necessary to induce the translation of this exigency in determined prescriptions. As there is a work of theoretical reason, constructing itself, there is a work of practical reason, constructing itself in the contingency of action. We find again here the role of judgment and the possibility of a certain form of justification from the situations. In a sense, it is from the principles that we interpret the situations. But in another sense, it is from the situations that we are judging the relevance and the adequacy of the principles. This is possible only, of course, if we are able to seize the situations in their ethical significance. This is what is done by the ethical judgment. It can be said that it expresses an intuition. Actually, it expresses the manner according to which the primary exigency, constitutive of action and of practical reason, meets the situation. Thus, by the intermediary of the situations, it is finally from this primary exigency that we judge the particular principles. But it is not a deductive way. It is much more an inductive way. Because the primary exigency is only a form in search of its content, the presence of the concrete situation is necessary to enable the virtue of the exigency to reveal its judging and inspiring force in this circumstance.

How then conceive the relationships between scientific rationality and ethical rationality? So far we have only presented an analogy. But we must wonder if there is not something more, more precisely if there is not an interiority of ethical rationality with respect to scientific rationality. If things are actually so, it will be possible to found an ethics of science in an intrinsic manner, that is to say from the very conditions which make possible the scientific enterprise. To be sure, we shall have to do there only with an ethics of science considered in itself, not with the ethical problems which are raised by the applications of science, and, more generally, by the interaction between science and other systems of action. But it is important to know if there is actually an ethics of science as such. In some cases, there are indeed conflicts between the demands of research and other demands, for example the respect of the human person. An ethics of science alone does not suffice to clarify such conflicts but appears in any case as a necessary condition for such a clarification.

There is undoubtedly in scientific practice a normative aspect. But it consists only in what could be called an espistemic normativity, relative to knowledge. There are in each discipline norms pertaining to the formation of concepts, to the construction of theories, to the processes of validation. Those norms are not given by advance and have never an absolute character. They are constructed in the course of research and are besides progressively transformed. A part of what is called foundational research bears on those epistemic norms. The construction of those norms is itself governed by a directing idea which constitutes the very project of science: the idea of a rational knowledge of reality. To be sure, this idea is effective only in the measure where it is expressed in relatively concrete norms. But in itself it is more than a program; it presents itself as the expression of a task and of a responsibility. Not perhaps for each individual as such, but in any case for the human community and, therefore, indirectly for everybody. This is the indication that the project of science is perceived as endowed with an ethical value. It is, it seems, what is true in the thesis of Jacques Monod according to which science is based upon a postulate of an ethical nature. The epistemic normativity can be thus understood as a conditional normativity and, therefore, as affected itself by an ethical coefficient: the epistemic norm acquires an ethical significance from the moment where it is recognized as a condition for the fulfiment of a project which is itself recognized as ethically desirable.

But the project of science is not sufficiently precise in order to make possible a deduction of the epistemic norms; on the contrary it is on the basis of a determination of those ones that the project of science receives a relatively precise content. There is thus no link of necessary implication between this project and the epistemic norms which are admitted at a given moment, and consequently a given norm will be able to take an ethical value only in a transitory manner. What is important, to ensure the realization of the project of science, is not so much the faithfulness to the norms which are accepted at a given moment as the effort of invention by virtue of which the system of the epistemic norms gains in efficacity. The relationship of the ethical normativity to the epistemic normativity could thus be formulated under the form of the following principle: the proper duty of science, considered in itself, is to ensure the free exercise of its epistemic normativity. We could see in such a principle the basis of an ethics internal to science.

But we have still to understand how the very project of science can acquire an ethical value. It is not sufficient for that to observe that scientific knowledge is a received value in a significant cultural area. It must be shown how the project of science is connected with the source of normativity, in other terms how theoretical reason is connected with practical reason. This will give besides the means in order to meet in a sufficiently fundamental manner the ethical questions which arise at the interface between scientific research and other systems of action.

The concept which helps us to understand the articulation between the domain of rational knowledge and the domain of ethics is, to be sure, the concept of action, considered as designating existence itself inasmuch as it has the responsibility of constructing itself. The enterprise of knowledge is a form of action. It represents one of the modalities by which existence makes itself capable of going outside itself, of opening itself, in principle without limitations, to the whole of the reality in which it is immersed, and by which it begins thus to reconquer itself on the determinations which define its situation. In trying to understand the world, existence discovers that it has in itself the power to overcome the particularity of the here and now of its situation; it begins thus to affirm itself as liberty. Knowledge is thus one of the moments by which action retakes in its actual initiative its proper conditions of functioning.

We must therefore find again in the enterprise of knowledge the general structure of action, that is to say the tension between a constituting dynamism, which is open on the infinite, and the effective procedures in which existence determines itself concretely and gives to itself its real figure. This is effectively what happens: there is indeed a tension between the project of science, which imposes to this one an immanent "telos", which is only a yet non-determined exigency of construction, and the actual norms according to which the real content of knowledge is constructed. Just as the general significance of action is to translate in effectivity the exigency which constitutes it, so the significance of the scientific enterprise is to translate in actual propositions the project which constitutes it. But if the effort of action in order to coincide with itself is what is at the source of the ethical normativity and of its rationality, then the effort of knowledge in order to fulfil its project is itself the manifestation, in its particular order, of the ethical exigency which permeates existence. There is thus indeed an ethical dimension in knowledge. And it is indeed inasmuch as it will be rational that knowledge inscribes itself in the integrating rationality of ethics. There is effectively an interiority of theoretical reason with respect to practical reason, but this does not suppress at all the difference. This relationship reveals itself only in the dimension of the foundations; it cannot find a translation in the order of deduction. It authorizes neither a deduction of science from ethics nor a deduction of ethics from science.

But it must be remembered that the ethical exigency is the demand of a concrete unity, that is to say of integrality and of integrity. And scientific knowledge is not the whole of existence. It does not even cover the totality of the domain of knowledge as such. If there is effectively a presence of ethics in science, and by this an insertion of the scientific enterprise in the total order of action, this very presence imposes on science to recognize the particularity of its position with respect to this total order of action. But this means two things: on the one hand, that science does not represent, indeed, but a stage in the development of action, defined precisely by the project of a rational knowledge, and on the other hand, that the ethical dimension which inhabits it inscribes in its very enterprise a reference to an exigency of integrity. This can lead to the formulation of two other principles, which could bring a complement to the purely internal principle of the ethics of science and could intervene there where science interferes with other domains of action: a principle of limitation and a principle of cooperation. The principle of limitation imposes on scientific research to recognize that it constitutes only a certain type of approach of reality, very important but not sufficient in order to judge completely the ethical import of a situation. The principle of cooperation imposes on scientific knowledge to understand itself as having to bring its contribution to an effort of integration, that is to say to the constructive effort by which existence endeavours to constitute itself as a totality. What does this concretely mean? Probably an appeal to ethical inventivity, to the work of practical reason. If science has an ethical value, in any case, it is to be sure because in itself rational knowledge is a constitutive dimension of action, but it is also because it is called to overcome itself in a movement of unification which is probably what is designated by the term of "wisdom". We find here again a very old idea, which had been retaken some years ago by the late Evert Beth: "Through knowledge to wisdom".

Barcan Marcus et al., eds., Logic, Methodology and Philosophy of Science VII

ETHICAL ASPECTS OF NON-ETHICAL THEORIES

MARIAN PRZEŁĘCKI

Inst. of Philosophy, Univ. of Warsaw, Warsaw, Poland

1. The deliberately loose notion of ethical aspect used in the title of my paper needs some explanation. Ethical aspects of scientific theories may be defined in various ways. It is one sense only of this notion that is going to be discussed in the present paper. Bernard D. Davis, speaking about "Alleged Threats from Genetics" at the last Congress in Hannover, 1979,[1] has distinguished three main kinds of such threats: dangerous products, dangerous powers, and dangerous ideas. It is the last of them — a threat characterized by the author as "socially dangerous knowledge", "knowledge that shakes the foundations of public morality" — that I want to concentrate upon in the following.

Let us first try to explain this notion somewhat more closely. Since the proper object of moral valuations seem to be human actions (and, derivatively, human agents), it is not theories themselves, but rather the acts of propounding a theory that may be morally evaluated. In the case of non-ethical theories, including all the scientific ones, the act of propounding such a theory seems to be, in itself, a morally neutral action. If it is morally judged, it is so in regard to its connection with some other kind of action of an explicitly moral character. In general, the connection is thought of as a causal one: the fact of propounding a given theory T brings about, as its effect, a morally relevant action A. The causal relation here involved is to be taken in a very broad meaning, including so-called "partial" and "probabilistic" causal connections. These may be regarded on individual as well as on social level. In the first case we have to do with psychological, in the second with social effects of the fact of advancing a given theory. An important type of that general relation is one in which

[1] Davis (1982), pp. 835, 840.

theory T is deliberately used in order to bring about action A. Here, as before, the whole connection may be treated either on individual or on social level. What is going to be the main object of the present discussion is a certain special case of that "intentional" relation: a case in which the connection between propounding theory T and action A is a mediate one, linking T with A through a moral valuation of the action A. The situation in question may then be defined as follows. Theory T is being used in order to justify a definite moral valuation of an action A, and this valuation, in turn, brings about (or is deliberately used for bringing about) the action A.

Now, what I consider decisive for a moral evaluation of the act of propounding theory T is the question whether theory T, being actually used for justifying a moral judgement E, does, indeed, justify that judgement. If judgment E amounts to a positive valuation, i.e. a moral approval, of action A, theory T which justifies the judgment may rightly be said to legitimize (support, or promote) the action A. Anyone who propounds such a theory bears thereby direct responsibility for the given action. In view of this, it is important to be able to tell whether a theory T justifies a moral judgment E. To do so, T must stand to E in a definite logical relation. The simplest version of it assumes that to be justified by T, E must logically follow from T. This is not assumed to necessarily be the case on some other, more liberal notions of justification. In the present discussion we shall restrict our concern to this strict notion, because it is on that conception that the problem of the scientist's responsibility for his theory's ethical import becomes especially urgent. It is to be noted that the looser concepts of justification also refer ultimately to the relation of logical consequence as holding, in some way or other, between statements of theory T and moral judgments. And this raises the general question of whether the concept of logical consequence may be said to be applicable at all to ethical statements.

The answer to this question given in the present paper is based on certain assumptions concerning the logical character of ethical statements. By the latter I mean, roughly speaking, sentences which express moral valuations, the simplest of them being of the kind: "action A is morally good", "action A is morally better than action B", and the like. Literally taken, they belong to the category of declarative sentences. There seems thus to be no difficulty in applying to them the concept of logical consequence. Accordingly, I shall assume that the relation of logical consequence is defined for all ethical statements of the kind considered. It is to be noted that this assumption is independent of whether ethical statements are assigned any truth value. I am inclined to do so and to regard (some, at least) ethical statements as, literally, true or false sentences. Some arguments for that claim have been provided by me elsewhere.[2] Basing on the semantic definition of truth in its model theoretic version, I have tried to show that the definition may well be applied to ethical statements in view of the fact that ethical predicates do not essentially differ from descriptive predicates as far as their model theoretic interpretation is concerned. I do not want, however, to prejudge that controversial question in the present context. And there is no need to. The concept of logical consequence I appeal to is a syntactic one. It is defined for all declarative sentences independently of their semantic status. And so, it is applicable to ethical statements even if they are said to lack any truth value.

Another assumption adopted in this paper asserts that ethical predicates form a separate class of expressions, having no elements in common with the class of purely descriptive predicates. This admittedly is an idealization of the actual situation obtaining in natural language, which abounds in predicates endowed with both descriptive and ethical meaning (words such as "cruel" may serve as examples). It seems possible, however, in principle, to separate these two components and to arrive at a language free from any ethical ingredients, such as languages characteristic of certain scientific theories. It is to languages so reconstructed that our assumption is intended to refer. On this assumption, then, there cannot be any essential logical connections between descriptive and ethical statements. In particular, no (simple) ethical statement can be said to logically follow from a (consistent) set of purely descriptive statements. If it is said to follow from such statements, it is not the strict concept of logical consequence that is meant in such a claim, but certain looser notions. One of them is the notion of so-called analytic consequence: E follows analytically from T if E follows logically from T and some analytic premise. According to the so-called naturalistic standpoint in the theory of value, any ethical value is assumed to be definable in terms of purely descriptive characteristics. The corresponding definition can then function as an analytic premise in deducing an ethical statement E from purely descriptive statements T. According to the non-naturalistic standpoint (which appears to me a more plausible one), the premise needed in such inference is to be always of a synthetic, i.e. factual, nature. But, whether analytic or synthetic, the premise is bound to contain some ethical predicates in an essential (i.e. non-eliminable) way. And this is what proves crucial for our problem.

[2] See, e.g., PRZEŁĘCKI (1974).

To the above assumptions concerning the logical character of ethical statements I should like to add some remarks on their methodological status. They reflect a certain general view of the structure of ethical systems. It is a view that treats ethical systems on the model of empirical theories, as far as their justification structure is concerned. The view has been succinctly put by D. Føllesdal, who writes: "Justification in ethics takes place from below ... A set of ethical principles is justified by showing that its consequences fit in whith the ethical reactions we have in concrete ethical situations, where e.g. a person is tortured in front of us, where one person helps another, etc."[3] According to that view, it is only elementary (i.e. simple and concrete) ethical statements such as: "this action is good (bad)", or "better (worse) than another", that are justifiable in a direct way (by appeal, say, to our moral intuition). All general, or abstract, ethical principles are justifiable only indirectly (by appeal to their directly justifiable consequences). This is a view which seems to me the most plausible one. But aware of its controversial nature, I will not consider it a necessary presupposition of the arguments adduced in this paper.

The main conclusion implied by the assumptions here adopted sounds quite obvious. Since a genuine ethical statement cannot logically follow from non-ethical premises, it is only ethical theories that bear an ethical import. Theories free from any ethical assumptions are, in this sense, ethically neutral. This seems to be true of all scientific theories, in particular. The undeniable fact is, however, that, contrary to our claim, some ethical import is actually ascribed to theories which seem quite free from any ethical ingredients. It is not infrequent that a scientific — and so, apparently non-ethical — theory *T* is being accused of legitimizing some kind of action *A*, otherwise felt as morally objectionable. Now, according to our explication, *T* does legitimize *A* only if *T* entails a definite moral valuation of *A* — either a positive one: "*A* is morally good", or the negation of a negative one: "*A* is not morally bad". But this can hold only if *T* itself contains some ethical assumptions. If this is not the case, *T* can entail such a valuation only when combined with a certain ethical principle supplied from without. Such an ethical premise may belong to our stock of common-sense opinions. It may even, in some philosophic views, bear the status of an analytic statement. This is why its presence can easily be overlooked. Nevertheless, a suitable ethical premise must always be present, if the alleged inference is to be logically valid.

[3] FØLLESDAL (1981), p. 406.

To show this in some detail, I will examine two examples of scientific theories which have been burdened with an ethical import. Both of them belong to well known and widely discussed cases, and so are not in need of any introduction.

2. One of the theories I have in mind is the notorious hypothesis of the heritability of intelligence, or — strictly speaking — of IQ test scores. Since its first formulation, the hypothesis has become the object of a heated controversy. I shall here restrict myself to mentioning one of its fairly recent critics, L.J. Kamin, who in his paper "Heredity, Intelligence, Politics, and Psychology"[4] has put the hypothesis to a severe criticism. The paper contains a thorough "examination of the empirical data purporting to support the idea that IQ test scores are highly heritable". The author is concerned with "two classes of studies: those involving separated identical twins, and those involving adopted children", since "these two types of studies by general agreement provide the most powerful evidence for the heritability of IQ". Bringing to light a number of serious methodological defects, the author discredits both kinds of studies and states, in conclusion: "I see no unambiguous evidence whatever in these studies for any heritability of IQ test scores".

The author's arguments sound fairly convincing, and I am far from questioning his conclusion (though, as far as I know, the problem continues to be highly controversial). Kamin's paper, however, as its title announces, is an essay on psychology as well as on politics. And what I am opposed to is the kind of connection that is supposed to hold between these two domains. The psychological hypothesis is being attacked so vigorously because of its political, and — generally — ethical, implications. As Kamin tries to show, "the intelligence test has been used more or less consciously as an instrument of oppression against the underprivileged — the poor, the foreign-born, and racial minorities". The genetic interpretation of differences in test scores has served to legitimate the discrimination of various ethnic groups. "The first major practical effect of the testing movement lay — according to the author — in its contribution to the passage and rationalization of the overtly racist immigration law of 1924". The data then available provided "the first large-scale evidence that blacks scored lower than whites". Likewise "the Latin and Slavic countries stood low"

---

[4] KAMIN (1973). All the following quotations, if not otherwise marked, are from this paper.

("The Poles, it was reported, did not score significantly higher than the blacks"). Basing on such data, "the Congress passed in 1924 a law not only restricting the total number of immigrants, but also assigning "national origin quotas" ... in order to curtail biologically inferior immigration". As the author contends, a similar political role has been played by the intelligence test studies in more recent times.

Now, let us take all these facts for granted and ask if they may be said to discredit the relevant psychological hypotheses. Are we really entitled to say that the hypotheses legitimize the political practices mentioned? Such, undoubtedly, was the claim of those politicians who spoke of the steps taken as "dictated by science". The psychological research was to make it possible "to take the national debate over immigration out of politics, and to place it on a scientific basis". But this clearly is a misunderstanding. A political decision can never be taken "out of politics". A scientific theory alone can never legitimize any kind of action. It can do it only when combined with suitable ethical (or specifically political) principles. Let our psychological hypothesis amount to the claim that a given ethnic group *G* shows a lower IQ than the rest of the population, the difference being genetically determined. Does this fact legitimize any discrimination of the members of group *G*? In other words, does the hypothesis entail a value judgment to the effect that such a discrimination is morally right (or, at least, not morally wrong)? Evidently, not. The hypothesis can imply some moral valuation of such a policy only on the basis of certain ethical principles. And it is those principles that determine whether the policy is to be judged right or wrong. It can be easily seen that they may well imply a moral condemnation of any discrimination policy with regard to the members of group *G*. This will be the case on the basis of any altruistic system of ethics with its supreme command: "Love thy neighbour as thyself". According to it, we ought to care about the members of group *G* certainly not less than about the rest of us. No factual hypothesis concerning the members of group *G* can annul that obligation. Whether and how much we ought to care about them depends on ethical principles, not factual hypotheses. The latter may decide only what exactly our proper care about them is to consist in. The psychological hypothesis discussed by us may imply that our care about the members of group *G* should take forms somewhat different from those appropriate for other people. But to care differently does not mean to care less. And it is he only who cares less about some kind of people that may rightly be accused of discrimination.

Discussing the connection between the hypothesis of the heritability of intelligence and the ethical ideal of social justice, Bernard Davis makes a similar point. He expresses the view that the hypothesis in question "might conflict with egalitarian preconceptions about the amount and the distribution of variation in genetic potentialities in our species", but it is difficult to see how it can itself "threaten justice". "On the contrary, as we deepen our understanding of the interaction of inborn and social factors that influence human behaviour we should be able to build more effective institutions of justice". "We must recognize that we can adapt our social institutions to our evolutionary legacy". He concludes, in effect, that "the assumption of an inherent conflict between genetics (or other areas of science) and justice seems philosophically unsound".[5] This is wholly in line with our argumentation.

What I want to stress emphatically in connection with the concrete case being considered is that no factual hypothesis can by itself justify any racist doctrine, since the latter is never restricted to a pure description but always involves certain valuations. The fact is that the psychological hypothesis criticized by Kamin and others is most probably false. But this is a matter of fact, not of necessity. The hypothesis, after all, might well turn out to be true. Hence it seems so important to realize that there is no logical connection between that hypothesis and any racist doctrine with its discrimination policy. This is how Bernard Davis puts the point: "We must rest the goal of racial justice on grounds of moral conviction, rather than on vulnerable assumptions about questions of fact".[6] Let me conclude these remarks by quoting a saying of another biologist: "Man ... cannot place responsibility for rightness and wrongness ... on nature".[7] I fully agree with this claim. It is unfair to blame facts for our moral faults. As moral beings, we are not at the mercy of the facts. No facts can prevent us from being good.

3. Another famous case of a scientific theory which from its very origin has been burdened with an ethical import is the theory of evolution. The situation here falls under somewhat different schema than in the case of the theory of intelligence. The latter theory has been said to establish certain facts, which may be exemplified by saying that a given ethnic group G is, so

[5] DAVIS (1982), pp. 840–841; DAVIS (1978), p. 390.

[6] DAVIS (1978), p. 390.

[7] SIMPSON (1949), p. 346.

to say, "biologically inferior" than the rest of the population. A fact like this has then been taken to legitimize a discrimination of group *G* through justifying a moral valuation to the effect that such a discrimination is morally right. As seen above, the moral valuation can be said to follow from the factual statement only in virtue of a certain general ethical principle — a principle, say, which claims that it is morally right to care more about "biologically superior" than about "biologically inferior" groups of people. Now, it is just that kind of general ethical principle that a scientific theory such as the theory of evolution is assumed to provide. The theory of evolution has notoriously been treated, by biologists as well as by philosophers, as a basis of some systems of ethics — the systems of so-called "evolutionary ethics". The theory has been supposed to supply a general answer to the question what kind of conduct is to be qualified as morally right. The common way to arrive at such an answer is to distinguish what may be called the general direction of evolution and to call morally good anything which is "in line" with that direction, which promotes the main trend of evolution.

The earlier attempts at constructing systems of evolutionary ethics were rather crude, and their proposals are quite discredited by now. Theories of a "Social Darwinist" kind involved straightforward application of such slogans as the struggle for existence or the survival of the fittest to human affairs and from them derived their ethical principles. This has rightly been criticized as unwarranted in view of the logical gap between the purely factual and the specifically ethical dimension. Even if the theories adequately described the factual course of evolution (which has been seriously questioned), this would provide, by itself, no sufficient reason to see in conformity with it a criterion of morally good behaviour. Why on earth ought we to promote a given process on that account only that it is shown to prevail in nature?

The most emphatic criticism of that kind of approach has been provided by the great evolutionist T.H. Huxley, who sees a fundamental contradiction between what he calls the ethical process and the cosmic process of evolution: "The ethical progress of society depends, not on imitating the cosmic process ... but in combating it ... The practice of that which is ethically best ... involves a course of conduct which, in all respects, is opposed to that which leads to success in the cosmic struggle for existence". Justifying this conclusion, he states what follows: "Moralists of all ages and of all faiths ... have agreed upon the "golden rule", "Do as you would be done by" ... Strictly observed, the "golden rule" involves ... the refusal to continue the struggle for existence ... The followers of the "golden rule" may indulge in hopes of heaven, but they must reckon with the certainty that other people will be masters of the earth".[8]

To avoid that kind of criticism, the later theories of "evolutionary ethics" have been resorting to a more refined argumentation. As their representatives we may mention scientists of such rank as Julian Huxley and C.H. Waddington. Now, in what respect do their views differ from the earlier ones? The main difference seems to lie in the fact that, in contrast to the former theories, the present ones refer not to "cosmic", but to specifically human evolution. This is how Julian Huxley criticizes the former attempts: "In the past, so-called evolutionary theories of ethics ... have been vitiated through their attempting to apply conclusions derived solely from the biological level of evolution to subjects like ethics which only come into existence on the social level". The contradiction which for T.H. Huxley existed between the ethical process and the cosmic process "can be resolved — according to the present writer — by extending the concept of evolution ... into the human domain".[9] It is to be realized that such an extension brings with it a radical extension of the very concept of evolution. As C.H. Waddington observes, "in addition to the biological mechanism of hereditary transmission, man has developed another system of passing information from one generation to its followers. This is the process of social teaching and learning, and it constitutes in effect a secondary mechanism by which evolution can be brought about".[10] Julian Huxley puts it even stronger: "Man's evolution is not biological but psychosocial: it operates by the mechanism of cultural tradition".[11] As it is seen, we have passed thereby far beyond what is usually identified as the theory of evolution. Let us see then if that radical extension of the concept of evolution allows us somehow to bridge the gap between evolutionary facts and evolutionary ethics.

In view of what has been said before about the relationship between factual and ethical statements, the answer seems quite straightforward. It clearly depends on how the so-called facts of human evolution are being understood. Do they or do they not include any ethical characteristics? If what is established by the theory of human evolution is taken to be free from any ethical qualifications, no ethical principles can follow from the

[8] T.H. Huxley and Julian Huxley (1947), pp. 51–52, 81–82.

[9] T.H. Huxley and Julian Huxley (1947), pp. 105, 148.

[10] Waddington (1960), p. 28.

[11] Huxley (1964), p. 76.

theory alone. They can so follow only if the very characteristic of the facts of human evolution involves — directly or indirectly — certain ethical valuations. And this is what actually happens in some cases being discussed now. The question is somewhat difficult to decide because of the notorious indefiniteness which characterizes the relevant conceptions. There are, however, cases where the presence of ethical components in the very identification of evolutionary facts is quite explicit. The conception of Julian Huxley may serve as an example (though its author is far from being clean-cut and consistent in his statements). The author contends that "the ultimate guarantees for the correctness of our labels of rightness and wrongness are to be sought for among the facts of evolutionary direction". But "there are more or less desirable ... directions in evolution". It is "the most desirable direction of evolution that provides the ... external standard for ethics". Now, how is such a direction to be defined? Being described by the author as a biological progress, it is partly characterized by such factors as "increase in control over the environment, increase in independence of the environment, and the capacity to continue further evolution in the same progressive direction". But these biological factors do not exhaust the characteristics. "They all make for progress ... in the biological sector of evolution. But in the human sector another criterion has been added — the understanding and attainment of intrinsic values": those which "are to be valued for their own sake". Into them the author explicitly includes "selfless morality". This makes it possible for him to assert that "progress is ... the desirable direction of change in the world, and desirable ethically as much as materially or intellectually".[12] The conclusion is nothing but a consequence of how the very concept of evolutionary progress has been defined.

Trying to avoid this kind of vicious circle in deriving ethics from the theory of evolution, C.H. Waddington abstains, in principle, from inserting ethical ideas into evolutionary framework. In summarizing his position, he writes what follows: "Observation of the world of living things reveals a general evolutionary direction, which has a philosophical status similar to that of healthy growth, in that both are manifestations of the immanent properties of the objective world ... Any particular set of ethical beliefs ... can be meaningfully judged according to their efficacy in furthering this general evolutionary direction".[13] This is claimed to provide a general

[12] T.H. HUXLEY and Julian HUXLEY (1947), pp. 32, 126, 142, 182, 214, 230, 234.
[13] WADDINGTON (1960), p. 7.

"supra-ethical" criterion for deciding between alternative ethical beliefs. The evolutionary direction referred to in the criterion, being identified with what is called anagenesis (i.e. progressive change) in the theory of evolution, is here described in an ethically neutral way: by means of such biological traits as those mentioned above. The criterion may then be said to assess ethical beliefs from some non-ethical point of view. This is why the author calls it a criterion of "biological wisdom". The criterion enables us to judge which ethical beliefs are "biologically wise". Now, is it a sufficient reason to prefer them to the "biologically unwise" ones? By doing so, one treats "biological wisdom" as a value overriding the ethical one, or straightforward identifies the latter with the former. And this seems quite unwarranted. He who is behaving "biologically unwisely" is said to be "out of step with nature". But is he thereby necessarily behaving in a morally evil way? There is no contradiction in supposing that an ethically good deed may turn out to be "biologically foolish". This is just what evangelical morality is sometimes blamed for.

As another outstanding biologist G.G. Simpson puts it, "survival, ... increase of life, integration of organic or social aggregations, or other such suggested ethical standards are characteristics which may be present in varying degrees, or absent, in organic evolution but they are not really ethical principles ... They become ethical principles only if man chooses to make them such ... Certainly it is "good" to survive, but we fall into a semantic trap if we think that this is an ethical good". The author concludes, in effect, that "an evolutionary ethic ... cannot be expected to arise automatically from the principles of evolution in general, nor yet, indeed from those of human evolution in particular".[14] And this sounds quite obvious on the assumptions adopted by us above. An evolutionary theory could entail an ethical principle only if it already involved some ethical valuations. Cases which appear to speak to the contrary result from the notorious confusion of descriptive and evaluative elements. The following argument (adduced, nota bene, by the same author) may serve as a good illustration. "One of the great ethical achievements of early Christianity ... was recognition, in principle, of the brotherhood of man ... Confirmation of the truth of evolution established this doctrine as a scientific fact".[15] Equivocation inherent in this argument is evident. The ethical principle of treating all men as brothers is here confused with the

---

[14] SIMPSON (1949), pp. 301, 308, 310.

[15] SIMPSON (1949), p. 281.

biological hypothesis of the common origin of all men. It is confusions like this that lend an air of plausibility to the evolutionary ethics. In fact, from the theory of evolution we can derive no more ethics than we have put into it before. Treated as a purely scientific theory, it is morally neutral.

There is, however, still another sense in which a scientific theory such as the theory of evolution is said to bear an ethical import. An ethically relevant consequence of a given theory need not amount to an explicit moral valuation of an action *A*, such e.g. as the object-statement "*A* is not morally bad"; it may express what can be called a meta-qualification of such a valuation, claiming e.g. that the sentence "*A* is morally bad" is deprived of any truth value, or that it cannot be said to be, straightforwardly, right or wrong, being essentially relative or subjective, and the like. Though, on the account of entailing a meta-statement like this, a theory cannot be said to legitimize action *A*, it can be said to make *A*'s condemnation theoretically unwarranted, or unjustified, and, in this sense, to bear, after all, some kind of ethical import. Now, the theory of evolution is typically regarded as a theory carrying that kind of meta-ethical import. Since human evolution comprises as its part the evolution of moral phenomena, the theory of evolution is expected to provide some account of the genesis and function of ethical beliefs. And by some authors such an account is taken to determine the proper interpretation of ethical statements: their literal sense and their semantic and epistemological status. Seeing in ethics a kind of "supporting mechanism", Julian Huxley declares, e.g., that all intuitive theories of ethics are thereby finally put out of court. But there does not seem to be any definite logical connection between the facts of evolution of ethical beliefs and their proper semantic status. Functioning as a kind of "supporting" or "adaptational" mechanisms, they may well have the cognitive status of true or false statements. There is nothing inconsistent in the conjunction of these two claims. Deducing from the function of ethical beliefs their semantic status, one commits what has been called a "nothing-but" fallacy: since ethical beliefs are supporting mechanisms, they are "nothing but" supporting mechanisms. The argument seems hardly cogent. To see this, let us compare ethics with science. The evolutionary function of scientific beliefs, whatever it is claimed to be, is not taken to prejudge their semantic status. Seeing in them a kind of "instruments", one need not consider them to be "nothing but" instruments: instrumentalism is not the only view consistent with the facts of their evolution. In that respect, the situation of ethical beliefs seems analogous. If so, the theory of evolution may be declared free not only from ethical, but also from any meta-ethical import.

4. The analysis of such cases as the two biological theories discussed by us above illustrates a kind of situation characterized in abstract terms at the beginning of our considerations: a scientific theory *T* is actually used for justifying a moral valuation *E*, though *T*, in fact, does not justify *E* in any definite sense of this term. In other words, there are actual attempts at justifying *E* by an appeal to *T*, but the justification attempted is not valid. It turns out, on a closer scrutiny, that *E* does not follow from *T*. Inferring it from *T*, either one reasons wrongly, or what he, in fact, is doing is inferring *E* not from *T* alone, but from *T* and some additional, tacitly assumed, ethical premises. And so, if *E* presents an objectionable ethical conclusion, it is not theory *T* that is to be blamed for it, but either he who misuses *T* in this way, or the additional ethical principles which are involved in the given inference.

I repeat these obvious truths because, in spite of their obviousness, they are not always fully realized and observed. It is a common practice to discredit — or, at least, to mistrust — a scientific theory on account of the uses it is put to — independently of whether those uses are legitimate or not. This is a practice which should be strongly opposed. If in his attempts to legitimize racial discrimination one appeals to some scientific theory, we must not take such "justification" at its face value and condemn the given theory. First of all, we have to submit any such procedure to a close logical scrutiny in order to expose all the logical gaps that it may contain. The essential question is whether the objectionable ethical conclusion follows from the given theory, or not, since a theory may be blamed only for its logical consequences, and not for what one wrongly infers from it. On the assumptions here adopted, such an inference can be valid only if its premises include some ethical statements. These need not be explicitly stated. In many cases they are only insinuated or otherwise "smuggled in", owing to the notorious vagueness and ambiguity inherent in the relevant expressions. Our task then is to bring those principles to light and to direct our moral criticism against them, since it is they that are responsible for the ethical import ascribed to the given theory. It is to be noted that, on the methodological conception of ethical systems as systems justifiable "from below", this is the proper way to criticize ethical principles: by criticizing the consequences they lead to in concrete situations.

In all our considerations we have regarded the difference between genuine and spurious justification as an essential circumstance. What has been decisive for our assessment of scientific theories is the question whether a theory, being used to justify a moral valuation, does indeed justify it in some definite sense. It might be argued, however, that from a practical point of view the distinction turns out to be unimportant. What is important is the question whether the attempt at justifying the valuation brings about the corresponding ethical belief as its — psychological or social — effect. What is said to be important, then, is not a logical relation of justification, but a factual relation of causation. Now, I do not want to deny the importance of that kind of aspect of scientific theories. I am far from suggesting that a scientist should be indifferent to the social effects of his theoretical activity, in particular — to the social uses his theories are being put to, even if these are based on spurious arguments. He ought to be concerned with them, as everybody ought to be concerned with all the effects of his activity. What I do want to emphasize is the difference between these two kinds of situations and between the obligations which they impose on the scientist. If I propound a theory which does entail a moral approval of some objectionable action, I am directly responsible for the action. What I am then obliged to do is to revise the theory in question. If, on the other hand, I propound a theory from which someone wrongly infers such an approval, the fault is with him, not with me. Being indirectly responsible for all foreseeable effects of my theoretical activity, I cannot ignore this fact, either. What I am, however, in this case obliged to is a criticism not of the theory itself, but of its actual misuse: a criticism of the attempt to legitimize an action, attitude, or policy, by an appeal to the given scientific theory. In this respect, the philosopher has a special duty. It is, first of all, his task to examine the questionable procedures, to brand all wrong inferences, to expose all hidden premises, and, in effect, to acquit the given scientific theory of its alleged ethical import. The fact that a scientific theory has been misused in a morally objectionable way does not discredit the theory, but only its misusers. Criticizing on that account the very theory is not only unwarranted, but harmful. It is bound to result in serious distortions of scientific inquiry by introducing into it preferences and biases motivated by extra-theoretical reasons. Many of us have witnessed deplorable effects of such practices in the past not so far remote, and this is what makes us so sensitive about this point now.

Let me conclude by removing a possible misunderstanding. The thesis of the ethical neutrality of scientific theories, advanced in the present paper, should not be taken as a claim denying those theories any ethical importance. The bearing of science on ethics is undeniable. It may be summarized by saying that scientific knowledge creates moral problems, but it does not determine their solutions. Providing us with new information and with, resulting from it, new powers, science brings to our consciousness difficult moral dilemmas, not realized by us before. Science alone, however, does not dictate, with regard to them, any definite moral decisions. Scientific knowledge may thus be said to be painful, but not vicious. Truth in science is morally innocent.

References

DAVIS, B.D., 1978, *The moralistic fallacy*, Nature 272, p. 390.

DAVIS, B.D., 1982, *Alleged threats from genetics*, in: Logic, Methodology and Philosophy of Science VI, Proc. Sixth International Congress of Logic, Methodology and Philosophy of Science (Hannover, 1979), L.J. Cohen, J. Łoś, H. Pfeiffer, K.-P. Podewski, eds. (PWN, Warszawa, North-Holland, Amsterdam–New York–Oxford), pp. 835–842.

FØLLESDAL, D., 1981, *Einige ethische Aspekte der DNS-Rekombination*, in: Philosophie als Wissenschaft/Essays in Scientific Philosophy, E. Morscher, O. Neumaier, G. Zecha, eds. (Comes, Bad Reichenhall), pp. 393–411.

HUXLEY, T.H. and Julian HUXLEY, 1947, *Evolution and Ethics 1893–1943* (The Pilot Press, London).

HUXLEY, Julian, 1964, *Essays of a Humanist* (Harper and Row, New York-Evanston).

KAMIN, L.J., 1973, *Heredity, intelligence, politics, and psychology*, a paper read at the EPA Meeting, Washington, 1973 (mimeographed).

PRZEŁĘCKI, M., 1974, *Some philosophical consequences of the semantic definition of truth*, Dialectics and Humanism 1, pp. 117–128.

SIMPSON, G.G., 1949, *The Meaning of Evolution* (Yale Univ. Press, New Haven, CT).

WADDINGTON, C.H., 1960, *The Ethical Animal* (George Allen and Unwin, London).

CONTRIBUTED PAPERS

Section 1: Proof Theory and Foundations of Mathematics

D.D. AUERBACH, Expressing Consistency
K. DOŠEN, Minimal Modal Systems in which Heyting and Classical Logic Can Be Embedded
W. DZIK, Lattices of Theories as Strongly Characteristic Models of the Intuitionistic Propositional Calculus
A. GIUCULESCU, Intensional versus Extensional Mathematics
L. GUMAŃSKI, On Decidability of the First-Order Functional Calculus
K. LEEB, The Busy Bee. Diagonalization in the Syntax of Well-Categories
P. MARTIN-LÖF, On the Distinction between Propositions and Judgements
P. PÄPPINGHAUS, A Typed λ-Calculus for Primitive Recursive Operations over Ordinals and Girard's Model of Ptykes
Ch. PARSONS, Intuition and the Concept of Number
H. PFEIFFER, A System of Ordinal Notations Containing a Symbol for the First Weakly Inaccessible Ordinal
T. PRUCNAL, Structural Completeness of Pure Implicative Intermediate Logics
W. SIEG, Reductions of Theories for Classical Analysis
E.W. WETTE, The Consistency-Critical Primitive Recursive Function and the Inconsistent Variable-Free Elementary Term within Formalized Peano Arithmetic
A. WROŃSKI, Maehara-Style Equational Interpolation Property

Section 2: Model Theory and its Applications

J.N. CROSSLEY, Recursive Categoricity and Recursive Stability
M. DROSTE, On the Lattice of Normal Subgroups of Transitive Automorphism Groups of Linearly Ordered Sets
M. DUGAS, Torsion Theories and Radicals of Abelian Groups in L
R. GÖBEL, Torsion Theories and Radicals of Abelian Groups in Models which May Contain Measurable Cardinals
R. KOSSAK, Inductive Satisfaction Classes with an Application
R. KURATA, On the Embedding of Grothendieck Topos into a Category of Sheaves on a *cHa*
G.-W. LEE, A Semantic Analysis of Decision Problem
A. MACINTYRE, Effective Galois Cohomology and the Siegel–Mahler Theorem
L.L. MAKSIMOVA, On Interpolation in Modal Logics containing $S4$
R. MURAWSKI, On A_2^--Expansions of Models of Peano Arithmetic
R. PALYUTIN, Complete Horn Theories
S. TAGHVA, On Models of Finite Direct Products of Theories
A. VINCENZI, Effective Logic L_e
V. WEISPFENNING, The Elementary Theory of Fields with Absolute Value
R. ZIVALJEVIĆ, A Homology and Cohomology Theory Based on Infinitesimal Chains of Hyperfinite Length

Section 3: Recursion Theory and Theory of Computation

B. APOLLONI, About a Formal Non-Probabilistic Theory of Choices
E. BÖRGER, Logical Decision Problems and Complexity of Computations
W. DEMOPOULOS, E.P. STABLER, JR., Count Relations and Regular Languages
S. GONCHAROV, Equivalences of Constructivizations and Computable Enumerations
Y. GUREVICH, S. SHELAH, The Decision Problem for Branching Time Logic
H.A. KIERSTEAD, G.F. MCNULTY, W.T. TROTTER, JR., A Theory of Dimension for Recursive Ordered Sets
H. LEVITZ, Some Decision Procedures Pertaining to Base 2 Exponential Diophantine Equations
S.-Ch. LIU, Multiple Recursive ZF-Provable Δ_1-Operations
A.A. PELIN, A Method of Solving Word Problems in Free Algebras by Computing Normal Forms
H. RASIOWA, Completeness in Logic of Nondeterministic Complex Algorithms
Th. A. SLAMAN, The Recursively Enumerable Branching Degrees Are Dense in the Recursively Enumerable Degrees
J. TALJA, Semantic Games on Finite Trees
H. THIELE, Propositional Computer-Tree Dynamic Logic
A. WASILEWSKA, Programs, Automata and Gentzen Type Formalizations

Section 4: Axiomatic Set Theory

N. BRUNNER, Weak Wellordering Theorems
Y. KAKUDA, The Role of Filter Quantifier in Set Theory
D.A. KUREPA, On the Formula $n!=1\cdot\ldots\cdot n$ for Transfinite Numbers
J.A. LARSON, Ordinal Graphs and Infinity Paths
R.M. MARTIN, On Mereological Mathematics and the Heroic Course
L.J. STANLEY, S. SHELAH, J. BURGESS, Some Applications of Morasses with Built-in Diamond
D. VELLEMAN, Morasses of Height ω
A. S. YESSENIN-VOLPIN, On the Use of Identifications and Tenses for Verifications of Deductoids

Section 5: Philosophical Logic

J. ALMOG, Reference and Modality: Two Quinean Dogmas
H.B. ANDERSEN, S.A. PEDERSEN, Natural Kinds and Essentialism
C.A. ANDERSON, Semantical Antinomies in the Logic of Sense and Denotation
A. ARNON, On Purely Relevant Logics
E. BENCIVENGA, Supervaluations and Theories
K. BERKA, Logic with or without Ontology: A Marxist Approach
J.M. BOCHEŃSKI, Mathematical Logic and Continental Philosophers
M. BOŽIĆ, A Semantics for Logics Weaker than **R**
W.S. CRODDY, Logic and Intensionalism
J.F. CROSBY, The Necessary Truth of the Principle of Contradiction
J. CZELAKOWSKI, Filter Distributive Sentential Logics
G. FORBES, On the Internal Structure of Possible Worlds
P. GÄRDENFORS, Epistemic Importance and Minimal Changes of Belief
A. GRIEDER, The Logic of Predicates and the Formalism of Linear Algebra

T. HAILPERIN, Probability Logic
A. HAUTAMÄKI, A Logical Analysis of Points of View
K. HAVAS, Logical Contradictions, Concepts, and the Dialectics of Objects
W. HEITSCH, Eine fragenlogische Interpretation von Fragebogengerüsten
G. HEYER, On the Notion of Generic Reference
J. HUMPHRIES, Is the Liar Paradox Unique?
A.A. JOHANSON, Nonstandard Deontic Logic
T. KAPITAN, Implication and Logical Form
D. KAPLAN, A Problem in Possible World Semantics
W. KISTNER, The Status of Ordinary Logic
S. KRAJEWSKI, The Relatedness Logic
A. KRON, Is the Concept of an oml Definable in Relevance Logic?
F. v. KUTSCHERA, A Logic of Vagueness
W.A. LABUSCHAGNE, Infinitary Logic and the Substitution Interpretation of Quantifiers
H. LEBLANC, Unary Probabilistic Semantics
W. LENZEN, Modal Value Logic
F. LEPAGE, Croyance et forme logique du contenu cognitif
D.L.C. MACLACHLAN, Conditional Statements and Material Implication
M. MCKINSEY, Causality and the Paradox of Names
P.K. MEYER, R. ROUTLEY, V. PLUMWOOD, A Farewell to Entailment
A. OBERSCHELP, Class-Theoretical Logic
E. ORLOWSKA, Semantics of Vague Concepts
P.L. PETERSON, Numerical Quantifiers and the Syllogism
J.F. POST, Correspondent Truth without Determinate Reference
C. PÜHRINGER, A Completeness Proof for a Certain First-Order Theory
W. RAUTENBERG, 2-Valued Algebraic Consequences
P. SCHROEDER-HEISTER, Inversion Principles and the Completeness of Intuitionistic Natural Deduction Systems
E.A. SIDORENKO, Semantic Approach to Construction of Relevant Conditional Logic
P.M. SIMONS, Embedding Classical and Free Logics in Extensions of Leśniewski's Ontology
V.A. SMIRNOV, Tense Logics with Non-Standard Interconnections between Past and Future
H.P. SMOLENOV, Paraconsistency and Some Prospects of Dialectical Logic
Z. STACHNIAK, Semantic Characterization of Structural Logics
E.-W. STACHOW, Constructive Temporal Logic
R. STUHLMANN-LAEISZ, Structures and Truth Systems for Many-Dimensional Modal Logics
H.C.M. de SWART, Gentzen-type Systems, Constructive Completeness Proofs and Practical Decision Procedures
M. TAMTHAI, Mathematical Logic and Philosophy of Mathematics
P.-B. TARNAY, Y-a-t'il une réalité sans l'individualisable?
R. TURNER, Semantic Theories of Nominalized Predicates
A. ULE, Z. KNAP, On Some Quantificative Interpretations of Modal Logic
D. ULRICH, Concerning Recursive Bounds on the Size of Countermodels for Propositional Calculi with the Finite Model Property
D. VANDERVEKEN, A Model-Theoretical Semantics for Illocutionary Forces
R. WACHBROIT, Logical Compulsion and Necessity
E. WALTHER-KLAUS, Zur Geschichte und zur Deutung des Reziprozitätsgesetzes
P. WEINGARTNER, J. CZERMAK, Identity Conditions for Propositions and Propositional Functions

H. WESSEL, Nichttraditionelle Prädikationstheorie und intuitionistische Aussagenlogik
D. WESTERSTÅHL, On Determiners
L. WIESENTHAL, True and Partly True
R. WOJCICKI, The Logic of Non-Deductive Arguments
B. WOLNIEWICZ, An Algebra of Situations
P.W. WOODRUFF, Approximate Semantics and Iterative Theories of Truth
J. ZYGMUNT, A Model-Theoretic Analysis of Multiple-Conclusion Consequence Relations

Section 6: General Methodology of Science

P. ACHINSTEIN, What is Retroduction?
M. AFTOWICZ-BIELECKI, Structure and Application of Physical Theories: Comments on three Methodological Approaches
A. BALTAS, Towards the Science of the History of Physics?
W. BALZER, The Problem of Theoretical Terms: A New Perspective
E.M. BARTH, Dialectical Fields and Transformations: Brouwer-Fields, Beth-Fields, and Naess-Transformations
I.V. BLAUBERG, The Concept of Wholeness and its Functions in Scientific Knowledge
R.P. BORN, General Methodology of Science versus Philosophy of Science?
M.K. CHYTIL, The Comparison of the Importance of Qualitative and Quantitative Predicates
F.J. CLENDINNEN, Intuition and Rationality
R. CREATH, The Pragmatics of Observation
U. D'AMBROSIO, Cultural Dynamics and the Transmission of Scientific Knowledge
A. DERECIN, E. ERRICO, Present Epistemology and the Notion of "Critical Discomfort"
N.E. ENEV, Methodology of Scientific Discovery
P. FEVRIER, A Way towards the New Methodologies and their Significance
M.A. FINOCCHIARO, Opportunism, Pluralism, and Judgment
A. FRANKLIN, Are Paradigms Incommensurable?
U. GÄHDE, A Formal Approach towards the Theory-Dependence of Measurement
D.P. GORSKI, On Difficulties of Application of the Theories with Idealizations
G. GRANBOULAN-EVEN, Is Popper's Conception of Reson Non-Falsifiable?
H. HAUFFE, Theoriendynamik und Wissensrepräsentation vom Standpunkt der Informationstheorie
R. HILPINEN, On the Principle of Information Retention
C.A. HOOKER, Surface Dazzle, Ghostly Depths: An Exposition and Critical Evaluation of van Fraassen's Vindication of Empiricism against Realism
W.B. JONES, Another Look at the Predictivist Thesis
R. KAUPPI, Wert, Wertung und Wissenschaft
J. KIM, Self-Understanding and Rationalizing Explanations
A. KOCHERGIN, "Simon Syndrome" as a Problem of Methodology of Science
M. KÜTTNER, Logical Conditions of D-N Predictions, Structural Identity Thesis, and Blau's Example
W.I. KUPZOW, Soziale Bedingtheit der wissenschaftlichen Entdeckung
N. LACHARITÉ, Quelques utilisations des catégories systémiques examinées en vue d'une épistémologie maximalement attentive
C.-S. LEE, On the Demarcation Problem in the Context of Discovery
E. MCMULLIN, Two Ideals of Scientific Explanation
U. MAJER, The Craig-Elimination: A Proof for the Indispensability of Theoretical Terms
C. MARE, Universality and Qualitative Specificity of Space-Time Structures

C.U. MOULINES, Links, Loops and the Global Structure of Science
G. MUNEVAR, In the Lion's Den: Taming Methodology into Oblivion
J.Z. NALIOTOV, Le développment de la science du point de vue de l'unité de l'abstrait et du concret
P.D. NICOLACOPOULOS, Science as Artifact, Knowledge as Praxis
I. PÂRVU, A Typological Approach to Scientific Theories
D. PEARCE, Theory Dynamics and Abstract Logic
S.A. PEDERSEN, Formation and Development of Scientific Concepts
B. PETKOFF, Kybernetisches Modell der wissenschaftlichen Forschung
A. POLIKAROV, On Possibilities and Restrictions in Science
V. RANTALA, Correspondence, Symmetries, and Continuity
N. RESCHER, Extraterrestrial Science
G. SCHURZ, Correct Explanatory Arguments and Understanding Why
M.L. SHAMES, Caveat Experimentor: On the Pitfalls of Experimentation, with Especial Reference to Control Groups
A. SIITONEN, Demarcation of Science from the Point of View of Problems and Problem Stating
M. SINTONEN, A New Look at Consilience
A.K. SUKHOTIN, Logical and Gnoseological Foundations of Extra-Empirical Criteria of Truth
K. SUNDARAM, Objectivity and the Language of Science
K. SZANIAWSKI, On Defining Information
J. VAN BRAKEL, Natural Kinds, Essentialism and Empiricism
J. VÁSQUEZ SÁNCHEZ, Theory and Experience
J. WETTERSTEN, The Rationality of Rationality
H. ZANDVOORT, Intrinsic Success and Extrinsic Success of Research Programs

Section 7: Foundations of Probability and Induction

L.E. BERTOSSI, Faktische Wahrscheinlichkeit und Brown'sche Bewegung
R.M. COOKE, Three Fallacies in Subjective Probability
A.I. DALE, An Urn Model for a Partially Exchangeable Markov Chain
R. FESTA, Epistemic Utilities, Verisimilitude, and Inductive Acceptance of Interval Hypotheses
W.L. HARPER, Consilience and Natural Kinds
J. HUMBURG, Foundations of a New System of Probability Theory
G.M.K. HUNT, A Neo-Bayesian Justification of Induction
P.-K. IP, Multiattribute Utility Theory and Epistemic Decisions
R.C. JEFFREY, Radical Probabilism
G.E. JONES, Conventionalism and Induction
A. MARGALIT, M. BAR-HILLEL, The Gideon Paradox or "Choose Irrationally, It's the Rational Way to Choose"
K. POPPER, The Calculus of Probability Forbids Ampliative Probabilistic Induction
M. VON RIMSCHA, A Logical Approach to Non-Numerical Probability
B. SKYRMS, A Bayesian Theory of Conditionals
L. SOFONEA, N. IONESCU-PALLAS, A Connection between the Variational Principles of Mechanics and the Concepts of Probability and Entropy (Epistemological Comments and Modelling Examples)
Z.G. SWIJTINK, Experimental Randomization and the Likelihood Principle
J. WATKINS, Is there a Popperian Solution for the Pragmatic Problem of Induction?

J. WILLIAMSON, Inductive Probability
Ch. ZWICKL-BERNHARD, W. KOENNE, F. ÖSTERREICHER, P. ZINTERHOF, Einige Bemerkungen zu Poppers Argumenten gegen induktives Schließen

Section 8: Foundations and Philosophy of the Physical Sciences

A. BARTELS, Modelle und physikalische Bedeutung — am Beispiel der allgemeinen Relativitätstheorie
M. ČAPEK, The Status of Future Events
A. CORDERO, Quantum Mechanics and the Ascription of Physical Properties
O. COSTA DE BEAUREGARD, Lorentz and CPT Invariances and the EPR Correlations
J.T. CUSHING, Is there Just *One* Possible World? Contingency vs. the Bootstrap
D. DIEKS, Simultaneity in Special and General Relativity
C. DILWORTH, Physical Theory and the Notion of a Mechanism
M.C. DUFFY, Ether Theory in the Late 20th Century
J. EARMAN, What is Locality? (A Sceptical Review of Some Philosophical Dogmas)
W. ESSLER, G. ZOUBEK, Some Remarks on the Philosophical Interpretation of the Quantum Mechanical Theory of Measurement
J. FANG, "Novum Organum" Redivivus: Towards a "Paraphysica"
D. FINKELSTEIN, Quantum Set Theory and Applications
M. FLONTA, Zwei wissenschaftstheoretische Betrachtungsweisen physikalischer Theorien
H.J. FOLSE, JR., Complementarity and Scientific Realism
M.R. FORSTER, Bell's Paradox: What's the Problem?
Y. GAUTHIER, L'Analyse syntaxique de la physique mathématique: Les procédures de renormalisation dans les théories quantiques des champs
K. GAVROGLU, Some Comments Concerning the Structure of Theories in Physics
P.F. GIBBINS, Quantum Logic as Sequent Calculi
V.S. GOTT, Fundamental Principle of Physical Science
H.H. GREIDANUS, Principles and Applications of a Psycho-Physical Theory
P. HAVAS, The Interrelation of Physical Theories of Increasing Universality
M. HELLER, Is Space-Time Manifold an Arena of Quantum Mechanics?
G. HELLMAN, Stochastic Einstein-Locality and the Bell Theorems
A.L. HISKES, Symmetry Groups and the Content of Physical Theories
P. HORWICH, Backwards Causation and Determinism
R.I.G. HUGHES, The *A Priori* in Quantum Theory
F. JENC, Principles of the CA (Conceptual Analysis)-Method
F. JENC, Application of the CA (Conceptual Analysis)-Method to Quantum Mechanics
A. KAMLAH, A Classification of Functions in Physics
W. KRAJEWSKI, Analysis of the Correspondence Principle
P.J. LAHTI, Complementary, Uncertainty, and the Popperian Concept of the Growth of Scientific Knowledge
K.V. LAURIKAINEN, Pauli's View on the Philosophical Implications of Quantum Theory
H. LENK, Antinaturalistische und antirealistische Fehlschlüsse
M. LORENTE, A Causal Interpretation of the Structure of Space and Time
A. MENNE, Zum Identitätsproblem chemischer Stoffe
J. MEURERS, The Silence of Physics
P. MITTELSTAEDT, Naming and Identity in Quantum Logic
F. MÜHLHÖLZER, A Non-Essentialist View of Special Relativity
H.P. NOYES, C. GEFWERT, M.J. MANTHEY, Toward a Constructive Physics

L.N. OAKLANDER, McTaggart, Schlesinger, and the Two-Dimensional Hypothesis
M. PAVICIĆ, The Einstein Locality without the Bell Inequality
J. PFARR, An Operational Approach to the Lorentz-Transformation
K.R. POPPER, A Realist's View of the Einstein–Podolski–Rosen Experiment
H.R. POST, The Nature of Experiments
C. RAJSKI, Note on Quantification of Formulae of Physics
M. REDHEAD, P. HEYWOOD, Nonlocality and the Kochen–Specker Paradox
M.C. ROBINSON, Probability, Determinism, and Quantum Mechanics
U. RÖSEBERG, Continuity and Discontinuity in Scientific Revolutions
L. ROPOLYI, On the Philosophy of Rational Thermodynamics
R.M. ROSENBERG, On the Concept of Force — A Remark on a Remark by Truesdell
S.F. SAVITT, J. COLLIER, Tachyons and Causal Theories of Space-Time
E. SCHEIBE, What Kind of Hidden Variables are Excluded by Bell's Inequality?
G.B. SCHMID, A New Approach to Physics Based upon Substance-Like Quantities and their Currents
S.D. SHARMA, Analogy a Tool in Mathematical Formalism
J.J.C. SMART, Are Fundamental Laws of Nature True?
L. SOFONEA, The Meta-Concept of "Many" in the Frame of Physical Thought
J.J. STACHEL, Special Relativity from Measuring Rods
E.-W. STACHOW, Application of Relativistic Quantum Language to the EPR-Gedanken-experiment
G.J. STAVENGA, Elementary Particles and the Third Limitation on Physical Experiments
I. STEIN, Relativity and Quantum Mechanics from Random Walks
M. STÖCKLER, EPR, Relativity and the Interpretation of Quantum Mechanics
P. SZEGEDI, The Second Wave of Deterministic Efforts in Quantum Mechanics (1952). Foundations and Philosophy
M.M. YANASE, On Aionity — between Time-Space and Eternity

Section 9: Foundations and Philosophy of Biology

W. BECHTEL, Building Interlevel Theories: the Development of the Embden–Meyerhof Pathway
F. ČIŽEK, The Analytical Method and Reduction in Biology
A.J. CLARK, The Intelligibility of Evolutionary Biology
B. COLEMAN, Methodology in Clinical Research
L. DARDEN, Reasoning in Theory Construction: Analogies, Interfield Connections, and Levels of Organization
J. JANKO, On the Initial Principles of Theoretical Biology: General Biology or Generalized Physics?
E.R. KRAEMER, On a "Platonic" Argument for Organismalism
Ch. KWA, Two Ecosystem Approaches and their Relation to the Control of Nature
W. LEINFELLNER, Foundations of the Theory of Evolution: Four Models of Evolution
A. LINDENMAYER, Formal Theories of Growth and Development
E.A. LLOYD, Mathematical Models in Evolutionary Theory and the Semantic Approach to Theory Structure
A.S. MAMZIN, The Integrative Function of the Evolution Theory in Modern Biology
L. NISSEN, Wright and Woodfield on Natural Functions and Reverse Causation
V.J.A. NOVÁK, The Biological Bases of the Development of Ethics and their Objective Essence
R.C. RICHARDSON, The Use of Models in Biological Explanation

A. ROSENBERG, Fitness
R.R. ROTH, The Principles of R.H. Francé's "Objective Philosophy"
V. SCHURIG, Klassifikationsprinzipien der Biologie
M. SEEL, J. LADIK, The Tragicomedy of Modern Theoretical Biology
K. SEN, The Genetic Explanation of Behaviour
N.P.L. SIMON, The Testability of Haeckel's Biogenetic Law
R. Wm. SMITH, A Critique of Impure Reasoning in Biological Sciences
S. ŠTRBÁŇOVÁ, Chemical and Biological Foundations of Biochemistry
N. TENNANT, Reflections on Reductionism
G.D. WASSERMANN, Evolutionary Adaptation by "TIMA"
M.B. WILLIAMS, The Units of Selection Controversy: Resolution by an Axiomatisation

Section 10: Foundations and Philosophy of Psychology

K. AAGARD, Psychology as a Cultural Phenomenon and the Need for a Hermeneutical Approach
W. BECHTEL, Functionalism and the Teleological Perspective in Psychology
J.I. BIRO, Grice on Utterance Meaning
N. BLOCK, The Photographic Fallacy
R.J. BOGDAN, Proposition Attitudes and Psychological Explanation
E. COHORS-FRESENBORG, I. SCHWANK, On the Modelling of Learning Processes by $\alpha\beta\gamma$-Automata
R.J. DOUGLAS, B.P. KEANEY, A Critique of Popper and Eccles' Psychophysical Interactionism
C.A. FIELDS, Computational and Ecological Approximations to Perceptual Systems
G.H. FISCHER, Consequences of the Principle of "Specific Objectivity" for Model Construction in Psychology
A.R. GILGEN, Prescriptions for the Conceptual Integration of Psychology
P.A. HEELAN, Is Visual Space Euclidean? A Study in the Hermeneutics of Perception
O. HUBER, Restricted Information-Processing Capacity and Rational Decisions
W. ICKES, A New Paradigm for the Study of Personality
W. KLIMESCH, Psychological Constraints on Thinking
K. KRZYZEWSKI, The Consciousness as a Source of the Present-Day Formulation of Emotions
J. KUNG, Method, Sense, and Virtue in Aristotle's Science
B. LINKE, Das Gehirn also Identitätsmodell
A. MARRAS, Cognitive Psychology and the Formality Condition
M.A. NESTER, M. COLBERG, Uses and Abuses of Logic in Psychometrics
J. PIETARINEN, On Mental Inconsciousness
S. ROSENBERG, M.A. GARA, Theoretical and Methodological Dimensions in Personality and Social Psychology
D. RUIMSCHOITEL, Theoretical Explanation: Some Philosophical Models Confronted with Examples from Psychological Reality
F. SCHRAG, Folk Psychology, Scientific Psychology and the Predictability of Human Behavior
A. STROLL, Gibson on Surfaces
P. THAGARD, Computer Programs as Psychological Theories
W.K. WANG, The Mapping between the Inner World (the Brain) and the Outer World
W.C. WATT, Divergence and Convergence in Iconic Analysis
R. WERTH, What is Perception?

K.V. WILKES, Logicalizing Psychological Functions
E. WÜST, Der konstruktiv-realistische Aspekt des Psychischen

Section 11: Foundations and Philosophy of the Social Sciences

M. BUNZL, Social Kinds
F. COLLIN, Reason Explanation and Deviant Causal Chains
A. DIAS DE CARVALHO, Educational Science and Social Science
S. FULLER, Theory and Practice Revisited
G. GEIGER, The Logic of Human Sociobiology
B. GRÜNEWALD, The Sense Structure of a Language as a Manifold System of Mental Processes
E. GUISÁN, Ethics as a Social Science: From Schlick to Stevenson
B. HAMMINGA, The Intuitive Goal of the Economist's Endeavours
A. HELMAN, Sciences versus Art Theory
F.D. HEYT, Karl Popper on Sociology
Y.-S. HO, The Planning Process
H. HÖRZ, Determinants of Science-Evolution
M. JANKOV, Ideologisches Paradigma und Ideologen-Gemeinschaft (Thesen)
H.S. KARIEL, Inquiry as Diversion from Finalities
V.Z. KELLE, Correlation of the Principles of Systematism and Determinism as a Problem of the Methodology of Social Knowledge
A. KOUTOUGOS, Meaning Relativism and the Possibility of Communication: The Basis for an Interdisciplinary Account of Scientific and Cultural Change as Adaption
E. KRAUSZ, In Defence of Methodological Collectivism
E. LAGERSPETZ, Money as a Social Contract
S. MAFFETTONE, Critical Theory and Social Justice
R. MATTESICH, Information Economics and Agency Theory: Philosophic Aspects
H. NURMI, Social Choice Theory and Democracy
P. PETTIT, The Holist Grail
K.E. ROBSON, Danto on Chronicles in History
C. SAVARY, On the Meaning of Interpretation for the Social Sciences
J. SEARLE, Intentionality and Explanation in the Social Sciences
I.S. STEINBERG, The Problem of General Theory
R. STRANZINGER, A Rational Principle of Justice
F. STUDNICKI, A Critical Note on von Wright's Nicomachean Paradox
P. SZTOMPKA, On the Change of Social Laws
R. TRIGG, The Social Sciences and Biology
R. TUOMELA, Social Action, Systems Theory, and Scientific Progress
G. ZECHA, Postulates for Value Freedom in the Social Sciences

Section 12: Foundations and Philosophy of Linguistics

E.M. BARTH, A Two-Role Problem-Analytic Analysis of Montague's PTQ
I. BELLERT, Interpretive Model for Linguistic Quantifiers
T.J. BLAKELEY, Dialectic, Logic, Computing
J. v. BRAKEL, Sentences and Naming Sentences
A. BRESSAN, Generalized Synonymy Notions, Corresponding Quasi-Senses, and a

General Theory of Quasi-Senses Capable to Deal with Iterated Belief Sentences
S.J. BRISON, Bilingualism and the Language of Thought
W. BUSZKOWSKI, An Algebraic Approach to Categorial Grammars
H.G. CALLAWAY, Meaning Sans Analyticity
M.J. CRESSWELL, A Highly Impossible Scene: The Semantics of Visual Contradictions
B. DIANKOV, A Spectral Model of Lexical Meaning and the Problem of the Semantic Structure of Propositional Contexts
G. DRACHMAN, The Empirical Status of Theories of the Mind: Example, Generative Grammar
M. EYTAN, A Doctrinal Semantics for Harris' Theory of Language
H. FESTINI, Dummett's Conception as Theory of Meaning for Hintikka's Type of Game-Theoretical Semantics
W. HEYDRICH, Synonymy and Necessity
H. HIZ, Information Semantics and Antinomies
J.N. KAUFMANN, Analyse fonctionelle des séquences d'actes de discours
A. LANGE-SEIDL, Sign-Constituting Factors
D. LAURIER, Signification et intention de communication
E. LEINFELLNER-RUPERTSBERGER, Semantic Nets: Their Linguistic Usefulness
F. LIU, The Functions of 'Yes' and 'No' in Chinese and English
R.J. MATTHEWS, Formal Learning-Theoretic Conditions on Linguistic Theory
A.G.B. TER MEULEN, Homogeneous and Individuated Quantifiers in Natural Language
K. MUDERSBACH, Proper Names as Properties
O. NEUMAIER, Is Use a Criterion for Meaning?
R.J. PAVILIONIS, Analysis of Language and the Relevance of Reference to Conceptual Systems (Belief-Sentences and Indexicals)
Yu.A. PETROV, The Logic and Methodology of Interrogative Language
U.T. PLACE, Behavioural Contingency Semantics
J. POGONOWSKI, Semantic Gaps
B.-O. QVARNSTRÖM, Indeterminacy of Empirical Meaning
M. SEYMOUR, Scope Distinctions and on Asserting That
P. SGALL, Linguistic Meaning and Intension
N. STEMMER, Frege's Notion of Intension
P. SWIGGERS, Some Reflections on the Epistemology of Linguistics
G. TODT, Ausdruckslogik LA — A Common Basis for Linguistic Theories
R.G. VAN DE VELDE, Inference in Text Interpretation
P. WEINGARTNER, Weak Relevant Logic for Natural Language
W. WENNING, Universals of Color Naming and the Neurobiology of Color Vision
R. ZUBER, Privative Oppositions and Intensional Equivalence

Section 13: History of Logic, Methodology and Philosophy of Science

W.R. ALBURY, The Methodology of Condillac's Logic and the Natural Sciences in France during the Revolutionary Period
I. ANGELELLI, Frege and Abstraction
M. BAAZ, Gödel's Justification of his Completeness Theorem
M. BLEGVAD, Adam Smith as a Philosopher of Science
I. BOH, Problems of Alethic and Epistemic Iteration in Later Medieval Logic
M. BRADIE, Science as Model and Metaphor
T.G. BUCHER, Cicero's and Augustine's Logical Arguments Against Scepticism

C.B. BURCH, Descartes and Huygens: Foundational Differences in the Mechanical Philosophy
T. BURGE, The Concept of Truth in Frege's Program
H. BURKHARDT, Duns Scotus, Leibniz und Crusius über mögliche Welten
R.E. BUTTS, Leibniz on Empirical Methodology
A. COFFA, Carnap's Route to Semantics
J. COUTURE, Les Classes comme symboles incomplets des *Principia Mathematica*
C. DAPUETO, P.L. FERRARI, On Some Recent Contributions on the Philosophy of Mathematics
M. DAVIS, Issues in Computer Science Anticipated in Gödel's Work on Undecidability
J.W. DAWSON, JR., Cataloguing the Gödel Nachlaß at the Institute for Advanced Studies
G.J.W. DORN, Poppers zwei Definitionsvarianten von 'falsifizierbar'
C. EISELE, Peirce's Application of Mathematical and Scientific Methodology to Problems in the History of Science
M.M. FARTOS, Polyadic Treatment of a Syllogism from Aristotle's Poetics
S. FEFERMAN, Conviction and Caution: A Scientific Portrait of Kurt Gödel
A. GODDU, A Realist Interpretation of the Hypothetical Reasoning of the Middle Ages — A Tentative Proposal
E. GROSHOLZ, Two Episodes in the Unification of Logic and Topology
M. HEIDELBERGER, Philosophische Konzeptionen der Messung im 19. Jahrhundert
E.N. HIEBERT, The Influence of Mach's Thought on Science
D.A. HOWARD, Realism and Conventionalism in Einstein' Philosophy of Science: The Einstein–Schlick Correspondence
N. IONESCU-PALLAS, L. SOFONEA, I. GOTTLIEB, T. TORO, Efficiency of Historical-Epistemological Models. Examples from the History of Modern Atomic Physics
K.L. KETNER, How Hintikka Misunderstood Peirce's Account of Theorematic Reasoning
R. KEVELSON, Charles S. Peirce's 'Method of Methods'
E. KÖHLER, Gödel and the Vienna Circle: Platonism versus Formalism
C.A. LERTORA MENDOZ, L'analyse logico-linguistique du terme 'infini' chez Ockham
P. LINDFORS, Reflections on the Concept of Invariance in Eino Kaila's Psychology of Perception and Philosophy of Science
J. LOSEE, Shapere's "Presuppositionless" Philosophy of Science and the Perils of Historicism
K. MAINZER, Plato and the Four-Color Problem. Philsophical Remarks on Logic, Mathematics and Computer Science
W. MARCISZEWSKI, The Comprehension Axiom as Cantor's Contribution to *Mathesis Universalis*
J. MITTELSTRASS, From Transcendentalism to Rational Reconstruction
J. MOUTON, Power and Therapy in Francis Bacon's Natural Philosophy
V. MUÑOZ DELGADO, La "Consequentia" y sus divisiones generales en Juan de Oria
A. MUSGRAVE, Constructive Empiricism and the Theory of Observation Dichotomy
U. NEEMANN, Über die strukturelle Verwandtschaft zwischen Nominalismus und Platonismus
N.J. NERSESSIAN, How are Scientific Concepts Formed?
T. NICKLES, Justification as Discoverability
M.H. OTERO, The Uproar of the Beotians: A Case Study in the Philosophy of Geometry
J. PURŠOVÁ, Mechanical and Organic Models in the History of Modern Science
Y. RAV, A Reassessment of the Work and Polemics of Ernst Zermelo
M.D. RESNIK, Frege's Proof of Referentiality: What's Right with It?

G.P. SCOTT, Nature Philosophy and Atomism in Popper's Reality Language
W.R. SHEA, Descartes' Science: Methodological Ideal and Actual Procedure
R. SMITH, Aristotle as Proof Theorist
L. TAIMINEN, Frege on Existence
H. TENNESSEN, Vexküllian Inspired Epistemology and Theories of Perception Revisited
B. TERRELL, Science, Design and the Science of Signs
C. THIEL, On Gödel's Involvement in the Debate on Behmann's Treatment of the Paradoxes
K. VASSILIS, On Lakatos' Theory of Rationality
P.R. WOLFSON, Models and Explanation in Mathematics
G. WOLTERS, Ernst Mach und die Relativitätstheorie. Ein neuer Anfang. Dem Andenken E. Machs zum 100-jährigen Jubiliäum des Erscheinens der "Mechanik"

Published selections of contributed papers

A selection of papers from sections 1, 2, 3, 4, 5, 7, and 12 has been published in: Georg Dorn and Paul Weingartner (eds.): *Foundations of Logic and Linguistics. Problems and their Solutions*, New York and London, Plenum Press, 1985.

Selected papers from sections 10 and 11 have been published in: Otto Neumaier (ed.): *Mind, Language and Society*, Vienna, Austria, VWGÖ, 1984.

A selection of papers from section 8 has been published in: Paul Weingartner and Georg Dorn (eds.): *Foundations of Physics*, Vienna, Austria, Hölder–Pichler–Tempsky, 1986.

Selected papers from section 9 have been published in: Paul Weingartner and Geory Dorn (eds.): *Foundations of Biology*, Vienna, Austria, Hölder–Pichler–Tempsky, 1986.

Finally, a selection of papers from sections 6 and 13 has been published in: Paul Weingartner and Christine Pühringer (eds.): *Philosophy of Science — History of Science*, Königstein, West Germany, Verlag Anton Hain Meisenheim, 1984.